DISCHARGE OF SEWAGE FROM SEA OUTFALLS

Supplement to Progress in Water Technology

DISCHARGE OF SEWAGE FROM SEA OUTFALLS

Proceedings of an International Symposium held at Church House, London, 27 August to 2 September 1974

Edited by
A. L. H. Gameson
Water Research Centre—Stevenage Laboratory

PERGAMON PRESS
OXFORD · NEW YORK
TORONTO · SYDNEY · PARIS · BRAUNSCHWEIG

Pergamon Press Offices:

U.K.	Pergamon Press Ltd., Headington Hill Hall, Oxford OX3 0BW England,
U.S.A.	Pergamon Press Inc., Maxwell House, Fairview Park, Elmsford, New York 10523, U.S.A.
CANADA	Pergamon of Canada Ltd., 207 Queen's Quay West, Toronto 1, Canada
AUSTRALIA	Pergamon Press (Aust.) Pty. Ltd., 19a Boundary Street, Rushcutters Bay, N.S.W. 2011, Australia
FRANCE	Pergamon Press SARL, 24 rue des Ecoles, 75240 Paris, Cedex 05. France
WEST GERMANY	Pergamon Press GmbH, 3300 Braunschweig, Postfach 2923, Burgplatz 1, West Germany

First edition 1975

Library of Congress Catalog Card No. 75-4188

Printed in Great Britain by William Clowes & Sons Limited, London, Colchester and Beccles

ISBN 0 08 018302 6

Contents

Foreword

Enjoyment of the various pleasures that are afforded by bathing beaches and neighbouring coastal waters is among the most popular means of relaxation of people the world over. Yet, as is all too well known, this enjoyment and other related amenities are often spoiled by pollution arising from unsuitably sited or overloaded sewage outfalls. In general there is no major technical difficulty in preventing such adverse effects, but opinions differ as to the best way to proceed. Without going into all details of the argument, controversy arises mainly on two issues.

Firstly, there are those who believe that discharge of sewage without biological treatment, even some considerable distance offshore, is an unsavoury practice, inconsistent with modern man's technological ability and resources, and potentially dangerous to health. Others feel that it is perfectly proper to do this, since natural processes of self-purification and dispersion in the sea will, given good outfall design, ensure that no adverse effects result and that some advantage will accrue from the lower costs; moreover, the need to sacrifice valuable land for the siting of treatment works becomes unnecessary.

Secondly, there are those who consider that, because sewage can contain pathogenic micro-organisms, disposal schemes should be designed to prevent the degree of microbial contamination exceeding a chosen standard, one expressed in terms of numbers of coliform bacteria being most often suggested. Others contend that because no causal exposure–effect relationship has been established between numbers of these organisms in bathing waters and incidence of disease, and because of various practical difficulties in monitoring contamination, such a requirement would be counter-productive.

In the early sixties the then Department of Scientific and Industrial Research recognized that to resolve arguments such as these a much better understanding was needed of the behaviour of sewage and its constituents when released to the sea. The Department therefore set up a Coastal Pollution Research Committee to advise on an appropriate programme to be carried out by the Water Pollution Research Laboratory. A great deal was learned as a result of this work and of parallel activities in other parts of the world and in 1972, after about 10 years' effort, those most closely involved at the Laboratory, in consultation with the Committee, decided that the time was opportune to see whether some agreement could be reached among the professionals involved in various countries, as to the technical facts underlying the political issues. The idea was thus conceived of holding an International Symposium in which the leading protagonists of particular view points would present the scientific grounds for their contentions. Once embarked on this idea it was but a short step to decide that other important issues, such as the assessment of bacterial mortality in the sea, should be included.

The reader must judge for himself the success of the venture. My personal impression is that the quality both of the papers and of the discussion is certainly sufficiently high to have made the exercise thoroughly worthwhile and to reflect credit on all those who contributed to it.

As to the organization of the Symposium and the preparation of these Proceedings, I know that my former colleagues at WPRL and other participants will heartily support my view that by his untiring efforts, unflagging enthusiasm, and inspiration, the lion's share of the credit must go to Mr. A. L. H. Gameson and with it our grateful thanks.

A. L. Downing
(formerly Director,
Water Pollution Research Laboratory)

Binnie & Partners
London

Introduction

The original aim of the Symposium was to get full discussion of papers by leading protagonists on three important but unresolved topics: microbial standards for bathing beaches, the treatment of sewage before discharge to the sea, and the various mechanisms to which microbial mortality in the marine environment has been attributed. The scope was then extended to cover almost all aspects of the discharge of sewage from sea outfalls, though (even with 42 papers) full justice could not be done to the subject; for instance, there was too little on ecological effects and virtually nothing on outfall construction. Nevertheless, it was felt better to cover some major aspects as thoroughly as possible, rather than to attempt to cover the whole field uniformly. Papers were by invitation—after seeking the advice of about 100 specialists of international repute.

The programme was approved by the Coastal Pollution Research Committee of the Water Pollution Research Laboratory, and organization of the Symposium was little affected by the transfer, on 1st April 1974, of the Laboratory from the Department of the Environment to become the Stevenage Laboratory of the new Water Research Centre (outside the Civil Service). The main financial burden for the Symposium was borne by the Department.

Despite various difficulties, such as the late submission of many of the manuscripts, it was possible —through the ready collaboration of her Majesty's Stationery Office—to circulate preprints of each paper prior to the holding of the Symposium, and it was gratifying that (with a single exception) each was presented by an author. These two factors undoubtedly contributed greatly to the success of the Symposium in which there were some very spirited and rewarding discussions.

I take this opportunity of thanking the Authors for their unfailing courtesy over the past 18 months, the Chairmen for ensuring that the sessions ran to time and that the discussions took place in an orderly fashion, Mrs. Lena Jeger for her stimulating Opening Address, and the rest of the 218 participants (from 24 countries, plus W.H.O. and the World Bank) for their gracious co-operation and exemplary attendance, no paper being given to an audience of less than 100.

Editorial comment

In editing the papers, an attempt was made to introduce some degree of uniformity in the lay-out of tables, diagrams, reference lists, etc., so as to lighten the task of participants in reading so many papers in advance of their presentation. English spellings are used except in papers by American authors. Metric units are employed almost exclusively, the main abbreviations being s(econds), min(utes), m—metres or meters, l.—litres or liters, g—gram(me)s, t(onnes) = 10^6 g, and the prefixes n, μ, m, c, and k denoting multiples of 10^{-9}, 10^{-6}, 10^{-3}, 10^{-2}, and 10^3 respectively.

References in each paper are numbered in order of their first occurrence in the text (or relevant tables or diagrams); where several occur together, they are usually presented in chronological order. In the reference lists they are expressed in the format laid down in the 4th edition of the *World List of Scientific Periodicals* (Butterworth, London, 1964). The help is gratefully acknowledged of Mrs. I. N. Ward in checking, so far as was practicable, all these references, and of Messrs. D. Munro, A. R. Agg, A. W. J. Bufton, G. Stanfield, and D. J. Gould in editing and proof reading. Mr. Gould was also responsible for the detailed organization of the arrangements in Church House and gave invaluable assistance throughout, and Mr. Munro prepared the Subject Index.

Authors were normally given 20 min in which to present their papers (which were taken as read), and an edited summary of the opening remarks will be found immediately following the text of each paper. Contributors to the discussions were invited to hand in their contributions in writing (and most did). These have been edited—sometimes severely, as an attempt has been made to keep individual contributions down to a maximum of 300 words, although several submissions exceeded 2000 words. The Authors subsequently provided written replies conforming with what they had said at the time. To ensure speedy publication, the contributors were not sent the edited versions of their submissions; although all the edited material relating to each paper was checked by the Authors, I naturally take full responsibility for any misinterpretation or omission of the discussion material submitted. Papers were generally presented and discussed in pairs, but the discussions have been separated wherever practicable; these will be found at the end of each paper and are followed by the Authors' replies. Reported speech and English spellings are used throughout. The discussion of Papers 5 and 6, 9–11, 17 and 18, and 32–34, however, could not be treated so simply, and the Authors' replies are included within the general discussions. The numbering of the many additional literature references arising from the discussions continues from the reference list of the relevant paper, or of the final paper in the case of the general discussions. These references, some of which I have added myself where they seemed appropriate, have not been checked as thoroughly as those in the main reference lists.

A. L. H. Gameson

Water Research Centre—Stevenage Laboratory
Elder Way
Stevenage
Herts., SG1 1TH
England

January 1975

Erratum

All references to Rapp. P.—V. Réun. Cons. int. Explor. Mer, 1975, **167** are incorrectly dated; although the volume was not published until 1975 it is dated December 1974.

Opening Address

Mrs. Lena Jeger, *M.P.*

In welcoming the Symposium participants, Mrs. Jeger said that the subject to be discussed was one very relevant to the United Kingdom because those countries (usually islands) with the longest coastlines in proportion to their areas were historically the most abusive of the sea, and this sea-girt isle was no exception. It was gratifying that there was an increasing understanding of the marvellous but finite self-cleansing powers of the oceans: legislation and public concern were accelerating in nearly every country, and there was a fashionable interest in the environment which would pass if it were only a fashion. The three main necessities for implementation for effective control of marine pollution were the right machinery for enforcement, scientific man-power (for research, monitoring, and policing), and public interest and support which would inspire and compel government policy—and pay for it.

There had recently been massive re-organization of the water industry in England and Wales, on the principles that water management needed a national council, and that there should be delegation to authorities based on watersheds and not on local government boundaries. (There would, of course, be greater problems where national boundaries were involved.) The present gathering showed no shortage of world experts but, to cross new frontiers of knowledge, many countries (and certainly the U.K.) needed a restructuring of faculties in the universities, with straddling of old disciplines, and this called for fragmentation and re-assembly. New career structures inside and outside laboratories were needed, but British government scientists were disgracefully underpaid and it was a miracle that so many talented and devoted scientists remained in the service. The same applied outside, right down and up the line: new dignity was needed for the sewage workers and managers—and a career structure, since no army could work without its non-commissioned officers.

Most democratic governments acted in response to pressures, and only rarely took initiatives. These pressures (rational and irrational) could often be generated by scientists like Bronowski and Cousteau with a gift for communication. Some did not try hard enough—leaving it to the science writer who often did a good job which many scientists were too busy or too inarticulate to do—but in the urgency of the present situation (particularly in the context of the Symposium) Mrs. Jeger believed the public wanted to know and understand, and was capable of digesting much more scientific sales-talk than it got, and turned to science fiction despite the abundance of interesting science fact. A scientific truth which could not be clearly stated might not even be true. Mrs. Jeger had been lucky, as an arts student struggling with the use of the subjunctive in Icelandic sonnets, to have been at a college (Birkbeck) where the science faculty included Blackett and Bernal who accepted (at the cost of leisure) the social duty of communication to students in other faculties—apparently without any diminution in their scientific achievement.

Too often the attempt to engage science with politics avoided a definition of either. This led to difficulty—unless politics were considered to mean the whole body of public opinion and understanding. The scientist and politician who wanted to get something done had sometimes, ironically, to keep clear of what C. P. Snow called 'the corridors of power', which swarmed with the ineffectually ambitious, and where much time was wasted. Politics had been described as the art of the possible—more often it was the art of the unexpected. Political decisions required innovation of ideas and education in them.

Lest her audience thought she was forgetting about the sea, Mrs. Jeger gave a relevant example. British politicians and civil servants—semi-literate about pollution—devised rules and standards (very variably implemented) for water quality in rivers, but these standards did not apply to estuaries or to coastal discharges, and the Working Party of which she had been Chairman had found* a move of polluting industry to the river mouths and the sea shore to escape the laws of the politicians. Sewage and trade waste from about 6 million people was discharged to the sea or estuaries with little or no treatment, about two-thirds of the coastal authorities discharging at least some untreated sewage to the sea. There was uncertainty about the long-term effects of highly toxic and persistent substances originating in industrial and chemical wastes and discharged with the sewage. The Working Party had concluded that crude sewage should be discharged only after screening or comminution, and only through diffusers on long outfalls. Seeking to agree about how to treat the oceans of the world—whether to pollute the sea or farm it, whether to enjoy it or abuse it—could well be the most immediate growing point of international co-operation, and any responsible scientist and politician must want this above all else.

* Ministry of Housing and Local Government and Welsh Office. 'Taken for Granted. Report of Working Party on Sewage Disposal'. H. M. Stationery Office, London, 1970.

Mrs. Jeger ended by referring to a non-scientific non-political extract from John Keats's last sonnet:

> The moving waters at their priest-like task
> Of pure ablution round earth's human shores . . . .

Oil slicks, dumping of chemical wastes with unknown results, anxieties about nuclear wastes and industrial rubbish, now made these lines seem idle nostalgia. Nevertheless she believed—especially at the present important gathering—that the poets, the scientists, and even the politicians could come together, and that the moving waters could resume their task of 'pure ablution round earth's human shores'.

Paper No. **1**

The Discharge of Sewage from Sea Outfalls into the North Sea

P. C. Wood

Ministry of Agriculture, Fisheries and Food, Fisheries Laboratory
Burnham-on-Crouch, England

This paper presents in general terms some of the major factors which have a bearing on the discharge of sewage from outfalls into the North Sea. A brief description will be given of the relevant topographic, climatic, and hydrographic features which may affect the distribution of sewage, and of those uses of the sea which may be directly or indirectly affected. The aim of the paper will be to summarize many of these factors briefly; more detailed considerations of some aspects are the subject of other papers presented later at this Symposium.

Topography

For the purposes of this paper, the North Sea is defined as that sea area lying south of latitude 59° N and bounded by the Orkney Islands and the east coast of Britain on one side, and the coast of Norway and continental Europe on the other. The southern boundary is taken as the Straits of Dover. The Baltic Sea is excluded from these considerations and for this purpose a line has been drawn across the Skagerrak from Kristiansand in southern Norway to Hanstholm in northern Denmark.

The main features of the bottom topography are shown in Fig. 1. Generally, the depth of the North Sea decreases from the north towards the south, although extensive shallow areas, such as the Dogger Bank, occur in relatively deep water. Depths greater than 50 fm (91 m) are found in the northern part,

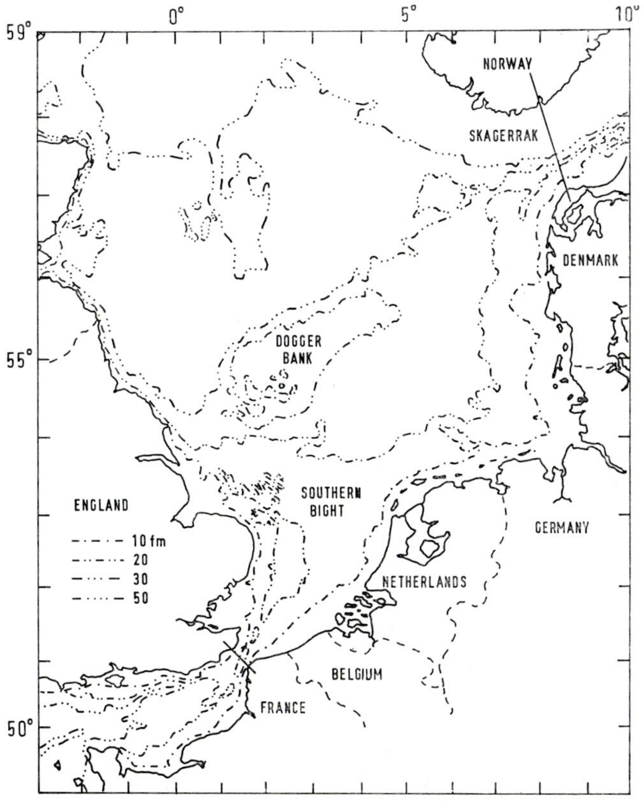

FIG. 1. Topography of the North Sea

including the area of the Norwegian Deeps adjacent to the south coast of Norway, where depths may exceed 150 fm (274 m). Extensive shallow areas of less than 10 fm (18 m) depth are found adjacent to the coasts of Denmark, Holland, Belgium, and the larger estuaries of Britain, such as the Firth of Forth, the Wash, and the River Thames. Most of the central northern part of the North Sea is between 20 and 30 fm (37–55 m) deep, the area of this depth approaching the coast of north-east Britain relatively closely.

The sea bed varies considerably, consisting mainly of sand, mud, or gravel or mixtures of them. Sand and gravel deposits are worked commercially in parts of the southern North Sea, and areas of mud occur in many of the estuaries as well as in well defined areas of deeper water[1]. In some places, usually near to rocky coasts, rocky outcrops occur. Sand is found on littoral beaches along many coasts of the North Sea, where it provides important recreational areas. The nature of the sea bed, and particularly its roughness, affects the turbulence of the water, which is important when considering the mixing characteristics of an area. Detailed consideration of this topic is outside the scope of this paper.

Climatic Conditions

Details of the climatic conditions prevailing in the North Sea are summarized by Hohn[2], whose data have been used extensively in the present paper. The North Sea is dominated by prevailing winds which bring humid conditions free from extremes of temperature. However, the North Sea is a transition area between oceanic and continental weather systems, the North Atlantic weather systems being modified by those of the continent. In winter, the lowest average temperatures occur on the eastern side of the North Sea (2–3°C) whilst those on the western side are about 5°C. In summer there is a gradient of average air temperature from the north (where it is about 13°C) to the south-east (where it reaches about 17°C). The mixed weather systems lead to a high frequency of winds, the average wind force and storm force frequency increasing from south-east to north-west. The average wind force is about 3 (Beaufort scale) in summer and 5–6 in winter.

Cloudiness and humidity are common in all seasons, but in coastal regions the effect of the land mass is to reduce cloud cover. In all months, the average frequency (days/month) of cloud cover of 8/10 or more in the southern and central North Sea is between 40 and 60 per cent, and in the northern North Sea between 50 and 70 per cent. The frequency of days with cloud cover of 2/10 or less is between 10 and 20 per cent. Clear weather occurs most frequently in the northern area from April to August, and in the remaining areas in April or May.

The frequency of precipitation is greatest in November and December and lowest in May to July. In the southern and eastern area the frequency (days/month) in the summer is about 8–10 per cent, but in many coastal areas precipitation is less as a result of the influence of the land mass. Eastern coastal regions of Britain are protected from the prevailing south-westerly wind and tend to have a lower rainfall.

Hydrography

The hydrographic features of an area play a predominant part in the fate of a waste released from a coastal discharge. The distribution of the waste will initially be influenced by (among other things) the tidal streams and in the long-term by the residual currents. Other hydrographic factors such as the effect of wind, the turbulence of the water mass, density gradients, and the differences in current velocity and direction at different depths all influence the dilution and transport of waste. Unfortunately, most hydrographic observations have until recently dealt mainly with those areas of the North Sea away from the coasts, although it is near the coasts that most discharges occur. The hydrographic characteristics of coastal waters vary widely, depending upon, for instance, topography, depth of water, and nature of the sea bed, and it is not possible to describe these features in any detail; some attention will, however, be given to this factor in subsequent contributions of a more specialized nature.

Much of the North Sea, except on part of the eastern side, is subjected to clearly defined tidal streams, in contrast with other seas such as the Baltic and Mediterranean, which, because of their restricted contact with the North Sea and the Atlantic Ocean, have only limited water movements associated with the tides. Little information is, however, available in a synoptic form illustrating the tidal streams in coastal waters, but some indication of the magnitude of the tidal streams throughout the North Sea, including coastal regions, can be inferred from consideration of co-tidal charts, from tables showing the mean range of tidal heights at selected coastal stations, and from current speeds shown at fixed stations on navigational charts.

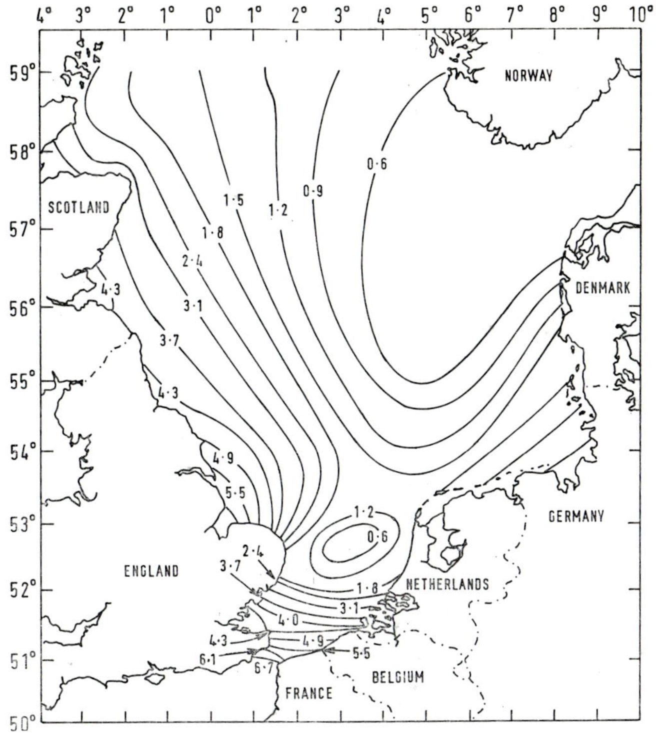

FIG. 2. Co-range chart of North Sea, showing range of mean spring tides (metres)

Co-tidal charts also show, by means of co-range lines, areas of the North Sea having the same range of tidal heights. Figure 2 shows the co-range lines for the North Sea. The salient features of this figure are the relatively small range of tides in the north and east, and the high range of tides in the south and west. Of particular note is the small range of tides adjacent to the entrance to the Skagerrak, and in the area between England and Holland. These areas surround the so-called amphidromic points. These observations are supported by examination of the range of tides predicted in tide tables at selected coastal stations (Fig. 3); for instance, the ranges of mean spring tides at stations in the Orkney Isles, in southern Norway, and in Denmark lie between 0·6 and 1·7 m, whereas on the coasts of Belgium, France, and the east coast of Britain the range of mean spring tides at 10 stations is between 1·9 and 6·6 m. The mean range of spring tides on the Dutch coast adjacent to the amphidromic point is about 1·7 m, but the range increases moving north-eastwards until, in the estuaries of the Elbe and the Weser, a tidal range of 3·2–4·0 m is recorded.

Where there is a large rise and fall of tide it may be inferred that there are strong tidal streams. This is borne out by examination of the tidal streams marked at selected stations on navigational charts (Fig. 3). The conditions of mixing and dispersal of wastes from coastal discharges are therefore likely to be more favourable on the western and southern coasts of the North Sea, than on the Norwegian, Danish, and Dutch coasts.

In addition to the immediate dilution and dispersal of a waste, broader hydrographic considerations have a bearing on the long-term distribution of wastes and their soluble components. This will be particularly important for conservative substances, i.e. metals, and nutrients such as nitrates and phosphates. The circulation pattern of the North Sea is complex[3–5], varying with depth and with the season, but its main characteristic is an anticlockwise pattern of movements. High-salinity water enters the North Sea from the Atlantic between the Orkneys and Norway and by the Straits of Dover. Run-off from the land produces estuarine and coastal water masses of lower salinity along the British coast and the coasts of the Low Countries, Germany, and Denmark. An outflow of low-salinity water from the Baltic enters the North Sea by the Skagerrak and flows northwards along the south Norwegian coast. It has been estimated[6] that about a third of the total volume of the North Sea is displaced each year by inflow from the Atlantic.

The layering of water in the North Sea, resulting from the density differences due to the salinity and/or temperature characteristics of each water mass, has an important effect on the biological processes of the sea, and on the rate of mixing and dispersal of wastes. The influence of air temperature on surface water temperature is greatest in the shallower coastal regions. According to Hohn[2], in winter the

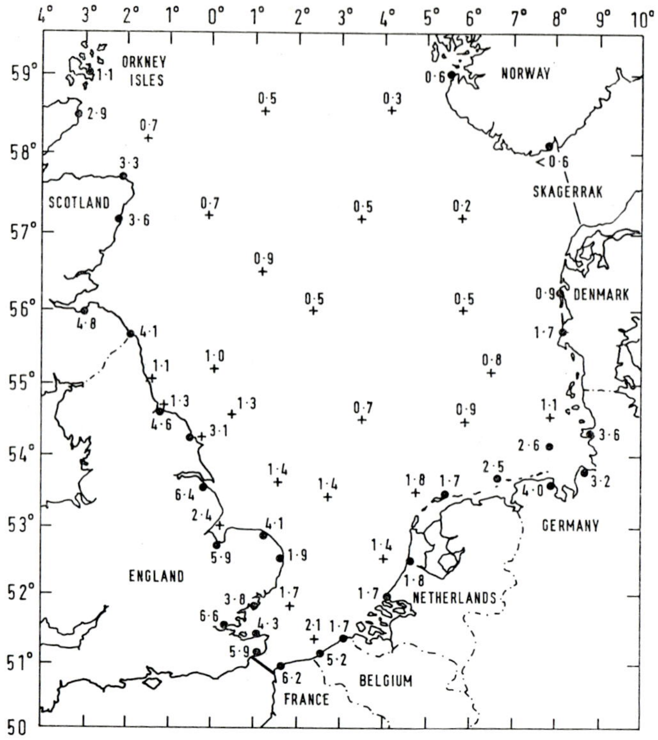

FIG. 3. Tidal range and current velocities in North Sea at mean spring tides. Range (metres) at coastal stations shown by circles, maximum velocity (knots) by crosses (*1 knot* = *0·515 m/s*)

temperature of the deeper northern parts is about 7–8°C, whilst those of the shallower eastern side range from 2 to 5°C. In summer, maximum temperatures in the eastern area reach 16–17°C, and in the western area 13–15°C, but in shallower coastal regions of the southern North Sea, particularly where there are exposed banks or restricted circulation, temperatures in excess of 20°C often occur.

The North Sea, which is heterogeneous with regard to density, has been divided into several regions by Dietrich[7]. Except close to estuaries, the coastal regions of the North Sea together with the whole of the Southern Bight and the German Bight are homogeneous with regard to temperature and salinity all the year round. Adjoining this region and extending north and west, the sea is layered for part or whole of the year as a result of its temperature structure.

Adjoining the Norwegian coast, in the Norwegian Deep, stratification due to salinity differences occurs, there being a strong bottom layer with regular changes in salinity as a result of the entry of deep Atlantic water. Between the Norwegian Deep and the thermally layered area is found a transition zone where the water column is only weakly stratified.

In addition to this general structure, marked salinity gradients occur in most of the larger estuaries and in adjoining waters. Where there is marked stratification, whether due to salinity or to temperature differences, vertical mixing is reduced, and the dilution and dispersal of low-density wastes discharged from pipelines are markedly reduced, and furthermore their transport may be influenced mainly by the movement of the surface water. The stability of the water column also has marked effects upon the productivity of the waters.

In considering the long-term effects of sewage disposal into the North Sea, reference must be made to the significance of mineral nutrients which are present in sewage. Nutrients are important in that they are required by phytoplankton at the bottom of the food web. In some circumstances excess nutrients are undesirable because they may lead to excessive growth of phytoplankton, which can lead to deoxygenation of sea water or to the production of undesirable toxic species of phytoplankton. The distribution of the major nutrients, nitrates and phosphates varies with time, position, and depth, but the best assessment can be made when the numbers and activities of planktonic organisms are at a minimum. According to Johnston[8] the North Sea may be divided into several clearly defined areas. The coastal waters of the United Kingdom, Holland, Belgium, Germany, and Denmark are relatively low in free inorganic nutrients, but there is evidence of enrichment from land run-off and sewage outfalls, particularly in the southern North Sea. The nutrient regimes of Norwegian waters are complex due to the influence of the Baltic, fresh-water run-off, the entry of deep Atlantic water, and the complex salinity-temperature structure. The north central area of the North Sea is rich in nutrients derived from the oceanic water, but, because of layering, only those nutrients in the surface waters are utilized. The south central North Sea is shallow and well mixed and has intermediate levels of nutrients.

In considering the nutrient input from rivers and coastal discharges, the high productivity of estuaries and coastal waters suggests that locally the quantities of nutrients thus discharged are probably significant. However, the total amount of nutrients entering the North Sea from these sources, when compared with that entering from other sources, appears to be of little significance except perhaps locally. It has been calculated[9] that of the total input of nitrogen, about 79 per cent enters with the inflow of Atlantic water, 17 per cent is derived from rivers and discharges, and the remainder comes from precipitation. In a more recent and probably more comprehensive survey of discharges made by the International Council for the Exploration of the Sea[10], sewage outfalls are shown to contribute only about 7 per cent of the total input. Of this amount, about half is discharged into the shallower southern part of the North Sea adjacent to dense areas of population, most of it through the larger river systems.

Effects of Discharges from Coastal Outfalls

The main effects that have to be considered when sewage is discharged into coastal waters are those influencing amenities and fisheries. Amenity considerations, which include those related to health, will be discussed in some detail later. Briefly, the coastal waters of the North Sea are used extensively for recreational purposes because of their close proximity to large areas of population. Bathing beaches are found widely distributed along all coasts of the North Sea, wherever conditions are suitable. However, because of the relatively low water temperatures and the incidence of wind, rain, and cloud in northern areas, the most important recreational areas are found in the southern part of the North Sea. Nevertheless, during the summer periods some beaches in all areas are subject to very heavy use. The main features of coastal discharges which may adversely affect these activities incude the presence of floating solids and surface slicks, the discoloration of water, and the psychological effect where discharge points are obvious. The public health aspects of sewage discharges and their effects on bathing must also be taken into account.

In addition to bathing, other water-sports such as skin-diving, boating, and water-skiing are important recreations for a large number of people, and the recreational value of beaches and other coastal areas to biologists, naturalists, bird-watchers, etc., are being increasingly recognized; all these activities can be affected by ill-sited outfalls.

According to the survey of discharges made by the International Council for the Exploration of the Sea[10], most of the sewage entering the North Sea is untreated before discharge, although there are regional variations; for instance, about 60 and 80 per cent respectively of discharges from the cities of Rotterdam and London are treated before discharge.

The main effects that sewage discharges may have on fisheries occur in coastal waters. Within recent years, trawl fisheries have moved into shallow coastal waters, and the physical presence of pipelines or discharge points, including any buoys used to mark them, are serious obstructions which must be avoided. However, probably the main effect of coastal sewage discharges is upon shellfisheries. Molluscan shellfisheries of high commercial value, for species such as oysters, mussels, cockles, clams, etc., occur mainly in the estuaries of eastern Britain, the Low Countries, and Denmark[11]. Many of these species become actively polluted by faecal bacteria at considerable distances from the points of discharge, and the risks of illness following consumption are particularly high because, in many countries, such shellfish are eaten raw or only lightly cooked. Where commercial shellfish beds are polluted by sewage, extra burdens are placed on the industry, for such shellfish have to be relaid in clean areas—which are not always available or convenient—or be sterilized or subjected to a process of purification in tanks before sale.

Molluscan shellfish may also become unacceptable as a result of the accumulation of biotoxins. These toxins are produced by certain well-defined species of dinoflagellates, a microscopic group of planktonic organisms. Although these organisms are widely distributed in small numbers, when conditions become suitable (high temperature and excessive nutrient) they may multiply rapidly, and the concentrations of toxin in filter-feeding molluscs may become unacceptable to man. Biotoxin accumulation has occurred in Norwegian fjords[12], off the north coast of Britain[13], and in Holland[14]. There is some evidence that, in the first of these cases, the development of dinoflagellates was associated with excessive nutrients derived from the discharge of sewage effluents. Although the precise conditions which precipitate development of large numbers of dinoflagellates are not known, the discharge of sewage effluents into restricted coastal waters adjacent to molluscan shellfish is a cause for concern.

Fisheries for crabs and lobsters are found in coastal regions of east Britain and of Norway. Such areas have hard bottoms of rock or mixtures of shell, sand, and gravel. The location of sewage discharges, particularly of untreated sewage, in positions which will allow the settlement of organic matter will clearly alter the nature of the bottom and make the habitat unsuitable for these species. Fisheries for shrimps occur in the shallower coastal regions where the bottom is softer. These fisheries may also be affected by gross pollution of coastal regions.

Juvenile stages of certain fish, notably the flatfish, sole and plaice, are found in very large numbers in many coastal and estuarine areas[11]. These areas provide shelter and a substrate bearing suitable food for small fish before they migrate into deeper waters. Two other important factors are temperature and salinity. The high water temperature which occurs in summer in many shallower coastal waters allows rapid growth of juvenile fish and encourages the production of benthic food organisms, whilst the low salinity, particularly in estuarine areas, cannot be tolerated by many of their predators. Thus coastal and estuarine waters, although yielding relatively few fish of commercial size, are nevertheless of importance for several valuable offshore fisheries.

Recent years have seen the development of long sea outfalls discharging untreated sewage at considerable distances from the coast. Economically such schemes are particularly attractive for large urban catchments, and very substantial flows may result. In addition to the interference that this may bring to fishing activities, and to the effect on the sea bed of any deposited material, fish and shellfish are very susceptible to pollutants contained in sewage. In northern Europe, sewage from large urban areas usually contains a substantial proportion of industrial wastes, and the quantities of metals contained in the effluent may in some localities be significant, leading to the accumulation of unacceptable levels in commercial species of fish and shellfish as a result of their incorporation in the food web.

Discussion

From this brief review, it is evident that the North Sea, and particularly its coastal region, has a wide range of uses, many of which can be affected by the discharge of sewage from coastal outfalls.

The use of coastal regions for recreation is particularly great in the southern North Sea, mainly because of the proximity of areas with a high density of population there. With increasing leisure, many more people spend their holidays in coastal areas, which are thus becoming of increasing economic value. It is therefore important that sewage discharge into coastal waters should be undertaken in such a way that the value of these areas is not reduced. The value of the fisheries of the North Sea is expanding each year, and, with a rising demand for fish from traditional fishing areas, increasing amounts of fish and shellfish are being taken from the shallower coastal waters. Shellfish of high value, some of which are of particular importance because they can be cultivated, are the species mainly at risk from coastal discharges. Coastal discharges may affect fisheries in many ways, including alteration of the habitat, pollution by faecal bacteria, and the risk of the development of species of plankton whose toxin may be transmitted to man. Further away from the coasts the long-term effects of increased nutrient supplies associated with sewage discharges cannot be ignored.

Despite the risks that sewage outfalls might pose, the observed effects in the North Sea are less than expected because of the tidal regime and the free exchange of water with the Atlantic Ocean. In those areas where there are vigorous tidal currents, i.e. on the British, French, and Belgian coasts, the dispersion of wastes from coastal outfalls is rapid, whereas on the eastern side, where tidal currents are relatively weak, the mixing and dispersal of wastes from coastal discharges is less effective.

Although the discharge of sewage can lead to undesirable effects near to outfalls, in view of the large exchange of water with the Atlantic it seems unlikely that sewage discharge will in the long run lead to a general deterioration of conditions throughout the North Sea. However, care is needed to ensure that the capacity of coastal waters to receive sewage wastes is not exceeded.

REFERENCES

1. McCave, I. N. Mud in the North Sea. In 'North Sea science' (ed. E. D. Goldberg). MIT Press, Cambridge, Mass., 1973, p. 75.
2. Hohn, R. On the climatology of the North Sea. In 'North Sea science' (ed. E. D. Goldberg). MIT Press, Cambridge, Mass., 1973, p. 183.
3. Lee, A. J. The currents and water masses of the North Sea. *Oceanogr. & mar. Biol.*, 1970, **8**, 33.

4. Lee, A. J., and Ramster, J. The hydrography of the North Sea. A review of our knowledge in relation to pollution problems. *Helgoländer wiss. Meeresunters.*, 1968, **17**, 44.
5. Hill, H. W. Currents and water masses. In 'North Sea science' (ed. E. D. Goldberg). MIT Press, Cambridge, Mass., 1973, p. 17.
6. Kalle, K. Der Einfluss des englischen Kustenwassers auf den Chemismus der Wasserkörper in der Sudlichen Nordsee. *Ber. dt. wiss. Kommn Meeresforsch.*, 1952, **13**, 130.
7. Dietrich, G. Die natürlichen Regionen von Nord- und Ostsee auf hydrographischer Grundlage. *Kieler Meeresforsch.*, 1950, **7**, 35.
8. Johnston, R. Nutrients and metals in the North Sea, In 'North Sea science' (ed. E. D. Goldberg). MIT Press, Cambridge, Mass., 1973, p. 293.
9. James, A., and Head, P. C. The discharge of nutrients from estuaries and their effect on primary productivity. In 'Marine pollution and sea life' (ed. M. Ruivo). Fishing News (Books) Ltd, West Byfleet, England, 1972, p. 163.
10. Report of Working Group for International Study of Pollution of the North Sea and its effects on living resources and their exploitation. International Council for Exploration of the Sea, Co-op. Res. Rep. Ser. A, No. 39, 1974.
11. Korringa, P. The edge of the North Sea as nursery ground and shellfish area. In 'North Sea science' (ed. E. D. Goldberg). MIT Press, Cambridge, Mass., 1973, p. 361.
12. Oftebro, T. Occurrence of paralytic shellfish poison in mussels (*Mytilus edulis* L.) from Norwegian waters 1964. *Nord. VetMed.*, 1965, **17**, 467.
13. Wood, P. C. Dinoflagellate crop in the North Sea. *Nature, Lond.*, 1968, **220**, 21.
14. Korringa, P., and Roscam, R. T. An unusual case of mussel poisoning. International Council for Exploration of the Sea, Shellfish Comm. Docum. 49, 1961.

Mr. Wood introduced his paper by saying that there was much difference of opinion about the effects of sewage disposal into the sea. He thought that this was due in part to failure by some to understand that many different conditions occurred in the areas of disposal. Furthermore, there were many different uses, or degree of use, to which these areas were put. He had therefore indicated in his paper those conditions occurring in the North Sea which had a major bearing on the long- and short-term effects of sewage disposal at sea. He outlined how these conditions were considerably more favourable than those occurring in the Mediterranean, off the western coast of North America, or in the Baltic; he showed that climatic conditions for bathing in the North Sea were considerably less favourable than elsewhere, so resulting in a significantly shorter immersion time.

DISCUSSION

Dr. J. F. Macqueen (U.K.) noted that Mr. Wood, Dr. Breittmayer, and Dr. Svannson had quoted the periods for water to be completely exchanged as 3 years for the North Sea, 100 years for the Mediterranean, and 36 years for the Baltic. He asked what information was available regarding the exchange period at points around the British coast, since such data could be of great importance in siting large discharges.

Mr. J. W. Talbot (U.K.) replied that there was little detailed information available, but investigations were in progress for some areas. The question was a difficult one since knowledge was needed of local circulations near the shore—these could be quite complex and might also exhibit considerable variability. Thus fairly extensive investigations might be necessary before a satisfactory understanding was reached.

Paper No. **2**

The Mediterranean

J. Aubert and J.-Ph. Breittmayer

C.E.R.B.O.M. (I.N.S.E.R.M.), Nice, France

The Mediterranean is a special field, distinct from other seas of the world (such as epicontinental, land-locked, and inland seas) by its geography and geology. In fact, it is the model of mediterranean seas according to the classical definition given by geographers: sea introduced inside the land and communicating with the oceans only by narrow and shallow straits or sounds. Its topography, hydrology, currents, and physico-chemical characteristics make the Mediterranean a special and wide receptacle. Warm and saline, almost tideless, and with a limited continental shelf, this sea is nearly without movement, so that the bacterial pollution remains there and is, more than anywhere else, a coastal phenomenon.

Before studying the bacteriological state of the Mediterranean shores, we shall examine its topography, hydrological structure, and water circulation.

Physical Characteristics

The Mediterranean Sea is formed by two basins:

(1) The western basin, bordered by old massifs, with coasts which are generally steep and rocky—except in the Lion Gulf and for the swampy coasts of Italy and North Africa.

(2) The eastern basin, which is more complex, with the Adriatic Sea in the north, the Black Sea and the Aegean Sea in the north-east, and the properly-called Mediterranean Sea extending between North Africa, Israel, Syria, and Turkey.

The water is renewed through the Strait of Gibraltar and, according to the oceanographers[1], this renewal occurs every century.

The hydrological structure makes it possible to distinguish a superficial layer, an intermediate one, and a deep one. The evolution of these bodies of water shows that the Atlantic water (superficial water) entering at Gibraltar has a general movement to the east, whereas the Levantine and Adriatic waters (intermediate occidental waters) have a general movement to the west.

In the superficial water, there is, during much of the year, an area of physical discontinuity of temperature (thermocline), generally at a depth of 20–50 m, and this acts as a real barrier to vertical dispersion. In addition, it should be pointed out that, with the exceptions of the Adriatic, the Lion Gulf, and the vicinities of Sicily and Tunisia, the continental shelf (where a low depth is generally favourable to an intensification of mixing of terrestrial waters with the marine body of waters) is virtually or completely absent. This is especially the case for the Spanish, French, and Italian coasts. All these hydrological characteristics result in the fact that the bacterial pollution in the Mediterranean Sea remains a phenomenon confined to neritic areas, rarely diffusing farther than a few kilometres off the coasts.

State of Bacterial Pollution in the Mediterranean Sea

By considering the littoral population (about 100 million) it is possible to estimate the sewage discharge as some 3×10^7 t/day; to this must be added about 2·5 km^3/day of polluting discharges from inland waters and rivers.

The evaluation established by the Group of Experts[2] in 1972 gives the data shown in Table 1 for the organic load from different countries.

Most of the waste waters from the Mediterranean coastal populations are discharged directly into the sea and cause a continual microbial contamination of the shores. We cannot, in a few pages, give a detailed report of the bacteriological state of the whole of the Mediterranean shores. Thus, we give only a general outline of this state, together with some precise data, taking into account not only microbiological data but also local meteorological, geographical, physico-chemical, and biological conditions.

TABLE 1. Organic load of domestic sewage discharged directly into the Mediterranean Sea or through rivers[2] (t/year)

	BOD[a]	Phosphorus	BOD per km coastline	P per km coastline
Spain	130 000	5 900	60	2·7
NW basin (France)	360 000	16 000	336	15·0
Italy	400 000	18 000	61	2·7
Yugoslavia	17 800	800	27	1·2
Malta	8 000	320	67	2·7
Greece	100 000	4 500	37	1·7
Turkey[b]	100 000	4 500	36	1·6
Cyprus	9 600	430	15	0·7
Lebanon	31 250	1 250	149	6·0
Israel	32 000	1 400	145	6·5

[a] Calculated on assumption that yearly BOD load in domestic sewage is 20–25 kg/person
[b] Excluding the Black Sea coast

Generally, the most polluted shores correspond to the most highly populated and industrialized areas, with worst conditions near large towns. Microbial pollution is less important in the eastern Mediterranean basin, but unfortunately we have little relevant information; the studies carried out concern essentially the human-activity zones which have important effects on the sea. According to Kouyoumjian[3] the Lebanese shores are quite polluted, the larger towns being mainly on the coast. Near Beirut from 30 000 to 1·6 million terrestrial bacteria per litre were found. The Israeli coast near Tel-Aviv is also very polluted, 30 000 m^3/day of sewage being discharged there as far back as 1963 and resulting in badly polluted beaches (several thousand to several million bacteria per litre). Shuval points out[4] that this state has obviously improved since the installation of a large outfall discharging urban waste waters 800 m offshore.

Hardly anything is known about the state of pollution of the Greek, Turkish, Egyptian, and Tunisian coasts.

For the western basin there are more complete data on the polluting discharges and their effects. Pollution is present in a few enclosed areas along the Algerian and Tunisian coasts, is of relatively little importance along the Spanish and Yugoslavian shores, but is sometimes very serious along the French and Italian coasts. Ambrosioni made a sanitary survey along the whole of the Italian shore in 1970 and his report[5] shows that, for 90 per cent of localities, the sewage is discharged to the sea without treatment.

The highest levels of pollution are found along the Ligurian coast, with an important maximum at Genoa—due, in part, to industrial wastes and the petroleum harbour. Conditions appear to be better along the coasts of Tuscany, Latium, and Campagna, except near the large towns (Leghorn, Piombino, Portoferraio, Fiumicino, Osti, Naples, and Salerno). The situation at Naples is particularly serious as the waters are considerably polluted. Paoletti[6] studied the coastline at Cumes, polluted by sewage from the main outfall which serves the whole of the suburbs of Naples. This outfall, discharging sewage containing 2×10^{10} bacteria per litre (including $9{\cdot}8 \times 10^8$ coliforms and an average of 4×10^4 salmonellae) with a flow of 5 m^3/s, pollutes the coastline of Cumes drastically and continuously over a distance of more than 6 km.

The coasts of Calabria, both on the Tyrrhenian and Ionian sides, are not very polluted because of the low population density and the absence of industry.

Along the Adriatic coast, intense bacterial pollution can again be found from the Yugoslavian border to the east of Romagnia. The situation is sometimes very serious there, especially in the vicinity of Romagnia which has shallow waters and a high degree of industrialization. However, many developments have been carried out in this part of Italy (as well as in others) to improve the situation.

The Gulf of Trieste is especially polluted: Majori[7] reports the presence of areas with more than 30 000 *E. coli* per litre, a bacterial density of faecal origin found again in mytiliculture beds, north of the Gulf, in which many salmonellae were found.

Conditions are far better along the coasts of Marches, Abruzzi, and Apuglia—with the exception of industrialized harbour areas (Pesaro, Ancona, Falconara, Bari, Polignano, and Gallipoli). In Bari the situation is singularly complicated by the presence of hydrocarbons resulting from the activity of the mineral-oil harbour.

This geographically localized aspect of bacterial pollution is found again, together with alternation of polluted areas and of portions of coast devoid of any contamination. The intensity of the factors affecting dispersion of bacteria is basic here, and the Strait of Messina provides an outstanding example of this aspect. Two hydrological phenomena occur simultaneously in the Strait: the waters are mixed by a strong northerly current and the deep waters well-up along the continental slope south of the Strait. Despite the strong pollution due to the sewage from Messina, Genovese[8] has shown the complete absence of enteric bacteria in the heart of the Strait, whereas he found more than 250 000 coliforms per litre in certain places on the coast near Messina. Furthermore, the Sicilian coast is generally polluted, especially near the main towns (Siracusa, Catania, and Messina). On the other hand, the situation is good, if not excellent, along the Sardinian coast, the only polluted areas being a few harbours, especially Cagliari, where the waters contain an average of 1000 to 1 million coliforms per litre, the bacterial density decreasing very rapidly in the marginal areas.

The coasts of Spain mainly face the open sea, and the eastern shores (except in areas of cliffs) have large sandy beaches. The bacterial contamination is generally slight, except near large cities such as Valencia, Tarragona, and particularly Barcelona[9] where the numbers of *E. coli* are very high (often 30 000 to 50 000 per 100 ml).

The French Mediterranean coasts have been carefully studied and checked over by our laboratory[10]: at many points on the coasts of Languedoc and Roussillon, with a flat coastline, the concentration of test organisms of faecal contamination is low (from a few hundred to a few thousand coliforms per litre). Even the presence of such towns as Sète does not much modify this state. Only a few places are really polluted, generally near the mouths of important rivers carrying domestic wastes—as is the case at Perpignan, Narbonne, Béziers, Montpellier, and Aigues-Mortes. Thus we find 950 000 coliforms (with 3000 *E. coli*) per litre near the mouth of the Aude, and 250 000 coliforms (with 4000 *E. coli*) per litre at that of the Hérault. The greatest pollution was found in Crau du Roi (8 million coliforms and 20 000 *E. coli* per litre). There are also some contaminated lagoons, such as the Etang de Thau (3000 coliforms per litre).

In the district of Marseilles, the pollution becomes more intense the closer we get to the city. The discharge area of the main sewer, between the Isle of Riou and the coast, south of the roadstead, and particularly the outfall area of the Huveaune are highly polluted zones. On either side of this area the pollution is slighter (coliforms often less than 100 per litre) though the pollution of marginal areas is still worrying—1000 coliforms per litre in Morgiou and 12 000 in Sormiou with 200 *E. coli*, and this pollution (caused by the Marseilles sewage) appreciably affects the ecology of the area. Bellan[11] reported large variations in the populations near the points of discharge of domestic sewage into the Mediterranean Sea: new colonies are established in vast concentric areas around the discharge point, the closest to it being semi-abiotic or with very few species among which are the annelids *Capitella capitata* and *Soolopsis fuliginosa* which are rightly considered as 'indicator species of pollution'. Because of the prevailing winds and currents, which are to the south and east, the edge of the Massif de l'Estaque is almost unpolluted. The north of the Etang de Berre is exceptionally polluted, the bacterial concentrations being higher than anywhere else along the French Mediterranean coast (40 million coliforms, 3 million *E. coli*, and 700 000 *Streptococcus faecalis* per litre); all the conditions of eutrophication and dystrophy are present, with the disappearance of vegetation and fauna. The water of this lagoon, discharging through the Caronte channel, maintains the high level of pollution found in Port de Bouc.

Around the city of Toulon, the bacteriological analyses showed that, despite generally satisfactory conditions, a few polluted areas can be found, usually indicating the discharge of waste waters in unacceptable conditions. These points of pollution become more and more numerous on approaching the Riviera, which can easily be explained by the steady increase eastwards in the density of housing along the coast as far down as the Italian border. Corsica is generally in a good sanitary state, except near cities and agglomerations which discharge sewage through very short pipes.

Up to 1969, the bacterial pollution of the French coast became more and more serious, the bacterial levels increasing regularly with the growing population. It was thus necessary to carry out a complete check of all the sewage installations. At Nice the whole of the sewage was discharged at a distance of 149 m offshore (east of the airport) with a flow of 3–4 m^3/s. In the Alpes Maritimes we often found more than 10^5 aerobic bacteria per ml.

From a very general point of view, the studies carried out before 1969 led to the following conclusions (which could be extended to all the Mediterranean coasts):

(1) The bacterial pollution is increasing with population along the coasts. Indeed, the sea water near large towns contains a very large enteric contamination extending, in the worst cases, for

6–7 km. Even areas in the proximity of short sewage outfalls from minor localities with summer tourist populations and those near estuaries are very polluted.

(2) The bacterial diffusion is always in the direction of the dominant currents; for example, the Ligurian current results in a constant drift of pollutants from east to west.

(3) Comparison of pollution of marine areas and the corresponding seashore always shows the highest contamination to be at the shore.

(4) The bacterial count is considerably higher in areas which are naturally more stable or artificially enclosed.

Changes in Pollution on the French Coast since 1969

Since 1969, the construction of new outfalls has influenced the degree of contamination of the French coast by domestic sewage. At the request of the Ministry of Equipment, the C.E.R.B.O.M. is studying the replacement of old outfalls. The problem submitted was to consider for each case the extent of the contaminated marine area which would result from a future outfall, and to determine the optimum point of future discharge, with or without treatment being provided, in such a manner that the salubrity of the beaches and the shellfish beds would be protected.

Different outlets were calculated with the mathematical schedule proposed by M. Aubert and N. Desirotte[12], taking into account the hydrological conditions of each case. This schedule is based on the equation

$$\log C = \log (C_0/h) - \frac{\beta\gamma}{1{\cdot}1} t \tag{1}$$

where C_0 is the bacterial concentration in the sewage before discharge, and C the concentration at time t (hours) after discharge from an outlet at a depth h (metres); β is a coefficient which depends on flow and is unity for a discharge of 1 m^3/s; γ is a coefficient of the bactericidal action of sea water and is unity for the Mediterranean (comparative measurements in other seas show that γ can vary, so that it is important for it to be included in the mathematical model).

To date, more than 30 new outfalls have been created (or have replaced old ones) taking the mathematical laws of bacterial diffusion into account in determining the discharge points. In particular we may mention the discharges from Banyuls, Toulon, Cannes, and Nice as the principal ones. Generally very long (at times several kilometres), they take the wastes far away from the coast, so that whatever the meteorological conditions the bacterial pollutants should not reach the shore alive. The discharge into the sea is sometimes under the thermocline—this is especially the case for the outlet at Banyuls—which may increase the possibility of dispersion of pollutant bacteria within the body of the water. This deep renewal of sewage outfalls has resulted in a great improvement in the sanitary condition of the French Mediterranean coasts, both continental and Corsican. As an example, we shall but give the results obtained in a study made by the C.E.R.B.O.M. during the summer of 1972 on the microbiological state of 36 seaside resorts located along the coasts of the Alpes Maritimes where many improvements were introduced to the network of waste-water discharges. This work made it possible to obtain a sanitary evaluation of this part of the coast and to compare the data with those collected in the same area before the setting up of new devices for improved sanitation or the installation of harbours or marinas.

The most representative point of the state of contamination being located 50 m offshore, our observations particularly concerned the variations in bacterial density at this position. We have found that:

(1) In only one area did the coliform count exceed 1000 per 100 ml; the exception was at Golfe Juan where part of the sewage is discharged at the shore and where the River Vallauris (draining this urban centre) enters the sea.

(2) The highest coliform counts were found at Roquebrune, Sainte Hélène, Saint Laurent du Var, Golfe Juan, and Madelieu.

(3) Comparison of the coliform counts observed during the 1972 campaign with those observed in 1966 brings out the following facts:

  (a) For the beaches between the Cap d'Antibes and the Cap de Nice, the coliform counts found in 1972 were substantially lower than in 1966; the setting up of new sanitation installations is especially significant around the Bay of Nice.

  (b) Between the Cap de Nice and the Italian border the coliform counts were higher in 1972 than in 1966, and an increase in contamination was apparent at certain other places—generally those coastal areas where pleasure harbours or coastal installations have been created, thus entailing a seasonal population increase. These are the areas of Menton Garavan, Roquebrune, Beaulieu, Saint Jean Cap Ferrat, Saint Laurent du Var, Golfe Juan, Mandelieu, and La Napoule. For some of them the pollution is due to a total lack of sanitation installations.

Thus, except at two points the water of the coasts of the Alpes Maritimes is of good or average quality according to the nomenclature of bacteriological criteria for quality of bathing waters as they are defined in France (*J. O.* July 1969)

Similar improvements are being made at the present time in all the countries bordering the Mediterranean Sea: either by treatment of sewage before discharge, or by drainage, collection, and discharge of sewage off the coast where mixing and dispersion are far more effective. Nevertheless, many points still remain very polluted.

Conclusion

It thus appears that bacterial contamination is present nearly everywhere in the Mediterranean coastal regions, but that it is not very high in relation to the size of the Mediterranean.

Areas near major urban centres which discharge sewage directly to the marine environment are responsible for the very substantial enteric contamination in some localities, and this contamination is increased by industrial wastes which can change the natural biological balance—as in the north-west and north Adriatic basin.

However, the disposal techniques which take into account the hydrology of the area and make full use of dispersion in the sea show that the existing conditions could be greatly improved.

REFERENCES

1. LACOMBE, H., and TCHERNIA, P. Hydrographie de la Méditerranée. In 'Consultation sur la protection des ressources biologiques et des pêches contre la pollution en Méditerranée'. Rome, February 1974, 12 pp.
2. GROUP OF EXPERTS ON MARINE POLLUTION (GFCM/ICSEM). Review on the state of pollution in the Mediterranean Sea. In 'Marine pollution and sea life' (ed. M. Ruivo). Fishing News (Books) Ltd, West Byfleet, England, 1972, pp. 28–32.
3. KOUYOUMJIAN, H. A preliminary study of bacterial pollution and physiography of Beirut coastal waters. *Revue int. Océanogr. méd.*, 1972, **26**, 5–26.
4. SHUVAL, H. I., COHEN, N., and FURER, Y. Y. The dispersion of bacterial pollution along the Tel-Aviv shore. *Revue int. Océanogr. méd.*, 1968, **9**, 107–121.
5. AMBROSIONI, P., BISBINI, P., and MARINELLI, M. Enquêtes sur l'état de pollution des eaux côtières de la péninsule italienne. *Revue int. Océanogr. méd.*, 1970, **18–19**, 225–234.
6. PAOLETTI, A. Pollution fécale du littoral de Cumes et considérations sur le pouvoir d'auto-épuration du milieu marin. *Revue int. Océanogr. méd.*, 1966, **1**, 44–55.
7. MAJORI, L. Research experiments on the pollution of the waters of the upper Adriatic. *Int. Conf. Oil Pollut. Sea, Rome, 1968*, pp. 1–32.
8. GENOVESE, S., BRUNI, V., and MACRI, G. Activité microbienne endogène et bactéries contaminates dans la mer. *Revue int. Océanogr. méd.*, 1968, **9**, 73–81.
9. SUÑER, J., and PIÑOL, J. Coliform bacteriophages and marine water contamination. *Proc. 3rd int. Conf. Wat. Pollut. Res., Munich, 1966*, Vol 3, pp. 105–120.
10. AUBERT, M., AUBERT, J., and DANIEL, S. Côtes de France. Inventaire national de la pollution bactérienne des eaux littorales. Vol 4—Méditerranée. *Revue int. Océanogr. méd.*, 1968, Suppl., 167 pp.
11. BELLAN, G. Pollution et peuplements benthiques sur substrat meuble dans la région de Marseille. I—Le secteur de Cortiou. II—L'ensemble portuaire marseillais. *Revue int. Océanogr. méd.*, 1967, **6–7**, 53–87, and **8**, 51–95.
12. AUBERT, M., and DESIROTTE, N. Théorie formalisée de la diffusion bactérienne. *Revue int. Océanogr. méd.*, 1968, **12**, 5–48.

Presenting the paper, Dr. Breittmayer read a shortened version of it. In so doing he emphasized the special characteristics of the Mediterranean—the virtual absence of tides, the existence of a thermocline for much of the year, and the narrow continental shelf.

Most of the waste waters from the Mediterranean coastal populations were discharged directly into the sea. The most polluted shores corresponded to the most populated and industrialized areas where the enteric contamination was increased by industrial wastes which could change the natural biological balance. The worst conditions appeared near large industrialized towns where the ecology of the areas around the discharge points was appreciably affected; new colonies were established in concentric areas, the closest to the centre being semi-abiotic. Indicator species of pollution, such as *Capitella capitata*, were found there. For a given amount of sewage the pollution was considerably higher in artificially enclosed areas.

In France, since 1969, the construction of new long outfalls had decreased the pollution of the nearby beaches, thus meeting the national criteria of good or average quality bathing water. Nevertheless, not all the necessary improvements had yet been made, and many points still remained very polluted.

DISCUSSION

Prof. G. Shelef (Israel), who was unable to attend the Symposium but submitted a written contribution, said that the Authors had correctly indicated the relative lack of data for the eastern basin of the Mediterranean and had thus emphasized the situation in the western basin. Israel provided a typical example of the eastern basin, particularly of the shoreline from the eastern part of North Africa to the southern part of Turkey. Geologically, this shoreline and its continental shelf were of the 'emerged' type—in contrast to those of many parts of the north-western Mediterranean as well as the Adriatic and Aegean seas which were of the 'submerged' type.

The emerged shoreline was typified by a moderately sloping continental shelf extending 15–25 km. The Israeli shelf had a slope of 0·6–1·2 per cent, so that any marine sewage outfall should be quite long to reach a depth which would give sufficient initial dilution. The emerged shelf was usually sandy or alluvial, which eased outfall laying but could create problems of outfall anchoring and diffuser clogging. Thermal and salinity stratification were minimal throughout the continental shelf. The absence of a marked thermocline for most of the year in Israeli waters[13] ruled out any significant trapping of sewage, and thus increased the risk of sewage-borne floatables being carried towards the shore.

Coliform die-off rates seemed to be quite high[14] and T_{90} values of $1\frac{1}{2}$–4 h were used in outfall design. These rapid rates were probably due to high temperature and the abundance of solar irradiation in this area.

In the past two decades there had been rapid increases in both residential and tourist populations along the coastal areas in most of the eastern Mediterranean countries. Discharge from the Tel-Aviv outfalls had increased from the 30 000 m^3/day (quoted in the paper) in 1963 to more than 160 000 m^3/day in 1973, and resulted in the closing of three of the seven Tel-Aviv beaches as the coliform standard of 2400 per 100 ml (80 percentile) had been surpassed[15]. The municipal authorities were now constructing a rather sophisticated sewage treatment plant to reclaim the waste water for re-use, thus eliminating the sewage discharge to the sea. The disposal of excess sludge by a long small-diameter outfall was being investigated, it being assumed that the sludge would rapidly be consumed by marine organisms.

Mr. J. A. Wakefield (U.K.) asked whether British authorities would be able to make use of the type of bacteriological survey referred to in the paper. In reply, Mr. A. L. H. Gameson (U.K.) said that the French surveys provided an extensive but rather superficial indication of bacterial pollution along the whole coast. Although such surveys might pin-point the areas of greatest pollution, they could not be used in the grading of British beaches (as he suspected that Mr. Wakefield would like), because it would not be practicable to analyse sufficient samples to obtain representative average or extreme values—owing to the enormous variability in bacterial counts which were found over periods of hours, days, or months, and which depended on the innumerable factors that unpredictably affected the bacterial level[16].

Mr. D. E. Cannon (U.K.) asked for enlightenment on certain aspects of the Aubert–Desirotte formula[12] which appeared to be somewhat unconventional. He asked for confirmation that the coefficient related to the flow from a single diffuser port rather than to the total flow. The formula did not appear to take cognizance of the work of Rawn, Bowerman, and Brooks[17] and others on initial dilution; the graphical representation of the results of their work and that of Abraham[18] showed a fairly complex relation between jet diameter, discharge rate, and depth. He understood that, in the formula, initial dilution was numerically equal to the depth, h (in metres), of the outlet below the sea surface. It might be that initial dilution and surface dispersion could not be considered as two separate stages in the formula and that initial dilution was something more than h. He also enquired whether the Authors had values of γ for seas other than the Mediterranean; he presumed that this was not exactly the bacterial mortality rate constant, but must bear some relation to it.

Finally, Mr. Cannon compared results obtained from the Aubert–Desirotte formula[12] with those from Abraham's curves for initial dilution[18], Brooks' equation for surface dispersion[19], and a T_{90} of $1\frac{1}{2}$ h. Other criteria being equal, the former method gave a predicted outfall length of 420m whereas the latter gave 800m.

Dr. J. A. French (U.S.A.) noted that, for much of the year, there was a thermocline barring vertical dispersion. This would effectively prevent the upward travel of sewage, with its bacterial and viral contaminants, but it would also impede the downward transport of the oxygen needed to satisfy the oxygen demand of the sewage. He asked if there was adequate dissolved oxygen in the ambient waters below the thermocline.

REPLY TO DISCUSSION

Dr. Breittmayer, replying to Mr. Cannon, confirmed that β was a function of the flow from a single port. He went on to point out that the formula was not a means of exact calculation of the diffusion of bacteria; it was simply used in predicting the outfall length required to give a bacterial count less than any particular value at the beach, and it had been derived statistically to take account of worst conditions rather than the actual conditions on a particular day. With regard to the magnitude of γ, he said that for other seas they compared the ratio of the decrease in bacterial counts to the decrease in Rhodamine concentration with the corresponding ratio in the Mediterranean.

Commenting on Dr. French's contribution, Dr. Breittmayer said there were several factors (such as planktonic populations, oxygen transfer rates, and diffusivity) which made the dispersion of bacteria less efficient under the thermocline. In many cases, however, because of the configuration of the Mediterranean shores, outlets would be too close to the beach if they discharged at depths of less than 20–30 m.

ADDITIONAL REFERENCES

13. Balashe-Yalon Consulting Engineers Ltd. Design of marine wastewater outfall for the city of Netaniya. 1974.

14. Gilath, C., Shuval, H. I., Yoshpe-Purer, Y., and Cohen, N. The use of radioisotope tracer in the study of disposal and inactivation of bacteria discharged in coastal waters. In 'Developments in water quality research' (ed. H. I. Shuval). Ann Arbor–Humphrey Science Press, Ann Arbor, Mich., 1970, p. 296.

15. Shelef, G. Criteria for marine waste disposal in Israel. In 'Marine pollution and marine waste disposal' (ed. E. A. Pearson and E. d. F. Frangipane). Pergamon Press, 1975 (*in press*).

16. Gameson, A. L. H. Experiences on the British coast. In 'Marine pollution and marine waste disposal' (ed. E. A. Pearson and E. d. F. Frangipane). Pergamon Press, 1975 (*in press*).

17. Rawn, A. M., Bowerman, F. R., and Brooks, N. H. Diffusers for disposal of sewage in sea water. *J. sanit. Engng Div. Am. Soc. civ. Engrs*, 1960, **86** (SA2), 65.

18. Abraham, G. Jet diffusion in stagnant ambient fluid. Delft Hydraulics Laboratory, Publ. No. 29, 1963.

19. Brooks, N. H. Diffusion of sewage effluent in an ocean current. *Proc. 1st int. Conf. Waste Disposal mar. Envir.*, Univ. Calif., Pergamon Press, 1960, p. 246.

Paper No. **3**

Regional Problems with Sea Outfall Disposal of Sewage on the Coasts of the United States

Gerald Berg
National Environmental Research Center
U.S. Environmental Protection Agency
Cincinnati, Ohio 45268

Almost half the 50 United States are contiguous to the sea. Within them, on the ocean's edge, are some of that nation's largest population centers, and within these, some of that nation's most difficult sewage-disposal problems. Most of the liquid wastes from the large coastal urban centers are discharged through outfalls to the oceans. Tonnes of solids, even from inland cities, are barged out to sea daily, and vast quantities of sewage are disposed of into rivers that eventually empty into the great seas. To what extent these waters that constitute most of the earth's surface can assimilate and recycle the garbage of man's existence may never be known completely.

By and large, the problems that outfall disposal of sewage to the seas brings about have been matters of major concern only in circumscribed areas. Sometimes important concern has been generated by economic impact as with the pollution of shellfish beds or recreational swimming areas, sometimes by the obvious unsightliness of pollution as in the appearance of floatables, sometimes by the crescendo of indignant environmentalists offended by man's onslaughts on the macrocosm.

Whatever the concern, investigations of effect usually are undertaken when the pollution is already extensive, often too late to determine the conditions that were normal before the assault on the environment began. We might still surmise what had been normal before pollution beset an area from what is normal elsewhere, but we may never surely know. Of course, we might ask if perhaps ocean water newly enriched by nutrients in sewage might make a finer menstruum for aquatic life than the sea itself. Sometimes, it seems to. Indeed, not all wastes are pollutants. Yet, the piper must be paid.

North East Coast

New England

In some of the coastal north-eastern United States, untreated municipal sewage is discharged through ocean outfalls to the sea. Although many larger communities are moving to secondary treatment, closed swimming beaches, even in sophisticated cities such as Boston, still occur.

Connecticut is the only New England state that treats all sewage discharged to the ocean. Here, all wastewaters receive primary treatment, and many receive secondary treatment before disposal. Massachusetts still discharges some raw sewage, but is increasing its commitment to primary treatment and secondary treatment. Rhode Island disposes some untreated sewage to the sea, but most of its discharge is primary effluent. Maine, however, is a state of isolated rural communities, many of which find treatment costs difficult to cope with. Here domestic discharges to the ocean, from small coastal islands and some mainland communities where terrain is unsuitable for septic tanks, aggravate the problem. More than 80 per cent of the terrain of Maine appears to be unsuitable for subsurface discharge.

With total coliform counts exceeding 700/100 ml, more than 60 estuaries in New England are closed to shellfishing. In all, more than 930 km^2 of New England coastal waters are closed to shellfish harvesting as a result of discharge to the Atlantic Ocean of untreated and partially treated municipal sewage, and the direct and indirect discharge to the ocean of untreated and poorly treated sewage from private disposal systems. Shellfish resources of proven and demonstrated areas of coastal waters (where tests have shown depuration will produce acceptable bacterial quality) can be harvested only with special approval from the health departments of the states. Thus, shellfish taken from approved areas with total coliform counts of 70–700/100 ml must be depurated before thay can be marketed.

Industrial waste discharge through outfalls to the ocean also constitutes an important pollution problem to the New England coast. Thermal discharges from increasing numbers of power plants have altered the aquatic ecology of some bay areas but, with little known about the aquatic life of these areas during the period preceding power-plant operation, the true extent of damage cannot be determined.

Power-plant discharges of chlorine also pose serious pollution problems. Chloramines, formed from the reaction of chlorine with ammonia nitrogen or with organic nitrogenous compounds, are highly toxic to fish and to the aquatic biota that constitutes important fish foods. Moreover, chlorinated biphenyls which may also form from chlorination of certain wastes, and other chlorinated organics similarly formed, may be carcinogenic or toxic to man and to other living forms.

Heavy metals also are discharged through outfalls into tidal rivers emptying into Long Island Sound by the metal-plating industry concentrated in southern New England. These discharges are afforded little treatment by the numerous companies, many small, that constitute much of the industry. To the north, there occur a few mines, some abandoned, from which zinc and lead leach into the bays and inlets along the coast of Maine. These metals are found in high concentrations in the biota and in the bottom sediments of the receiving waters.

Paper-mill wastes also constitute an important marine problem, especially in isolated estuaries along the coast of Maine where the high organic loads discharged by the mills may serve as rich menstruums for bacterial pathogens such as *Klebsiella pneumoniae* and as sludge blankets that sometimes smother beds of shellfish. Discharge permits, now required by Federal law, demand 85–90 per cent solids and BOD removals from paper-mill wastes prior to ocean outfall disposal and may eventually mitigate this problem.

New York Bight

New York's access to the Atlantic Ocean is limited to the area of the New York Bight. Here metropolitan New York–New Jersey discharges $7{\cdot}3 \times 10^6$ m^3 of sewage daily into New York Harbor and its adjacent bays. Of these 7·3 million m^3, 1·8 are untreated, 2·0 undergo primary treatment, and 3·5 undergo biological treatment before discharge to the sea. During heavy rains, 90 per cent of the sewage may bypass treatment, washing directly into the bight through the combined sewer systems. Most of the sludge solids settled from the treated domestic wastes are barged out to sea and disposed of into the ocean 19 km from the entrance to the harbor of New York. But, with an anticipated tripling of sludge volume when all treatment plants have inaugurated secondary treatment (as they must by mid-1977 to comply with Federal law), alternative methods for sludge disposal are under intensive consideration.

The raw sewage discharged directly to upper New York Bay deposits $1{\cdot}7 \times 10^6$ m^3 of sludge in the harbor area annually, and this sludge is dredged and deposited at a site just 8 km from the recreation beaches along the New Jersey coast. Moreover, along this stretch of New Jersey coastline, with its numerous sandy swimming beaches, 32 municipal sewage treatment plants discharge 150 000 m^3 of wastes daily 300 m or less into the Atlantic Ocean. Many of these treatment plants are primary. Until enjoined by the Federal courts in February 1972, these communities collected their treatment sludges in settling and storage tanks during and preceding the warmer resort-season months and then discharged the collected sludge with effluent through the effluent outfalls during the winter season. Today, the sludge is barged 19 km out to sea to approved discharge sites.

Within the environs of the New York Bight, there is discharged 70 per cent of all the sludge dumped by the United States. Almost 60 per cent of the industrial wastes disposed to the seas by this country are discharged at these approved sites. More than 90 per cent of these wastes are acids disposed of 24 km out to sea. The remainder of the industrial wastes is today barged 170 km to the edge of the continental shelf. Alternative methods for disposal of industrial wastes are being intensively sought.

Central East Coast

Along the Atlantic coast from southern New Jersey to North Carolina and south to Florida, population and industrial intensities diminish and the pollution problems resulting from ocean outfalls appear to be less intense, even in the waters of the Chesapeake and Delaware Bays.

The waters of the heavily polluted Potomac River empty into the Chesapeake Bay, however, and with some partially treated domestic and industrial wastes discharged into the bay through outfalls, this pollution has closed some shellfish harvest areas such as those in Cape Harbor. Other bay areas, solely because of their proximity to sewage outfalls, are also closed to commercial fishing and shellfishing. Occasionally, domestic sewage pollution closes Virginia Beach and other beach areas.

South East Coast

Decades of a continuous inflow of vacationers and retirees to the Atlantic shore of southern Florida have created a serious waste-disposal problem in this area. From a rapidly urbanizing 16-km-wide strip of coast from Palm Beach to the Florida Keys, bordered on the west by the Everglades, increasing volumes of sewage are being discharged through 10 outfalls to the Atlantic Ocean. Most of this sewage is raw. Discharge from only two of these outfalls is secondary effluent, and discharge from only one other is primary effluent. Most of these outfalls approach or extend beyond the continental shelf, which ends only about 2 km east of the economically important recreational beaches and fishing waters of southern Florida.

Dispersion of sewage discharged into the sea along this narrow strip of coastline is determined primarily by the Florida current (that part of the Gulf Stream that flows through the Straits of Florida) but is also influenced somewhat by winds and tides. Toward-shore eddies off the powerful Gulf Stream (which usually carries away much of the discharge that reaches into it) occasionally push plumes of sewage towards the beaches.

The discharges from the outfalls are augmented by drainage from numerous canals that carry domestic and small amounts of industrial wastes, together with herbicides and pesticides, from the rural and inland urban areas to the sea.

Along this stretch of coast at the outfall serving the City of Hollywood, studies indicate that marine plankton have been somewhat affected by the discharge of primary effluent. Some reduction in numbers of plankton has occurred in the boil. Preliminary studies indicate that some changes apparently have occurred in bottom biota at the point of discharge also, but that none have occurred beyond 30 m from the outfall. At the cities of Pompano Beach and Delray, where raw sewage is discharged, increased numbers of benthic species normally associated with sewage have occurred at the outfalls.

Fin-rot affects many fish at the Hollywood outfall, but the disease occurs in fish found in cleaner waters as well. It has not been possible to establish cause and effect. Fish often congregate at the outfall, apparently finding sewage a good source of nutrients. Spiny sea urchins also seem to seek the nutrients of the outfall.

Large quantitites of viruses have been recovered from the primary effluent of the Hollywood treatment plant, from the secondary effluent at the Port Everglades treatment plant, and from other sites as well. Viruses always constitute a potential hazard to recreationalists exposed to them, but they occur in virtually all domestic sewage and survive in some numbers even after secondary treatment. Moreover, viruses survive for long periods even in sea water which generally constitutes a noxious environment for them.

The Gulf Coast

The Gulf Coast's major single pollution source is the Houston Ship Channel and Galveston Bay through which the channel passes. Sewage disposal to coastal waters by the State of Texas is largely indirect. Most municipal and industrial sewage from the State's coastal communities is discharged into ship channels, bays, and rivers that empty into the Gulf of Mexico. The 80-km-long Houston Ship Channel receives partially treated sewage from Houston and from other industrial communities along its path through Galveston Bay as it meanders to the Gulf. The ship channel and the bay are surrounded by one of the heaviest industrial concentrations in the United States. These waters have been described as among the most polluted in this country, but, according to the Texas Water Quality Board, have experienced a reduced loading of 75 per cent in the past 5 years. Houston itself is still responsible for half of the wastes discharged into the channel today.

Along with the heavy industry concentrated in this area, shrimp and oyster harvesting is also a principal industry. This industry had been somewhat suppressed, but it is not clear whether pollutants were responsible.

The treated domestic sewage discharged into the waterway contains viruses, and viruses have been recovered repeatedly from the waters of the channel.

Several industrial discharge problems—industrial effluents containing chlorine disposed of into the Brazos River emptying into the Gulf, a cyanide discharge into the channel by a coke manufacturer, and a discharge into the channel of an industrial waste containing mercury—have recently been abated.

All industrial plants on the Houston Ship Channel are expected to install secondary treatment facilities by the end of 1974.

The Southern California Bight

The Pacific Ocean along the warm southern California coast is an important water-contact recreation area throughout most of the year, and the tidal basins along this expanse of coastline serve an important shellfish-harvesting industry.

The Southern California Bight accepts discharges from 174 municipal sewage treatment plants that discharge $4{\cdot}2 \times 10^6$ m^3 of sewage daily into the Pacific Ocean. Twenty-five of these treatment plants discharge $3{\cdot}8 \times 10^6$ m^3 of this daily total, and 95 per cent of this discharge is contributed by 5 of the plants. Of the total daily discharge to the coastal waters, $3{\cdot}29 \times 10^6$ m^3 are primary effluent and $0{\cdot}48 \times 10^6$ m^3 are secondary effluent. Industrial wastes contribute only a small fraction of the pollution in the bight. About 1000 m^3 of such wastes, about 70 per cent from the petroleum industry, are disposed of daily into San Pedro Bay which empties into the ocean. Metal processing and fish canning wastes and 20×10^6 m^3 of cooling-water from 15 power-generating stations along the bays and estuaries of the bight also are discharged to coastal waters.

Bottom sediments in the vicinities of major municipal outfalls contain higher concentrations of metals than sediments collected some distances from the outfalls. The highest metal concentrations are found in the vicinity of the large-volume Los Angeles County outfall. Some of these metals are toxic. Moreover, municipal discharges are a major source of the DDT and PCB found in coastal waters. As with the metals, sediments taken near major outfalls contain higher levels of DDT and PCB than sediments taken some distance from the outfalls. The Sanitation Districts of the City of Los Angeles and the Counties of Los Angeles and Orange are currently under court cease-and-desist orders to remove solids from their sewage before discharge.

Outfalls, of course, are sources of nutrients and many fish are attracted to outfall discharges to the bight. Dover sole apparently have increased in size and number in the environs of new discharges and have decreased in abundance in areas where discharges were discontinued. Levels of DDT are several orders of magnitude higher in such fish captured in the vicinity of major outfalls than in the same species trapped in non-polluted areas. Sixty per cent of the sole captured near Los Angeles Harbor contain DDT levels higher than those permitted by the United States Food and Drug Administration for commercial sale. Many white croakers contain such high DDT levels also. Pesticide levels in fish and mussels have correlated well with the pesticide concentrations in the bottom sediments at the locations where the fish were taken. Fin-erosion disease in dover sole and in white croakers also seems to be associated with outfall discharges.

Benthic populations are also affected by the nutrients supplied in sewage. Such populations decreased near outfalls in the bight that ceased to discharge, and increased near new outfalls after they began to discharge. To a lesser degree, similar changes in plankton populations have also been reported.

North West Coast

Puget Sound

Puget Sound serves Seattle and its adjacent communities. The waters of the sound accommodate hundreds of pleasure boats and sports fishermen, and the beaches of the sound are used extensively by swimmers and sunbathers. Some commercial fishing and shellfishing occur offshore, but inshore shellfishing has been somewhat restricted by beach pollution.

Studies undertaken in 1957 and in 1964 apparently showed that sludge deposits in the vicinity of Seattle's north trunk outfall, one of the most polluted regions in Puget Sound, did not change extensively in quality or quantity between those years. The solids, deposited by the raw sewage discharged from this outfall, were usually only a few centimeters thick within 300 m of the boil. The same kinds of microflora, perhaps fewer in number, were found in 1964 as were found in 1957. There appeared to have been little smothering of biota by sludge deposits. Some kelp and eelgrass beds thrived close to the outfall, but no clams, snails, or starfish were found in the areas closely contiguous to the boil.

Treatment plants constructed in the 1960's have replaced the more than 20 smaller facilities that served this area. The largest of these, the West Point plant, provides primary treatment and chlorination for 470 000 m^3 of sewage daily and discharges 1200 m into Puget Sound through a 180-m diffuser 73 m beneath the surface. The Renton plant treats 180 000 m^3 of sewage per day by an activated-sludge process and discharges to the Duwamish River which empties into the sound. The remaining plants discharge a total of less than 80 000 m^3 of chlorinated primary effluent daily directly into the sound. Industrial wastes are pretreated at the source before discharge when this is necessary.

Median total coliform counts (per 100 ml) in Puget Bay have dropped from about 2000 in 1966 to about 25 in 1973, apparently as a result of improved sewage treatment. Dechlorinated primary effluent from the West Point outfall has almost no acute toxicity for fingerling coho salmon, but chronic toxicities have yet to be determined. Chlorinated effluent from this outfall has some toxicity, however, presumably chloramine induced.

Alaska

Alaska, sparsely settled though it is, suffers several important pollution problems, all of industrial origin. In Silver Bay at Sitka and in Ward Cove at Ketchikan, pulp mills degrade the waters with sulfite waste liquor and solids. In St. Paul Harbor and Gibson Cove, 15 seafood-processing plants degrade these waters with partially treated wastes.

Studies in 1970 and 1971 showed that outfall discharge from the pulp industry at Silver Bay had polluted 28 km^2 of the bay area and Eastern Channel (which opens into Sitka Sound) with sulfite waste liquor. This discharge, which extended to Sitka Harbor, reached concentrations of more than 300 mg/l. in some places. Throughout most of this area, the waste-liquor concentrations exceeded 5 mg/l., a level toxic to benthic life, phytoplankton, and clam larvae and to food chain organisms consumed by salmon. Sludge deposits composed of pulp fibers, wood chips, and other anaerobic solids layered the bottom of the receiving cove and generated hydrogen sulfide gas which occasionally floated bottom debris. In September 1970, the gas reached a concentration sufficient to kill 100 000 fish.

Early in 1973, in order to meet Alaska's water quality standards, the pulp mill installed new clarifiers and implemented better chemical recovery methods. New outfalls that should give better effluent dispersion have not yet been constructed.

In 1968 and 1969, studies in Ward Cove and Tongass Narrows at Ketchikan showed that outfall discharges from pulp-mill wastes were producing pollution problems similar to those in the Silver Bay area. About 3·6 km^2 of the cove and narrows had been affected by the solids and sulfite waste liquor discharge. Phytoplankton and microscopic food organisms upon which salmon feed were not found in the affected area. Fish and shellfish, once indigenous to the cove and narrows, were not found in the affected area either. Both the fish and shellfish larvae are sensitive to even 1 mg/l. of the waste liquor. Benthic forms normally present in such areas were not found in the sludge blanket that covered much of the bottom of the cove, and blue mussels were absent from many parts of the polluted area.

Early in 1973, clarifiers improved the effluent quality somewhat at Ketchikan as they did at Silver Bay. New outfalls to improve effluent dispersion have not yet been installed here either.

The seafood industry is one of Alaska's oldest and largest, and Kodiak is its major processing site. In 1970 and 1971, studies in Kodiak showed that 15 seafood-processing plants were discharging 33 000 t of wastes annually as untreated sewage into St. Paul Harbor and Gibson Cove. The wastes created an anaerobic, hydrogen sulfide-producing sludge in those waters. Large quantities of floating fish parts defiled the harbor and cove waters. Recently installed screening devices have reduced the quantities of floatables, and abated the pollution significantly.

Hawaii

The most important ocean outfall disposal problem in Hawaii centers around Pearl Harbor and the City of Honolulu along Mamala Bay, the State's most heavily populated area. Although Pearl Harbor serves a military purpose primarily, the recreational beaches and waters along the ocean shores of the city and beyond are of major economic importance. Fisheries also comprise an important economic factor in this area.

The city's major outfall, located to the east of Pearl Harbor on Sand Island, discharges 210 000 m^3 of raw sewage daily into the bay 1100 m off the island. An additional 50 000 m^3 of treated sewage ranging from poorly treated primary to extended aeration effluents are discharged daily directly into the harbor.

Fecal coliform counts in some areas of Pearl Harbor average 100 000/100 ml. Bacteria-contaminated oyster beds are closed to harvesting, and shrimps, crabs, and sponges are less prevalent in polluted areas of the harbor than in the cleaner channel entering it. Important bait-fish populations for the tuna fisheries exist in the lochs that comprise the harbor. Fish kills which destroyed thousands of fish in this area in 1960 and 1965 have been attributed to domestic pollution and insecticide poisoning, respectively.

East of the harbor in Mamala Bay, Keehi Lagoon is impacted by the major outfall on Sand Island and by a small Army outfall both of which discharge just outside of the lagoon. The reefed lagoon is used for boating, commercial bait fishing, and other commercial and industrial purposes. Inside the reef where high levels of turbidity and organics occur, corals and other species requiring clean water do not exist. Worms and other species that thrive in polluted areas occur here in large numbers, however. The benthic algae now in the lagoon are species that thrive in highly organic waters, and these algae competitively crowd out reef-building algal species.

Fecal coliform counts in Honolulu Harbor are also high. The bottom of this harbor is thick with anaerobic silt which suppresses the growth of desirable benthic flora. The silt also contains high concentrations of heavy metals and pesticides. Two streams carrying urban drainage and industrial discharges drain into this harbor and add to the impact upon the harbor of the Sand Island ocean outfall.

The quality of the water south-east to Waikiki seems to be good generally, although winds and currents occasionally direct plumes of sewage towards the recreational beaches. Some illegal boat pollution plagues the smaller harbors, also. Moreover, heavy-metal concentrations have been detected in some areas and at least one fish kill of unknown origin has occurred in recent years.

In Kaneoha Bay, on the island's north-east shore, water quality is poor and continues to deteriorate even though two major plants afford activated-sludge and trickling-filter treatment and a third plant affords tertiary treatment. The deterioration, brought about by a combination of circumstances centering about poor water circulation, an abundance of nutrients from the treatment plants, and sedimentation caused by erosion from poor land development practices, is producing areas devoid of coral and inhabited by noxious algae.

At the outfall discharge site off Sand Island, 40 000 m^2 of ocean bottom are severely affected, and lesser damage affects an area of about 1 km^2. In the affected area, reduced numbers of species are evident, and of those that occur, many are of varieties that feed on detritus. Sewage solids seem to be an important nutrient source to these organisms.

Two independent activated-sludge treatment operations, intended to serve the urban and the military needs of the Honolulu area, are presently under consideration as a means for alleviating the pollution problems of the region. Both plants would discharge treated effluents through extended outfalls with diffusers to depths 67 m or deeper into the ocean.

Summary

By and large, nutrients discharged with domestic sewage appear to attract fish which, along with the microflora that constitutes a part of the food chain, often thrive in such waters, but some species do disappear from such areas. Even if the oceans do assimilate many non-toxic wastes, however, toxic effluents are clearly another matter that will always need attention. Diseases such as fin-rot (and possibly tumors) seen in certain fish, even in some that thrive in polluted waters, are also a matter of concern. And, of course, the risk to the recreationalist constituted by the discharge of viruses and other pathogens, even within sewage that has undergone secondary treatment, is ubiquitous. Moreover, there must be concern with the possibility of subtle, long-term, untoward effects that may not be easily detected.

In any event, in the United States, many of the pollution problems of today may well dissipate tomorrow. The Federal Water Pollution Control Act Amendments that became law in 1972 require that all sewage treatment plants must provide secondary treatment by mid-1977; they proclaim as goals the attainment, by 1983 wherever possible, of waters clean enough for swimming and the propagation of fish and other wildlife, and, by 1985, zero discharge of pollutants into the nation's waters.

There may well be reason on the side of those who argue that the assimilative capacity of the seas can cope with at least the nutrient content of sewage, but it is difficult to quarrel with the enshrouding wisdom of the 1972 Amendments in their greater design to retain for man the pureness of the earth he so wilfully sacrifices in his quest for the better life.

ACKNOWLEDGEMENT

The Author gratefully acknowledges personal communications from J. Farris (Texas Water Quality Board, Austin, Texas) and the following personnel of the U.S. Environmental Protection Agency: H. M. McCammon (Region I, Boston, Mass.), E. F. M. Wong (Region I, Needham Heights, Mass.), A. Montague and F. L. Ross (Region III, Philadelphia, Pa), and J. E. Osborn and S. G. Provant (Region X, Seattle, Wash.) without whose help it would have been impossible to prepare an up-to-date account of regional problems in the United States.

BIBLIOGRAPHY

The following documents have been heavily drawn upon, and, although it has not been found practicable to reference the text, obvious associations can be made between the titles in the bibliography and the section headings in the paper.

1. Dewling, R. T. Statement: ocean disposal. U.S. Envir. Protect. Ag., Reg. II, Public Relations Release, New York, 1974.
2. Dugan, G. L., and Young, R. H. F. Effects of coastal waste disposal in Hawaii. *J. envir. Engng Div. Am. Soc. civ. Engrs*, 1973, **99** (EE5), 691–701.
3. Gibbs, C. V. Testimony presented to Subcommittee on Environmental Pollution. U.S. Senate Public Works Committee, 93rd Congress, 2nd Session, Honolulu, 1974.
4. Metcalf, T. G., Wallis, C., and Melnick, J. L. Environmental factors influencing isolation of enteroviruses from polluted surface waters. *Appl. Microbiol*, 1974, **27**, 920–926.
5. Municipality of Metropolitan Seattle. Disposal of digested sludge to Puget Sound—the engineering and water quality aspects, July 1965. Seattle, 1965.
6. Provant, S. G., and Stewart, R. K. September 1971 survey of the effects of pulp mill wastes on receiving waters at Silver Bay, Alaska. U.S. Envir. Protect. Ag., Reg. X, Anchorage, Alaska, 1972.
7. Sigel, M. M. Miami-metro offshore sampling program. Interim report to H. J. Ross Associates, Inc., Miami, Fla, 1972.
8. Southern California Coastal Water Research Project. The ecology of the Southern California Bight: implications for water quality management. SCCWRP TR 104. El Segundo, Calif., 1973.
9. U.S. Environmental Protection Agency. Ocean dumping in the New York Bight—facts and figures. Reg. II, Surveillance and Analysis Division, New York, 1973.
10. U.S. Environmental Protection Agency. Ocean outfalls and other methods of treated wastewater disposal in southeast Florida. Reg. IV, Atlanta, Ga, 1973.
11. U.S. Environmental Protection Agency. Mamala Bay wastewater treatment and disposal system, Oahu, Hawaii. Environmental Statement D-EPA-24002-HI. Reg. IX, San Francisco, Calif., 1973.
12. U.S. Environmental Protection Agency. Effects of pulp mill wastes on receiving waters at Silver Bay, Alaska. Reg. X, Anchorage, Alaska, 1971.
13. U.S. Federal Water Quality Administration. Effects of pulp mill wastes on receiving waters at Ward Cove, Alaska. Reg. X, Alaska Operations Office, Anchorage, Alaska, 1970.

Prof. Berg, introducing his paper, said that the great seas were the final repository for most of man's liquid wastes. Much of those wastes was discharged directly through outfalls to the oceans, sometimes taxing the assimilative capacities of these greatest of all ecosystems. With currents often washing wastes back to the shores, even an unlimited ocean capability for regeneration would not mitigate completely the economic and recreational damage inflicted by floatables, tars, oils, toxicants, and infectious agents.

Man's existence on earth was measured in millennia, but his capability for seriously disrupting the earth's ecosystems and seriously affecting his own existence went back barely a few score years. Even if he decided he had to live with some of his pollution, he could not live with all of it. In the next decade or two, man had little alternative but to alter his ways, or pay the piper a measure and beyond.

DISCUSSION

Mr. R. G. Ludwig (Brazil) observed that recommendations were currently being made to the U.S. Congress by the Environmental Protection Agency to (1) remove the requirement of secondary treatment for all sewage discharges by 1977, and (2) remove the requirement of secondary treatment prior to ocean disposal where suitable long outfalls and diffusers discharged to open ocean waters.

Mr. H. L. Jones (Australia) asked whether engineers and scientists were taking seriously the 'no discharge' proposals of the U.S. Government.

Mr. A. J. O'Sullivan (U.K.) noted that Prof. Cronin had suggested (in 1971)[14] that the capacity of the Atlantic shelf of the Eastern seaboard of the United States to receive pollutants would be exceeded in some 40 years—this was based on the current increase in waste disposal, population projections, industrial predictions, hydrographic data, and information available on biological tolerance limits. He invited Prof. Berg's comments on this prediction, particularly with regard to whether the recent improvements referred to in his paper would put the critical date further away.

Mr. J. T. Calvert (U.K.) noted that several of the cases cited by Prof. Berg—such as the Potomac River and the Houston ship canal—were those of pollution of inland waters. He asked for examples where discharges of sewage to the 'ocean' at reasonable distances from the low-water mark had created serious problems.

Dr. G. E. B. Kullenberg (Denmark) enquired if Prof. Berg had considered the relative importance of the discharges he had discussed and the atmospheric fallout. In the North Sea, for example, it had been suggested that atmospheric fallout was the major source of some pollutants.

Mr. P. C. Wood (U.K.) referred to the Author's statement that virus had been found in coliform-free sea water, and asked if there was any epidemiological evidence that coliform-free shellfish had caused virus infections following consumption by man. Mr. J. A. Wakefield (U.K.) asked Prof. Berg to enlarge on his reference to virological surveys, and to give the cost of analysing a sample for virus as compared with analysing for coliform or other bacteria.

REPLY TO DISCUSSION

In reply to Mr. Ludwig, Prof. Berg said that the degree of treatment to be required was being carefully re-evaluated by the Environmental Protection Agency and also by the National Water Quality Commission which would make its recommendations to Congress in about a year. To Mr. Jones he replied that Congress had expressed its intent in law (the Amendments of 1972 to the Water Pollution Control Act of 1965) that there should be no discharge of pollutants into the waters of the United States by 1985. This was a goal intended to protect the environment of the United States and of the entire world which shared with the United States the oceans and the air. Prof. Berg thought this intention would be taken seriously unless Congress, in its wisdom, saw fit to alter the goal.

Replying to Mr. O'Sullivan, he said that if universal secondary treatment became a reality by 1977—as it would in the United States under current law—the Atlantic shelf would, of course, be better able to cope with the better quality discharge. If zero discharge became a reality in 1985, then a problem should no longer exist for outfall effluent discharge, but sludge was another matter to be dealt with.

The key to Mr. Calvert's question was in the definition of 'serious'. It was a serious matter, for example, when outfall pollutants were washed up on the bathing beaches of south-eastern Florida or southern California. Some might consider such pollution less of a problem (at least not of economic impact) on the frigid beaches of northern countries.

He told Dr. Kullenberg that he had indeed considered the pollution impact of atmospheric fallout on the oceans, but since such pollution was not an outfall problem he had thought discussion of the matter to be inappropriate in his paper.

In reply to Mr. Wood, he doubted if such evidence existed, because of the impracticability of gathering it—to obtain such data would require an excessively expensive effort. He pointed out also that shellfish could be harvested legally from waters containing some faecal coliforms. The cost of analysing a sample for viruses (about which Mr. Wakefield had enquired) varied with the kind of sample and the degree of analysis required. It might vary from \$100 to \$500. On the same basis, a test for faecal coliforms might vary from \$5 to \$15.

ADDITIONAL REFERENCE

14. Cronin, E. L. A discussion on biological effects of pollutants in the sea. IV—Pollution prevention. *Proc. R. Soc.*, B, 1971, **177**, 439–450.

Paper No. **4a**

Some Problems in Hong Kong

J. R. Preston

J. D. & D. M. Watson, High Wycombe, England

Background scientific information is likely to be available in urbanized and industrialized countries where long-standing pollution problems have been experienced and the social consequences are most pressing. Elsewhere in the world, community and industrial revolutions have frequently taken place within the last two or three decades and the potential dangers are only just beginning to be defined. In such places, background information is most hard to come by and assessments have to be founded more on contemporary than on past research.

A further common primary difference between highly-developed and developing countries is in the assimilative capacity and reaction of receiving waters, and the consequent behaviour of discharged effluents. The seas around the more developed zones are usually temperate whereas those surrounding less developed communities are often tropical or sub-tropical. Particular differences which are apparent in the tropics as compared with temperate zones are a generally higher biological reaction rate and accelerated bacterial die-off, arising most probably from more intense ultra-violet radiation and plankton activity. Dissolved-oxygen patterns are probably more varied in tropical areas: saturation levels are lower, and concentrations can be more significantly affected by vigorous algal activity, contributing dissolved oxygen by photosynthesis often to the point of supersaturation, and depleting it when the death of undernourished populations exerts a decay demand. Both of the two main aspects just described have been manifested in investigations carried out in Victoria and Tolo Harbours in Hong Kong, between latitudes 22° and 23° north of the equator.

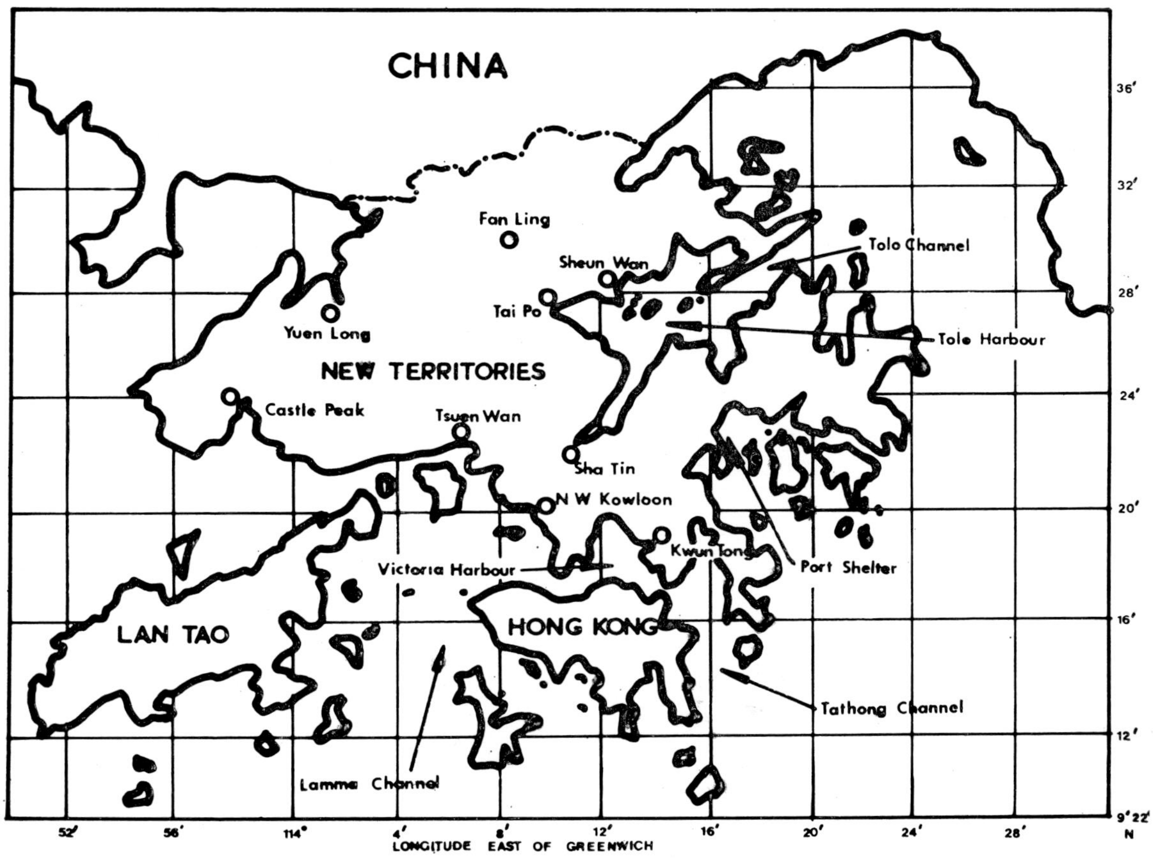

FIG. 1. Principal geographic features of Hong Kong

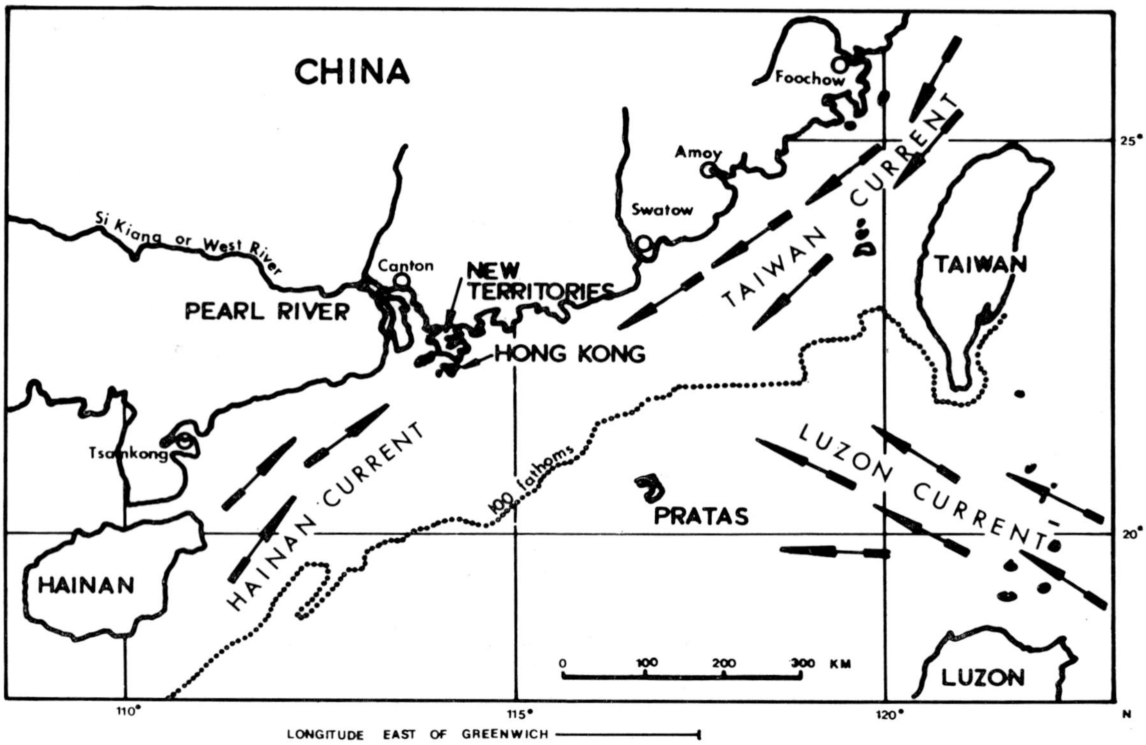

FIG. 2. Principal sub-oceanic currents in the China Sea

In Hong Kong (Fig. 1) a total colony population of about 4 million is expected to increase to 6·2 million by the year 2000. The urban centre located about Victoria Harbour and adjacent waterways contains 3·3 million people which will probably grow by 1 million. Satellite towns are being developed in the New Territories, and the land-locked Tolo Harbour area will eventually support some 750 000 people.

Sewage from the present urban centres is discharged raw into the main harbour area through 14 principal outfalls. Screenable material is first removed from the larger discharges. It was feared that the situation in the main harbour might be reaching a critical point and it was generally accepted that Tolo Harbour, an area of considerable natural beauty, should be protected from the harmful effects of pollution. An investigation was therefore ordered into the measures necessary to maintain tolerable circumstances in both waterways.

A two-directional monsoon exists in Hong Kong, with seasonal variations in climate. The north-easterly winds are cool and the south-westerlies warm and humid. The monsoonal currents are shown on Fig. 2. The Pearl River to the west of Hong Kong floods in the spring, bringing down large quantities of silt and reducing the salinity of local waters to as low as 0·5 per cent for two or three months.

It was decided that quantitative standards for individual discharges into Victoria Harbour were inappropriate and that the accent would be better placed on overall receiving-water condition, with dilution and dissolved oxygen as the primary parameters. For both harbours minimum initial dilutions of 50 and 85 for settled and screened sewage respectively were proposed to safeguard against smell nuisance, and a general minimum dissolved oxygen concentration of 50 per cent saturation was suggested. Solids-size and toxic-substances limits were recommended, but a coliform standard was discounted although check measurements were carried out throughout the study with the following result:

Percentage of samples with counts above	Victoria Harbour	Tolo Harbour	Castle Peak
10 000 per 100 ml	1	0	9
2400 per 100 ml	6	3	15

The Victoria Harbour and the westerly Castle Peak results were affected by the large seasonal increases coincident with Pearl River flooding when some very high counts of up to 10^7 per 100 ml were observed.

The quantitative evidence from the considerable numbers of dissolved-oxygen measurements taken is summarized on isovariate plots in Figs. 3 and 4. Worst overall circumstances observed were 20–25 per cent saturation in North West Kowloon and 40 per cent in Central Harbour. Tolo Harbour waters were generally healthy at the surface but deteriorated towards the bottom with values of 50 per cent saturation in early mid-summer.

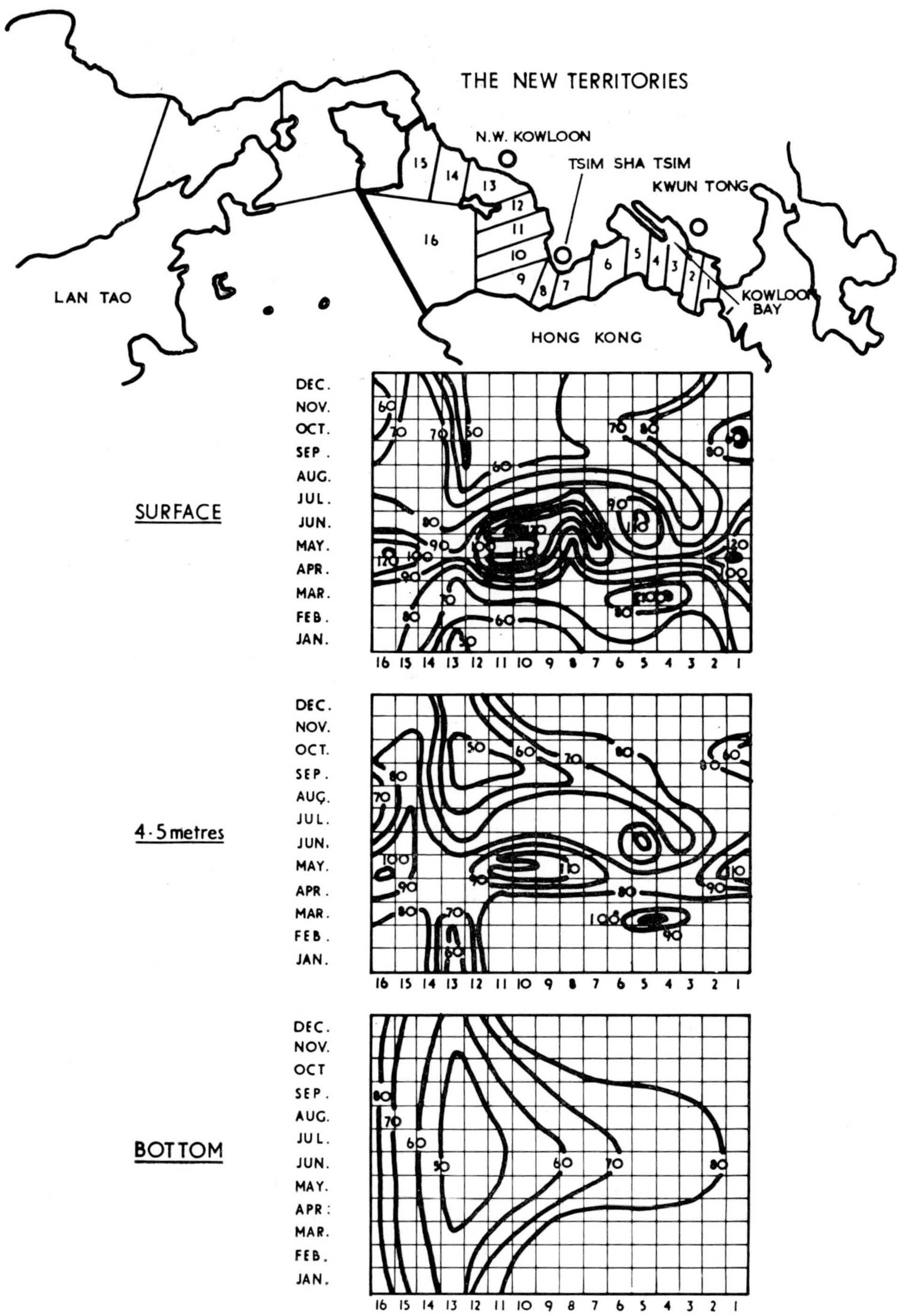

FIG. 3. Dissolved-oxygen isovariates (per cent saturation) in Victoria Harbour, Hong Kong

The daily pollution load into Victoria Harbour in terms of BOD was estimated at 185 t and the average diluting volumes through the harbour at 45 $\times$ 10^6 t at neap tides and 80 $\times$ 10^6 t at springs, at each turn of tide. It was estimated that the BOD load could increase ultimately to 2·3 times the 1971 level. It was considered that the load into the central harbour could be increased by 50 per cent whereas in the North West Kowloon area it was already at a critical level.

Tentative recommendations were made to introduce staged treatment facilities at two principal locations. Something short of conventional rate treatment was thought necessary at North West Kowloon and possibly full treatment at Kwun Tong at the east end of the harbour. The rest of the central area's sewage was to continue to be discharged untreated except for screening.

Prediction of short-term and mass movement of receiving waters in Tolo Harbour proved very difficult, and resort was taken to salt-balance determinations. In a total harbour volume of 500 $\times$ 10^6 m^3,

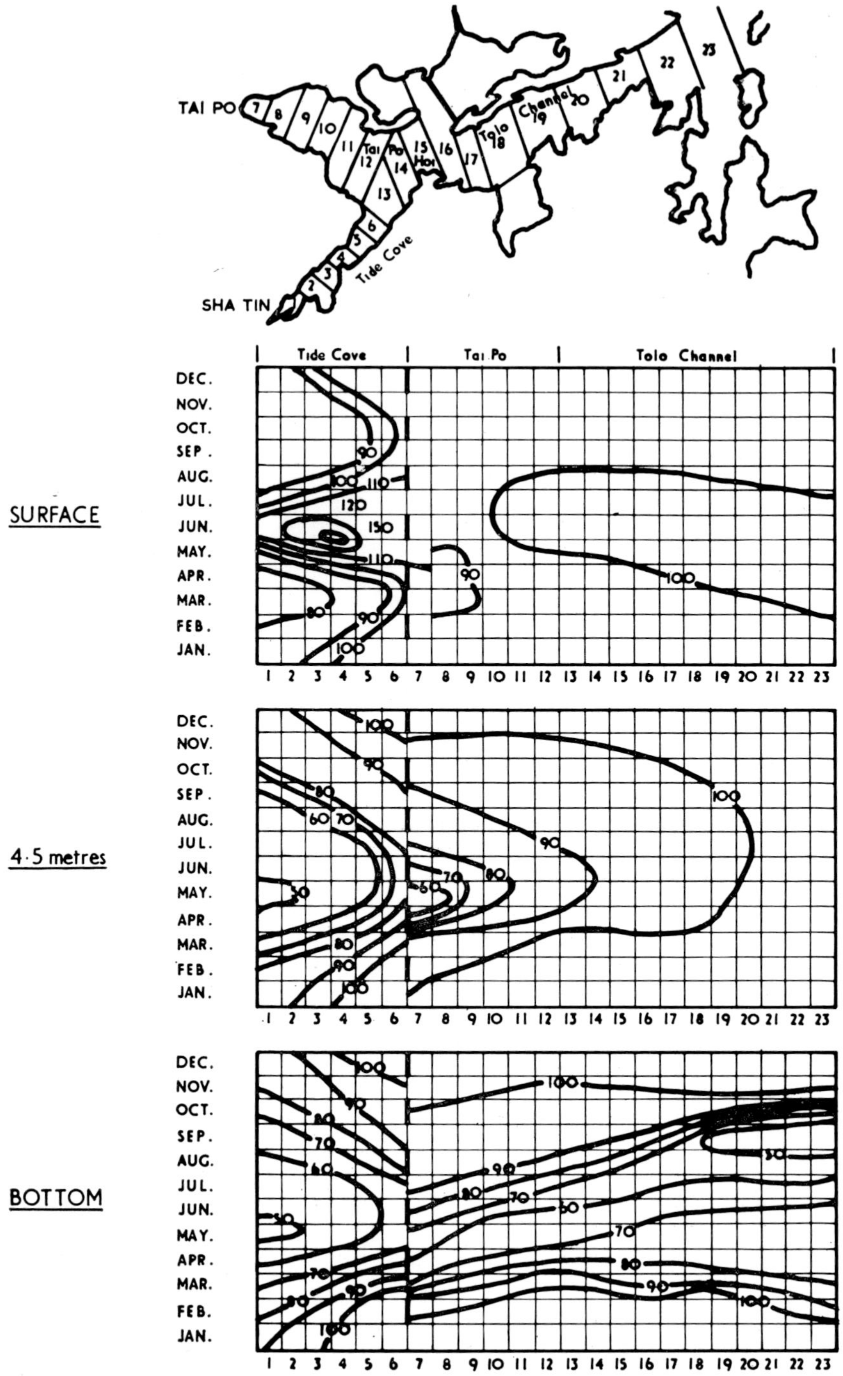

FIG. 4. Dissolved-oxygen isovariates in Tolo Harbour, Hong Kong

a residence period of 28 days was estimated for discharge into the central part of the harbour and up to 42 days from the remotest point in the waterway. It was estimated that the sewage from a then planned 280 000 population could be satisfactorily dispersed through a long outfall after primary settling or through a shorter pipe after full treatment. Consideration was given to the possible dangers of eutrophication and the risks were thought low.

Development plans have recently been revised and a shoreside population of at least 750 000 is foreseen around Tolo Harbour. Recommendations have been made for 'full' treatment with nitrification of ammonia for the sewage from 600 000 of this ultimate population. The risk of intolerable nutrient build-up has increased proportionately but there is insufficient evidence for the risk to be quantified. The precaution is to be taken of designing the treatment plant to accommodate nutrient removal facilities should regular monitoring show them to be necessary.

Mr. Preston, introducing his paper, said that it emphasized two important differences in the examination of marine receiving waters, between developed temperate regions and developing tropical or sub-tropical areas: behavioural differences were likely in biological reaction rates and assimilation phenomena, and specific background information was usually less easy to come by in less developed areas. This was not to imply that scientific effort was absent in Hong Kong. Valuable data on fluctuations in temperature, salinity, and dissolved oxygen had been reported by Chau and Abesser[1] in 1958, and Watts[2] had worked in the continental shelf waters off the coast of China. The Hong Kong Government's Fisheries Research Division was currently developing a programme to examine use of Hong Kong's extensive coastline for mariculture.

Statistical evidence of direct human contamination from ingestion of polluted sea water in the amenity areas was non-existent. An arbitrary coliform standard could have been adopted, but experience in Hong Kong and elsewhere in S.E. Asia was that die-off was extremely rapid, with T_{90} values probably less than 30 min. Because of interference from numerous adjacent outfalls, an attempt was made by the use of bacterial tracer, *Serratia marcescens*, to identify worst offenders and examine die-off. Reductions of 98 per cent were indicated in the receiving water in the immediate vicinity of the outfalls shortly after discharge.

The treatment needs had therefore been assessed by reference to levels of dissolved oxygen in the receiving waters. An overall lower limit of 50 per cent saturation was considered reasonable. Given screening and efficient diffusion of all untreated discharges, local blemishes would be removed and receiving waters generally maintained in their existing condition which was aesthetically acceptable. It might be argued that a lower limit could be tolerated. An absolute limit of 10 per cent had been proposed for the Thames Estuary[3] as providing some margin of safety against odour nuisance, and life chains could probably be maintained in estuarine conditions with a level of 30 per cent saturation. It was considered, however, that reduction of the limit, and hence the margin of safety against deoxygenation, was not advisable in Hong Kong circumstances.

DISCUSSION

Mr. J. A. Wakefield (U.K.) requested Mr. Preston to enlarge on the background leading to the disposal scheme described in the paper. He also asked if there was any bathing in the area and if there was an anti-sewage-in-the-sea campaign.

Mr. E. Wilmot-Morgan (U.K.) spoke as a former Chief Engineer Drainage in Hong Kong. He referred to the wide seasonal variations in rainfall (497 mm in October–April and 1690 mm in May–September on a 44-year average) and to the fact that Hong Kong had always been sewered on the separate system, the foul sewage discharging mainly to the harbour waters in the residential areas. As Mr. Preston had mentioned, the tidal flow through the harbour had prevented any serious problem except in the land-locked Yau Ma Tei Typhoon Shelter where pollution had been mitigated by a requirement for septic tanks in the contributing properties. In the late 1940's an intercepting sewer with a screening plant and a deep outfall into the harbour had been constructed, the screenings being dumped well outside the harbour limits. The population explosion of 20 years later had not been foreseen and he asked Mr. Preston if the Yau Ma Tei sewerage system had been able to cope with the increased flows and whether there were now frequent overflows from the two pumping stations.

In the early 1960's it had been accepted that discharge of untreated sewage to the harbour might not be desirable, although the interception of the multitude of outfalls would have been very expensive and land was not available for treatment plant; in addition, the increasing use of sea water for flushing systems might present treatment problems. Furthermore, it was not possible to control pollution from the large population living on junks and sampans. Finally, he asked whether submarine outfalls with screening plants were cheaper to construct than contact stabilization plants at the outfalls where land might be made available on, say, the new Wanchai reclamation.

Mr. R. G. Ludwig (Brazil) remarked that he believed the rapid rate of coliform disappearance in Hong Kong to be due primarily to the high water temperature; T_{90} values reported for tropical waters were invariably higher than those for colder waters. He asked Mr. Preston what was his definition of screening.

Dr. G. E. B. Kullenberg (Denmark) asked whether Mr. Preston had investigated the dispersion conditions in Hong Kong; he thought the tidal currents there were extremely complicated and that this might imply good dispersion.

REPLY TO DISCUSSION

In reply to Mr. Wakefield, Mr. Preston said that the Hong Kong Government had had sewage treatment needs in mind for some years. In 1964 J. D. & D. M. Watson were instructed to report on sewerage and sewage treatment for the Northern New Territories in connection with the Government's proposal to develop several satellite townships, two of which (Sha Tin and Tai Po) would contribute sewage effluent to land-locked Tolo Harbour, and in the following year it was recommended that marine investigations be carried out there. These investigations, which were extended to include the main Victoria Harbour and adjacent waterways, were completed in 1970 and detailed engineering feasibility studies had since been carried out into sewage treatment at Sha Tin and North West Kowloon. Public interest had been aroused during the investigatory periods, but he thought it would be correct to say that at no time had there been an anti-sewage-in-the-sea campaign although strong criticism had been voiced about the condition of the harbours in the areas for which improvement works were proposed.

In response to Mr. Wilmot-Morgan's query about the sufficiency of the Yau Ma Tei sewer, Mr. Preston undertook to make enquiries in Hong Kong. He subsequently reported that several large sewers existed in that district, and that it had been found generally difficult in Kowloon to separate completely the foul from the surface-water discharges. Foul sewers became overloaded in wet weather, and overflow to storm nullahs was common. In Hong Kong, where outfalls were relatively short, simple screening and discharge was undoubtedly cheaper than full treatment, but the opportunities for dispersion of polluting loads were limited and the objective of reducing pollution concentrations could not always be met.

Answering Mr. Ludwig, he said that the screen openings were 5 mm. In reply to Dr. Kullenberg, Mr. Preston said that, although he could confirm that there was rapid dispersion in the vicinity of the outfalls studied, no precise determinations of dispersion coefficients had been made; efforts had been concentrated on estimation of mass removal of polluted water.

REFERENCES

1. CHAU, Y. K., and ABESSER, R. A preliminary study of the hydrology of Hong Kong territorial waters. *Hong Kong Univ. Fish. J.*, 1958, **2**, 43–58.
2. WATTS, J. C. D. The hydrology of the continental shelf area south of Hong Kong. I—Oceanographic observations for the year 1969. *Hong Kong Fish. Bull.*, 1971, **2**, 51–57.
3. DEPARTMENT OF SCIENTIFIC AND INDUSTRIAL RESEARCH. Effects of polluting discharges on the Thames Estuary. Water Pollution Research Technical Paper No. 11. H.M. Stationery Office, London, 1964, p. 527.

Paper No. **4b**

Some Problems in the Baltic

Artur Svansson

Fishery Board of Sweden, Göteborg

The Baltic is sometimes assumed to be the most polluted sea area in the world, but such a statement is easy to misunderstand and clarification is necessary. It will be shown that the Baltic is like a large estuary with a very long residence time, and that it is also like a large fiord with a shallow sill and relatively deep parts where stagnation is a regular phenomenon. The condition has deteriorated during the past few years, and many of the coastal waters are now seriously polluted; therc is a ban on the selling of cod liver due to high amounts of DDT, but (on the other hand) fishing has increased.

The Baltic is a kind of sill estuary, the narrow and shallow passages of the Belt Sea and the Kattegat (Fig. 1) constituting the 'sill area'. (These two seas are sometimes included in the concept of the Baltic, but usually, and so in this paper, the Baltic is considered to consist of the Baltic proper, the Gulf of Finland, the Bothnian Sea, and the Bothnian Bay.) The mean salinity of the Baltic is low (7 g/kg) and

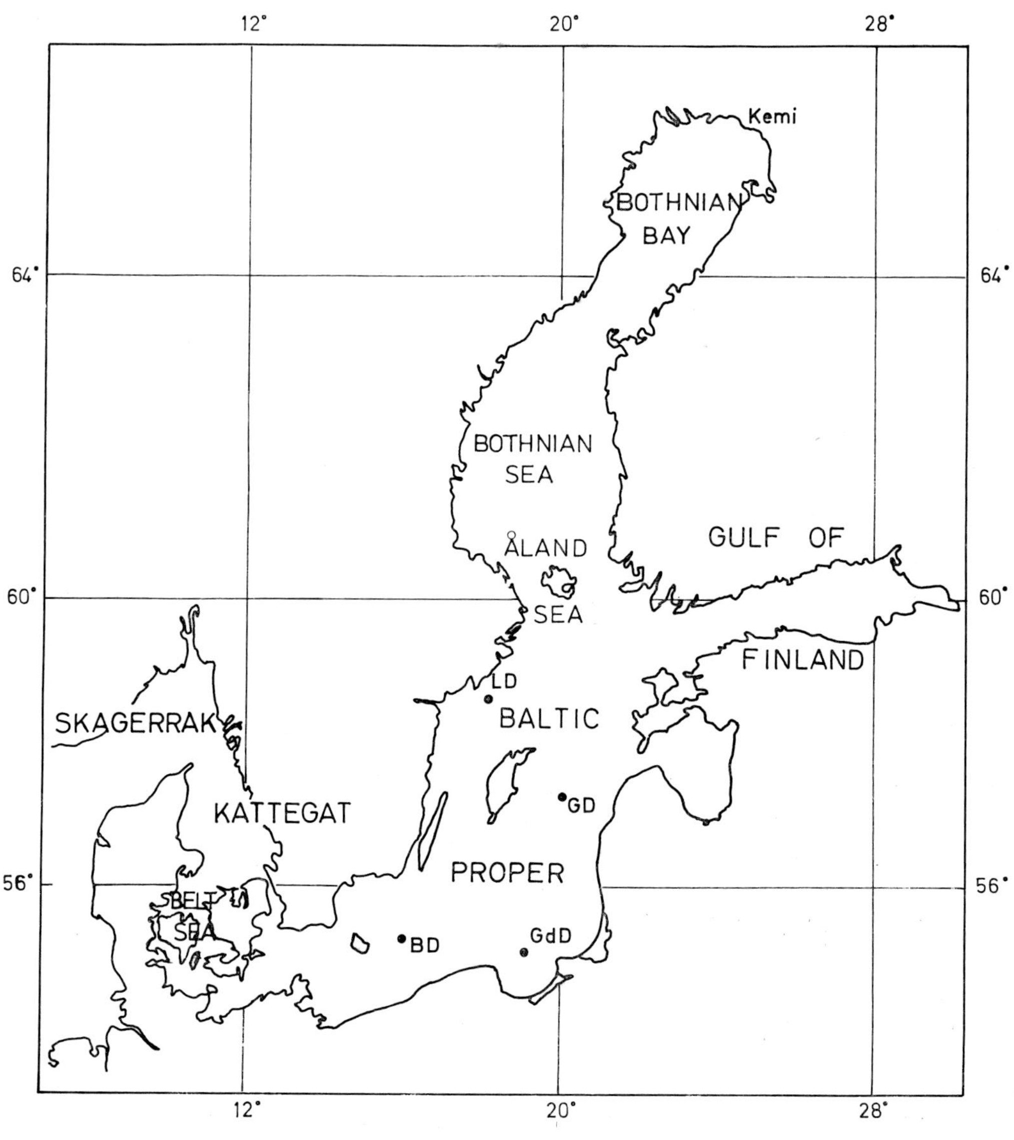

FIG. 1. The seas of the Baltic

so is the exchange with the ocean. On average, the yearly precipitation on the Baltic is balanced by evaporation and left only is the fresh-water supply of approximately 15 000 m^3/s to balance the inflow of ocean water. The following very rough calculation has been made: for a conservative dissolved substance (such as salinity) which by definition is not involved in biological cycles or lost in sediments, the Baltic can accumulate 36 years' supply before there is balance between supply and outflow. Thus, if 10 000 t is the annual supply, the conservative substance will increase to 360 000 t before equilibrium is reached.

The salinity of 7 g/kg is a mean value. In the innermost parts the salinities are lower; in the south they are higher. This is seen in Fig. 2 which also shows that the deeper salinities are higher than the corresponding surface values—particularly in the Baltic proper. The Bornholm Deep (see BD in Fig. 1; bottom depth 89 m), the Gotland Deep (GD, 259 m), and the Landsort Deep (LD, 459 m) are the deepest parts of three basins. In the deeps the water is usually stagnant, by which we mean that it has very restricted exchange with the overlying water. The dissolved-oxygen content there is usually low and the nutrient concentration high. Stagnation is probably a normal feature of the deeps, but when Fonselius[1] plotted all available values of oxygen in the near-bottom water it was obvious that there had

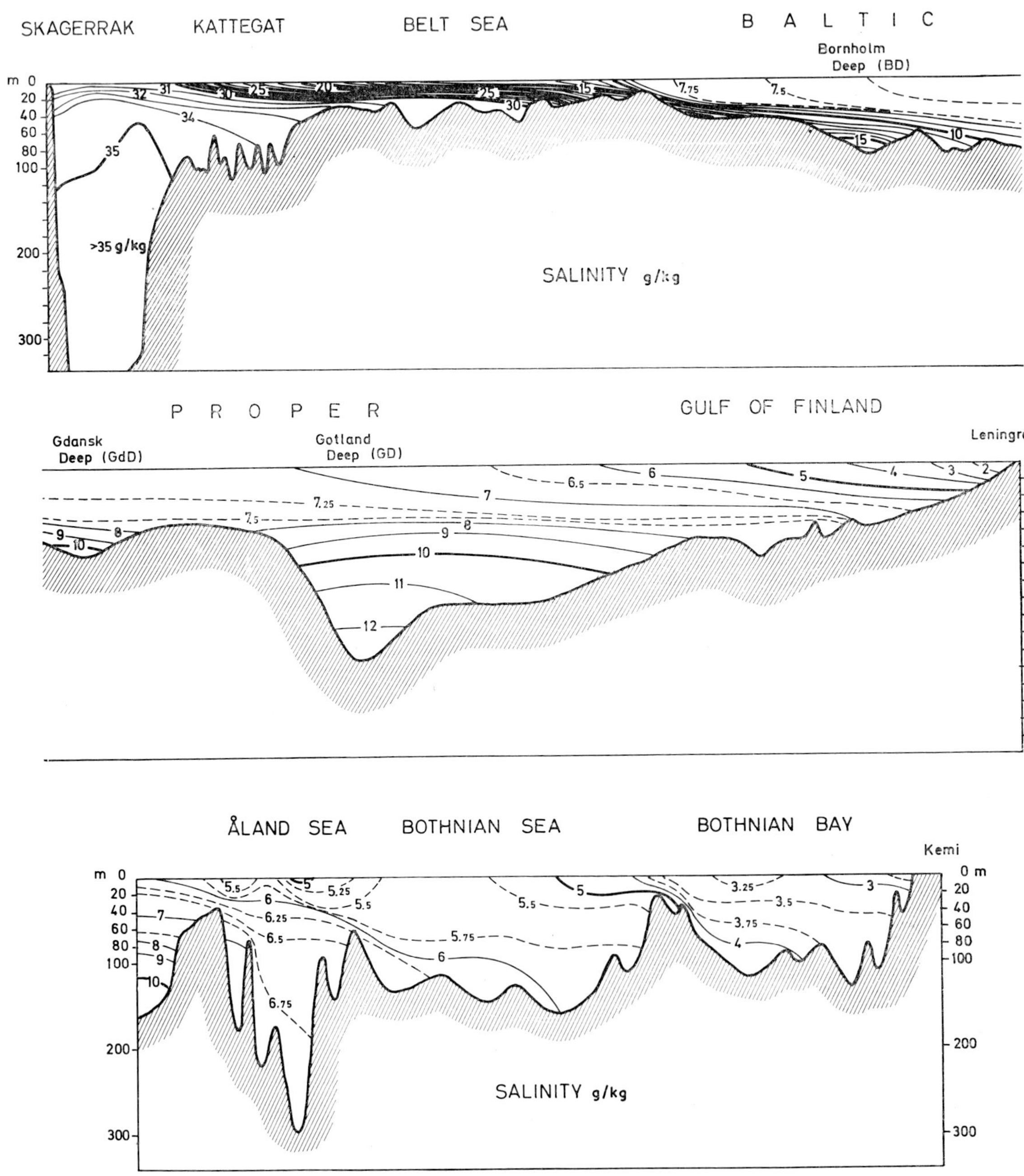

FIG. 2. Vertical salinity profiles through the Baltic

been a long-term trend of decrease in oxygen content. People began to ask if this change had been caused by the increased load of organic material—mostly due to eutrophication following increased nutrient supply. It has been estimated[2] that the annual sewage phosphorus supply is approximately 15 000 t; in this case the industrial contribution is probably negligible, but if we talk of organic substances, there is the large discharge from forest industries to be taken into account. Thus, in terms of BOD, the polluting load is 300 000 t/year for sewage and 800 000 t/year for industry. The latter figure, however, probably does not well express the real oxygen demand because the products from the forest industries are sparingly degradable, and some people think that the long-term oxygen decrease in the deeps is mainly due to the slow degradation of these products.

However, it is also quite clear that deep oxygen may change naturally due to climatic variations. Thus, Fig. 3 shows long-term variations in (a) fresh-water supply, (b) and (c) salinity, and (d) oxygen content (hydrogen sulphide being counted as negative oxygen). Soskin[3] presented the theory that surface salinities increase when fresh-water supply diminishes. That deep salinities follow suit seems to be obvious, even if it is not proven. But when the fresh-water supply is low, saline water is more available in the Kattegat and the Belt Sea. Inflow of water of relatively high salinity to the deep basins of the Baltic, normally associated with a high level of the Baltic, is thereby favoured by a small fresh-water supply. Because the increase of salinity in the bottom water is larger than that in the surface water the stagnation is reinforced and the oxygen conditions may deteriorate accordingly.

So the deep oxygen can apparently be reduced by (1) increased load of organic material and (2) climatic variations. Unfortunately we do not know which of these two factors is dominant. A substantial part of future Baltic research will aim at clarifying this problem. An important tool will be a theoretical model; the first trials have been carried out, but there is much to be done before a conclusive answer can be presented.

A higher amount of nutrients has quite obviously caused a higher primary production of phytoplankton[4]. Also the catch of fish has increased, although the fishery biologists are very cautious in explaining the larger catches. In the 1930's the total catch of herring and cod was approximately 50 000 t. In 1970 the corresponding figure was nearly 450 000 t, and if the catches of sprat and flounder are added

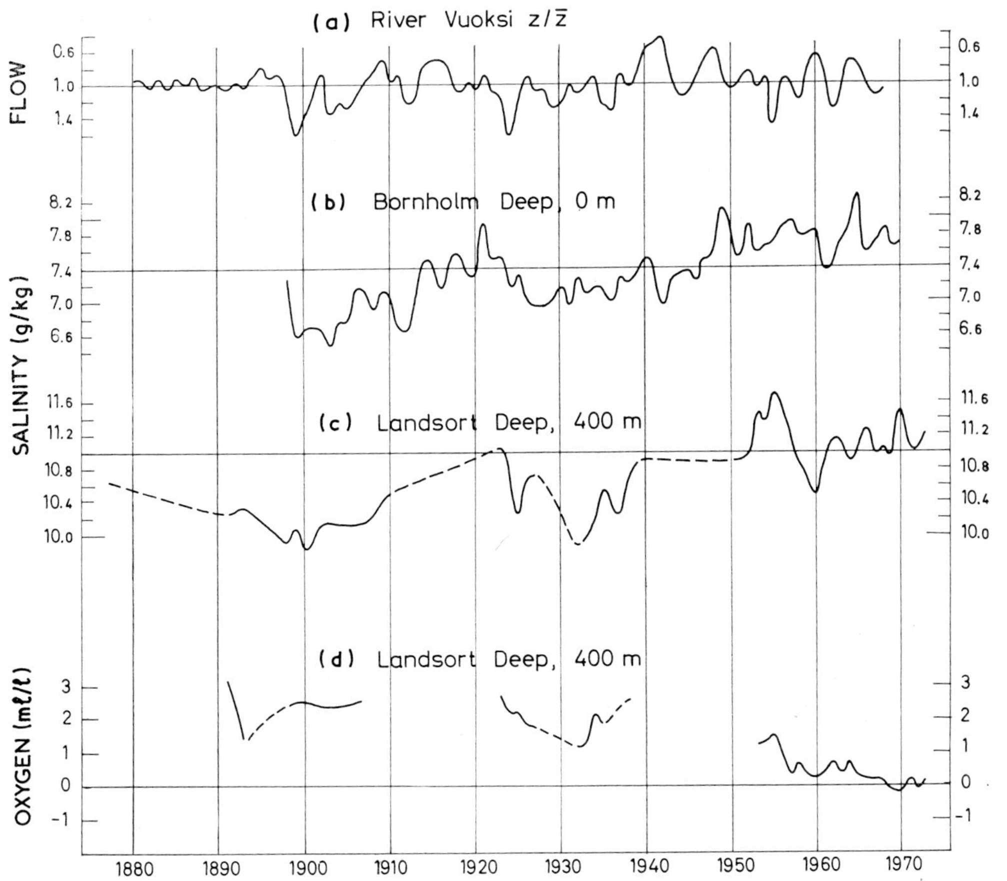

FIG. 3. Long-term variations in (*a*) flow of River Vuoksi (discharging to Gulf of Finland via Lake Ladoga) expressed as ratio to long-term average, (*b*) and (*c*) salinity, and (*d*) dissolved oxygen

we may end up with a catch of 700 000 t. It appears that these increased catches could well have resulted from increased fishing effort and improved fishing methods. For the cod fishery, Otterlind[5] considers that only the variations in catches since the 1950's reflect hydrographic variations. Most critical for cod are the spawning conditions, the necessary ones being (1) salinity >10 g/kg (or, better, density $>1{\cdot}008$ g/cm^3), (2) temperature 4–6°C, and (3) oxygen content >1 ml/l. These conditions seem to be fulfilled only during certain years.

Whereas eutrophication may be hazardous to cod (too low deep oxygen values) the opposite is probably true for herring and sprat. There may be risks of poisonous effects on herring because the grounds for the very sensitive spawning of the herring are situated rather near land, but so far there have been no indications to suggest this. A real sprat fishery in the Baltic did not start until 1950, but has since increased every year. Since 1973 there has been a convention between the countries round the Baltic to prevent overfishing.

Returning to pesticides, mercury pollution has so far been assumed to be only a coastal problem in the Baltic: the sale of some fish is prohibited in three parts of the archipelago owing to high concentrations of mercury, but some measurements in sea birds show that the influence may have reached the open sea. The mercury content of feathers of osprey (a bird which consumes fish almost exclusively) showed values around 4 μg/g in determinations on museum specimens dating from 1840–1940, but increased by a factor of three thereafter[6], whereas the mercury content of guillemot eggs decreased from 0·5 μg/g in 1969 to 0·3 μg/g in 1973, during which time the use of mercury was discontinued in both Swedish and Finnish pulp mills[7].

The situation for DDT and PCB is worse than that for mercury. As measured in herring (extractable fat), DDT was highest in the central and southern parts of the Baltic proper—about 50 μg/g—whereas concentrations of 15 μg/g were found in the Bothnian Sea and 2–3 μg/g in the Skagerrak[8]. It was also found that DDT in pike increased from lakes, through rivers, to the Baltic. Although PCB is more evenly distributed, the concentrations are of the same order of magnitude. Guillemot eggs which in 1969 (as already mentioned) contained 0·5 μg/g of mercury in the fresh tissue, had 600 μg/g of DDT and its metabolites, and 250 μg/g of PCB, in the extractable fat[9].

Sedimentation in the open Baltic is probably at a maximum in the deepest parts where the current velocity is usually least, but conditions in stagnant basins may be complicated by the different geochemical reactions in oxidized and reduced environments. Hallberg has shown[10] that the concentrations of copper and molybdenum are highly sensitive to the presence or absence of hydrogen sulphide. Zinc, on the other hand, seems to be independent of stagnation, and a marked increase in the sediment surface at the bottom of Gotland Deep is believed to be an effect of pollution.

Nearly 200 sub-surface samples were analysed for non-polar hydrocarbons (oil) during 1970–73. In over 80 per cent of the samples the concentration was below the detectable limit of 0·05 mg/l, and no concentration exceeding 0·10 mg/l was found[11]. Nevertheless, it is quite clear that 'the Baltic area is a very sensitive one and that a few disasters like the Torrey Canyon accident outside southern England in March 1967 could annihilate the marine life and fisheries'[2].

There is some evidence that during the last few years there has been an increase in fish diseases, particularly of eel and pike but also of cod. The problem, which is mainly a coastal one, is thought to have been aggravated by pollution from oil, heavy metals, etc.

Recently a convention for Baltic pollution was agreed upon. When ratified, the convention will put a ban on the discharge of PCB, DDT, many other pesticides, and heavy metals.

REFERENCES

1. FONSELIUS, S. H. Hydrography of the Baltic deep basins III. Fishery Board of Sweden, Hydrography Ser. Rep. No. 23, 1969.
2. Report of the ICES Working Group on Pollution of the Baltic Sea. International Council for Exploration of the Sea Coop. Res. Rep., Ser. A, No. 15, 1970.
3. SOSKIN, I. M. Long term changes in the hydrological characteristics of the Baltic. Hydrometeorological Press, Leningrad, 1963, 159 pp. (in Russian).
4. FONSELIUS, S. H. On biogenic elements and organic matter in the Baltic. Ambio Spec. Rep. No. 1 (1st Soviet–Swedish Symposium on Pollution of the Baltic), 1971, p. 29.
5. OTTERLIND, G. Fish stocks and fish migration in the Baltic Sea environment. 2nd Swedish–Soviet Symposium on Control of Baltic Sea Pollution, Riga, 1973 (*to be published*).
6. JOHNELS, A. G., EDELSTAM, C., OLSSON, M., and WESTERMARK, T. *Fauna Flora, Upps.*, 1968, **63**, 323.
7. OLSSON, M. Personal communication.

8. Jensen, S., Johnels, A. G., Olsson, M., and Otterlind, G. DDT and PCB in herring and cod from the Baltic, the Kattegat and the Skagerrak. Ambio Spec. Rep. No. 1 (1st Soviet–Swedish Symposium on Pollution of the Baltic), 1971, p. 78.
9. Jensen, S., and Olsson, M. DDT and PCB in the Baltic. 2nd Swedish–Soviet Symposium on Control of Baltic Sea Pollution, Riga, 1973 (*to be published*).
10. Hallberg, R. O. A geochemical method for investigation of paleoredox conditions. 2nd Swedish–Soviet Symposium on Control of Baltic Sea Pollution, Riga, 1973 (*to be published*).

Dr. Svansson, when presenting his paper, said that after studying the preprints for the Symposium, he was aware of the fact that his paper dealt much less with sea outfalls than did most of the others. Nevertheless, it stressed the consequences of discharges of treated or untreated sewage from such outfalls on the condition of the whole Baltic in terms of phosphorus, deep oxygen, etc. Whereas the winter phosphate level in the 1950's had been approximately 9 μg/l. it now seemed to have doubled. Primary production had thereby increased, and this had probably enhanced the pelagic fishery but worsened oxygen conditions in the deteriorating deep cod fishery; conditions in the stagnant deeps, however, were also subject to climatic variations.

DISCUSSION

Dr. J. A. French (U.S.A.) asked how the broken part of the curve at the left of Fig. 3(c) had been obtained; this ndicated that salinities might have been higher at earlier times and that salinity variations might be periodic (with a period of the order of a century) rather than monotonically increasing, and he invited the Author's comments. The curves in Fig. 3(b) and (c) showed definite increases in salinity, and the text suggested a correlation with decreasing flow, yet the Vuoksi flow did not appear to be decreasing. He asked whether there had been a measurable decrease in rainfall or in the flows of the other rivers.

Mr. H. M. Cotton (New Zealand) referred to Dr. Svansson's remarks on climatic changes in the Baltic, and asked him to indicate the types of change considered.

REPLY TO DISCUSSION

In reply, Dr. Svansson explained that the broken line in Fig. 3(c) started from a single measurement in 1877 during a large pioneer cruise. The salinity variations could, of course, be explained in terms of periods of different lengths, e.g. 5, 10, 30, or 80 years. He agreed that whereas for periods up to 30 years there was correlation with flow, this was not true for longer periods. One hypothesis of the mechanism involved was an influence of change in flow (as outlined in the paper), another had been presented by Dickson[11] who argued that the inflow from the ocean was changed. Both hypotheses, however, originated from the same source: variations in atmospheric activity.

ADDITIONAL REFERENCE

11. Dickson, R. R. A recurrent and persistent pressure-anomaly pattern as the principal cause of intermediate-scale hydrographic variation in the European shelf seas. *Dt. hydrogr. Z.*, 1971, **24**, 97–119.

Bacterial Indicators of Sewage Pollution

G. J. Bonde
Aarhus University, Denmark

Indicator organisms may be grouped functionally into three general categories with, however, no sharp demarcations[1]: those used primarily to identify environmental changes or factors which may be unknown, those used primarily to quantify pollution levels, and those which can be studied under controlled conditions in the laboratory to interpret and further evaluate the importance of field data.

Bacteria are typical monitors of faecal pollution in the aquatic environment. Many are restricted in their usefulness to areas near the source of pollution, e.g. *Escherichia coli*, whereas spore-formers such as *Clostridium perfringens* are wide-ranging and can also identify old and remote pollution. Generally, bacteria are non-specific in identification, although, for instance, some *Salmonella* types[2] and faecal streptococci[3] may indicate specific industrial wastes or distinguish human from animal pollution. One of the advantages of bacteria as indicators is that they may be identified by simple techniques, so that a large number of samples can be handled at one time with greater sensitivity than that of physico–chemical methods.

When monitoring aquatic sites, large samples must be collected frequently. Bacterial populations in sediments are more stable and need to be checked less frequently.

The literature dealing with indicator bacteria is comprehensive. Generally, reference may be made to text-books and standard methods[4–10]. The ideal requirements for indicators have been given by the Author[8,11]. For the examination of sewage, only a quantitative estimation is relevant. The fundamental methods of water and sewage examination were established at the beginning of the century—mainly inspired by Houston[12,13] who maintained the superiority of microbiological over chemical methods.

Rationale of Bacteriological Examination

Water is not harmful to bacteria, viruses, or parasites, but may be an ideal vehicle for transport. However, few pathogens can multiply in water (exceptions that can be mentioned are *Vibrio parahaemolyticus* and some types of *Salmonella*), and bacteria are seldom free-floating in water, but tied up in heaps and clumps of debris, so that it is possible by ingesting one such particle to consume the infective dose even when this is large.

For most diseases, water is only an occasional source of infection, and for others only a rare one. Reference may be made to Bonde[8] and to Craun and McCabe[14] regarding water-borne diseases.

Investigations in recent years[15–19] have demonstrated some relationship between pollution and disease in various fresh-water areas, but not, however, in coastal areas[20–22]. Another epidemiological investigation conducted in England[23] concluded that bathing carried only a negligible health risk and that the risk, where present, was probably associated with chance contact with infectious faecal particles.

Minimal Infective Dose

The available data suggest a wide range of levels among the water-borne pathogens, and there is no generally accepted minimal concentration[24,25]. It is usually considered that for *S. typhi* and *Vibrio cholera* quite low doses are necessary (3–5 organisms) whereas other *Salmonella* types and enteropathogenic *E. coli* require considerably higher doses[26]. Besides the dose of infection, the amount of water swallowed during bathing is also important, but not generally agreed upon.

Demonstration of Pathogens in Aqueous Environments

Although it is often difficult to prove the transmission of water-borne diseases by epidemiological methods, the techniques for demonstrating pathogenic micro-organisms have improved greatly in recent years, and pathogens of the *Salmonella* genus as well as different types of enteroviruses have often been demonstrated, also by quantitative methods[2,5,10,27–47]. Sewage treatment processes have often been found to reduce the content of pathogens to a very slight degree or not at all[35,38]. In other cases, reductions of 80–99 per cent were found for *Salmonella*[48,49]; enteroviruses might be reduced by 40–99 per cent[50,51], the latter reduction, however, depending entirely on the proper operation of the purification plant.

Another recently demonstrated problem is the transfer of R-factors (entailing resistance to antibiotics) from *E. coli* to *Salmonella* and *Shigella*[52,53].

Presentation of Indicators

Bacteria of the coliform group must be considered the primary, most frequently applied indicators. This is, however, very much due to tradition and not to the fact that the coli group is more easy to handle as regards efficiency or precision, nor regarding methods of demonstration.

The coliform group

To the coli group belong, beside the indicator organisms usually named 'coliforms' or 'bacteria of the coli-aerogenes group', several species with slight or dubious sanitary importance, as well as numerous species of no interest to hygiene at all (*Serratia*, *Erwinia*).

No single test is found that will at the same time permit an easy and unambiguous demonstration of these different groups of Gram-negative rods, and this fact is an essential obstacle to a numerical estimation. A majority of strains of *Escherichia coli* are able to ferment lactose with the production of acid and gas, and this property is found of use in presumptive, quantitative determinations. Not all strains of *E. coli* are, however, able to ferment lactose, just as this property is not specific for *E. coli* but is also found in other related bacteria that are often present in the same samples. Micro-organisms that ferment lactose with formation of acid and gas in the presumptive media and cannot be identified as *E. coli* are named false presumptives.

On the other hand, some organisms that belong in other families and can never be called 'coliforms' because of their different morphology may, by virtue of their vigorous fermentative breakdown of carbohydrates, appear as false presumptives, e.g. *Bacillus polymyxa* and *B. macerans*, and clostridia, in particular *Cl. perfringens*. Many authors apply the name 'false presumptives' only to these organisms that are not 'coliforms' but still give false fermentation of the presumptive media (see, for example, Bonde[8a]).

In addition, special problems arise in connection with the quantitative estimation of bacteria by inoculation of samples of water, soil, and sewage, as the organisms under investigation may be attenuated by their stay in water and may grow slowly or may even have lost some of their characteristics.

Classification of lactose-fermenting bacteria in water examination

A review of the lengthy history of coliforms in water examinations has been given by the Author[8b].

Within water bacteriology a division must so far cut across all the classifications mentioned, as only strains producing visible acid and gas within 48 h in lactose media are accepted as belonging to the coliform group.

Increasing evidence is given that *E. coli* (type I) is the most reliable indicator of fresh faecal pollution in contrast to the count of total coliforms[8,54–56], and tests of thermotolerance gain more and more attention[57]. It is thus questionable whether coliforms can be regarded as true indicators of faecal pollution at all.

The pseudomonad group

As most species of the genus *Pseudomonas* are widely distributed in nature, living as saprophytes in fresh surface water and in sea water as well as in soil, on plants, etc., the bacilli of this genus clearly do not lend themselves to an application as indicators of pollution. Some of the pseudomonads are plant pathogens, a smaller number are pathogenic for amphibians, reptiles, and fish, but so far only *Ps. aeruginosa* is known as a pathogen for man and warm-blooded animals. *Ps. aeruginosa* was described as an indicator of pollution by Houston[13], later on by Ringen and Drake[58], Taylor[6a], Buttiaux[4a], and Reitler and Seligman[59].

Determinations of *Ps. aeruginosa* are important particularly in estuarine waters when high water temperatures and available nutrients might allow growth of this organism. Generally the concentration of *Ps. aeruginosa* in sewage and receiving waters is quite low: 0–30 per 100 ml[8,60].

Aeromonas spp. as indicators

Aeromonas species have recently attracted considerable interest through the work of Schubert[61,62] who reclassifies the group and also gives evidence that these bacteria, in particular *A. hydrophila*, may indicate pollution. Further information on this group is given later in the present paper and by Hansen and Bonde[63].

Faecal streptococci

Faecal streptococci are supposed to indicate recent faecal pollution[3,64–68]. The most valuable application of the faecal streptococcus relates to the ratio of faecal coliforms to faecal streptococci, a ratio of 4 or greater indicating a discharge of municipal wastes, and one of 0·6 or less indicating surface-water run-off[3].

Unfortunately, the faecal streptococcus group also includes several biotypes that are of limited sanitary significance[3]. Therefore, until better methodology is available, the use of faecal streptococcus limits below 100 organisms per 100 ml in recreational water may be unrealistic unless confirmed by parallel faecal coliform examinations[69].

Clostridium perfringens

Bonde[8] gives a review of the classification, production of toxins, and pathogenicity, as well as occurrence in soil, water, and sewage, of this organism. The most controversial point is whether it is really ubiquitous and mainly found as spores in nature. Klein and Houston[12], Houston[13], and McGowan, and others, considered *Cl. perfringens* to be a primary faecal and pathogenic organism, which as an indicator bacterium was of indisputable and greater significance than the coli bacteria. Wilson and Blair[70] considered the demonstration of cells in the vegetative form to be particularly valuable, and stressed that *Cl. perfringens* is an organism indisputably of faecal origin and is thus of great importance for the detection of intermittent and occasional pollution. Many other workers such as Miege-Cellot[29a], Buttiaux[4b], and Taylor[6b] considered the search for *Cl. perfringens* to be a valuable supplement to other examination methods. A more negative attitude towards the value of the cultivation of anaerobes is adopted by American workers[28].

Indicators in extreme climates

In hot climates the differential tests generally used for the identification of *E. coli* cannot be taken advantage of, as other coliform types are thermotolerant and even indole positive without belonging in the *E. coli* group[71].

In cold climates, on the other hand, man and warm-blooded animals do not always harbour *E. coli* in their intestines, but other coliforms[72]. As a substitute for the examination of *E. coli* and coliforms, the application of *Cl. perfringens* counts was recommended by Bonde[8]. Evison and James[73] recommend the application of *Bifidobacterium* (*Lactobacillus bifidus*).

Other indicators

Bacteria of other genera, when they occur in higher numbers correlated to pollution, or unexpectedly occur in localities because of pollution, may also be taken into consideration as indicators, but will generally not be applicable as monitors since they are demonstrable only by qualitative methods[55,56]. Among these are sulphite-reducing bacteria and *Thiobacillus* species. *Bacillus* species may by their formation of antibiotics in presumptive media disturb water examinations. Additionally, *Bacillus* species by the special preference of certain species for certain environments (soil, sewage, fresh water) also arouse suspicion as to the mode of pollution of a body of water[56]. Strains of *B. licheniformis* occur with a frequency above expectation in fresh and saline waters (as well as in sewage), whereas treated waters and filter sand have a higher representation of strains with bulging spores, mostly *B. brevis*.

Sediments near the coast show an increased frequency of *B. cereus*, *B. pantothenticus*, and *B. sphaericus* strains, and the samples of strains from faeces are dominated by *B. subtilis*.

Bacteriophages are sometimes used as indicators of the presence of *E. coli*[74], and of specific disease-producing organisms[75].

Finally it may be mentioned that a demonstration of all *Salmonella* spp. has been advocated. However, these organisms are widely different as to origin and sanitary significance, and they occur in quite small numbers, probably also intermittently.

Some Remarks on Methodology

Space does not permit a detailed account of methods for demonstration of indicators in sewage, but a few essential items should be stressed. As quantitative methods are particularly important for this purpose, the distribution followed by such counts should be ascertained. In some instances a simple Poisson law is found to apply[8,76], but in most cases more complicated laws—exponential or negative binomial—were better descriptions of the data[77,8], much depending upon the kind and treatment of the sample.

The computation of Most Probable Numbers by the dilution method is based upon the assumption of a simple Poisson law. Much criticism has been raised against this method, but it has become a tool of wide utility, and the loss of precision following from the use of it is counter-balanced by the methodological deficiencies and great variability of colony counting[78].

Membrane filtration, likewise, is no ultimate solution as this procedure will have all the disadvantages of counts on solid media (many false presumptives, confluent colonies), and the necessity of making an unknown number of dilutions[79,54,69]. Another essential point is the choice of the right medium for the special purposes. Media fitted to the investigation of drinking water will not automatically be applicable in the examination of sewage, sea water, or sediments. The glutamic acid medium was thus found less useful for a primary incubation at 44°C of sewage samples[80], and simple lactose medium not at all applicable for sewage or use at higher temperatures[8].

Relation Between Indicators

A functional dependence of the occurrence of one kind of organism on the other could not be expected, and the word 'correlation' should be used with caution.

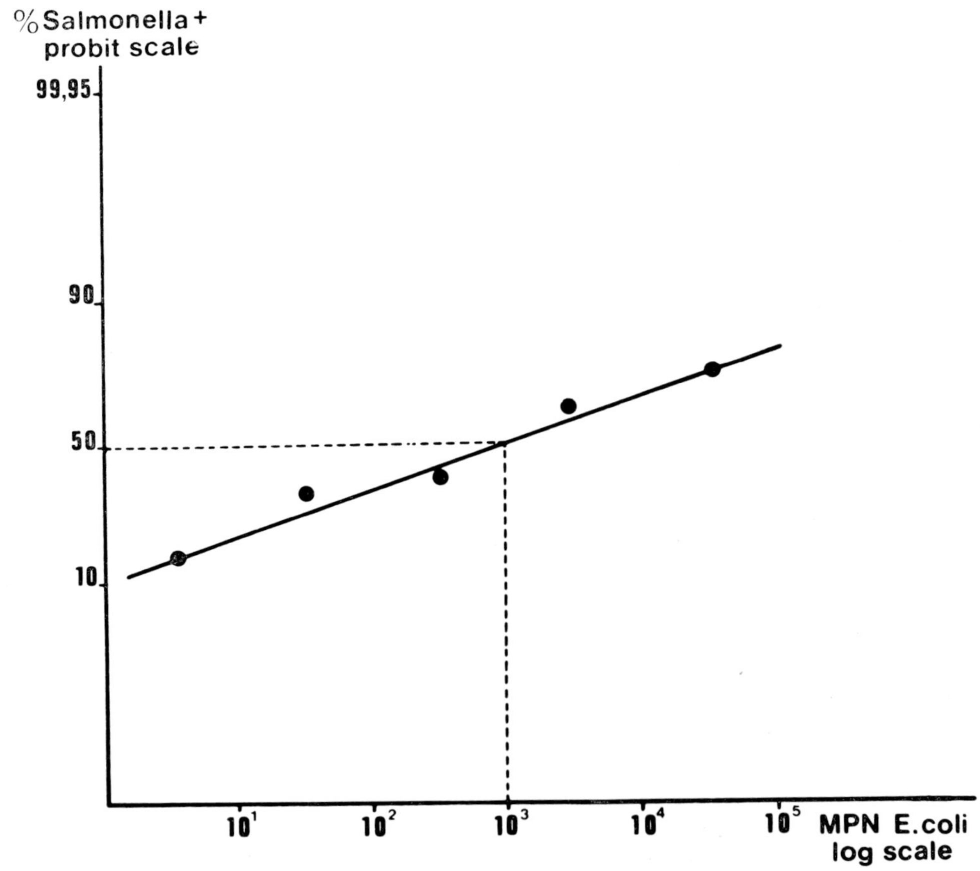

FIG. 1 Probit transformation of data from Grunnet *et al.*[43] showing probability of finding *Salmonella* at different levels of *Escherichia coli*

Association of pathogenic organisms to faecal coliforms

From what was previously said, one could not expect such relationships between human pathogens and total coliforms as the latter may be of varied origin and may multiply in nature. For the same reason, one could not even expect a constant relationship between *E. coli* and total coliform counts, just as it is not possible to make use of different types of enterococci for distinguishing completely between animal and human pollution.

As the methods for quantitative estimation of *Salmonella* are, still, less advanced and of unknown precision, the frequency of demonstration of *Salmonella* following an enrichment of filters or gauze pads is compared to the counts of faecal coliforms.

The graph of the data of Grunnet *et al.*[43] from estuarine waters is slightly S-shaped and has a certain resemblance to a dose-response curve, and following a probit transformation the shape is quite rectilinear (Fig. 1). A count of 1000 faecal coliforms corresponds to an average of 50 per cent probability of demonstrating *Salmonella*.

The probability of recovering viruses is not so well established. By means of data from Lund[46], Shuval *et al.*[47], and Taylor[81] it is found that viruses were recovered from 53 per cent of the samples for which the *E. coli* count exceeded 1000 per 100 ml, and from 18 per cent of those with lower counts.

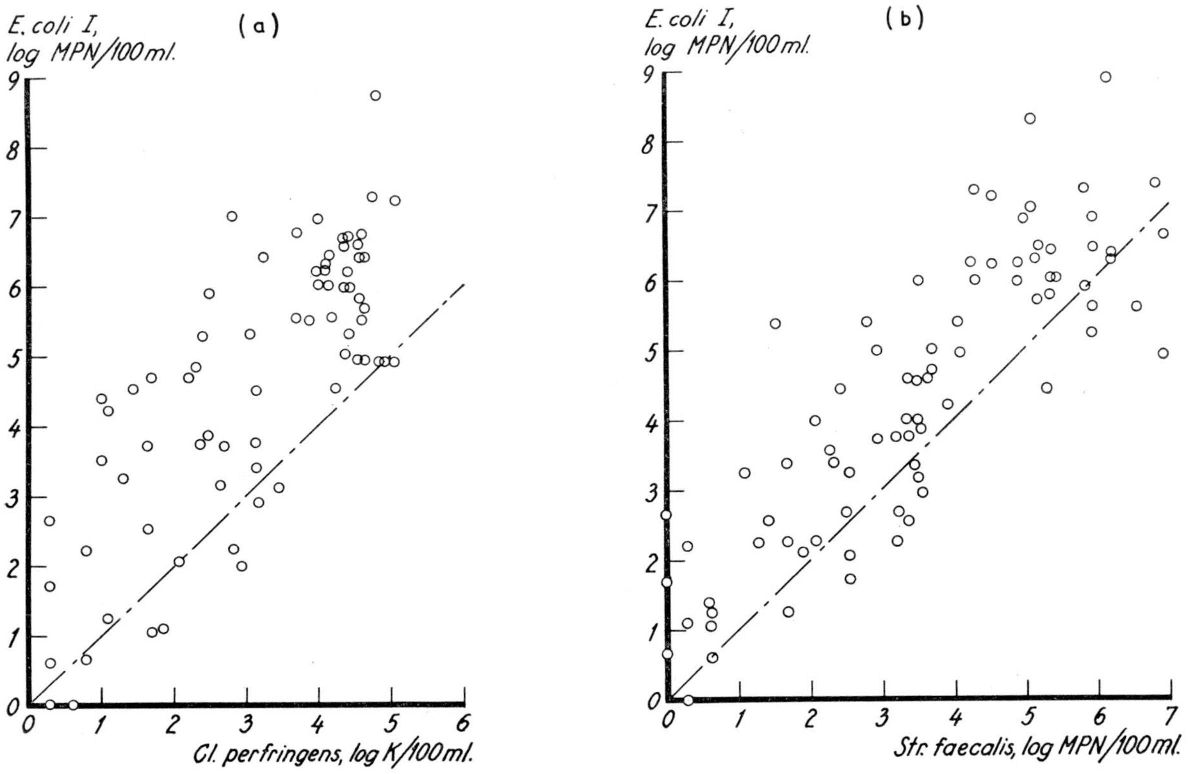

FIG. 2 Relation of log counts of *Escherichia coli* I to those of (*a*) *Clostridium perfringens* and (*b*) *Streptococcus faecalis*

Association of *Cl. perfringens* to faecal coliforms

In a suspension of faecal matter the ratio of *Cl. perfringens* to *E. coli* may differ from that in the effluent from a biological filtration plant, in sea water, or in sludge. Furthermore, the two sets of estimates are based on different methods of determination: the *Cl. perfringens* estimate on a colony count, the *E. coli* estimate on a multiple-tube method.

Figure 2(a) shows the relation between counts of faecal coliforms per 100 ml and of *Cl. perfringens* in non-pasteurized samples. It is seen that the counts of faecal coliforms are often about 100 times as great as those of *Cl. perfringens*.

Comparison of *Streptococcus faecalis* and faecal coliforms

These two organisms show a slightly better relationship, the counts of faecal coliforms being generally ten times as great as those of faecal streptococci (Fig. 2(b)). Occasionally the faecal streptococci give the higher counts.

Occurrence of *Pseudomonas* spp. related to coliform and faecal coliform counts

The distribution of samples with or without *Ps. aeruginosa* on the different groups according to the numbers of thermo-tolerant coliforms is given in Table 1. It is seen that *Ps. aeruginosa* were found in far smaller numbers than *E. coli*. However, they seem to be able to survive for a long time in dried digested sludge. If they occur in greater numbers they will indicate faecal pollution. Nevertheless, *Ps. aeruginosa* occurs too sporadically for it to be considered an ideal indicator organism.

TABLE 1. Relations between coliform counts and presence of *Pseudomonas*
(Numbers in parentheses are percentages for coliform ranges)
(a) Faecal coliforms and *Ps. aeruginosa*, 89 samples

Faecal coliforms (per 100 ml)	<920	920–9200	>9200
Ps. aeruginosa present	2 (4)	7 (26)	10 (83)
Ps. aeruginosa absent	48 (96)	20 (74)	2 (17)

(b) Coliforms (MPN) and *Pseudomonas* spp., 220 samples

Coliforms (per 100 ml)	0	2–95	110–920	>920
Pseudomonas spp. present	4 (31)	43 (52)	47 (70)	53 (90)
Pseudomonas spp. absent	9 (69)	40 (48)	18 (30)	6 (10)

In groups with less than 100 faecal coliforms per 100 ml the demonstration of *Pseudomonas spp.* succeeds or fails with equal frequency; however, when the more polluted samples are considered the frequency of success increases rapidly, and that of failure decreases correspondingly. No consistent relation has been found between the frequency of occurrence of fluorescent colonies and the degree of pollution.

Ecological Investigations

Indicators in sewage and sewage effluents

Table 2 gives results of a survey of counts of *Cl. perfringens* and Table 3 of *E. coli* in sewage. The rates at which the two organisms diminish through the plants are different. The number of *Cl. perfringens* is reduced both by sedimentation and in the biological filter; the number of *E. coli* is only slightly reduced by sedimentation, but on the other hand the reduction in the filter is greater, the reduction in number throughout the whole plant being greater than that of *Cl. perfringens*. It is evident that the ratio of *E. coli* to *Cl. perfringens* will be different in different kinds of samples, and the ratio of 100–1000:1 often cited cannot be expected to exist. Sometimes receiving waters contain just as many or more *Cl. perfringens* than *E. coli*. In certain industrial wastes all coliform bacteria may be killed while *Cl. perfringens* spores generally remain intact.

TABLE 2. *Clostridium perfringens* (counts of black colonies) per ml of sewage

	No. of samples	Min.	Max.	Mean		Mean percentage reduction
				Arith.	Geom.	
Raw sewage	27	298	3312	1323	1000	0
Settled sewage	43	80	1160	471	380	62–64
Biologically filtered sewage	23	51	515	224	162	83–84

TABLE 3. *Escherichia coli* (MPN values) per ml of sewage

	No. of samples	Min.	Max.	Geom. mean	Percentage reduction
Raw sewage	8	11 000	160 000	30 000	0
Settled sewage	15	2 400	160 000	27 000	10
Biologically filtered sewage	14	280	9 500	1 600	95

Indicators in receiving waters

Laboratory investigations, where sedimentation and dilution can be disregarded, may give an estimate of the isolated mortality factor. However useful for elucidating one or a few factors at a time such laboratory experiments may be, they are rather confusing and leave us only the laborious way of studying what takes place in nature on the spot by two procedures: (i) studies of the distribution of different taxonomic units in the sea and sediments at different distances from the source of pollution by different conditions of weather, currents, depth, etc., and (ii) the controlled quantitative study of the decline in numbers of one species parallelled by an examination of the rate of dilution in areas with only one predominant source of pollution. It must be stressed that there is an essential difference in respect to reduction of enteric bacteria between primary and secondary effluent[82,83]. After 5–6 h for primary effluent, the mortality curves approach an almost horizontal, asymptotic course.

Further discussion of microbial mortality would be out of place in the present paper, since a whole day of this Symposium is devoted to the subject.

An analysis of variance of the results of more than 19 000 individual *E. coli* counts collected under varied conditions of current, wind, week, time of the year, and distance from source of pollution in the Sound[55,11] during a five-year period has shown the effects of (1) current, (2) wind, (3) distance from source, (4) week, and (5) year to be of importance in the order mentioned. The factor 'week' seems to be identical with weekly fluctuations of the water temperature and rainfall. The counts show a definite maximum in July–August. At coastal stations the amount of pollution and distance from its source were the absolutely dominating factors[11]. A consequence of these investigations is that the value of *E. coli* as an indicator is quite restricted geographically. Another more resistant indicator is needed to trace the distant and old pollution[55,56].

Dispersion of indicators in sediments

Sedimental bacteria form an interstitial flora between grains of sand etc. and may cover enormous surfaces and exert a corresponding influence on the total environment. Because of its stationary character the sedimental flora is of importance in the estimation of long-term effects and variations over long periods.

TABLE 4. Percentage distributions of *Clostridium perfringens* in waters from six areas

Cl. perfringens (per 100 ml)	1	2	3	4	5	6
0	2	0	0	1	0	2
1–10	0	0	0	8	0	20
10–100	2	0	5	43	63	24
100–1 000	15	39	45	38	37	45
1 000–10 000	64	46	41	11	0	9
10 000–100 000	17	15	9	0	0	0

1. Limfjorden. 2. Svendborg. 3. Oeresund. 4. Gothenburg. 5. Kungsbacka. 6. Lillebaelt.

TABLE 5. Percentage distribution of species in sediments from same areas as in Table 4, and also (7) Sargasso

	1	2	3	4	5	6	7
E. coli I	14	1·4	0	0	0	8	0
Coliforms	14	21·4	0	28·6	1·2	8	16
Proteus	0	0	0	0	0	0	2
Aeromonas-Vibrio	18·4	5·7	6	17·9	2·4	0	7
Pseudomonas	24·2	24·3	12	35·7	0	31	12
Comamonas	8·8	7·1	4	0	0	0	0
Achromobacter	0	0	1	7·1	0	0	0
Acinetobacter	0	0	10	0	8	8	0
Alcaligenes	0	7·1	2	0	0	0	2
Coryneforms	0	2·9	4	3·6	3·5	0	0
Cytophaga-Flavobact.	1·2	2·9	3	0	2·3	0	0
Micrococci	5	1·4	12	0	2·3	0	0
Bacillus	14·4	24·3	46	7·1	80	46	61
Spirillum	0	1·4	0	0	0	0	0

The ideal indicator of pollution in sediments is the persistent anaerobic spore-former *Cl. perfringens*, but the composition of the total heterotrophic flora is also of importance as an overall estimate, whereas Gram-negative rods of enteric origin like *E. coli*, according to the studies of decay in water, might be expected only in recently polluted areas[55,56]—see Tables 4 and 5.

REFERENCES

1. Butler, P., Andrén, L. E., Bonde, G. J., Jernelöv, A. D., and Reish, D. J. Test, monitoring, and indicator organisms. In 'A guide to marine pollution' (ed. E. D. Goldberg), Gordon and Breach, New York, 1972, p. 147.
2. Grunnet, K., and Nielsen, B. B. *Salmonella* types isolated from the Gulf of Aarhus. *Appl. Microbiol.*, 1969, **18**, 985.
3. Geldreich, E. E., and Kenner, B. A. Concepts of fecal streptococci in stream pollution. *J. Wat. Pollut. Control Fed.*, 1969, **41**, R336.
4. Buttiaux, R. L'analyse bactériologique des eaux de consommation. Paris, 1951, *a* p. 180, *b* p. 131.
5. Buttiaux, R. Surveillance et contrôle des eaux d'alimentation. *Revue Hyg. Méd. soc.*, 1958, **6** (2), 170.
6. Taylor, E. W. The examination of waters and water supplies. Churchill, London, 7th Edn, 1958, *a* p. 462, *b* p. 468.
7. Buttiaux, R., and Mossel, D. A. A. The significance of various organisms of faecal origin in foods and drinking water. *J. appl. Bact.*, 1961, **24**, 353.
8. Bonde, G. J. Bacterial indicators of water pollution. Teknisk Forlag, Copenhagen, 1962, *a* p. 221 *b* p. 226.
9. World Health Organization. European standards for drinking water. WHO Regional Office for Europe, Copenhagen, 1970, p. 4.
10. Scarpino, P. V. Bacterial and viral analysis of water and waste water. In 'Water and water pollution handbook', Vol 2 (ed. L. L. Ciaccio). Dekker, New York, 1971, p. 639.
11. Bonde, G. J., and Thomsen, A. M. Computer analysis of a bacteriological monitoring system for pollution control. *Proc. 5th int. Colloquium med. Oceanogr., Messina*, 1973, p. 73.
12. Klein, E., and Houston, A. C. Further reports on bacteriological evidence of recent and therefore dangerous sewage pollution. In 'Metropolitan water supply', 28th Annual Report of Local Government Board, Supplementary Report of Medical Officer, Appendix B, London, 1898–99, p. 498.
13. Houston, A. C., and McGowan, G. In '2nd Report of Royal Commission on Sewage Disposal', Interim Rep., Vol 2, Wyman, London, 1902, p. 326.
14. Craun, G. F., and McCabe, L. J. Review of the causes of water-borne disease outbreaks. *J. Am. Wat. Wks Ass.*, 1973, **65**, 74.
15. Smith, R. S., Woolsey, T. D., and Stevenson, A. H. Bathing water quality and health. I—Great Lakes. U.S. Public Health Service, Environmental Health Center, Cincinnati, 1951.
16. Smith, R. S., and Woolsey, T. D. Bathing water quality and health. II—Inland river. U.S. Public Health Service, Environmental Health Center, Cincinnati, 1952.
17. Smith, R. S., Woolsey, T. D., and McCabe, L. J. Bathing water quality and health. III—Coastal water. U.S. Public Health Service, Robert A. Taft Sanitary Engineering Center, Cincinnati, 1961.
18. Conn, N. K., Heymann, C. S., Jamieson, A., McWilliam, J. M., and Scott, T. G. Water-borne typhoid fever caused by an unusual Vi-phage. *J. Hyg., Camb.*, 1972, **70**, 245.
19. Geldreich, E. E. Principles of developing criteria for coastal bathing waters. Working paper for WHO–GESAMP. Conference on Scientific Aspects of Marine Pollution, Vienna, 1973.
20. Stevenson, A. H. Studies on bathing water quality and health. *Am. J. publ. Hlth*, 1953, **43**, 529.
21. Moore, B. The present status of diseases connected with marine pollution. *Revue int. Océanogr. méd.*, 1970, **18–19**, 193.
22. World Health Organization. Health criteria for the quality of recreational waters with special reference to coastal waters and beaches. WHO/W.Poll/72.10, Geneva, 1972.
23. Public Health Laboratory Service. Sewage contamination of coastal bathing water in England and Wales. *J. Hyg., Camb.*, 1959, **57**, 435.
24. Plotkin, S. A., and Katz, M. Minimal infective doses of viruses for man by the oral route, In 'Transmission of viruses by the water route' (ed. G. Berg), Wiley-Interscience, New York, 1967, p. 151.
25. Du Pont, H. L. The response of man to virulent *Shigella flexneri*. *J. infect. Dis.*, 1969, **119**, 296.
26. McCullough, N. B., and Eisele, C. B. Experimental human salmonellosis. *J. infect. Dis.*, 1951, **88**, 278.
27. Kiribayashi, S., and Aida, T. A study in the fate of cholera vibrio in the sea water of Keeling Port, Formosa. *J. publ. Hlth Ass. Japan*, 1933, **9** (11), 1.
28. Prescott, S. C., Winslow, C. E. A., and McCrady, M. H. Water bacteriology with special reference to sanitary water analysis. Wiley, New York, 6th Edn, 1946.

29. Miege-Cellot, J. Étude bactériologique des eaux superficielles de la poche de Dunkerque. Thesis, Lille, 1947, *a* p. 81.
30. Moore, B. The detection of paratyphoid carriers in towns by means of sewage examination. *Mon. Bull. Minist. Hlth*, 1948, **5** (7), 241.
31. Stryszak, A. Behaviour of microorganisms of the *Salmonella* group in the sea water of the Gulf of Gdansk. II. *Bull. Inst. mar. trop. Med. Gdansk*, 1949, **2**, 213.
32. Steiniger, F. On shifts of frequency of freely living *Salmonella* forms at the coasts of Heligoland. *Zentbl. Bakt. ParasitKde*, I, Orig., 1955, **162**, 440.
33. Greenberg, A. E., Wickenden, R. W., and Lee, T. W. Tracing typhoid carriers by means of sewage. *Sew. ind. Wastes*, 1957, **29**, 1237.
34. McCoy, J. H. Salmonellae in crude sewage, sewage effluent and sewage-polluted natural waters. *Proc. 1st int. Conf. Wat. Pollut. Res., London, 1962*, Vol 1, p. 205.
35. Berg, G. (ed.) Transmission of viruses by the water route. Wiley-Interscience, New York, 1967.
36. Shuval, H. I., Cymbalista, S., Wachs, A., Zohar, Y., and Goldblum, N. The inactivation of enteroviruses in sewage by chlorination. *Proc. 3rd. int. Conf. Wat. Pollut. Res., Munich, 1966*, Vol. 2, p. 37.
37. Brisou, J. Viral and parasitic pollution of littoral waters and its consequences on public health. *Bull. Wld Hlth Org.*, 1968, **38**, 79.
38. Grabow, W. O. K. The virology of waste water treatment. *Wat. Res.*, 1968, **2**, 675.
39. Metcalf, T. G., and Stiles, W. C. Viral pollution of shellfish in estuary waters. *J. sanit. Engng Div. Am. Soc. civ. Engrs*, 1968, **94**, 595.
40. Slanetz, L. W., Bartley, C. H., and Stanley, K. W. Coliforms, fecal streptococci and *Salmonella* in sea water and shellfish. *Hlth Lab. Sci.*, 1968, **5**, 66.
41. Taylor, E. W. Forty-second report of Director of Water Examination, 1965–66, Metropolitan Water Board, London, 1968, p. 117.
42. Brezenski, F. T., and Russomano, R. The detection and use of salmonellae in studying polluted tidal estuaries. *J. Wat. Pollut. Control. Fed.*, 1969, **41**, 725.
43. Grunnet, K., Gundstrup, A. S. P., and Bonde, G. J. Isolation of *Salmonella* from a polluted marine environment. *Revue int. Océanogr. méd.*, 1970, **17**, 165.
44. Brezenski, F. T. Estuary water quality and *Salmonella. Proc. Nat. spec. Conf. Disinfect., Am. Soc. civ. Engrs, New York*, 1971.
45. Kristensen, K. K. Kvalitative og kvantitative Salmonella-underøgelser af spildevand og Øresundsvand. In 'Report on the investigations of the Swedo–Danish Committee on Pollution in the Sound 1965–1970'. Statens Trykningskontor, Lund, 1971, p. 226.
46. Lund, E. Undersøgelser over forekomst af virus i Øresundsvand. In 'Report on the investigations of the Swedo–Danish Committee on Pollution of the Sound 1965–1970'. Statens Trykningskontor, Lund, 1971, p. 292.
47. Shuval, H. I., Thompson, A., Fattal, B., Cymbalista, S., and Wiener, Y. Natural viruses inactivation processes in sea water. *Proc. Nat. spec. Conf. Disinfect., Am. Soc. civ. Engrs, New York*, 1971.
48. Mathews, W. W. Annual report on sewage treatment works for 1956. Gary Sanitary District, Gary, Indiana.
49. McCoy, J. H. The presence and importance of *Salmonella* in sewage. *Proc. Soc. Wat. Treat. Exam.*, 1957, **6**, 81.
50. Clarke, N. A., Stevenson, R. E., Chang, S. L., and Kabler, P. W. Removal of enteric viruses from sewage by activated sludge treatment. *Am. J. publ. Hlth*, 1961, **51**, 1118.
51. Gilcreas, F. W., and Kelly, S. M. Significance of the coliform test in relation to intestinal virus pollution of water. *J. New Engl. Wat. Wks Ass.*, 1954, **68**, 255.
52. Smith, H. W. Incidence in river water of *Escherichia coli* containing R-factors. *Nature, Lond.*, 1970, **228**, 1286.
53. Smith, H. W. Incidence of R^+ *E. coli* in coastal bathing water of Britain. *Nature, Lond.*, 1970, **234**, 155.
54. Bonde, G. J. Bacteriological methods for the estimation of water pollution. *Hlth Lab. Sci.*, 1966, **3**, 124.
55. Bonde, G. J. Heterotrophic bacteria in a polluted marine environment. *Proc. 3rd int. Conf. Wat. Pollut. Res., Munich, 1966*, Vol. 3, p. 87.
56. Bonde, G. J. Studies on the dispersion and disappearance phenomena of enteric bacteria in the marine environment. *Revue int. Océanogr. méd.*, 1968, **9**, 17.
57. Geldreich, E. E. Sanitary significance of fecal coliforms in the environment. U.S. Federal Water Pollution Control Administration, *Wat. Pollut. Control Res. Ser.* Publ. No. WP-20-3, 1966, p. 122.
58. Ringen, L. M., and Drake, C. H. A study of the incidence of *Pseudomonas aeruginosa* from various natural sources. *J. Bact.*, 1952, **64**, 841.
59. Reitler, R., and Seligman, R. *Pseudomonas aeruginosa* in drinking water. *J. appl. Bact.*, 1957, **20**, 145.
60. Grunnet, K., Gundstrup, A. S. P., and Bonde, G. J. Tetrationate broth as a medium for simultaneous demonstration of *Salmonella* and *Ps. aeruginosa. Nord VetMed.*, 1974, **26**, 239.
61. Schubert, R. H. W. Die Systematik der *Aeromonaden. Arch. Hyg. Bakt.*, 1967, **150**, 682.

62. SCHUBERT, R. H. W. Das Vorkommen der *Aeromonaden* in oberirdischen Gewässern. *Arch. Hyg. Bakt.*, 1967, **150**, 689.

63. HANSEN, J. C., and BONDE, G. J. *Aeromonas hydrophila* as a possible cause of furunculosis in eel. *Nord. VetMed.*, 1973, **25**, 121.

64. MALLMANN, W. L., and SELIGMAN, E. B. A comparison study of media for the detection of streptococci in water and sewage. *Am. J. publ. Hlth*, 1950, **40**, 286.

65. LITSKY, W., MALLMANN, W. L., and FIFIELD, C. W. Comparison of the most probable numbers of *E. coli* and enterococci in river water. *Am. J. publ. Hlth*, 1955, **45**, 1049.

66. SLANETZ, L. W., and BARTLEY, C. H. Numbers of enterococci in water, sewage and feces determined by the membrane filter technique with an improved medium. *J. Bact.*, 1957, **74**, 591.

67. BURMAN, N. P. Some observations on coli-aerogenes bacteria and streptococci in water. *J. appl. Bact.*, 1961, **24**, 368.

68. KENNER, B. A., CLARK, H. F., and KABLER, P. W. Cultivation and enumeration of streptococci in surface waters. *Appl. Microbiol.*, 1961, **9**, 15.

69. GELDREICH, E. E., JETER, H. L., and WINTER, J. A. Technical considerations in applying the membrane filter procedure. *Hlth Lab. Sci.*, 1967, **4**, 113.

70. WILSON, W. J., and BLAIR, E. M. M. Correlation of the sulphite reduction test with other tests in the bacteriological examination of water. *J. Hyg., Camb.*, 1925, **24**, 111.

71. RAGHAVACHARI, T. N. S., and IYER, P. V. S. The coli-aerogenes index of pollution used in the bacteriological analysis of water. *Indian. J. med. Res.*, 1939, **26**, 867.

72. HENRIKSEN, S. D. Comparison of coliform organisms from feces and from water in Norway. *Acta path. microbiol. scand.*, 1954, **35**, 75.

73. EVISON, L. M., and JAMES, A. A comparison of the distribution of intestinal bacteria in British and East African water sources. *J. appl. Bact.*, 1973, **36**, 109.

74. KOTT, Y., ARI, H. B., and BURAS, N. The fate of viruses in a marine environment. *Proc. 4th int. Conf. Wat. Pollut. Res., Prague, 1969*, p. 823.

75. GERNER-RIEUX, C., BUTTIAUX, R., and MUCHEMBLÉ, G. Epidemiological aspect of research on bacteriophages in water. *Hydrobiologica*, 1949, **1**, 105.

76. ROBERTS, E. A., and COOTE, C. G. The estimation of concentration of viruses and bacteria from dilution counts. *Biometrics*, 1965, **21**, 600.

77. BLISS, C. I. The analysis of insect counts as negative binomial distributions. *Proc. 10th int. Congr. Ent.*, 1956, Vol. 2, 1015.

78. COCHRAN, W. G. Estimation of bacterial densities by means of the MPN. *Biometrics*, 1950, **6**, 105.

79. BONDE, G. J. Om hygiejnisk-bakteriologisk vurdering af vand. *Nord. VetMed.*, 1963, **15**, 696.

80. GRUNNET, K., and BONDE, G. J. Methods for estimation of *E. coli* in receiving waters and sewage. *Proc. 5th int. Colloquium med. Oceanogr., Messina*, 1973, p. 291.

81. TAYLOR, E. W. Thirty-ninth report of Director of Water Examination, 1959–60, Metropolitan Water Board, London.

82. GUNNERSON, C. G. Sewage disposal in Santa Monica Bay. *J. sanit. Engng Div. Am. Soc. civ. Engrs*, 1958, **84** (SA1), Paper 1534.

83. WHEATLAND, A. B., AGG, A. R., and BRUCE, A. M. Some observations on the dispersion of sewage from sea outfalls. *J. Proc. Inst. Sew. Purif.*, 1965, 291.

Prof. Bonde, introducing his paper, said that after more than 20 years' work in the field of bacterial indicators in sewage he considered that the problems divided quite naturally into four groups:

1. Elaboration of methods and comparison between methods. The most decisive changes had been the introduction of membrane filtration, specific synthetic media, and dehydrated preparations.
2. Study of occurrence and distribution of indicator bacteria in relation to ecological factors, and setting up of mathematical models as a basis for computation and comparison.
3. Establishing of relationships between the occurrence of indicators and selected pathogens, with the purpose of evaluating the indicator systems.
4. Relevant grouping of micro-organisms, and their importance to human health or industrial processes. Taxonomy was not a static field: whole classes of organisms could disappear overnight by re-grouping or re-naming.

It had been possible to arrive at reasonable and reliable results within the first three fields. There was general agreement on the demands put on methods and media; mathematical models, based on reasonable approximations and assumptions had been set up, and some had proved useful in practical work. In his opinion, the greatest challenge lay in the instruction of the relevant procedures, and he emphasized the difficulty in the correct application of Most Probable Number tables.

Many organisms hitherto considered non-pathogenic had gained new importance with the advent of antibiotics, immuno-suppressive treatment, and the complicated equipment of synthetic materials. Related to the problems of possible pathogenicity were those of infective dose, invasive power, formation of antigens, and the possibility of growth in tissues and cavities—though these problems were perhaps more important in the bacteriology of drinking water.

He considered the organisms most closely associated with fresh human faecal pollution were *E. coli* or faecal coliforms, but that examination of more remote pollution (such as sediments in the sea) could preferably be made by counts of sulphite-reducing *Clostridia* at 48°C.

[*The discussion of this paper starts on p. 60*]

Human Enteric Viruses and Bacteriophages as Indicators of Sewage Pollution

Pasquale V. Scarpino
Professor of Environmental Engineering
Department of Civil and Environmental Engineering
University of Cincinnati, Ohio 45221

Microbial Indicators of Sewage Pollution

A myriad of pathogenic and non-pathogenic intestinal microorganisms are contributed by humans and other warm-blooded animals to sewage which enters rivers, streams, and large bodies of fresh and saline water, and contributes to the pollution of these environments. Therefore, the question posed by the early water bacteriologists was "Which bacterium would be the most useful to indicate sewage contamination of water and thus the presence or absence of enteric pathogens?" Today we can expand their search for a bacterial indicator to one that reflects also the presence or absence of human enteric viruses. Thus, the properties that an ideal indicator microbe should possess include[1] applicability in all types of water, presence whenever pathogens are present, and a reasonably longer survival time than the hardiest enteric pathogen; it should not reproduce in the contaminated water (thereby leading to an inflated value), and it should be harmless to man—although pathogens should usually pose no problem to a trained microbiologist. The density of the indicator in contaminated water should have some direct relationship to the degree of fecal pollution. The test procedure should have high specificity and sensitivity, and the method should be easy to perform.

It would be desirable to find one microbe to fulfil the above criteria, but such a fortuitous circumstance has not as yet occurred. With the exception of the free-living amoeba *Naegleria gruberi*, which causes amoebic meningoencephalitis among swimmers[2], the normal microbiota of natural waters is not of sanitary significance: it is not involved in the actual transmission of human or animal diseases, and has no known association with human or animal wastes.

Various microbiological determinations have been used to assess the extent of fecal pollution of water (and therefore the presence or absence of enteric pathogens) and include standard plate count, total coliform-group density, fecal coliform density, fecal streptococci density, ratio of fecal coliforms to fecal streptococci, anaerobic spore-forming bacteria (i.e. *Clostridium perfringens* or *C. welchii*), pathogenic bacteria (i.e. *Salmonella* and *Shigella* species), lactobacilli, *Pseudomonas aeruginosa* density, enteric viruses, and density of bacteriophages (e.g. coliphages — the viruses of coliform bacteria).

Each determination has its intrinsic limitations and advantages which make it impossible to name the one ideal indicator of fecal pollution that fulfils all the previously stated criteria. In the United States, the total coliform group and fecal coliform density are considered to be reliable indicators of bacterial pathogen presence in potable water and other types of water and waste water, respectively. However, their usefulness in assessing viral presence in water and waste water is seriously open to question[3–5].

Human Enteric Viruses as Indicators of Sewage Pollution

Types found in feces, sewage, and polluted surface water

Over 100 new human enteric viruses have been described since the investigations of Enders *et al.*[6] on viral propagation techniques using tissue cultures. This group includes all viruses known to be excreted in quantity in the feces of man; these are listed in Table 1 along with their associated diseases. Thus, the enteric viruses found in feces and eventually in sewage and polluted surface water consist of the enteroviruses (poliovirus, coxsackievirus, and echovirus), infectious hepatitis, adenoviruses, and reoviruses. Other viruses may be swallowed by humans (e.g. influenza, mumps, and cold- or fever-sore viruses) and may be isolated later in feces, but are not particularly significant in disease transfer via contaminated water. Since viruses do not multiply outside of living susceptible cells, human enteric viruses can be expected to decrease in numbers with time even in that nutritional environment constituted by domestic waste[11].

TABLE 1. Human enteric viruses and their associated diseases[7–10]

Major sub-group	Number of types	Associated disease
Poliovirus	3	Paralytic poliomyelitis, aseptic meningitis
Coxsackievirus		
Group A	23[a]	Herpangina, aseptic meningitis, paralytic disease
Group B	6	Pleurodynia, aseptic meningitis, and infantile myocarditis
Echovirus	31[b]	Aseptic meningitis, fever and rash, diarrheal disease, respiratory infections
Infectious hepatitis	1(?)[c]	Infectious hepatitis
Adenovirus	31	Respiratory and eye infections
Reovirus	3	Fever, respiratory infections, and diarrhea

[a] Type 23 was found to be identical with echovirus type 9, and A23 has been dropped and the number is unused[10]
[b] Echovirus serotypes 10 and 28 have now been reclassified, and these numbers are now unused[10]
[c] Isolation uncertain

The presence in water and sewage of viruses of non-human animal origin, and of bacterial, insect, and plant origins, has been documented, but the importance of their presence in water to man has not been fully elucidated. These non-human animal, plant, insect, and bacterial viruses find their way into natural waters through urban and rural storm-water run-off from land, by direct fecal excretions of non-human animals, and may be native to fish and other aquatic life, including microbes, found in natural waters.

Origin and importance in water

From feces

While essentially everyone discharges coliform bacteria in feces, the excretion of enteric viruses in feces by apparently healthy individuals is largely confined[12] to children under age 15. According to several studies, about 10 per cent of these children on the average will be shedding viruses in their feces at any given time[12–14]. According to the 1960 United States census, about 30 per cent of the population fall into this age group. Clarke *et al.*[7] then made the assumption that juvenile virus-excretors discharge enteric virus in amounts comparable to poliovirus excreted by human volunteers after infection[15]; i.e. an average concentration discharged in feces during the carrier state of about 10^4 $TCID_{50}$ (tissue-culture infective dose) per gram. The enteric virus density in feces was then computed on a *per capita* basis to be about 300 $TCID_{50}$ per gram, or about 200 virus units per gram of feces. The relative enteric virus to coliform density in human feces was then determined to be 15 virus units for every million coliforms, or a 1 to 65 000 ratio, as shown in Table 2.

TABLE 2. Calculated enteric virus to coliform ratios (From Clarke *et al.*[7]; modified by Geldreich and Clarke[3])

Source	Contents in: Virus units	Contents in: Coliform density	Virus to coliform ratios
Feces	200/g	13×10^6/g	1:65 000
Sewage	500/100 ml	46×10^6/100 ml	1:92 000
Polluted surface water	1/100 ml	5×10^4/100 ml	1:50 000

From sewage and in polluted fresh surface water

Clarke *et al.*[7] also calculated the average enteric virus density in domestic raw sewage at 500 virus units per 100 ml of sewage, and not more than 1 virus unit per 100 ml of polluted surface water (see Table 2). Using information on the density of coliforms in sewage and also in surface waters obtained in their laboratory, the enteric virus to coliform ratio in sewage can be calculated to be about 1 to

92 000, and in polluted surface water to be about 1 to 50 000 coliforms (see Table 2). Shuval[16] found in actual field studies that the enteric virus density in sewage of five communities in Israel averaged 100 per 100 ml, while the virus concentration in polluted river water, due to physical dilution alone of the incoming sewage, would range from 0.1 to 1 virus infectious unit per 100 ml (Shuval and Katzenelson[4]). Based on actual recoveries, an enterovirus to coliform ratio of roughly 1 to 10^7 (or about 1 to 10^6 based on the maximum virus concentrations detected) was determined to be present in Israel raw sewage[16]. This ratio is much larger than the previously reported 1 to 65 000 theoretical ratio of Clarke *et al.*[7]. More recently, Buras[17] reported enterovirus to coliform ratios of 1 to 20 000 for undiluted sewage, and 1 to 2000 for a sewage effluent.

England[18] pointed out that the virus content in sewage fluctuates widely, and is influenced not only by viral infection rates in the community at any given time, but also by the hour of the day, the season of the year, the ratio of industrial to domestic waste, and the extent of sewage dilution from seepage or surface run-off into the sewerage system. For example, Kelly and Sanderson[19] reported seasonal fluctuations in the occurrence of enteric viruses in sewage, finding maximum virus density during warm months and minimal amounts during cold months. However, although Shuval[16] noted higher average virus recoveries in the sewage of Israeli communities inhabited by lower than by higher socioeconomic levels—1677 and 480 average plaque-forming units (PFU) per liter, respectively—no clear seasonal variation in the enterovirus levels in raw sewage from four cities was observed.

In estuarine and ocean waters

Many coastal communities discharge their wastes into adjacent estuaries, bays, harbors, and inshore coastal waters of the oceans. As populations concentrate in urban areas, marine outfalls and ocean dumping of wastes have increased. The oceans receive both treated and untreated domestic and industrial wastes, including sewage effluents and sludges, not only from numerous coastal but also from inland sources. When the ocean is employed as a convenient receptacle for such wastes, there is constant danger of pollution of shellfish areas and bathing beaches, and such occurrences constitute a potential threat to public health. The presence of enteric viruses in estuarine and ocean waters has been documented[20,21]. For example, Shuval[16], using the phase-separation method, reported the detection of enteroviruses in 14 out of 48 sea-water samples obtained at varying distances from a sewage outfall off the coast of Tel-Aviv, Israel. Persistence of enteroviruses in the marine environment was demonstrated in two samples taken just opposite a bathing beach that was about 1.5 km north of the marine outfall. One of these samples showed a virus density of 32 PFU per liter, whereas a sea-water sample obtained on the same day in the boil over the end of the sewage outfall showed an enterovirus concentration of 60 PFU per liter. Radioisotope tracer studies revealed that the outfall sewage had been diluted by a factor of 1000 in sea water by the time it had reached the sampling point opposite the bathing beach.

Studies have also been made on the survival characteristics of enteric viruses in estuarine and ocean waters. The dependency of enteric virus survival in estuary and ocean waters on temperature, biotic flora, degree of pollution, and virus type was recognized by Metcalf and Stiles[22]. Enteroviruses were first found by them to survive equally well in sterile and non-sterile sea water, thus negating the possible influence of predators present in the water on viral survival. Temperature was found, however, to be influential. For example, in their studies, echovirus 6 and poliovirus 1 were both present in sea water stored at 5°C for 130 days, but at 37°C these same viruses failed to survive more than 28 days. Virus survival was also less in sea water contained in cellulose dialysis bags immersed in estuary water than in the same water stored in the laboratory, although estuary temperatures hovered around 0°C. This was explained as perhaps being due to water-soluble viral-interfering substances produced by the abundant biotic flora present in the estuary, that could pass through the pores of the dialysis membrane and affect the enteroviruses inside. In similar studies with viruses in dialysis tubing immersed in estuary water, enteroviruses suspended in primary sewage effluents survived longer than those in dialysis tubing containing sea water. Enhanced survival in sewage was attributed to the organic content of the sewage acting as a protective colloid. Finally, viral survival depended upon the type of virus present; i.e. coxsackievirus survived longer than echovirus, while poliovirus exhibited the least survival ability.

A virucidal activity in sea water has been detected by several investigators in as widely separated sampling areas as the Mediterranean and Red Sea[16] and the Atlantic Ocean[23]. This factor has not been clearly elucidated, and it may have only a minor role in inactivating enteric viruses in estuarine and ocean waters. In any case, as pointed out by Mitchell[23], the frequent infectious-hepatitis outbreaks associated with the consumption of shellfish indicate that a percentage of viruses evidently survive this inactivation factor in sea water.

Importance of enteric viruses in water

The importance of enteric viruses in water resides not in their numbers but in their infectivity. The importance of low-level transmission of enteric viruses to man is evident when consideration is given in Table 3 to the small quantity of virus units that are capable of producing an infection in man by the oral route. Plotkin and Katz[24] reviewed the available literature concerned with the minimal dose of viruses that would be infective for man via the oral route, and concluded that one tissue-culture dose

TABLE 3. Infective doses of viruses for man[24,25]

Virus	Virus dose[a]	Route of inoculation	Number of persons inoculated	Per cent of persons infected
Poliovirus 1 (SM strain)	2 PFU	Oral (gelatin capsule)	3	67
Poliovirus 3 (Fox strain)	1 TCD_{50}	Gavage	10	30

[a] As plaque-forming units (PFU), or as the quantity of virus that will infect 50 per cent of the tissue cultures inoculated (TCD_{50})

(1 TCD_{50}) was sufficient to infect a human if it came in contact with susceptible cells. Later experimentation by these workers[25] with the attenuated poliovirus demonstrated that one tissue-culture dose constituted an infectious dose. Thus, since enteric viruses are found in feces, sewage, and fresh and saline polluted surface waters, and only a few virus particles are needed to initiate an infection in a susceptible host, enteric viruses may constitute a health hazard to humans. Disease outbreaks due to waterborne viruses have been documented. For example, the explosive New Delhi outbreak of infectious hepatitis in 1955–56, responsible for some 30 000 overt cases of the disease before subsiding, was traced to inadequate treatment of the heavily polluted Jumna River water used as a raw source of the municipal water system[26]. The United States has experienced several major outbreaks of infectious hepatitis that were shown to be waterborne, including the Worcester, Mass., outbreak involving 90 per cent of the members of a football team[27]. In addition, during the period 1961–1970 there were 30 waterborne outbreaks of infectious hepatitis reported in the United States, and 39 outbreaks of gastroenteritis of undetermined etiology which may have been viral in origin[28]. However, as pointed out by Clarke *et al.*[7] and McDermott[27], the number of recognized waterborne outbreaks of enteric virus disease is not large, which indicates that many may not be reported or understood to be viral in origin.

Use of enteric viruses as indicators of sewage pollution

Since enteroviruses of human origin are found in water and waste water and may remain infectious for a significant period of time[29], depending upon environmental factors and outside interventions, Cliver[30] suggested that enteroviruses may be themselves the most valid model agents to study the effect of water treatment on virus survival. Other investigators[31] have suggested the use of one of the enteroviruses, poliovirus, as an indicator of enteric virus pollution, especially that by the infectious hepatitis virus (hepatitis A), because of the frequent isolation of poliovirus from surface water polluted by domestic sewage effluents. Metcalf[32], however, pointed out that such usage is in error because of a paucity of information concerning the fecal excretion rates of hepatitis A virus, its survival in water and during waste treatment, and its inactivation by chlorine. Thus, there would be no assurance that hepatitis A would be present every time poliovirus was found, or conversely for that matter.

In recent years a multitude of methods has been proposed for virus detection; these have been reviewed by Scarpino[1], Hill *et al.*[33,34], and more recently by Berg[35]. It is definitely becoming easier to recover enteric viruses from a variety of waters and waste waters with a great deal more efficiency. Thus, a direct measurement of the virus hazard in a water or waste water can now reasonably be made, and the presence of sewage pollution thereby indirectly determined. However, as pointed out by Berg[35] the cell-culture systems used for detecting enteric viruses are sensitive to less than 50 per cent of the viruses excreted by man, and there are probably viruses in sewage that have not yet been detected and identified. Thus, the number of viruses detected is several orders of magnitude below the quantities actually present[35].

Finally, I might speculate that since the discovery of the Australian antigen, which is a marker for hepatitis B virus and may be itself a hepatitis virus, detection of hepatitis A virus itself may be within our reach. It is instructive to note that the Australian antigen has already been isolated from clams contaminated by drainage of untreated sewage from a coastal hospital[36]. During the 3 months preceding this isolation, individuals with Australian-positive hepatitis were patients in this hospital. Hepatitis B virus, it should be emphasized, can also be transmitted by the oral route[10], although its usual route is parenteral. Thus, hepatitis B may also pose a danger in sewage-contaminated water.

Bacteriophages as Indicators of Sewage Pollution

Detection of levels of bacteriophages (i.e. bacterial viruses) in water and waste water has been recommended by some investigators as being a sensitive, convenient, economic, and reliable index of water contamination by sewage containing pathogenic enteric bacteria and viruses, and as a useful epidemiological tool in tracing the origins of waterborne microbial diseases. Although reports have stressed the advantages of bacterial viruses in the assessment of fecal contamination of water, there is disagreement as to the ultimate merit of such an indicator model. I will attempt in this section, therefore, to discuss the origins of bacteriophages found in water, as well as their possible utility as indicators of bacterial and viral contamination of water, and their usefulness in epidemiological investigations.

Origin of bacteriophages found in water

Bacteriophages are a constant inhabitant of the intestinal tract of man, and may be encountered wheresoever fecal contamination has occurred[37]. Thus, they are found not only in human feces but also in domestic sewage and in sewage-polluted waters. According to studies by Ware and Mellon[38], raw sewage contained 37 600 coliphages (i.e. bacterial viruses or bacteriophages specific for strains cf *Escherichia coli*) per 100 ml of sewage that were active against *E. coli* B, whereas Pretorius[39] reported higher coliphage levels in raw sewage of 265 000 per 100 ml. The coliphage to fecal coliform (i.e. *E. coli*) ratio in raw sewage was found by Pretorius to be 1 to 52, whereas Ware and Mellon reported a coliphage to total coliform ratio of 1 to 358. Dias and Bhat in a comprehensive report[40] found coliphage numbers in sewage sampled over a period of time to fluctuate from a low of 1800 to a high of 960 000 per 100 ml. Sewage from diverse rural and urban sources in Hong Kong was also examined[41] for coliphage density, which ranged from 3600 to 1 590 000 PFU per 100 ml. Urban sites yielded higher plaque counts than rural. Average recoveries of 1300 PFU per 100 ml were found by Dhillon *et al.*[41] for five rural samples and 4800 PFU per 100 ml for six urban samples. Fifty per cent of their tested bacteriophage isolates were able to grow on the three indicator strains of *E. coli* used, and these non-specific bacteriophage strains were found to be more widely distributed in their samples. In a recent communication, Dhillon and Dhillon[42] were able to isolate specifically from sewage RNA-containing bacteriophages MS2 and Q-Beta. The MS2 bacteriophages were more widely distributed than the Q-Beta ones, although most of the habitats they sampled yielded only one or the other. Single-stranded DNA-containing F-specific bacteriophages were not detected, and this lack of recovery might have been due to their low efficiency of plaquing on the bacterial indicators used, the presence in this type of bacteriophage of a modification that caused them to be restricted by the bacterial host, or they simply were not present.

Enteric bacterial pollution

The literature pertaining to the utility of bacteriophages in determining the presence of enteric bacteria in water is quite extensive, and has been reviewed by Scarpino[1] and more recently by Kenard and Valentine[43]. This literature tends to indicate that direct correlations can exist between bacteriophage content of fresh and marine waters and the degree of fecal contamination by bacterial pathogens such as *Salmonella* and *Shigella* species, and indicator bacteria such as *E. coli*.

For example, a number of investigations by Guélin established the public health significance of bacteriophages in water, and tended to show that the numbers of bacteriophages infective for *E. coli* correlated with the density of coliforms in polluted river water (see Scarpino[1]). In studies investigating the detection in waters of bacteriophages infective for *Salmonella typhi*, Guélin[44,45] found that the levels of Vi bacteriophage, specific for *S. typhi* organisms that have the Vi antigen, varied directly with the degree of fecal pollution. These bacteriophages were practically absent from waters that were slightly polluted. The presence of bacteriophages in sea water active against the coli–typhoid–paratyphoid–dysentery groups of bacteria was later reported by Guélin[46] from sampling points near Roscoff, France. He found that the number of bacteriophages varied directly with the degree of sewage contamination, and that they survived in sea water for a longer time than bacteria. Bacteriophages active against

Clostridium welchii were also isolated from polluted water in quantities that were proportional to the degree of sewage pollution[47]. However, it was found that in the same water the numbers of *C. welchii* and its bacteriophage were often lower than the numbers of *E. coli*.

Diénert's work[48,49] supported the contention that bacteriophages were useful indicators of the degree of pollution of water by pathogenic bacteria. For example, bacteriophages active against *E. coli* and the typhoid and paratyphoid fever organisms were recovered from polluted surface waters but not from unpolluted ones. He concluded that the bacteriophage test was more reliable than the *E. coli* determination for detecting pollution of water. This was believed to be especially true when the source of pollution was several kilometers from the bathing area, since bacteriophages, unlike bacteria, were easy to detect when present in small numbers. A bathing water was considered satisfactory if no bacteriophages were present in 20 cm^3 of water[48].

Carlucci and Pramer[50] described the occurrence, persistence, and activity of bacteriophages active against the coliforms *E. coli* and *Enterobacter aerogenes* and against a marine bacterium, *Serratia marinorubra*, in sea water collected *ca.* 200 m offshore at Long Branch, New Jersey. These authors noted no difficulty in isolating bacteriophages active against the above three species. Coliphages infective for *E. coli*, however, were inactivated rapidly in natural sea water containing no added organic matter, with less than 2 per cent remaining after 10 days, and 0.01 per cent after 20 days. They concluded that the ability of coliphages to multiply in sea water was dependent upon the presence of sufficient nutrients to support the growth and multiplication of host cells. In areas of pollution, the organic content of sea water may reach levels that support bacterial growth, and multiplication of bacteriophages may then occur. This may be of significance in reducing bacterial levels in sea water. Little dissolved organic matter, on the other hand, is found in sea water free of pollution. Extensive growth or rapid multiplication of bacteria therefore does not occur, and as a consequence bacteriophage multiplication ceases. This research thus tends to support the use of bacteriophage (coliphage) levels to indicate the degree of bacterial pollution of water.

Reports by Kott and co-workers[51,52] indicated that a definite relationship existed between total coliform levels and *E. coli* bacteriophages in shellfish and in polluted waters. For instance, it was found that the coliform numbers in oysters from three sampling stations were ten times as numerous as *E. coli* bacteriophages[51]. In addition, in raw sewage a ratio of coliphages to total coliforms of 1 to 100 was noted[52] whereas in the effluent from the Haifa treatment plant in Israel the ratio was 1 bacteriophage to 10 coliform bacteria. The ratio of *E. coli* bacteriophage to *Enterobacter aerogenes* bacteriophage in the effluent, on the other hand, was found to be 100 to 1. It was concluded that the coliphages were rapid and reliable indicators of water pollution.

Studies by Suñer and Piñol[53] showed a bacteriophage to coliform ratio of 1:1 in polluted sea water collected along the Barcelona coast of Spain. These authors used a C-5 strain of *E. coli* for the detection of bacteriophages, and pointed out that their most-probable-number method of coliphage assay was an indirect means of arriving at the *E. coli* pollution level in water and perhaps was not really useful, since *E. coli* could be detected easily anyhow. Nevertheless, it was emphasized, bacteriophage assay might be useful when dealing with pathogens, such as *Salmonella* species that are more difficult to isolate. However, as pointed out by Slanetz[54], recent advances dealing with the direct isolation of pathogenic bacteria such as salmonellae from water and waste water have definitely affected the ultimate worth of this indirect bacteriophage assay approach for *Salmonella* species. Since microbiological techniques for pathogens have been perfected, Slanetz believed that indicator organisms (i.e. coliforms, fecal coliforms, fecal streptococci, or bacteriophages) were not entirely needed to establish the health hazard or the safety of sea water.

In a recent communication, Kenard and Valentine[43] put forth persuasive arguments for the detection of bacteria in water by the isolation of specific bacteriophage, using a test procedure that can be completed after 6–8 h of incubation. Fecal coliform, total coliform, and total coliphage numbers were determined in over 150 water samples from different geographical locations in the Washington, D.C., and northern Virginia areas over a period of 2 years. A high degree of correlation was found in the samples between the fecal coliforms and the coliphages, and presumptive coliform counts based on coliphage recoveries were found to agree closely with actual bacterial counts. The ratio of coliphages to fecal coliforms was about 0.7 to 1, regardless of the contamination level, whereas the coliphage to total coliform ratio was 1 to 25.

Some investigators, however, don't agree concerning the potential usefulness of bacteriophages as indicators of bacterial pollution of water, and have reported little or no correlation between bacteriophage levels and water potability[55,56]. In addition, the bacteriophage technique was found by others[57] not to be superior to the usual bacteriological techniques used for evaluating water pollution from fecal sources, since one bacterial culture could not be attacked by all strains of bacteriophages found in polluted waters. Finally, Zieminska[58] compared the levels of *E. coli* in river water over a period of a year to the levels of bacteriophages infective for *Salmonella typhi* Vi, *Shigella flexneri*, and *Salmonella paratyphi* B. He could not find any correlation between the *E. coli* titer and the titers of any of the above bacteriophages. It was concluded that the determination of bacteriophage levels was not a suitable procedure for the routine examination of water to detect contamination by enteric bacterial pathogens.

Thus, although this bacteriophage criterion for determination of fecal pollution of water should be considered by researchers, it is obvious that additional studies must be made before it can replace (a) those criteria—such as the total and fecal coliforms—that are presently accepted as indicators of fecal pollution of drinking or other types of water and waste water, respectively, or (b) the direct isolation of bacterial pathogens from polluted waters.

Certainly, the complexity of the bacteriophage method and the time required before final results are available do not allow for a more convenient, economical, rapid, or reliable view of bacterial pathogens in water when compared with existing methods, such as the membrane filter, for detection of presently accepted bacterial indicator organisms in water.

Enteric viral pollution

Several bacteriophage models have been proposed for the detection of enteric viral pollution of water. For example, bacteriophages of *Serratia marcescens* have been used by South African researchers[59] as indicators for tracing the reduction during sewage treatment processes of not only pathogenic bacteria but also animal viruses. It was reported that there was a similarity between the inactivation curve for the bacteriophage infective for *S. marcescens* and for the bacterium *E. coli.* In addition, the *S. marcescens* bacteriophages were inactivated at a slower rate than the Mahoney strain of poliovirus 1, thus providing a margin of safety. Later reports by Carstens *et al.*[60] and Coetzee[61] demonstrated that bacteriophage infective for *S. marcescens* could be useful in evaluating the removal of poliovirus 1 during sewage treatment processes.

Coliphage models of enteric virus presence in water and waste water have attracted considerable attention in recent years. As pointed out by Metcalf[32], justification for the use of coliphages instead of enteroviruses was based upon alleged similarity in behavior under various environmental influences, the convenience and lower costs in working with bacteriophages, the shorter time required for assay, and the ability to enumerate bacteriophages in a quantitative fashion. Little actual information, however, is available comparing the relative survival time of coliphages to that of enteric viruses of pathogenic significance to man. Several studies in this latter regard can be mentioned.

Joyce and Weiser[62], in laboratory studies, compared the survival of enterovirus (Sabin poliovirus 1, coxsackievirus, and echovirus) to that exhibited by T-2 coliphage in untreated farm-pond water held at 20° and 4°C. It was found that the coliphages did not survive beyond 7 days in the absence of their bacterial hosts, and disappeared from the water sometime beyond 14 days when their hosts were present. On the other hand, the enterovirus survival was considerably longer in similar water, poliovirus being recoverable up to 84 days at 20°–25°C, and present at 91 days at 4°C, while coxsackievirus and echovirus were recovered up to 13 weeks at both holding temperatures.

Kott *et al.*[63] carried out studies on the fate of bacterial viruses in a marine environment using a coliphage model that was felt by them to be similar in behavior to enteroviruses. The model included the T-1 to T-7 coliphages that were infective for *E. coli* B. From their results (and those obtained by Shuval *et al.*[64] in separate studies using the same sewage source) they assumed that coliphages were as resistant as enteroviruses to the marine environment. A rough, tentative estimation of enterovirus pollution of sea water was made based upon the presence of the following coliphage levels: less than 10/100 ml of sea water indicated slight and not recent pollution; 100/100 ml indicated recent, medium-sized pollution; and 1000/100 ml indicated recent, heavy pollution. Actual correlations, however, between the coliphage densities and enteroviruses were never made. Berg[65] pointed out that coliphage counts often will not reflect enterovirus levels in water, since the latter will occur in human feces at incidence levels ranging from 0 to 100 per cent, depending upon season of the year, sanitary conditions, population age, and other factors that exist in a community. Obviously, if coliphages are present in fresh domestic sewage when enteroviruses are present and, assuming an equal resistance to the marine environment, the absence of coliphages would indicate under the circumstances the absence of enteroviruses. Thus, a margin of safety would be provided when coliphages were more numerous than enteroviruses.

A recent study by Kott *et al.*[66] focused on the coliphage to human enteric virus ratios found in tap,

TABLE 4. Coliphage and enterovirus relationships in waste-water influents and effluents during different seasons of the year[66]

Waste-water type	Virus examined	Virus particles (per 100 ml) in			
		Winter	Spring	Summer	Autumn
Trickling-filter influent	Coliphage	$1{\cdot}8 \times 10^6$	$7{\cdot}6 \times 10^6$	$8{\cdot}1 \times 10^6$	$3{\cdot}8 \times 10^6$
	Enterovirus	57	54	54	30
Trickling-filter effluent	Coliphage	$9{\cdot}8 \times 10^5$	$4{\cdot}6 \times 10^6$	$2{\cdot}6 \times 10^5$	$1{\cdot}8 \times 10^5$
	Enterovirus	71	51	60	41
Oxidation-pond effluent	Coliphage	$6{\cdot}6 \times 10^4$	$1{\cdot}6 \times 10^5$	$1{\cdot}9 \times 10^4$	$5{\cdot}2 \times 10^2$
	Enterovirus	53	46	42	45

flood, and waste waters. Their observations on the numbers of coliphages and enteric viruses in a waste-water treatment plant during various seasons of the year are summarized in Table 4. The bacteriophages recovered were those capable of attacking the bacterial host *E. coli* B. Although both coliform and coliphage densities were large, resulting in a 1 to 100 coliphage to coliform ratio, enterovirus numbers were low, giving a 1 to 1000 enterovirus to coliphage ratio. Flood-water samples contained between 10^2 and 10^5 coliforms/100 ml, whereas the coliphage numbers ranged from 2 to 5400 per 100 ml. About two-fifths of the coliphages in the flood water were within the range 2–100/100 ml, compared with 10^3–10^5/100 ml for coliforms, or a 1 to 10^3 ratio for the coliphage to coliform numbers. Enteroviruses ranged from 1 to 15/100 ml, while more than half of the coliphages recovered ranged from 10^2 to 10^3/100 ml. About half of the coliforms were present in numbers exceeding 10^5/100 ml. The ratio of enteroviruses to coliphages to coliforms was thus 1 to 10^{2-3} to 10^5. Coliphages and enteroviruses were also detected by Kott *et al.*[66] in some tap-water samples at the time when enteric disease outbreaks were observed in the community, resulting in a 1 to 10 ratio of enteroviruses to coliphage.

An experimental oxidation pond over a two-month period demonstrated no decrease in either added poliovirus 1 or f2 coliphages[66]. A comparison of the relative resistances to chlorination of several coliphages (coliphages active against *E. coli* B and K_{12}, MS2, and f2) and poliovirus 1 was also made. It was found that all coliphages examined were generally more resistant than poliovirus 1 to various doses of chlorine at pH 6 and 8.3 after 1 or 2 hours' contact time in oxidation-pond effluents. The

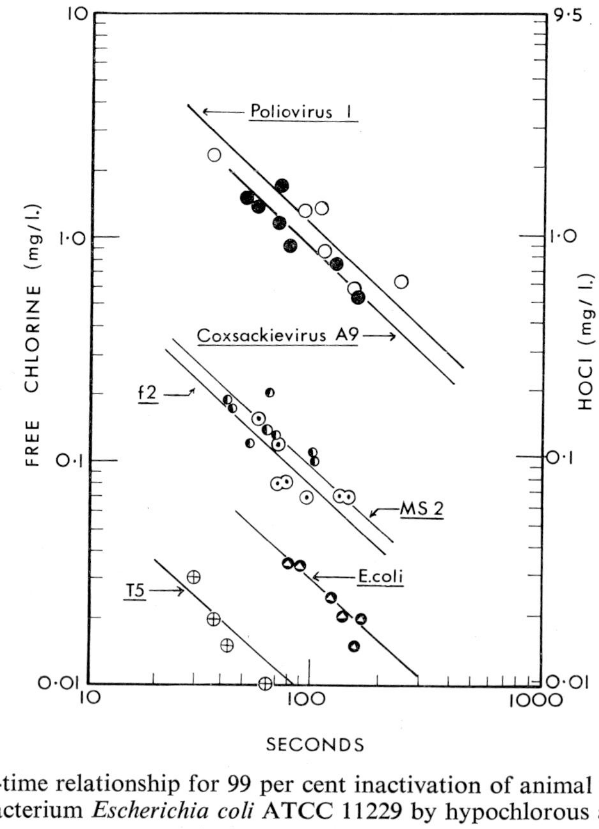

FIG. 1. Concentration-time relationship for 99 per cent inactivation of animal viruses, bacterial viruses, and the reference bacterium *Escherichia coli* ATCC 11229 by hypochlorous acid at 5°C and pH 6

resistance of f2, however, was always greater than poliovirus 1, which is the same as found by Shah and McCamish[67]. The present Author has found that although the f2 coliphage exhibited greater resistance than both poliovirus 1 and coxsackievirus A9 to monochloramine, this was not the case with hypochlorous acid.

There is little doubt that Dr. Kott's studies have provided new insight into our understanding of the use of coliphages as indicators of viral contamination of water. The simultaneous presence of coliphages in large numbers and one of the enteroviruses, poliovirus, in tap, flood, and waste waters is certainly quite convincing argument for the use of coliphages as an indicator of viral presence in water. Additional studies, however, are definitely needed not only with poliovirus but also with other enteric viruses to determine if the coliphage to enteric virus indicator relationship holds true in other situations as well. One such study that was done by Scarpino *et al.*[68] is at variance with this use of coliphages as enteric viral indicators. Precise comparisons were made by us in carefully controlled laboratory studies of the survival characteristics after chlorination at pH 6 and 5°C of several coliphages (T5, f2, and MS2), *E. coli*, and two animal viruses (poliovirus 1 and coxsackievirus A9). At pH 6 more than 95 per cent of the free chlorine present was in the HOCl form. The results in Fig. 1 show a greater resistance to hypochlorous acid by the two animal viruses, compared to the faster death of the three coliphages and the test bacterium, *E. coli*. However, other (unpublished) studies by us indicate that in the presence of monochloramine the coliphage f2 is indeed more resistant than poliovirus 1. Thus, it is probable that in certain environmental circumstances a coliphage model as f2 can be an effective indicator of enteric virus presence and inactivation, but perhaps not in all situations.

Epidemiological use of bacteriophages

Finally, I would like to mention briefly the epidemiological use of bacteriophages. Several investigators have noted the use of bacteriophages in tracing the origin and the extent of a bacterial disease. Gell *et al.*[69] reported the usefulness of bacteriophages in epidemiological studies, while Peso[70] investigated the origin of waterborne diseases by looking for bacteriophage tracers. The focus of infection can be determined by examination of cases of the disease (as typhoid) and healthy carriers for specific bacteriophages. Although Hruby[71] could not isolate typhoid organisms from the Shušnica River in Yugoslavia, bacteriophages infective for *Salmonella typhi* and *paratyphi* B, and for *Shigella dysenteriae* I and *flexneri*, were detected. Thus, the water from the river was implicated in a previous disease outbreak. Bacteriophages were also used by Gerner-Rieux *et al.*[72] to trace sources of infection, especially from waterborne diseases, while Takeya *et al.*[73] developed a sensitive method of diagnosing the carriers of the Celebes type El Tor vibrios by detection of the vibrio bacteriophages in stool specimens. The phage-detection method was found by Takeya to be more sensitive than, and as reliable as, the actual isolation of the vibrios. Understanding of the origins and extent of waterborne diseases can therefore be accomplished using bacteriophages, especially when the pathogenic bacteria initiating the problem are present in low numbers or are not amenable to present recovery procedures.

REFERENCES

1. SCARPINO, P. V. Bacterial and viral analysis of water and waste water. In 'Water and water pollution handbook', Vol 2 (ed. L. L. Ciaccio). Dekker, New York, 1971, p. 639.
2. CHANG, S. L. Pathogenic free-living amoebae and recreational waters. *Proc. 6th int. Conf. Wat. Pollut. Res., Jerusalem, 1972*, p. 85.
3. GELDREICH, E. E., and CLARKE, N. A. The coliform test: a criterion for the viral safety of water. *Proc. 13th Wat. Qual. Conf.*, Univ. Illinois, Urbana, 1971, p. 103.
4. SHUVAL, H. I., and KATZENELSON, E. The detection of enteric viruses in the water environment. In 'Water pollution microbiology' (ed. R. Mitchell). Wiley-Interscience, New York, 1972, p. 347.
5. DUTKA, B. J. Coliforms are an inadequate index of water quality. *J. envir. Hlth*, 1973, **36**, 39.
6. ENDERS, J. F., WELLER, T. H., and ROBBINS, F. C. Cultivation of the Lansing strain of poliomyelitis virus in cultures of various human embryonic tissues. *Science, N.Y.*, 1949, **109**, 85.
7. CLARKE, N. A., BERG, G., KABLER, P. W., and CHANG, S. L. Human enteric viruses in water: source, survival and removability. *Proc. 1st int. Conf. Wat. Pollut. Res., London, 1962*, Vol 2, p. 523.
8. HORSFALL, F. L., and TAMM, I. (eds.) Viral and rickettsial infections of man. Lippincott, Philadelphia, 4th Edn, 1965.
9. RHODES, A. J., and ROOYEN, E. C. v. Textbook of virology. Williams & Wilkins, Baltimore, Md, 5th Edn, 1968.
10. DAVIS, B. D., DULBECCO, R., EISEN, H. N., GINSBERG, H. S., WOOD, W. B., and MCCARTY, M. Microbiology. Harper and Row, Hagerstown, Md, 2nd Edn, 1973.

11. Berg, G. Integrated approach to problem of viruses in water. *J. sanit. Engng Div. Am. Soc. civ. Engrs*, 1971, **97** (SA6), 867.
12. Ramos-Alverez, M., and Sabin, A. B. Intestinal viral flora of healthy children demonstrated by monkey kidney tissue culture. *Am. J. publ. Hlth*, 1956, **46**, 295.
13. Honig, E. I., Melnick, J. L., Isacson, P., Parr, R., Myers, I. L., and Walton, M. An epidemiological study of enteric virus infections. *J. exp. Med.*, 1956, **103**, 247.
14. Gelfand, H. M., Fox, J. P., and LeBlang, D. R. The enteric viral flora of normal children in southern Louisiana. *Am. J. trop. Med. Hyg.*, 1957, **6**, 521.
15. Sabin, A. B. Behavior of chimpanzee-avirulent poliomyelitis viruses in experimentally infected human volunteers. *Am. J. med. Sci.*, 1955, **230**, 1.
16. Shuval, H. I. The detection and control of enteroviruses in the water environment. In 'Developments in water quality research' (ed. H. I. Shuval). Ann Arbor–Humphrey Science Publishers, Ann Arbor, Mich., 1970, p. 47.
17. Buras, N. Recovery of viruses from waste water and effluent by the direct inoculation method. *Wat. Res.*, 1974, **8**, 19.
18. England, B. Concentration of reovirus and adenovirus from sewage and effluents by protamine sulfate (Salmine) treatment. *Appl. Microbiol.*, 1972, **24**, 510.
19. Kelly, S. M., and Sanderson, W. W. Density of enteroviruses in sewage. *J. Wat. Pollut. Control Fed.*, 1960, **32**, 1269.
20. Metcalf, T. G., and Stiles, W. C. The accumulation of enteric viruses by the oyster, *Crassostrea virginica*. *J. infect. Dis.*, 1965, **115**, 68.
21. Metcalf, T. G., and Stiles, W. C. Enteroviruses within an estuarine environment. *Am J. Epidemiol.*, 1968, **88**, 379.
22. Metcalf, T. G., and Stiles, W. C. Survival of enteric viruses in estuary waters and shellfish. In 'Transmission of viruses by the water route' (ed. G. Berg). Wiley-Interscience, New York, 1967, p. 439.
23. Mitchell, R. Destruction of bacteria and viruses in seawater. *J. sanit. Engng. Div. Am. Soc. civ. Engrs.* 1971, **97** (SA4), 425.
24. Plotkin, S. A., and Katz, M. Minimal infective doses of viruses for man by the oral route. In 'Transmission of viruses by the water route' (ed. G. Berg). Wiley-Interscience, New York, 1967, p. 151.
25. Katz, M., and Plotkin, S. A. Minimal infective dose of attenuated poliovirus for man. *Am. J. publ. Hlth.* 1967, **57**, 1837.
26. Dennis, J. M. 1955–56 infectious hepatitis epidemic in Delhi, India. *J. Am. Wat. Wks Ass.*, 1959, **51**, 1288.
27. McDermott, J. H. Virus and water quality: occurrence and control—Conference summary. *Proc. 13th Wat. Qual. Conf.*, Univ. Illinois, Urbana, 1971, p. 211.
28. Taylor, A., Craun, G. F., Faich, G. A., McCabe, L. J., and Gangorosa, E. J. Outbreaks of waterborne diseases in the United States, 1961–1970. *J. infect. Dis.*, 1972, **125**, 329.
29. Akin, E. W., Benton, W. H., and Hill, W. F. Enteric viruses in ground and surface waters: a review of their occurrence and survival. *Proc. 13th Wat. Qual. Conf.*, Univ. Illinois, Urbana, 1971, p. 59.
30. Cliver, D. O. Viruses in water and wastewater: effects of some treatment methods. *Proc. 13th Wat. Qual. Conf.*, Univ. Illinois, Urbana, 1971, p. 149.
31. Coin, L., Menetrier, M. L., Labonde, J., and Hannoun, M. C. Modern microbiological and virological aspects of water pollution. *Proc. 2nd int. Conf. Wat. Pollut. Res., Tokyo, 1964*, Vol 1, p. 1.
32. Metcalf, T. G. Biologic parameters in water transmission of viruses. *Proc. 13th Wat. Qual. Conf.*, Univ. Illinois, Urbana, 1971, p. 1.
33. Hill, W. F., Akin, E. W., and Benton, W. H. Detection of viruses in water: a review of methods and application. *Wat. Res.*, 1971, **5**, 967.
34. Hill, W. F., Akin, E. W., Benton, W. H., and Metcalf, T. G. Virus in water. II—Evaluation of membrane cartridge filters for recovering low multiplicities of poliovirus from water. *Appl. Microbiol.*, 1972, **23**, 880.
35. Berg, G. Methods for detecting viruses in environmental waters—a status report. *Proc. 1st Microbiol. Seminar Standardization of Methods.* U.S. Environmental Protection Agency, Publ. No. EPA–R4–73–022, Washington, D. C., 1973, p. 72.
36. Mahoney, P., Fleischner, G., Millman, I., London, W. T., Blumberg, B. S., and Arias, I. M. Australia antigen: detection and transmission in shellfish. *Science, N.Y.*, 1974, **183**, 80.
37. d'Herelle, F. The bacteriophage and its behavior. Williams & Wilkins, Baltimore, Md, 1926.
38. Ware, G. C., and Mellon, M. A. Some observations on the coli/coliphage relationship in sewage. *J. Hyg., Camb.*, 1956, **54**, 99.
39. Pretorius, W. A. Some observations on the role of the coliphages in the number of *Escherichia coli* in oxidation ponds. *J. Hyg., Camb.*, 1962, **60**, 279.
40. Dias, F. F., and Bhat, J. V. Microbial ecology of activated sludge. *Appl. Microbiol.*, 1965, **13**, 257.

41. DHILLON, T. S., CHAN, Y. S., SUN, S. M., and CHAU, W. S. Distribution of coliphages in Hong Kong sewage. *Appl. Microbiol.*, 1970, **20**, 187.
42. DHILLON, E. K. S., and DHILLON, T. S. Synthesis of indicator strains and density of ribonucleic acid-containing coliphages in sewage. *Appl. Microbiol.*, 1974, **27**, 640.
43. KENARD, R. P., and VALENTINE, R. S. Rapid determination of the presence of enteric bacteria in water. *Appl. Microbiol.*, 1974, **27**, 484.
44. GUÉLIN, A. Étude des bactériophages typhiques. VI—Dans les eaux. *Annls Inst. Pasteur, Paris*, 1948, **75**, 485.
45. GUÉLIN, A. Sur le choix des souches étalons pour la détection du bacille typhique dans les eaux par la recherche des bactériophages spécifiques. *Annls Inst. Pasteur, Paris*, 1950, **79**, 186.
46. GUÉLIN, A. Étude quantitative de bactériophage de la mer. *Annls Inst. Pasteur, Paris*, 1948, **74**, 104.
47. GUÉLIN, A. Sur la présence du bactériophage *perfringens* dans les eaux et son rôle dans l'épuration des eaux stagnantes. *Annls Inst. Pasteur, Paris*, 1950, **79**, 447.
48. DIÉNERT, F. Hygiene of bathing places. *Bull. Acad. Méd.*, 1944, **128**, 660.
49. DIÉNERT, F. Investigation and estimation of bacteriophages in water. *C. R. Acad. Sci., Paris*, 1945, **221**, 574.
50. CARLUCCI, A. F., and PRAMER, D. An evaluation of factors affecting the survival of *Escherichia coli* in sea water. IV—Bacteriophages. *Appl. Microbiol.*, 1960, **8**, 254.
51. KOTT, Y., and GLOYNA, E. F. Correlating coliform bacteria with *E. coli* bacteriophages in shellfish. *Wat. Sewage Wks*, 1965, **112**, 424.
52. BURAS, N., and KOTT, Y. Bacteriophage as an indicator for the estimation of water pollution. *Israel J. med. Sci.*, 1966, **2**, 660.
53. SUÑER, J., and PIÑOL, J. Coliform bacteriophages and marine water contamination. *Proc. 3rd int. Conf. Wat. Pollut. Res., Munich, 1966*, Vol 3, p. 105.
54. SLANETZ, L. W. Discussion of paper by Suñer and Piñol[53]. *Proc. 3rd int. Conf. Wat. Pollut. Res., Munich, 1966*, Vol 3, p. 115.
55. COUTURE, E. The presence of bacteriophage in waters as a method of judging their quality. *Revue Hyg. Méd. prév.*, 1936, **58**, 371.
56. EMILIANOWICZ, W. Viability of the anti-Vi bacteriophage in natural aquatic habitats in comparison with viability of homologous typhoid bacteria. *Bull Inst. mar. trop. Med. Gdansk*, 1952, **4**, 342.
57. PESO, O. A., and MIGNONE, R. M. The determination of bacteriophages as an indicator of water pollution. *Revta Obr. sanit. Nac., B. Aires*, 1955, **34**, 433.
58. ZIEMINSKA, S. Comparative studies on the presence of bacteria of the *E. coli* group and bacteriophages of some pathogenic intestinal bacteria in river water. *Roczn. pánst. Zakl. Hig.*, 1958, **9**, 225.
59. SOUTH AFRICAN COUNCIL FOR SCIENTIFIC AND INDUSTRIAL RESEARCH, NATIONAL INSTITUTE FOR WATER RESEARCH. Report of the Director for 1961. C.S.I.R. Spec. Rep. W20, 1963.
60. CARSTENS, E. M. J., COETZEE, O. J., MALHERBE, H. H., and HARWIN, R. M. Bacteriophage of *Serratia marcescens* as an index of human virus survival during sewage purification. C.S.I.R. Res. Rep. 241, 1965.
61. COETZEE, O. J. Bacteriophage as an index of faecal pollution. *Publ. Hlth, Johannesburg*, 1966, **66**, 45.
62. JOYCE, C., and WEISER, H. H. Survival of enteroviruses and bacteriophage in farm pond waters. *J. Am. Wat. Wks Ass.*, 1967, **59**, 491.
63. KOTT, Y., ARI, B. H., and BURAS, N. The fate of viruses in a marine environment. *Proc. 4th int. Conf. Wat. Pollut. Res., Prague, 1969*, p. 823.
64. SHUVAL, H. I., CYMBALISTA, S., WACHS, A., ZOHAR, Y., and GOLDBLUM, N. The inactivation of enteroviruses in sewage by chlorination. *Proc. 3rd int. Conf. Wat. Pollut. Res., Munich, 1966*, Vol 2, p. 37.
65. BERG, G. Discussion of paper by Kott *et al.*[63]. *Proc. 4th int. Conf. Wat. Pollut. Res., Prague, 1969*, p. 833.
66. KOTT, Y., ROZE, N., SPERBER, S., and BETZER, N. Bacteriophages as viral pollution indicators. *Wat. Res.*, 1974, **8**, 165.
67. SHAH, P. C., and MCCAMISH, J. Relative chlorine resistance of poliovirus 1 and coliphages f2 and T_2 in water. *Appl. Microbiol.*, 1972, **24**, 658.
68. SCARPINO, P. V., LUCAS, M., DAHLING, D. R., BERG, G., and CHANG, S. L. Effectiveness of hypochlorous acid and hypochlorite ion in destruction of viruses and bacteria. In 'Chemistry of water supply, treatment, and distribution' (ed. A. J. Rubin). Ann Arbor Science Publishers, Ann Arbor, Mich., 1974, p. 359.
69. GELL, P. G. H., HOBBS, B. C., and ALLISON, V. D. An outbreak of water-borne typhoid investigated by bacteriophage typing and 'selective' sewage examination. *J. Hyg., Camb.*, 1945, **44**, 120.
70. PESO, O. A. Investigation of the origin of water-borne disease by bacteriophage. *Revta Obr. sanit. Nac., B. Aires*, 1948, **22**, 337.
71. HRUBY, J. An epidemic of typhoid fever in Slunj in 1951–1952. *Higijena*, 1953, **5**, 1.
72. GERNER-RIEUX, C., BUTTIAUX, R., and MUCHEMBLÉ, G. Intérêt épidémiologique de la recherche des bactériophages dans les eaux. *Hydrobiologia*, 1949, **1**, 105.
73. TAKEYA, K., SHIMODORI, S., and ZINNAKA, Y. New method for early diagnosis of Celebes type El Tor vibrio carriers. *J. Bact.*, 1965, **90**, 824.

Prof. Scarpino introduced his paper by saying that in it he had attempted to put into proper focus the utility of human enteric viruses and bacteriophages as indicators of sewage pollution. The enteric viruses were those excreted in the faeces of man and potentially transmissible to man via ingested contaminated water, whereas bacteriophages were environmentally ubiquitous organisms that specifically infected bacterial cells. He had reviewed the relationship and relevance of both enteric viruses and bacteriophages to other microbial indicator organisms or groups, and discussed the types, origins, and distribution of enteric viruses and bacteriophages, along with their possible roles as indicators of sewage pollution. The controversy concerning the utility of bacteriophages as indicators of bacterial and viral contamination of water was explored, and the usefulness of bacteriophages in epidemiological investigations was briefly presented.

GENERAL DISCUSSION ON PAPERS 5 AND 6

Dr. K. Grunnet (Denmark) described a quantitative method of determining *Salmonella* sp. in receiving waters by means of a 'filter buoy' which gave a summation of concentrations present at all times over a period. The buoy weighed 150 kg including equipment. The 24-V supply from two accumulators in the bottom of the buoy (transformed to 220 V by a converter) operated a peristaltic pump which drew water continuously through three series of filters at relative rates of 1:2:4. A total volume of about 30 l. could be filtered in a 24-h period, and the Most Probable Number was estimated from the number of positive and negative titres. When this method was compared with the swab method in the Gulf of Aarhus it was found that a swab isolation frequency of 50 per cent corresponded to 1 *Salmonella* per 15 l. of sea water, and that when the concentration was as high as 2–3 per l. the frequency was nearly 100 per cent.

Prof. R. R. Colwell (U.S.A.) said that the problem of drug-resistance factors, both the transfer of plasmid-mediated drug resistance occuring *in situ* and the influx of drug-resistant bacteria into the marine environment, had to be given more attention. Reports were beginning to accumulate concerning the presence of such bacteria in rivers and coastal waters. She and her colleague Dr. Sizemore[74] had isolated multiple-drug resistant bacteria carrying plasmids from the deep ocean. The disturbing point was that it appeared that such plasmids might be transferred to autochthonous marine bacteria, thus creating a new environmental problem with respect to public health and mariculture. It might well be that other characters, such as toxin production and invasive properties could similarly be transmitted to marine and estuarine bacteria. In reply, Prof. Scarpino said that Prof. Colwell's point was well taken and her comments were very appropriate. Prof. Bonde said that in his paper he had referred[52, 53] to the transfer of R-factors; he believed the exchange of genetic material could take place only in treatment plants where the necessary proximity existed.

Dr. M. Waldichuk (Canada) also took up Prof. Colwell's point and asked what risk there was in the transmittal of viruses through fish as a vector—say in the handling or consumption of raw fish taken in sewage-contaminated waters. Prof. T. G. Metcalf (U.S.A.) said that the potential enteric virus transmission within the food chain was currently being examined. Experimental transmission of virus from shellfish to worms feeding on shellfish faeces had been demonstrated, and consumption of such worms by detritophagous fin-fish extended the transmission to fish. These results suggested a method by which enteric viruses might find their way into detritophagous species of fish. Dr. V. J. Cabelli (U.S.A.) added that, in the case of bacteria, the genera *Pseudomonas-Aeromonas*, and *Vibrio* included species pathogenic to both humans and fish.

Dr. Waldichuk, speaking as a non-microbiologist, said that the impression had been given that viruses were indeed important in sewage-polluted waters, but were not identifiable by standard techniques used for monitorng sewage pollution. There was public concern regarding the risk of infection by water-borne viruses about which little was known. It had been implied by Prof. Bonde that *E. coli* was a good indicator and correlated well with *Salmonella*, and by Prof. Scarpino that bacteriophages might have a good potential for epidemiological use. He asked whether more emphasis should not be put on development of a routine and non-hazardous method for measuring viruses in sea water, rather than perpetuating the less diagnostic techniques. In reply, Prof. Bonde stressed that he had compared existing methods and chosen the one best suited to the demands of the task. The intermittent occurrence of viruses and the great cost of viral assays had hitherto precluded the application of routine viral tests. Prof. Scarpino said he had not suggested the exclusive use of bacteriophages as enteric viral pollution indicators. He believed they had potential usefulness under certain circumstances and in certain situations. As he had emphasized in his paper, considerably more research was needed on relationships between coliphage and enteric viruses, in as many different situations as possible, before he would completely subscribe to the coliphage indicator concept. Non-hazardous methods for detecting animal viruses could be very expensive. He would rather look directly for animal viruses in sea water than use an indirect indicator system, though he recognized that routine viral analyses were beyond the reach of many environmental laboratories; a central assay laboratory might be the answer.

Dr. Cabelli agreed with a statement by Prof. Bonde that the real threat of salmonellosis from water recreational use lay in multicell particulates. Only in this way could the data on human ID_{50}, amount of water ingested, and density of organism in water, be reconciled to provide a reasonable probability of infection. As was recorded in Paper 7, his organization had found MPN values for salmonellae not exceeding 22 per litre.

Dr. L. J. Guarraia (U.S.A.) asked how one took into account the pathogenic populations which might reside in sediments, and what was their relationship to the indicator species in the water column which were far more susceptible to die-away and dispersion; such sediments could be re-suspended after the indicator species had disappeared. Prof. Bonde said in reply that *Clostridium perfringens*, as an indicator of old and remote pollution, was excellently fitted to demonstrate the presence of pathogens in sediments. Prof. Scarpino added that (as had been pointed out earlier by Prof. Berg) animal viruses were recoverable from sediments, and so could be a potential problem in the marine environment.

Dr. J. J. Barnard (South Africa) enquired whether chemical methods could not be used for detecting sewage pollution, and asked if the ultimate aim was a sterile environment and what would be achieved by sterilizing sewage discharged to the sea. Prof. Scarpino replied that chemical methods were used to detect sewage pollution, but that they were inadequate in the detection of bacterial and viral pollution. He went on to say that they were not attempting to obtain a sterile environment, but only to minimize microbial and chemical contamination of the sea, particularly in areas near human concerns, such as shellfish harvesting areas and bathing beaches.

Dr. G. E. B. Kullenberg (Denmark) brought the physical conditions into the discussion. He asked the Authors for guidance to the physicists on the best dispersion situation: high initial dilution, better dispersion in the early stages, or suspension for a long time in the water. This was particularly relevant to the dumping of sewage sludge. Prof. Scarpino, in reply, said the dispersion of sewage (and microbes) was best covered by the engineering papers later in the Symposium.

ADDITIONAL REFERENCE

74. Colwell, R. R., and Sizemore, R. K. Drug-resistant bacteria in the marine environment. *Proc. a. Meet. mar. technol. Soc., Washington, D.C., 1974*, pp. 427–430.

Paper No. 7

The Development of Criteria for Recreational Waters

V. J. Cabelli,* M. A. Levin,* A. P. Dufour,* and L. J. McCabe

United States Environmental Protection Agency, Recreational Water Criteria Program, National Environmental Research Center, Cincinnati, Ohio

Untoward effects on human health arising from the pollution of estuarine and coastal waters are obtained primarily through man's use of the marine environment as a source of recreation and food. Polluted shellfish as a potential route of disease transmission will be discussed later in this Symposium by Dr. Metcalf[1]. The present report considers water-quality criteria for marine recreational waters, health-effects information needed as the basis for such criteria, past and present efforts to obtain this information, and the impact of these criteria on the treatment of domestic and industrial wastes discharged into estuarine and coastal waters.

Unfortunately, one of the more visible manifestations of criteria and standards for recreational waters has been beach closures or their posting as unsafe and, hence, the denial of much needed recreational facilities to the public. With specific reference to public health hazards associated with the use of such recreational waters, competent authorities have questioned the validity of such closures (except in extreme cases, i.e. beaches in close proximity to outfalls for untreated sanitary wastes) along with the standards used to effect them. However, most of these same authorities would agree that water-quality criteria and standards are essential in defining the type and extent of treatment required for sanitary and industrial wastes whose effluents are carried to recreational waters. Information on the risk of disease, categorized by type, severity, and economic impact and associated with some indicator(s) of water quality, are needed by local regulatory agencies for the cost versus 'acceptable risk' decisions required in the planning of treatment facilities. Storm-water run-off is frequently the cause of excessive bacterial concentrations in recreational waters, and control of these point sources (and in some cases non-point sources) is just being developed.

Past considerations of water-quality criteria[2–4] have been limited by the paucity of data from disease outbreaks or epidemiological studies which clearly show that individual cases of disease, or the rate of illness associated with the use of marine recreational waters, is correlated to a given level of pollution as measured by any of several potential indicators of water quality. There have been a number of case reports in which disease was attributable to swimming in recreational waters[5]; however, in most of these instances, the association with some measured indicator of water quality was vague. This situation has its origin in the fact that many of the infectious diseases which can be transmitted by this route are neither fatal nor 'reportable' by most public health authorities. In addition, most of the diseases in question can be transmitted by routes other than recreational water use. A notable exception to the above is 'swimmers ear', an otitis externa whose major etiological agent is *Pseudomonas aeruginosa*[6]. However, even with this well-defined clinical entity, which has been clearly associated with swimming, it is not known how frequently the source of the organism is endogenous as opposed to exogenous in origin.

The consequences of the aforementioned situation are two-fold. First of all, published mortality statistics cannot be expected to provide a basis for the establishment of criteria. Some experts would suggest that, in the absence of 'reportable' or fatal disease, standards and even criteria are not required[2]. However, as noted above, this course of action provides no guidance for environmental control. The second consequence is the difficulty in designing well-controlled studies whose objective is to relate the incidence of illness among swimmers at a beach subject to sanitary and other forms of pollution to some health-effects indicator(s) of water quality. Thus, it is not surprising that there have been only two such epidemiological studies—as contrasted to scattered observations—and that the findings have not resolved the question of criteria and standards for recreational waters. The results and conclusions from these studies undoubtedly will be discussed later in this symposium by Drs. Mosley[7], Shuval[8], and Moore[9]. Since Stevenson's investigation[10] provided the background for the work to be described, it should be noted that in his fresh-water studies, one comparing two beaches on Lake Michigan and the other comparing a swimming pool to a beach on the Ohio River, he reported two instances suggesting

* Present address: National Marine Water Quality Laboratory, West Kingston, Rhode Island 02892.

health effects associated with swimming in waters having a coliform density of about 2000 per 100 ml. Neither Stevenson[10], from his comparison of illness rates at two bathing beaches in Westchester, New York, nor Moore[5], from his study of poliomyelitis among children residing near the seashore in England and Wales, produced data suggestive of an increased incidence of illness associated with swimming in 'polluted' marine waters as defined by the coliform densities therein. The approaches taken by Moore and Stevenson were completely different. In Stevenson's study, the known factor was the swimming experience—bathing at a more polluted versus a less polluted beach—and the unknown factor was the rate of illness. Moore, starting with known cases of poliomyelitis and paired controls, compared the frequency of swimming in the case histories of the two populations.

In the design of the present investigation, we rejected Moore's approach for several reasons. First of all, he presupposed—quite possibly with justification from the public health point of view—which diseases are or are not significant in establishing criteria. Secondly, the reliability of his data may be questioned because of the protracted interval of time between the bathing experience itself and the inquiries made concerning its nature and circumstances. Finally, because of day-to-day variability in pollution due to hydrographic and meteorological factors such as tide, rainfall and wind, the 'bathing' experiences cannot be accurately associated with the quality of the water.

Stevenson's experimental design and findings[10] will be examined in more detail since they are essential in understanding the design of our study. Fortunately, advances in the state of the art with regard to epidemiological and microbiological methodology as well as data analysis have removed many of the constraints placed on Stevenson in the design of his study.

Definition of swimming

Neither Stevenson, in defining his bathers as opposed to his non-bathing controls, nor Moore, in his inquiries concerning bathers, appear to have defined 'swimming' such that individuals actually 'at risk'—those whose upper body orifices were significantly in contact with the water—were isolated and examined. Thus, if swimming is not defined precisely, it is possible that, the water quality at the two beaches notwithstanding, differences in pollution-associated illness may be sought between two populations in which many individuals never were appreciably at risk.

Multiple exposures

The day-to-day variability in pollution levels requires that, ideally, the study group be limited to individuals who have had a single (one-day) swimming experience during the observation interval associated with a given trial. In both the fresh-water (Lake Michigan) and salt-water (Westchester) studies, the day-to-day variability as measured by coliform indicators was considerable; in fact the pollution levels at each pair of beaches appreciably overlapped each other. Furthermore, in both these studies, the use of 'calendars' to record illness made it necessary to limit the study to seashore residents. This maximized the probability that multiple exposures would occur. Stevenson, in comparing the incidence of illness during 'high' and 'low' pollution days, obviated only part of this difficulty.

Non-bathing controls

Stevenson's non-bathing controls were individuals who did not go to the beach. Thereby, beach-going but not swimming-associated illnesses, such as gastroenteritis due to food ingested from restaurants, stands, etc., could be erroneously included in calculating the illness rates of swimming as opposed to non-swimming populations. This could affect illness-rate comparisons between 'high' and 'low' days as well as between beaches.

Demographic considerations

Stevenson analyzed his data with consideration to age and sex but not to ethnic or socioeconomic (SES) factors. However, especially in the salt-water study, the test beaches appear to have been paired with reference to ethnic and SES factors of the resident populations. Susceptibility to disease, background rate of illness, nature of the swimming experience, and even the reliability of the respondents' information concerning illness and the swimming experience could vary by ethnic or social class.

Tidal effects

Hourly variability in the pollution levels due to tide, wind, rainfall, etc., can present a problem in the interpretation of findings from epidemiological-microbiological trials. In Stevenson's study this was uncontrolled. Except in those instances where a 'captive' study population is available, such as institutionalized individuals or organized groups, there is little that can be done to mediate such effects since individuals at the beach during a given day can be expected to swim on several occasions during a half tidal cycle.

Indicators of pollution

At the time of Stevenson's study, the state of the art was such that only two microbiological parameters were measured. Coliform determinations were made in accordance with confirmed test procedures described in the 9th Edition of Standard Methods. Enterococcus levels were examined. It was subsequently determined that, although the media gave results that were specific for enterococci, the quantitative estimates were too low for use in the study. This points out the need to evaluate the methods to be used with waters from the study areas. The coliforms and enterococci were used as indicators of the threat of enteric disease from pathogenic microorganisms associated with fecal wastes. It is of interest that in his study Stevenson found that upper-respiratory disturbances were more prevalent than gastrointestinal complaints.

Marine Recreational Water Criteria Program

The first phase of the program to develop criteria for marine recreational waters consisted of a pre-test of the epidemiological and available microbiological methodology and verification of the suitability of the test and control beaches tentatively selected. After screening a number of salt-water beaches in the vicinity of large urban population centers, those in the vicinity of 20th Street, Coney Island, and 67th Street, Rockaways, in New York City were selected as the 'barely acceptable' and 'relatively unpolluted' beaches respectively (Fig. 1). In order to avoid 'human volunteer experiments' or the experimental bias which could occur by conducting the trials at beaches which local authorities post as unsafe, the test beach must be one which is 'barely acceptable' according to existing criteria and standards; thereby the initial trials are, in fact, a test of those standards. It is known that people do swim at beaches posted as unsafe; but these individuals may be a unique portion of the overall population and, generally, are too few in numbers for the conduct of the trials.

The second phase of the research will compare the rates of illness at the 'relatively unpolluted' beach and a 'barely acceptable' test beach using study populations large enough for significant results. In the absence of differences in illness rates between the relatively unpolluted and barely acceptable beaches, a 'more polluted' test beach must be obtained by the relaxation of local standards, the conduct of the trials during periods of peak pollution, or the relocation of the study to an area governed by less stringent standards. Thus, the second phase may be an iterative process.

The third phase of the study, which is designed to confirm and extend the second phase results, will be conducted at a series of beaches on a pollution gradient. Thereby, it is hoped to obtain a mathematical expression between the incidence of illness and some microbiological or chemical indicator(s) of water

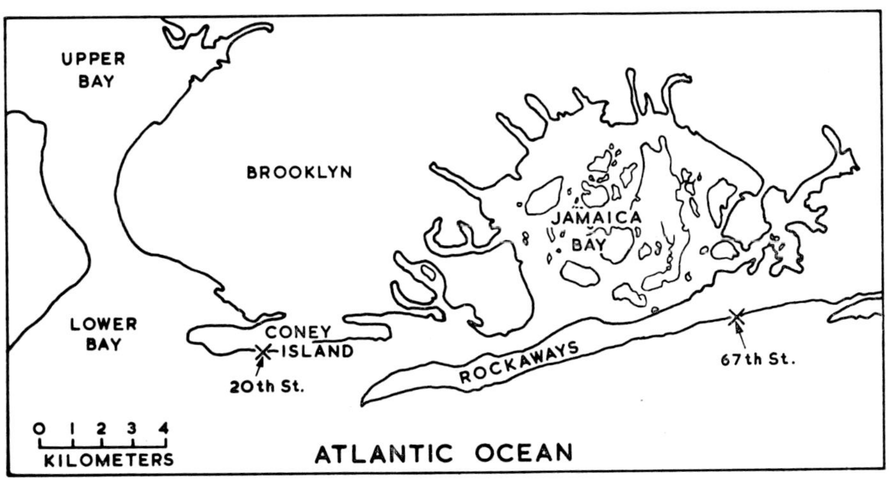

FIG. 1. Location of 'barely acceptable' (20th St., Coney Island) and 'relatively unpolluted' (67th St., Rockaways) beaches designated in Phase I trials

quality. An additional series of trials will be conducted at subtropical beaches. Studies to evaluate, refine or develop microbial and epidemiological methods were conducted concurrently with the first phase of the study and have continued along with the second phase this past summer.

Epidemiological methods

Features of the epidemiological-microbiological research, incorporated into the experimental design in response to problems encountered by Stevenson, are as follows:

1. Since swimming is defined as significant exposure of the upper-body orifices to the water, subjects are queried on the nature and duration of swimming activity. The validity of this information was pre-tested by observing family groups over a day at the beach and comparing these observations with information obtained from a representative of the group. The more rigorous definition of swimming will allow for a beach-going but non-swimming control group and thereby eliminate the bias from non-swimming associated illnesses.
2. Beach interviews are conducted only at weekends. Thus, by eliminating from the study individuals who swim during the midweeks before and after the weekend trial, subjects are limited to exposure to the water on a single day, or at most on two successive days. The use of weekends maximizes the size of the study population but limits the illness observation period to 8–10 days. This feature of the study will facilitate the analysis of the data 'by days', thereby obviating the effect of day-to-day variability in pollution levels.
3. The impact of within-day variability in pollution, primarily attributable to tidal effects, cannot be eliminated. It has been minimized, however, by choosing test and control beaches which are markedly different in the pollution levels reaching them and by maximizing the data-gathering effort at those times when minimal tidal effects (high or low tide) coincide with peak beach usage periods (usually 11 A.M. to 4 P.M.).
4. Demographic effects, which could assert themselves as differences in susceptibility to infection, in swimming activity and in the reliability of respondent information, have been minimized by selecting test and control beaches whose populations are demographically similar and by obtaining age, sex, ethnic, and SES information that can be used in isolating and identifying the influence of these factors.
5. An illness-severity and economic-impact index has been obtained from information collected on appropriate factors, such as remain home, stay in bed, visit a physician.
6. The validity of the illness information can be tested by providing the name of a physician in the reminder letter sent on the Monday following a trial and by requesting the names of other physicians consulted during the observation period. This also will provide follow-up medical and, possibly, laboratory data.
7. The illness questionnaire solicited information on irritations and disturbances of the skin, upper respiratory tract, eyes, and ears not only with regard to pollution-associated infectious processes but also against the possibility of toxic and hypersensitive conditions attributable to chemical pollution and to pollution-associated changes in marine biota.

The sequence of events during and subsequent to the beach interview is shown in Table 1.

TABLE 1. Sequence of events for epidemiological-microbiological trials

Day of week	Day number	Activity	Function
Saturday	1	Beach interview, water sampling	(a) Obtain name, address, phone, etc. (b) Reject pre-trial midweek swimmers (c) Query on beach activity (d) Assay of water samples
Sunday	2	Same as Saturday	As above
Monday	3	Reminder letter	(a) Provide name of physician (b) Reminder to note illness
Monday	10	Phone or mail interview	(a) Obtain illness information (b) Reject post-trial midweek swimmers (c) Obtain demographic information

Microbiological and chemical methods

A major deficiency in previous studies has been the limited spectrum of microbial or chemical health effectors/indicators for which quantitative estimates were obtained. This was due, in part, to the state of the art concerning enumerative methods for use with marine water samples. Stevenson[10] determined total coliform densities. Moore[5] measured coliforms and fecal coli quantitatively and *Salmonella* qualitatively. Dr. Bonde[11], who spoke earlier in this Symposium[12], and other workers[2,13] have discussed the inadequacy of coliforms and even fecal coliforms as the sole indices of water quality, the need to measure a more restrictive portion of the coliform population, and the possibility of including other pathogens and indicators in classifying recreational waters with regard to health effects in populations using them. In addition to enteropathogenic organisms found in feces and indicators of the contamination of the water therewith, the spectrum of health-effects organisms which could be enumerated in recreational waters includes pathogens which multiply in sewage effluents during the course of treatment, those discharged in certain industrial effluents[14,15], and aquatic microorganisms which are pathogenic for man and which may multiply in estuarine or marine environments subject to nutrient pollution and thermal effects. Table 2 lists those organisms whose enumeration was considered for this study, the sources of these organisms and the availability of methodology for their enumeration. Quantitative methods for estimating enterovirus densities in estuarine and coastal waters are considered too primitive, time consuming, and expensive for inclusion in this study at the present time. In addition to the microbial health-effect indicators listed in Table 2, coprostanol will be examined as an indicator of fecal pollution. Inclusion of a given microbial pathogen in the list of organisms to be enumerated does not necessarily imply that there are definitive data linking it to recreational waterborne disease but, rather, that it is a known human pathogen and has been recovered in large numbers from coastal or estuarine waters. Sanitary surveys are performed and relevant hydrographic and meteorological data obtained.

It can be seen from Table 2 that a number of methods have been developed or are undergoing development in our laboratory. At least one of the methods, that for *P. aeruginosa*, has been published[16] and reports describing several others are in preparation. In each instance where an enumerative method was developed, it may be assumed that no method was available, that one or more existing methods were evaluated and found unsatisfactory *for the purposes of this investigation* or that the existing method was a most probable number procedure with the attendant problems of precision and logistics. It is beyond the scope of this report to describe all the methods which have been or will be used in our studies.

TABLE 2. Source, role, and status of enumerative methodology of candidate health-effects indicators

Organism	Significant source[a]	Possible role[b]	Methods development[c]
Enteroviruses	F S	P	U
Salmonella	F S	P	D (S–HVS)
Shigella	F S	P	Z
Pseudomonas aeruginosa	S I R A	P	D (mPA)
Aeromonas hydrophila	S R A	P	D (mA)
Coliforms	F S I R A	W[e]	D (TC MPN), (mC)
Escherichia coli	F S	P W	D (mC)
Klebsiella	F S I R A	P W N	D (mC), (mK)
Enterobacter	S I R A		D (mC)
Citrobacter	R A		D (mC)
Fecal coliforms	F S I	W	D (EC MPN), (mC)
Fecal streptococci	F S R	W	D (mSD)
Vibrio parahemolyticus	A	P[e] N	U
Clostridium perfringens	F S	W	U
Candida albicans	F S[d]	P W	Z

[a] Relative to other sources: F, feces of warm-blooded animals; S, sewage; I, industrial wastes; R, run-off from uncontaminated soils; A, estuarine or coastal waters

[b] Role: P, pathogen; W, indicator of sanitary wastes; N, indicator of nutrient pollution

[c] Status of method (quantitative development): D, developed; U, under-development; Z, no method

[d] Insufficient information

[e] Questionable

However, as an illustration of the direction that our methods research has taken, we have included a flow diagram for the millipore-filter procedure for coliforms (mC) which can be used to obtain total and fecal coliform densities as well as the distribution of the component genera of the coliform population (*Citrobacter*, *Enterobacter*, *Escherichia* and *Klebsiella*). Like our British colleagues and a number of investigators in the United States[13], we feel that *E. coli* is the most specific indicator of contamination with the fecal wastes of warm-blooded animals. However, since sources of pollution in addition to the fecal wastes of warm-blooded animals must be considered in establishing water quality criteria, information on the distribution of coliform biotypes should be more instructive. The flow diagram for the mC procedure is presented in Table 3. As shown, the procedure utilizes an *in situ* substrate (urea) test for the identification of *Klebsiella* but requires the picking of colonies for biochemical identification in the differentiation of *Escherichia*, *Enterobacter*, and *Citrobacter*. An oxidase test from a nutrient agar plate is used to eliminate *A. hydrophila*. A modification which markedly simplifies this procedure has been developed and evaluated. Incubation of the membrane on a selective-differential medium is followed by three successive *in situ* tests, a urease test for *Klebsiella*, an oxidase test to eliminate *A. hydrophila*, and an indole test to identify *Escherichia*. The remaining coliforms are either *Enterobacter* or *Citrobacter*. The medium appears to be sufficiently differential so that typical blue colonies regularly confirm as coliforms, thereby obviating the necessity for confirmation. If a fecal coliform estimate is desired, an appropriate number of colonies has to be picked to lactose broth and then to EC broth.

TABLE 3. Flow diagram of mC procedure for total coliforms and coliform biotypes

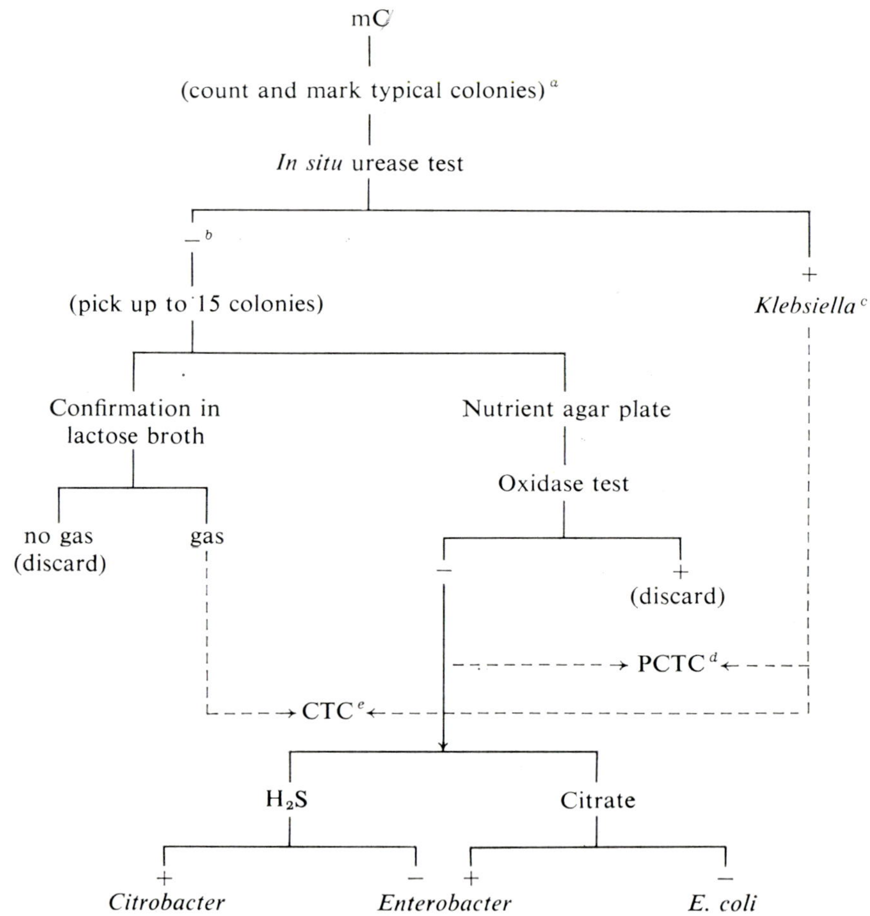

[a] Presumptive total coliforms
[b] Urease-negative coliforms (UNC)
[c] Typical urease-positive colonies, rarely do not confirm

[d] Partially confirmed total coliforms $= \text{UNC}\,\dfrac{\text{Oxidase neg.}}{\text{Total picked}} + \textit{Klebsiella}$

[e] Confirmed total coliforms $= \text{UNC}\,\dfrac{\text{Oxid. neg., lactose pos.}}{\text{Total picked}} + \textit{Klebsiella}$

– – – for calculating coliform densities

Program status

In cooperation with Mr. Paul Habermann and Dr. Jack Elinson of the Center for Policy Research and the Division of Socio-Medical Sciences, Columbia University School of Public Health, a pre-test of the epidemiological and microbiological methodology was completed during the summer of 1973. The details of this phase of the overall program will be published at a later date since, as of this writing, the data have not been completely analyzed. However, it seems quite certain that both the epidemiological and the microbiological methodology which was evaluated will be appropriate for the second phase of this study.

TABLE 4. Comparison of log mean indicator and pathogen densities (per 100 ml) at the 'barely acceptable' and 'relatively unpolluted' beaches

Organism(s)	Method	Coney Island (20th St.)	Rockaways (67th St.)
Total coliforms	mC	983*	39·8
Fecal coliforms	mC	165*	21·5
E. coli	mC	174*	24·8
Klebsiella	mC	122*	13·7
Enterobacter	mC	530*	11·1
Fecal streptococci	mSD	91·2	21·8
P. aeruginosa	mPA	30·4*	6·5
A. hydrophila	mA	25·3	26·5

* Significantly different at 95 per cent confidence level

With the exception of *A. hydrophila* and possibly fecal streptococci, the differences between the two beaches with respect to the densities of the candidate health-effects indicators were both significant and appreciable (Table 4), this in spite of within and between trial variability (Fig. 2). Although the Coney Island beach was 'barely acceptable' in that the adjacent beach to the west was posted as unsafe, the mean total coliform density during the period of the trials was about half the local standard of 2400 total coliforms per 100 ml. That this was not due to the assay method used can be seen from Table 5

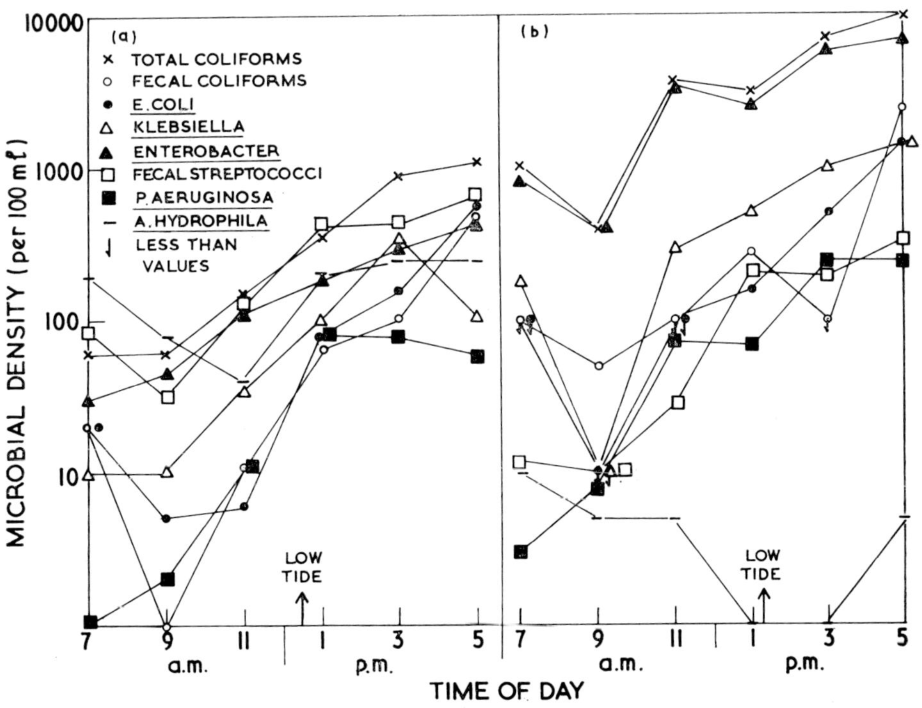

FIG. 2. Day-to-day and within-day variability in log mean microbial densities at Coney Island beach, on (*a*) 28 July and (*b*) 12 August 1973
Period of significant swimming activity was from 11 A.M. to 5 P.M.

TABLE 5. Comparison of total coliform recoveries (per 100 ml) by MPN and mC procedures in matched samples
Samples '1' and '2' collected at opposite ends of beach, about 300 m apart

Date of Trial[a] (1973)	MPN 1	MPN 2	mC 1	mC 2
14 July	490	78	160	130
22 July	330	2400	1200	600
28 July	220	1300	120	600
29 July	130	170	170	110
11 Aug.	3500	3300	15 000	14 000
12 Aug.	400	7000	3100	3500
18 Aug.	4900	4600	1900	1200
Median	900		900	
Log mean	870		850	

[a] Data for 19 Aug. omitted because mC values not available

which compares mC and MPN total coliform estimates. Rather, the disparity in the total coliform densities used to classify the beaches and the data obtained in this study appears to reflect the influence of day-to-day and within-day variability. During the trials, the Coney Island beach did meet the National Technical Advisory Committee recreational water requirement[17] of a mean of 200 fecal coliforms per 100 ml. However, more than 10 per cent of the samples had a fecal coliform density in excess of 400 per 100 ml, the second aspect of the criterion.

The pre-test data did produce two interesting observations associated with the choice of the test and control beaches. From a comparison of the *E. coli* and *Enterobacter* ratios (Table 6), it can be seen that the very low levels of pollution (as measured by coliforms) reaching the Rockaways beach are not a further dilution of those at Coney Island. Rather, the data suggest an immediate source of raw fecal wastes. Furthermore, the *Salmonella* recoveries obtained by the *Salmonella* high-volume sampling (S–HVS) method[18] (Table 7), do not reflect the disparity in the total coliform densities at the two beaches. This should not affect the choice of the Rockaways as the control beach. It does, however, point out the value of examining the distribution of coliform biotypes in determining sources of pollution.

The second relates to the analysis by trial (day) of the illness rates at the Coney Island beach. The trials (days) were ordered according to the mean coliform or fecal coliform density (Table 8) and segregated into four groups of two trials each. The illness rates for the four groups were determined by type (respiratory, gastrointestinal and other) for swimmers and the non-swimming controls. This analysis revealed no increase in illness associated with swimming during the 'high' as compared to the 'low' pollution days. On the other hand, there was a strong suggestion that the incidence of gastro-intestinal illness among swimmers, when corrected for the rate among non-swimmers, was higher at the 'barely acceptable' (Coney Island) than at the 'relatively unpolluted' (Rockaways) beach. As noted previously, since this was a pre-test of methodology, the number of individuals queried was too small to draw any firm conclusions relative to water quality criteria. However, these findings reaffirm the recognized need to extend the range of pathogens and indicators enumerated in the course of such studies.

TABLE 6. *E. coli*/*Enterobacter* ratios at Coney Island (20th St.) and Rockaways (67th St.) beaches by trial

Trial date	Mean recovery (per 100 ml)				*E. coli*/*Enter.* ratio	
	Coney Island		Rockaways		Coney Island	Rockaways
	E. coli	*Enter.*	*E. coli*	*Enter.*		
14 July	59	162	13	5	0·36	2·6
22 July	197	324	16	7	0·61	2·3
28 July	79	226	24	4	0·35	6·0
29 July	88	239	20	27	0·37	0·74
11 Aug.	1099	6998	12	2	0·16	6·0
12 Aug.	493	4487	17	14	0·11	1·2
18 Aug.	240	279	41	16	0·86	2·6
19 Aug.	78	249	56	14	0·31	4·0
Median ratios					0·35	2·6

TABLE 7. Comparison of *Salmonella* and total coliform densities (per 100 ml) at Coney Island and Rockaway beaches
Arranged in descending order of coliform values

Coney Island			Rockaways		
Date	Total coliforms[a]	*Salmonella*[b]	Date	Total coliforms[a]	*Salmonella*[b]
11 Aug.	14 500	0.00020	18 Aug.	350	<0.00018
12 Aug.	3300	0·00045	22 July	205	<0·00018
19 Aug.	1850[c]	0·00020	29 July	185	0·00040
18 Aug.	1550	0·00020	19 Aug.	90	<0.00018
22 July	900	0·00040	12 Aug.	70[c]	0·00020
29 July	435	0·00020	14 July	30	<0·00018
28 July	360	0·00020			
14 July	145	0·00020			

[a] mC estimate of total coliforms from low-tide samples collected concurrently with those for the *Salmonella* assays
[b] Obtained from examination of 55·5 l. by S-HVS method[18]
[c] Estimate obtained by MPN method

TABLE 8. Mean total and fecal coliform densities (per 100 ml) at Coney Island beach by trial
Log mean recoveries from samples collected over four tide periods at two locations

Trial date	Total	Fecal
11 Aug.	9268	565
12 Aug.	5668	859
22 July	794	149
18 Aug.	640	201
29 July	476	77
28 July	467	77
19 Aug.	491	113
14 July	300	58

REFERENCES

1. METCALF, T. G. Evaluation of shellfish sanitary quality by indicators of sewage pollution. This volume, pp. 75–84.
2. HENDERSON, J. M. Enteric disease criteria for recreational waters. *J. sanit. Engng Div. Am. Soc. civ. Engrs*, 1968, **94**(SA6), 1253–1276.
3. SENN, C. L. Coliform standards for recreational waters. *J. sanit. Engng Div. Am. Soc. civ. Engrs*, 1963, **89**(SA4), 57–94.
4. FOSTER, D. H., HANES, N. B., and SABIN, M. L. A critical examination of bathing water quality standards. *J. Wat. Pollut. Control Fed.*, 1971, **43**, 2229–2241.
5. PUBLIC HEALTH LABORATORY SERVICE. Sewage contamination of coastal bathing waters in England and Wales. A bacteriological and epidemiological study. *J. Hyg., Camb.*, 1959, **57**, 435–472.
6. JONES, E. H. External otitis, diagnosis and treatment. C. C. Thomas Publications, Springfield, Illinois, 1965.
7. MOSLEY J. W. Epidemiological aspects of microbial standards for bathing beaches. This volume, pp. 85–93.
8. SHUVAL, H. I. The case for microbial standards for bathing beaches. This volume, pp. 95–101.
9. MOORE, B. The case against microbial standards for bathing waters. This volume, pp. 103–110.
10. STEVENSON, A. H. Studies on bathing water quality and health. *Am. J. publ. Hlth*, 1953, **43**, 529–538.
11. BONDE, G. J. Bacteriological methods for estimation of water pollution. *Hlth Lab. Sci.*, 1966, **3**, 124–128.
12. BONDE, G. J. Bacterial indicators of sewage pollution. This volume, pp. 37–47.
13. DUTKA, B. J. Coliforms are an inadequate index of water quality. *J. environ. Hlth*, 1973, **36**, 39–46.
14. DUNCAN, D. W., and RAZZELL, W. E. Occurrence of *Klebsiella pneumoniae* in common environmental situations. B.C. Research and Vancouver Laboratory Fisheries Service Publications, Vancouver, 1970.

15. DUFOUR, A. P., CABELLI, V. J., and LEVIN, M. A. Occurrence of *Klebsiella* species in wastes from a textile finishing plant. *Bact. Proc.*, 1973, p. 3.
16. LEVIN, M. A., and CABELLI, V. J. Membrane filter technique for enumeration of *Pseudomonas aeruginosa*. *Appl. Microbiol.*, 1972, **6**, 864–870.
17. NATIONAL TECHNICAL ADVISORY COMMITTEE. Water Quality Criteria. Federal Water Pollution Control Administration, Dept. of Interior, Washington, D.C., 1968, pp. 7–14.
18. LEVIN, M. A., CABELLI, V. J., and FISCHER, J. R. A quantitative large volume sampling technique (*in preparation*).

In presenting the paper Dr. Cabelli said that a modification of the approach used by Stevenson[10] in the 1950's appeared to be the one most likely to yield the positive results required for establishing microbiological criteria for marine recreational waters. The three most essential modifications seemed to be (1) a more restricted definition of swimming, (2) a decrease in the time span over which a trial was conducted, and (3) the elimination of multiple exposures to the water. Furthermore, in a programme adopted to develop the criteria, three basic tenets should be observed, namely that (a) not only the numerical value but also the appropriate indicators of recreational water quality should be treated as unknowns, (b) the design of the programme should be minimally (if at all) influenced by existing criteria at the local or national level, and (c) the criteria developed should provide a relationship—between the incidence of illness among swimmers and some microbiological or chemical measure of water quality—from which numerical standards could be derived.

The United States Environmental Protection Agency was carrying out investigations in consonance with these guide lines and with emphasis on epidemiological-microbiological trials conducted at marine recreational beaches. Membrane-filtration methods had been developed for the enumeration of a number of potential water quality indicators. Both the microbiological and epidemiological methodologies were found to be satisfactory in a pre-test conducted at New York City beaches during the summer of 1973. In fact, despite the small numbers of individuals questioned in the pre-test, significant differences in illness rates between swimmers and beach-going but non-swimming controls were obtained at a 'barely acceptable' as compared to a 'relatively unpolluted' beach.

DISCUSSION

Mr. H. M. Cotton (New Zealand) said that there were some small beaches in New Zealand which were intensively used for swimming. He asked what was the impact of the swimmers themselves due to the bacteria, etc. they shed to the waters, and if the Authors had considered the number of bathers per unit area as a relevant variable. Shellfish (such as cockles) were often taken by visitors at three beaches in New Zealand; he thought this practice unwise and invited Dr. Cabelli's comments on it.

Mr. J. A. Wakefield (U.K.) asked if Dr. Cabelli sometimes had the impression that, due to his and other scientists' inability to prove a strong relationship between pollution and ill health, his work was counter-productive, and whether it would not be better to quantify the need to spend money on anti-pollution measures in terms of human revulsion rather than of poor health or mortality.

Mr. J. C. T. Arnold (U.K.) said that, as a consulting engineer concerned with the design of sea outfalls for the discharge of sewage, it seemed to him very important that a 'standard' should be agreed upon. It probably did not matter much to the engineer what that standard was—provided it was attainable. Prior to 1 April 1974, U.K. designers found themselves producing schemes, with no precise idea of the criteria to be met, for approval by an authority which could reject the proposals on the grounds that they did not satisfy requirements which it did not clearly define. This was a deplorable waste of time and resources. The Royal Commission standard for discharge of treated sewage to watercourses was a fairly crude criterion, but it had worked satisfactorily in most cases. He therefore asked if a similar approach could not be adopted for discharges to the sea.

Dr. M. J. Suess (W.H.O.) took the opportunity of informing participants that the World Health Organization was currently developing a Manual on recreational waters; the data on criteria reported in the paper were, of course, of great interest to the W.H.O. He also pointed out the importance of chemical substances, such as the carcinogenic polynuclear aromatic hydrocarbons which were present in sewage, had been demonstrated to be present in shellfish, and should definitely be considered in conjunction with standards for shellfish.

REPLY TO DISCUSSION

Dr. Cabelli, in response to Mr. Cotton, said he thought that the problem of bather-to-bather transmission of infective agents was improbable except in small bodies of water, such as swimming pools and ponds, which were not subject to much water exchange. He considered that it would be injudicious to allow the harvesting of shellfish from areas used for swimming, because of the accumulation of organisms by shellfish and because a faecal indicator had no value when dealing with such sources.

He disagreed with Mr. Wakefield's contention. For reasons stated in the paper, he did not believe that the British experience with 'pollution associated' recreational-borne illness was representative. Furthermore, the Authors' preliminary findings in this context had suggested that it might, in fact, be possible to obtain a meaningful relationship which could be used by the appropriate regulatory agencies.

Dr. Suess's point was well taken. With regard to recreational water quality criteria, the Authors were currently measuring coprostanol as an indicator of sewage pollution; in time, they hoped to isolate and examine such 'skin manifestations' as contact dermatitis as possible effects of chemical pollution. Similarly, pollution of shellfish by both inorganic and organic chemicals (including carcinogens) was receiving considerable attention.

Paper No. **8**

Evaluation of Shellfish Sanitary Quality by Indicators of Sewage Pollution

T. G. Metcalf

Department of Microbiology, University of New Hampshire
Durham, New Hampshire 03824

Restriction of shellfish to an estuarine abode inevitably leads to exposure to pollutants in industrial and domestic wastes discharged into estuarine waters. The sanitary status of shellfish exposed to domestic waste discharges has been a matter of concern since the beginning of the nineteenth century. A causal relationship between shellfish and outbreaks of enteric infections was reported in 1895, when a third of the typhoid fever cases in Brighton, England, were estimated to result from consumption of sewage-polluted oysters or mussels[1]. Since that time many outbreaks of bacterial enteric infections resulting from consumption of sewage-polluted shellfish have been recognized around the world. Transmission of viral-induced enteric infections by shellfish was shown to occur with recognition of infectious hepatitis epidemics following ingestion of raw shellfish[2–4]. Paralytic shellfish poisoning (PSP) arising from accumulation of poisonous species of dinoflagellates is another type of public health problem for 'shellfish eaters'[5,6].

The edible bivalve mollusks of the class Pelecypoda (oysters, clams, mussels) are the only molluskan shellfish of commercial importance for which sanitary controls are required. The microbiological basis of these tests is not applicable to tests for PSP where mouse bioassay is the only practical way of detecting poison. The outline of a shellfish certification program used subsequently in the United States and Canada was formulated in 1925[7]. While the program involved cooperation between the shellfish industry and local and federal governments to maintain sanitary conditions during processing and shipping stages, the predominant public health control measure was directed at the water in which shellfish grow.

The value of coliform tests as the principle microbiologic criterion for assessing the sanitary quality of shellfish and shellfish growing waters is one part of a more general controversy over the validity of such tests as indicators of the biological safety of water[8]. Procedures currently used in the United States for examination of sea water and shellfish include tests for index microorganisms indicative of fecal contamination and, accordingly, indicate possible danger of transmission of enteric diseases[9]. Specific methods for the detection and enumeration of salmonella, shigella, other bacterial pathogens, and enteric viruses have not been recommended because laboratory procedures are not considered sufficiently established to justify routine use. Questions of sensitivity, accuracy, reliability, and epidemiologic significance of these tests are involved.

Adoption of coliphage as an indicator of fecal pollution was first advocated in 1948[10]. Advantages cited on behalf of the use of coliphage included convenience, economy, and quantitative enumerative capability[11]. More recently use of coliphage f2 as an indicator of enteric virus in water was recommended[12]. Properties of chlorine resistance, natural occurrence in large numbers in polluted water, and a size and structure comparable to the enteric viruses were cited in favor of this indicator virus.

The following study was concerned with use of the coliform test to assess the danger of pollution by bacterial or viral pathogens, and its significance as a criterion of the sanitary quality of shellfish. Public health aspects of the sanitary status indicated for shellfish growing in waters polluted by sewage discharges are considered.

Methods

Shellfish examinations were conducted in New Hampshire and Galveston Bay estuaries. The eastern oyster, *Crassostrea virginica,* was used throughout these examinations. Information on the sanitary quality of shellfish and shellfish waters was obtained from the results of coliform, salmonella, and enteric virus tests.

Water samples were collected by catch, gauze-pad, or virus-concentrator procedures. Samples examined for virus were concentrated to a final volume of 10–20 ml. Concentrates were prepared from gauze-pad extracts following pH adjustments, differential centrifugation, and ether treatment. Details of these procedures have been reported previously[13]. Reconcentrates of virus-concentrator eluates from adsorptive filters were made by successive adsorption and elution utilizing membrane filters. Details of these methods have been described previously[14].

Virus isolations or assays were made using primary monkey kidney monolayers, with examinations for plaque or cytopathic effects. Isolates were identified by serum neutralization tests[15].

Salmonella samples were obtained by passing up to 2 l. of water through a celite-coated membrane filter, followed by addition of the filter to enrichment broth. Isolation and characterization procedures have been described previously[16].

Oysters were collected by tonging or dredging. Salmonella examinations were made on samples derived from homogenates of pools of ten or more oysters; 5–10 ml amounts were added to enrichment media, with subsequent isolation and characterization procedures the same as described for water samples.

Virus recovery from shellfish in studies comparing coliform indices, salmonella, and virus occurrence was made from samples derived from a common homogenate source. Final reconcentrates for use in virus isolations were obtained by either of two methods. The first made use of low and high speed centrifugation, with ether treatment of resuspended sediments. Details of the method have been reported previously[17]. The second involved separation of tissue debris at pH 4·5, with recovery of virus from sediment or supernate, depending upon low or high conductivity values. Sediment elution with 0·05 M glycine, pH 11·5, led to recovery of virus. Supernates or neutralized eluates were concentrated further by membrane adsorption procedures[18]. Removal of membrane coating components (MCC), if present, was accomplished by mixture, at pH 8·5, of eluate with a washed cationic resin (Ionac C-249, Ionac Chemical Co., Sybron Corp., Birmingham, New Jersey 08011). Recovery of any virus attached to resin followed washing with 0·05 M glycine, pH 8·5. Natural virus isolations were made using cell cultures, and isolates were identified as described previously[15].

Water and shellfish samples were examined for total and fecal coliform values in accordance with Standard Methods[8,9]. Five-tube fermentation tests with lauryl tryptose broth were used for presumptive tests; MPN coliform counts were based on confirmation of positive presumptive results in brilliant green lactose bile, 2 per cent. Circulating constant-temperature water baths at $44·5 \pm 0·1$°C were used for fecal coliform tests. Presumptive and confirmed tests were read at 24 ± 2 h and 48 ± 3 h. Fecal coliform tests were read at 24 ± 2 h.

Fecal streptococci in water and oysters were enumerated on M-enterococcus medium. The membrane filter procedure was used with water samples, 1-ml amounts of homogenate being plated for oyster samples. Details of the procedures used have been described[19].

Coliphage isolations were made from samples derived from water or shellfish homogenates following differential centrifugation and overnight treatment with chloroform. Quantitative enumeration was made by mixing 0·5-ml samples with 0·7 per cent agar overlays containing the appropriate *Escherichia coli* host cell. Qualitative determinations using chloroform-treated supernates after overnight enrichment with *E. coli* were made in the same manner.

Coliphage used in experimental studies was isolated from New Hampshire estuary waters[20]. It was a single-stranded RNA virus, 25 nm in diameter, and of icosahedral symmetry.

Results

A New Hampshire estuary subject to pollution from domestic waste discharges provided a natural laboratory within which the relationship of pollution to sanitary quality of shellfish and shellfish-growing waters was studied intensively. It was demonstrated that salmonellae and enteric viruses do occur in shellfish and shellfish-growing waters judged to be sanitarily acceptable by fecal coliform indices.

Shellfish waters shown by pollution indices to be of approved or conditionally approved classification were examined first for the presence of bacterial indicators of pollution and salmonellae and viruses in parallel tests. Examples of representative findings are shown in Table 1.

Median fecal coliform numbers less than 70 per 100 ml of sea water indicated sanitarily acceptable water at four of five stations. At first glance the validity of these findings seemed to be challenged by the salmonella and virus isolations. Members of both of these groups were found at Station 1, and virus isolations were made at all stations. Bacterial indicator and salmonella isolations were made from catch samples, while virus isolations came from gauze-pad samples. Thus bacterial results were indicative only of the sanitary status prevailing at the time of sample collection, while virus isolations were made from samples collected over a time span of 4–72 h. Under these conditions, ample opportunity was provided

TABLE 1. Actual and indicated sanitary values of shellfish waters subject to sewage pollution

Station	No. of samples	Median numbers per 100 ml sea water: Fecal coliforms (EC-MPN)	Median numbers per 100 ml sea water: Fecal streptococci (MF)	Isolates: Salmonella[a] No.	Isolates: Salmonella[a] Per cent	Isolates: Enteric virus[b] No.	Isolates: Enteric virus[b] Per cent
1	13	49	15	3	23	4	31
2	14	170	17	4	28	4	28
3	10	2	2	0	0	5	50
4	25	2	1	0	0	4	16
5	24	3	2	0	0	4	17

[a] Salmonella isolates were made from grab samples
[b] Enteric virus isolates were made from gauze pads immersed 3 days in estuary water

for introduction of pollutants not measured and therefore not recognized by bacterial indicator analysis. Agreement between indicated and actual sanitary status at three of four stations was considered a better measure of the usefulness of bacterial indicators for bacterial pathogens. Nevertheless salmonellae were isolated from 23 per cent of samples at one station considered safe from a fecal coliform viewpoint.

Examinations were made in oysters where indicator tests and salmonella and virus isolations all used the same sample. Results showed the actual sanitary status of oysters could not always be forecast correctly by the results of indicator tests. Data obtained from these examinations are shown in Table 2.

TABLE 2. Actual and indicated sanitary status of oysters from waters subject to sewage pollution

Station	No. of samples	Median numbers per 100 g oysters: Fecal coliforms (EC-MPN)	Median numbers per 100 g oysters: Fecal streptococci (MF)	Isolates: Salmonella No.	Isolates: Salmonella Per cent	Isolates: Enteric virus No.	Isolates: Enteric virus Per cent
1	10	25	0	0	0	2	20
2	8	78	0	2	25	2	25
3	30	52	40	3	10	2	7
4	13	130	66	1	8	3	23

Each shellfish pool from each collecting station was of acceptable sanitary status according to median coliform numbers. Yet salmonella isolations were made from these pools at three stations, and virus isolations were made at all stations. These data showed both salmonella and virus could be present in oysters judged by bacterial indicator test results to be of acceptable sanitary quality. The greater number of virus isolations made paralleled the results obtained in sea water and led to an impression of a more widespread dissemination of virus than salmonella throughout the estuary.

Oyster beds and overlaying sea water were monitored daily for fecal coliforms, salmonellae, and viruses. Evaluation of oyster sanitary value indicated by coliform numbers was made only where salmonellae and viruses were recovered from water. The results are shown in Table 3.

TABLE 3. Fecal coliform numbers found in parallel examinations of oyster and overlaying sea-water samples positive for salmonellae and enteric viruses

Station	No. of samples	Fecal coliforms (median EC-MPN): Numbers per 100 g oysters	Fecal coliforms (median EC-MPN): Numbers per 100 ml sea water	Isolates: Salmonella[a] Oysters	Isolates: Salmonella[a] Water	Isolates: Enteric virus[b] Oysters	Isolates: Enteric virus[b] Water
1	15	120	19	0	2	1	3
2	13	330	33	3	2	7	6
3	10	25 000	790	5	7	8	7
4	6	6750	131	1	4	6	4

[a] Salmonella isolates from water were made from grab samples
[b] Enteric virus isolates from water were made from gauze pads immersed 3 days in estuary water

One station produced acceptable quality oysters and three stations acceptable quality water. Yet, with one exception, salmonellae and viruses were found at least once in both types of samples at each station.

The data in Tables 1–3 showed bacterial indices to be useful for gauging the likelihood of pathogen presence resulting from significant levels of fecal pollution. Their usefulness was open to question when applied to sanitary assessments involving shellfish subjected to low levels of fecal pollution. To gain further insight into indicator usefulness, tests were conducted only in shellfish-growing waters with a conditionally approved classification. Data were collected on bacterial numbers found in parallel examinations of oysters and overlaying sea water where the oyster samples were shown by direct tests to harbor salmonellae and viruses. The data are given in Table 4.

TABLE 4. Fecal coliform and fecal streptococci numbers in oysters and overlaying sea water at time of salmonella and enteric virus isolations from oysters

Trial	Fecal coliforms: Oysters (EC-MPN/100 g)	Fecal coliforms: Sea water (EC-MPN/100 ml)	Fecal streptococci: Oysters (MF/100 g)	Fecal streptococci: Sea water (MF/100 ml)
1	0	0	0	0
2	4·5	4	200	1
3	700	46	0	16
4	0	7·8	0	0
5	45	0	200	2
6	0	0	0	1
7	130	33	200	4

Salmonellae and viruses were isolated from oysters on seven different occasions. Each time the sanitary quality of sea water overlaying the oyster bed was acceptable according to fecal coliform numbers. Oysters were unacceptable only once. Fecal streptococci numbers were in agreement with coliform values and supported the findings of acceptable sanitary quality. This study emphasized the inability of bacterial indices to insure the absence of enteric pathogens or indicate a potential health hazard for oysters harvested from conditionally approved waters.

The appearance of bacteria or virus in shellfish requires introduction of these forms into the environmental sea water. The flushing action of tidal exchanges promotes water movement and an opportunity for pollutant introduction. Fecal coliform numbers and salmonella and virus occurrence were studied during tidal exchanges. Parallel examinations of oysters and overlaying sea water were made to compare indicator assessment of sanitary quality with the results of direct tests for salmonellae and viruses. The comparisons found when salmonellae and viruses were shown to occur are given in Table 5.

An acceptable sanitary quality for sea water was indicated by fecal coliform numbers before, during, and after two sets of tidal exchanges. Oysters examined at high and low tide were of acceptable quality. Yet water was positive for salmonellae and viruses halfway through the exchanges, and oysters at either half or low tide. The data showed the pollutants were introduced during tidal exchanges. Salmonellae

TABLE 5. Fecal coliform numbers in oysters and overlaying sea water positive for salmonella and enteric virus during tidal exchanges

Trial	Tidal exchange	Oysters[a]: Fecal coliforms (EC-MPN/100 g)	Oysters[a]: Isolates[b] Salmonella	Oysters[a]: Isolates[b] Virus	Water: Fecal coliforms (EC-MPN/100 ml)	Water: Isolates[b] Salmonella	Water: Isolates[b] Virus
1	High	78	0	0	0	0	0
	½		0	+	6	+	+
	Low	230	+	+	0	0	0
2	High	130	0	0	0	0	0
	½		0	0	12	+	+
	Low	110	+	+	0	0	0

[a] Oyster pools with 10 oysters per pool

[b] Salmonella isolates were made from grab samples, virus isolates from gauze pads immersed 3 h in estuary water

and viruses were accumulated during the 6-h exchange period, and retained after disappearance of these forms from environmental sea water. Retention of these forms held great significance for assessment of oyster sanitary quality. In order to detect oyster carriage of bacteria or viruses at low tide it was necessary to test directly for these pathogens. Assessment of oyster sanitary quality through fecal coliform indices was not possible, either by examination of overlay water or oysters themselves.

Introduction of new equipment and methods for direct enumeration of virus in water has provided an important new tool for monitoring virus occurrence in shellfish-growing waters. Concomitantly it is possible simultaneously to test the same water for its coliform content, with removal of samples for coliform tests at frequent intervals. Examples of fecal coliform numbers found in sewage effluents and river water sampled every 5 min for 1 h are given in Table 6.

TABLE 6. Fecal coliform numbers (per 100 ml) in sewage effluents and river water during continuous monitoring

Sample (min)	Waste-treatment plant effluents[a]			River water
	No. 1	No. 2	No. 3	
0	490	3300	1 100 000	210
5	1110	170	790 000	—
10	2600	50	1 300 000	70
15	3300	50	3 300 000	—
20	80	50	300 000	<20
25	7900	80	2 300 000	—
30	50	20	2 300 000	50
35	330	490	3 300 000	—
40	700	140	4 900 000	50
45	—	—	—	
50	—	—	—	110

[a] Suspended-solids contents of effluents (mg/l): No. 1—54, No. 2—56, No. 3—72. Plant capacities (m^3/s): No. 1—3·1, No. 2—0·7, No. 3—1·7

Fluctuations in coliform numbers made it virtually impossible to gain a true picture of coliform density from the result of a single test. The fluctuating values were indicative of an uneven distribution of coliform bacteria and suggested that a similar uneven distribution of pathogenic forms might be expected. The results raised a question about the way by which shellfish-water sanitary quality should be assessed. As a criterion of sanitary quality, what significance should be attached to the results of single fecal coliform tests on catch samples at given test sites? The data in Table 6 offered a possibility of rapidly fluctuating coliform numbers in shellfish waters exposed to fecal pollution. These data indicated no single coliform value would adequately describe the extent or measure the degree of pollution experienced over a period of time. The soundest basis for evaluating coliform numbers at a given site would be provided by the results of frequent sampling over a period of time.

Direct virus monitoring of shellfish waters was combined with direct examinations of oysters for natural virus in Galveston Bay. Parallel coliform and virus examinations of oysters and overlaying sea water were made on the same sample. The results of a series of examinations are given in Table 7.

Fecal coliform numbers indicated a uniformly acceptable sanitary quality of water and oysters, with the exception of one water sample. Virus monitoring of overlay sea water failed to detect virus in sample volumes of 91–397 l. examined during a 3-month period. Despite the agreement of all test criteria approving the sanitary quality of overlay water, direct virus testing of oyster pools led to three natural virus isolations. The one pool retested led to recovery of the same virus type isolated initially. Recovery of *Vibrio parahemolyticus* on three occasions suggested little if any parallelism existed between occurrence of this species and fecal coliform indices of acceptable sanitary quality.

Results of the Galveston Bay study provided important additional evidence of an inadequacy of fecal coliform indices. The most serious though not the only offenders in all studies were viruses. The number of times virus was detected in coliform-acceptable shellfish or overlay water exceeded greatly the incidence of salmonellae or other bacterial pathogenic forms.

The possible use of coliphage as an indicator of enteric virus was examined by comparing frequency of coliphage occurrence in shellfish and shellfish waters, and coliphage survival and growth-potential characteristics. The distribution of coliphage and enteric virus in oysters, oyster-bed mud, and estuary water was studied to determine coliphage dissemination. The results of this study are given in Table 8.

TABLE 7. Fecal coliform numbers found in parallel examinations of oysters and overlaying sea water at one collecting station, with virus and *Vibrio parahemolyticus* recoveries from oysters only September–December 1972

Samples		Fecal coliforms		Isolations				
				Virus			*Vibrio parahemolyticus*	
					Oysters			
Trial	(l.)	Water[a] (EC-MPN/100 ml)	Oysters (MPN/100 g)	Water[b]	Type	PFU/100 g	Water	Oysters (MPN/g)
1	91	49	45	0	—	0	—	0
2	95	7	20	0	P1	14, 11	—	0
3	189	17	45	0	—	0	—	2100
4	151	9	79	0	P2	26	—	0
5	167	33	130	0	—	0	—	43
6	231	< 18	45	0	—	0	—	0
7	163	330	78	0	—	0	—	15
8	397	< 2	20	0	—	0	—	0

[a] Collecting-station fecal coliforms (EC-MPN per 100 ml) were also determined (1) by Civil Engineering Department, Texas A & M University, during the same period, giving median values of 43 (at surface, 8 m, and 11 m) and 23 (at 4 m), and (2) by Texas Water Quality Board, giving values of 9 (15 June), 15 (3 Aug.), and 43 (20 Sept.)

[b] Examinations using a virus concentrator to sample the volumes indicated

Coliphage was isolated whenever virus was found in water. In most cases coliphage was detected in oysters when virus was recovered. The frequency of coliphage isolations indicated a population density greatly in excess of virus in oysters and seawater. A coliphage indicator would have signaled the presence of virus in all water samples, and the majority of oyster samples from which virus was recovered. It also would have indicated the presence of virus many times in both types of samples where virus was not detected. Its use with mud samples where virus isolations exceeded coliphage by a factor of five did not seem justified.

Coliphage isolations from estuary waters were made with the help of three host *E. coli* strains. The influence of host cell upon isolation success is shown in Table 9. Three coliphage 'types' were found during the study. The isolates were qualitatively different. Cross-reactivity with a host strain not used for coliphage isolation was not detected. Shifts in the distribution of coliphage 'types' suggested no single type could be counted on to occur exclusively. These data indicated both successful detection and enumeration of coliphage numbers would depend upon *E. coli* host bacteria used.

The possibility of an increase in coliphage numbers resulting from growth in host *E. coli* bacteria in estuary waters was studied using dialyzing membranes immersed in estuary waters. A doubling of

TABLE 8. Fecal coliform numbers (EC-MPN per 100 g oysters or 100 ml water), and coliphage and enteric virus isolations (number of isolations/samples tested) from oysters, overlaying water, and oyster-bed mud

Sample	Oysters[a]			Water[b]			Mud[c]	
	Fecal coliforms	Coliphage	Virus	Fecal coliforms	Coliphage	Virus	Coliphage	Virus
1	130	10/24	3/24	27	8/30	0/30	0/6	1/6
2	200	8/11	1/11	52	7/10	0/10	0/6	1/6
3	350	7/12	2/12	260	2/11	0/11	0/3	0/3
4	892	10/12	4/12	1300	2/10	0/10	0/3	0/3
5	91	24/58	2/58	33	20/139	3/139	1/6	1/6
6	145	16/28	3/28	15	10/32	0/32	0/8	2/8
Totals		75/145	15/145		49/232	3/232	1/32	5/32

[a] Isolates made from pools of 10 oysters

[b] Fecal coliform and coliphage tests were made on catch samples, virus tests on samples obtained from gauze pads immersed 3 days in water at collection station

[c] Isolates made from 1000 g samples

TABLE 9. Numbers of coliphage isolations made from estuary samples with three host strains of *Escherichia coli*

Trial	No. of samples	UNH[a]	ATCC 11303	ATCC 9637
1	138	9	52	16
2	100	3	11	39
3	115	0	4	17
Totals	353	12	67	72

[a] Isolated from raw sewage

coliphage members occurred within 6 h. It was concluded coliphage multiplication theoretically could occur in estuary waters containing host *E. coli* bacteria. Increases in coliphage numbers leading to increasing disparities between numbers of coliphage and enteric viruses present in estuary waters increasingly distorted original ratios existing at the sewage outfall.

Discussion

Microbiological evaluation of shellfish sanitary quality has been regarded in some quarters as an imperfect art executed in an imperfect way. The use of a fecal coliform index for detection of fecal pollution is central to any discussion on microbiological aspects of shellfish sanitation[21]. The studies described focused on extensive field examinations of the usefulness of fecal coliform indices as a measure of the presence of salmonellae and viruses in shellfish. The fecal coliform index was found to provide useful information when the fecal pollution measured was moderate to excessive. Correlation between coliform-indicated sanitary status and the actual status determined by direct examination was quite good. Inadequacies of the index were detected increasingly with declining levels of pollution. Correlation between indicated and actual microbiological quality became erratic in sea water at numbers less than 70 per 100 ml, or less than 230 per 100 g in shellfish. An acceptable coliform index of overlay water had little significance for the actual microbiologic quality of oysters. The most pronounced inadequacies occurred in the presence of enteric viruses.

Study results offered a possible explanation for inaccuracies of coliform indices at low incidences of fecal pollution, and suggested remedial measures. Continual monitoring of water showed significant fluctuations of coliform numbers to be possible at rapid intervals. The probable irregular periodicity of pollutant introduction into shellfish waters could only be detected by continuous monitoring. Initiation of a continuous monitoring policy would serve three important purposes. It would permit recognition of fluctuating coliform numbers, provide a truer picture of microbiologic quality, and might help to explain current inadequacies associated with sanitary interpretations drawn from coliform indices.

The discrepancies between enteric virus occurrence in shellfish and the coliform-indicated sanitary quality, represent another parameter of shellfish sanitary quality determinations. Continuous monitoring of fecal coliform numbers in sea water resulting in a truer expression of sanitary quality might allow a more satisfactory indication of virus occurrence. On the other hand, greater resistance of virus to environmental factors, leading to selective survival of virus only, would result in exposure of shellfish to virus in the absence of fecal coliform bacteria. In such circumstances the negative or minimal coliform findings gathered for water or shellfish would bear no relationship to the actual existing sanitary status. Recovery of poliovirus from Galveston Bay oysters shown by direct tests to possess acceptable coliform indices, and to be growing in water of uniformly acceptable microbiologic sanitary quality, may have been an example of such circumstances. The only sure way of resolving this situation would be direct enteric virus testing of shellfish.

Another possible avenue of approach, and an alternative to enteric virus testing, entails the use of a coliphage index. Objections to a coliphage index were raised by study results. An increase in coliphage numbers was found theoretically possible through contact and interaction of coliphage with host *E. coli* bacteria in estuary waters. Any increase in coliphage numbers by such means would distort the original numerical relationship between coliphage and enteric virus and hopelessly complicate interpretation of indicator significance. The occurrence of three coliphage 'types' based upon the host bacterial cell necessary for detection indicated use of multiple *E. coli* strains could prove necessary for the successful detection and enumeration of the coliphage population of an estuary. The disparity between coliphage and enteric virus isolations in estuary samples made it difficult to determine what significance coliphage numbers had as an indication of possible enteric virus occurrence.

Results of the study support the following recommendations. As part of the sanitary survey of shellfish growing waters, microbiological evaluations should be made by continuous monitoring of shellfish waters for fecal coliform numbers. For example, monitoring might be carried out for appropriate intervals on each of three occasions: before, during, and after tidal exchanges. The sanitary acceptability of shellfish should be determined by direct tests conducted for potential bacterial and virus pathogens. Consideration should be given to further examinations of the usefulness of a coliphage indicator such as f2 for assessing the possible presence of virus.

ACKNOWLEDGEMENTS

The participation of Dr. L. W. Slanetz and Dr. C. H. Bartley, Department of Microbiology, University of New Hampshire, and Dr. R. Wende and Mr. Claire Gothard, Houston, Texas Public Health Laboratories, in bacterial indicator studies is gratefully acknowledged.

Dr. J. M. Vaughn, Woods Hole Oceanographic Institution, participated in the conduct of coliphage studies.

The cooperation of Office of Sea Grant, Texas A&M University, in furnishing boat support is gratefully acknowledged.

The studies were made possible by grants from the U.S. Public Health Service, Food and Drug Administration, FD 00351 and FD 00170.

REFERENCES

1. Newsholm, A. *Publ. Hlth, Lond.*, Sept. 1898.
2. Roos, B. Hepatitis epidemic conveyed by oysters. *Svenska Läkartidn.*, 1956, **53**, 989–1003.
3. Mason, J. O., and McLean, W. R. Infectious hepatitis traced to the consumption of raw oysters. *Am. J. Hyg.*, 1962, **75**, 90–111.
4. Dougherty, W. J., and Altman, R. Viral hepatitis in New Jersey. *Am. J. Med.*, 1962, **32**, 704–716.
5. Sommer, H., and Meyer, K. F. Paralytic shellfish poisoning. *Archs Path.*, 1937, **24**, 560.
6. Sommer, H., Whedon, W. F., Kofoid, C. A., and Stohler, R. Relation of paralytic shellfish poisoning to certain plankton organisms of the genus *Gonyaulax*. *Archs Path.*, 1937, **24**, 537.
7. Report, Committee on Sanitary Control of the Shellfish Industry in the U.S. *Publ. Hlth Rep., Wash.*, 1925, Suppl. No. 53, 6 November.
8. Detection of pathogenic microorganisms in water and wastewater. In 'Standard methods for the examination of water and wastewater'. American Public Health Association, American Water Works Association, and Water Pollution Control Federation, New York, 1971, 13th Edn, pp. 697–704.
9. Recommended procedures for the examination of seawater and shellfish. American Public Health Association, New York, 4th Edn, 1970.
10. Guélin, A. Étude quantitative de bactériophage de la mer. *Annls Inst. Pasteur, Paris*, 1948, **74**, 104–112.
11. Carstens, E. M. J. Bacteriophages and their possible use in sewage purification. *J. Proc. Inst. Sew. Purif.*, 1963, 467–468.
12. Shah, P. C., and McCamish, J. Relative chlorine resistance of poliovirus 1 and coliphages f2 and T_2 in water. *Appl. Microbiol.*, 1972, **24**, 658–659.
13. Metcalf, T. G., Slanetz, L. W., and Bartley, C. H. Enteric pathogens in shellfish. In 'Microbial safety of fishery products' (ed. C. O. Chichester and H. D. Graham), 1973, pp. 215–234.
14. Wallis, C., Homma, A., and Melnick, J. L. Apparatus for concentrating viruses from large volumes. *J. Am. Wat. Wks Ass.*, 1972, **64**, 189–196.
15. Wallis, C., Parks, W., Sakurada, N., and Melnick, J. L. A rapid plaque method using vertical tube cultures for titration of viruses and neutralizing antibodies. *Bull. Wld Hlth Org.*, 1965, **33**, 795–801.
16. Metcalf, T. G., and Stiles, W. C. Enteroviruses within an estuarine environment. *Am. J. Epidemiol.*, 1968, **88**, 379–391.
17. Metcalf, T. G., and Stiles, W. C. The accumulation of enteric viruses by the oyster, *Crassostrea virginica*. *J. infect. Dis.*, 1965, **115**, 68–76.
18. Wallis, C., Henderson, M., and Melnick, J. L. Enterovirus concentration on cellulose membranes. *Appl. Microbiol.*, 1972, **23**, 476–480.
19. Slanetz, L. W., Bartley, C. H., and Stanley, K. W. Coliforms, fecal streptococci and salmonella in seawater and shellfish. *Hlth Lab. Sci.*, 1968, **5**, 66–78.
20. Vaughn, J. M. The use of coliphage as an index of human enterovirus pollution in an estuarine environment. Doctoral dissertation, University of New Hampshire, 1972.
21. Geldreich, E. E. Sanitary significance of fecal coliforms in the environment. U.S. Dept. of Interior, Fedl Wat. Pollut. Control Adm., Wat. Pollut. Control Res. Ser. Publ. No. WP-20-3, 1966.

Presenting his paper, Prof. Metcalf said that field experiments had been made in order to determine the correlation between faecal coliform indices and the occurrence of bacterial and viral pathogens in shellfish growing in sewage-polluted waters. Agreement between the coliform indices and pathogen presence was best with moderate to excessive degrees of pollution, and the accuracy of the coliform indices became marginal when pollution reached minimal levels. The greatest discrepancy between coliform-indicated and actual sanitary values of shellfish determined directly occurred with enteric viruses. Objections to the use of a coliphage indicator of virus were based upon disparities in occurrence of the two forms, the occurrence of coliphage of differing host-cell specificities, and implications arising from possible increases in coliphage numbers in estuary waters. Significant fluctuations of faecal coliform numbers at frequent intervals were found in continuously monitored sewage and river samples. Recommendations for improving the accuracy and microbiological evaluation of shellfish sanitary quality were presented in the paper.

DISCUSSION

Mr. P. C. Wood (U.K.) said that in Europe the suitability of shellfish for consumption was assessed by examining the shellfish themselves, and asked what was the current thinking in the U.S.A. with regard to the use of shellfish quality as a means of regulating the shellfish industry. He also asked if Prof. Metcalf knew of any instance where virus infection had occurred following consumption of shellfish which either satisfied coliform standards or were taken from approved areas; to his own knowledge virus hepatitis had occurred only following consumption of grossly polluted shellfish.

Prof. H. I. Shuval (Israel) observed that it was not really surprising that shellfish often showed viral contamination in cases where grab samples of overlying water showed low levels of microbial contamination; the shellfish served as integrators of water quality over a period of weeks, whereas most water tests were for much shorter periods. Monitoring methods were required for integration of exposure levels over such long periods. [See also contribution by Dr. Grunnet on p. 60—*Ed.*] The gauze pad could be effective for this purpose despite its low efficiency—his own studies had shown that although the pads picked up only about 1 per cent of the water passing through them they could lead to detection of viruses in sea water where grab-sample tests were negative. As to an 'indicator virus', he suggested that in countries where routine polio inoculation of all infants was practised, the sewage would invariably contain relatively high levels of poliovirus which could thus fulfil many of the accepted functions of an indicator organism. His colleague Dr. Katzenelson had developed a rapid test for poliovirus in water using a fluorescent technique. This type of approach could place virus monitoring on a practical basis, with results becoming available as rapidly as those of conventional coliform tests which could no longer be considered reliable where viral contamination was concerned.

Prof. Y. Kott (Israel) referred to a study made about 10 years previously in Texas in which oysters were placed in baskets at various distances from a sewage outfall. It was found that the number of coliphages isolated from the oysters decreased with increasing distance from the outfall, none being found at distances exceeding 200 m.

Prof. R. Mitchell (U.S.A.) said there was growing evidence that chemotactic responses were involved in bacterial activities in the sea. He asked Prof. Metcalf to comment on chemotaxis of virus to shellfish, and to say if there was any specificity or whether attraction was affected by shellfish physiology. Bacterial chemotaxis had been found in Prof. Mitchell's laboratory, and he asked if virus accumulated less in shellfish in chemically polluted areas.

Prof. R. R. Colwell (U.S.A.) pointed out that specificity of shellfish for given species of bacteria could be clearly shown—as had been done at her laboratory with radioactive mercury labelling of micro-organisms. One could also show loading of heavy metals into higher trophic levels via bacteria using this methodology[22]. She asked if anyone had shown an effect of oil or any other hydrocarbon on retention or uptake of viruses by shellfish.

REPLY TO DISCUSSION

In reply to Mr. Wood's first question, Prof. Metcalf said that U.S. shellfish sanitation requirements stipulated exercise of strict control over the sanitary quality of shellfish entering the market place. The necessity of gaining public approval and consumer confidence in the safety of the product was regarded generally as a legitimate basis for imposition of strict quality controls. No satisfactory answer could be given to Mr. Wood's second question; outbreaks of sufficient size and severity to gain attention had been associated with a degree of pollution permitting laboratory identification of faecal bacteria. This raised the questions of how many isolated non-epidemic cases were subject to laboratory or physician scrutiny, and what epidemiological or other information had been gathered in such cases.

In answer to Prof. Mitchell he said that minute quantities of arabinose-like substances natural to sea water had been reported to increase shellfish feeding activities, but no chemotactic responses influencing attraction of virus to shellfish had been recognized. Virus accumulation proceeded as a direct function of its occurrence in sea water passed through shellfish gill structures during filter-feeding activities. In at least one estuary study area, virus accumulation and subsequent assumption of virus carriage state had been greater in the presence of industrial pollution than in its absence.

With regard to Prof. Colwell's question, Prof. Metcalf said that apart from the interference exerted upon the accumulation process resulting from a direct inhibition of shellfish feeding activity, no effect of hydrocarbons on uptake or retention had been shown.

ADDITIONAL REFERENCE

22. Colwell, R. R., and Nelson, J. D. EPA final report 1974. Environmental Protection Agency, Washington, D.C. (*in press*).

Paper No. **9**

Epidemiological Aspects of Microbial Standards for Bathing Beaches

James W. Mosley

Associate Professor of Medicine, University of Southern California School of Medicine
Los Angeles, California 90007

The discharge of sewage into the sea has several potentially adverse consequences, one of which is the causation of human infections. Sewage in which the agents of human disease can be demonstrated with regularity is increasingly contaminating the sea because of the continued growth of human populations and present methods of disposal. We must ask, therefore, if there is evidence that this contamination has already produced, or is producing, a sufficient number of thalassogenic (Greek *thalassa* the sea + *genesis* source) infections to justify the expense of altering present practices in handling of human wastes. Whatever our answer to that question, we must also try to determine what may happen in the future if the level of contamination continues to increase.

The point at which we have no choice but to take remedial action is reached whenever marine pollution sustainedly results in as many or more infections than the number in the previous generation from which that contamination was derived. Ecologically, the organism has then taken advantage of contamination of the sea as a circumstance permitting perpetuation or augmentation of its numbers. I may say now that I know of no agent of disease for which this criterion is presently fulfilled, but I believe that it was fulfilled by mollusk-associated typhoid fever in some areas of the British Isles in the 1890's. For Brighton, Newsholme[1] in 1896 estimated that 30 per cent of cases resulted from ingestion of oysters and other mollusks, which was probably a conservative figure in view of data subsequently provided by Nash[2]. For Belfast, Mair[3] presented evidence that opening of a sewage outfall into the estuary in 1897 resulted in typhoid fever at epidemic levels until harvesting of mussels and cockles from contaminated water was restricted in 1902. Despite a probable increase in Belfast's carrier reservoir during that five-year interval, mortality rates during the subsequent five years fell by 70 per cent.

At lesser levels of morbidity due to diseases caused directly or indirectly by marine pollution, we must decide how many such infections can be tolerated. As an example, we may cite the endemic cases of shellfish-associated type A hepatitis in the north-eastern United States. There was adequate evidence for this occurrence in the mid-1960's[4,5], and there has been no change in harvesting or sanitation practices to give us reason to believe that cases have disappeared. Should we continue to ignore this problem as has been done for the last decade?

For some potentially thalassogenic infections there is no definite information at present that any cases are being produced because of contamination of the sea. Enteroviral disease, for example, would be in this category. We can be reassured about the apparent lack of any problem only if the possibility that cases are being missed has been reasonably excluded, and that there is a mechanism for surveillance of sufficient sensitivity to detect a significant increase associated with one or more mechanisms of exposure to contaminated water.

As background to the discussion by Professor Shuval and Dr. Moore of microbial standards, I shall review the kind of evidence we now have, as well as the new evidence we will need, concerning the occurrence of disease as a result of discharge of sewage into the sea. My comments will be focused upon the following questions: (1) By what mechanisms do infections occur? (2) Which pathogenic organisms and diseases can be used as indices of disease related to pollution? (3) Which cases should we attribute to marine contamination? (4) Which populations are suitable for deriving conclusions about health hazards? (5) How can we monitor the extent of the problem on a continuing basis?

Mechanisms of Infection

The circumstances under which one or more persons may be exposed at any given time to pathogens in the sea can be conveniently characterized as: (1) non-recreational immersion, (2) recreational immersion, and (3) consumption of mollusks (and perhaps other shellfish) harvested from contaminated waters.

Non-recreational immersion

There is a variety of circumstances under which an individual is accidentally immersed in the sea. These include falls from piers, bridges, causeways, ships, and boats. In this category of non-recreational immersion we may also place those incidents in which persons deliberately throw themselves into water in a suicide gesture or unsuccessful attempt. Under either circumstance, submersion of the head is likely, with contamination of the mucous membranes of the conjunctivae and nasopharynx. In addition, there is undoubtedly ingestion of at least some water in many cases. In a number of instances, this type of non-recreational immersion occurs in waters very heavily contaminated with sewage.

A systematic search of the medical literature would probably disclose at least occasional cases in which infection was attributed to non-recreational immersion in heavily contaminated water. The frequency, however, in relation to the total number of such accidents must be very low. I am aware of no attempt to carry out a systematic investigation of persons with this type of exposure. Such lack of attention is an unfortunate omission, because the degree and intensity of exposure seem unparalleled by some of the other situations about which there has been concern (e.g. swimming along contaminated beaches). If accidental immersion in heavily contaminated water seldom results in infection, then there is probably little reason to worry about recreational use of water contaminated to a lesser extent.

Recreational immersion

Recreational immersion has been of major concern to public health workers because they are charged with the responsibility of deciding whether beaches and other areas for aquatic sports are safe or unsafe. Thus far, they have had to make decisions on a largely empirical basis. No adequate data to relate the epidemiologic consequences of anything other than gross pollution are yet available.

From a public health viewpoint, it is worthwhile to distinguish between swimming and incidental immersion. The latter is a convenient designation for a dousing that is not an inherent part of the sport, but not an unexpected occurrence. This category is particularly pertinent for water-skiing, and, to a lesser extent, fishing and sailing. Its importance is that water-skiing, fishing, and sailing all take the individual from circumscribed areas that may be definable as safe for swimming into broad expanses of water that are either marginally or heavily contaminated. There has been particular concern about water-skiing, but exclusion of participants from areas recognized to be polluted is generally difficult to enforce. To my knowledge, there has been no investigation of infection among water-skiers, and the health hazard the sport entails is a subject of conjecture. Because immersion is expected at least occasionally, most participants are unlikely to remember specific incidents or to be able to identify their location. This fact presents formidable difficulties to any meaningful study.

The term 'swimming' implies that the individual propels himself through the water. His face is thereby in sufficient proximity to the water for the mucous membranes to become contaminated, and for water to be actually ingested. The hygienic aspects of swimming, both in fresh and salt water, have long been a matter of concern, and there have been a number of investigations[6–8]. Most attention has been focused upon swimming pools, for which conditions can be reasonably well defined and to which sanitation measures can be applied. Lakes and rivers are less satisfactorily studied, and marine swimming poses the greatest methodologic problems of all.

The likelihood of infection in recreational waters depends upon whether a pathogen is in sewage at any given time, its numbers, its ability to survive, its infective dose, and the degree of dispersal. Even adding outbreaks and sporadic cases resulting from swimming in fresh water to those associated with the sea, there is little evidence that recreational immersion is frequently a means by which infection is acquired. Only leptospirosis has been regularly associated with infections of this type[9–12]. This exception is notable in relation to the stipulations mentioned above. Leptospira are able to multiply outside of an animal host if environmental circumstances are favorable, thereby greatly increasing their numbers. They also have the capacity to penetrate intact mucous membranes, and probably to infect at low multiplicities of organisms. Furthermore, most outbreaks have been associated with stagnant ponds in which there was little if any dispersal of contaminating drainage.

Enteric agents seem seldom to be transmitted by swimming. The early literature[13–16] records four outbreaks of typhoid fever acquired in grossly polluted water. There are also occasional instances in which sporadic cases of typhoid or paratyphoid disease were associated with swimming[14,17]. The enteroviruses, which would seem likely to be transmitted by contaminated water, have been little implicated as water-borne agents despite much attention.

Strenuous efforts were made to associate poliomyelitis with swimming when that disease was a major problem in the United States. Unfortunately, the few published studies alleging a relationship have been inconclusive, as exemplified by a recently reported investigation[18] of an epidemic of Coxsackie B5 infection at a boys' summer camp. The virus was detected in one of two water samples from the swimming area, but the study was carried out after the epidemic was already in progress. No bacteriologic studies were made to exclude seepage of sewage from nearby septic tanks, nor any effort to demonstrate entrance of an indicator dye or organism from toilets in the camp into the water. Such incomplete investigations are of little help.

There is no reason to doubt the potential for infectivity of enteric organisms in sea water. It is presumably dilution of the agents, the small quantity of water ingested during swimming, and a requirement for a relatively large inoculum that have prevented any significant problem thus far. It is still possible, however, that more infections are acquired by swimmers than we presently appreciate. The epidemiologic association may be missed in overt cases, and an unusually large proportion of infections may be subclinical because of a small infecting dose.

Shellfish-associated infection

If dilution of pathogens in human waste saves swimmers from infection at our bathing beaches, their reconcentration by edible mollusks does visit upon man retribution for dumping sewage into the sea. About enteric infections resulting from consumption of uncooked or undercooked oysters, clams, mussels, and cockles, there can be no question. About the present extent of the problem and its future management, however, there are many questions.

Although one epidemic of oyster-associated typhoid fever was reported from France[1] as early as 1816, it was not until the 1890's and early 1900's that the importance of this mechanism was recognized. As mentioned above, the large proportion of cases in some parts of Britain at that time were found to be due to consumption of contaminated oysters, mussels, and cockles. Epidemics of shellfish-associated typhoid were also described in the United States[19–22], although they do not appear to have been so numerous or extensive as in Britain because shellfish were less widely consumed. The recognition of shellfish as a vehicle for enteric agents resulted in some modifications of sewage disposal, with lessening of marine pollution, and there was an inception of shellfish sanitation as a public health activity. The impact of relatively simple measures upon the incidence of typhoid in Brighton between 1895 and 1907 is illustrated in Fig. 1.

One approach to shellfish-associated infection was to move the mollusks from contaminated to clean waters where their natural pumping action cleanses them of pathogens[23,24]. This approach was gradually elaborated into the system that Americans have recently called 'depuration'[25]. It is upon this procedure, adapted to holding tanks rather than natural waters, that the British have placed great reliance. In the United States, given impetus by its major epidemic in 1924–25[26,27], the shellfish sanitation program has been based largely upon a system of 'approved' waters, with policing to prevent

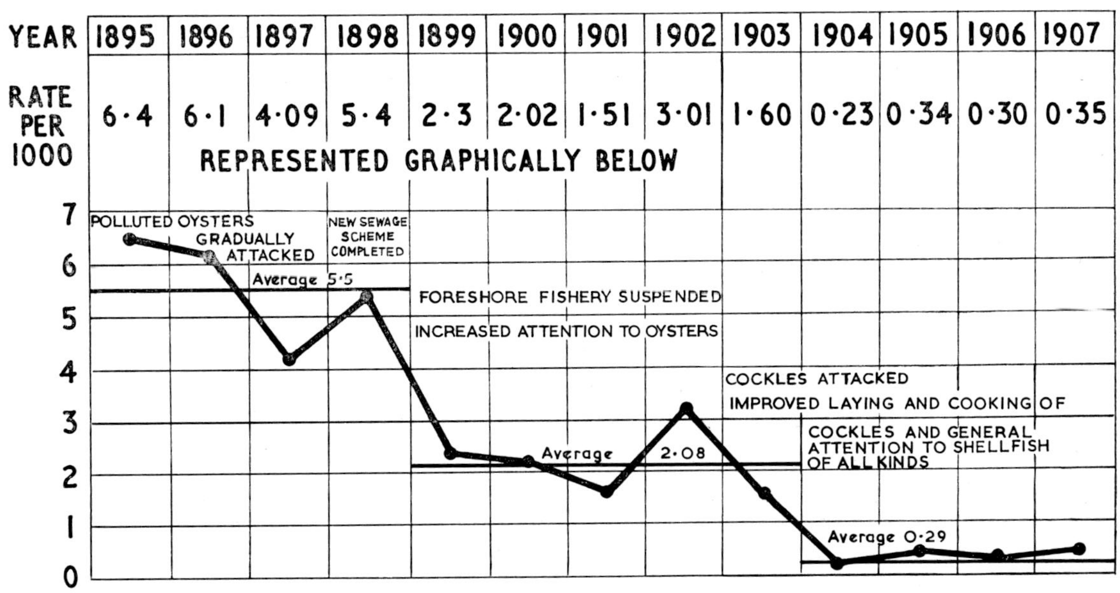

FIG. 1. Incidence of typhoid fever in Brighton, England, between 1895 and 1907 (prepared by Nash[2] in 1909) showing impact of various shellfish sanitation measures

illegal harvesting. Both programs met with apparent success, and it is probable that the measures did contribute to the subsequently decreased incidence of typhoidal disease in both countries. Their use, however, coincided with a variety of other improvements in general sanitation, including the surveillance of food handlers and supervision of known typhoid carriers. Whether the British program continues to be a success is not known, although one Medical Officer of Health has suggested [28] on the basis of his personal experience with gastroenteritis that the Mosaic prohibition concerning shellfish be more generally adopted. There is no question that the American program was insufficient to prevent the problem of shellfish-associated hepatitis.

The Swedish oyster-associated epidemic of 1955–56 [29] was regarded in the United States as an epidemiologic curiosity until 1962. In that year an oyster-associated outbreak in the southern United States [30] and a clam-associated epidemic centered in New Jersey [31] made it evident that the absence of shellfish-associated typhoid fever was not synonymous with absence of contamination by sewage. Two additional clam-associated epidemics involving hundreds of cases followed in 1964 [32,33], and a new outbreak of major proportion has been recently recognized [34]. In addition, there is an 'endemic' background of sporadic cases that may total as many as several hundred each year [4].

Outside of the United States, there has been little interest in mollusk-associated type A hepatitis. Since the Swedish report in 1965, only one investigation has been reported in the literature. Stille and co-workers [35] indicated that consumption of contaminated mollusks accounted for an estimated 19 per cent of type A hepatitis in Frankfurt. The German cases were chiefly attributable to eating oysters and mussels on the Mediterranean littoral, especially southern France and Italy.

The occurrence of shellfish-associated hepatitis is not confined to mollusks eaten raw, because steaming as usually practised fails to raise the internal temperature sufficiently to inactivate any viral agent [36].

A recent question is that of transmission of type B hepatitis by accumulation of hepatitis B virus in oysters and clams [37]. The only demonstration of molluskan contamination, however, was in samples taken from a bed adjacent to a hospital's sewage outfall. Many other samples along the coast were negative when tested for the virus-induced antigen that serves as a marker of hepatitis B virus. It is not probable that this is a significant route of transmission of type B hepatitis. The occurrence of the agent in feces is doubtful, and oral infectivity of serum is low. Disposal by pouring down a sink of one or several units of contaminated blood at a hospital, however, could conceivably result in an occasional infection.

One category of infection not yet associated with mollusks is that produced by enteroviruses. Metcalf and Stiles [38] demonstrated these agents in oysters harvested from contaminated waters closed to harvesting, but Denis [39] recently reported recovery of Coxsackie A viruses in market samples taken in France. There is no question but that further investigation is warranted.

This designation 'shellfish' includes not only mollusks but also crustaceans. Mollusks are frequently eaten raw and the entire animal is consumed, including the gastrointestinal tract. This fact, in addition to their ability to concentrate particulate matter, is very relevant to their frequent implication as vehicles of enteric infection. Crustaceans, on the other hand, are commonly eaten after removal of the gastrointestinal tract and other viscera, and cooked to a degree that would seem adequate to kill any organisms that might be present. Crustaceans also do not have the same efficiency of concentrating small organic particles from sea water as do mollusks. Recently, however, DiGirolamo and co-workers [40] have demonstrated that West Coast (United States) shore crabs accumulated poliovirus both in artificially contaminated sea water and when allowed to feed on virus-contaminated mussels. In another experiment [41], they demonstrated phage uptake into the edible flesh of the crab and its persistence after cooking in boiling water for as long as 20 minutes.

Pathogens and Diseases for Potential Study

For recognition of the hazard of exposure to contaminated sea water, the agent (and its disease) most suitable for study would be one continuously present in sewage; able to survive well in the marine environment; infective when ingested at low doses; unlikely to be transmitted other than by water-borne contamination; and causative in the majority of infected persons of a distinctive clinical syndrome of sufficient severity to make medical attention likely. Although no such organism or disease is now known, these characteristics provide criteria in our search for epidemiologic indices. The organisms in which we are most likely to be interested are those for which the fecal-oral route of infection is usual. These include the typhoid bacillus, salmonella other than the typhoid bacillus, hepatitis A virus, and enteroviruses.

In the 1890's we would, of course, have studied typhoid fever because of the prevalence of cases and carriers, the ability of the typhoid bacillus to survive in salt water, and the likelihood that most cases would come to medical attention. It is still possible to study typhoid fever in areas where large numbers of cases and carriers exist. The other salmonella, aside from the paratyphoid organisms that cause enteric fever, do not produce a distinctive clinical syndrome. In addition, it is recognized that these agents are easily acquired at seaside resorts through contaminated food or drinking water rather than bathing in the ocean.

Hepatitis A virus produces a distinctive clinical syndrome with an inapparent-to-overt case ratio that is low enough to permit clinical detection of an adequate proportion of infections for study. The agent can be acquired by a variety of mechanisms, but comparison of the frequency of any given background with that in a control group does test the significance of that exposure. Hepatitis A virus, however, is by no means continuously present in sewage, as evidenced by many water-borne outbreaks of gastroenteritis not followed by simultaneously transmitted hepatitis[42]. In addition, the incidence of type A disease is declining in the United States[43] as well as in some European countries[44].

Enteroviruses of one or more serotypes are almost always present in urban sewage, survive well in a marine environment, and may not require large doses to be infective[45]. Their major disadvantage is that overt cases present a variety of clinical syndromes, and the inapparent-to-overt case ratio is usually high. The availability of simple laboratory procedures for their detection and identification, however, makes it possible to monitor exposed persons for evidence of subclinical infection in prospective or even cross-sectional studies. Furthermore, their prevalence in the United States, as indicated by reported morbidity from aseptic meningitis, appears to be increasing (Fig. 2). They are, therefore, the agents most suitable for detailed investigation at the present time.

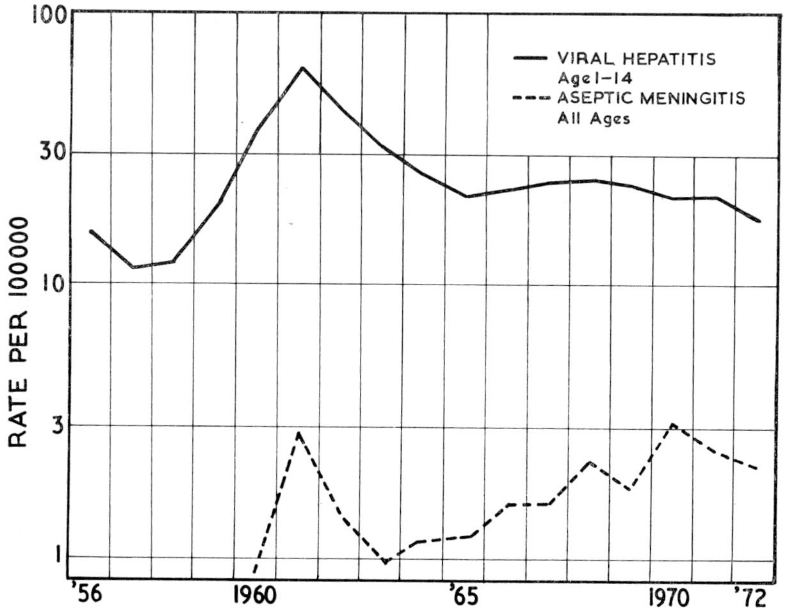

FIG. 2. Incidence of reported viral hepatitis (an index of morbidity due to type A hepatitis) and aseptic meningitis (index of morbidity caused by enteroviruses) in United States between 1956 and 1972

The Cases to be Counted

The epidemiologist is likely to establish the following criteria for cases to be counted as thalassogenic: (1) The person had definite exposure prior to onset of symptoms at an interval appropriate to the incubation period of the agent in question. The exposure may have been by immersion of the head in waters where contamination is known, or by ingestion of a marine product with the potential for serving as a vehicle. (2) The individual had no exposure to any other likely source or vehicle of infection, at least under the particular circumstances prevailing for that case. (3) The person became clinically ill, because subclinical infection represents no burden of morbidity for either that individual or the community. An additional circumstance favoring epidemiologic implication of marine exposure would be the occurrence of other cases with a similar background, but stipulation of this point as a requirement would preclude any definition of sporadic cases.

Although the epidemiologist may be satisfied with these criteria, many sanitary engineers, aware of the extent to which coastal waters are polluted, believe that their use results in a vast understatement of the problem. They hold that the morbidity attributed to contaminated water as a vehicle should include not only cases resulting from primary exposure, but also secondary and later generations of cases, even when contaminated water has no subsequent role in transmission. If this view is adopted, then subclinical infection caused by marine contamination assumes an importance proportional to the inapparent-to-overt case ratio.

Secondary cases may be reasonably attributed to the thalassogenic mechanism whenever there is no great likelihood of acquisition from some other person who becomes infected by a different mechanism. Thus, cases of typhoid fever in the United States secondary to a household member who ate raw oysters could at present be counted as indirectly due to the mechanism of infection for the primary case on the basis that exposure to the typhoid bacillus would otherwise be improbable. This attribution would not be appropriate, however, in a country still having a high endemic burden of typhoid.

For cases subsequent to the primary and secondary, there is similar justification for the engineers' view *if* the agent is one not likely to be introduced into the community otherwise, *and* marine contamination represented a frequent rather than a happenstance mechanism for transmission to the primary case(s). For example, three or four generations of cases of typhoid fever in an Icelandic community free of typhoid and typhoid carriers, introduced by an inhabitant who acquired his infection by consumption of raw mussels in Italy, could represent a situation in which the later cases could also be considered thalassogenic. If the disease is one just as likely introduced by a person infected in some other way (e.g. contact transmission), or is due to an agent already present in the community (e.g. hepatitis A virus), then the argument that subsequent cases should be attributed to the one case acquired from contaminated water collapses.

By these considerations, we can neither categorically restrict ourselves to primary cases of overt disease nor automatically include all subsequent cases anytime infection is acquired through water as a vehicle. *The ultimate criterion is the number of cases we would prevent on a long-term basis if contamination of the sea were abolished.*

Because the total burden for the community needs to be defined, it does seem worthwhile to include inapparent cases in any detailed investigations of thalassogenic infections. These cases also have public health importance because they amplify the agent, increasing thereby the possibility that sewage will cause additional infections when it contaminates the sea.

The Populations to be Studied

The ideal population to investigate is one with a high rate of susceptibility to the infection; a random distribution of persons having a marked affinity for, or dislike of the activities by which they may be exposed; and having little or no background activity of the disease due to other mechanisms of transmission. For epidemiologic purposes, it would be very helpful if the chance for participating in the activity were confined to a few occasions more than one incubation period apart, and provided such pleasure that the date would be subsequently remembered for weeks to months.

For populations living in a seaside resort where there is easy and frequent access to the sea, it may be very difficult to distinguish cases that are acquired because of contamination of the sea from cases acquired by other mechanisms. The latter must obviously exist or the organism would not be in sewage contaminating the sea in that particular area. Also, unless the advent of pollution were recent or the organism had been absent from the community for a number of years, most individuals are likely to have had many opportunities to be exposed and thereby be immune. Finally, in at least some instances, biases will be introduced. Thus, the more polluted beaches may be avoided by an upper socio-economic group able to travel longer distances, or to purchase waterfront property to which access can be restricted. Because of the inverse correlation of enteric infections with socio-economic status, a spuriously low incidence may then be found. All of these difficulties could have been in part responsible for the failure to implicate contaminated beaches in the transmission of poliomyelitis[7,46].

Preferable would be a population living within one to two hours' drive of a beach, thereby not being a part of the resort city but also not likely to stay overnight. The latter minimizes (but does not obviate entirely) the exposure to drinking water and food in the resort.

Finally, the population at true risk on beaches is that which swims rather than merely 'bathes'. This distinction may or may not have been made in past studies.

Methods of Study to be Used

It is useful to distinguish between investigative and surveillance methods[47]. The differences inherent in these two epidemiologic techniques are summarized in Table 1.

TABLE 1. Comparison of investigation and surveillance as epidemiologic activities

	Investigation	Surveillance
Stimulus	To answer the question whether occurrence of disease can be related to a specific etiology or mechanism of exposure	To monitor the frequency of diseases in relation to one or several mechanisms of exposure
Data collection	Organized to be applicable to a specific situation	Organized to cover broad population groups
Quality of data collected	An effort is made to collect accurate data not otherwise available	Crude data are acceptable if they indicate when a significant increase occurs in frequency of the disease
Character of data utilized	Non-routinely available	Routinely available
Duration	Limited to the time necessary to answer the question posed	A continuing activity as long as the disease remains a threat

The investigative approach is appropriate whenever data must be obtained that cannot be collected on a routine basis. Thus, to find out if infections occur as a result of swimming along beaches where salmonella or enteroviruses can be recovered from the water, it will be necessary to identify a susceptible population, confirm its exposure, and carry out follow-up—including collection of specimens. For validation of that exposure as the one responsible for any infections observed, contemporaneously followed controls are necessary. Clearly, such an elaborate program can be applied only to a few selected situations, and cannot be envisaged as a continuing activity even if a positive answer is obtained.

Surveillance is a technique that relies upon routinely collected data known to be adequate to signal a significant change in the level of a disease. That level and the disease must be of sufficient public health importance to warrant the effort. Ordinarily, surveillance is applied to broad population groups that are at potential risk, and it continues as long as the disease has the potential for being important. Surveillance must rely, therefore, upon non-investigative personnel. Its applicability to thalassogenic infections is illustrated by its use in maintaining shellfish-associated hepatitis. Cases reported through routine morbidity notifications are surveyed, utilizing a brief questionnaire covering consumption of raw or undercooked shellfish. Response is also sought concerning other significant mechanisms of potential exposure. The percentage of positive answers concerning mollusk consumption will vary with the cultural habits of the population, but should be reasonably constant in the absence of a shellfish-associated outbreak. Any endemic level of transmission can be established by internal comparisons—if shellfish are not responsible for cases, consumers should have the same frequency of exposure as non-consumers[4].

Microbial Standards

Use of microbial standards is basically a surveillance procedure. They can be helpful for monitoring beaches and shellfish sanitation *if* they predict occurrence or non-occurrence of human infection. To establish their usefulness, it is necessary to show that they are predictive. The mere occurrence of pathogens in water used for swimming does not seem to be necessarily significant[48], but there is undoubtedly a level at which they become so. On the other hand, recovery of any human enterovirus from shellfish may indicate exposure to sufficient pollution to produce human infections. Under any circumstance, the frame of reference for the value of microbial standards is always the epidemiologic one.

REFERENCES

1. Cited by MAXCY, K. F. Food sanitation: meat, fish, eggs and plants. In 'Rosenau preventative medicine and hygiene' (ed. K. F. Maxcy). Appleton-Century-Crofts, New York, 7th Edn, 1951, p. 909.
2. NASH, J. T. C. Discussion of paper by L. W. D. Mair[3].
3. MAIR, L. W. D. The aetiology of enteric fever in Belfast in relation to water supply, sanitary circumstances, and shellfish. *Proc. R. Soc. Med.*, 1909, **2**, 187.
4. HEPATITIS SURVEILLANCE UNIT. Non-epidemic shellfish-associated hepatitis. Communicable Disease Center, Atlanta, Hepatitis Surveillance Rep. No. 23, 1965, p. 18.

5. Koff, R. S., Grady, G. F., Chalmers, T. C., Mosley, J. W., Swartz, B. L., and Boston Inter-Hospital Liver Group. Viral hepatitis in a group of Boston hospitals. III—Importance of exposure to shellfish in a non-epidemic period. *New Engl. J. Med.*, 1967, **276**, 703.
6. Stevenson, A. H. Studies of bathing water quality and health. *Am. J. publ. Hlth*, 1953, **43**, 529.
7. Public Health Laboratory Service. Sewage contamination of coastal bathing waters in England and Wales. A bacteriological and epidemiological study. *J. Hyg., Camb.*, 1959, **57**, 435.
8. McLean, D. M. Transmission of viral infections by recreational water. In 'Transmission of viruses by the water route' (ed. G. Berg). Wiley-Interscience, New York, 1967, p. 25.
9. Schaeffer, M. Leptospiral meningitis. An investigation of a water-borne epidemic due to *L. pomona*. *J. clin. Invest.*, 1951, **30**, 670.
10. Cockburn, T. A., Vavra, J. D., Spencer, S. S., Dann, J. R., Peterson, L. J., and Reinhard, K. R. Human leptospirosis associated with a swimming pool, diagnosed after eleven years. *Am. J. Hyg.*, 1954, **60**, 1.
11. Coggins, W. J. Leptospirosis due to *Leptospira pomona*—an outbreak of nine cases. *J. Am. med. Ass.*, 1962, **181**, 137.
12. Nelson, K. E., Ager, E. A., Galton, M. M., Gillespie, R. W. H., and Sulzer, C. R. An outbreak of leptospirosis in Washington State. *Am. J. Epidemiol.*, 1973, **98**, 336.
13. Reece, R. J. Discussion of paper by L. W. D. Mair[3].
14. American Public Health Association. Report of Committee on Bathing Places. *Am. J. publ. Hlth*, 1922, **12**, 121.
15. Winslow, C. E. A., and Moxon, D. Bacterial pollution of bathing beach waters in New Haven Harbor. *Am. J. Hyg.*, 1928, **8**, 299.
16. Typhoid fever from bathing in polluted waters. *Wkly Bull. N.Y. Cy Dep. Hlth*, 1932, **21**, 257.
17. Moore, B. A survey of beach pollution at a seaside resort. *J. Hyg., Camb.*, 1954, **52**, 71.
18. Hawley, H. B., Morin, D. P., Geraghty, M. E., Tomkow, J., and Phillips, C. A. Coxsackievirus B epidemic at a boys' summer camp: isolation. *J. Am. med. Ass.*, 1973, **226**, 33.
19. Conn, H. W. The outbreak of typhoid fever at Wesleyan University. Connecticut State Board of Health Report, 1894, p. 243.
20. Soper, G. A. Typhoid fever at Lawrence, N.Y., due to oysters. *Med. News*, 1905, **86**, 241.
21. Stiles, G. W. Sewage-polluted oysters as a cause of typhoid and GI disturbances. *Chem. Bull.*, 1912, **156**, 16.
22. Brooks, P. B. Outbreak of typhoid attributed to infected oysters. *J. Am. med. Ass.*, 1916, **66**, 1445.
23. Phelps, E. B. Some experiments upon the removal of oysters from polluted to unpolluted waters. *J. Am. publ. Hlth Ass.*, 1911, **1**, 305.
24. Dodgson, R. W. Report on mussel purification. Ministry of Agriculture and Fisheries, Fish. Invest., Ser. II, Vol 10, No. 1, 1928.
25. Liu, O. C., Seraichekas, H. R., and Murphy, B. L. Viral depuration of the Northern Quahang. *Appl. Microbiol.*, 1967, **15**, 307.
26. Lundsen, L. L., Hasseltine, H. E., Leaky, J. P., and Velder, M. V. A typhoid fever epidemic caused by oyster borne infection (1924–25). *Publ. Hlth Rep., Wash.*, Suppl. No. 50.
27. Bundesen, H. N. Typhoid epidemic in Chicago apparently due to oysters. *J. Am. med. Ass.*, 1925, **84**, 641.
28. Vickery, K. Coxsackie group A in oysters and mussels. *Lancet*, 1973, i, 1385.
29. Roos, B. Hepatitis epidemic conveyed by oysters. *Svenska Läkartidn.*, 1956, **53**, 989.
30. Mason, J. O., and McLean, W. R. Infectious hepatitis traced to the consumption of raw oysters. *Am. J. Hyg.*, 1962, **75**, 90.
31. Dougherty, W. J., and Altman, R. Viral hepatitis in New Jersey. *Am. J. Med.*, 1962, **32**, 704.
32. Ruddy, S. J., Johnson, R. F., Mosley, J. W., Atwater, J. B., Rossetti, M. A., and Hart, J. C. An epidemic of clam-associated hepatitis. *J. Am. med. Ass.*, 1969, **208**, 649.
33. Hepatitis Surveillance Unit. Clam-associated epidemics of infectious hepatitis. Communicable Disease Center, Atlanta, Hepatitis Surveillance Rep. No. 18, 1964, p. 14.
34. Center for Disease Control. Follow-up on shellfish-associated hepatitis. *Morbid. Mortal. wkly Rep.*, 1973, **22**, 388.
35. Stille, W., Kunkel, B., and Nerger, K. Austern-Hepatitis. *Dt. med. Wschr.*, 1972, **97**, 145.
36. Koff, R. S., and Sear, H. S. Internal temperature of steamed clams. *New Engl. J. Med.*, 1967, **276**, 737.
37. Mahoney, P., Fleischner, G., Blumberg, B. S., Millman, I., London, W. T., and Arias, I. M. Australia antigen: detection and transmission in shellfish. *Science, N.Y.*, 1974, **183**, 80.
38. Metcalf, T. G., and Stiles, W. C. Viral pollution of shellfish in estuary waters. *Am. J. Epidemiol.*, 1967, **88**, 3.
39. Denis, F. A. Coxsackie group A in oysters and mussels. *Lancet*, 1973, i, 1262.
40. DiGirolamo, R., Wiczynski, L., Daley, M., and Miranda, F. Preliminary observations on the uptake of poliovirus by West Coast shore crabs. *Appl. Microbiol.*, 1972, **23**, 170.
41. DiGirolamo, R., Wiczynski, L., Daley, M., Miranda, F., and Viehweger, C. Uptake of bacterio-

phage and their subsequent survival in edible West Coast crabs after processing. *Appl. Microbiol.*, 1972, **23**, 1073.

42. MOSLEY, J. W. Transmission of viral diseases by drinking water. In 'Transmission of viruses by the water route' (ed. G. Berg). Wiley-Interscience, New York, 1967, p. 5.
43. MOSLEY, J. W. Epidemiological implications of changing trends in type A and type B hepatitis. In 'Hepatitis and blood transfusion' (ed. G. N. Vyas, H. A. Perkins, and R. Schmid). Grune and Stratton, New York, 1972, p. 23.
44. WORLD HEALTH ORGANIZATION. Infectious hepatitis, 1959–1972. *Wld Hlth Statists Rep.*, 1973, **26**, 686.
45. PLOTKIN, S. A., and KATZ, M. Minimal infective doses of viruses for man by the oral route. In 'Transmission of viruses by the water route' (ed. G. Berg). Wiley-Interscience, New York, 1967, p. 151.
46. THOMPSON, A. W. S. Poliomyelitis in Auckland, 1947–49: an epidemiological study. *J. Hyg., Camb.*, 1950, **48**, 96.
47. MOSLEY, J. W. Epidemiology. In 'Infectious diseases' (ed. P. D. Hoeprich). Harper and Row, Hagerstown, Md, 1972, p. 17.
48. STEINIGER, F. Paratyphus-B-Bakterien im Nordseewasser. *Zbl. Bakt.*, Orig., 1951, **157**, 52.

Introducing his paper, Prof. Mosley said that thalassogenic infections of humans might be produced by nonrecreational (accidental) immersion, recreational immersion, or consumption of contaminated shellfish. Illness potentially caused by swimming had received the greatest attention, but without evidence for its being a significant problem at the present time. In contrast, epidemic and endemic shellfish-associated infections had been documented, but seemed to excite very little interest among public health workers or engineers. The question whether swimming at contaminated beaches produced infection with enteric organisms was one still requiring an investigative approach. Consideration of the characteristics of pathogenic agents in sewage, and of human populations that might be exposed through swimming, suggested that enteroviral infection in a community 1–2 hours' drive from a popular beach would be the most suitable model. An attempt ought to be made, therefore, to identify those cases of enteroviral disease that would be prevented if contamination of the beach were ended. For infections acquired by consumption of molluscs, surveillance was now appropriate. Shellfish-associated type-A hepatitis was a disease convenient to monitor by this technique.

[*The discussion of this paper starts on p. 110*]

Paper No. **10**

The Case for Microbial Standards for Bathing Beaches

Hillel I. Shuval

Associate Professor, Director Environmental Health Laboratory
Hebrew University—Hadassah Medical School
Jerusalem, Israel

Early Microbial Standards and Engineering Feasibility

The fact that water can serve as a highly effective vector of enteric disease agents was known even before there was a full understanding of the nature of the causative pathogenic microorganisms themselves. Methods of treating water by sand filtration to remove the causative agents were also practised prior to the time that the efficiency of such processes could be controlled by evaluating the degree of removal of enteric microorganisms. However, as time went on it became the practice to check the effectiveness of treating polluted water by bacterial testing.

In 1900 the slow sand filter at Lawrence, Mass., treating the highly polluted water of the Merrimac River, had been in operation for several years and its installation had been followed by an immediate clear-cut drop in the typhoid fever death rate in the city. Based on their slow sand filter performance and epidemiological insight they were able to set one of the first microbial standards for drinking water which in today's terms would be equivalent to 69 coliform bacteria per 100 ml[1]. With the introduction of water disinfection technology by chlorine it was possible in 1910 for the U.S. Public Health Service to establish its first statutory standard for drinking water at 2 coliform bacteria per 100 ml of water. In both these cases the standard set was essentially a function of *engineering feasibility* based on the assumption that the less enteric organisms ingested in drinking water the safer it would be.

Since that time, health authorities in many countries throughout the world have set similar numerical standards for the microbial quality of drinking water as one of the first administrative aids in the monitoring and control of the safety of water supplies. There can be little debate today as to the wisdom and foresight of those who established that early environmental quality standard which has made a major contribution in the prevention and control of water-borne enteric disease and protecting the health of hundreds of millions of persons.

The bacterial standard for drinking water was established and put into practice long before there was a full scientific evaluation of such factors as: the minimal infectious dose, the ratio of pathogens to coliforms, the dose-response relationship, the reliability of coliforms as an indicator under various conditions in the natural environment, and the relative resistance of coliforms and pathogens in different water treatment processes.

In presenting this brief historical background of early microbial water standards it is not to be implied that, since the almost universally accepted drinking water standard was established in its day based mainly on engineering feasibility criteria rather than epidemiological and scientific data, we should automatically do likewise today.

There is nevertheless something to be said for the basic logic and direct approach of the early public health pioneers. They had good reason to believe that by ingesting pathogenic bacteria, carried by water, people could become infected and they therefore strived within the limits of the technology available to reduce the potential risk to a reasonable minimum. They then set standards based on engineering feasibility so that others would have goals to meet which had been proved to be obtainable. The standards set also enabled them to monitor and control water treatment plants once they had been built and put into operation so as to assure their continued high level of performance. In the case of drinking water, where treatment costs were relatively small and benefits demonstratively large, the decision to set a microbial standard based on the lowest level obtainable under reasonable engineering conditions may in light of today's knowledge and experience be considered obvious and simple enough. However, 70 years ago there were many who opposed spending the money needed to build the water treatment plants required to meet those early standards. What may be obvious today was not so obvious to those who held the purse strings at the time.

The case for establishing microbial standards for bathing water is much more complex and a lot less obvious.

Epidemiological Data

The association between bathing in sewage-polluted sea water and enteric disease has been reviewed in detail by others[2,3] and it is apparent that there is not the same strong, clear-cut case that exists for enteric disease such as typhoid fever and contaminated drinking water.

It has been correctly pointed out that even where such an association does indeed exist there are serious methodological obstacles in studying population groups exposed to contaminated bathing water. For example, resort cities often have large transient populations who may stay for a few days or weeks at most and then return home, while the incubation period of some water-borne diseases such as infectious hepatitis may extend for a month or more. Exposure to risk varies greatly among visitors at bathing beaches since some may swim energetically for hours ingesting considerable volumes of water while others may barely wet their ankles.

Another factor is that the probability of ingesting a pathogen even among those who do actually ingest sea water is a function of the endemic disease prevalence in the community whose wastewater is contaminating the bathing areas. Since the few well-designed systematic epidemiological studies have been carried out in countries such as the U.S.A. and the U.K. where high levels of community hygiene have prevailed for years, enteric disease prevalence in the normal population may be so low as to make it difficult to detect differences in disease rates among those exposed to risk unless very large study samples are employed.

Despite these serious methodological obstacles there have nevertheless been a number of studies which have indicated that certain enteric disease outbreaks could best be explained by exposure of bathers to sewage-contaminated sea water at beaches[4,5]. The evidence is not massive by any means but it is there and cannot be overlooked. There is ample evidence, however, that massive outbreaks of typhoid fever and infectious hepatitis have been caused by consuming raw clams grown in sewage-polluted sea water. There can be no question of the ability of sea water to be an effective vector of enteric pathogens under certain circumstances.

The consumption of polluted drinking water from community water supplies tends to lead to universal exposure of the total population who all are then at risk. This makes epidemiological studies much more clear cut. The percentage of the population involved in sea bathing has in the past been relatively small. This is changing rapidly, however, as standards of living increase and free time for recreation becomes widespread, many millions of people are turning to the bathing beaches as a popular pastime. Likewise as urban populations grow rapidly in developed as well as developing countries and modern water-carried waste disposal systems are introduced and expanded, the volume of wastewater liable to be dumped in coastal waters grows continuously. We are therefore obliged to evaluate the public health risks associated with this new and growing situation where more and more people are exposed to bathing waters contaminated with higher and higher concentrations of fecal wastes. The population at risk is growing as is the potential degree of risk.

Is it sufficient to assume that, since the epidemiological evidence to date is not strong, we can continue to proceed on the premise so forcefully stated in the well-known report of the U.K. Medical Research Council Committee headed by Dr. B. Moore[2] which concluded that "the risk to health of bathing in sewage contaminated sea-water can, for all practical purposes, be ignored" while further stating that under circumstances where no gross pollution of beaches occurs "the need for extensive improvements in the quality of sewage effluents discharged into the sea does not arise"?

A careful evaluation of the potential risks involved in bathing in contaminated sea water is now required, particularly in light of the tremendous increase in ocean bathing throughout the world which is paralleled by increases in the degree of pollution at such beaches.

Analytical Approach

In light of the dearth of reliable epidemiological data, an analytical approach may prove to be useful in evaluating the risks associated with bathing in sewage-contaminated sea water. Some of the variables that must be considered are discussed in the following sections.

Minimum infectious dose

Kehr and Butterfield in their classical studies[6] concluded that 1 per cent of the individuals ingesting a single typhoid bacterium become infected. Plotkin[7] has shown that for poliovirus one tissue-culture infectious dose is sufficient to be a human infectious dose. It may be reasonable to assume that this is correct for the agent of infectious hepatitis as well. While the concept of a single pathogenic organism being sufficient for human infection is contested by some, there appears to be sufficient experimental and epidemiological evidence to support the hypothesis, at least in the case of some of the enteric diseases that are potential candidates for transmission by bathing water.

For the purposes of this discussion we feel it is not unreasonable to assume that at least in certain cases a swimmer who ingests a single pathogen can potentially become infected, say one out of every ten. We can further assume that of those infected only one in ten will develop a detectable case of clinical disease.

Ratio of pathogens to coliforms

Here, too, Kehr and Butterfield[6] provided the pioneering studies indicating that the ratio of pathogens to coliforms in contaminated water is a function of the prevalence of the disease in the community whose wastes are polluting the water. They found, for example, that in areas with a typhoid morbidity rate of 10 per 100 000 the ratio of typhoid bacteria to coliforms was 8:1 000 000. They showed that, as typhoid rates increase, the number of typhoid organisms per million coliforms goes up, since coliform excretion per person remains constant in the population while the number of persons shedding typhoid organisms increases. It is reasonable to expect that the risk of a non-immune person becoming infected when ingesting polluted water should increase measurably as he goes from the U.K. with a typhoid rate of 0·45 cases/100 000 population to certain Mediterranean countries with 10 times that rate or to some South American or Asian countries with 100 times that rate[8].

In our studies of sewage in various communities in Israel we have found[9] a ratio of enteric viruses to coliforms of about 10:1 000 000 with enterovirus concentration in sewage averaging some 1000 PFU per 100 ml.

From these findings we can conclude that the risk of ingesting a pathogen in any given volume of water ingested at a given coliform standard increases as the enteric disease rates increase.

A further complicating factor that must be considered in the case of sea water is the differential die-away rates of coliforms and pathogens. We have shown that in the Mediterranean coastal areas of Israel coliforms die away rapidly with a T_{90} of 1 h or even less while polioviruses under similar conditions have a T_{90} of 24 h or more[9]. This may mean that, under actual field conditions of a bathing beach exposed to sewage contamination from an outfall placed so that the travel time of the sewage field is 2 h, the coliform count will be reduced by a factor of 100 by die-away alone while there will be no equivalent reduction in virus count. Thus if the initial virus to coliform ratio is 1:100 000 it may be increased to 1:1000 at the bathing beach due to the differential die-away rates. In our studies we have actually detected poliovirus at a bathing beach showing coliform counts of 1000 per 100 ml.

In summation, the ratio of pathogens to coliforms in sewage-contaminated sea water is a function both of the enteric disease rates and the differential die-away rates of coliforms and pathogens.

Volume of water ingested by swimmers

While it can be safely assumed that most individuals will consume about 1 l. of drinking water per day, if not more, in hot dry climates, no such simple assumption can be made as to bathers. Mechalas *et al.*[10] suggest that 10 ml is the average amount ingested by swimmers per exposure period while Steiniger[11] reports on a survey made among physical education physicians who found 50 ml to be a good estimate for active swimmers in high quality fresh water. Of course, there will be individual differences and differences due to climate. For example, a child spending the day at a warm Mediterranean beach resort may spend as many as 2–3 full hours in the water and ingest 100 ml of sea water during the day, while a bather at a colder Scandinavian beach will spend much less time in the water and can thus be expected to ingest much less water. For the purposes of this discussion it is assumed that bathers ingest about 10 ml per bathing day. It is felt that this assumption is a conservative one.

Hypothetical risk of becoming sick

Based on the above assumptions let us attempt to calculate the risk of becoming sick bathing in sewage-polluted water, showing a coliform count of 10 000 per 100 ml. For the purposes of this estimate let us assume that 1 per cent of the enteric virus count in sewage of 1000 PFU/100 ml the agent is of a virulent undefined viral gastroenteritis, with a minimum infectious dose of one, similar to that of polio-virus. The coliform count of the raw sewage is assumed to be 10^8 per 100 ml.

The total virus to coliform ratio in sewage would then be $1:10^5$ while the specific viral gastroenteritis pathogens we are considering would have a ratio of $1:10^7$ coliforms. The concentration of virus in the raw sewage would be 10 PFU/100 ml.

A coliform count of 10 000/100 ml at the bathing area might result from a physical dilution of the sewage of 1:100 and a coliform die-away factor of 100. This would reduce the raw sewage coliform count of 10^8/100 ml by a factor of 10^4 but would increase the virus to coliform ratio, assuming no virus die away in the sea in 2 h to $1:10^4$. That is, one virus infectious dose per 1000 ml of sea water.

Assume that each bather ingests 10 ml of sea water per day of exposure, so that only 1 in every 100 bathers per day ingests a virus, and that of those infected only 1 in 100 shows clinical symptoms of the disease. If these assumptions are correct, the chances are that an individual will become sick under these conditions once in 10 000 bathing days. This may be considered a small risk for an individual. Stated in other terms, however, one out of every 10 000 bathers on a given day may become sick at this particular level of contamination. If the sewage contamination is decreased to 1000 coliforms/100 ml then we might expect to see one person in 100 000 sick with the particular viral gastroenteritis under discussion. With a coliform standard of 100/100 ml, the risk of infection might be expected to be reduced accordingly.

If, however, an epidemic of infectious hepatitis breaks out in the community, or in another community with a higher endemic disease prevalence, then one would expect the percentage of persons shedding pathogens to the sewerage system to increase measurably, thus increasing the potential risk of transmitting that disease through contact with contaminated sea water even though the coliform standard achieved remains constant.

The Need for Standards

These are, of course, only hypothetical calculations. It is difficult to say what the real risks actually are at this time. However, we have no reason to question the assumption that people do indeed ingest water while bathing and that pathogens do enter the sea in large numbers and survive long enough to reach bathing beaches in detectable quantities. The popular belief that the sea possesses powerful therapeutic qualities may have some justice to it, but there is ample evidence today to show that when insulted with massive doses of sewage even the sea cannot rid itself instantly of the heavy doses of pathogens which can be detected in considerable numbers at most polluted bathing beaches. We also believe that some percentage of the people who swallow a single pathogen do actually become ill. This must lead to the unavoidable conclusion reached 70 years ago when drinking-water standards were first set, that by decreasing fecal contamination in water ingested by man we will reduce the risk of becoming infected.

Lacking as we do a firm epidemiological basis and dose-response data, we find it difficult to establish a rational bacterial standard to eliminate the health risks involved. We can, however, revert in part to the classical engineering feasibility concept adopted by the early pioneers in environmental standards.

It appears that within the limits of reasonable engineering technology today, a partial treatment of sewage can remove objectionable floatable solids and reduce coliforms by some 90 per cent without chlorination. Good diffuser design can usually provide initial dilution of about 100 while a reasonably long outfall sewer can often provide a further physical dilution factor of 10 or more. This might provide a coliform and virus reduction of a factor of 10^4 without counting on natural die-away. We can assume that a minimal further reduction in coliforms, some 90 per cent, is often obtained by sedimentation and/or natural die-away in the sea. This suggests that a standard or goal of 1000 coliforms per 100 ml can be feasibly attained.

Some authorities prefer to formulate the standard or criteria for bathing waters in terms of fecal coliform bacteria as a more specific and reliable indicator of fecal contamination. This approach has much to be said for it and there is considerable laboratory and field evidence to support this contention. Chlorination would provide even better results as far as coliforms are concerned but its efficacy against enteroviruses is open to question and the problem of coliform regrowth after chlorination still presents a problem.

Another way of stating the above performance criteria, would be to require that partially treated sewage be discharged to the sea in such a way as to achieve a physical dilution approaching 1000 at adjacent bathing beaches. The advantage of such a formulation is that it does not take into account coliform reductions due to chlorination or natural die-away neither of which can be assured of providing an equivalent removal of such critical pathogens as viruses.

A standard based on effective physical dilution of 1000 and the removal of floatables, difficult as it may be to achieve in many situations, has much to be said for it on aesthetic grounds alone. One bathing beach we studied was exposed to raw sewage contamination with a physical dilution at the beach of 100 but showed low coliform counts due to rapid die-away in the sea. Nevertheless, swimming in such water was essentially equivalent to taking a bath with several cups of sewage added to the tub. I wonder how many people find such an experience an acceptable health risk or aesthetically tolerable, no matter how sparse the epidemiological evidence.

A more attractive possibility would be to establish a direct virus standard. This is feasible today since virus monitoring techniques are available for detecting very low concentrations of viruses in large volumes of water[9]. Today one can test 100–500 l. of water to determine whether viruses are present or not. A microbial standard for bathing beaches based on the absence of enteric viruses in say 100 l. is in many ways the most logical standard in light of the higher degree of resistance of such pathogens to natural inactivation in the sea and their known low minimal infectious dose.

Discussion and Conclusions

The claim[2] that "the risk of bathing in sewage contaminated sea-water can, for all practical purposes, be ignored" is not consistent with a more fail-safe public health approach which insists on identifying *potential* risks and taking preventive action even before major outbreaks occur. It is further claimed by some that the present-day widespread and recognized presence of pathogens in sewage-contaminated sea water at bathing beaches is of importance only to the extent that it demonstrates that those particular diseases are present in the community. It has been suggested that the correct approach would be to deal primarily with the prevention and control of such disease within the community rather than to take any specific measures to reduce the contamination at the bathing beaches.

What would have happened if this approach had been followed in the case of sewage-contaminated drinking water, or milk contaminated by unhealthy cows, or polluted farm facilities? I doubt whether we would have succeeded as we have without setting rigorous standards for the control of water treatment and milk pasteurization.

There is very good epidemiological evidence to show that, in those areas where milk pasteurization was not introduced and control was based solely on attempts to improve the health of the animals or the sanitation on the farms, milk-borne disease transmission was not successfully stayed. The modern public health approach calls for the establishment of *multiple barriers of protection* to block the transmission of disease by environmental vectors.

It is certainly unfounded to imply that the control of the transmission of infectious hepatitis by the water route be based solely on the eradication of this disease in the community when at this time no known method of treatment or immunization as yet exists. Preventing sewage from getting into water ingested by man, or inactivating the virus by chemical means, are to date the only known effective means of preventing the transmission of this disease by water. This holds for drinking water and I suggest for contaminated bathing water as well.

There are a number of arguments presented by those who propose that no microbial standards are needed. Will they stand the test of universal application as well as the test of time, particularly in light of ever increasing sewage pollution at bathing beaches and ever increasing numbers of persons exposed to risk?

There is ample evidence that pathogens, particularly viruses, survive for sufficient periods of time in sewage-contaminated sea water to appear at bathing beaches in detectable concentrations. There is good reason to believe that ingesting a single pathogenic bacterium or virus can in certain cases lead to infection and disease. There is also evidence that bathers do indeed ingest 10–50 ml of sea water during a single day of exposure. Taken together, one must conclude that there is a risk of becoming infected if one bathes in sewage-contaminated water and that the risk would be greater as the level of sewage contamination increases and in areas of high enteric disease rates.

It would appear from the foregoing that a prudent policy of providing reasonable protection to the millions of bathers at seaside beaches would be to call for the reduction of fecal microorganisms of sewage origin to the lowest feasible level even prior to more definitive epidemiological evidence becoming available. The classical approach of basing tentative environmental standards on engineering feasibility leaves much to be desired but can in this case, as was true for the early drinking-water standards, provide a yardstick for evaluating existing sea-pollution levels at various beaches exposed to sewage contamination and provide an obtainable engineering objective for reducing fecal contamination to a relatively low level.

Hopefully, in time it will be possible to refine such a standard as more scientific evidence is gathered. An empirical standard based on the concept of engineering feasibility should be applied with flexibility and judgement since we are unable at this time to estimate the increased or decreased risk associated with some slight deviation from it. It should not be considered as a sacred limit to be rigidly adhered to or unchallengeable in the light of additional evidence but rather as a guide and tentative goal to be strived for.

We are attempting to establish an optimal environment for man, whose spectrum of responses to detrimental stress is so wide and so variable that he at times defies the ultrasensitive laboratory measurements designed to evaluate these stresses. The many indeterminate variables of the environment and of human biology will require the continued use of sound professional judgement based on experience and understanding in the flexible application of health standards which are vital tools to: provide uniform goals, evaluate improvement, and compare program effectiveness in different areas.

This need for quantitative standards on the one hand and flexibility on the other can be summed up by the epitaph: A yard stick is needed even if it is made of rubber[12].

REFERENCES

1. PHELPS, E. B. Public health engineering. John Wiley, New York, 1948, p. 278.
2. PUBLIC HEALTH LABORATORY SERVICE. Memorandum No. 37, Sewage contamination of bathing beaches in England and Wales, H.M. Stationery Office, London, 1959, p. 23.
3. WORLD HEALTH ORGANIZATION. Health criteria for the quality of recreational waters with specific reference to coastal waters and beaches. WHO/W.Poll/72.10, Geneva, 1972.
4. SNOW, D. J. R. Typhoid and the city beach. Report for the year 1958. Commissioner of Public Health, Western Australia, 1958.
5. SHUVAL, H. I. World Health Organization, Document EP/WP 72.4, 1972.
6. KEHR, R. W., and BUTTERFIELD, C. T. Notes on the relationship between coliform and enteric pathogens. *Publ. Hlth Rep., Wash.*, 1943, **58**, 589–607.
7. PLOTKIN, S. A., and KATZ, M. Minimal infective doses of viruses for man by the oral route. In 'Transmission of viruses by the water route' (ed. G. Berg). Wiley-Interscience, New York, 1967, pp. 151–166.
8. WORLD HEALTH ORGANIZATION. Statistical Report 1968, pp. 2 and 448.
9. SHUVAL, H. I. Detection and control of enteroviruses in the water environment. In 'Developments in water quality research' (ed. H. I. Shuval). Ann Arbor-Humphrey Science Press, Ann Arbor, Mich., 1970, pp. 47–71.
10. MECHALAS, B. J., HEKIMIAN, K. K., SCHINAZI, L. A., and DUDLEY, R. H. An investigation into recreational water quality. Office of Research and Monitoring 'Water quality criteria data book'. U.S. Environmental Protection Agency, Washington, D.C., Vol 4, 1972, 256 pp.
11. STEINIGER, F. Aus der Freihandbiologie der Salmonellen. *Dt. med. Wschr*, 1954, **79**, 1118–1120.
12. SHUVAL, H. I. The uses and misuses of environmental health standards. *Am. J. publ. Hlth*, 1964, **54**, 1319–1329.

In presenting his paper Prof. Shuval said there was ample evidence that pathogens, particularly viruses, survived for long periods in sewage-contaminated sea water to appear at bathing beaches in detectable concentrations, and there was good reason to believe that ingesting a single pathogen could, in certain cases, lead to infection and disease. There was also evidence that bathers ingested 10–50ml of sea water in a single day. Taken together, one had to conclude that there was a risk of becoming infected if one bathed in sewage-contaminated water and that the risk would be greater as the level of contamination increased and in areas of high enteric disease rates. In addition, aesthetic considerations alone provided ample justification for reduction of faecal contamination of bathing beaches. Public officials, including health authorities, should weigh both potential health risks and aesthetic factors when deciding public policy.

It would thus appear that a prudent policy for providing reasonable protection to the millions of bathers at seaside beaches would be the reduction of faecal micro-organisms of sewage origin to the lowest feasible level—even before more definitive epidemiological evidence was available. The classical approach of basing tentative environmental standards on engineering feasibility left much to be desired, but could in this case (as it had for the early drinking-water standards) provide a yardstick for evaluating existing pollution levels at various beaches exposed to sewage contamination and provide an obtainable engineering objective for reducing faecal contamination to a relatively low level.

Hopefully, in time it would be possible to refine such a standard as more scientific evidence was gathered. An empirical standard based on engineering feasibility should be applied with flexibility and judgement, since it was not yet possible to estimate the change in risk associated with a slight deviation from it. It should not be considered as a sacred limit to be rigidly adhered to, or unchallengeable in the light of additional evidence, but rather as a guide and tentative goal to be strived for.

[*The discussion of this paper starts on p. 110*]

Paper No. **11**

The Case against Microbial Standards for Bathing Beaches

B. Moore

Public Health Laboratory, Exeter, England

It is common ground for those who argue either for or against microbial standards for bathing waters that the progressive elimination of hazards of the environment is an important goal of preventive medicine. At issue here is whether the adoption of microbial standards is relevant to this goal. A call for standards is familiar to the public health bacteriologist in a number of different contexts. Commonest perhaps is the demand from medical or environmental health administrators that maximum microbial counts be specified for different foods, or that surface swabs or washings from catering establishments should be cultured quantitatively by the laboratory and be judged as satisfactory or unsatisfactory, this furnishing a basis for further administrative action.

The application of bacterial standards to sea water can be traced back to the cautious suggestion by Winslow and Moxon[1] in a report on New Haven Harbor pollution that the coliform count of samples from bathing waters should not exceed 100 per 100 ml. They gave no logical basis for this figure, but their study was part of a campaign to convince the New Haven authorities of the need for a sewage treatment scheme to replace the discharge of crude sewage into the harbour. A generation later, Garber[2] reported that different public health authorities in the United States had bathing water standards ranging from a requirement that no coliform organisms be present to a permitted median count of up to 2400 coliforms per 100 ml. At every level these standards were justified by those who had imposed them on the grounds that with their use epidemiological experience had been satisfactory. Unquestionably, the demand for bathing water standards in recent years has been greatly reinforced by worldwide concern about environmental conservation. That this demand also has a strong aesthetic and emotional component is indicated by the contrasting public apathy on the far greater hazards posed by the spread of say an influenza epidemic in the crowded environments of places of public entertainment.

The case against microbial standards for bathing waters stated below rests on two main lines of argument. First, the classification of bathing waters in terms of permitted median and 90-percentile levels is simplistic to the point of being spurious, given the complexity of pollution patterns that detailed studies have consistently demonstrated. Secondly, no satisfactory evidence linking the risk of communicable disease with the degree of sewage contamination of coastal or other bathing waters has yet been found, and to that extent microbial standards for bathing water are irrelevant to the public health.

Grading of Beaches by Coliform or Faecal Coli Count

To sharpen the focus of discussion, a recent proposed classification of marine waters in the State of Maine[3a] may be cited. It is a good example of the kind of legislation resulting from present policies in the United States on water pollution control. Four grades are specified, as shown in Table 1 in order of diminishing excellence. The figures are for areas in which there are no shellfish beds.

TABLE 1. Maximum permitted counts (per 100 ml) for four classes of water[3a]

Class		Median	90-percentile
SA	Coliform	70	230
	Faecal coli	15	50
SB_1	Coliform	240	500
	Faecal coli	50	150
SB_2	Coliform	500	1000
	Faecal coli	100	200
SC	Coliform	700	2300
	Faecal coli	150	500

Waters of classes SA, SB_1, and SB_2 are deemed suitable for bathing; waters of Class SC are not. The median counts are to be calculated on "a series of samples representative of the waters", the number of samples not being specified.

The justification for discriminating between beaches of Class SB_2 and Class SC on health grounds is questioned later. The more immediate question here is whether a given beach can on the findings of bacteriological survey be confidently allocated to one or other of the categories shown in Table 1. In 1959, a Public Health Laboratory Service committee[4], in its report on a six-year bacteriological and epidemiological study of British coastal bathing waters, expressed the view that, health grounds apart, it could see no logical justification for the application of rigid bacteriological standards because "by a judicious selection of sampling points and times, a considerable upgrading or downgrading of a given beach could easily be arranged at will."

During the past twelve years, the studies of Water Pollution Research Laboratory workers at a number of coastal beaches around the English coast have reinforced these doubts as to the validity of bacteriological grading, at least for beaches in the United Kingdom. Gameson, Bufton, and Gould[5] reported the first of these intensive studies, done at Sidmouth on the south Devon coast. Coliform counts obtained on sea-water samples taken at fixed sampling points along Sidmouth beach were found to be lognormally distributed, both within and between days. During May–June and September–November 1965, routine samples were collected six times a day from eight sampling stations on a total of 86 days in all. The daily median coliform counts on the beach as a whole ranged between 3 and 2860 per 100 ml. The following year, six samples over the course of a day were collected from one of the fixed sampling stations over the period April–October. On 10 of the 86 occasions when sampling was done on successive days at this point, the median count changed by a factor of at least 10 from one day to the next, a range that spans all four classes in Table 1. The overall 90-percentile coliform count for this sampling point was 757 per 100 ml and the median value 82 per 100 ml. The authors cited pointed out that this nine-fold ratio of 90-percentile to median counts was comparable to similar ratios found in other surveys, including two of the surveys done by the Cincinnati workers referred to below. With a ratio of this magnitude, however, compliance with the 90-percentile levels shown in Table 1 would require the setting of a very much lower permitted level for the median count.

Recently, Gameson[6] has reviewed the main findings of the Water Pollution Research Laboratory's field studies since 1961 at seven sites in British coastal waters. These provide a consistent picture that confirms and supplements the Sidmouth findings cited above. The coliform counts on samples from single sampling stations on five different bathing beaches all gave an approximately lognormal distribution. The most extensive long-term surveys were done at Bridport, Dorset, in a series of sampling periods that lasted about two weeks on each occasion. For all sampling stations along Bridport beach, the maximum median coliform count recorded during the successive sampling periods exceeded the minimum median counts for the same sampling stations by roughly one order of magnitude. Gameson also tabulates the 90-percentile/median ratios for single sampling stations on five of the beaches studied. The values obtained, with the number of samples on which they were based alongside in brackets, were 2·6 (60), 3·4 (43), 8·5 (1197), 10·5 (169), and 10·7 (515). Not surprisingly, the ratio tended to increase the more extensive the survey done at a given sampling station. Gameson also discusses at some length the various factors that accounted for the considerable variability in the coliform counts recorded, including tidal state, wind velocity, sea turbulence, solar radiation, and other factors, but these are not the direct concern of this paper.

These findings indicate that the bacteriological contour map of coliform contamination on any one bathing beach is highly variable and cannot be realistically accommodated within the confines of the kind of classification that is being advocated in many countries at the present time. This is not to say that a specified bathing area subject to sewage contamination cannot be adequately described in statistical terms based on faecal coli counts of sea-water samples from the area in question. The results of large surveys have suggested that such counts may be lognormally distributed. If therefore a given bathing area were randomly sampled throughout and the number of samples was adequate, estimates of the mean count and of its variability would be available and the probability of a random sample exceeding a given value could be calculated. To make allowance for the effects of tides, winds, solar radiation, and other factors, extensive sampling over a lengthy period of time would in fact be needed to characterize a given beach with reasonable precision. If at that point the beach was made subject to long-term monitoring, a later judgment that its quality had significantly deteriorated over a period would again require an extensive sampling programme to substantiate it.

So much for any one beach. Grading a series of beaches each with its own distinctive pattern of pollution presents additional problems, given the implication that a beach of class SA, for instance, is 'better' than one of a lower class. The ranking of a series of beaches according to 90-percentile levels, for example, would not necessarily give the same order as one based on 95-percentiles. Add to this finally the well-known technical error of the coliform test. The practical conclusion would seem to be that microbial standards for bathing waters as usually proposed have little scientific validity.

Relationship between Microbial Standards and Health Hazards

To allow discussion of the second leg of the case against standards, assume that bathing beaches can in fact be suitably classified according to the degree to which they are contaminated with sewage. The next question that arises is whether the bacteriological grading can be matched by a corresponding scale of health hazards, to the point where a beach of a given class can be declared suitable for bathing and one of a lower class unsuitable.

What health hazards are to be considered here? Clearly, a bacterial standard based on organisms indicative of sewage contamination can only be expected to relate to the risk of contracting diseases the causal agents of which are present in sewage. It cannot be expected to measure the risk of say leptospiral infection or of amoebic meningoencephalitis, neither of which is a sewage-borne disease. To avoid confusion, however, the point should be made here that the presence of a pathogenic agent in sewage-contaminated water does not necessarily imply that bathing in that water carries a significant health hazard. This is so for a number of reasons. Some organisms, for instance, are carried by a high proportion of the normal population but cause disease only in a small specially susceptible minority. Individual susceptibility is therefore the determinant of disease for these organisms and not simply contact with the organism itself in a polluted environment. Other agents, of which rubella virus is an example, may be present in the excreta of persons with the clinical disease and could therefore in principle be recovered from sewage-contaminated water if suitably delicate techniques for the purpose were available; nevertheless, no one has ever suggested that rubella should be looked on as a potentially water-borne disease. Even for an organism like the typhoid bacillus, the likelihood of a non-immune person contracting the disease from bathing will depend on the number of typhoid bacilli ingested and not simply on whether the organism can be recovered by highly sensitive procedures from a given volume of bathing water.

The risk to health of bathing in sewage contaminated water

The linkage of a bacteriological classification of bathing waters with health risks, which is implicit in Table 1, may be scrutinized by seeking answers to three questions. First, how is one to link a polluted environment with a particular degree of associated health hazard? Secondly, what is the present evidence that bathing in sewage-contaminated water does carry a health risk? Thirdly, where a health related microbial standard has been proposed, is it based on satisfactory evidence?

Linking the environment with disease

Take a hypothetical coastal resort with a single beach subject to contamination from an adjacent sewer outfall and falling into a particular class in bacteriological terms. The first problem is to identify those persons who have been taken ill through having picked up a disease-producing organism while bathing on this beach. The practical difficulty of finding them in a shifting holiday population need not concern us here. If, however, all bathers during the relevant period could be found and a list of their ailments made, this would provide a maximum figure for the disease incidence that could possibly be ascribed to bathing. Certain illnesses like the common cold would be struck off the list, because, although they might well follow bathing, nothing that is known of the biology of the common cold would point to the causal virus being acquired from water rather than through personal contact with another person. A somewhat shorter list of illnesses worth further inquiry would remain. It would include conditions known to be spread by sewage-polluted drinking water, such as typhoid fever or less specifically gastro-enteritis, i.e. diarrhoea and/or vomiting. By analogy it would be reasonable to suppose that these might also be caused by bathing in polluted water.

The next problem would be to decide whether persons reporting the various illnesses had in fact contracted them while bathing. Typhoid fever would present a particular problem here because the incubation period of water-borne typhoid fever may be several weeks in duration. Moreover, the presence of typhoid bacilli in the bathing water would indicate that carriers of the organism must be living in the sewered population in question and that food-borne typhoid fever was therefore also a possibility. Retrospective linkage of the disease with a history of bathing would thus be difficult.

Take, however, a more common condition at seaside resorts, namely gastro-enteritis, and a holiday-maker suffering from this minor illness and with a history of having bathed at least once on the beach concerned. The first difficulty here is that gastro-enteritis may be caused by a whole range of different circumstances and different causal agents. When due to food poisoning, the cause might be established by circumstantial evidence that a group of persons of whom the particular patient was one had eaten a specific meal at their hotel and been taken ill a few hours later, while those not partaking of the meal had escaped. Similarly, a higher incidence of gastro-enteritis in bathers by comparison with non-bathers would point to the possibility that bathing was the cause, particularly if the incidence of symptoms could be shown to be greater in persons who had bathed more frequently or for longer periods.

What emerges at this stage of the discussion is that the first pointer to bathing as a cause of disease would usually be the demonstration of an excess of the disease in question in bathers by comparison with non-bathers. A significant association does not however necessarily prove causation. As a trivial example, a paratyphoid B outbreak at a seaside resort investigated many years ago by the writer and ascribed at the time to bathing in polluted sea water, was shown later to have been caused by ice-cream from a van parked on the beach in question at the relevant time. The criteria for postulating causation of an illness by an environmental hazard as against an association that might be fortuitous or the result of other unknown factors have been clearly set down by Bradford Hill[7] and have been previously discussed at some length by the writer[8] in the context of sea-water pollution. Once a significant association has been demonstrated between bathing and a particular illness, these criteria must be considered before a direct cause-and-effect relationship can be accepted.

The literature

A full review of the literature on bathing as a cause of communicable disease would be outside the terms of reference of the present paper. This literature has been summarized in the following sentence: "Thus, apart from a modest tally of enteric fever cases ascribed to bathing during the past half-century or so, little evidence of a health risk associated with bathing in polluted water has so far been found"[3b]. The Public Health Laboratory Service committee[4] already referred to above could only find four cases of paratyphoid fever, out of a total of over three thousand notified during the years 1953–59, where the evidence pointed to infection having been contracted through bathing. The affected children had played on two grossly polluted beaches of the 40-odd beaches examined during the study. This was the basis of the committee's conclusion that "bathing in sewage-polluted sea water carries only a negligible risk to health, even on beaches that are aesthetically very unsatisfactory". The committee expressed the opinion that such risk as was demonstrated was probably associated with chance contact with intact aggregates of faecal material that happened to have come from infected persons.

Because of the relatively low incidence of enteric fever, even in countries where the disease is endemic, a grading of beaches in terms of the likelihood of contracting enteric fever has not proved practicable. The American standards to be discussed in the next section were based on the risk of minor illness. It has recently been possible, through the courtesy of the Consumers' Association in the United Kingdom, to seek more up-to-date evidence in this country that might indicate a greater risk of minor illnesses in bathers. Just over a thousand families belonging to the Association completed a questionnaire about their experience of holidays on the British coast during 1972. The information, after deletion of the names of those concerned, was examined in the writer's laboratory. The findings are summarized in Table 2, which gives the incidence of gastro-intestinal illness, upper respiratory tract related illness, and other miscellaneous illnesses in bathers and non-bathers of the group in question.

TABLE 2. Number of individuals taken ill while on holiday or immediately afterwards (Figures in parentheses are percentages of corresponding totals)

	Total	Gastro-enteritis etc.	Upper respiratory tract infection	Other illnesses	Total No. of illnesses
Non-bathers	1686	49 (2·9)	46 (2·7)	36 (2·1)	131 (7·8)
Bathers	1892	49 (2·6)	34 (1·8)	53 (2·8)	136 (7·2)
TOTAL	3578	98 (2·7)	80 (2·2)	89 (2·5)	267 (7·5)

In this table, 'other illnesses' included conditions like sunstroke and crab bites, to which bathers are more likely to be exposed than non-bathers. Clearly, however, there is no pointer here to a higher incidence of minor illness among bathers as compared to non-bathers. In the absence of any indication of a significant excess of disease among the bathers, the establishment of cause-and-effect relationships between bathing and illness does not arise.

A critique of American health-related standards

The American studies purporting to have linked the risk of disease contracted through bathing with a particular level of coliform contamination of the bathing water call for detailed discussion at this point, because they provide the only basis so far available for safety standards not only in the U.S.A. but also in other countries. Three prospective surveys aimed at assessing the risk to health of bathing in polluted water were carried out during the years 1948–50 by the United States Public Health Service. The findings were summarized and discussed by Stevenson[9]. These field studies were organized by the Environmental Health Centre at Cincinnati, Ohio. The 1948 survey was done on Lake Michigan at Chicago, the 1949 study on the Ohio River at Dayton, Kentucky, and the 1950 survey on coastal populations at New Rochelle and Mamaroneck, N.Y.

Similar methods were used throughout. In each study, the intention was to take two bathing areas, one with low average bacterial counts, the other more polluted. In the event, the median coliform counts of the two Chicago beaches were similar. In the Ohio River study, the only convenient good quality bathing area available for comparison with river bathing was a fresh-water swimming pool with recirculation and chlorination. During the studies, regular water samples were collected from all the beaches chosen and examined by the coliform test.

For evidence of illness and relevant bathing histories, the investigators relied on calendar record forms filled in by co-operating families. These families were instructed to note swimming and illness experience daily and to record eye, ear, nose, and throat ailments, gastro-intestinal disturbances, and skin irritations. The forms were collected for scrutiny at the end of each bathing season. About 80 per cent of the records contributed useful information, although there was evidence that in some instances records had not been maintained on a day-to-day basis.

The main finding of the surveys as a whole was an appreciably higher overall incidence of illness in swimmers than among non-swimmers irrespective of the quality of the bathing water. Stevenson's comment on this was that "this was expected inasmuch as water is an abnormal habitat for man regardless of its bacterial quality". Ailments of eye, ear, nose, and throat accounted for more than half the illnesses recorded, and gastro-intestinal disturbances for about one-fifth.

Preliminary scrutiny of all the records showed no clear association between illness incidence and the degree of pollution of the bathing waters. The investigators then examined the various tabulations in detail and found indications in two studies of a possible relationship between illness incidence and bathing in water when the logarithmic mean coliform count was about 2500 per 100 ml. As these two findings are the logical basis on which present American bacteriological standards rest, they are worth citing briefly. In each case the figures of illness incidence for the various sub-groups were appropriately adjusted before significance tests were applied to take account of differences in age distribution of the persons in each of the sub-groups concerned. The two positive findings were as follows:

1. For one of the two Chicago beaches studied, three days were selected from the records, on which the mean coliform density of the bathing water was 2300 per 100 ml, and three other days when the mean count was 43 per 100 ml. A comparison was then made between the three-day illness incidence rates per 100 swimmers, during the week following a swim, of those who bathed on the days in question. The figures so obtained were 12·2 for those who swam on the more polluted days and 8·5 for those who swam in cleaner water. This observation, based on a significance test against the hypothesis that pollution caused an increase in illness incidence was one that could occur by chance once in 100 trials. No information is given in Stevenson's paper as to what was the nature of the illnesses in both groups. The findings of the particular analysis cited were, however, not borne out by similar comparisons made in the other surveys.
2. In the Ohio River study, the group of river swimmers, who were exposed to water with a median coliform count of 2700 per 100 ml, totalled 977. The expected number of gastro-intestinal disturbances in this group was 40 assuming no relationship between bathing and illness incidence and basing the figure on the overall incidence of illness in the study population of some 7500 persons in all. The actual number so affected was 53. In the significance test used, the difference could occur by chance once in twenty trials.

The original investigators were extremely cautious in their interpretation of these findings and cannot be blamed for the retrospective weight placed on them nearly twenty years later when pollution became an issue of greater popular concern. The Chicago results can certainly not stand up to serious scrutiny. On the evidence given, the survey tables were inspected and two groups of swimmers picked out because their three-day illness incidence rates of 12·2 and 8·5 per 100 swimmers respectively showed an apparent excess of illness in those who had bathed in the more contaminated bathing water. Opposite findings in the other surveys were ignored. Secondly, all the illnesses reported were lumped together. It would be manifestly absurd to compare the incidence of skin irritations in bathers on the cleaner days with the frequency of diarrhoea in those who swam on the other days, but there is no way of knowing to what extent this might have been done. Thirdly, the comparability of the two swimming groups in other material respects is not discussed. Fourthly, the inclusion of all illnesses reported during the week after a bathe makes the association of these ailments with the bathing episode more tenuous. With all these doubts, the emphasis placed on a rather marginal level of statistical significance as an index, not only of association between bathing and illness but of a cause-and-effect relationship between the two, is difficult to justify.

The Ohio River study findings are equally unsatisfactory in this respect. Significance is attached to an incidence of gastro-intestinal disease in Ohio River swimmers in excess of that suffered by swimmers in a fresh-water pool with recirculation and filtration by a margin that could occur by chance once in twenty trials. The difference was accounted for by gastro-intestinal complaints reported by 13 river swimmers out of a total of nearly a thousand. Here again, the multiple aetiology of minor gastro-intestinal complaints makes it difficult to assess to what extent the comparisons made between the two swimming groups and tested by statistical tests of significance have any real validity.

The writer's conclusion is that classifying beaches of classes SA, SB_1, and SB_2 in Table 1 as being suitable for bathing and Class SC as being unsuitable on health grounds cannot be justified on the evidence cited. One final point on attempts to relate bacterial standards to health risks calls for brief comment. This is the claim that such standards also reflect the relative risks of contracting enterovirus infections. Quite apart from the virtual absence of evidence that enterovirus infections can be water-borne, the classic work of Melnick[10] and others has shown that enteroviruses are present in polluted water only when infections with these viruses are already circulating in the population. Polluted water cannot therefore be satisfactorily monitored for enteroviruses except as a part of a wider epidemiological study of diseases caused by these agents. Equally, bacterial indicators of sewage contamination cannot be assumed to mirror the likelihood of enteroviruses being present except during the periods of prevalence of enterovirus infections in the community.

Value of Microbial Standards from Public Health Standpoint

The argument against microbial standards in the foregoing pages rests on their inadequacy as a basis for classifying bathing areas and on the lack until now of any satisfactory evidence linking a differential incidence of communicable disease with increasing contamination of bathing water. Suppose, however, that with further study a standard that conformed to both criteria was in fact established. The further question would then arise of the value of such a standard from the public health standpoint, given the manpower and laboratory resources required to exercise appropriate surveillance over all bathing areas. This question has many facets. First, pollutionists tend to forget that when a pathogenic micro-organism is present in sewage, this indicates that the corresponding disease is already circulating in the population. The public health worker will consider the epidemiology of communicable disease in this wider context. To take a classic water-borne disease, a bathing water standard in the United Kingdom that claimed to protect bathers from contracting typhoid fever would be quite out of tune with the realities of the epidemiology of this disease in the United Kingdom at the present time. The annual incidence is about a hundred cases, of which two-thirds to three-quarters are contracted abroad. Again, salmonellae are often present in sea water but the prevention of salmonella infection in man is overwhelmingly a matter of surveillance of foods rather than of environmental measures. It would be wrong for a public health authority to concern itself with the risk of acquiring a salmonella infection during bathing to the detriment of active campaigns to improve catering standards and the quality of the various foods that are known to be liable to salmonella contamination.

More important perhaps is the broad question whether there may not be alternative philosophies that do in fact safeguard the bather without recourse to the imposition of microbial standards. In the United Kingdom, there has been general acceptance of the Public Health Laboratory Service committee's conclusion that the health risk of bathing in sewage-contaminated sea water is trivial even on aesthetically unsatisfactory beaches. There is also general agreement that in an age of increasing recreational opportunities public beaches must be made aesthetically above reproach. As mentioned above, the committee argued against the imposition of microbial standards. The practical philosophy that has emerged from this field of debate is that new sewerage schemes should be so planned as to ensure that nearby beaches do not become polluted. The stimulus to better disposal schemes is here the spur of public interest rather than action based on an unreliable microbiological parameter that is itself unrelated to health risks.

This subject may also be viewed from another standpoint. As mentioned in the introductory paragraph of this paper, the public health bacteriologist is familiar with the demand for standards in a number of other fields, e.g. in food hygiene or in the preparation of safe pharmaceutical products. In both fields, microbiologists stress that process control is more important than product control. It is quite impracticable, for example, to test all bottles of intravenous fluids turned out by a given manufacturer or to devise any realistic sampling schedule that can be guaranteed to detect a small proportion of bottles that are not sterile. What can, however, be done is to devise a process with built-in safety precautions that can virtually guarantee that a sterile product will result. The analogous approach to problems of environmental pollution is that for bathing areas control should be exercised over the siting and other details of sewerage schemes for sewage disposal into the sea. If these are satisfactory, the microbiology of bathing areas need cause no concern. Similarly, factory effluents are a more important focus for attention in preventing river pollution than too great an emphasis on chemical examinations of river water for traces of various toxic compounds.

In summary, the case against bathing water standards argued in this paper is (1) that bathing beaches cannot be satisfactorily graded in terms of coliform or faecal coli counts, (2) that no good evidence of a quantitative relationship between coli counts and health risks is yet available, and (3) that the allocation of resources needed to check conformity to an imposed standard cannot be justified on public health grounds.

REFERENCES

1. Winslow, C. E. A., and Moxon, D. Bacterial pollution of bathing beach waters in New Haven Harbor. *Am. J. Hyg.*, 1928, **8**, 299.
2. Garber, W. F. Bacteriological standards for bathing waters. *Sewage ind. Wastes*, 1956, **28**, 795.
3. World Health Organization. Health criteria for the quality of recreational waters with special reference to coastal waters and beaches. WHO/W.Poll/72.10, Geneva, 1972, *a* Annex II, p. 22, *b* p. 5.
4. Public Health Laboratory Service. Sewage contamination of coastal bathing waters in England and Wales. *J. Hyg., Camb.*, 1959, **57**, 435.
5. Gameson, A. L. H., Bufton, A. W. J., and Gould, D. J. Studies of the coastal distribution of coliform bacteria in the vicinity of a sea outfall. *Wat. Pollut. Control*, 1967, **66**, 501.
6. Gameson, A. L. H. Experiences on the British coast. In 'Marine pollution and marine waste disposal' (ed. E. A. Pearson and E. d. F. Frangipane). Pergamon Press, 1975 (*in press*).
7. Hill, A. B. The environment and disease: association or causation? *Proc. R. Soc. Med.*, 1965, **58**, 295.
8. Moore, B. The present status of diseases connected with marine pollution. *Revue int. Océanogr. méd.*, 1970, **18–19**, 193.
9. Stevenson, A. H. Studies of bathing water quality and health. *Am. J. publ. Hlth*, 1953, **43**, 529.
10. Melnick, J. L. Poliomyelitis virus in urban sewage in epidemic and non-epidemic times. *Am. J. Hyg.*, 1947, **45**, 240.

Dr. Moore introduced his paper with a definition of the word 'standard' as a definite level of excellence viewed as the measure of what was adequate for some purpose. His arguments against microbial standards for bathing beaches were (1) that they were, as so far proposed, anything but definite with regard to categorizing bathing beaches into the good, the bad, or the indifferent, and (2) that they were inadequate for their avowed purpose of improving the public health, because the existence of a significant public health problem in this area had not been demonstrated, and the minimal risk entailed had certainly not been satisfactorily collated with microbial counts.

Taking as a basis a proposed American classification shown in Table 1 of his paper, he showed that individual beaches studied in detail on the south coast of England spanned all four grades of the American classification according to which batch of samples one relied on for this grading.

As regards the health aspects, he had suggested in his paper that three questions needed answers. First, how did one spot diseases caused by bathing?—this was a question of epidemiological method. Secondly, was there good reason for believing that bathing in sewage-contaminated water was a significant cause of disease? Thirdly, where a health-related microbial standard had been proposed, was it based on satisfactory evidence? The literature amounted to a record of a very moderate tally of enteric fever cases in the past half-century. For this reason, prospective studies of minor illness had been made, notably by American workers; the Cincinnati studies in this field had been criticized at some length in his paper as being unsatisfactory.

In his final comments, Dr. Moore tried to put the bathing water problem into the wider context of public health in the field of communicable disease as a whole.

GENERAL DISCUSSION ON PAPERS 9–11

Mr. R. G. Ludwig (Brazil) said he had followed with considerable interest the Israeli work on virus disappearance and the status of coliform standards for bathing beaches, and he wished to raise some questions he had asked when that work had been presented[11] at the end of 1973. Prof. Shuval had said in his paper that they had shown coliforms died away rapidly in the Mediterranean coastal waters of Israel, with a T_{90} value of 1 h or even less, while polioviruses under similar conditions had a T_{90} value of 24 h or more. In Mr. Ludwig's opinion there was no evidence for the claim that the latter value was applicable in the sea. Thus, whereas the value for coliforms had been determined from *in situ* studies, it appeared that that for viruses had been found in the laboratory—but laboratory investigations of disappearance rates bore little resemblance to rates found from *in situ* studies and coliform values of 24 h or more had been found in the laboratory.

He believed that although microbial standards served a useful purpose for guiding the design and monitoring the operation of sea disposal systems, they could not be justified on the basis of disease transmission. Experience had fully proved that sewage-polluted drinking water and contaminated milk represented great health risks, and there was ample evidence to support rigid microbial standards for shellfish-harvesting areas, but there was no such evidence regarding the risk of bathing in heavily polluted sea water.

Mr. Ludwig then referred to the situation at Rio de Janeiro, where he believed the population made more recreational use of the beaches than anywhere else in the world. None of the beaches met the Californian bathing water standards, yet his colleague Dr. Fausto P. Guimarães (after many years of study) had advised him that there was no evidence to indicate disease transmission of any type resulting from this recreational use. During Sundays and holidays in the summer season more than 30 000 persons swam in the grossly polluted waters at one of these beaches, at Ramos, where coliform MPN values of 10^6–10^7 per 100 ml had been recorded; according to Prof. Shuval's estimate there should be 3000 to 30 000 persons becoming sick at Ramos beach on a summer Sunday.

It had been because of this total lack of evidence concerning disease transmission from bathing in polluted waters that the well-known Californian standard of 1000 coliforms per 100 ml had been developed essentially on the basis that below this level the aesthetic condition of the beaches remained satisfactory with no evidence of sewage pollution. From long-term studies it had been concluded[12] that a reasonable standard for the beaches at Rio de Janeiro was 10 000/ 100 ml, based on similar observations of aesthetic acceptability. One might criticize coliform standards as being unscientific, but they had served their purpose in the planning and design of submarine outfall systems for bathing-beach protection over several decades. He knew of no system designed on the basis of proper coliform standards that had not resulted in an aesthetically acceptable beach, and of parallel importance was the fact that the use of such standards resulted in designs of reasonable cost.

Replying to Mr. Ludwig, Prof. Shuval agreed that the *in situ* T_{90} values of about 1 h for coliforms in the Mediterranean were not strictly comparable with the laboratory values of 24–48 h for poliovirus. Nevertheless, similar results had been obtained from die-away experiments in the sea, using dialysis bags. Even so, he still felt that the definitive answer would be provided only by *in situ* virus die-away studies. As to the situation in Brazil, he thought it would be valuable if the results of such a study were to be published. However, he doubted if the study included tourists, who were the people most likely to become infected in such a situation.

Mr. Ludwig then added that in the population of about 5 million people in the Rio de Janeiro area there were currently less than 60 reported cases of typhoid per year, and that while both population and pollution had been steadily increasing the number of cases continued to fall. There was, however, evidence of a slight increase in hepatitis cases, with 500–600 being reported per year, but it was known that there were other contributory factors such as contaminated vegetables. All attempts to correlate cases of disease with the effects of swimming in the most heavily polluted waters had given negative results. Regarding tourist exposure, most of the tourists came from Argentina and other Latin American countries where conditions were similar to those of Brazil, and they fortunately frequented the less polluted beaches.

Dr. D. D. Mara (U.K.) said that there would be no risk to health by bathing in sewage diluted with 10^x parts of sea water if x were sufficiently large. How large x should be had not been resolved (though Prof. Shuval believed it should be at least 3), and it was not simple to make accurate routine measurements of dilution. The best answer to what could be done was to do what Dr. Moore had done—undertake an epidemiological survey. If the survey showed there was no significant excess of disease among bathers in the area under consideration, then there was no problem and no need for microbial standards. Such was the case in England and Wales (as Dr. Moore had shown), but it did not necessarily follow that the same situation existed in, for instance, the Mediterranean where times of immersion tended to be much longer. However, if the survey showed there was a significant health risk to bathers, the responsible authorities had a duty to see that the risk was removed; closing beaches was generally impracticable, and the proper solution was to change either the character of the sewage discharged (by treatment) or its location (by outfall siting). It was then necessary to demonstrate that the alteration had been successful. A further epidemiological survey would provide information on whether the alterations had been sufficient, but would not show if the degree of improvement had been unnecessarily high (and this knowledge was of no small economic importance). Follow-up studies of the original survey could be used to correlate the degree of risk with some parameter of water quality; a microbial parameter was appropriate, and most authorities would use the coliform count—at least initially and despite its shortcomings. Such studies might show that incidence of excess disease among bathers in a particular region was associated with a coliform count exceeding a particular value which might vary from place to place. To quantify this relationship was one of the major challenges facing engineers and epidemiologists concerned with coastal pollution.

It was in the absence of numerical knowledge that the hypothetical calculations such as those presented by Prof. Shuval and previously by Coetzee[13] (who had shown that a person swimming daily for 30 days in South African coastal waters had a 1 in 28·5 chance of contracting typhoid) were so valuable. The standards resulting from Prof. Shuval's calculations were orders of magnitude, and thus escaped the ridicule which Dr. Moore rightly levelled at the Maine standards. If a standard were set in terms of the 90-percentile count, then a good designer would automatically aim for a median count an order of magnitude lower.

In reply, Dr. Moore said he had no quarrel with the logic of Dr. Mara's approach. He agreed that the results of the epidemiological surveys done in England and Wales[4] certainly did not necessarily apply to Mediterranean countries: the onus was on those countries to produce satisfactory epidemiological evidence, or otherwise to stop claiming that bathing on their beaches was a danger to health. So far as he personally was concerned, Dr. Moore would be very much more worried about eating raw salads or drinking contaminated water on a Mediterranean holiday than about bathing hazards. With regard to the situation if a survey were to show a significant health risk to bathers, this did not necessarily indicate the need for a radical change in the methods of sewage disposal into the sea. The health risk of contracting a particular communicable disease should be looked at against the broader epidemiological background of the disease in the community as a whole, and blocking the bathing route might not be the most effective way of controlling the disease.

Dr. Mara had approved of the hypothetical calculations of Shuval and Coetzee indicating the magnitude of the health risk from bathing in the absence of definite epidemiological evidence. On this, Dr. Moore had a general and a specific comment. The general one was that, with the present problems of hunger and poverty in the world, health priorities were only too clear already. He could see little justification for an attempt to establish a health hazard on hypothetical grounds when there was no factual evidence at the present time that the hazard existed. His specific comment referred to the paper that Dr. Mara had cited by Coetzee[13] which was written to discredit the Public Health Laboratory Service Committee's report[4] and even included a hint that adverse comments on this report were being systematically sought in the U.K. The calculation cited by Dr. Mara purported to be based on one which Coetzee claimed had been made by Kehr and Butterfield[14] to the effect that the chance of a daily swimmer contracting typhoid fever during a 90-day bathing season in a population with a typhoid morbidity of 4 per 100 000 was 1 in 950. This was quite simply untrue—a discussion of swimming in general, and this supposed calculation in particular, was not to be found in Kehr and Butterfield's paper.

Prof. Shuval said he could not agree with Dr. Moore's policy of waiting until an outbreak of disease occurred at Mediterranean beaches before setting standards. Public health policy had traditionally aimed at reducing potential risks to a minimum even if no outbreak had occurred. In many areas of public health there were standards for toxic chemicals in water, food, and air which were based on laboratory evidence alone. In other cases, standards were set only after accidents had occurred. It would be wrong to wait for outbreaks of disease at polluted beaches when there was already enough information on which to base a tentative standard from sound laboratory evidence coupled with legitimate aesthetic considerations. He did not know when or where would occur the outbreaks that Dr. Moore suggested waiting for, but he felt that they would indeed occur if preventaive action was not taken on the knowledge already available.

Mr. J. A. Wakefield (U.K.) took the opportunity of outlining the activities of the Coastal Anti-Pollution League (of which he was Chairman), and asked Dr. Moore if he was satisfied with the situation in the U.K. where 85 per cent of the outfalls discharged less than 100 m beyond the low-water mark[15] and, if not, what method (other than establishing a standard) he could suggest for correcting this unsatisfactory situation. Dr. Moore agreed that some outfalls were unsatisfactory and needed improvement, but said that this was an amenity issue and not one of health. Prof. Shuval took up this last point and disputed the existence of a strict dividing line between aesthetic considerations and potential health hazards when talking about faecal contamination of bathing beaches. In his view, a public health official had to take account of both when deciding public policy. Bathing beach standards could certainly be justified on aesthetic grounds alone, and he doubted if anyone would find it acceptable to take a bath in a tub with a few cupfuls of sewage in it—even if a very reputable public committee assured him that it had, to date, found no evidence that this would lead to sickness. He could not agree that there was absolutely no risk in bathing in such water, since it was not known what pathogens might be present and because some small amount of water would be ingested. Policy should therefore be based on aesthetic factors in addition to the small, but nevertheless possible, risk to health.

Prof. Mosley said that there might well be reasons for establishing water quality standards for bathing beaches that had nothing to do with health. Those who argued for improving the aesthetics of water quality, however were reluctant to cease admonishing their opponents about health effects of contamination, but it was time to put aside the question of health if an adverse effect could not be substantiated. So far as the aesthetic desirability of controlling contamination of bathing beaches was concerned, he assumed that the basis for judgement was essentially personal. Despite his scepticism about a demonstrable injury to health, he would still prefer not being immersed in polluted sea water, and he imagined that most other people had a similar preference. The question for debate then became one of cost. Unfortunately, it was likely that the use of existing standards would require an enormous expenditure of resources, so that the aesthetic issue had to be restated in terms of how much money was available to pay for how much of an improvement in water quality.

Dr. L. J. Guarraia (U.S.A.) thought that microbial water quality standards served at least three basic functions: protection of human health, protection of the environmental microbial ecology, and provision of a uniform water quality for a mobile population using water for recreational purposes. He invited Dr. Moore's comments on these aspects in relation to his statement for the lack of need for microbial standards. Dr. Moore replied that there might well be a case for standards on grounds apart from health; ecologists and others should then try to formulate standards appropriate to their particular requirements. He had been concerned only with the health aspects and with the mathematical absurdities of some of the microbial standards currently in use. Prof. Y. Kott (Israel) added that there was a need to establish standards of aesthetic acceptability.

Prof. R. R. Colwell (U.S.A.) took up the reference to ecological standards and said it was imperative to recognize that there was a characteristic natural microbial flora of sea water, shellfish, fish, marine plants, etc., just as there was in man. The influx of sewage bacteria, with all its effects on the natural flora, should be measured to assess the stress imposed on the environment. Hence, here was a standard for the environmental impact. The coliform index served a useful purpose since it was generally accepted that faecal coliforms were part of the normal flora of neither the ocean waters and sediments nor of the inhabitants thereof. Changes in the natural flora might well contribute to chronic or sub-lethal effects—such as interference with set of shellfish larvae[16]—rendering the biota more susceptible to disease or incidence of new diseases or pathological states, as well as affecting the normal assimilation–dissimilation activities of micro-organisms in their role of cycling matter.

Dr. V. J. Cabelli (U.S.A.) remarked that the points of view of the three Authors were not really as divergent as it would seem. When faced with the design of a sea outfall they would invariably consider health effects and effluent standards in developing the design and operating criteria.

Mr. B. Grover (World Bank) said that the World Bank was frequently asked for financial assistance in pollution-control projects for developing countries where receiving-water standards did not exist. The European or North American standards were inappropriate in such countries with their warmer climates, lower levels of public health, and greater scarcity of capital for expenditure in the public sector. While accepting Dr. Moore's contention that microbial standards for bathing waters had not yet been justified on grounds of public health, he shared Prof. Shuval's view that performance criteria were nevertheless desirable—though Dr. Moore seemed to have concurred indirectly by alluding to project planning stimulated by the spur of public interest to ensure that nearby beaches did not become polluted. Mr. Grover thought the main problem was how best to formulate rational bathing-water criteria (for regions where shellfish criteria did not apply) which took due account of the local environment, public health, aesthetics, and economics: the promotion of standards which resulted in total project costs exceeding total project benefits (however difficult these might be to measure) was to be deplored. Until there were better quantification techniques, reliance should be put on the best available professional judgement in the light of this basic economic concept. Dr. Moore commented that until epidemiological evidence of a health hazard was obtained, aesthetic considerations were probably the best yardstick: beaches should be made visually attractive and should have no visible evidence of sewage contamination.

Prof. Mosley said that with regard to the potential results of dumping untreated sewage into natural waters he found it useful to compare the engineer with the theologian. The theologian looked at the amount of sin in the world and became convinced it would bring down upon mankind a variety of plagues. The engineer looked at the amount of sewage pollution permitted and became convinced that it must be producing disease. Without question, both sin and pollution could have health consequences. Sin often seemed to go unpunished (at least in this world) and it was quite possible that the same was true for pollution. Finally, even if sin and pollution coincided with an excess of disease in the community, the scientist would still prefer direct to indirect evidence of causal relationships.

On the subject of causal relationships, Dr. Moore said he was sceptical of the validity of Plotkin and Katz's evidence[17] to the effect that one poliovirus particle could cause disease. In vaccination campaigns using the kind of attenuated virus on which the statement was based, 10^5–10^6 times this amount was given to each person immunized. A more detailed discussion of this point would take account of the shape of the dose–response curve. Prof. G. Berg (U.S.A.) agreed that there were insufficient data available to answer several critical questions. However, there had been sufficient independent demonstration that different viruses were capable of infecting susceptible hosts when inoculated at the lowest levels measurable in cell cultures which were the usual detection system; there was little likelihood that cross-infection among inoculated individuals could consistently confuse the picture. Prof. Shuval said that the work of Kehr and Butterfield[14] on minimal infective dose might not have been as rigorous as it ought, but there now appeared to be ample evidence that a single tissue-culture infective dose was sufficient to infect a percentage of exposed humans. This had been confirmed by several people subsequent to Plotkin's original work. He felt there were no grounds for complacency in situations where individuals might ingest only a few viruses in drinking water, or, for that matter, in bathing water.

Mr. E. W. Flaxman (U.K.) said that on reading the papers by Prof. Shuval and Dr. Moore he had been struck by the complete conflict between the figures on which their arguments were based: Dr. Moore's earlier findings[4] appeared to show with reasonable certainty that the risks to health were insignificant in the U.K., whereas Prof. Shuval's numerical example seemed to show that there ought to be a detectable incidence of disease if people bathed in grossly polluted sea water. However, doubts had been raised by several speakers as to the validity of conclusions on the minimum infective doses for typhoid and viruses, and Prof. Shuval's numerical example depended entirely on his assumptions of minimum infective dose. Before this matter could be resolved, it seemed essential that further evidence be obtained.

Mr. F. M. W. King (U.K.) said that he was a civil engineer but had listened intently to the debate among the scientists on the question of standards. In his view no case had been made for any kind of microbial formula governing discharges of sewage from sea outfalls. He was glad that Dr. Moore (whose work he had studied with keen interest for nearly 30 years) had not been moved by the weight of international argument. Speaker after speaker had asked for criteria to work on from a microbial viewpoint, but Mr. King felt this to be unnecessary. He saw in the published work of people like Rawn and Palmer, Abraham, Gameson, and Agg on each of the major factors—such as initial dilution, T_{90}, and subsequent dispersion—the essential design criteria that engineers required. The adoption of fundamental 'rules' of this sort would provide adequate bacteriological safeguards and take care of aesthetic considerations at the same time. There was an economic limit to what could and should reasonably be done.

Mr. D. G. Davis (U.K.) pointed out that the application of legally enforceable water-quality standards in the U.K. invariably applied to the discharge and not to the receiving water, but the discussion had centred on standards applicable to the latter; he had serious doubts on the practicability of enforcing such standards by the usual legal procedures. He asked if it were out of the question to control bacterial contamination by treating the discharge with chemical sterilizing agents, confining such treatment to bathing beaches known to be seriously contaminated by sewage [see pp. 155–163—*Ed.*]. Several references had been made to the possible necessity of closing beaches if a standard were exceeded, and he asked Dr. Moore if he was aware of any beaches in the U.K. ever having been closed on the grounds of hazard to health. Dr. Moore replied that he knew of no such beach that had been closed to bathers because of pollution.

Dr. G. E. B. Kullenberg (Denmark) said that the main advantage of a sea outfall was that with proper design, and by using the physical conditions at a well chosen site, high initial dilution and good dispersion were obtained, so that sewage treatment could be minimized.

Mr. B. Sterregaard (Denmark) remarked that while treated sewage might be aesthetically satisfactory it still had a high content of micro-organisms. He asked if it was acceptable for such an effluent to be discharged right onto a bathing beach, or should some minimum dilution at the beach be required.

Mr. T. J. Leonard (U.K.) emphasized that there was as yet no absolute knowledge of the threshold of microbial density which was hazardous to health, that an acceptable amenity standard was distinctly subjective and would vary from one community to another, and that ecological considerations would also vary. It should be accepted that it was the duty of scientists to advance their particular knowledge to which the engineers (such as himself) should look, and that it required informed professional investigation in each case to advise the Authority concerned with a sea discharge. It was then ultimately for that Authority to take the politico-economic decision, as Mrs. Jeger had said.

Mr. J. T. Calvert (U.K.) said the application of a microbial standard for a bathing beach presented almost insuperable problems. The assessment of beach conditions depended on wind, currents, microbial mortality, and atmospheric conditions, with the result that mathematical calculation became almost impossible.

Dr. J. J. Barnard (South Africa) said that a health authority should implement standards set for recreational waters. If it was not possible to meet the standard the beaches must be closed—with serious economic consequences. He questioned the relevance of epidemiological studies, since polluted sea water was but one vehicle of infection by pathogens. The whole environment contributed millions of these organisms. However, there was a case for establishing standards: if the quality of sewage effluent discharged to the sea were controlled, the responsible contributor might be persuaded to reclaim the effluent for irrigation or industrial re-use rather than discharge it into the sea.

Dr. M. Waldichuk (Canada) referred to conditions in Canada where attempts to maintain microbiologically clean waters, based on bathing requirements, had been abandoned long ago. Nevertheless, there was a sound basis for a stringent standard on coliform counts in shellfish areas. In situations where clean beaches were wanted for recreational purposes, the requirements were thus simplified if there was an oyster bed nearby. Similarly, pollution-free sea water was required by coastal fish-processing plants. However, it was quite clear from the discussion on the three papers presented, that apart from the microbiological requirements for growing uncontaminated shellfish, there was little case for microbial standards in bathing waters.

Prof. Mosley, ranging over a number of the topics that had been discussed, said that (as stated in his paper) a microbial standard was generally used as a mechanism for public health surveillance. It was helpful if, and only if, it alerted public health workers that measures were needed because contamination had reached a point at which undesirable effects on health became likely. So far as contamination of the sea was concerned, there was a way (admittedly not an easy one) of deciding whether any particular microbial standard was appropriate when applied to shellfish. Prof. Metcalf[18] had demonstrated that bacteriologically contaminated growing-waters were likely to produce virologically contaminated oysters. Unfortunately, he had also shown that virologically contaminated shellfish could come from growing-waters apparently satisfactory at the time of sampling.

However, in relation to swimming, it was still not possible to identify any increase in human disease occurring in conjunction with an increase in pollution along bathing beaches. This was not to say that such increases did not, or could not, occur—only that they had not been sufficiently large or obvious for them to be identified. There was therefore no public health basis for deciding on any particular microbial standard.

Adverse consequences upon health were not the only reason for concluding that sewage contamination of beaches was undesirable. There could also be ecological or aesthetic reasons, but it was questionable whether a microbial standard designed to protect health was the most suitable one for these purposes. From the aesthetic point of view, particle size and water turbidity would seem more appropriate parameters.

Prof. Shuval, in his closing remarks, said he would like to suggest that Dr. Moore had, in his paper, advocated the need for standards since he had said that new sewerage schemes should be planned so as to ensure that nearby beaches did not become polluted. This was, in essence, calling for a set of engineering criteria to achieve an effective reduction of faecal matter at bathing beaches, and in that they were in complete agreement. It was not important whether the criterion was a microbial standard or a physical dilution standard—they would both achieve the same ends.

ADDITIONAL REFERENCES

11. KATZENELSON, E., and SHUVAL, H. I. Viral pollution considerations in marine waste disposal. In 'Marine pollution and marine waste disposal' (ed. E. A. Pearson and E. d. F. Frangipane). Pergamon Press, 1975 (*in press*).

12. KOBLITZ, A. V. Sugesto para criterio bacteriologico das aguas de recreaco. *VI Congresso Brasileiro de Engenharia Sanitaria, São Paulo, Brazil, January 1971.*

13. COETZEE, O. J. Comments on sewage contamination of coastal bathing waters. *S. Afr. med. J.*, 1961, **35**, 261.

14. KEHR, R. W., and BUTTERFIELD, C. T. Notes on the relationship between coliform and enteric pathogens. *Publ. Hlth Rep., Wash.*, 1943, **58**, 589–607.

15. DEPARTMENT OF THE ENVIRONMENT and WELSH OFFICE. Report of a survey of the discharges of foul sewage to the coastal waters of England and Wales. H.M. Stationery Office, London, 1973, p. 3.

16. TUBIASH, H. S., COLWELL, R. R., and SAKAZAKI, R. Marine vibrios associated with bacillary necrosis, a disease of larval and juvenile bivalve mollusks. *J. Bact.*, 1970, **103**, 272–273.

17. PLOTKIN, S. A., and KATZ, M. Minimal infective doses of viruses for man by the oral route. In 'Transmission of viruses by the water route' (ed. G. Berg). Wiley-Interscience, New York, 1967, pp. 151–166.

18. METCALF, T. G. Evaluation of shellfish sanitary quality by indicators of sewage pollution. This volume, pp. 75–84.

Paper No. **12**

Heavy Metals in the Marine Environment

A. Jernelöv

Swedish Water and Air Pollution Research Laboratory, Stockholm

Some of the heavy metals, especially mercury and lead, undoubtedly belong to those that have been most extensively studied as marine pollutants. Despite this it is evident, when making an attempt to summarize present knowledge in the field, that existing data form scattered bits and pieces and that the assumptions and hypotheses required to connect and bring the pieces together are at best unverified. Frequently the data contradict each other and little objective information exists that can be used to confirm or refute the results of other investigations.

In the present paper few individual figures for metal concentrations are quoted, but some examples are given to indicate typical values or ranges, or to provide bases for mass-balance calculations. Naturally attempts are made to identify and find patterns in distribution and occurrence.

The term 'heavy metals' has been used in a very wide (in fact in a non-chemical) sense, and includes metalloids. The information is presented as case studies.

Mercury

The first incident that drew attention to mercury as an environmental pollutant occurred in the coastal marine environment of Minamata Bay, Japan. Although this tragic event had the positive consequence that we became aware of the hazards that could be connected with mercury pollution, we now know that from most points of view it was not typical of the marine mercury contamination problems.

The Minamata case was very localized, methyl mercury being discharged directly to a river and the bay. Around the coasts of the world we have many areas with local mercury pollution, but in most cases mercury is or has been discharged in inorganic form or as organo-mercurials other than methyl mercury. This means that before the mercury becomes a hazardous contaminant in fish or shellfish, it has to undergo methylation. Compared to bio-accumulation, biological methylation is a slow process. Some conditions do exist, however, under which the rate of biological methylation of mercury is considerably enhanced, and they will be discussed below.

For the open ocean the mercury problem is quite different. No local sources can affect the mercury levels in organisms in the open sea. The mercury content in them results from natural processes or man-affected changes of global significance.

Of the total mercury present in an aquatic ecosystem, only a small proportion is generally found in the methylated form. This fraction, however, is the important one from a fish-contamination point of view, as the biological half-life of methyl mercury in fish can be measured in years, while that for inorganic mercury is a matter of weeks. Furthermore, for man as a consumer of marine organisms, methyl mercury is far more dangerous than inorganic mercury. From what has been said above, it is evident that there are two principal ways in which methyl mercury levels in marine organisms, notably fish, can be increased: (1) through an increase in the total amount of mercury, and (2) through an increase in the proportion of the total mercury that exists in the form of methyl mercury, which would be the result of an increased rate of biological methylation of mercury.

Mercury both from man-made and natural sources may reach the ocean in two principal ways: through surface run-off and through the atmosphere. Mercury in fresh water is generally found attached to particles of organic or inorganic origin. The size of these particles can naturally show a wide range, some of them being extremely small. As their mercury-binding ability is more dependent on surface area than on mass, the smaller particles play a more important role than should be expected from their relative weight.

When entering the sea, the formation of soluble tetrachloride complexes results in an extraction of mercury from some of the inorganic particles such as clay, while the bond strengths between most organic substances and mercury allow persistence of the complexes. The net result of this is a relocation of particle-bound mercury into association with organic ligands. Methylation of mercury takes place as a detoxification when the micro-organisms, while consuming organic substances, happen to come into contact with the mercury ions.

When the organic particles settle, predominantly in nearshore areas, the attached mercury is incorporated in the deposits. Should the sediments become anaerobic during the degradation of the organics with the production of sulphide ions, mercury will be bound as sulphide. The result of these processes is that mercury transported to the marine environment by surface run-off will have but a small impact on the open sea. Most of it will be retained in the coastal zone.

The mercury that reaches the upper mixed layer of the open oceans through rivers or through the atmosphere will eventually be accumulated by organisms. Attached to or incorporated in dead organisms or faecal pellets, it will then be transported vertically toward the deeper parts of the oceans. In connection with the final decomposition of the carrier material, the mercury is released again into a soluble form, from which it may either be transported back up to the surface with upwelling currents or incorporated in the authigenic minerals forming on the sea bottom.

In the coastal zone itself, industrial discharges frequently give rise to elevated concentrations of mercury in marine organisms. In some instances, where methyl mercury is discharged as such, or where mercury-rich organic sediments are partially exposed to air as a result of tidal motions, very high levels of methyl mercury ($> 10\ \mu g/g$) may occur in marine organisms.

Volatile mercury, such as its elemental form, or dimethyl mercury, will evaporate into the atmosphere. Industrial sources handling mercury as such, the burning of mercury-containing fuels, or the washing of mercury-containing minerals, will contribute to the atmospheric load. Dimethyl mercury is unstable in the atmosphere. In contact with acids it will be degraded to water-soluble monomethyl mercury and in contact with ultra-violet light to elementary (inorganic) mercury.

The world production of mercury[1] is in the order of 10^{10} g/year. To losses from the subsequent usage should be added direct release of mercury in connection with burning of fossil fuels, heating of mercury-containing minerals, and so on. At the very most, these sources could account for a release of mercury of the same order of magnitude as the 'intentional' production[1].

Estimates of mercury transports[2] to the sea indicate that the rivers may contribute about 10^{10} g/year while the washout with precipitation from the atmosphere could be $(2{\cdot}5\text{–}15) \times 10^{10}$ g/year. The last figure is based on an estimated total atmospheric load of 4×10^9 g, a close to complete washout in connection with precipitation, and an average time between rain of 10 days. The estimate of the atmospheric burden of mercury has been questioned and it has been suggested that it represents a tenfold or greater overestimate. The alternative figure ($2{\cdot}5 \times 10^{10}$ g/year) is arrived at by multiplying the amount of rain falling over the oceans by a conservative estimate of the average mercury concentration in precipitation of $0{\cdot}06\ \mu g/l$. derived from mercury content in apparently uncontaminated glacial ice[2]. Accepting the concept that $1{\cdot}5 \times 10^{11}$ g/year represents an overestimate, a compromise of 5×10^{10} g/year may be a reasonable estimate for calculation purposes.

Analyses of the Greenland ice sheet have demonstrated a tendency to an increase in mercury concentrations during the last decades. When the estimated atmospheric flux of mercury is related to the human production—of which not all is released to the atmosphere—it is apparent that the observed increase, perhaps with a factor of two, that seems to have occurred during the last decades cannot be accounted for by the mercury released from mercury-handling industries, fossil-fuel burning, and so on.

The 'natural' flux of mercury—'degassing from the earth crust'—appears to be of dominating importance. Weiss *et al.*[3] have estimated it to be in the order of $(2{\cdot}5\text{–}44) \times 10^{10}$ g/year. It is possible that man, through his activities, has affected this degassing during the last decades. This enhancement of the process may have occurred through the changed agricultural practice involving more effective ploughing, the use of fertilizers, and so forth. Should this be the case, it is possible that the degassing involves a biological dimethylation of mercury forming a volatile mercury compound which, after introduction to the atmosphere, will be degraded as described above. If that is the case, the mercury case study would have provided us with another example of ecological interaction of a hard-predicted type.

Recently, however, Carr and Wilkniss[4] have presented new data questioning whether the mercury content of Greenland glacial ice has increased at all drastically in recent years. They suggest that the observed pattern in mercury concentration could rather be due to an inhomogeneous mercury distribution through the ice sheet.

Surface water in the oceans has been found to contain $\sim 0{\cdot}13$ ng/g of mercury while the few available analyses from deep water from both the Atlantic and Pacific Oceans indicate higher figures ($\sim 0{\cdot}30$ ng/g). In their classic work from the 1930's, Stock and Cucuel[5] reported concentrations in sea surface water of 0·03 to 0·04 ng/g. Several later reports have also indicated similar figures. The total amount of mercury in sea water would thus amount to $\sim 2{\cdot}1 \times 10^{14}$ g. Evidently, man has not so far been able to affect this amount measurably.

Marine plankton frequently contain mercury levels of 2–10 ng/g, most of which is in the inorganic form, while fish concentrations vary considerably from 0·01 to 2 μg/g with a substantial fraction of the mercury in the form of methyl mercury. The (methyl) mercury in the fish is bound to the protein and is not dissolved in the fat as are the chlorinated hydrocarbons. Unlike them, the mercury content in fish is consequently not very fat-dependent but much more related to trophic level, size, and age of the fish. The highest mercury content is found in large predators like swordfish and tuna. The fact that many of these fish have mercury levels above the permissible levels in North America (0·5 μg/g) has during the last years had a clearly adverse effect on the fishing industry.

Analyses of museum specimens of tuna and swordfish have not indicated any notable change in mercury content during the last 150 years. A surprising finding during these analyses was the large variation in mercury content of these fishes, in both the old specimens and the recent ones, in view of the common understanding among ecologists that these species should be among the best integrators over large areas that we have. If the existing data are closely examined to find a pattern in these variations, it appears as if the tuna caught in areas with upwelling currents—the most productive fishing areas—tended to have a higher mercury level than those from other parts of the oceans. A possible explanation of this could be that the higher mercury content in the deep water in connection with the high (micro) biological activity in these super-productive zones results in a higher rate of biological methylation of mercury. The phenomenon could thus be a natural one. Some marine mammals, notably seals, have been found to contain very high amounts of mercury, especially in the liver. Despite the fact that most of the mercury they consume in fish is in the form of methyl mercury, the liver contains predominantly inorganic mercury. It has been suggested that seals may have an enzyme that breaks the covalent bonds between the mercury and carbon atoms in methyl mercury.

From laboratory experiments with methyl mercury chloride or other salts, effects on photosynthesis of algae and on behaviour of fresh-water organisms have been reported at very low concentrations of methyl mercury. Acute toxicity on fish has been found to occur at methyl mercury concentrations in the tissues of ~ 20 μg/g. Except for a few incidents such as Minamata Bay, however, there are no field reports from the marine environment describing effects of existing mercury levels on the organisms. The mercury problem is commonly considered as a problem for the fish consumer, not as a problem for fish production. In some instances, the national standards on permissible levels of mercury in fish—enforced to protect public health—prevent the use of certain fishing resources. Specialized fish eaters—such as the seals mentioned above—may have developed certain protective mechanisms through evolution.

Recently it has been suggested that selenium may have an antagonistic effect to the toxicity of mercury, and that there may be an uptake-regulating process leading to stoichiometric relations between mercury and selenium in marine organisms that render both elements less toxic than if they had occurred independently. So far, this is a matter for further research.

Cadmium

High concentrations of cadmium have been found in marine organisms in some coastal areas. Cadmium concentration in sea water near the coast has been reported to be within the range of 0·05 to 1 ng/g. In the open sea the levels are reported to be from 0·01 to 0·05 ng/g. Probably the true range is much narrower. Analyses of cadmium contents in North Atlantic deep-sea sediment have shown an average concentration of 225 ng/g with somewhat lower levels in areas rich in shell debris.

In fresh water, cadmium has been found to bind strongly to organic substances, especially in chlorophyll-containing plants. In marine water, however, cadmium appears in complexes to a higher degree than mercury does. In comparative studies, about 80 per cent of cadmium has been found to go through a 0·22-μm membrane that will retain 80–90 per cent of the mercury.

The concentration factors from sea water to marine organisms for cadmium are commonly 100–1000. Certain marine mammalians (sea otters) have been reported to contain up to 500 μg/g of cadmium in their kidneys. Corresponding concentrations in man would cause severe kidney damage. The biological effects of cadmium in these sea otters and of the levels found in other marine organisms are unknown.

Lead

The behaviour of lead in the marine environment is to a high degree dependent on the properties of the individual lead species in question. Like most other elements and compounds described as case studies here, lead enters the oceans both through run-off water from land and through the atmosphere. It is believed that in prehistoric times the rivers were the main source of lead for the oceans. The lead thus transported to the sea could be as insoluble forms attached to organic or inorganic particles, as soluble chelates, or as soluble inorganic complexes. Naturally there was some exchange taking place between the soluble and insoluble phases. The particle-bound lead settled in the near-coast areas similar to the way described for mercury, with smaller fractions entering the deeper parts of the oceans. Of the lead entering the oceans in a soluble form, a large part was accumulated in organisms of the photic zone. It is believed that in prehistoric times only about $1{\cdot}8 \times 10^{10}$ g reached the open sea each year. The residence time for lead in the upper mixed layers of the ocean has been estimated to be only a few years.

The smelting of lead and production of alkyl leads from 1750 to 1966 is presented in Table 1. The amounts of lead aerosols produced per year during four years within the time interval of Table 1, namely 1753, 1815, 1933, and 1966, are estimated to have been 2, 4, 10, and 100 $\times$ 10^9 g respectively[7]. A substantial part of the $\sim 1{\cdot}5 \times 10^{11}$ g/year of lead aerosols that are now produced will fall out over the ocean.

TABLE 1. Lead smelted or burned as alkyl lead (10^9 g/year)
From Murozumi *et al.*[6]

Year	Northern hemisphere			Southern hemisphere
	Primary smelting	Secondary smelting	Burned alkyl lead	Primary smelting
1750	60			40
1800	90			50
1860	220			
1880	400			
1890	520			40
1900	750	0		80
1910	940	60		100
1920	880	200	0	110
1930	1200	400	4	170
1940	1300	400	36	230
1950	1300	550	110	240
1960	1900	600	180	360
1966	2400	700	310	350

The present injection of lead to the sea through surface run-off has been estimated to be 5×10^{11} g/year. A very large part of this is estimated to be retained in the coastal zone.

Analyses of lead in the Greenland ice sheet have shown an increase in lead fallout of about 500 times over the last 2800 years. A substantial part of this increase occurred after the introduction of ethyl lead in gasoline. Analyses in sea water support the idea of a recent increase due to fallout. 'Young' surface water contains variable and high concentrations of lead ranging from 0·07 to 0·35 ng/g while 'old' deep sea water contains 0·02 to 0·04 ng/g.

When this picture is compared with that found for chemically similar metals such as barium and radium, which tend to have relatively high concentrations in 'old' deep sea water and relatively low ones in 'young' surface water, the indications of a recent increase due to higher fallout are further strengthened. This is even more likely as the lead distribution is similar to those of radioactive debris from nuclear explosions.

The levels of lead in marine organisms and the biological and ecological effects of the environmental contamination are practically unknown. There are, however, reasons for considerable concern.

Copper

Extremely high concentrations of copper have been found in water, sediment, and biota in some coastal areas. To a very large extent, however, the copper appears to remain in the nearshore area. Studies of long-term changes in copper content of seaweed in a coastal area without any known direct discharge have not revealed any notable change during the last decade.

The suggestion that atmospheric transport of copper to the open ocean should occur to a significant extent has yet to be verified and quantified. There are at present no indications of a food-chain magnification of copper.

Zinc

In connection with substantial local discharge, highly elevated levels of zinc have been found in sediment and biota in some coastal waters. Nothing is known about zinc contamination in the open sea. From heavily contaminated coastal areas, it is known that zinc content generally decreases when going up the food chain.

Selenium

Selenium is found in a variety of marine organisms. Especially in livers of seals and penguins, the reported concentrations are very high ($\sim 100\ \mu g/g$). To what extent this is due to man's impact is unknown. The theory of interdependence of mercury and selenium concentrations in marine organisms is treated in the mercury section.

Arsenic

Reports of high concentrations of arsenic in the biota have generally been confined to coastal areas with known discharge. In the Baltic, the concentrations of arsenic in fish have been found to be well correlated with salinity. Even extremely high local contamination in the northern low-salinity part does not render the arsenic concentration in fish much higher than those commonly found in the oceans.

Some marine invertebrates such as crabs and lobsters have long been known to contain high concentrations of arsenic. As stated[3] already in the 1930's, "the concentrations of arsenic in some marine organisms are so high that if it had been arsenic as As^{3+} one single meal would have been toxic to man. As no effects are observed, it has to occur in the pentavalent form (As^{5+})."

During the 1960's, some further light was thrown on the form of arsenic in marine organisms. In fish oil, arsenic was observed to occur as an organo-arsenic compound. Further investigations showed that marine algae accumulate arsenic, probably mistaking it for phosphates and converting it into organic compounds, one type of which seems to be lipids with phosphorus replaced by arsenic. Some other organo-arsenic compounds have not yet been identified in the form in which they occur in marine biota. After conversion into organic compounds in the algae, arsenic is further accumulated along the food chain.

Vanadium

In connection with burning of some petroleums, large amounts of vanadium are entering the atmosphere. Open-ocean surface water has been found to contain more vanadium than does sub-surface water. This could be due to atmospheric fallout but also to surface contamination from oil spills or to a natural accumulation process in the surface film.

Vanadium is known to have very high accumulation factors from sea water to some marine organisms. The extent of man's impact on vanadium concentrations in marine organisms is unknown and so are the biological and ecological effects of vanadium contamination.

Manganese and Iron

In connection with mining operations, large amounts of iron and manganese are discharged to the marine environment. Compared to the natural flux, however, the amounts released and discharged by man are small. Elevated concentrations of iron and manganese are only found locally in coastal areas.

The natural transport of iron and manganese involves release from minerals on the continents, transport with surface run-off to the oceans, solubilization as complex ions in contact with saline water or precipitation with mineral particles on the continental shelves.

Bio-accumulation provides a mode for vertical transport in dead organisms or faecal pellets. In the bottom water, iron and manganese are released and precipitated as oxides. In this process of precipitation of iron and manganese, heavy metals such as mercury, cadmium, copper, and zinc may be incorporated.

Manganese and iron are mentioned here not because of their biological significance as man-made contaminants, which is believed to be small, but because they demonstrate a special problem of transport and because of their impact on mercury and other heavy metals during the formation of ferromanganese minerals.

Discussion

In the paper the behaviour of some heavy metals and metalloids in the environment, and specially in the marine ecosystem, is discussed. An attempt is made to relate special patterns in their turnover to chemical and physical properties of the individual substances. For the evaluation of hazards connecting environmental contamination to the substance in question, it is concluded that the following environmental characteristics and properties are relevant:

1. Human production and release in relation to natural flux.
2. Persistency in the environment (residence time in waters, sediment or soil, atmosphere, and organisms).
3. Bio-accumulation—
   (a) the relation between fat-solubility and water-solubility may give an indication of passive accumulation,
   (b) if 'active' accumulation is known to occur, the concentration factor should be estimated.
4. Properties affecting the dispersion of the compound (e.g. volatility, tendency to adhere to or form complexes with organic or inorganic particles, tendency to form soluble complexes with chlorides or other abundant ions in sea water, tendency to form oxides or sulphides with low solubility).
5. Toxicity to aquatic organisms.
6. Toxicity to man or other mammals.
7. Biological effects such as selective enhancement or retardation of metabolic activities of certain organisms that may have long-term effects on the balance of the ecosystem.
8. Transformation reactions that may change the properties of the compound; most such reactions performed by organisms increase the water-solubility of the compounds, thereby making it possible to excrete them, but there are some notable exceptions such as the biological formation of methyl mercury and organo-arsenic compounds.

REFERENCES

1. Man's impact on the global environment. Report of the Study of Critical Environmental Problem SCEP. MIT Press, Cambridge, Mass., 1970.
2. Baseline studies of pollutants in the marine environment and research recommendations. International Decade of Ocean Exploration Baseline Conference, New York, May 1972.
3. WEISS, H. V., KOIDE, M., and GOLDBERG, E. D. Mercury in a Greenland ice sheet: evidence of recent input by man. *Science, N.Y.*, 1971, **174**, 692.
4. CARR, R. A., and WILKNISS, P. E. Mercury in the Greenland ice sheet: further data. *Science, N.Y.*, 1973, **181**, 843.
5. STOCK, A., and CUCUEL, F. Die Verbreitung des Quecksilbers. *Naturwissenschaften*, 1934, **22**, 390.
6. MUROZUMI, M., CHOW, T. J., and PATTERSON, C. Chemical concentrations of pollutant lead aerosols, terrestrial dusts and sea salts in Greenland and Antarctic snow strata. *Geochim. cosmochim. Acta*, 1969, **33**, 1247.
7. PATTERSON, C. Lead. In 'Impingement of man on the oceans' (ed. D. W. Hood). Wiley-Interscience, New York, 1971, p. 245.

In the absence of Dr. Jernelöv his paper was read by Mrs. A. L. Martin (Sweden). The paper summarized the known biological effects of heavy metals contaminating ocean water. The total amount of mercury in the hydrosphere was estimated, effects of bio-accumulation of methyl mercury in plankton and fish were described, and possible interdependence between mercury and selenium was indicated. The formation of strong complexes of cadmium in sea water gave total cadmium concentrations of 0·01–0·05 ng/l. in the open sea, but biological transformations and effects were not known. The inflow of lead into the oceans (including formerly clean areas such as the Arctic) had been estimated. Levels of copper and zinc, although locally very high, had not given evidence of accumulation in the food chain. Vanadium showed magnification in the food chain, but no biological effects had been reported. Recent findings suggested that arsenic magnification was due to the resemblance of arsenic oxides to phosphates. Manganese and iron were themselves not toxicologically significant; they incorporated other heavy metals into ferromanganese minerals on precipitation as oxides onto the ocean floor. The paper ended with a list of the environmental characteristics and properties relevant to the evaluation of hazards from heavy metals and metalloids.

DISCUSSION

Dr. M. Waldichuk (Canada) said that much attention had been given to the human health aspects of metals in coastal waters and marine organisms. This was understandable in view of the tragic incidents in Japan with poisoning by mercury and cadmium. However, there were also problems due to effects of metals on aquatic organisms and ecosystems, and he thought more emphasis should perhaps be placed on these problems[8]. He referred particularly to sub-lethal effects of metals on aquatic organisms. Authorities in Canada and the U.S.A. had imposed an administrative maximum permissible mercury concentration of 0·5 mg/l. in fish products, but work in Sweden[9, 10] had shown that fish containing concentrations as low as 0·1 mg/l. in their tissues could not react effectively to an imposed torque in a rotating cylinder. The more important effects, however, were perhaps on the sensitive embryonic stages. Thus it had been shown[11] that exposure of pre-eyed eggs of sockeye salmon and pink salmon to mercury concentrations of 10–20 μg/l. led to teratogenic effects in the alevins and fry; even if the deformed offspring survived to adulthood they would not reproduce. Cadmium, on the other hand, was not apparently toxic to the eggs, but was highly toxic to the fry—especially when they became free-swimming. Recent work[12] had also shown sub-lethal effects of cadmium on eggs and larvae of Pacific herring, giving rise to serious behavioural defects; the cadmium apparently interfered with the calcium metabolism. The osmotic balance of the eggs was modified and their buoyancy altered so that they might sink or rise, with possible population losses if they moved into unfavourable currents. In the larvae, there appeared to be replacement of calcium by cadmium in the otolith, thereby adversely affecting the balance and orientation. There was also evidence of important sub-lethal effects of cadmium on the reproductive capacity of certain fish species[13].

Endocrinological effects of metals in fish were also known. At copper levels as low as 0·1 mg/l. there was a sharp increase in corticosteroids in sockeye salmon[14]—a manifestation of stress in the fish. He thought it would be prudent to examine more closely the endocrinological effects of metals entering the marine environment through man's activities.

Finally, Dr. Waldichuk referred to the effect of mixtures of metals, which related to the real world in terms of sewage sludge and sewage generally. Although some antagonism might occur between such constituents as mercury and selenium (as noted by Dr. Jernelöv in his paper) there was also known synergism between such metals as zinc, copper, and cadmium[15]. The bio-accumulation of these metals further complicated their acute and chronic toxicity. The problems of different sensitivities of different trophic levels, and of the various life stages of individual species in an ecosystem, added to the complexity. It was obvious that this area of interaction of different elements and of numerous organisms required further intensive investigation before waste disposal into the marine environment could be managed adequately.

On the theme of subtle effects of low levels of pollutants, Dr. A. D. McIntyre (U.K.) added that for some years his laboratory had been studying such effects over periods of at least 6 months in simplified three-stage food chains (primary producers, herbivores, and carnivores) in large outdoor tanks. They had shown, for example, that each stage was adversely affected by levels of copper 2 orders of magnitude below the LD_{50} and only 2–3 times as great as the normal background levels in the sea. Although these experiments had been conducted over long periods under semi-natural conditions, it was not clear to what extent the results were applicable to the very complex conditions in the natural environment. Dr. D. Taylor (U.K.) took up this point and remarked that in trying to estimate the chronic or sub-lethal effects of heavy metals it was essential to remember that the test media might not be in any way similar to the natural environment in terms of the ionic species present. In his own laboratory it had been shown that with copper sulphate solutions of about 1 ng/l. (similar to natural sea-water levels) detrimental effects could be produced in the burrowing activities of marine bivalves that patently did not occur in the natural environment. Much more work was needed in the area of ionic speciation if erroneous conclusions were not to be reached in sub-lethal metal-toxicity studies.

Mr. J. W. Gunster (Netherlands) said that, for reasons given in the paper, it had been stated that most of the mercury from surface run-off was retained in the coastal zone. He thought it was an important conclusion that mercury (and possibly other heavy metals) would not be available for influences on the oceanic ecosystems. He asked from which regions Dr. Jernelöv had obtained his data, whether they could be quantified, and if any literature references could be given.

Prof. G. E. Jones (U.S.A.) remarked that the interaction of complexing between organic nutrients and toxic cations in laboratory media affected the expression of toxicity on bacterial cultures. Using pulse polarography at the University of Liverpool, three major peaks were found in a sea-water medium with tryptone and yeast extract. As the concentrations of metals, such as copper, were increased, the three peaks of organo-metallic complexes were saturated with the metal ions. In addition, the inorganic ions in the medium formed speciation with some metal ions. When the complexing of both the organic and inorganic complexes was exceeded, free metal ions immediately caused bacterial toxicity.

The Chairman, Prof. R. R. Colwell (U.S.A.) said that she and her colleagues had shown that mercury mobilization via bacteria involved release of elementary mercury into the water column[16] where it could be methylated via trimethyl tin[17]. It appeared that, in the water–sediment interface, this metabolic pathway was most significant. Furthermore, it had been observed by other workers that even if input of mercury to the environment were stopped, it could be released into the water column from older mercury-laden sediments via methylation, and this could occur months (or even years) after the mercury input[18]. Thus, the role of micro-organisms in the cycling of mercury and probably other heavy metals in the aquatic ecosystem was highly important. Concentration of mercury in higher levels of the food chain, such as in oysters, was enhanced in the presence of mercury-metabolizing bacteria[19].

REPLY TO DISCUSSION

Although Mrs. Martin was able to deal with most of the points raised in the discussion, the formal replies were subsequently provided by Dr. Jernelöv.

He agreed with Drs. Waldichuk and McIntyre that it was most important to study the sub-lethal effects of physiological, biological, genetic, and teratogenic types. A modest start had been made in this direction within the framework of a WHO project on sub-lethal effects of pollutants in marine organisms. The study currently involved institutes around Oslo Fjord and the Wadden Sea, but he hoped it would later include the Firth of Clyde. He also fully agreed with Dr. Taylor's comment, and he gave Mr. Gunster the reference[20] for which he had asked.

Dr. Jernelöv regretted that from Prof. Jones's discussion and paper[21] he failed to understand that the rapid death of non-marine bacteria was attributable primarily to heavy-metal toxicity; he was therefore unable to comment further.

He found Prof. Colwell's theory of biological methylation and further transalkylation of mercury most interesting. However, the data he had seen had not helped him to discriminate between this possible process and that suggested by Wood[22] and Landner[23] with direct methylation of mercury. He added that he was eager to hear further about the progress of her work with methylation and transmethylation of other heavy metals.

ADDITIONAL REFERENCES

8. Waldichuk, M. Some biological concerns in heavy metal pollution. In 'Proc. Symposium on Pollution and the Physiological Ecology of Estuarine and Coastal Water Organisms', University of South Carolina, Georgetown (*in press*).

9. Lindahl, P. E., and Schwanbom, E. A method for the detection and quantitative estimation of sublethal poisoning in living fish. *Oikos*, 1971, **22**, 210–214.

10. Lindahl, P. E., and Schwanbom, E. Rotary-flow technique as a means of detecting sublethal poisoning in fish populations. *Oikos*, 1971, **22**, 354–357.

11. Servizi, J. A., and Gordon, R. W. Effects of selected heavy metals on early life of sockeye and pink salmon. *In preparation.*

12. Alderdice, D. F., and Rosenthal, H. Sub-lethal effects of pollutants on marine fish eggs and larvae. To be presented at 13th Pacific Science Congress, Vancouver, B.C., 1975. *In preparation.*

13. Sangalang, G. B., and O'Halloran, M. J. Cadmium-induced testicular injury and alterations of androgen synthesis in brook trout. *Nature, Lond.*, 1972, **240**, 470–471.

14. Donaldson, E. M., and Dye, H. M. Corticosteroid concentrations in sockeye salmon (*Oncohynchus nerka*) exposed to low concentrations of copper. *J. Fish. Res. Bd Can.* (*submitted*).

15. LaRoche, G. Biological effects of short-term exposures to hazardous materials. In 'Control of hazardous material spills', Environmental Protection Agency, University of Houston, Texas, 1972, pp. 199–206.

16. Nelson, J. D., Blair, W., Brinkman, F. E., Colwell, R. R., and Iverson, W. P. Biodegradation of phenylmercuric acetate by mercury-resistant bacteria. *Appl. Microbiol.*, 1973, **26**, 321–326.

17. Jewett, K. L., and Brinkman, F. E. Paper in *Proc. 167th nat. Meet. Am. chem. Soc., Los Angeles, April 1974.*

18. Proceedings of Conference on Persistent Chemicals in the Environment, Ottawa, 1974. National Research Council of Canada.

19. Colwell, R. R., and Nelson, J. D. EPA final report 1974. Environmental Protection Agency, Washington, D.C. (*in press*).

20. Jernelöv, A. Case study for mercury. In 'Changing chemistry of the oceans', 20th Nobel Symposium, Gothenburg, 1971.

21. Jones, G. E. Heavy metal ions as the principal bactericidal agent in Caribbean sea water. This volume, pp. 199–208.

22. Wood, J. M., Kennedy, S. F., and Rosén, C. G. Synthesis of methyl mercury compounds by extract of a methylogenic bacterium. *Nature, Lond.*, 1968, **220**, 173–174.

23. Landner, L. Biochemical models of the biological methylation of mercury suggested from methylation studies *in vivo Neorospora crassa. Nature, Lond.*, 1971, **230**, 452–453.

Paper No. **13**

Persistent Organic Residues

J. E. Portmann

Ministry of Agriculture, Fisheries and Food, Fisheries Laboratory
Burnham-on-Crouch, England

Urban sewage as discharged to the sea, at least from north European countries, does not generally receive more than primary treatment to remove solid matter. A recent survey of sewage discharges to the North Sea conducted by ICES[1] revealed that less than 1 per cent of the total amount (estimated at 7345 × 10^3 m^3/day) received tertiary treatment, 35 per cent received secondary treatment, and 44 per cent received no treatment at all.

Treatment of sewage in a biological treatment plant is aimed mainly at reducing the organic content of sewage, and simple compounds such as the 'soft' detergents are largely degraded during passage through a sewage works. Other compounds are not degraded at all, or only slightly so. These substances therefore pass out in the liquid effluent from the works or are adsorbed on to the sewage sludge and retained. However, it is common practice to discharge sewage sludge periodically, either by pipe or by dumping vessel, and the net result with regard to persistent chemicals will be much the same whichever method is adopted. For this reason, in this paper the relationship between these substances and whole sewage will not be considered, although comments will be made as appropriate where there is a strong measure of partition between the sludge and liquid phases.

The majority of urban sewerage systems carry both domestic and industrial wastes. The proportion of the latter may be quite small; the ICES survey[1] indicated this to be less than 1 per cent in some areas, whereas in others it may be quite large (e.g. 83 per cent for the Bremen area of Germany). Such trade discharges are likely to introduce substantial quantities of organic chemicals to the sewerage system. Most will probably be readily degradable, a few will not be, and certain substances (although degradable) may create short-term problems before they are decomposed.

Unfortunately the composition of industrial effluents is usually very complex and bears little relation to the raw materials used in the manufacturing process. For this reason, limits imposed on the quality of an effluent accepted into a town sewer are usually of a general nature; for example, total BOD, COD, extractable matter, loss on ignition. However, a few substances, such as PCB's, chlorinated solvents, and phenols, are recognized as being liable to create problems, and in the United Kingdom limits are often set for these substances.

Comparatively few organic chemicals are recognized as being regularly present in sewage, and as this paper deals with those which are persistent, only a few substances will be discussed in detail. The precise meaning of 'persistent' or of other similar terms is one which has caused considerable discussion in international fora recently. For the purpose of this paper, 'persistent' will be taken to include 'hard' substances and others which although degraded to some extent in a treatment works or in the sea may cause problems, albeit in the short term.

Groups of Substances Presenting Special Problems

General

Most of the organic compounds present in sewage are degradable and yield simple materials such as ammonia, phosphate, water, and hydrochloric acid on decomposition. With the possible exception of the nutrients ammonia, nitrate, and phosphate, which will be considered elsewhere in this Symposium, these ultimate degradation products are unlikely to cause any problems in the marine environment. However, in some cases the degradation process may be slow, and intermediate compounds may be more toxic or may be bio-accumulated; some of these are considered below.

Organochlorine pesticides

The presence of organochlorine pesticide residues in sewage has been recognized for some considerable time. Holden and Marsden[2] reported in 1966 that in Scotland sewage effluents of largely domestic origin contained at least 100 ng/l. of dieldrin and rather less than 100 ng/l. of DDT. Where the sewage was treated in admixture with certain trade wastes, particularly those from industries using wool, the dieldrin value was as high as 10 μg/l. and for DDT up to 500 ng/l. Similar levels were reported by Lowden *et al.*[3] in an examination of 20 sewage effluents from England which were found to contain an average of 57 ng/l. of dieldrin and 95 ng/l. of DDT. These early figures may have been slightly high due to interference by PCB's, but values recently reported for a number of marine sewage outfalls in California[4] are (with a single exception, see later) of a similar order, ranging from 20 to 98 ng/l. These outfalls were estimated to carry a total of 49 kg/year of DDT in a total daily flow of 172 $\times$ 10^3 m^3/day. This compares with the estimated annual input of DDT into the North Sea by sewage of 656 kg in a total flow of 7345 $\times$ 10^3 m^3/day. Although this indicates a somewhat higher input of DDT per unit volume of sewage (by a factor of *ca* 3·2) in California compared with north Europe, the pattern of usage of DDT has been much greater in North America than in north Europe.

The source of organochlorine pesticides in sewage of a purely domestic origin is not precisely known, but since these compounds are ubiquitous in both water supplies and food it is probable that at least a proportion enters via the human diet. Additional quantities of DDT are probably added to sewage following its use in the home but the amount so introduced is probably small and, as a result of the gradual prohibition of home use of DDT, is likely to become even smaller. Dieldrin has in the past found extensive use as a moth-proofing agent in woollen garments and carpets; these may well be the source of at least some of the dieldrin reported in domestic sewage. As reported by Holden and Marsden[2], use of dieldrin on trade premises engaged in moth-proofing of woollen goods could introduce very large quantities of dieldrin to a sewage effluent. Similar findings were reported by Croll[5] in a survey of organochlorine pesticide residues in English fresh-water rivers, where by far the highest concentrations were found in rivers passing through wool-processing areas of Yorkshire and in the River Severn. The enormous importance of industrial input is best illustrated by the White Point (California) discharge, which was found to contain between 59 and 97 μg/l. of DDT, i.e. *ca* 1000 times that normally found in sewage effluents in that area; this largely originated from a DDT manufacturing plant.

Organochlorine pesticides such as DDT are slowly degradable, the major routes of decomposition for DDT being via DDE or TDE (DDD). Under the anaerobic conditions in which sludge is treated in a sewage works, DDT is rapidly decomposed and the main decomposition products are likely to be TDE, DDMU, DBP, and DDCN[6,7]. Aerobic degradation routes have also been reported. Under the normal retention time found in a sewage works, the extent of degradation of DDT seems likely to be small. However, Schmidt *et al.*[4] reported that with the exception of the White Point effluent only DDE was found in other Californian sewage effluents analysed.

There is a considerable body of evidence to suggest that organochlorine pesticides are readily adsorbed on to sewage sludge particles and are present in much higher concentrations in the sludge than in the sewage effluent. Schmidt *et al.*[4] reported that the waste water discharged through the Hyperion outfall contained 400 ng/l. of DDT, whereas the sludge contained four times that amount. Work at our own laboratory suggests that typical DDT concentrations in sewage sludge from industrialized urban areas are of the order of 0·1 to 1 mg/l. which, assuming that the earlier figures of Lowden[3] and Holden and Marsden[2] apply, indicate a concentration or enhancement factor of between 40 and 400 relative to sewage effluent.

This concentration of the pesticide residues by sewage-sludge particles is not a unique phenomenon, for there is considerable evidence that organochlorine pesticides are adsorbed on to other particulate matter in the sea and removed from the water column by deposition to the sediments. Work off the Keratsini (Athens) sewage outfall in Greece[8] showed DDT concentrations of up to 1·5 mg/kg in the sediments immediately adjacent to the outfall, with concentrations decreasing with distance from the outfall to less than the detectable level of about 0·01 mg/kg total DDT.

Although studies by the Burnham laboratory in two major sewage-sludge disposal areas, the Thames Estuary and Liverpool Bay[9], have failed to reveal any build-up of organochlorine pesticide residues in these areas, studies elsewhere—for example off California[10]—showed appreciable accumulation of organochlorine residues in the sediments of the dumping areas. The reason for the lack of accumulation of organochlorine residues in the dumping areas off the United Kingdom is undoubtedly linked with the good dispersion characteristics in those areas, the reasons for which are discussed more fully by Wood[11].

The presence of organochlorine pesticide residues, especially DDT, in marine flora and fauna is well documented and there have been many reports in recent years of monitoring studies[12,13]; DDT has been reported to be present in concentrations of up to 27 mg/kg carbon in phytoplankton samples from the Pacific[14]. Although it is unclear whether the pesticide residues found in plankton samples are truly absorbed or merely adsorbed on the surface, laboratory tests have indicated that, at the higher concentrations recorded, photosynthesis is likely to be affected[14]. Of all the organochlorine pesticides reported, DDT is usually the one present in the highest concentration, and it is generally accepted that DDT and/or its metabolite DDE has been responsible for the decline in breeding success of several avian species. The levels commonly occurring in fish do not appear to have any ill effects, although in a few areas of the U.S.A. where levels were exceptionally high, breeding success has been affected in certain fish species[15].

The degradation of accumulated DDT by metabolic activities of fish is also well documented[16,17] and, even in the areas where DDT has affected breeding success, removal of the source of pollution has been followed by recovery of the fish population. The major risk posed by organochlorine pesticides present in sewage discharged into the marine environment therefore seems likely to be dependent upon whether benthic animals (the food of demersal fish) can absorb these materials from the sediments. The facts are at present uncertain but in the long term there seems little doubt that most of the organochlorine pesticide compounds entering the sea are either rapidly degraded or immobilized in the sediments and not available to marine life[18].

Polychlorinated biphenyls (PCB's)

Although of industrial origin and usually determined with DDT, etc. by gas-liquid chromatography, PCB's are mixtures of compounds made up of several different levels of chlorination of a biphenyl molecule. Both classes of compound are lipophilic and behave in similar ways in the marine environment; much of what has been written regarding the risks and fate of organochlorine pesticides applies equally to PCB residues. Although PCB's have slight pesticidal activity, they are not used as pesticides. However, although they have a wide range of industrial uses, the number of practical applications has been restricted in recent years in order to reduce environmental pollution.

The presence of PCB in sewage effluents was first reported by Holden[19] who found an average concentration of 1 mg/kg in sewage sludge dumped in the Clyde Estuary from Glasgow. This was estimated to introduce approximately 1 t/year to the Clyde. He also reported that the Burnham laboratory had found PCB concentrations of 0·1–5 mg/kg in sewage sludge from Manchester and 0·2 mg/kg in sewage sludge from London. In each case, the sludge introduces about 1 t/year of PCB to the marine environment. More recent work has shown that the concentration found in sludge from London was fairly typical: 11 sludges ranged in content from <0·004 to 2·4 mg/kg dry weight[20]. Similar concentrations have been reported in sewage sludge discharged from the Hyperion outfall in California[4], i.e. 0·08–0·09 mg/kg. These concentrations are considerably higher than those reported for sewage effluents and it appears that, in common with the organochlorine pesticides, PCB's are adsorbed on particulate matter: the equivalent concentration of PCB in the Hyperion outfall waste water was only 0·27 μg/kg. In sewage effluents which receive only primary treatment the concentrations of PCB were reported to be about 10 times that found in the Hyperion effluent, which received secondary treatment.

The exact route of entry by which PCB arises in urban sewage is uncertain. Like the organochlorine pesticides, PCB's are ubiquitous in the environment, but as they occur in most animal tissues it is therefore probable that at least a proportion of the total input into sewage enters via the human diet. Until recently, when curbs were imposed on the use of PCB's, there were many ways in which they could enter industrial discharges to town sewers. In the U.K. and the U.S.A. the use of PCB's is now generally confined to closed-system applications and it is difficult to see how they can continue to occur so widely in sewage discharges. Nevertheless, there has been little evidence of a decline in the concentration of PCB found in sewage sludges in the two or three years since the restrictions were imposed. It has recently been reported that under certain circumstances DDT might decompose to produce PCB but the evidence for this appears to be limited, particularly in relation to quantitative aspects. Almost certainly the major sources of PCB in sewage are likely to be relatively small users of PCB formulations or materials containing PCB. The quantities lost on an annual basis are not large; for example, the input to sea from the Hyperion outfalls is only 2 kg/day (sludge and waste water) and that to the North Sea from sewage has been estimated to total a mere 6·6 t/year from all countries. A typical commercial formulation of a PCB may contain up to 100 or more compounds; gas chromatographic traces usually show at least 14 well-defined peaks. There is a steadily increasing amount of evidence to suggest that the lower-chlorinated compounds can be degraded by loss of chlorine atoms, either in sewage works or in the environment. This is supported by the general finding in U.K. environmental samples of residues which appear similar to Aroclor 1254, Clophen 50, or similar formulations, whereas the greatest use has always been of the lower-chlorinated formulations, e.g. Aroclor 1242.

Although it has been demonstrated[21,22] that PCB's are toxic to marine life, including phytoplankton, at concentrations in the range 0·005–0·1 mg/l., the highest concentrations reported in open-ocean waters are considerably lower than this[23], and although PCB is accumulated by marine animals (including fish and seals), there is no evidence that the levels normally encountered are detrimental to marine life. However, in certain restricted areas, for example off California, levels in marine animals and especially fish-eating birds are high enough to be implicated in the decline in numbers.

As with organochlorine pesticides, PCB's are accumulated on particulate matter, and in areas of discharge of sewage or sewage sludge they are likely to be found in the sediments. Concentrations of up to 2·4 mg/kg dry weight have been found in the sediments near the major Californian outfalls[10] and it has recently been reported[8] that the sediments off the Athens sewer outfall contained up to 0·7 mg/kg. There is some doubt about the availability of the PCB to marine organisms once it is adsorbed on the sediments; this will depend to some extent on the nature of the sediment and the strength of adsorption. Nimmo *et al.*[24] have, however, reported that PCB was absorbed from sediments by fiddler crabs and pink shrimps, although they were not able to establish whether this was directly from the ingested sediment particles or indirectly from water containing leached PCB.

Chlorinated solvents

There have been a number of reports of the presence of chlorinated aliphatic hydrocarbons in the marine environment, both in sea water and in marine organisms[25,26]. These compounds are produced in very large amounts, on a global basis, probably[27] in excess of 3×10^6 t/year. Some are used as intermediates, for example in the manufacture of PVC, but large quantities are used as degreasing agents and in dry-cleaning. As a rough approximation, the annual sale for these uses can be equated to the loss to the environment. Since many of the simple aliphatic solvents (e.g. carbon tetrachloride, trichloroethylene, perchloroethylene, methylene chloride) are relatively volatile, much of the transfer to the sea is probably via the atmosphere and precipitation.

It is however likely that significant quantities of chlorinated solvents enter the sea via sewage. This was recognized by Montgomery and Conlon[28] in the course of an investigation into possible causes of sewage works failures; they reported concentrations of carbon tetrachloride, trichloroethylene, chloroform, etc. in two crude sludges, a digested sludge, and the digester overspill. There was little difference between the crude and the digested sludge, but the concentrations in the digester overspill were about half those in the digested sludge; for example, for carbon tetrachloride the respective concentrations were 0·07 mg/l. and 0·2 mg/l. These concentrations are only about 1000 times as great as those found in Atlantic water, and it has been suggested[26] that certain of the simple aliphatic solvents can be produced by natural inorganic chemical reactions in the troposphere.

Although small quantities of solvents might be expected to enter purely domestic discharges, the biggest sources must undoubtedly be industrial, especially degreasing operations followed by the discharge of wash water into sewers. It might therefore be expected that much higher concentrations than those reported by Montgomery and Conlon[28] might be found in sewage effluents containing a high proportion of industrial effluent. This is supported by early results of analyses of sewage sludges from a few industrial towns, which were found to contain, for example, 4 mg/kg chloroform, 8 mg/kg trichloroethylene and 6 mg/kg perchloroethylene.

There is considerable controversy as to whether the simpler halogenated hydrocarbons are accumulated by marine organisms. Generally, the concentration factors are not large, being usually 2–100 times. Uptake appears to be a rapid but entirely passive process, probably based on relative solubilities in animal lipid and in sea water, and as soon as clean sea water is encountered loss takes place almost as rapidly as uptake. Degradation is probably through successive dechlorination steps, but as the compounds in question are relatively simple such processes probably take place reasonably rapidly. Certain compounds are in fact rapidly decomposed at the pH of sea water; for instance, pentachloroethane is said to change to tetrachloroethene. Examination of sediments in sewage-dumping areas supports the theory of limited persistence, because concentrations generally do not exceed 0·5 μg/kg.

Detergents and detergent additives

The use of synthetic detergents as substitutes for soap products created many problems in sewage works and in rivers, in terms of foaming, toxicity to aquatic life, and interference with biological processes in sewage works. Since the late 1960's there has been a gradual change to the use of linear alkyl benzene sulphonate type detergents. These are much more readily degradable in sewage works processes and presumably also in the marine environment. A detailed discussion of these materials is therefore hardly relevant in this paper. However, for certain industrial processes, especially the textile industry, the switch to 'soft' detergents has been much slower and it is possible that significant concentrations of 'hard' detergent enter the marine environment wherever particular industries discharge to town sewers.

Generally the content of anionic detergent entering a sewage works ranges from 8 to 25 mg/l., and removal during treatment is usually *ca* 90 per cent. In the U.K., undigested sludge normally contains somewhat less than 1000 mg/l. and humus sludges about 150 mg/l. Removal of non-ionic 'hard' detergents in a sewage works is relatively small, and typical effluents from textile industries are likely to contain[29] about 2 mg/l. If it is assumed that the concentration factor on sewage sludge is the same for non-ionics as for anionic detergents, then concentrations in crude sewage sludge and humus sludge might be expected to be 100 and 15 mg/l. respectively. In the main, detergent compounds will be associated with the solid matter and not the liquid fraction, resulting in conditions which will be most favourable for bacterial attack. Persistence of even 'hard' detergents in the marine environment is therefore unlikely to be observed, and as the bulk of evidence suggests that the toxic concentration for fish is usually greater than 1 mg/l., it seems unlikely that the quantities of detergents present in sewage or sewage sludges will affect marine life anywhere except in the immediate discharge zone.

Synthetic detergents as sold contain a number of other compounds and one, an optical brightener (2,5-di-(benzoxole-2-yl) thiophene), has been reported to occur in concentrations of up to 40 mg/kg dry weight in sewage sludge from certain Swedish towns[30]. This compound was estimated to have a half-life in water of 30–60 days, and fish were found to accumulate the compound from water by a factor of about 1000. Loss in clean water was relatively rapid but concentrations in fish liver reached 2 mg/kg. Although the compound has a low acute toxicity, doubts have been expressed about its long-term effects.

Owing to the problems which might be caused in fresh water by discharges of large quantities of nutrients, there have been attempts to replace the use of phosphates in synthetic detergents. One of the compounds which was used on a pilot scale was nitrilotriacetic acid (NTA). This has now been withdrawn from use in the U.S.A. and its use was never permitted in the U.K. Although readily degradable[29,31] at temperatures above 20°C, degradability is markedly dependent on temperature, and removal may be as low as 3 per cent at 5°C. It is not acutely toxic to aquatic life (96-h LC_{50} values for diatoms and bluegills range from 185 to 487 mg/l.) and toxicity decreases with increasing hardness[32]. Direct toxic effects in sea water are therefore unlikely. Since NTA is a chelating agent, it might be argued that it is likely to render metal ions more soluble and hence more available to marine life, and thereby increase their toxicity. Work by Sprague[33] suggests that this is unlikely, presumably because NTA has a very strong chelating action, which precludes release of the metal to the animal.

Phenol

Phenol does not occur in significant amounts in sewage of purely domestic origin; however, it does occur in the wastes from a wide range of industrial processes. Phenol is degradable in sewage works, although minimum amounts of nutrients are necessary to ensure completion of the process. Toxicity to aquatic life is variously reported within the range 1 to 100 mg/l., depending on species, temperature, and salinity.

In view of the fact that controls are usually imposed on discharges of phenol at about 0·5 mg/l., the risk of acute toxic action on marine life is not very great. Since phenol is degradable it is unlikely to be persistent in sea water or sediments, and there do not appear to be any reports of it being detected in the marine environment. However, phenol is accumulated by fish and shellfish, and tainting of marketable species is a very real danger. This danger is emphasized if the phenol is mixed with active chlorine, which might happen if sewage is chlorinated before discharge; oysters were found[34] to be appreciably tainted when exposed to monochlorophenol in concentrations in water of only 0·1 μg/l.

Summary

Sewage of urban origin contains a wide variety of organic chemicals. Some of these, e.g. organochlorine pesticides, occur quite fortuitously and enter via the human diet; others are of industrial origin. Whereas most organic chemicals appear to be fairly readily degradable during passage through a sewage works and are probably also degradable in the sea, a few are recognized as being relatively persistent; these include the organochlorine pesticides, PCB's, certain chlorinated aliphatic hydrocarbons, 'hard' detergents, and phenols. In most cases the discharge of sewage containing such substances, whether treated or not, does not cause damage to the marine environment. Local accumulation can, however, occur in areas where dispersion is poor and, if the substance is either highly toxic or bio-accumulated, damage to marine life can result. It should be added, as a further cautionary note, that at present only certain classes of persistent organic substances are recognized, and that as analytical techniques improve, and manpower is extended to allow a greater range of investigations, other potentially damaging substances are likely to be revealed. However, at the present time, despite the ubiquitous distribution of some of the more persistent substances, with a few localized exceptions, there is little evidence of damage to marine organisms, and no risk to man through accumulation in food organisms.

REFERENCES

1. Report of the ICES Working Group for International Study of Pollution of the North Sea and its effects on living resources and their exploitation. International Council for Exploration of the Sea, Co-op. Res. Rep. Ser. A, No. 39, 1974, 191 pp.
2. Holden, A. V., and Marsden, K. The examination of surface waters and sewage effluents for organo-chlorine pesticides. *J. Proc. Inst. Sew. Purif.*, 1966, 295–299.
3. Lowden, G. F., Saunders, C. L., and Edwards, R. W. Organo-chlorine insecticides in water. Part II. *J. Soc. Wat. Treat. Exam.*, 1969, **18**, 275–294.
4. Schmidt, T. T., Risebrough, R. W., and Gress, F. Input of polychlorinated biphenyls into California coastal waters from urban sewage outfalls. *Bull. envir. Contam. Toxicol.*, 1971, **6**, 235–243.
5. Croll, B. T. Organo-chlorine insecticides in water. Part I. *J. Soc. Wat. Treat. Exam.*, 1969, **18**, 255–274.
6. Albone, E. S., Eglinton, G., Evans, N. C., and Rhead, M. M. Formation of bis(*p*-chlorophenyl)-acetonitrile (*p,p′*-DDCN) from *p,p′*-DDT in anaerobic sewage sludge. *Nature, Lond.*, 1972, **240**, 420–421.
7. Jensen, S., Göthe, R., and Kindstedt, M.-O. Bis(*p*-chlorophenyl)acetonitrile (DDN), a new DDT derivative formed in anaerobic digested sewage sludge and lake sediment. *Nature, Lond.*, 1972, **240**, 421–422.
8. Dexter, R. N., and Pavlou, S. P. Chlorinated hydrocarbons in sediments from southern Greece. *Mar. Pollut. Bull.*, 1973, **4**, 188.
9. Wood, P. C., and Portmann, J. E. Toxicity of sludge. In 'Out of sight out of mind. Report of Working Party on Sludge Disposal in Liverpool Bay', Vol 2. H.M. Stationery Office, London, 1972, pp. 345–355.
10. The ecology of the southern California Bight: implications for water quality management. Southern California Coastal Water Research Project, California, TR 104, 1973, p. 180.
11. Wood, P. C. Disposal of sludge to sea. National Symposium on disposal of municipal and industrial sludges and solid toxic wastes, London, 1973. Institute of Water Pollution Control. Paper 8.
12. Butler, P. A. Organochlorine residues in estuarine mollusks, 1965–72—National Pesticide Monitoring Program. *Pestic. Monit. J.*, 1973, **6**, 238–362.
13. Jensen, S., Johnels, A. G., Olsson, M., and Otterlind, G. DDT and PCB in marine animals from Swedish waters. *Nature, Lond.*, 1969, **224**, 247–250.
14. Cox, J. L. DDT residues in marine phytoplankton. *Residue Rev.*, 1972, **44**, 23.
15. Butler, P. A., Childress, R., and Wilson, A. J. The association of DDT residues with losses in marine productivity. In 'Marine pollution and sea life' (ed. M. Ruivo). Fishing News (Books) Ltd, West Byfleet, England, 1972, pp. 262–266.
16. Mendel, J. L., Klein, A. K., Chen, J. T., and Walton, M. S. Metabolism of DDT and some other chlorinated organic compounds by *Aerobacter aerogenes*. *J. Ass. off. analyt. Chem.*, 1967, **50**, 897.
17. Young, R. G., St. John, L., and Lisk, D. J. Degradation of DDT by goldfish. *Bull. envir. Contam. Toxicol.*, 1971, **6**, 351–354.
18. Woodwell, G. M., Craig, P. P., and Johnson, H. A. DDT in the biosphere: where does it go? *Science, N.Y.*, 1971, **174**, 1101–1107.
19. Holden, A. V. Source of polychlorinated biphenyl contamination in the marine environment. *Nature, Lond.*, 1970, **228**, 1220–1221.
20. Department of the Environment. Water pollution research 1972. H.M. Stationery Office, London, 1973, p. 105.
21. Keil, J. E., Priester, L. E., and Sandifer, S. H. Polychlorinated biphenyl (Aroclor 1242): effects of uptake on growth, nucleic acids, and chlorophyll of a marine diatom. *Bull. envir. Contam. Toxicol.*, 1971, **6**, 156–159.
22. Hansen, D. J., Parrish, P. R., Lowe, J. I., Wilson, A. J., and Wilson, P. D. Chronic toxicity, uptake, and retention of Aroclor 1254 in two estuarine fishes. *Bull. envir. Contam. Toxicol.*, 1971, **6**, 113–119.
23. Harvey, G. R., Steinhauer, W. G., and Teal, J. M. Polychlorobiphenyls in North Atlantic ocean water. *Science, N.Y.*, 1973, **180**, 643–644.
24. Nimmo, D. R., Wilson, P. D., Blackman, R. R., and Wilson, A. J. Polychlorinated biphenyl absorbed from sediments by fiddler crabs and pink shrimp. *Nature, Lond.*, 1971, **231**, 50–52.
25. Murray, A. J., and Riley, J. P. Occurrence of some chlorinated aliphatic hydrocarbons in the environment. *Nature, Lond.*, 1973, **242**, 37–38.

26. LOVELOCK, J. E., MAGGS, R. J., and WADE, R. J. Halogenated hydrocarbons in and over the Atlantic. *Nature, Lond.*, 1973, **241**, 194–196.
27. GOLDBERG, E. D. Atmospheric transport. In 'Impingement of man on the oceans' (ed. D. W. Hood). Wiley-Interscience, New York, 1971, p. 75.
28. MONTGOMERY, H. A. C., and CONLON, M. The detection of chlorinated solvents in sewage sludge. *Wat. Pollut. Control*, 1967, **66**, 190–192.
29. DEPARTMENT OF THE ENVIRONMENT. Thirteenth progress report of Standing Technical Committee on Synthetic Detergents. H.M. Stationery Office, London, 1972.
30. JENSEN, S., and PETTERSSON, O. 2,5-di-(benzoxazole-2-yl)thiophene, an optical brightener contaminating sludge and fish. *Envir. Pollut.*, 1971, **2**, 145–155.
31. RUDD, J. W. M., and HAMILTON, R. D. Biodegradation of trisodium nitrilotriacetate in a model aerated sewage lagoon. *J. Fish. Res. Bd Can.*, 1972, **29**, 1203–1208.
32. STURM, R. N., and PAYNE, A. G. Environmental testing of trisodium nitrilotriacetate: bioassays for aquatic safety and algal stimulation. In 'Bioassay techniques and environmental chemistry' (ed. G. E. Glass). Ann Arbor Science Publishers, Ann Arbor, Mich., 1973, pp. 403–424.
33. SPRAGUE, J. B. Promising anti-pollutant: chelating agent NTA protects fish from copper and zinc. *Nature, Lond.*, 1968, **220**, 1345–1346.
34. BOËTIUS, J. Foul taste of fish and oysters caused by chlorophenol. *Medd. Danm. Fisk. Havundersøg.*, 1954, **1**.

Dr. Portmann introduced his paper by outlining the difficulties encountered in preparing a paper on persistent organic residues, and by indicating that his rationale for the inclusion of so many chemicals which were clearly of industrial origin was based on the facts that this was the source of major concentration and that most sewerage systems in northern Europe accepted a large proportion of trade wastes. He then briefly summarized the content of the paper, section by section, and included some new information on the oil content of U.K. sewage sludges which could range from 600 to 3000 mg/l. The proportion of oil which could be described as of neither animal nor vegetable origin was variable, but (on average) at least a third fell into the category of persistent petroleum hydrocarbons. The sources were also variable, but a significant proportion was believed to be from used motor engine oils discarded from home maintenance activities.

In conclusion he stated that sewage contained a wide variety of organic chemicals, many of which were degradable in sewage works and probably also in the sea. However, although the dangers of a few organic substances were now well recognized, there were probably many more such substances of equal persistence as yet undetected.

DISCUSSION

Mr. C. Chamberlin (U.S.A.) noted that Dr. Portmann had mentioned in his paper that the PCB levels shown to be toxic to phytoplankton were much higher than those generally reported for open-ocean waters. As Prof. Colwell had pointed out earlier in the Symposium, important effects need not be so drastic and could be extremely subtle. Remson, Woodwell, and co-workers at the Woods Hole Oceanographic Institution had shown that, at PCB levels commonly reported for coastal waters, the affinity of phytoplankton for nutrients was affected, and this might produce changes in the dominance patterns of phytoplankton. If these changes were from diatom dominance to dominance by dinoflagellates of the red-tide type this would certainly be an important effect, and he invited Dr. Portmann's comments. Prof. R. Mitchell (U.S.A.) added that chronic sub-lethal concentrations of chemical pollutants often adversely affected chemical communication both between and among some animal species, causing behavioural aberrations. He and his colleagues had observed similar inhibition in marine micro-organisms—preventing predation, changing species dominance, and slowing decomposition of organic matter. Their studies of Red Sea coral reefs had shown that low concentrations of hydrocarbons caused subtle changes in the surface microflora of the reefs, resulting in slow death of the corals. There was a need for more detailed analysis of such effects of low concentrations of pollutants on the biota of the oceans.

Prof. R. R. Colwell (U.S.A.) asked whether the route of entry of an organic compound was via particulate matter ingested by detritus feeders.

Mr. D. W. Mackay (U.K.) pointed out that in the Clyde Estuary the dumping of sewage sludge had resulted in enhanced PCB levels both in the sediments and in the benthic fauna. He added that, in recent years, the PCB output in the sludge had been reduced by 60 per cent due to the action of the pollution control authorities, but that no reduction in PCB concentrations in the sediment had yet been observed. Prof. Colwell remarked that this suggested very little degradation, and so raised the question of what would be the long-term effects of layers of such material on the sea bed.

REPLY TO DISCUSSION

Dr. Portmann thanked Mr. Chamberlin and Prof. Mitchell for their informative contributions. There was, of course, some dispute on the 'normal' levels of PCB in sea water, and he thought it likely that the values found by the Woods Hole Institution, and used by them in their experiments with phytoplankton cultures, were rather higher than the true sea-water values.

In reply to Prof. Colwell he said that ingestion of particulate matter was certainly one (though not the exclusive) route of entry, and there was probably a dependence on compound. Thus both uptake of DDT across the gill surface and uptake via food could readily be demonstrated.

He was glad that Mr. Mackay had provided information on the Clyde Estuary where accumulation was more likely to occur than in areas such as the Thames Estuary where water movements were greater.

Paper No. **14**

Effects of Nutrient Enrichment from Sewage in the Sea

A. D. McIntyre and R. Johnston

Department of Agriculture and Fisheries for Scotland, Marine Laboratory, Aberdeen

Nutrient enrichment from sewage in the sea is here taken to mean the addition of biologically available forms of nitrogen, phosphorus, and organic carbon. This addition may take place in three ways—by disposal of sludge, by piped effluent, or by run-off from the land, mainly via rivers. Field observations indicate that significant effects can be detected in all these cases.

The effects of nitrogen and phosphorus addition normally operate mainly through the primary producers (phytoplankton or macro-algae), and secondary effects on the environment or on the rest of the food chain may follow. Additions of organic matter, on the other hand, can affect the consumer organisms more directly. The additional nutrients are clearly measurable in winter when plant life is inactive but in summer they may be rapidly consumed by vigorous plant growth within a few daylight

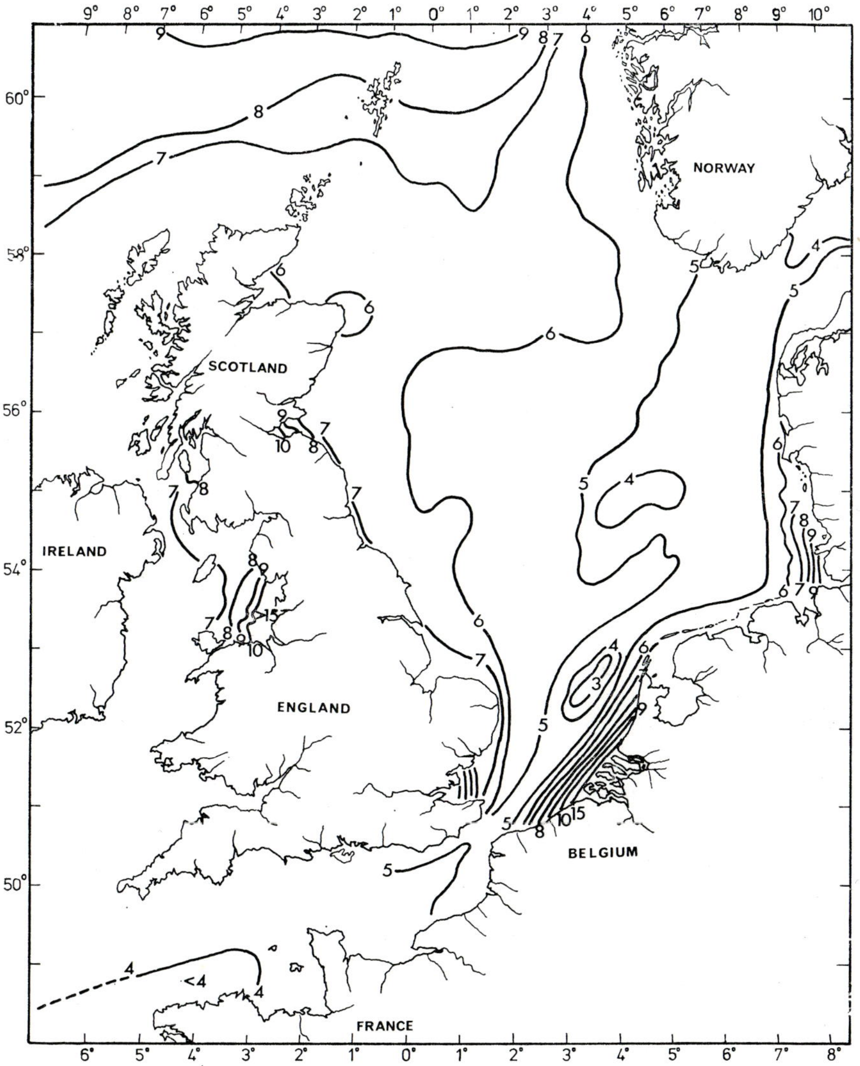

FIG. 1. Winter phosphate (10^{-7}g-at PO_4–P per l.) in North Sea
(Multiply by 3·1 to convert to μg P/l.)

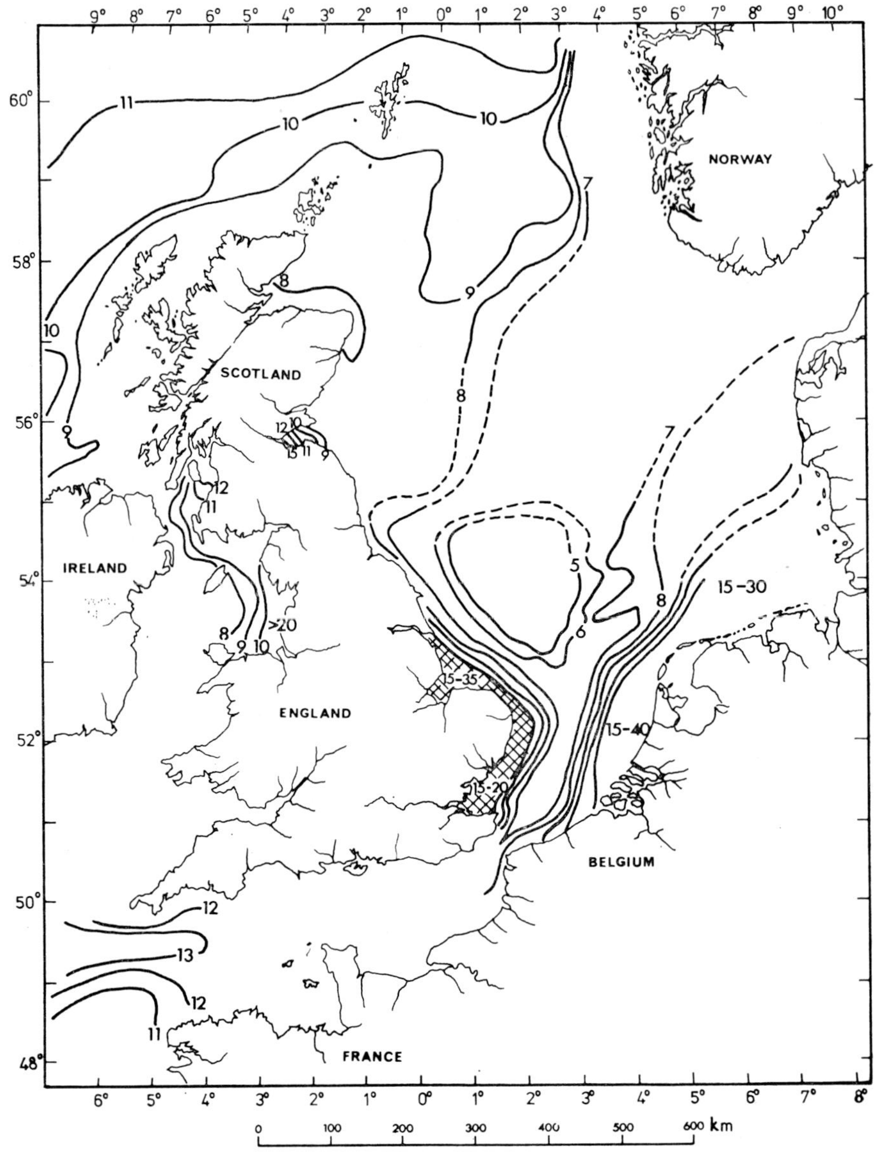

FIG. 2. Winter nitrate (μg-at NO_3–N per l.) in North Sea
(Multiply by 14 to convert to μg N/l.)

hours of discharge. Direct sewage discharges to the sea are usually untreated except for those from the major cities which may receive, at best, primary treatment and separation of solids. Almost all discharges to rivers are treated, and further breakdown of residual sewage material takes place en route to the sea. One of the problems of evaluating any given situation is the difficulty of separating nutrient effects from those of contaminating substances. The combined impact of agricultural, industrial, and domestic wastes on the nutrients of coastal waters can be clearly shown (Figs. 1 and 2), but it is almost impossible to determine accurately the varying contributions of the domestic sewage component, although for the northern part of the North Sea this may be estimated at about a third of the total and for the southern part (and the Irish Sea) about half of the total[1].

Sewage Sludge Dumping

Since marine sewage sludge disposal usually involves known quantities of material of approximately known composition being dumped at a specific site, this is perhaps the most straightforward situation. The effects will be determined not only by the amount dumped, but also to a large extent by the location of the site, and broadly speaking there are two extreme possibilities—calm or turbulent sites. In calm situations, which may be sheltered or relatively deep with little water movement and little stirring of the bottom deposit, the deposited material will tend to accumulate in a limited area, where monitoring of changes can be straightforward. On turbulent sites, on the other hand, which may be exposed and perhaps also shallow, with stronger tides or currents, and where even wave action may disturb the bottom, dumped material will tend to be rapidly dispersed, and the detection of effects will be much more difficult. While no site may be fully in one or other of these categories, it can be said that of the presently well documented dumping grounds, those for New York[2] and Glasgow[3] are 'accumulating' grounds, while those for London[4] and Manchester[5] are 'dispersing' grounds.

Effects on the water column

Considering 'accumulating' grounds in New York Bight[2], where in 1972 four million m^3 of sludge from primary and secondary treatment were dumped at 37 m depth on grounds which had been used for over 40 years, orthophosphorus levels in winter at the sea surface averaged 42 μg/l. and total phosphorus 86 μg/l., but values as high as 165 and 232 for ortho and total phosphorus respectively have been measured at other times of the year. Although these values are higher than would be expected from unpolluted waters in the area (where the range throughout the year is 6–28 μg/l.) and are clearly attributable to sludge dumping, they are exceeded in nearby coastal areas where pollution from rivers is even more marked.

The direct effect will be on primary production and on most dumping grounds this means on phytoplankton. Summer levels of chlorophyll *a* over the New York Bight sludge grounds rose to 15 μg/l., a high but not excessive level, and the low N:P ratio suggests that a nitrogen-limiting situation exists. This supports earlier experimental work[6] from which it was shown that nitrogen enrichment of water from this dumping ground produced substantial growth of diatoms. However the situation appears to be rather complex since even with a low N:P ratio and not excessively high chlorophyll levels, quite large quantities of unused nitrate and phosphate were present in some years. One explanation of this might be the inhibition of primary production by some component of the sludge, and such a possibility has been discussed[7]. The effect of increased turbidity may also be a factor. However, a two-year study of the area concluded[2] that no obvious effect of sludge dumping could be detected in the phytoplankton of New York Bight.

In the Clyde area, sewage sludge from Glasgow (about 10^6 t/year) is dumped in water of about 80 m depth 5 km off the coast on a ground with little current movement. Surface levels of phosphate and nitrate on the dumping ground in winter were 40 and 171 μg/l. respectively, again values higher than expected from non-polluted areas, but substantially less than in some nearshore regions of the Clyde. In summer, nitrate concentrations sometimes drop to zero in the surface waters over the dumping area, but unusual phytoplankton blooms are not encountered and chlorophyll *a* values considerably less than 5 μg/l. are usual[8].

On the exposed grounds where sludge does not accumulate, water-column data are available from Liverpool Bay[5] where about 10^6 t of sludge are dumped annually, but the studies of the effects are unfortunately rather inconclusive. In 1970 the mean winter levels at the surface over the dumping ground were 172 μg/l. for nitric nitrogen and 11 μg/l. for nitrous nitrogen, while ammoniacal nitrogen was extremely high at 114 μg/l. The corresponding summer values for these three forms of nitrogen were 22, 3·2, and 83 μg/l. respectively. The mean surface ortho-phosphate level in winter was 34 μg/l. (total P 54) and in summer 18 (total P 31). These levels are about twice the offshore concentrations, but they are not excessive and are substantially lower than those found farther east in Liverpool Bay where the influence of the River Mersey is stronger.

There appeared to be no exhaustion of these nutrients in the summer of 1970, suggesting that they would not be limiting for plant populations. In general, apart from concentrations of the flagellate *Phaeocystis* (which seems to depress the growth of other phytoplankton species), no large crops of phytoplankton were detected in the area, and the mean levels of chlorophyll *a* in June and July 1970 were 0·36 and 0·87 μg/l., indicating poor populations. Growth of phytoplankton experimentally in water from the dumping ground compared with that in clean offshore water did not indicate any consistent difference in performance[5].

Data on zooplankton from sludge dumping grounds are even sparser, but a report of the two-year study of the New York Bight area[2] concludes that no effect could be detected in the zooplankton population of the area, either in species composition or in distribution. In the Clyde, however, the centre of the dumping ground was rich in large zooplankton to such an extent that they could be seen on the underwater television screen. This observation was confirmed by plankton net hauls which showed that numbers of such organisms as copepods, which are normally abundant in the Clyde, were low in the dumping area, but that larger species (euphausiids and *Sagitta*) were sometimes comparatively high[9].

Effects on the bottom

The main effect on benthos from sewage sludge arises from the organic matter dumped on the bottom, and any indirect effects from enhanced phytoplankton growth and settlement would tend to be obscured by this.

On the 'non-accumulation' grounds, effects on the benthos are negligible. Sewage sludge from London is deposited on such grounds in the Thames, at present in quantities of over 5×10^6 t/year (wet weight). The area is only 13–20 m deep, and tidal velocities of 1 m/s and sometimes more occur, so that bed conditions are unstable and the fauna is generally poor. Surveys of this ground[4] show that in the organically richest mud the only effect is at most a slight increase in numbers and density of polychaete worms. In Liverpool Bay the mobility of the bottom deposits again appears to be the main factor controlling the benthos, so the fauna is also poor[5], and no general benthic effects can be attributed to the sludge.

It is on the accumulation grounds that significant effects can be detected. In the Clyde, for example, the muds in clean water off the Garrochhead dumping ground have between 0·3 and 2·0 per cent organic carbon by dry weight[10]. This value increases in the vicinity of the dumping location until, for about 10 km² around the central dumping site, values between 3·0 and 8·2 per cent are encountered. In this area a marked change in the fauna was detectable, and evidence of this could be seen on both the underwater television screen and in bottom deposit collections[11]. In the mud with less than 3 per cent organic carbon there was a mixed fauna of molluscs, crustaceans, echinoderms, and polychaetes. The diversity of the fauna became progressively reduced as the dumping ground was approached, until in the centre of the ground, where organic carbon values of around 8 per cent were recorded, only a few species of polychaetes and a single species of oligochaete worm made up the bulk of the population. However the biomass of the infauna was high—much higher than on the clean grounds, and the total number of individuals present was also very high.

A similar picture is reported for the New York Bight[12]. Dumping on this ground had resulted, as expected from the substantial quantities of sludge involved, in a sediment which is in places 3 or 4 times higher in organic content than that found in the Clyde, and again as expected a considerable modification of the benthos was found. However, even here there were no substantial areas devoid of life, and the total area affected was relatively small—some 35 km².

The possible impact of enrichment of sludge-dumping grounds on species of commercial importance must also be considered. Fish do not seem to favour as food the invertebrate benthos which dominates the high organic areas at the centre of the dumping grounds, but at the periphery, where the effect is enhancement rather than gross pollution and species reduction, fish do actively feed. In the New York Bight, 22 species of fish were reported from the dumping ground, and of these 6 were common, while several others such as herring and mackerel were occasionally numerous[2]. The fish were shown to be actively feeding at the edges of the ground. In the Clyde, although bottom feeding fish were not found on the bottom itself, the television camera showed very substantial concentrations of fish just off the bottom at the very centre of the dumping area and these were obviously feeding on the abundant crustacean plankton already referred to. These observations do not suggest any serious adverse effect on fish generally.

But the most significant change for fisheries must be expected from those species which actually inhabit the deposit, and for muddy grounds in the British Isles the most important commercial species in this category is *Nephrops norvegicus*. The Clyde television survey showed a clear transition from ground suitable for *Nephrops* to black mud of a fibrous texture which would inhibit normal burrowing and in which, as indicated above, the normal fauna was absent. The relatively confined nature of the sludge dumping situation in the Clyde gives us an opportunity to quantify the effects. Our surveys suggest that about 10 km² of former *Nephrops* ground had been made unsuitable for this species, and we may therefore calculate the value of the stock displaced. The average density of commercial-sized *Nephrops* in the area may be estimated from underwater television pictures and from divers' counts as between 1 and 2 individuals per m². If we accept an average of 1·5 individuals, this gives a total of about $1{\cdot}5 \times 10^6$ *Nephrops* per km². The mean weight of a single *Nephrops* from landings in the Clyde in 1970 was 32·58 g and the mean value of the first sale in the market was £0·31 per kg. Using these figures we arrive at a value of the standing stock of *Nephrops* per km² in the Clyde as about £15 000. The yield in this area represents at most half the standing stock, indicating a figure of roughly £7500 per km² as the annual value of these grounds in terms of *Nephrops*. We can further suggest that the 10 km² of ground lost here through dumping represents a loss in sterling of about £75 000 per year. While this type of calculation can give no more than an 'order of magnitude' impression, it is one way of quantifying the effects of sewage sludge dumping.

Coastal Input

Pipe outfalls

A recent review[13] on effects of marine outfalls showed that influence on primary production in open waters is slight and that the major effect is on benthos in the area of deposition in the vicinity of outfalls where a reduction of species diversity occurs. Later observations confirm this. The Pacific coast of America for example, with deep water nearshore, is perhaps the ideal situation for ocean outfalls, and the San Diego system has been studied in detail[14]. It discharges 3 km offshore at a depth of over 60 m where the physical characteristics of the water column are such that the discharge does not reach the surface. The changes in the receiving water are slight, and sludge accumulation round the diffuser is the main effect. There is, however, evidence of some effect in the water column in this region of southern California. One of the problems is to distinguish between enrichment due to sewage and that caused by upwelling, which is a seasonal feature of this coast, so studies must be timed for a period of no upwelling. One such survey[15] showed that primary productivity round two outfalls was 2–3 times the normal seasonal average for that coast, and was associated with corresponding levels of chlorophyll *a* and particulate organic carbon and nitrogen, indicating high phytoplankton standing crops. However, the levels of nutrients and dissolved organic constituents were not greatly different from the normal coastal background, suggesting rapid uptake of nutrients and/or very effective dispersal from the outfall. Since a modern sewage effluent pipe will usually be designed and positioned to achieve the maximum dilution and dispersion, and since improved techniques of construction and laying now make possible pipe lines several kilometres in length, it may be expected that the situation would not deteriorate in spite of increasing populations and more complete sewerage systems.

But to give some appreciation of the effects of sewage outfalls even on open coasts, it is instructive to compare the inputs to the bottom from such effluents with that which might be expected from natural processes. Considering sedimentation and the normal yield to the bottom, very few measurements have been made of the input of natural detritus. One method of studying this is deploying sediment traps, and a major problem in interpreting results from these is in deciding the extent of the resuspension of particles which have already reached the bottom at least once. The upper limit of deposition must be the total primary production for the year (any subsequent production stages will usually be included in this) irrespective of indications of spot measurements.

Attempts at evaluating such data for the Scottish west coast sea lochs by Steele and Baird[16] suggest that the amount of organic carbon produced in the water column which finds its way to the bottom is about 80 mg/m² day and this value may be applied to the entire year giving an estimate which is about 30 per cent of the annual primary production. Rates of production are sufficiently similar round Scotland to allow these data on natural inputs to the bottom to be applied to east coast areas.

Turning to inputs from sewage, we can calculate that if sewage solids discharged to Scottish east coast waters are uniformly distributed within a 1-km coastal belt, the input considered in various

TABLE 1. Domestic wastes entering North Sea from Scotland

	Population (thousands)	Coastline (km)	BOD of liquid fraction			Dry weight of solid fraction	
				t/km day			
			t/day	mean	peak[a]	t/day	t/km day
Cromarty Firth	17	50	0·75	0·015	0·015	1·2	0·02
Beauly, Inverness, and Moray Firths	160	190	4·7	0·025	0·055	10·9	0·06
Kinnaird Head to Bridge of Don	109	65	4·1	0·063	0·095	7·5	0·11
Aberdeen, Kincardine, and Angus	303	95	11·0	0·116	3·75	20·7	0·22
Firth of Tay	330	95	11·5	0·121	3·75	22·6	0·24
Firth of Forth to Dunbar	1271	190	55·2	0·291	9·35	87·0	0·46
Rest (extreme north and south)	160	120	5·3	0·044	0·178	10·9	0·09

[a] Maximum value for 20-km sections

sections ranges from 0·02 t/km^2 day (dry weight) for sparsely populated highland areas to 0·46 t/km^2 day on average for densely populated areas (Table 1). These conditions are reasonably well met in practice since at present most of the coastal sewage is untreated and is discharged through numerous short sea outfalls. Assuming 70 per cent of this dry weight is organic matter with 50 per cent carbon content, the inputs given in Table 1 average 16 mg/m^2 day for the more sparsely populated coastal zone and 106 mg/m^2 day for the region from Aberdeen to the south shore of the Firth of Forth. In the region of densest population near Edinburgh, an area of 20 km^2 is known to be grossly affected by sewage pollution[17] which, on the basis of the calculations above, would contribute 273 mg/m^2 day to the bottom. In addition there is perhaps another 40 per cent loading due to mixed industrial inputs which increases this figure to 382 mg/m^2 day. To this should be added the natural input which in this eutrophic area must be substantially more than the 80 mg/m^2 day estimated for non-industrialized waters. The overall loading for this region of the Firth of Forth is thus close on six times background, with even greater amounts near the outfalls, where the bottom is mainly of muddy sand and is predominantly of the accumulating type, although it is quite shallow.

River input

Effects in the water column

Since sewage input to rivers is more carefully controlled than to the sea, and since in Britain treatment is required before discharge to fresh waters, we may expect that the main effect of rivers in the present context will be measured in terms of inorganic salts, but here again it is difficult to separate the sewage component from that produced by agricultural run-off or by industry which frequently uses rivers for effluent discharge. Further, the river input may be compounded with the effects from offshore sewage sludge dumping. Thus in Liverpool Bay the main nutrient influence is from the River Mersey, whose effect at least in summer seems to extend as far out as the sludge dumping ground[5]. The mean winter surface levels in 1970 of nitric, nitrous, and ammoniacal nitrogen were 396, 26·6 and 155 μg/l. respectively and the ortho and total phosphate values were 40·6 and 62·0 μg/l. Decreases in the level of nutrients in the area during 1970 seem to be correlated as much with low river inflow as with plant production, which was low in that year except for *Phaeocystis*. The mean N:P ratio in the area most affected by the river in February was 32:1 and even in the more distant sludge dumping area this was 25:1, and although these ratios are unusually high it was not thought that phosphate was low enough to cause the low standing crop of phytoplankton observed. The unusually high ammoniacal nitrogen might possibly have a general adverse influence. While there appears to be much to explain about these 1970 data, the situation seemed to be more normal in 1971. Although sampling was less extensive, the pattern of water circulation and mixing was different from that of the previous year, and much higher standing crops of phytoplankton were observed which utilized the nutrients more completely.

In New York Bight a similar effect of river outflow can be demonstrated at the point where the Raritan River, New Jersey, enters the Bight. Here surface organic carbon levels in September reach 13 mg/l. and inorganic nitrogen almost 700 μg/l. with phosphate-P at 220–250 μg/l. These levels are substantially higher than normally found even at the sludge dumping sites and evidence of the effect is given by an observation of bright apple-green colour of the water which contained an almost pure culture of the alga *Didymocystis*[6].

In general, for a polluting effect in the water column to be anything more than a localized or transitory phytoplankton bloom, the area affected must, as discussed by Steele[18], be large enough to allow the phytoplankton patch to grow more quickly than the forces of turbulent diffusion and zooplankton grazing can act to disperse it. Steele suggests that on open coasts, length scales of at least 10 km must be involved in enrichment before such effects would occur and, even then, that some reduction in herbivore grazing would be necessary. This supports the observations that effects from individual outfalls tend to be slight but that large blooms may occur at river mouths especially where polluted water or unpalatable *Phaeocystis*-type species inhibit grazing. The situation which perhaps deserves more attention, however, is the one where increasing population or industrialization results in a gradual multiplication of pipelines or other points of effluent, any one of which by itself would not be significant but which could eventually on a stretch of open coast produce an input exceeding the actual length scale postulated by Steele.

Effects on the bottom

On open coasts the river influence on the water column can sometimes be separated from that of individual outfalls but this effect tends to be smoothed out by the time the bottom is reached. It is of interest to look at the overall effect on the benthos of nutrient enrichment on a stretch of comparatively open coast in Ayrshire in Scotland.

The coast in question is shown in Fig. 3. In the north part of the area, factory effluent containing among other things considerable quantities of nitrate is discharged, and from the joint mouth of two rivers in the central section there is a substantial input of domestic sewage as well as industrial effluent and agricultural run-off. At intervals along the beach and in particular at both the north and south ends there are substantial inputs of untreated sewage on to the beach not far from the low water mark. Figure 4 showing measurements of nitrate in the area indicates two distinct plumes, one associated with the river outflow and one with the factory effluent. Apart from these local hot spots the general background level of nutrients is significantly above that of coastal waters in unpolluted parts of Britain, and chlorophyll levels are correspondingly elevated (Fig. 3). In examining the benthos of the beach and shallow water of this area, grain-size analyses were made as well as measurements of carbon and chlorophyll in the sand. A quantitative sampling programme for both meio and macrofauna was carried out at intervals along the beach at selected tidal levels and, in the subtidal zone, meiofauna samples were collected by divers, and grabs were used at selected positions out to the 10-m line for macrofauna. Finally, beam trawl hauls provided estimates of the density of large epifauna and juvenile flatfish. Particulate carbon and chlorophyll were measured at the north and south ends of the beach every 2 cm down sand cores of 20-cm length. There was no consistent vertical trend and average values per core for carbon ranged from 923 to 1357 μg per gram of sand, and chlorophyll from 0·78 to 2·61 μg/g. For meiofauna (small metazoans and larger protozoa which pass a 0·5-mm sieve), most of the individuals were found in the top 12 cm of sand. A survey in the spring of 1972 showed that the meiofauna population in the lower part of the intertidal zone averaged over $4{\cdot}8 \times 10^6$ individuals per square metre and that 83 per cent of these consisted of nematodes. For macrofauna (the animals retained on a 0·5-mm screen) the surveys indicated a well developed *Tellina tenuis* community with 34 species of macrofauna

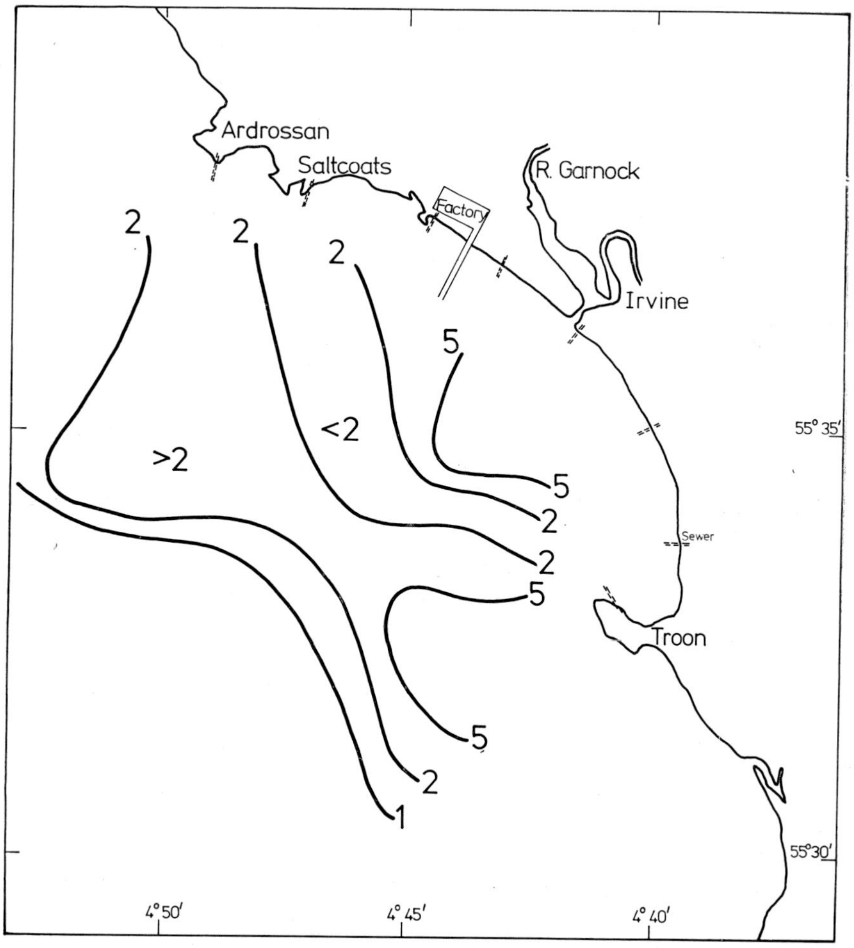

FIG. 3. Irvine Bay, Ayrshire, showing positions of sewers, factory outfall, and rivers
Contours indicate distribution of chlorophyll a *(μg/l.) at 4-m depth, July 1972*

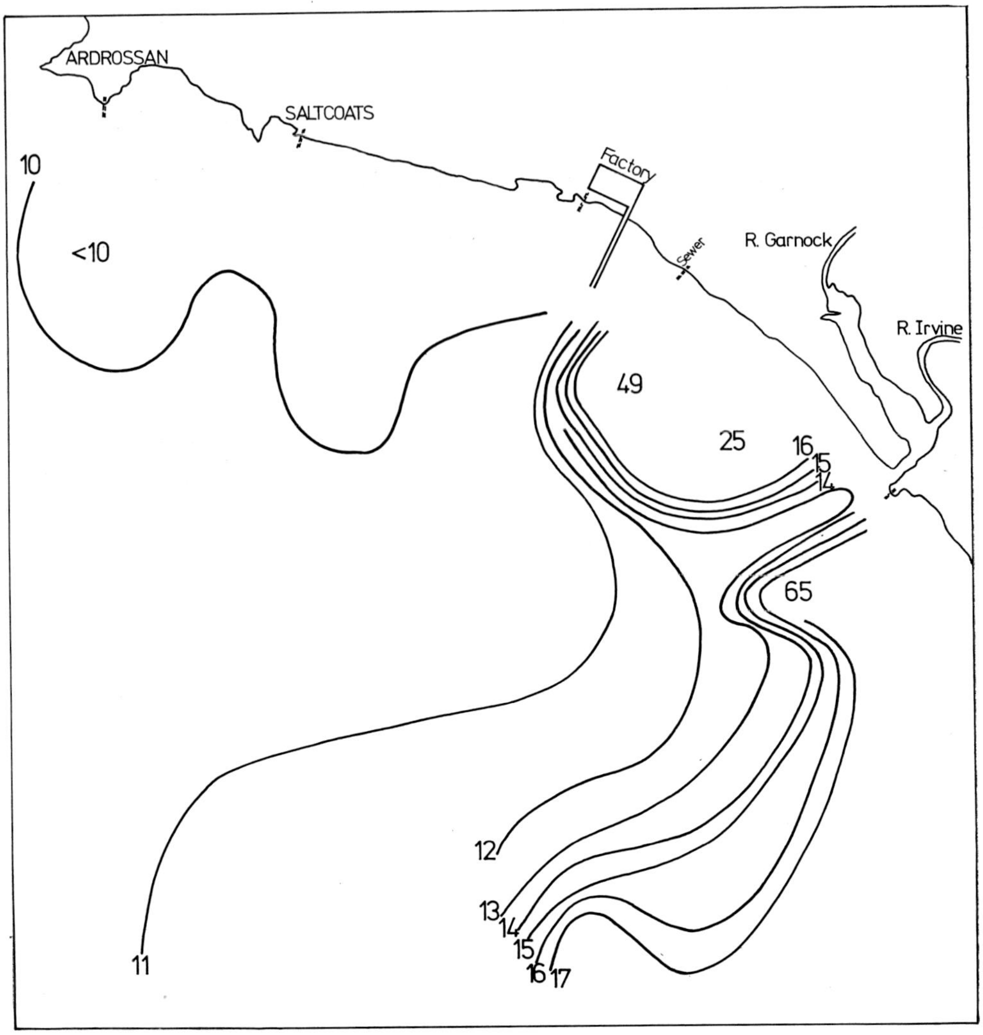

FIG. 4. Distribution of nitrate (μg-at NO_3–N per l.) at 3-m depth in Irvine Bay, April 1970 (*Multiply by 14 to convert to μg N/l.*)

invertebrates reaching a density of over $8000/m^2$ and a dry-weight biomass of more than 20 g/m^2. For epifauna from a series of hauls made at right-angles to the shoreline using a 2-m beam trawl the most common invertebrates collected were the heart urchin, *Echinocardium cordatum*, which on some occasions completely filled the net. Shrimps, *Crangon* species, were next in abundance (maximum 1200 per 10-min haul) and hermit crabs were also present in large numbers. The main catches in the hauls however were juvenile flatfish, mainly plaice and dabs, the count of which sometimes reached 1800 per 10-min haul.

These results suggest that the high nutrient levels already referred to are in fact reflected in the benthos. While it is difficult to make accurate comparisons between polluted and unpolluted beaches because of such variables as general topography, grain size, run-off, and exposure, data from Loch Ewe on the north-west Scottish coast[19], which contains a number of unpolluted beaches of various degrees of exposure, do provide a basis for comparison. The macrofauna off the Ayrshire coast has a greater species diversity and a higher density of individuals than even the most sheltered of the Loch Ewe beaches, and the biomass is more than three times as great. The meiofauna is also richer particularly in numbers of nematodes. Finally, comparison of the 1972 juvenile flatfish populations shows that the density in Irvine Bay is higher, by substantially more than an order of magnitude, than that of the unpolluted Firemore beach in Loch Ewe.

These levels indicate that even in an area where the nutrient levels are 5–10 times higher than normal and where the presence of untreated sewage on the beach is beginning to affect amenity values, the invertebrate and vertebrate populations may not be obviously affected adversely and in fact may be enhanced. The only suggestion of a possible adverse effect rises from a detailed study of the meiofauna. A striking feature of the Irvine Bay populations is the complete absence of interstitial copepods at many of the intertidal stations and a marked reduction in their species diversity at those stations where they do occur. Observations on a large number of unpolluted Scottish sandy beaches show that copepods are normally an important component of the meiofauna and their absence from a large part of Irvine Bay together with the reduction of the species diversity on the rest of the beach may be the first sign of adverse effects of effluent in this area. Here again we are faced with the impossibility of disentangling the effects of a wide range of different pollutants, but this is the situation with which we most frequently have to deal, and it is perhaps useful to look at a substantial section of the potentially threatened ecosystem rather than at single indicator species.

Conclusions and Summary

While sewage effluent may cause dramatic changes in enclosed regions, the influence of nutrients from sewage in open marine areas is usually much less significant.

The effects of sludge dumping tend to be confined to the bottom, and to sites where water movement is slight. On such sites the diversity of benthos is reduced, and if a bottom fishery is involved the most important consequence may be the loss of a measurable section of fishing ground. This loss is roughly quantifiable, and may well be offset in the broad view, by the convenience and economy of disposal at sea.

The major effects from sea outfalls also tend to be on the bottom, and an attempt to examine these effects in the perspective of natural input suggests that from one coastal area carrying a population of over half a million, the input of carbon to the bottom may be up to six times normal. Changes in plant and animal communities are therefore not unexpected.

The major influence on the water column comes from river input, where the scale of enrichment may be sufficient for the formation of more than short-lived plankton patches. Although the input from individual point sources of nutrients would normally be of too small a scale for the maintenance of such patches, the multiplication of these points with increasing population on an open coast may eventually cause the threshold to be exceeded, especially where industrial effluent is also present, and may result in gross algal blooms.

In such areas where nutrient enrichment from diverse sources mingles with industrial and agricultural effluent along a stretch of coast, the obvious result may be general enhancement of the benthic ecosystem, but signs of imbalance between the various components of the community may indicate the beginning of a change from 'eutrophication' to 'pollution'.

REFERENCES

1. Report of the ICES Working Group on Pollution of the North Sea. International Council for Exploration of the Sea, Co-op. Res. Rep., Ser. A, No. 13. Charlottenlund, Denmark, 1969.
2. Effects of waste disposal in the New York Bight. National Marine Fisheries Service (AD 739539), 1972 (9 sections).
3. MACKAY, D. W., and TOPPING, G. Preliminary report on the effects of sludge disposal at sea. *Effl. Wat. Treat. J.*, 1970, **10**, 641.
4. SHELTON, R. G. J. Some effects of dumped solid wastes on marine life and fisheries. In 'North Sea science' (ed. E. D. Goldberg). MIT Press, Cambridge, Mass., 1973, p. 415.
5. DEPARTMENT OF THE ENVIRONMENT. Out of sight out of mind. Report of Working Party on Sludge Disposal in Liverpool Bay. Vol 2. H.M. Stationery Office, London, 1972, 485 pp.
6. RYTHER, J. H., and DUNSTAN, W. M. Nitrogen, phosphorus, and eutrophication in the coastal marine environment. *Science, N.Y.*, 1971, **171**, 1008.
7. BARBER, R. T., and KRIEGER, D. Growth of phytoplankton in waters from the New York City sludge dumping grounds. Paper presented at 33rd Annual Meeting of American Society of Limnology and Oceanography, 1970.
8. Unpublished survey data. Department of Agriculture and Fisheries for Scotland, Marine Laboratory, Aberdeen.
9. CONWAY, D. Unpublished survey data. Department of Agriculture and Fisheries for Scotland, Marine Laboratory, Aberdeen.
10. MACKAY, D. W., HALCROW, W., and THORNTON, I. Sludge dumping in the Firth of Clyde. *Mar. Pollut. Bull.*, 1972, **3**, 7.
11. TOPPING, G., and MCINTYRE, A. D. Benthic observations on a sewage sludge dumping ground. International Council for Exploration of the Sea, C.M. 1972/E:30.
12. PEARCE, J. B. The effects of solid waste disposal on benthic communities in the New York Bight. In 'Marine pollution and sea life' (ed. M. Ruivo). Fishing News (Books) Ltd, West Byfleet, England, 1972, p. 404.

13. Pike, E. B., and Gameson, A. L. H. Effects of marine sewage disposal. *Wat. Pollut. Control*, 1970, **69**, 355.

14. Chen, C. W., and Orlob, G. T. The accumulation and significance of sludge near San Diego outfall. *J. Wat. Pollut. Control Fed.*, 1972, **44**, 1362.

15. Eppley, R. W., Carlucci, A. F., Holm-Hansen, O., Kiefer, D., McCarthy, J. J., and Williams, P. M. Evidence for eutrophication in the sea near southern California coastal sewage outfalls. Calif. Co-op. Oceanic Fish. Invest. Rep. 16, 1972, p. 74.

16. Steele, J. H., and Baird, I. E. Sedimentation of organic matter in a Scottish sea loch. *Memorie Ist. ital. Idrobiol.*, 1972, **29** (Suppl.), 3.

17. Covill, R. W. The quality of the Forth Estuary (2). *Proc. R. Soc. Edinb.*, B, 1971/72, **71**, 143.

18. Steele, J. H. Patchiness. In 'Ecology of the seas' (ed. D. Cushing and J. Walsh). Blackwell, Oxford (*in press*).

19. McIntyre, A. D. The range of biomass in intertidal sand, with special reference to the bivalve, *Tellina tenuis*. *J. mar. biol. Ass. U.K.*, 1970, **50**, 561.

In his opening remarks Dr. McIntyre said that the effects of added nitrogen, phosphorus, and organic carbon in the sea were discussed in the paper, and because of the difficulty of separating the input of sewage from those of rivers and land run-off, all major sources of nutrients had been considered.

The dumping of sewage sludge seemed to have little effect in the water column, but in deep areas, where the current movements were slight, the bottom sediment was altered and species diversity of the benthos was greatly reduced. This tended to inhibit feeding of commercial fish species on the bottom near the centre of the dumping ground, although these species might congregate off the bottom and on the periphery of the ground. Burrowing commercial species were usually eliminated from the centre of the ground. The quantification of such effects was illustrated in the paper.

Piped sea outfalls also tended to have their main effect on the bottom. Calculation suggested that the natural input of organic carbon to the sea bed round the Scottish coast was about 5 times what might be expected from sewage in sparsely populated regions, but that in industrialized areas the sewage input might be 4–6 times as great as that for primary production—indicating that changes in the bottom communities might be expected in such areas.

The nutrient input to the sea from rivers could be very much greater than from sludge dumping or from outfalls, and would tend to mask effects from these sources. Input from outfalls was usually on too small a scale to produce stable phytoplankton patches, but multiplication of outfalls in a limited area coupled with river inflow could produce gross algal blooms in the water column. The effect on the bottom would initially be enhancement of the benthic communities, though further addition could lead to changes in the population structure; a situation where this might be taking place, shown first by changes in the meiofauna, was discussed in the paper.

DISCUSSION

Dr. A. J. Lacey (U.K.) asked if the Authors had measured the effect of organic load on oxygen concentration, and how sensitive were marine fish and other organisms to oxygen levels below saturation. He also enquired if Dr. McIntyre had observed felt-like mats near outfalls, either of silt bound by the filamentous red alga *Rhodocorton* and developing on rocky shores, or of diatoms and red algae occurring on sand surfaces. Arising from Dr. McIntyre's impression of the different sensitivities of copepods to pollution, he asked if he thought there was any possibility of developing a biotic index of quality as had been done for fresh-water situations.

Mr. J. W. Gunster (Netherlands) noted that the Authors had referred to nitrogen-limiting situations during the productive season as a consequence of sludge dumping in both the Clyde and New York Bight areas, but no mention had been made of the effects where there were nutrient discharges from rivers to areas where nutrients limited production. He then drew attention to the situation in the Dutch coastal waters north of the Rhine Estuary where phosphorus had been found to be a growth-limiting nutrient for spring *Phaeocystis* blooms (N:P ratios of over 100 had been found). This was a matter of increasing concern since the total phosphate content of these waters had nearly trebled during the past decade as a result of increased phosphate discharge from the Rhine. Certain effects of algal blooms had been noticed in recent years, such as foam blankets and odour nuisance along the coast; there was probably also shellfish toxicity in some estuarial and tidal-flat regions, and there was concern about possible undesirable changes in species composition. However, there was, as yet, no clear evidence that the high nutrient content of the Rhine was the sole cause of these effects. The findings of phosphorus-limiting situations were thought to be indicative of possible eutrophication due to human activity. Research by the Netherlands Institute of Sea Research at Texel was continuing, with monitoring at more frequent intervals.

Dr. H. A. Cole (U.K.) asked if Dr. McIntyre had any definite evidence associating *Phaeocystis* blooms with nutrient enrichment from sewage. In addition to creating amenity problems, *Phaeocystis* seemed to be biologically undesirable—work at Conway had shown that, following *Phaeocystis* blooms, substances were present in the sea which inhibited the development of some marine larvae.

REPLY TO DISCUSSION

In reply to Dr. Lacey, Dr. McIntyre said they had measured the oxygen concentration of bottom water in the Clyde sludge dumping ground and found values of 94·5 and 97·5 per cent saturation, which were at least as high as in clean locations in the Clyde. On the other hand, depressed oxygen levels had been reported over the dumping grounds in New York Bight. Regarding sensitivity to low oxygen levels, there was a very wide range of response among marine organisms so that it was difficult to generalize, but around most of the outfalls with which he was familiar the water movement and mixing were such that oxygen deficiency was not one of the major problems.

The mats to which Dr. Lacey had referred were not uncommon where local shelter allowed silt to form in the rocks. On sand, his experience had been that outfalls were usually in relatively exposed areas too unstable for the formation of sand mats, but he had measured increased growth of attached episammic microalgae.

The Authors were examining the applicability of copepods as a biotic index in the Clyde where the few species of harpacticoid copepod surviving were just those which survived longest in their stress experiments.

In reply to Dr. Cole he said that although *Phaeocystis* blooms often seemed to be associated with sewage-enriched waters, this was not always so. During recent work in a sea loch on the north-west coast of Scotland they had recorded a substantial *Phaeocystis* bloom which was clearly not connected with pollution. The situations producing such blooms obviously required further study.

Paper No. **15**

The Significance of Surface Pollution in Coastal Waters

Robert E. Selleck

Professor of Sanitary Engineering, University of California, Berkeley, Calif.

The objectives of this work were to develop sampling and analytical techniques suitable for the evaluation of flotage at the surface of marine waters, and the measurement of flotage found in the vicinity of marine outfalls and in regions free of water-surface pollution. Two sampling techniques were evaluated: surface trawling for particulates $>0\cdot5$ mm in size, and surface screening for coliform bacteria, small particulates, and surface film materials.

General Description of Ocean Surveys

The ocean surveys were conducted in the San Pedro Channel and Santa Monica Bay regions of the Southern California Bight as shown in Fig. 1. Sampling Stations 1 and 4 were located over the marine outfalls of the Hyperion Plant of the City of Los Angeles and the Joint Water Pollution Control Plant (JWPCP) of the County of Los Angeles, respectively. Those outfalls had about equivalent flows and contributed together about 71 per cent of the total wastewater flow released to the California Bight n 1971[1]. Their dimensions are presented in Table 1 together with their 1971 average flows and effluent quality characteristics.

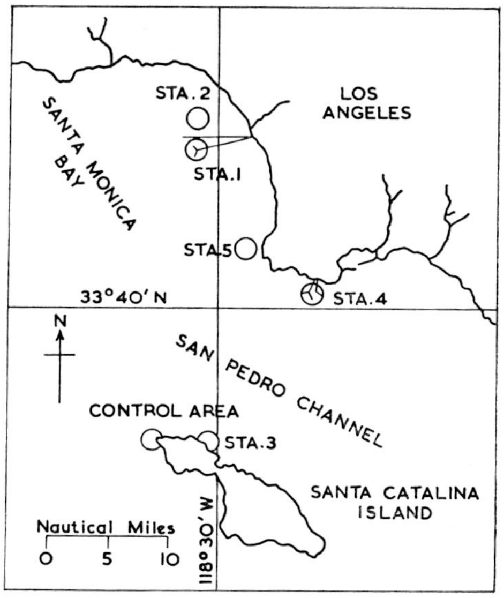

FIG. 1. Study area

The JWPCP effluent was a primary effluent containing a significant proportion of industrial wastewater[1]. The Hyperion effluent, on the other hand, stemmed almost entirely from domestic sources and a portion of it received secondary treatment (activated sludge). Its quality was better than that of a normal primary effluent, reflecting its domestic nature and the influence of the secondary treatment. No chlorine was added to the effluents during 1971, but it was being added to the JWPCP effluent in July 1973.

Three separate study-area surveys were made in April 1971, August–September 1971, and July 1973. Extensive ocean stratification existed during those periods, typical examples being shown in Fig. 2. The wastewater discharge plumes were always submerged beneath the ocean pycnocline at depths ranging from 12 m to approximately 30 m.

TABLE 1. Characteristics of outfalls sampled

	Hyperion 144-in.	JWPCP outfalls 90-in.	JWPCP outfalls 120-in.
Outfall			
Diameter (m)	3·66	2·29	3·05
Length (m)	8000	2600	3600
Diffuser			
Length (m)	2414	732	1353
Shape	Y	Y	L
Average depth (m)	59	63	56
No. of ports	168	100	740
Effluent (average, 1971)			
Flow (m^3/s)	14·7	5·26	11·0
HEM (mg/l.)	19	70	
BOD (mg/l.)	120	320	
SS (mg/l.)	73	330	
VSS (mg/l.)	55	210	

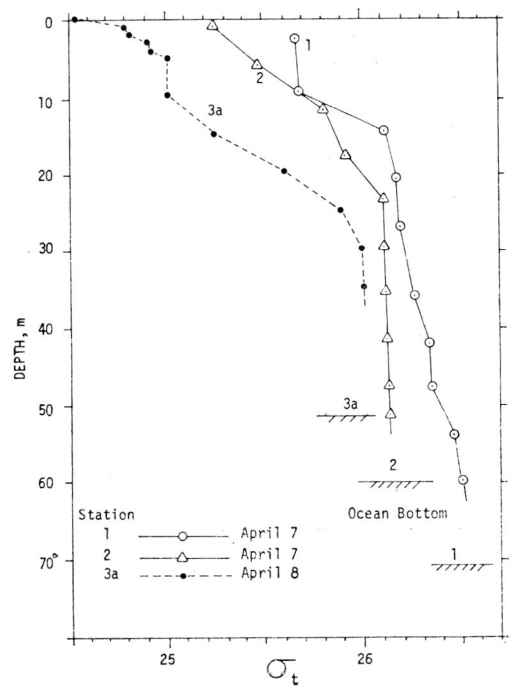

FIG. 2. Water-density profiles in April 1971
Density = 1 + 0.001 σ_t g/cm^3

Trawled Particulates

Floating particulates greater than 0.5 mm in size were collected by straining an ocean surface layer approximately 3 cm deep and 0.87 m wide through a 500-μ pore-size nylon screen fastened behind a trawling device. The analyses conducted on the particulates varied from survey to survey but included at times: (1) type, size, and number of particulates captured, (2) total dry weight, (3) hexane-extractable materials (HEM) and the composition of the HEM in terms of fatty acids (FA), pesticides, and polychlorinated biphenyls (PCB), and (4) coliform bacteria.

Type, size, and density

Some of the particulates appeared to be of wastewater origin; consequently, they were called 'sewage types'. Included in this category were such objects as pieces of grease and wax, seeds, plastic, fiber, and rubber. The grease and wax particulates were always predominant, and the seeds and plastic were usually prevalent. Other types of particulates were obviously of non-wastewater origin. Pieces of kelp and tar were the most common in that category. The average concentrations of the prevalent types of particulates captured at the surface in the surveys are shown in Table 2. The grease and wax particulates were predominant in the samples collected over the outfalls and north of the Hyperion outfall.

TABLE 2. Concentration (number per 100 m²) and size (mm) of prevalent trawled particulates in 1971

Particulate type	Hyperion outfall		JWPCP outfalls	Hyperion, 3.7 km N		Control areas		Particulate size (Summer only)		
	7 Apr.	31 July	3 Sept.	7 Apr.	1 Sept.	8 Apr.	2 Sept.	Median	Mean	Max.
Sewage types										
Grease & wax	573	60	234	16	52	0·48	<0·08	1·0	1·3	10
Seeds	85	7·4	11	7·0	20	0·29	0·08	(All 2 to 3 mm)		
Plastic	4·4	4·7	88	0·3	1·0	0·10	0·25	(Most 2 to 5 mm)		50
Other	54	9	9	8·4	4	1·8	0·53			
Natural										
Tar	32	4.0	13	1.7	3.5	1·8	0·84	1·5	1·9	35
Kelp	22	0.3	23	1.7	16	14	0·51	2·5[a]	4·0[a]	150[a]
								7·5[b]	16[b]	150[b]

[a] At Hyperion, 3.7 km N for 1 Sept.
[b] At JWPCP outfalls

The sizes of the particulates collected in the August–September 1971 survey were measured, with the results also being shown in Table 2. The mean sizes of the more prevalent types showed some tendency to vary from location to location but the trends were not pronounced except in the case of the kelp. The mean size of the grease and wax particulates was 1·3 mm, and the largest particulate had a length of 10 mm. The grease and wax particulates correlated fairly well with other non-hexane extractable particulates classed as 'sewage types' as shown in Fig. 3(a).

Hexane-extractable materials (HEM)

The HEM of the samples correlated closely with the numbers of grease and wax particulates found in the samples in all cases where tar was not excessive, as shown in Fig. 3(b). The correspondence was almost 1:1, i.e. one grease and wax particulate equaled 1 mg of HEM. This would arise only if most of those particulates were extracted with hexane because their mean size was only 1·3 mm.

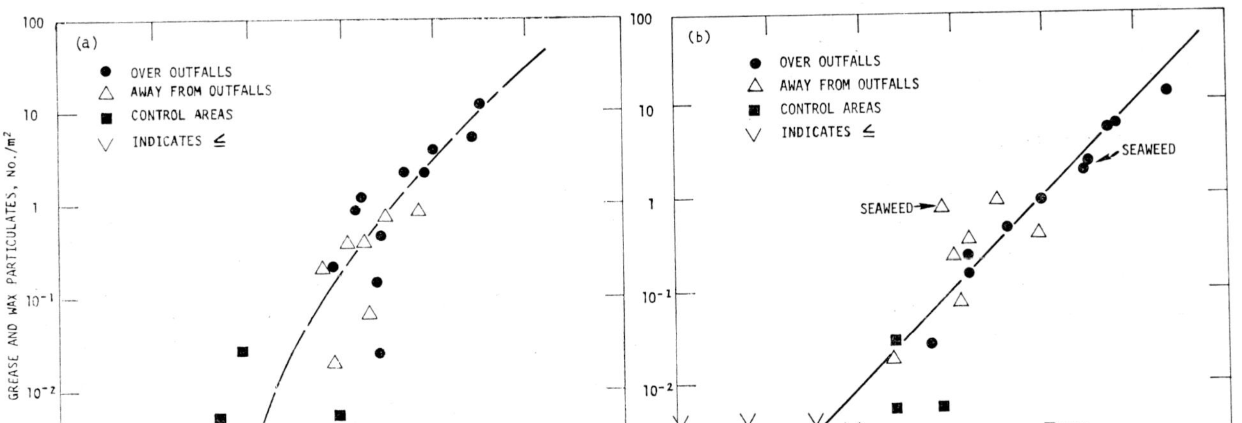

FIG. 3. Comparison of grease and wax particulates with (*a*) non-extractable sewage types and (*b*) hexane-extractable materials

Long-chain fatty acids (C_{12}–C_{22})

The fatty acids were determined directly on the unacidified or unsaponified HEM; consequently, they were free fatty acids. They comprised, on average, about 10 per cent of the total HEM with no trends being noted. Their average composition did vary appreciably with distance from the outfalls, however, as indicated in Fig. 4 which shows a preponderance of unsaturated C_{18}-acids in the particulates captured over the JWPCP outfalls in July 1973. This matched the composition of the free fatty acids found in particulates screened directly from the JWPCP effluent at the same time.

The composition of the HEM derived from one trawl sample collected over the JWPCP outfalls in 1973 was analyzed in more detail. The free fatty acids comprised approximately 12 per cent of the HEM, and an additional 18 per cent were saponifiable, composed mainly of saturated C_{16} and C_{15} acids. Carbon-hydrogen-oxygen analyses indicated that 38 per cent of the remaining HEM were bound water, 21 per cent alkanes, and 10 per cent other unidentified long-chain organic compounds. Thus the visual description of grease and wax particulates was apt, they were composed primarily of grease and waxes.

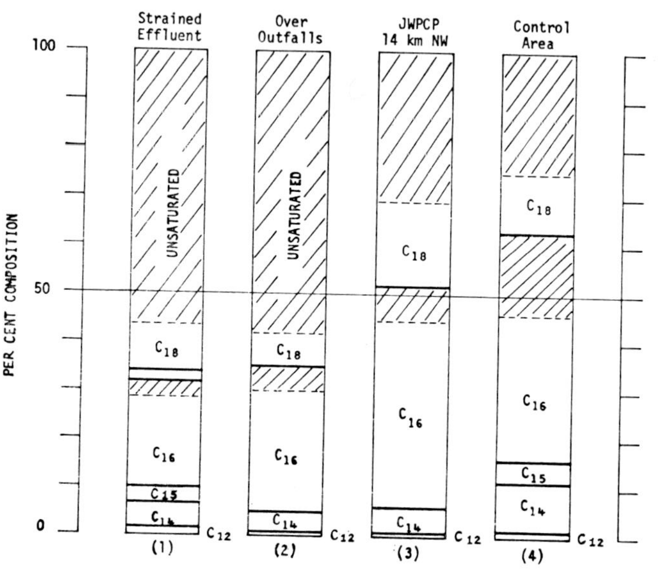

FIG. 4. Fatty-acid composition of large particulates, July 1973

PCB and pesticides

A few samples collected in the August–September 1971 survey were analyzed for PCB and some pesticides. Aroclor 1254 was detected in one sample collected over the Hyperion outfall, but it amounted to only 0·13 per cent of the sample's HEM. The pesticides were not detected to any significant degree in any of the samples.

Coliform bacteria

Aside from the observations made over the Hyperion outfall in August 1971, the number of organisms found per mg of HEM was remarkably constant in the samples taken over or near the marine outfalls as shown in Table 3. No bacteria were present in 100-ml samples taken 10 cm underneath the surface.

The significance of the data presented in Table 3 is difficult to judge in terms of pollutional significance. A commonly accepted standard for ocean bathing waters in the United States is less than 1000 total coliforms/100 ml for 80 per cent of the time, but a surface concentration cannot be expressed conveniently in terms of a volumetric concentration. For example, the surface trawl strained the surface water to a depth of approximately 3 cm, but the same particulates would have been captured at a trawl depth of 2 cm, or less. Approximately 99 per cent of the grease and wax particulates captured in 1971, however, had sizes of less than 5 mm. Assuming this was the sampling depth of significance, and that 1 mg of HEM contained 12 000 organisms, then 4 mg/m^2 of HEM would be equivalent to approximately 1000 organisms/100 ml. The observed median surface concentrations of coliform bacteria were about half the 80 percentile, however, so a median of about 2 mg/m^2 or less of HEM would be required to meet the ocean bathing standard.

TABLE 3. Comparison of coliform bacteria with HEM

Location	Date (1971)	Average HEM (mg/m²)	Coliform bacteria (median per m²)	Bacteria per mg of HEM
Hyperion	7 Apr. 31 Aug.	10·8 2·0	150 000 3 000	1.4×10^4 1.5×10^3
JWPCP	3 Sept.	2·2	63 000	2.9×10^4
2·7 km N of Hyperion	7 Apr. 1 Sept.	0·34 0·13	6 450 2 500	1.9×10^4 1.9×10^4
Controls Sta. 3(a) Sta. 3(b)	 8 Aug. 2 Sept.	 0·025 0·0012	 <0.0043 <0.00070	 <0.2 <0.6

Aesthetic impact

In the 1973 survey it was possible to evaluate the aesthetic impact of the particulates with a team of 6 or 7 observers composed of sampling personnel and the ship's crew. A rating sheet was used and the average response was determined from a scale ranging from 0 to 3 points, with the points increasing as the surface appearance deteriorated. The results are summarized in Table 4 which indicates that a relatively high HEM concentration was required to produce an offensive condition unless the particulates were contained in sea slicks. It might be appropriate, however, to maintain an HEM concentration of less than 3 mg/m² to be on the safe side at all times.

TABLE 4. Comparison of aesthetic impact with HEM

HEM (mg/m²)	Average rating
0·005	0 —Water surface in virgin state
0·050	1 —Some flotage evident but insignificant
3	3 —Offensive (in standing slick over Hyperion outfall, April 1971)
9	1·5—Flotage obvious and somewhat unsightly
32·5	2·3—Flotage obvious and unsightly, offensive to some

HEM mass balances

Different methods were employed to determine the concentration of the large particulates in the treatment-plant effluents. The best approach was to pump 18 l. of effluent through a screen identical to that used in the surface trawl. The average of five such samples collected from the JWPCP effluent on 13 July 1973 was 0·65 mg/l. as HEM.

The mass-emission rates of HEM from the treatment plants were estimated, with the results shown in Table 5. The results indicated that a tonne or so of HEM was being released daily to the study area.

TABLE 5. Mass-emission rates of particulate HEM from treatment plants

Plant	Date	HEM (kg/day)
Hyperion	7 Apr. '71	190
Hyperion	31 Aug. '71	390
JWPCP	11 July '73	890

To determine if all of the HEM in the plant effluents actually attained the ocean surface, a local mass balance was made about the outfalls using the steady-state continuity equation in the form

$$Q_w c_w = cUb \tag{1}$$

where Q_w is the effluent flow rate, c_w is the concentration of HEM in the effluent, c the concentration of HEM on the ocean surface, U the surface drift, and b the width of the diffuser projected normal to the direction of U. The product $Q_w c_w$ is the plant emission rate given in Table 5.

Estimation of concentrations from Equation 1 was necessarily crude because the surface drift was estimated with a drogue set at a 2-m depth, but the agreement between computed and observed concentrations of HEM over the outfalls was relatively good, and the differences were in the right direction (the observed being less than the computed) as shown in Table 6.

TABLE 6. Local mass balance on HEM of trawled particulates

Outfalls	Date	Concentration of HEM (mg/m²)	
		Calculated, c	Observed
Hyperion	7 Apr. '71	12	10·8
Hyperion	31 Aug. '71	14	6·6 (eastern-most trawl)
JWPCP	11 July '73	114	32·5 (L diffuser)
			5·4 (Y diffuser)

The data on surface concentration of HEM were much too few to estimate accurately the total mass of HEM of wastewater origin present on the ocean surface at any given time. Still, it is believed that even a crude estimate is better than none at all, so an average concentration distribution of HEM was estimated and the total mass of HEM computed as shown in Fig. 5. The total was estimated to be

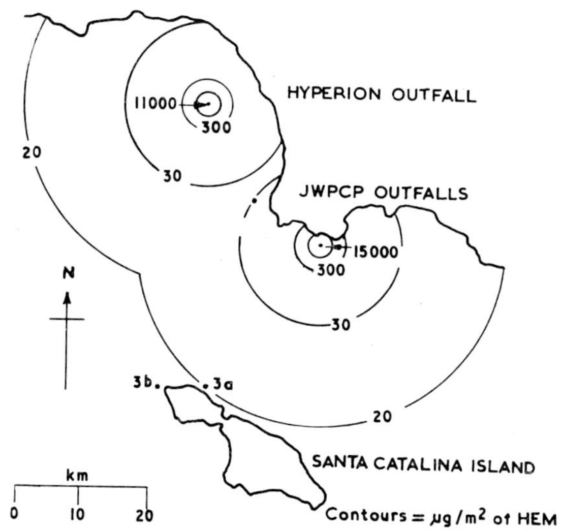

Calculations

HEM concentration (μg/m²)	Area (km²)	Average end area (km²)	Incremental concentration (μg/m²)	Incremental weight (kg)
0	2300			
		2300	20	46
20	2300			
		1500	10	15
30	700			
		400	270	108
300	90			
		55	12 200	671
12 500	20			
			Total	840

FIG. 5. Estimate of mass of wastewater HEM present on ocean surface, trawl samples

only 840 kg, about the same order of magnitude as the daily mass emission of HEM to the study area from the wastewater outfalls. Either the residence time of the particulates on the ocean surface was very short or they had concentrated in regions not sampled.

Surface-Screen Samples

Sampling accuracy

Cloth screens, first used in this study, could be extracted relatively easily with hexane by folding and inserting a soxhlet apparatus. They also tended to conform to the ocean surface and could be handled easily at sea. Their major disadvantage was that most fabrics could not be stripped of their background of hexane extractables, but glass fabrics could be cleaned relatively easily and were used in this study. They were cut to an area of 0·1 m² for sampling at sea and collected a layer 0·13 mm deep.

It is believed that the glass-cloth sampler collected insoluble duplex and stable monolayer surface films accurately. Experiments using the classic approaches of Langmuir[2] and Adam[3] indicated that the cloths removed 95 per cent of a palmitic acid monolayer from clean distilled water with a coefficient of variation of only 3·7 per cent. Heavy duplex films of vegetable fats were also collected from the water surface with the same order of accuracy.

The analyses conducted on the samples paralleled those made on the trawled particulates; i.e. HEM, free fatty acids (FFA), PCB and some pesticides, and coliform bacteria. (A coarsely woven nylon cloth was used to collect the bacterial samples because the organisms did not wash free from the glass cloth.)

Results

Figure 6 shows typical results obtained in the control area in July 1973. The median HEM concentration was significantly greater in slick than in non-slick areas but the results were extremely variable and there was a considerable overlap in the data. Also the HEM concentrations generally were extremely low throughout, approaching at times the background level of HEM present in the glass sampling cloths (10 μg or less). Even this modest difference between slick and non-slick areas was not apparent in samples taken over the JWPCP outfalls, although the average concentration was appreciably greater as shown in Table 7.

TABLE 7. Fatty-acid and PCB concentrations (μg/m²) in cloth-screen samples, July 1973
(Figures in parentheses are percentages of HEM values)

Location	HEM	Fatty acid	Stearic acid	PCB (as Aroclor 1254)
JWPCP outfalls	920	88 (9·6)	13 (1·4)	4·8 (0·52)
14 km NW of JWPCP	410	13·6 (3·3)	1·9 (0·46)	2·1 (0·51)
Control area	210	5·9 (2·8)	0·69 (0·33)	1·2 (0·57)

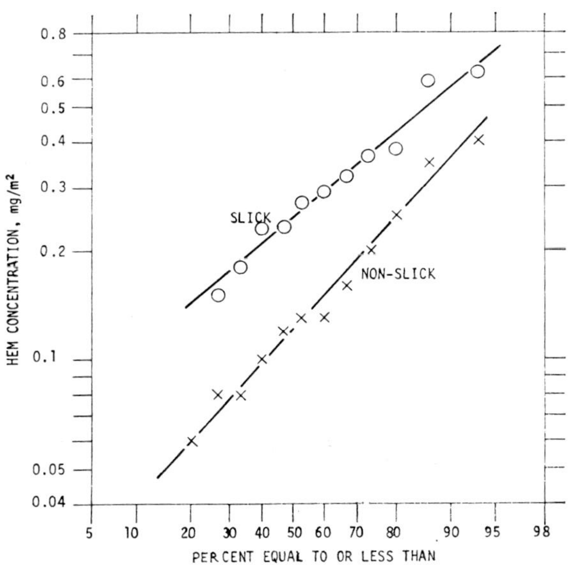

FIG. 6. Cumulative frequency distribution of HEM found in cloth-screen samples in control area, July 1973

The concentrations of the fatty acids also did not vary greatly between slick and non-slick areas. Table 7 shows that their proportion of the total HEM decreased from an average of approximately 10 per cent over the outfalls to 3 per cent in the control area. The fatty acids analyzed were free fatty acids because the glass-cloth samples were not acidified or saponified before extraction with hexane. No assurance can be made that the concentrations measured were those actually existing at the air–water interface because the methods of sampling and analysis may have altered the equilibria established between the acids and their salts.

The free fatty acids form salts readily with sea water. Two-liter grab samples of JWPCP effluent were mixed with 34 l. of clean water in a flat pan. Typical results showed that the acids in the bulk water (acidified prior to extraction) decreased from 40 to 20 per cent of the HEM with sea-water dilution as shown in Table 8. The acids must have formed salts which migrated to the pan and air interfaces. In addition, the free fatty acids in the cloth-screen samples taken from the surface of the pan represented only about 7 per cent of the HEM. The similarity between the effluent cloth samples (Table 8) and those taken over the outfalls (Table 7) was striking in terms of the saturated fatty acids such as stearic acid.

TABLE 8. Fatty-acid and PCB concentrations in pan data for JWPCP effluent grab sample, 13 July 1973 (Figures in parentheses are percentages of HEM values)

Type of sample	HEM	Fatty acid	Stearic acid	PCB (as Aroclor 1254)
Bulk effluent (mg/l.)	71·4	29 (41)	5·8 (8·1)	
Diluted bulk in pan (mg/l.)	2·36	0·48 (20)	0·15 (6·4)	0·000117 (0·005)
Glass-cloth samples (μg/m²)	1490	100 (6·7)	30 (2·0)	2·1 (0·14)

PCB compounds were found in appreciable quantity in the HEM derived from the cloth-screen samples, but the pesticides occurred in only trace amounts. The PCB portion of the HEM was nearly constant throughout the study area as shown in Table 7.

Coliform bacteria were not analyzed in the 1973 survey, but large numbers of the bacteria were found at times in the cloth-screen samples collected over the outfalls in 1971. The results were highly erratic, however, with either large numbers being recovered or none at all. The nylon cloth screen collected a surface layer 0·1 mm deep, so the number of bacteria/m² may have been approximately equal to the number of bacteria/100 ml. A cumulative frequency diagram of the bacteria in terms of number/100 ml is shown in Fig. 7. The distribution very nearly exceeded a bathing water standard of 1000 organisms/100 ml in 80 per cent of the samples.

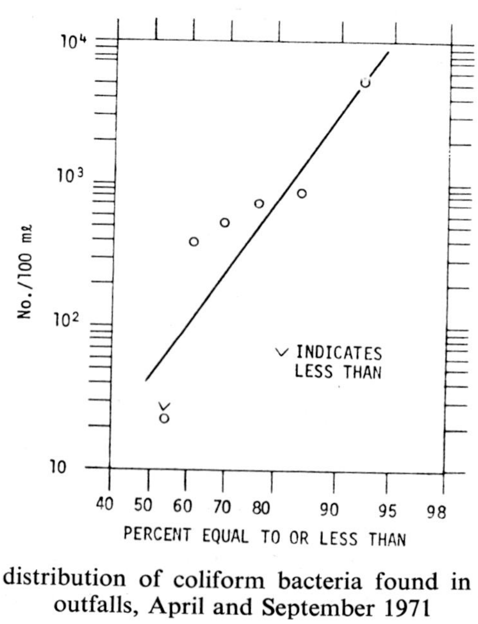

FIG. 7. Cumulative frequency distribution of coliform bacteria found in screen samples taken over the outfalls, April and September 1971

Bulk water samples

Eighteen liters of bulk ocean water were collected 0.5 m beneath the ocean surface at each station sampled in July 1973. The samples were acidified when collected, with the results of the analyses being summarized in Table 9.

TABLE 9. Fatty-acid and PCB concentrations (μg/l.) in bulk ocean samples, July 1973 (Figures in parentheses are percentages of HEM values)

Location	HEM	Fatty acid	Stearic acid	PCB (as Aroclor 1254)
JWPCP outfalls (L diffuser only)	54	6·4 (12)	0·92 (1·7)	0·103 (1·9)
14 km NW of JWPCP	51	9·1 (18)	0·92 (1·8)	0·025 (0·49)
Control area	31	3·3 (11)	1·1 (3·6)	0·018 (0·58)

The table shows that the PCB fraction of the HEM remained relatively constant at approximately 0·5 per cent except over the JWPCP outfalls where it was unusually high. Stearic acid, on the other hand, had a nearly constant concentration of approximately 1 μg/l. throughout the entire study area. This differed appreciably from the distribution of the acid on the surface ocean, which decreased grossly with distance from the outfalls as shown in Table 7.

HEM mass balance

Some of the cloth-screen samples (data not shown) indicated that particulates less than 0·1 mm in size were extremely numerous on the sea surface, sometimes exceeding a concentration of $10^6/m^2$. They were, of course, collected in addition to the surface films by the glass-cloth sampler. Considering the methods of analysis used in this study and the results discussed above, it appears that most of the free saturated fatty acids were derived from the microparticulates rather than the surface film proper; consequently, free stearic acid was selected as a tracer of such particulates, being the most likely to salt-out in the marine environment and the most conservative of the acids analyzed.

Tables 7 and 8 show that the stearic acid proportion of the HEM decreased from 2·0 per cent in the effluent pan samples to 0·33 per cent in the control area. Assuming that the stearic acid proportion would have been 2·0 per cent over the JWPCP outfalls if all of the HEM had been derived from wastewater microparticulates, the ratios of wastewater HEM to total HEM shown in Table 10 were obtained.

TABLE 10. Estimated concentration of HEM of wastewater (WW) origin in cloth-screen samples

Location	Proportion of WW origin	HEM (μg/m²)	
		WW origins	Background
Over JWPCP outfalls	0·75	690	230
14 km NW of JWPCP	0·22	90	320
Control area	0·16	34	177

Using this information and the computation methods shown in Fig. 5, the mass of HEM in the study was estimated to be 312 kg, or about the same order of magnitude as the HEM derived from the trawled particulates.

The results of the study were inconclusive regarding the appropriate control of surface pollution as collected with the glass-cloth samplers: HEM is appropriate only if the pollution is so gross as to produce heavy duplex surface films; PCB could have served as an excellent tracer of such pollution but unfortunately other sources of this material existed in the study area (atmospheric fallout)[1]. Restrictions could be placed on the coliform bacteria but it is not known what impact such requirements might have on the practice of marine wastewater disposal without further study.

Conclusions and Recommendations

1. Both the cloth-screening and surface-trawling techniques demonstrated clearly the presence of surface pollution derived from submerged primary wastewater effluent plumes.

2. Hexane-extractable materials (HEM) provide a convenient means of gauging surface pollution derived from particulates greater than 0·5 mm in size. Aesthetic and bacterial considerations indicated a reasonable median concentration of such material in the range of 1–2 mg of HEM/m^2.
3. Free fatty acids, PCB compounds, and coliform bacteria proved that surface pollution of the type collected by glass-cloth screens existed on the ocean surface. Additional work is required, however, to determine the behavior of fatty acid films on the ocean surface and appropriate means of analyzing them. It appears that the polluting substances found were derived from very fine floating particulates (<0·10 mm in size), and HEM proved not to be a reliable means of gauging such pollution.

ACKNOWLEDGEMENTS

This study was funded primarily by the United States Environmental Protection Agency, Contract No. R-800373, with Mr. Walter F. Rittall being the project officer. Also, the Southern California Coastal Water Research Project under the direction of Dr. George Hlavka provided the means for collecting the ocean samples in the 1971 surveys.

REFERENCES

1. Southern California Coastal Water Research Project. The ecology of the Southern California Bight: implication for water quality management. SCCWRP TR 104, El Segundo, Calif., 1973.
2. Langmuir, I. Fundamental properties of solids and liquids. II. *J. Am. chem. Soc.*, 1917, **38**, 1858.
3. Adam, N. K. The physics and chemistry of surfaces. Oxford University Press, Dover edn, New York, 1968, p. 27.

Introducing his paper Prof. Selleck said that the purpose of the study reported had been to ascertain the nature and significance of those portions of a sewage effluent reaching the surface when released in coastal regions. The effluents studied had all received at least primary treatment, were discharged at depths around 60 m, and were then trapped below the pycnocline at depths of 12–30 m. Care had to be taken not to ascribe flotage of extraneous origins to these discharges.

Development of methods of sampling and analysis had been necessary, as few attempts had previously been made to measure flotage quantitatively. Large particulates (>0·5 mm) were collected by a surface trawl, and small particulates (<0·1 mm) and surface films by glass-cloth screens which collected stable monolayers and oil duplex films relatively efficiently.

It was found that the large particulates did penetrate the thermocline and reach the surface; they remained in the study area for less than a day. In the absence of tar, the concentration of these particulates was measured simply and accurately as hexane-extractable materials (HEM) by the Soxhlet extraction method.

Characterization of the cloth-screen samples proved much more difficult. The unassociated (free) fatty acids (C_{12}–C_{22}) in primary effluent, when mixed with sea water, formed salts which rapidly migrated to all interfaces The free fatty acid concentrations found were relatively small, but nevertheless significant; it was concluded that they were attributable to small debris also collected by the sampler. The measured concentrations of the fatty acids, together with those of coliform bacteria and PCB's, indicated that a substantial portion of the cloth-screen HEM had been derived from the sewage discharges. The effects of such discharges on slick formation and prevalence was not ascertained.

More recently, flotage samples had been collected from *surfacing* primary and *secondary* effluent plumes. Preliminary results indicated that secondary effluents could also yield large particulate flotage containing appreciable numbers of coliform bacteria, even when the effluent had been chlorinated before discharge.

DISCUSSION

Mr. R. G. Ludwig (Brazil) asked for an estimate of the effect on the nearby beaches due to the flotage from the outfalls studied.

Prof. J. E. Foxworthy (U.S.A.) commented on the significance of the paper to the design engineer and treatment-plant operator. In California there were many ocean standards set by the state, one of which pertained to the quantity of floatables allowed in the vicinity of an outfall. There had previously been no efficient, reproducible means of sampling and analysing *in situ* samples; Prof. Selleck's technique filled that void.

Prof. A. G. Occhipinti (Brazil) pointed out that the coliform concentrations in Fig. 7 were similar to the bathing-water standard of 1000 per 100ml, and asked why such long outfalls had been constructed.

Mr. A. R. Agg (U.K.) asked if Prof. Selleck had any evidence for the mechanism of formation of the grease and wax particulates. He also drew attention to the observations of surface slick formation reported by Newton[4] who considered the formation of a slick to be the aesthetic criterion most amenable to quantitative description. (Other factors resulting in amenity deterioration considered had included discoloration, turbidity, presence of floating or stranded sewage solids, and odour.) Newton had observed that the frequency of occurrence of slicks decreased with increasing wave height—when the height was greater than 0·3 m—with a corresponding decrease in the area of slick formed by a continuous sewage discharge. For a given outfall configuration, the frequency of slicks increased as the sewage flow increased and the current decreased. It was found at Bridport that a slick usually formed when the initial dilution was less than 100, but only infrequently when it was more than about 400. The substances in sewage most likely to cause slick formation were higher fatty acids and their esters whose total concentration in untreated sewage was of the order of 140 mg/l., this figure being considered consistent with surface-tension theory and with the observation that 1 m^3 of sewage resulted in 70–800 m^2 of slick in the sea.

REPLY TO DISCUSSION

Prof. Selleck, replying to Mr. Ludwig, said that no surf zone or beach samples had been collected during the field studies, but he believed (from evidence not reported in the paper) that at least the large particulates would move shorewards with the internal waves during periods with wind speeds less than 5 m/s.

In reply to Prof. Occhipinti he said that he considered the coliform concentrations in Fig. 7 to be very high, since no coliforms had been found at a depth of 10 cm. He pointed out that the effluent plumes had been trapped some 12–30 m below the surface at the time of sampling, whereas the plumes sometimes surfaced during the winter months and this was why the outfalls were made so long. Also, new federal regulations made it mandatory to exhibit no more than 200 faecal coliforms per 100 ml.

He regretted that he had no evidence on which to answer Mr. Agg's question. Grease and wax particulates tended to fall visually into two or three distinct categories—such as hollow spheres showing layering of different materials around the shell. It could be that some of these particulates were formed in the long outfall pipes rather than in the treatment plant. They had been unable to observe slicks derived specifically from the outfalls because of the depth of the effluent plumes, with the exception of one occasion when a large 'standing' slick outlined almost perfectly the Hyperion outfall diffusers 59 m below the surface. On that day the plume was very nearly surfacing, being trapped at a depth of only 12 m. Unfortunately, the glass-cloth technique had not then been fully developed so the slick had not been sampled. It was not known if this slick had been formed by film substances from the effluent or if it resulted from a slight surface compression created by some hydraulic phenomenon.

ADDITIONAL REFERENCE

4. NEWTON, J. R. Factors affecting slick formation at marine sewage outfalls. In 'Pollution criteria for estuaries' (ed. P. R. Helliwell and J. Bosanyi). Pentech Press, London, 1975, Chap. 12 (*in press*).

Effluent Quality of Chlorinated Sewage Discharged from Sea Outfalls

Yehuda Kott
Environmental Engineering Laboratories, Technion
Israel Institute of Technology, Haifa

From the time that chlorine was discovered in 1774 by Karl W. Scheele, its uses have been increasing. The chemical curiosity which had no valuable application in the eighteenth and nineteenth centuries has changed, and today chlorine is used as a basic element for the production of many inorganic and organic derivatives such as methyl chloride, chloroform, and chlorinated phenols[1].

The use of chlorine in the disinfection of water started at the end of the last century, when in 1897 the first application of chlorine at Maidstone, England, stopped an outbreak of enteric fever[2,3].

The application techniques of chlorine in water supplies have been described in detail in various sources[4,5]. The reasons for the use of chlorine in waste water could be mentioned as follows:

1. To ensure proper disinfection of waste water.
2. To oxidize harmful and dangerous materials that might be discharged in the waste water.
3. To oxidize gases that might cause an odour problem.
4. To reduce the BOD of waste water.
5. To oxidize hydrogen sulphide which might cause corrosion.

From the above mentioned factors it can be learned that chlorination for waste water was applied only when the wastes were to be discharged to a river or for restricted irrigation. The literature shows that it is common for sewage to be chlorinated before discharge through marine outfalls, particularly when the outfall is situated in the vicinity of public beaches[6] or when other reasons apply—such as medical, aesthetic, and ecological hazards. It is therefore necessary to investigate the theoretical hazard potential of sewage before and after chlorination, in order to evaluate more precisely the advantages gained by the discharge of chlorinated, as compared with unchlorinated, sewage from marine outfalls.

The purpose of the present paper is to provide data concerning effluent quality after chlorination, in order to be able to assess the microbial quality of sea water close to bathing beaches. It is hoped that this will contribute positively to future criteria for the microbial quality of sea water and bathing beaches.

Materials and Methods

Most Probable Numbers (MPN's) were determined for coliform bacteria at 35°C and for faecal coliforms at 44°C using the EC medium and the multiple-tube fermentation technique recommended in APHA Standard Methods[7]. A 5-day 20°C BOD test[7] of effluents was also performed.

Vibrio cholerae (El Tor strain) was inoculated to test tubes to obtain an MPN using the standard 15-tube technique[7] with tubes containing peptone broth (pH 8·6). After 5 h growth in these tubes at 37°C, a loopful from each tube was inoculated to another set of peptone-broth tubes, followed by 5 h incubation at 35°C. A loopful from each of these tubes was inoculated to TCBS cholerae medium (Oxoid code No. CM333) to obtain isolated colonies which were then transferred to agar slants and incubated for 18 h at 37°C. The slants were checked serologically with polyvalent antisera, and results were considered positive when a clear agglutination was observed. Samples showing autoagglutination were discarded; positive results were recorded as *Vibrio cholerae* bacteria found in 100 ml.

Enteroviruses were isolated and counted using the ultrafilter method proposed by Gartner[8], and the concentrated viruses were inoculated on primary MK cells as described in detail in previous work[9]. Results were expressed as plaque-forming units (PFU) in 100 ml of original sample.

Samples for enumeration of *Entamoeba histolytica* were taken from the Haifa percolating filter sewage treatment plant. Up to 5 l. of influent and 5–100 l. of effluent were collected for cyst counting; 100 l. of effluent were collected for chlorination experiments. Concentration, flotation, growth, and chlorination experiments were done by the methods previously published[10].

Chlorination tests

Two-litre volumes of effluents were placed in a thermostatic bath, and sodium hypochlorite solution was added in amounts resulting in chlorine concentrations ranging from 5 to 80 mg/l. A magnetic stirrer was used to ensure good mixing. Samples were de-chlorinated after 1 or 2 h contact, and the above mentioned tests were performed. For continuous chlorination a 350-l. vessel divided by four baffle plates was used. Chlorination was carried out by adding sodium hypochlorite solution to give a final concentration of 10–80 mg/l. Contact times of 1 and 2 h were used.

Results

Table 1 summarizes the results of chlorination experiments on secondary treatment plant effluents. When the number of coliforms in the effluent (N_0) was approximately 10^7 per 100 ml, the concentration after chlorination (N_t) showed a decrease of about 5 orders of magnitude—as is shown by the survival ratio (N_t/N_0). It may be noted that the residual chlorine concentration measured at the end of each experiment was relatively high.

Table 2 shows the results achieved with various contact times. It can be seen that the survival ratio with short contact was low, whereas with long contact six to seven orders of magnitude reduction was achieved. It is interesting to mention that because of the specific type of the effluents the pH was high; it is well known that at such pH values very few hypochlorous ions are available for chlorination, most of the chlorine being in the form of hypochlorite ions.

When carbon dioxide was bubbled through the sewage to reduce the pH value to 5·6–6·0, fewer coliforms survived the chlorination. Table 3 shows results obtained using oxidation pond effluents. These effluents showed that, with good mixing conditions, the number of coliforms could decrease to less than 2 per 100 ml, whereas in other conditions the decrease was only three orders of magnitude. It is interesting to note that, whilst the numbers of *Vibrio cholerae* decreased considerably in these adverse conditions and disappeared completely after chlorination, the number of human enteric viruses decreased significantly but with fairly high numbers surviving chlorination. It may also be noted that *Entamoeba histolytica* cysts, which are generally considered to be very resistant, were present in the domestic sewage and proved to be sensitive to chlorination.

TABLE 1. Batch chlorination of Haifa treatment plant effluent

Serial No.	Chlorine concentration (mg/l.)	Contact time (min)	Coliform MPN (per 100 ml): Before chlorination, N_0	After chlorination, N_t	Survival ratio, N_t/N_0	Residual chlorine (mg/l.)
1	5	30	$2·3 \times 10^7$	9 300	$4·0 \times 10^{-4}$	3·7
2			$2·1 \times 10^8$	700	$3·3 \times 10^{-6}$	3·4
3			$1·3 \times 10^7$	2 800	$2·2 \times 10^{-4}$	3·5
4			$3·3 \times 10^7$	1 300	$3·9 \times 10^{-5}$	3·4
5			$7·9 \times 10^6$	17 000	$2·1 \times 10^{-3}$	4·0
6			$7·9 \times 10^7$	2 200	$2·8 \times 10^{-5}$	—
7	8	30	$3·3 \times 10^7$	56	$1·7 \times 10^{-6}$	6·3
8			$7·9 \times 10^7$	130	$1·6 \times 10^{-6}$	6·6
9			$1·3 \times 10^7$	32	$2·5 \times 10^{-6}$	6·4
10			$7·9 \times 10^6$	33	$4·2 \times 10^{-6}$	6·5
11			$7·9 \times 10^7$	70	$8·8 \times 10^{-7}$	6·7
12			$9·3 \times 10^7$	230	$2·5 \times 10^{-6}$	4·3
13			$2·3 \times 10^7$	230	$1·0 \times 10^{-5}$	—
14			$2·1 \times 10^8$	1 100	$5·2 \times 10^{-6}$	5·0
15		120	$7·9 \times 10^7$	46	$5·8 \times 10^{-7}$	5·4
16			$1·3 \times 10^7$	49	$3·8 \times 10^{-6}$	5·4
17			$3·3 \times 10^7$	17	$5·1 \times 10^{-7}$	4·5
18			$7·9 \times 10^6$	6·8	$8·6 \times 10^{-7}$	5·8
19			$7·9 \times 10^7$	17	$2·1 \times 10^{-7}$	—
20			$9·3 \times 10^7$	64	$6·9 \times 10^{-7}$	3·6
21			$2·1 \times 10^8$	1 100	$5·2 \times 10^{-6}$	4·4

TABLE 2. Batch chlorination of oxidation pond effluents
Applied chlorine: 8 mg/l.

Serial No.	Contact time (min)	Effluents before chlorination		Effluents after chlorination				Effluents after CO_2-bubbling and chlorination			
		pH	Coliform MPN (per 100 ml) N_0	pH	Coliform MPN (per 100 ml) N_{t1}	Residual chlorine (mg/l.)	Coliform survival, N_{t1}/N_0	pH	Coliform MPN (per 100 ml) N_{t2}	Residual chlorine (mg/l.)	Coliform survival, N_{t2}/N_0
1	5	7·7	$1{\cdot}7 \times 10^7$	7·8	$3{\cdot}3 \times 10^5$	5·8	$1{\cdot}9 \times 10^{-2}$	5·7	$3{\cdot}3 \times 10^4$	4·4	$1{\cdot}9 \times 10^{-3}$
2		8·0	$4{\cdot}9 \times 10^5$	8·1	$4{\cdot}9 \times 10^3$	7·1	$1{\cdot}0 \times 10^{-2}$	5·7	$1{\cdot}1 \times 10^2$	4·0	$2{\cdot}2 \times 10^{-2}$
3	10	7·7	$1{\cdot}7 \times 10^7$	7·8	$4{\cdot}9 \times 10^4$	4·9	$2{\cdot}9 \times 10^{-3}$	5·7	$2{\cdot}4 \times 10^4$	4·2	$1{\cdot}4 \times 10^{-3}$
4		8·0	$9{\cdot}3 \times 10^5$	8·0	$1{\cdot}1 \times 10^3$	6·6	$1{\cdot}2 \times 10^{-3}$	5·7	31	5·3	$3{\cdot}3 \times 10^{-5}$
5		8·0	$4{\cdot}9 \times 10^5$	8·1	$9{\cdot}3 \times 10^2$	6·6	$1{\cdot}9 \times 10^{-3}$	5·9	$2{\cdot}4 \times 10^2$	3·8	$4{\cdot}9 \times 10^{-4}$
6	15	7·7	$1{\cdot}7 \times 10^7$	7·8	$2{\cdot}4 \times 10^4$	4·4	$1{\cdot}4 \times 10^{-3}$	5·8	$2{\cdot}4 \times 10^4$	3·5	$1{\cdot}4 \times 10^{-3}$
7		8·0	$4{\cdot}9 \times 10^5$	8·1	79	6·4	$1{\cdot}6 \times 10^{-4}$	6·0	2	4·0	$4{\cdot}1 \times 10^{-6}$
8	20	8·0	$9{\cdot}3 \times 10^5$	8·0	23	6·6	$2{\cdot}5 \times 10^{-5}$	5·7	18	4·9	$1{\cdot}9 \times 10^{-5}$
9	30	7·4	$7{\cdot}9 \times 10^6$	—	$2{\cdot}4 \times 10^2$	3·8	$3{\cdot}0 \times 10^{-5}$	5·6	$4{\cdot}6 \times 10^2$	2·9	$5{\cdot}8 \times 10^{-5}$
10		8·0	$9{\cdot}3 \times 10^5$	8·0	23	6·6	$2{\cdot}5 \times 10^{-5}$	5·9	23	4·6	$2{\cdot}5 \times 10^{-5}$
11	60	7·4	$7{\cdot}9 \times 10^6$	—	33	3·1	$4{\cdot}2 \times 10^{-6}$	—	7·8	2·2	$9{\cdot}9 \times 10^{-7}$
12	120	7·4	$7{\cdot}9 \times 10^6$	—	2	2·7	$2{\cdot}0 \times 10^{-7}$	—	79	2·0	10^{-5}

TABLE 3. Survival of enteropathogens in chlorinated oxidation pond effluents in more than 20 experiments
Experimental conditions: 20°C, 1 h contact time, 8 mg/l. chlorine in form of sodium hypochlorite

	Per 100 ml				In 10 l.
	Coliforms (confirmed)	Faecal coliforms	*Vibrio cholerae* (El Tor)[a]	Enteroviruses	*Entamoeba hystolytica* cysts[b]
Before chlorination					
Minimum	$1\cdot3 \times 10^4$	$4\cdot9 \times 10^4$	0	30	2
Median	$7\cdot0 \times 10^5$	$4\cdot6 \times 10^5$	2·2	85	4
Maximum	$1\cdot1 \times 10^7$	$1\cdot3 \times 10^6$	2·2	290	28
After chlorination					
Minimum	<2	<2	0	40	0
Median	4·5	<2	0	57	0
Maximum	$2\cdot3 \times 10^4$	49	0	100	0

[a] *V. cholerae* added to raw sewage (10^3 per 100 ml) introduced daily to pond[11]
[b] Haifa percolating filter effluent

Table 4 shows results of poliovirus 1 (attenuated) compared with several bacteriophages. It is clearly seen that more than 20 mg/l. of applied chlorine was needed to achieve a very significant decrease, in fact total elimination of the poliovirus required at least 60 mg/l. It may be noted that removal of both virus and phage was favoured by the lower pH value and the longer contact time[12].

TABLE 4. Percentage survival of bacteriophages and poliovirus 1 (attenuated) after chlorination in oxidation pond effluents

Chlorine dosage (mg/l.)		20		40		60		80	
Virus tested	Chlorine contact time (h)	pH 8·3	pH 6·0	pH 8·3	pH 6·0	pH 8·3	pH 6·0	pH 8·3	pH 6·0
Coliphages on	1	56	13·5	5·8	1·05	31	0·2	15	0
E. coli B	2	36	4·5	22	0·3	10	0·06	2·7	0·06
Coliphages on	1	52·1	19·1	17·6	0·5	28·9	0·3	13·1	0·01
E. coli K_{12}	2	7·8	5·8	7·8	0·2	8·1	0·1	3·6	0·002
f_2	1	43·5	46·5	64	21·5	30·6	38	42·6	13
	2	19·7	28·5	52	14·5	24·3	38	43·2	38
MS_2	1	25·5	21·2	22·6	9·5	7·4	2·1	10·6	3·4
	2	10·6	10	15	7·4	10·6	5·6	15	3·4
Poliovirus 1	1	35	10·2	7·0	14·2	0·7	0·2	0·4	0
(attenuated)	2	12·2	14·2	12·2	13	0·4	0	0	0

Table 5 summarizes results from a continuous-flow chlorination at very high chlorine dosages, and shows that under these specific conditions negative viral isolations were recorded, coinciding with the high concentrations used in batch experiments.

TABLE 5. Effect of continuous chlorination on coliform bacteria and human enteric viruses (HEV) during continuous flow of oxidation pond effluents in vessel giving 2 h contact

Chlorine applied (mg/l.)	Before chlorination		After chlorination	
	Coliforms (MPN/100 ml)	HEV (PFU/100 ml)	Coliforms (MPN/100 ml)	HEV (PFU/100 ml)
40	$4\cdot6 \times 10^6$	100	14	0
50	$2\cdot4 \times 10^6$	150	11	0
60	$7\cdot9 \times 10^6$	200	2	0

One of the most polluted sea-water coastlines in Israel is that of Greater Tel-Aviv. These public beaches are exposed to sewage pollution, and frequently in the past 10 years have not satisfied the standard which is a coliform count not exceeding 2400 per 100 ml. The restrictions imposed on one of the most important facilities on the one hand, and the knowledge accumulated on chlorination dynamics on the other, resulted in the Tel-Aviv Municipality's decision to start chlorination experiments on raw sewage flowing through a marine outfall.

In the summer of 1973, experiments were performed with various concentrations of sodium hypochlorite applied to the sewage. It was noticed that there was a significant difference between the number of coliforms found when the sea was calm or rough. In the absence of chlorination, about 18 per cent of the samples taken under calm conditions showed more than the upper limit of 2400 per 100 ml allowed for bathing, but about a third exceeded this limit when the sea was rough. When chlorine was added, none of the samples taken under calm conditions exceeded the limit; however, the proportion was again about a third when the sea was rough.

The effectiveness of chlorination in the Tel-Aviv outfall pipe is illustrated in Table 6, from which it can be seen that the best results were achieved with the highest concentration (30 mg/l.) of chlorine, although reductions of two orders of magnitude were obtained with somewhat lower concentrations[13].

TABLE 6. Eighty-percentile coliform values obtained at various chlorine dosages in Tel-Aviv sewage
All data taken from Zohar[13]

Applied chlorine (mg/l.)	0	20	25	30
80-percentile coliform value (per 100 ml) after contact time of				
15 min	$5{\cdot}0 \times 10^8$	$1{\cdot}0 \times 10^7$	$2{\cdot}0 \times 10^6$	$2{\cdot}0 \times 10^5$
25 min	$5{\cdot}0 \times 10^8$	$1{\cdot}5 \times 10^6$	—	$4{\cdot}0 \times 10^5$
Reduction, orders of magnitude	0	1–2	2	3

It may be noted that because of dilution, dispersion, and natural die-off the number of coliform bacteria reaching the bathing area was probably a tenth of the maximal counts allowed.

In order to achieve the maximum effect of chlorine applied to sewage it must be remembered that there are many inorganic and organic materials that can be oxidized by chlorine, such as H_2S and aminoacids, which are present in sewage and cause some reduction in the chlorine oxidizing capacity[5,14], but it was often demonstrated that the efficiency of chlorination is dependent on the close contact of the chlorine applied to the sewage by good mixing[15–18].

When chlorinated sewage reaches the sea, there is usually little or no available chlorine left, especially in the case of raw sewage. The chlorinated organic materials discharged to the sea may influence marine life.

In recent years investigations, including life cycle studies with aquatic organisms, have shown that chlorinated organics can be toxic and cause changes in the species diversity of fishes[19]; in addition there is indication of interference in sensitive mechanisms like the nerve centre controlling respiration[20]. It is often stated that residual chlorine and also free residual chlorine has little effect on various fishes[21,22]. In various places raw sewage from big cities flows to the marine environment without any treatment[23]. However, in other places primary treatment that includes chlorination is a routine practice. Moreover the Special District No. 1 in East Bay Municipal Utility District, Oakland, California, primary treatment plant uses a pre-chlorination dosage of 10 mg/l., and 20 mg/l. chlorine injected to the effluent sewer, which makes a total of 33 t/day[24,25]. This enormous quantity of chlorinated sewage is discharged 1·6 km offshore in San Francisco Bay. It is therefore understandable that a toxicity study of the effect of this sewage on marine life in the area was carried out[26]; it was found that there was a significant toxicity on fish. De-chlorination of chlorinated sewage reduced toxicity.

Discussion

Chlorination of sewage before discharge through sea outfalls is practised in few cities in the world. In most cities sewage flows either raw or after treatment to the marine environment without chlorination. The question of whether to chlorinate or not can be dealt with only after taking into consideration the benefit that comes to the public bathing beach from chlorination, and to balance this with the possible damage to marine life when chlorinated effluents are discharged to the sea.

In the present paper data were presented to demonstrate the possibility of reducing coliform bacteria, killing *Entamoeba* cysts, and inactivating human enteric viruses. From the various results it could be seen that reductions of three to six orders of magnitude can be achieved by efficient chlorination. *Entamoeba histolytica* cysts that are found in raw or treated sewage were shown to be sensitive to chlorine. It should be mentioned that cysts placed in sea water died within 2 days[27]. While *V. cholerae* bacteria might migrate from an endemic area to other countries, it was observed that very slight chlorination had a lethal effect on the vibrio. It is often stated that viruses are very resistant to chlorination. In fact very little is known on the mechanism of action of chlorine on micro-organisms. Some published studies mention the effect of chlorine on enzymatic action[28,29] in the sensitive and essential respiration pathways.

In recent years many studies have been published dealing with viral inactivation. Viruses seem to be much more resistant than other micro-organisms to the action of chlorine because of their different metabolism. The results presented in this study show that, with practicable concentrations, the reduction in viruses is small, and that in order to achieve a complete inactivation of viral particles it is necessary to apply more than 40 mg/l. of chlorine. It is possible to achieve better results if the pH of the sewage is acidic.

It should be mentioned that studies published recently have shown that poliovirus 1 (attenuated) and others, and some bacteriophage were found to be more sensitive at pH 10 than at pH 6 in a chlorine-demand-free system. It could be that penetration of chlorine into the virion is more effective at pH10[30,31].

This phenomenon was investigated using poliovirus 2 (attenuated) under slightly different conditions in a chlorine-demand-free system and it was found that[32] at pH 10 the viruses were less sensitive than at pH 6·0, chlorination at pH 6·0 causing greater and more rapid kill than at pH 10. The results obtained on survival of coliforms after bubbling CO_2 through effluents confirmed this, using bacteriophages and human enteric viruses. These results might require some re-evaluation of the importance of finding a way to lower the pH of sewage in order to achieve higher yield in disinfection, and of the necessity to reduce the chlorine concentration and thus prevent some of the toxic effects that result when chlorinated compounds are formed.

It seems that not enough attention has been given to constructing contact chambers for efficient mixing and contact of chlorine. The money invested in an excellent mixing chamber could save, in the long run, money spent in chlorine consumed.

Results given in Table 6 showed that the Tel-Aviv marine outfall did not prove to be very efficient in reducing coliform numbers, although if a chlorination tank had been built, or some of the chlorination had been done in the outfall, then probably a better reduction would have been shown. In addition there are a few more ways to achieve satisfactory chlorination, and the crucial point is what we expect from chlorination—reduction of number of micro-organisms, safeguard from infectious diseases, or just the application of traditional treatment as done routinely in sewage treatment plants. It seems that if the target of chlorination is known, the application of chlorine is efficient when concentration, contact time, pH, and mixing are calculated beforehand and maintained afterwards.

REFERENCES

1. Chlorine facts. The Chlorine Institute, New York.
2. Brown, D. F. Chlorination—a historical survey. *Effl. Wat. Treat. J.*, 1969, **9**, 203.
3. Whitlock, E. A. The chlorination of water. *J. Instn. sanit. Engrs*, 1954, **53**, 61.
4. Weber, W. J. Physicochemical processes for water quality control. Wiley-Interscience, London, 1972, p. 427.
5. White, G. C. Handbook of chlorination. Van Nostrand Reinhold, New York, 1972.
6. Pike, E. B., and Gameson, A. L. H. Effects of marine sewage disposal. *Wat. Pollut. Control*, 1970, **69**, 355.
7. A.P.H.A., A.W.W.A., and W.P.C.F. Standard methods for the examination of water and wastewater. American Public Health Association, Washington, 13th Edn, 1971.
8. Gartner, H. Detection and recovery of polioviruses on a soluble ultrafilter. In 'Transmission of viruses by the water route' (ed. G. Berg). Wiley-Interscience, New York, 1967, p. 121.
9. Kott, Y. Coliphages as virus indicators in water and wastewater. Second Annual Report. Technion Research & Development Foundation, Haifa, 1971.
10. Kott, Y. Chlorination dynamics in wastewater effluents. *J. sanit. Engng Div. Am. Soc. civ. Engrs*, 1971, **97** (SA5), 647.

11. KOTT, Y., and BETZER, N. The fate of *Vibrio cholerae* (El Tor) in oxidation pond effluents. *Israel J. med. Sci.*, 1973, **8**, 191.
12. KOTT, Y., ROZE, N., SPERBER, S., and BETZER, N. Bacteriophage as viral pollution indicators. *Wat. Res.*, 1974, **8**, 165.
13. ZOHAR, Y. Report on chlorination of the Reading sewage pumping station. Ministry of Health, May 1973 (in Hebrew).
14. FAIR, G. M., GEYER, J. C., and OKUN, D. A. Water and wastewater engineering. Vol 2—Water purification and wastewater treatment and disposal. John Wiley, New York, 1968, Chap. 31.
15. WHITE, G. C. Disinfection practices in the San Francisco Bay area. *J. Wat. Pollut. Control Fed.*, 1974, **46**, 89.
16. COLLINS, H. F., SELLECK, R. E., and WHITE, G. C. Problems in obtaining adequate sewage disinfection. *J. sanit. Engng Div. Am. Soc. civ. Engrs*, 1971, **97** (SA5), 549.
17. KOKOROPOULOS, P. Designing post chlorination by chemical reactor approach. *J. Wat. Pollut. Control Fed.*, 1973, **45**, 2155.
18. DOTHENDARAMAN, V., SOUTHERLAND, H. L., and EVANS, A. L. Performance characteristics of chlorine tanks. *J. Wat. Pollut. Control Fed.*, 1973, **45**, 611.
19. TSAI, C.-F. Effects of chlorinated sewage effluents on fishes in upper Patuxent River, Maryland. *Chesapeake Sci.*, 1968, **9** (2), 83.
20. INGOLS, R. S., GAFFNEY, P. E., and STEVENSON, P. C. Biological activity of halophenols. *J. Wat. Pollut. Control Fed.*, 1966, **38**, 629.
21. MERKENS, J. C. Studies on the toxicity of chlorine and chloramines to the rainbow trout. *Wat. Waste Treat. J.*, 1958, **7**, 150.
22. BRUNGS, W. A. Effect of residual chlorine on aquatic life. *J. Wat. Pollut. Control Fed.*, 1973, **45**, 2180.
23. SUÑER, J., and PIÑOL, J. Coliform bacteriophages and marine water contamination. *Proc. 3rd int. Conf. Wat. Pollut. Res., Munich, 1966*, Vol 3, p. 105.
24. Water Pollution Control Division Special District No. 1, East Bay Municipal Utility District Annual Report 1965, California, 1966.
25. DAVIS, G. H. Personal communication, 1973.
26. ESVELT, L. A., KAUFMAN, W. J., and SELLECK, R. E. Toxicity assessment of treated municipal wastewaters. *J. Wat. Pollut. Control Fed.*, 1973, **45**, 1558.
27. KOTT, H., and KOTT, Y. Viability of *Entamoeba histolytica* cysts exposed to sea water. *Revue int. Océanogr. méd.*, 1970, **18–19**, 85.
28. KNOX, W. E., STUMPF, P. K., GREEN, D. E., and AUERBACH, V. H. The inhibition of sulphydryl enzymes as the basis of the bactericidal action of chlorine. *J. Bact.*, 1948, **55**, 451.
29. MILBAUER, R., and GROSSOWICZ, N. Reactivation of chlorine-inactivated *Escherichia coli. Appl. Microbiol.*, 1959, **7**, 67.
30. SCARPINO, P. V., BERG, G., CHANG, S. L., DAHLING, D. R., and LUCAS, M. A comparative study of the inactivation of viruses in water by chlorine. *Wat. Res.*, 1972, **6**, 959.
31. SCARPINO, P. V., LUCAS, M., DAHLING, D. R., BERG, G., and CHANG, S. L. Effects of hypochlorous acid and hypochlorite ion in destruction of viruses and bacteria. In 'Chemistry of water supply, treatment, and distribution' (ed. A. J. Rubin). Ann Arbor Science Publishers, Ann Arbor, Mich., 1974, p. 359.
32. NUPEN, E. M., ROSS, W. R., and KOTT, Y. The importance of pH as a factor in future chlorination practice. Unpublished data, 1973.

Prof. Kott, introducing his paper, said that in Israel—where over 90 per cent of the water resources were utilized—the discharge of waste waters to the sea was itself a waste. Although efforts at reclamation were in progress, sewage was currently discharged to the sea.

Chlorination of sewage with about 10 mg/l. of applied chlorine had given reductions of up to 5 orders of magnitude in numbers of faecal coliforms, complete destruction of *Vibrio cholerae* bacteria and *Entamoeba histolytica* cysts, and 30 per cent reduction in enteroviruses. Recent experiments with a dose of 20 mg/l. and 2–4 hours' contact time showed devitalization of all enteroviruses in the Haifa percolating-filter effluents—destruction was not confirmed owing to insufficient knowledge on viral existence and survival. A bacteriophage had been isolated which was extremely resistant to chlorine; when examined under the electron microscope, it was found that the tails of this phage were detached from the heads, thus suggesting this to be the weak point of the phage leading to its inactivation.

Complete mixing and close contact in chlorination chambers were essential for good results, and pH was an important factor. Some authors considered poliovirus 1 to be less resistant at pH 10 than at pH 6, using the PFU method in a chlorine-demand-free system. However, when the $TCID_{50}$ method was used as a comparison in South African studies, the viruses were shown to be more sensitive at pH 6. In field experiments with oxidation pond effluents it was found that attenuated poliovirus 1 was more sensitive to a high chlorine dose at pH 6 than at pH 8. These results were very important and would have to be considered when operating future water-reclamation plants, and the application of chlorine with or without pH adjustment (after lime treatment) would also become important. This problem had arisen on a small scale in the Dan Region reclamation area where an oxidation pond effluent by-pass was installed to discharge (in case of breakdown) through a 30-m sea outfall; the benefit of pH adjustment had to be weighed against its cost.

It might sound Utopian, but the diversion of acidic industrial wastes to an on-site chlorine generator placed and operating at a marine outfall would help to achieve the ultimate goal of safe and pleasant beaches.

DISCUSSION

Prof. P. Harremoës (Denmark) asked what had been the extent of nitrification of the secondary and oxidation pond effluents. This was important because of the well-known break-point phenomena which might give free chlorine in the latter but not in a non-nitrified trickling filter effluent.

Dr. J. J. Barnard (South Africa) said that an electrolytic oxidation process (in which chlorine was produced by electrolysing sea water) had been developed by Prof. Mendia in Naples and Prof. Føyn in Oslo and investigated at Portsmouth[33]. The merit of this process appeared to be that the chlorine was produced relatively cheaply. He asked what had happened to this process.

Prof. H. I. Shuval (Israel) referred to a study in which he had been concerned with the effectiveness of chlorination of sewage discharged 800 m offshore at Tel-Aviv. They had detected coliform regrowth in the sewage plume, and this presented a problem if ambient water standards were to be met. Prof. J. E. Foxworthy (U.S.A.) pointed out that, in about half the 40 studies of surface sewage fields off California, coliform regrowth had been found in the immediate vicinity of outfalls discharging unchlorinated effluents; this phenomenon seemed to be related to water temperature and/or season[34].

Prof. Shuval added that chlorination alone was of limited effect on raw sewage because of the solid particles. Even if it were effective, he doubted if the public would accept faecal material on the beaches. Removal of floatable solids and sufficient dilution of sewage were still required. Prof. G. Berg (U.S.A.) said it was possible to achieve extensive reduction or complete removal of bacteria from effluents, and even sewage, using relatively small amounts of chlorine if mixing was sufficiently rapid to allow the halogen to reach and inactivate the bacteria before it was dissipated by formation of chloramines and other inert substances. Moreover, if chlorination was at pH values outside the range of about 4·5–10·5, chloramines formed relatively slowly and this allowed perhaps 1–2 h for HOCl (the lethal form of chlorine) to destroy pathogens before being dissipated. However, it was still not known whether viruses embedded within solids could be reached by any chemical disinfectant.

Prof. Foxworthy asked if Prof. Kott had taken account of the toxic effects of continuously discharged chlorinated effluent on the ecology of the local marine environment. It was known that the residuals of free or combined chlorine could be highly toxic. In the event of such toxicity resulting in seriously adverse conditions, it might be necessary to de-chlorinate effluents before discharge—as was done in some Californian treatment plants. The total cost would then be increased substantially, and another form of disinfection might then be economically preferable. Prof. Shuval added that the toxic by-products of chlorination might indeed be the limiting factor in sewage treatment for certain aquatic environments. Another problem was the accidental release of chlorine gas from chlorination plants to nearby urban areas. Ozonization might overcome both these problems and have the added advantage of being a powerful virucidal agent. His studies had shown that with an ozone residual of only about 0·1–0·2 mg/l. a poliovirus reduction of 99·5 per cent could be achieved in 10 s, while the same concentration of chlorine took about 10 times as long.

Dr. P. K. Baay (Netherlands) asked about the possibility of organochloro compounds being formed during chlorination; these compounds might be persistent or even toxic.

Prof. P. V. Scarpino (U.S.A.) said that he had repeatedly demonstrated poliovirus 2 and coxsackievirus A9 to be more sensitive at pH 10 than at pH 6 (whereas Prof. Kott had found the reverse for poliovirus 1). More rapid inactivation of both the test viruses occurred in a borate buffer solution at pH 10, where the chlorine was in the form of OCl^-, than in a phosphate buffer solution at pH 10 where it was in the form of HOCl. As he had indicated in a recent paper[31], incorporation of KCl into the borate buffer influenced the OCl^- and HOCl virucidal relationships. Removal of the KCl shifted the OCl^- virus inactivation curve. He was currently assessing this phenomenon more completely for a subsequent publication.

REPLY TO DISCUSSION

Prof. Kott said in reply to Prof. Harremoës that the Haifa plant operated on a loading which reached 80–90 per cent above the design value. The effluent BOD was often above 60 mg/l., and typical figures for organic and total nitrogen (during 1971) were 18 and 60 mg/l. respectively. In recent experiments to reach break-point chlorination, a dose of over 40 mg/l. had been needed to provide a free residual. The oxidation ponds studied at a continuous flow of 250 l./h showed very good mixing, and a chlorine dose of 20 mg/l. with 2 hours' contact resulted in almost zero coliform counts and total absence of enteroviruses. Most of the nitrogen was present as ammoniacal nitrogen—for instance, 62 mg/l. out of a total of 88 mg/l. The applied chlorine was usually well below the break point.

He was unable to give Dr. Barnard any information on the electrolytic process, but agreed that using sea water at the discharge point should be efficient, convenient, and cheap. [An electrolytic plant is in operation in Guernsey[35]—*Ed.*]

On the subject of regrowth of coliform bacteria, he said that this had been studied at the Technion since 1966 and that he was confident it would not occur in the sea because of the adverse conditions there. Moreover, regrowth had never been observed in San Francisco Bay which received chlorinated primary effluents.

He had referred in his paper to the toxic effects of chlorinated organic materials in the sea, and agreed that chlorination followed by de-chlorination would involve unjustifiable cost. Other means might help in reducing the concentration of halogen applied—a combination of bromine and chlorine which acts synergistically, iodination, or replacement by other chemical disinfectants which might be found in the future.

He considered that the storage arrangements for chlorine were adequate, the risk of serious escapes to the atmosphere being minimal. Moreover, one-ton cylinders were used routinely in water and sewage treatment plants, and 50-ton containers were also used with safety. Regarding ozonization, he feared that this would produce the same kind of problems as in chlorination, since the resulting synthesized oxidized organics might be even more toxic than the chlorinated organics.

ADDITIONAL REFERENCES

33. MARSON, H. W. Electrolytic sewage treatment; the modern process. *Wat. Pollut. Control*, 1967, **66**, 109.

34. FOXWORTHY, J. E., and KNEELING, H. R. Eddy diffusion and bacterial reduction in waste fields in the ocean. University of Southern California, Los Angeles, 1969.

35. FRAMPTON, B. G. The Guernsey electrolytic sewage treatment plant. *J. Instn. munic. Engrs*, 1969, **96**, 91.

Paper No. **17**

The Case for Treatment

Kjell Baalsrud
Norwegian Institute for Water Research, Blindern, Oslo

For nearly a decade scientists have acknowledged that the sea is a resource of limited size. It has also been acknowledged that substances which enter the sea have no further place to go. Either they will be disintegrated and returned to the elementary substances found in the sea, or they will accumulate in the water or in the sediments. The water and the sediments are habitats for a great variety of organisms which exist in a delicate ecological balance. No major changes in the balance between marine organisms in the marine ecosystem can take place without a direct impact on man; yet we use the sea as our final receiving water, risking, and in some cases actually causing, deleterious changes to local or regional ecosystems.

Man has for centuries exploited nature, harvested from nature's abundance, cultivated new land, extracted raw materials, and developed urban and industrialized areas. Since Malthus (1803) we have heard warnings that our exploitation of nature could not proceed without limits. It has not been until the last two decades that man has realized the truth of such warnings. It has become clear that we may no longer depend on simply harvesting the products of the earth, but we must cultivate the earth and make use of renewable resources as the farmer on his land always has done.

As a result of our technological development, proper disposal of all kinds of wastes has become a major problem. If the quality of air, water, and soil is to be maintained, waste products discharged in the form of gases, liquids, or solids must come under more effective control. We must know the path of disposed waste until it is finally assimilated, stabilized, or inactivated. In some cases discharges of wastes into the environment can be accepted, but to permit this we must have a full understanding of the primary, secondary, and final effects of the discharges. Even in cases where nature is able to accept pollution, man himself may have special views. People may feel that the discharge is an offence to them and state their rights to freedom from pollution.

The oceans are vast. Can it be correct that they are threatened by pollution? Or is the volume of the ocean so great that waste will be highly diluted and that the self-purifying capacity in time may remove the small amounts of pollution that are left after dilution? This is the key issue of this paper.

Emphasis is usually put on mineral and oil resources in discussions on the use of the natural resources. However, in the use of various mineral resources and production of energy, man has some choice. But he has a limited choice with regard to how to produce his food and how to live in an acceptable environment. Man must be regarded from a biological point of view. Like all other animals, man is totally dependent on photosynthesis by the primary producers on land, in fresh water, and in sea water. He either eats plant material directly or he eats the secondary and succeeding producers. Man is as dependent on the living environment as a whole, as is any other biological species. Preservation of the quality of air, water, and soil is for these reasons a major issue for the individual government and for the intergovernmental organizations.

In discussing the use of the ocean for ultimate disposal of wastes we consider the use or misuse of the greatest and the most important of all resources on this planet. In order to clarify my position let me list four basic concepts for discussion:

1. We do not know the fate of many chemicals discharged to sea water.
2. We do not know the tolerances of marine organisms and ecosystems to artificial changes.
3. We do not know the compositions of pollutants presently discharged to the sea.
4. We do not know all future uses of marine environments, particularly the coastal areas, fjords, and continental shelves.

As is indicated by the title of this paper I am convinced that we must have full control of substances disposed of in the sea and that we must therefore apply treatment to polluted waste water before it is discharged. In arguing the case for treatment I will discuss the real danger of sea-water pollution on the one hand, and the protection that can be obtained through waste-water treatment on the other.

Much information about the chemistry of the oceans has been collected. But in spite of all the data that are registered and the many efforts to characterize the oceans from them, we must admit that on the whole the chemistry of the oceans is not too well understood. The concentrations and distribution of the major ions were established a long time ago and led to the important observation of a remarkably constant relationship between the major ions. The nutritive elements form a group of sea-water constituents which has been subject to extensive measurements. These elements may vary considerably in occurrence according to the biological balance and the primary production through photosynthesis, and according to depth, location, currents, and time of year. During periods of high primary production several of these elements are almost entirely removed from sea water in their inorganic form. Only in the deep waters of the main oceans do we find a fairly constant and abundant supply of nutrients.

Another group of sea-water constituents is made up of a large number of elements which occur in very small concentrations. The significance of these elements for marine organisms is not fully known. So far some 76 elements have been identified in sea water, and it is not unlikely that in reality all stable elements are present there.

Marine organisms in the open oceans live under quite stable conditions. No other ecosystem is developed under similarly stable conditions. Physical, chemical, and mechanical factors vary between narrow limits. Only along the shores, in estuaries and certain enclosed sea-water areas do we find more varying habitats. In such places we therefore find specialized organisms and ecosystems which have adapted to these particular conditions.

A phenomenon that may be important in our discussion is the pronounced ability of certain organisms to accumulate trace elements. In the case of phosphorus, which is an important nutrient, this ability is a necessity for life. The same holds for other nutrients. But in other instances, such as with heavy metals, the cause of such bio-accumulations is not known. Not only do some organisms accumulate substances directly from sea water into their tissues or shells, but also one organism preys on another. These substances may be transferred from one organism to another. In some cases the transfer results in further accumulation so that the top consumer in a food chain may be very rich in certain substances.

Sea-Water Pollution

Mineral pollution of sea water may be discussed for each of the three overlapping groups of elements: the major ions, the nutritive elements and the micro-constituents.

Major elements

It is entirely unlikely that we may alter the composition of the major elements in open-sea water. The amounts of these elements in the oceans are so great that even man with his tremendous powers and large-scale activities cannot cause changes in their concentrations. The same holds for the pH due to the high contents of bicarbonate and the great buffer capacity of the oceans. In local areas or in bays, fjords, and estuaries, these factors may be changed and local effects may be quite pronounced. The composition of open-sea water, however, is a result of the compositions of the entire crust of the world. Therefore if we should be able to change the composition of the major elements or pH of sea water the present situation would re-establish itself as a result of interaction between sediments and other geological structures and the sea water itself. In practical terms this means that the discharge to sea water of sodium chloride, calcium sulphate, chalk, or mineral acids is not of great concern from a pollutional point of view. Such discharges do occur in great quantities in several places and experience so far is in agreement with the theoretical considerations that can be made.

Nutritive elements

The occurrence of nutritive substances in sea water can be altered through pollution. A number of estuaries, fjords, and other confined sea-water areas have received so much additional nutrients, particularly phosphates and nitrates, that new biological conditions have developed in the form of heavy growths of planktonic and attached algae. This phenomenon, called eutrophication, has been of great concern in many places. Nutrients, which are considered to be a necessary addition for any top productive soil, are a nuisance and cause damage when they reach the water. Today eutrophication is one of the most pronounced effects of pollution of the sea. Although it may occur in fairly large bodies of sea water, like the Baltic, it is not very likely that the main oceans can be affected in this way.

The volume of the ocean is estimated to be $1{\cdot}4 \times 10^{18}$ m^3. If we assume an average phosphorus content of 50 μg/l., we have a total of 7×10^{10} t dissolved in sea water. A world population of 4×10^9

persons would produce a total of $3{\cdot}65 \times 10^6$ t/year (on the basis of 2·5 g phosphorus per person per day), which is about 1/20 000 of the contents of the ocean. Only a fraction of the phosphorus produced by humans may reach the sea. A similar evaluation could be made for nitrogen.

As a result of the rapid fixation of phosphorus and nitrogen in plants, coastal areas may be significantly influenced, but it is very unlikely that we are able to affect the total phosphorus situation of the seas. Most open-ocean surface waters are very poor areas as far as productivity is concerned, and have been called marine deserts. It is not likely that we can fertilize these areas.

It stands to reason, however, that because of the many local and regional instances of eutrophication, control of the amounts of nutrients which are discharged to the sea is highly necessary.

Micro-elements

This group consists of a large number of elements which occur in small quantities in sea water. Marine life has developed in the presence of these substances and the organisms tolerate their presence. Some of these elements are contained in waste water and are transferred to the sea through discharges. It is possible that some are added in such quantities, compared to the contents of sea water, that a significant increase in their concentrations in sea water may result. Consequences of changes in sea-water composition of the minor elements are most difficult to predict. We must expect, however, that some of the changes which we impose on sea water may have biological effects and distort the eco-balance of marine life. This seems fairly evident when considering the complex structure of marine life, and different growth stages of the individual organisms may have very particular requirements and tolerances. In the absence of detailed information, the conclusions to be drawn about this very serious problem are, unfortunately, at present a matter of personal judgement.

Organic micro-pollutants

The organic world is considered to be subject to continuous synthesis and breakdown. Except for geological preservation of such materials as oil or coal, we have learned that organic material produced by nature is also broken down by nature. At least this was so until man was able to produce new organic substances with molecular structures never seen in nature before. Some of these new substances, halogenated hydrocarbons in particular, are extremely stable. It seems that physical, chemical, and microbiological breakdown of some of these substances is very slow. For example, DDT has a half-life of 10–15 years. In addition, substances like DDT and PCB's have properties which are unfortunate from a pollution-control point of view and with respect to their environmental distribution and effects. They are fat-soluble and thus accumulate through, and are concentrated by, food chains. They are spread through the air and water and by migratory animals. Short-term effects are fairly well understood for these substances as for the individual elements, and long-term effects on life, reproduction, and behaviour are becoming increasingly well known. For example, it has been found that PCB's can cause severe disruption of hormonal systems in some animals. Very low concentrations of PCB's and other chlorinated hydrocarbons appear to have caused the elimination of at least one bird species from the entire East Coast of North America. In another case an initial environmental concentration of 3×10^{-12} of DDT in sea water was brought to 25 million times its original concentration by a four-step food chain, resulting in reproductive failure for the top carnivore.

Treatment Will Protect the Sea

Extensive research on treatment processes is going on in many countries. These efforts are concerned with the conventional types of treatment like mechanical and biological treatment, and with the more advanced treatment processes such as chemical precipitation, filtration, and carbon adsorption. A full account of the improvements of water quality which each of these processes can give is beyond the scope of this paper. It may, however, be satisfactory for the present argument in favour of applying treatment to recapitulate the major items in this connection.

Mechanical treatment

This will protect the local areas against accumulation of sediments at the bottom, and floatables and oily substances at the surface and the shores. In some instances it is expected that the main trouble from the particulate matter can be eliminated by disintegration of these substances rather than removal by mechanical treatment.

Biological treatment

This is usually applied in order to remove degradable organic matter. In those cases where it is necessary to protect the oxygen balance of sea water, biological treatment may be imperative. It is also considered by many to be an essential part of any extensive waste-treatment process. Biological treatment may also be considered as an indirect control measure for receiving waters. It is well known that sewage which is to be subject to biological treatment should not contain toxic substances, or otherwise be unfavourable to the growth of micro-organisms. The mere fact that waste water can pass through biological treatment without poisoning the system is a guarantee that the waste at least is tolerable to some living organisms. Usually mechanical and biological treatment will remove only a small fraction of the nutrients and very little of the micro-pollutants of organic or inorganic nature.

Chemical treatment

A very promising treatment process, which in recent years has been used to treat waste water, is the application of chemicals which precipitate and flocculate materials in the water, and thereby remove particles as well as certain chemical substances. The most conventional method in Scandinavia is to use aluminium salts as a precipitant. Usually the main purpose is to remove phosphates. Iron salts and calcium hydroxide are also being used. Although the major object of this treatment has been to achieve control of eutrophication through removal of phosphates, it is well known that the main part of the organics and a number of other substances are removed at the same time. Research in this field is progressing rapidly and there is already sufficient knowledge for design and application of the process. As an example, the cleaning up of Oslofjord will be based on a combination of mechanical and chemical treatment.

Filtration

Much of the remaining organics after biological treatment, and most of the remaining phosphorus after chemical treatment, will exist in the form of small particles and can be removed quite efficiently by filtration through conventional rapid sand filters or more modern modifications. Filtration has long been used for the treatment of drinking water, and the technology is available and well developed.

The integration of filtration technique with the other unit processes in a waste-water treatment plant may cause difficulties. It should be possible to master these problems, for example through the use of hydraulic equalization and modern techniques of automated control.

Activated carbon

Certain constituents of polluted water such as organohalogens and some micro-constituents are difficult to remove even when chemical precipitation and filtration are applied. Many of the substances have great affinity to activated carbon which seems to have a great potential both as a final treatment step (preceded by mechanical, biological, and chemical treatment) and as an alternative to biological treatment. In the latter case the activated-carbon step will normally be preceded by mechanical and chemical treatment (physical-chemical treatment). It seems that the application of active carbon as a final step in the treatment process will prove to be quite feasible, and may give extensive protection to receiving waters.

Other methods

There are several more processes which are under investigation, and may be useful as part of a complete treatment scheme. Examples are polishing in retention ponds, nitrogen removal, reverse osmosis, and electro-dialysis. Application of these steps must of course be considered carefully, because every process needs raw materials and energy to be performed and a sound balance must be found between the costs and the effects of waste-water treatment.

Treatment is not the only answer to pollution. Much of the pollution can be stopped at the source. Wastes can be recycled or turned into new products. Reduced consumption of pure water will usually give less water pollution and also make it easier to apply extensive treatment.

Monitoring

The basis for a choice between several alternative solutions is usually their economic consequences. In disposal of waste water from coastal areas, we often have a choice between treatment of sewage and a long outfall system. Managers will usually apply the most economic solution. The economic consideration must be based on the cost of construction and the cost of operation.

It should be required that any discharge of polluting material be supervised and monitored. In the case of treatment plants, monitoring is easy to achieve so it usually forms a part of the running operation. For ocean outfalls monitoring is more difficult and supervision is often neglected. When a long outfall system is constructed to discharge wastes at some depth, the manager may feel that the problem as such is solved as long as the sewage will flow through the pipe and the outfall is in operation ion.

It is essential that the deep-outfall solution is accompanied by an adequate monitoring programme. The State of California started to apply strict conditions for a monitoring programme before an outfall scheme was approved. The monitoring has given valuable scientific results and has at the same time been a guarantee to the public that any damage to nature which the outfall might cause would be detected. The cost of monitoring ocean outfall systems may be high, since it is necessary to monitor the adjacent bottom areas, the free sea water itself, and a number of selected organisms. If a persistent current is passing by the outfall system, it may be necessary to monitor the waters at great distances from the point of discharge. The monitoring programme may be so expensive that it directly affects the choice of a technical solution for the outfall system.

A monitoring programme will always have an additional value as a description of the quality of the environment at a certain time. Monitoring will give reference data for the evaluation of time-dependent changes, whether they are of a natural or man-made nature.

Coastal Water Pollution

It is, of course, important to point out that man is particularly dependent upon near-shore areas of the sea. These are the areas which we most frequently visit and use for different purposes, be it recreation, commercial fishing, or other activities. Waters above the continental shelves, along the shores, and in the fjords are the most productive areas of the sea. At the same time they are the waters which are most readily contaminated through man's activities.

The conflicts of different interests in the use of coastal waters have been subject to much debate. Conflicts have arisen between different types of users in a specific region and between regions or nations. These conflicts have been real ones in the sense that there have been a number of serious incidents of coastal water pollution with fish kills and other damage and hazards, including human deaths, as a result.

It has been stated that if disposal arrangements include a long pipeline to a sea outfall at great depth, pollution incidences can be avoided. Although this may be true in some cases, it may in others be just a matter of how good we are at observing the results of our activities. There is an old saying that 'out of sight is out of mind', and although this philosophy is not used deliberately, it may still be true that because of lack of direct observations it is assumed that results are satisfactory.

The continental shelves are now considered to be the location of vast untapped resources, for instance for extraction of special raw materials like oil and metals. In particular, offshore oil-drilling has led to discussion of the conflicts between these new activities in relation to the sediments and geological structures of the shelf, and to the use of water and the sea bed for harvesting food. Technical structures such as pipelines for the transfer of waste waters to a point of discharge, or to bring oil into shore, are features which may cause conflicts with other users, for instance trawling for bottom fish. It is often a question of economy how to lay these pipelines. It may be possible, but difficult and expensive, to have them put into the sea bed. Conclusions on this point are important for the discussion of use of the shelf area, and are important for the economic comparison of the alternative methods of waste disposal.

TABLE 1. Acceptability of sea-water pollution

Polluting material	Open sea—high dilution	Coastal water—low dilution
Particles	partly acceptable	not acceptable
Oil, floatables	not acceptable	not acceptable
Organic matter	acceptable	not acceptable
Nutrients	acceptable	not acceptable
Heavy metals	not acceptable	not acceptable
Organic micro-pollutants	not acceptable	not acceptable
Pathogenic organisms	acceptable	partly acceptable

Pollution is an open term and covers a number of different effects of waste materials. According to the interests and to the nature and value of the resources, different types of pollution may be considered to be particularly important to evaluate. A short description of some categories may be useful. See also Table 1.

Visible pollution

Visible pollution will, for the community at large, probably be considered to be a severe effect of water pollution. Although it may appeal mainly to the aesthetic values and emotions of people, a disposal scheme which leads to visible pollution is a quite unsatisfactory solution. Why should we flush our wastes away from our homes and industries, only to meet them again when we are somewhere else? Any visible signs of polluting materials in the sea will cause a reaction from the public, and hence be the source of public opinion and pressure directed towards abatement of the nuisance. Visible pollution may be avoided through conventional, mechanical, and biological treatment in most cases.

Eutrophication

In several fjords, estuaries, and coastal areas in the developed parts of the world, a marked eutrophication of the waters has developed. Although a slight increase in productivity of certain waters may be desired, the effects of eutrophication are nearly always of such a degree that they are entirely negative. Eutrophication in sea water usually leads to the excessive growth of undesirable types of attached algae, and to the formation of large blooms of free-floating phytoplanktonic organisms which may colour the water in shades of green, brown, or red. Eutrophication is one of those phenomena which may be caused by the fact that a resource is put in a wrong place. When used in agriculture or forestry, nutrients may be beneficial to man, but when discharged to confined water areas, they will cause an amount of primary production of algae which the ecosystems of these waters cannot assimilate. If 100 t of phytoplankton may be transferred to 1 t of fish, 500 t of phytoplankton at the same location may lead to less than 1 t of fish or even a fish kill.

Artificial eutrophication can be avoided through combinations with chemical treatment in most cases.

Organic matter

Domestic waste will usually be rich in organic matter which will be broken down through microorganisms. Industrial waste may be very rich in organics. The breakdown of the organic material will deoxygenate the water, and may change the biology and chemistry of the water drastically. Harmful effects caused by discharge of organics are a well-known water pollution problem and the usual control measure is to apply biological treatment to the waste water.

Bio-negative effects

A number of toxic substances are discharged in waste water. Some of these are organic in nature and some are inorganic. Some of the organic substances, particularly halogenated organics, are quite resistant to breakdown and may easily be absorbed into organisms and transferred in increasing concentrations via the food chain. The same holds for some of the inorganic substances (heavy metals). The nature of the processes involved for the various bio-negative substances is not too well known, and only those which are currently considered to be particularly dangerous have been subject to intensive studies.

Although removal of bio-negative pollutants is often difficult, it is believed that they can be removed or greatly reduced through advanced waste-water treatment.

Sediments

Where currents are not too strong, most of the particulate matter will settle in the ocean close to the place of discharge. Discharged particles will, however, in many cases change the structure of the bottom surface, chemically or physically, in such a way that it distorts the bottom fauna. When particulate matter of an organic nature is deposited, it is subject to slow microbiological breakdown. This will turn the top layer of the sediments anaerobic, and quite often remove oxygen from the waters which are in contact with the bottom. This kills all aerobic bottom organisms. In many cases sewage sludge has been dumped in places where renewal of bottom water is frequent. Marked changes are observed, but it seems that negative effects are limited to fairly small areas. An increase in the number of certain bottom organisms is usually observed.

Pathogenic organisms

Waste of human or animal origin will usually contain a great variety of micro-organisms, some of which are pathogenic. These pathogens may find their way to man either by direct use of the contaminated water or by accumulation in shellfish or other edible organisms. In several countries there is much concern for the bacteriological quality of coastal waters which are used for recreational purposes. The marine environment will usually have a high self-purifying capacity for this type of pollution, and micro-organisms or their spores will usually not persist for any length of time. However, a discharge may spread rapidly over fairly large areas depending on current and wind activities. The removal of pathogenic organisms can be achieved with several kinds of treatment processes and a final treatment of the effluent with substances like chlorine may make the discharge free of this type of pollution.

Discussion

The following set of arguments as discussed above should, in the Author's opinion, in most cases lead to the installation of waste-treatment facilities when waste water is to be discharged to the oceans:

1. The ocean is the ultimate sink of the world and therefore needs a maximum of protection.
2. The discharge of untreated waste waters does cause nuisance and harm.
3. Long outfall structures along the sea bed and ocean floor may be in conflict with other uses and interests in these areas.
4. Treatment processes do remove a number of polluting substances.

In considering the use of the marine environment as a recipient of wastes from modern society, it is extremely important to stress that we are discussing a problem of great complexity, and that our knowledge is far too limited to provide conclusive answers. It may take a long time for research and experience to produce the necessary information. In this situation we have two attitudes and two courses to follow:

A. We can go ahead and use the sea as a sink and use sea outfall systems for waste discharge whenever it is economical and feasible. If damage and nuisance result, we may change our plans and methods according to the best experience and knowledge gathered at the time. This, really, is the general practice for all kinds of waste disposal, be it solids, liquids, or gases. It has led to a lot of experience, a lot of knowledge about environment, and a lot of trouble and damage including serious threats to human health. As long as the harmful effects were of a local nature, the method of trial and error has been fairly successful in the long run. We destroyed some estuaries and fjords and learned how to protect others.

B. We can on the other hand say that the oceans are the most sensitive, the largest, and the most important of all natural resources, and that they need a maximum of protection. Any disposal of wastes in the ocean should be avoided. Used water should be subject to the best practicable means of treatment before it is returned to a river or directly to sea.

Any choice between these approaches should be based on facts, knowledge, experience, and the size of the problem. It is questionable whether we are allowed to follow Alternative A for a resource that covers 71 per cent of the earth's surface. On the other hand, on a short-term basis, it will be stated that we cannot afford to use the second alternative. Technology and trade have been allowed to develop according to the trial-and-error concept for so long that no major change can be expected to take place at very short notice.

Conclusion

It is the duty of the scientific community to study available information and draw conclusions which we can report to the leaders of our respective societies. The presentations and discussions at this Symposium will be a useful way for re-examination of the problems concerned and hopefully be an important step towards sound management.

It is tempting for a manager to apply a wait-and-see philosophy. If something is going wrong due to pollutional discharges, we may observe it and then we can correct for it. This philosophy has been used on several other occasions, particularly in the case of pollution of surface waters. When full-scale experiments are done in small areas, it may, from a global point of view, be a satisfactory approach to the problem. However, when we discuss the oceans, we are not allowed to do a full-scale experiment. We must at any time be sure that we have done our utmost to protect the sea.

We should stop pollution at the source, and when elimination of pollution is not to be achieved, we must apply treatment before discharge. Dilution is not the answer to pollution.

Mr. Baalsrud took the opportunity, when introducing his paper, to emphasize several relevant aspects of it. He proposed that by 'treatment' was meant a technical process giving at least primary or mechanical treatment. The need for treatment, even of discharges to the ocean, ought to be based on the large volume of valuable evidence of marine pollution that had been published in recent years—despite which he still felt there was a lack of knowledge on the character of polluting material and its degradation and stabilization in nature, as well as on the tolerances of single species and whole ecosystems in the marine environment. Also, little was known of the future uses of marine resources.

In the context of environmental protection engineering he focused attention on three factors: (1) The time factor. The world population was expected to rise from 4000 to 6500 million between the years 1974 and 2000; it was to be hoped and expected that ever more people would enjoy a higher standard of living, and this meant technological development and more pollution per individual. (2) The cost factor. He considered that the importance of the cost of environmental protection was greatly overstressed, and was convinced that the cost could be afforded if ways could be found of including it in the economic mechanisms. (3) The need for choices. Engineers should be more eager to present alternative plans on every scheme affecting natural resources. Each alternative should be evaluated carefully with regard to costs, benefits, and threats, thus making it easier for the decision-makers to reach environmentally sound decisions.

Personally, he thought the only acceptable goal to be that of zero pollution discharge—or of stopping pollution at the source. It was the responsibility of the environmental engineer to push development in this direction on every possible occasion.

In conclusion, he suggested that the 1974 approach to environmental problems should be: no argument was ever needed *for* treatment; a very good argument was needed for an ocean outfall *without* treatment.

[*The discussion of this paper starts on p. 177*]

The Case against Treatment

John T. Calvert
Senior Partner, John Taylor & Sons, London

As so often happens when an effort is made to condense a great deal of meaning into a few cryptic words, the title of this paper is most misleading. In fact no case can be made against treatment because it is impossible to prevent treatment occurring by natural processes. It is only in the very exceptional circumstances where sterile or highly toxic wastes are involved that degradation can be prevented and, after all, the process of treatment is merely the degradation of organic materials.

The argument between Dr. Baalsrud's paper[1] and mine is not a question of whether treatment should take place but rather where it should take place. Consequently if the title of my paper were to be expanded to indicate the views I hold it would read 'Whether it is desirable to use valuable land area for the biological treatment and purification of sewage and organic wastes or whether the process should be allowed to take place naturally in the wide expanse of the ocean'.

Natural Treatment and Purification

The fallacy that when sewage is discharged to sea no treatment or purification occurs, is to a large extent responsible for the opposition encountered to this method of disposal not only from the lay population but also from many scientists who one would have expected to have been better informed. Many of the opponents of sea disposal seem to assume that the sewage remains unchanged in the coastal waters, and that so far as the United Kingdom is concerned eventually our islands will be surrounded by a sea of sewage. With this thought in mind it is wise to consider the changes which take place when crude sewage is discharged to sea. For the moment I shall exclude from the discussion the coarser visible material with which I shall deal later.

When sewage is diluted one hundred times with clean sea water it is from a bacteriological point of view indistinguishable from a fully treated secondary effluent and so far as suspended solids and biochemical oxygen demand are concerned it is not only indistinguishable but, in fact, is of a superior quality. Subsequent to the dilution the natural chemical and biological processes of purification occur with the result that no change in the character and composition of the sea can be detected. In an inland river the prime indicator of pollution is the depletion of the oxygen content, but in the ocean there are very few places—except in the immediate vicinity of a large discharge—where any oxygen depletion can be recognized.

My case, therefore, is that nature will do the work of purification in an efficient and economic way without requiring the construction of elaborate, expensive, and land-consuming treatment works. If one is satisfied that sewage can be discharged to sea without causing harm to the composition of the water, to the life which exists in the ocean, or to the amenities of those enjoying coastal pastimes then, in my opinion, the strongest argument for the adoption of this method is the elimination of the profligate use of land if treatment is adopted.

Land Use

Land is a very precious commodity which can rarely be manufactured or extended, and its use for sewage treatment appears most wasteful when it can, in fact, be avoided. In seaside resorts the selection of suitable land for sewage treatment poses particularly difficult problems. Firstly there is the problem of geography which is concerned with the configuration of the ground. On the coast the land slopes to the sea so that sewage from any development naturally gravitates to the bottom of a valley leading to the sea. Discharge through a pipe outfall to the sea is a natural extension of this process. If, however, a significant area of land is to be dedicated to the treatment of the sewage it is impossible to utilize an area on the coastal strip. This follows from the fact that most development on the coast takes the form of holiday resorts where obviously a sewage-treatment works would not be welcome on the front or esplanade. The sewage, therefore, has to be collected in a pumping station (or stations) situated on the coast and lifted back inland, usually for a considerable distance. It is obviously an expensive matter.

Amenity

Secondly there is the problem of amenity which inevitably arises when it is necessary to select the site for sewage treatment. Inland, where there is no possibility of discharge to the sea, the arguments are always put forward that the selected site is the best agricultural land in the county and in any case it is far too near residential development. It is very rare for an inland sewage-treatment works at all times and under all conditions to be entirely free from nuisance and, in my opinion, the interference with amenities by a discharge to sea is far less than that caused by the establishment of an inland sewage-treatment works.

Even though it may be possible at the present time and with the present degree of development to select a site which is acceptable to the majority of the population, there is no case in my experience without some objectors. Furthermore, the amenity effect spreads much wider than the actual site occupied by the treatment works as, for the reasons stated, it is undesirable to build immediately adjacent to such a works. Yet in spite of this undesirability people will persist in building right up to the boundary of a works and then complain that their living standards are spoiled by the works' existence.

Cost

Thirdly there is the overall question of cost, both of capital construction and annual operating costs. In my experience the discharge of sewage to sea is far cheaper than the alternative of providing inland treatment works even when primary treatment only is taken into account. The capital cost of discharge to sea depends on the length of the outfall, which in turn depends on the configuration of the coast and the strength of the tidal currents. With a very long outfall, contrasted with the provision of primary treatment and ultimate discharge through a very short outfall, the capital cost of the two schemes can in general be very similar. However, when the annual operating cost is taken into account, sea discharge in all cases becomes more economical. The operating cost of a sewage-treatment works in the United Kingdom can vary between 50p and £1 per head of population per annum and this additional operating cost invariably throws the financial balance in favour of sea discharge.

In order to illustrate this point I would like to give figures for a scheme which has been constructed in the Wirral in Cheshire on the west coast of England which incorporates a very long sea outfall. The scheme as existing prior to the construction of the new works included a joint sewer for two District Councils, discharging to the sea through a very short outfall which was recognized as being totally unsatisfactory and quite inadequate in length. The joint sewer carried sewage from both of the Districts and at this stage politics entered into the argument. One of the Councils took the view that they would not permit discharge of untreated sewage into their coastal waters, and the other Council adopted the attitude that they would not allow the use of land for the treatment of their sewage. Obviously the sewage from each Authority in the joint sewer could not be separated and it appeared that an impasse had been reached. This was only resolved by agreeing to discharge the sewage 5 km out to sea which just satisfied the District Council objecting to sea discharge. This decision was put into practice and the scheme has recently been commissioned.

Detailed estimates were made of the cost of the alternative proposals and, despite the great length of the outfall, Table 1 (taken from a description of the scheme) shows that sea discharge was financially advantageous.

TABLE 1. Comparison of cost (£'000) of six alternative schemes for disposal of North Wirral sewage[2]

Scheme	A Full treatment at Dove Point outfall	B Primary treatment at Dove Point outfall	C Semi-full treatment at Dove Point outfall	D Long sea outfall	E Works and lagoon treatment outfall to River Birkett	F Sea outfall to Dee Estuary (Hilbre outfall)
Capital cost	2677	2281	2512	1933	2600	2182
Annual operating cost *plus* loan repayment	264	222	248	156	264	174

Five kilometres to sea is a terrific length for the discharge of what is solely a domestic sewage. It is quite impossible at this distance to see any visible slick, and float experiments indicate that if there were any possibility of material being washed back to the shore it could only reach the beach after sufficient time for virtually complete die-off of any bacteriological contamination.

Sludge Disposal and Agriculture

Another considerable advantage of locating treatment works in the middle of the sea is that it alleviates the problem of sludge treatment and disposal. With inland treatment works the disposal of sludge is not only one of the most expensive sections of the works, but also presents the most difficult technical problems and is the most likely to create a nuisance. The system of drying on open beds is extravagant with land and almost inevitably creates a smell in the vicinity unless the sludge is very thoroughly digested before being put on to the beds. This applies particularly to summer conditions with high temperatures. Similarly, mechanical methods of sludge treatment and disposal can create a nuisance around the works, although the area occupied by the sludge-processing equipment is considerably reduced.

Probably the cheapest and most satisfactory method of sludge disposal is to discharge it in liquid form after digestion on to agricultural land. This, however, suffers from the disadvantage that at certain seasons of the year it is impracticable to put wet sludge on to agricultural land and consequently some alternative method of disposal must be provided.

It has been argued that it is wasteful to discharge sewage and sewage sludge to sea as this represents the loss of valuable fertilizer. This argument, however, must be put in its proper perspective. In 1970 it was estimated, as quoted in the Report of the Jeger Working Party on Sewage Disposal[3], that the total dry solid matter recoverable from a contributing population of 43 million is probably around 1.2×10^6 t/year. Using figures from the Agricultural Research Council this total of dry solids from sewage would include 29 300 t of nitrogen, 15 500 t of phosphorus, and 3650 t of potassium. This represents the contribution to agricultural needs if all the recoverable sewage sludge now treated inland were used on the land. In contrast to these figures the consumption of artificial fertilizers in England and Wales in 1968 amounted to over 600 000 t of nitrogen and nearly 400 000 t of both phosphorus and potash. Thus utilization of all the available sewage sludge represents only about 4.5 per cent of our present annual nitrogen and phosphate consumption and only 1 per cent of the potash consumption. When one considers the cost of recovery of these elements from the sewage for use in agriculture this small loss cannot be considered reprehensible.

It cannot also be contended that the discharge of the sludge into the sea creates in any way undesirable conditions. This, I think, has been more than adequately proved by the extended and detailed investigation carried out on behalf of the Merseyside Authorities[4], which showed beyond doubt that the sea has an enormous capacity to absorb and treat quite satisfactorily very large quantities of domestic sewage sludge. Manchester and Salford discharge at the present time about 40 000 t of dry material yearly as sludge (from a population of about a million) off the west coast outside the Mersey Estuary, and the Report suggested that no harm would accrue if the quantities of sludge were increased sixfold. I have no doubt whatever that the discharge of sewage to sea is the best method of solving the sludge-disposal problem.

Furthermore, even where for any reason it is not possible to discharge all the sewage to the sea, sea disposal of the sludge is an economical and satisfactory solution of this problem wherever shipping has access to the site of works. In such cases the sludge can either be carried out to sea by pipeline or by barging or in sludge-vessels. In addition to Manchester and Salford many other local authorities in the U.K. including London, Glasgow, and Exeter have very satisfactorily adopted this means of sludge disposal.

Essential Requirements

In my view all the arguments based on land use, amenity, cost, and agriculture favour the discharge of sewage to sea. Nevertheless in order to obtain these conditions it is essential that certain requirements should be observed and it is not proper to discharge crude sewage at low-water mark as has often been done in the past. Most of the complaints of beach pollution arise not from high bacteriological counts but from coarse visible solids which can be washed up on to a bathing beach. Many of these solids which are easily recognizable as being of sewage origin are not readily disintegrated and could exist unchanged for a very long period. It is, therefore, essential that coarse solids should either be broken down or removed from the sewage before it is discharged to sea however far from the coast such discharge takes place.

It is debatable whether it is preferable to break down these solids by comminutors or similar machines or whether they should be removed by fine screens. Neither of these processes is completely effective, but in my view the removal of the larger solids by screening is generally more effective in removing a cause of complaint than the introduction of comminutors. If this comminution or screening is to be regarded as treatment then my case falls to the ground. I contend, however, that it does not conflict with the requirements of land use, amenity, and cost which are involved when primary and secondary treatment of the sewage is carried out inland. In particular, the machines can be housed in a very limited area attached to a pumping station and can be sited quite close to the coast without detracting from the amenities.

The other essential requirement is that the discharge should take place at a 'proper' distance from the shore and the determination of what is a 'proper' distance is the objective not only of this Symposium but of all the research work currently in progress. The so-called 'long' outfalls have become fashionable in recent years and although 'long' is not precisely defined, they seem to have varied from about 1 km to about 10 km beyond low-water mark. In my view many of these have been designed without any true design criteria and in many cases, judging from the work of the Water Pollution Research Laboratory, they would appear to me to be longer than absolutely necessary. If, as I suspect, these outfalls can be made shorter with equally efficient disposal, the economic argument in favour of discharge to sea is noticeably strengthened.

Conclusion

There are exceptions to every rule, and of course there are some situations even in coastal towns where treatment works are desirable or preferable, but such cases in my experience are few and far between. They arise where the configuration of the coast is such that discharge into an adequate depth of water cannot be achieved or where the currents are such that there is a danger of contamination of a widely used bathing beach. However the most cogent reason for the adoption of treatment works always seems to me to consist of the sentimental objection of those ill-informed people who imagine either that sewage is not purified in the sea or that contact with it is dangerous however well it is purified. They are perhaps not aware that a large proportion of London's drinking water consists of purified sewage.

In my opinion, and I hope in that of many participants of this Symposium, the proper method of sewage disposal in a coastal town is generally by discharge to sea. The arguments of land use, amenity, and cost are overwhelming.

REFERENCES

1. Baalsrud, K. The case for treatment. This volume, pp. 165–172.
2. Roberts, D. G. M., Summerton, J. D., Palmann, O., and Thompson, V. G. North Wirral Authorities and Hoylake UDC long sea outfall. *Proc. Instn civ. Engrs*, 1972, Suppl. (v), 101–126.
3. Ministry of Housing and Local Government and Welsh Office. Taken for granted. Report of Working Party on Sewage Disposal. H.M. Stationery Office, London, 1970, p. 11.
4. Department of the Environment. Out of sight out of mind. Report of Working Party on Sludge Disposal in Liverpool Bay, Vol. 1. H.M. Stationery Office, London, 1972, pp. 12 and 32.

In his introductory remarks Mr. Calvert said that his paper was intended to deal with discharge of largely domestic and generally biodegradable sewage to the open sea; it was not intended to cover highly toxic wastes, or discharges to estuaries or to coastal waters close to the shore. With these limitations in mind, he then outlined the four arguments on which he had based his conclusion that the discharge of untreated sewage to the sea was a proper method of sewage disposal.

First, he could see no reason why the natural process of biodegradation which took place in the sea should not be used in preference to artificial processes involving complicated engineering construction.

Secondly, land was one of the most precious commodities, and its use for treatment works (when other satisfactory means were available) seemed to him to squander that commodity uselessly. Land could not be created, and should therefore be preserved for essential uses or for recreational purposes.

Thirdly, in his experience, the alternative of an inland treatment works resulted in far more interference with amenity than did a proper discharge to the ocean. Anyone who had attempted to establish new treatment works (particularly in Britain) must know of the terrific weight of objections invariably received on amenity grounds.

Fourthly, he maintained that it was almost always cheaper to discharge the sewage to the sea than to build inland treatment works—in view of the costs of collection, land, and treatment works. In his paper he had discounted any suggestion that sewage should be treated inland for the recovery of natural fertilizers; even if the whole of the fertilizers were recovered from the sewage in the United Kingdom it would represent only a very small proportion of the total fertilizers used.

In conclusion, he said there were three principles he thought it important to bear in mind in the practical solution of pollution problems: (1) most actions were, in effect, some form of compromise, (2) virtually everything done involved some risk, and (3) the public purse, even that of the World Bank, was not limitless.

GENERAL DISCUSSION ON PAPERS 17 AND 18

Mr. F. M. W. King (U.K.) considered the argument in favour of treatment to have been put very well, but that the case against treating domestic sewage from seaside towns by inland works was irrefutable. The Division of Wessex Water Authority for which he was responsible had a sea-board of some 35 km along the English Channel and a summer population of about $1\frac{1}{4}$ million. In addition to some ninety-three inland sewage works, there were ten sea outfalls, two of which discharged into about 19 m of water at distances of 1370 m and rather more than 1 km offshore; the remaining eight outfalls were shorter than was to be desired and it was hoped that six of them would eventually be replaced by one main outfall—possibly in tunnel.

The long sea outfalls had given no trouble at all. One of them (with its 30-m predecessor) at Bridport had been the subject of intensive investigations by the Water Pollution Research Laboratory from 1967 to 1973. Some of the results had been published in detail[5–8], but the conclusions that made the biggest impact on him were that any pollution of the nearby bathing beaches was more likely to come from two small rivers (receiving treated effluent from inland works) than from the long sea outfall, and that a sizeable colony of gulls nesting on cliffs near the shore was also considered to be a significant source of bacterial pollution of the nearshore waters. Results of the Laboratory's investigations with the sewage discharging through experimental outlets at 430 m and 680 m offshore—as alternatives to the design diffuser at 1370 m—supported Mr. Calvert's contention that there was perhaps a tendency to take sewage further out than was absolutely necessary.

In Mr. King's opinion, the work at Bridport firmly endorsed the case against treatment when conditions could be shown to be suitable for discharge to the sea by a properly sited, correctly designed, and well-constructed outfall; the economic advantages, particularly in running costs, were very marked. Capital costs, however, were rising for all forms of sewage disposal, and he wondered if Mr. Calvert could say what progress had been made in the development of less expensive techniques for outfall construction, and whether, for example, thick-walled polyethylene pipe might eventually supersede steel for long sea outfalls. Mr. Calvert replied that there was greater danger of damage to a plastic pipe and that the cost of an outfall was more in the laying than in the pipe material.

Mr. D. E. Cannon (U.K.) did not think the question was one of treatment or non-treatment but one of degree of treatment; most sea outfalls included some preliminary treatment in the form of screens. It would be wrong to make a general statement that all sewage could be discharged to the sea without treatment, since both the composition of the sewage and the ability of the receiving waters to absorb the pollution effect varied from place to place. Land-locked seas did not have the absorbing capacity of open ocean waters, and heavy metals in industrial wastes could have disastrous effects when discharged to the marine environment. On the other hand, the preoccupation of some politicians with cleaning up the environment needed to be kept within reasonable bounds. Present concern was with the concentration of waste products which could be discharged into the seas at certain points along coastlines. He felt that each case should be considered on its merits. Finally, he pointed out that Mr. Baalsrud's argument on the cost of monitoring a sewage outfall was incomplete, since monitoring would also be required (though not to the same extent) for the sea discharge of a treated effluent.

Prof. P. Harremoës (Denmark) said that everyone could agree on certain cases for treatment: where local conditions gave rise to significant problems of oxygen depletion, or where discharges of nutrients resulted in eutrophication. The agreement ended when neither of these phenomena was likely to occur. The question was thus whether there was a baseline with respect to treatment when discharging to the ocean from a straight coastline with currents which would rapidly disperse the pollutants. Mr. Calvert had given the definite answer that no treatment was required, his principal arguments being based on land use, amenity, and cost—though Prof. Harremoës had some difficulty in understanding that land was so precious as to be a significant factor. Mr. Baalsrud's statement for treatment was a less clear recommendation—the best practicable means—but from Table 1 of his paper it appeared that his main argument rested on the unacceptability of heavy metals and organic micro-pollutants for discharge into the ocean. There was general agreement that these pollutants should be stopped at source wherever possible, but this was no argument for treatment of domestic sewage. Although the industrial contribution of heavy metals could be stopped at source, it would be very difficult to make significant reductions in the contributions to domestic sewage. He therefore asked if treatment processes were needed for removal of metals, but pointed out that this removal could be detrimental if the resulting sludge was then disposed of on to farm land. There was evidence that both biological treatment and chemical treatment using aluminium and iron removed 50–90 per cent, depending on the particular metal. He asked if this was sufficient or if lime treatment was also needed, and where should the line be drawn. In his opinion, the removal of micropollutants required activated-carbon treatment. All these processes, to which even reverse osmosis might be added, were (in Mr. Baalsrud's terms) practicable means of treatment which provided the best protection of the sea.

It was clear that there was no scientific basis for deciding which degree of treatment should be favoured, ranging from no treatment to the decrease in the concentrations of polluting constituents by several orders of magnitude. This left plenty of room for emotions, and the decision would be made by some politician who was (in this respect) ignorant enough to hold a firm opinion. In Denmark the regulatory agency had chosen mechanical treatment as the baseline (except for Copenhagen where biological treatment would be required), and in England no treatment was apparently the baseline. This was so far from the concept of 'maximum protection' and 'best practicable treatment' that the concept became meaningless. He did not consider it likely that mechanical–biological–chemical–activated-carbon treatment and disinfection would become a standard procedure for ocean disposal even in the United States under the zero-discharge concept, which was equally meaningless. Whatever level of treatment was chosen there would always be a residual, and this called for a wait-and-see philosophy Awareness of this was the basis for the design of monitoring programmes which would make the scientific community better prepared for pertinent questions next time.

Mr. Calvert commented that he had advocated the use of mechanical treatment to remove coarse visible solids (in the section of his paper that dealt with essential requirements); his point was that such treatment involved neither the collection of sewage at an inland site nor the use of a large area of land for biological treatment. With regard to land availability, he said that from his experience of attempting to locate sewage treatment works in the United Kingdom, this always presented serious problems: any selected site was invariably 'of the highest amenity value' or 'the best agricultural land in the county'. Mr. Baalsrud admitted that Prof. Harremoës had found the weak spot in his arguments. The question was how could just one or two stages of treatment suffice when much more comprehensive treatment would be required for full protection of marine resources. The answer was in the words 'best practicable means of treatment'. At any time the treatment processes which had become generally available should be applied, and these would be significant steps towards the final goal. Mechanical treatment was likely to be part of any combination of processes, and was the reasonable first stage. Thus mechanical and chemical treatments had been proposed as the first stages in the Oslo area, and would remove 85 per cent of the particulate matter, 85 per cent of the phosphorus, and 70 per cent of the organic matter. As the proposed outfall would discharge to the inner fjord, there would be full control of the receiving system and further treatment steps could be proposed as they became available.

Mr. E. W. Flaxman (U.K.) took up Prof. Harremoës's point on sludge disposal, and said that Mr. Baalsrud had apparently not considered the disadvantages that might accompany treatment. The sludge produced during treatment could contain many of the contaminants which were a cause for concern, but it had to be disposed of somewhere. There were only three basic places it could go: the sea, the land, or the air. Mr. Baalsrud had expressed doubts on disposal at sea, but the alternatives were not entirely unobjectionable. If sludge were dumped or spread on land, some of the contaminants might become bound in the soil, but others would be taken up in crops and yet others would be leached out and eventually find their way to the ocean. The effects of the ash and flue gases produced by sludge incineration were not fully known. The point he wished to make was that if attention was concentrated solely on safeguarding the oceans, without considering the impact of the basic pollutants on the environment as a whole, there was a risk of adopting unsound policies.

Dr. D. D. Mara (U.K.) pointed out that, in many parts of the world, a realistic economic appraisal of sewage disposal schemes should include considerations of the benefits to be gained by re-use of sewage or effluent[9]. The agricultural use had a long history, and with modern techniques it had a profitable future[10]. Discharge of sewage, whether treated or not, into the open sea (indeed, into *any* watercourse) might therefore represent an irresponsible waste of a valuable resource. Prof. R. Mitchell (U.S.A.) added that it was time to stop looking at wastes in a negative way, by striving for means of neutralizing their harmful properties before disposing of them in the sea. Both the water and the faeces in domestic sewage were valuable resources. Advanced treatment could recover the water, and sophisticated fermentation technology was available to convert the wastes into single-cell protein and other useful products.

Mr. J. A. Wakefield (U.K.) said that, in his view, both Authors were right, but that Mr. Baalsrud's recommendations were not practicable in an imperfect world. He was sure that it was a matter of priorities and economics, and that it would be better to re-site or lengthen existing outfalls, to monitor the effects of the discharges, and then to add treatment works as and when necessary. In reply, both Authors essentially agreed with Mr. Wakefield's remarks, though Mr. Baalsrud insisted that the practical choice of priorities was no argument when discussing the *need* for treatment.

Mr. B. Grover (World Bank) said that Mr. Calvert had correctly observed (in his opening remarks) that even the World Bank did not have enough money to finance the complete treatment of all sewage from coastal populations. In fact, if the Bank were to support Mr. Baalsrud's simplistic criteria—used water subjected to the best practicable means of treatment and an ultimate goal of zero pollution discharge—the amount of money available for projects of all kinds would soon dry up. This was because the Bank's funds were borrowed on the world's capital markets; bonds could be sold at the best commercial rates since the Bank insisted on its projects being well conceived and economically efficient, but there would be a loss of confidence, and funds would cease to be provided, if it were to finance projects not meeting these sensible criteria. Responsible authorities, particularly in developing countries, were keenly aware of the need to determine priorities in the allocation of scarce resources, and would not spend money to achieve an unpolluted ocean when people lacked hospitals, schools, safe water supplies, or even the food necessary for survival.

Mr. Baalsrud replied that he could not accept that the cost of treatment was too high. The total technical system from water source to consumer should be considered. People were apparently willing to pay the costs of extraction, treatment, and conveyance of water, of water installations such as toilet, bath, washing machine, and dish washer, and of the drainage system from house connections, through the streets to intercepting sewers, to the point of discharge. For these expenses there seemed to be little choice and few protests, but for the one stage, treatment, which was that part of the whole technical structure making the extensive use of water acceptable to the environment, there was no money. The cost of waste-water treatment in Norway had recently been estimated to be less than 10 per cent of the total cost. The argument that waste treatment could not be required in poor areas or in developing countries was untenable, for the simple reason that water pollution was a product of technical development and it was essentially only the rich people and rich industries which caused pollution. He maintained that treatment of all wastes was an obligatory service of the well-to-do to the rest of the world.

Dr. M. J. Suess (W.H.O.) said he believed that mechanical treatment should be obligatory, but that each case had to be examined in the light of local constraints and economic factors before deciding on the degree of further treatment required. He emphasized, however, that in many cases, where there was a lack of appropriately trained personnel, simple treatment and a well-designed outfall were preferable to a complicated treatment plant which could not be operated properly. Again, in some cases a sufficient length of outfall into a strong current—such as the Gulf Stream off Miami or the outflow from the Black Sea to the Mediterranean—could be satisfactory and would require no more than mechanical treatment.

In his concluding remarks, Mr. Baalsrud said he was afraid that several speakers had missed his main point. Experts responsible for the good use and protection of one of the most important natural resources should set an ultimate and ideal goal. The world was imperfect, but everyone should strive for perfection in his own field. If the goal could be agreed upon, there were innumerable things to be considered in attempting to reach it; but the best should not be the enemy of the good. None of the practical difficulties should be used as an argument for not trying to achieve the final worthy goal.

Mr. Calvert concluded by saying that he found himself to a considerable extent in agreement with Mr. Baalsrud. Thus he agreed that objectionable visible solids should be removed, that toxic materials should not be discharged, that eutrophication should be avoided, and that bathing beaches should be protected. Where he disagreed was in Mr. Baalsrud's attitude of a Jeremiah predicting catastrophe with very little evidence, and his adoption of Malthus as his guide when Mr. Calvert had always understood that Malthus had been discredited before the end of the last century. Mr. Baalsrud's fear of toxic metals must apply equally (or perhaps with even greater force) to the establishment of inland treatment works, and the proper treatment of those materials—whether for treatment or for sea discharge—was to keep them out of the sewer. Furthermore, those substances at present unknown which were going to poison the sea were more likely to cause damage if discharged with a treated effluent into inland waters or tidal estuaries.

ADDITIONAL REFERENCES

5. Gameson, A. L. H., Munro, D., and Pike, E. B. Effects of certain parameters on bacterial pollution at a coastal site. In 'Water pollution control in coastal areas'. Institute of Water Pollution Control, London, 1970, pp. 34–54.
6. Newton, J. R. Factors affecting slick formation at marine sewage outfalls. In 'Pollution criteria for estuaries' (ed. P. R. Helliwell and J. Bosanyi). Techno Press, London, 1975, Paper 12 (*in press*).
7. Gameson, A. L. H. Experiences on the British Coast. 'In Marine pollution and marine waste disposal' (ed. E. A. Pearson and E. d. F. Frangipane). Pergamon Press, 1975 (*in press*).
8. Munro, D. Observed and predicted coliform distributions near a sea outfall. This volume, pp. 353–362.
9. World Health Organization. Reuse of effluents: methods of wastewater treatment and health safeguards. *Tech. Rep. Ser. Wld Hlth Org.*, 1973, No. 517.
10. Chaiken, E. U., Poloncski, S., and Wilson, C. D. Muskegon sprays sewage effluent on land. *Civ. Engng, Easton, Pa*, 1973, **43**(5), 49–53.

Paper No. **19**

In Situ Methods for Determination of Microbial Disappearance in Sea Water

Poul Harremoës

Professor, Sanitary Engineering Department
Technical University of Denmark, Copenhagen

Although the rationale for the significance of bacterial indicators of pollution may be questionable, it remains that bacterial standards are often the determining factor in the choice of purification scheme and disposal arrangements for domestic sewage discharged to the sea; the traditional pollutants, such as BOD and nutrients, are seldom restrictive in the choice of engineering solutions. Knowledge is thus required of the decrease in bacterial concentration during transport from the outfall to a beach or a shellfish bed. For a given bacterial standard and a known concentration of the bacterial indicator in the sewage to be discharged, statistical relations for the path of transport and the reduction in bacterial concentration during transport then become the crucial factors in the choice of outfall location. Using these relations, and taking account of the different engineering solutions (including the degree of treatment) an economic optimum meeting the standard may then be found [1]. A prerequisite for such an approach is that the processes to which the sewage is subject after discharge are known and can be simulated with sufficient accuracy. In general terms this is far from being the case today—thus putting emphasis on gaining more insight into the phenomena. A fundamental part of this is to solve the problems related to methodologies by which the phenomena can be determined.

A General Statistical Approach

One approach is to consider the receiving water as a box with an input equal to the discharged sewage and an output consisting of bacterial concentrations at several locations. By statistical means it is then possible to find empirical relations between bacterial concentration and environmental factors, such as distance from outfall to point of interest, current velocity, and wind velocity. This can be carried so far that the remaining variance can be accounted for as inherent to the methods by which bacteria are enumerated [2–4]. For a given locality this approach can contribute significant information, but the general applicability of information on the phenomena, such as reduction with distance, is slight. The general reason for this is that macro-environmental factors, as exemplified above, will dominate. These factors are very local in nature and no transfer of data on parameters characterizing the phenomena can be relied on.

Micro-environmental factors, e.g. temperature of water, chemical composition, radiation, turbulence, and biological factors, influence the reduction of bacterial concentrations, but in order to identify the effect of these factors it is necessary to eliminate the reduction caused by dilution in the sea. This is not possible by the general statistical approach, because no specific information on phenomena within the box is available.

Clear distinction has to be made between the reduction caused by dilution and the reduction caused by processes affecting the bacteria in the water. It is reasonable to believe that only micro-environmental factors affect the processes. There is no doubt that, in order to obtain data of more general value on parameters characterizing the processes, they have to be related specifically to micro-environmental factors. The aim is that a set of easily measured data on local micro-environmental parameters will make possible a not-too-inaccurate prediction of the process.

The various factors, other than dilution, thought to influence the reduction of bacterial concentrations in sea water, are discussed in Papers 20–25 of this Symposium.

Different concepts regarding the dominance of the different factors is the reason for some confusion as to the term used to characterize the process: die-away [5], mortality [6], reduction [7], and disappearance [8]. The first two terms imply a process connected with death of bacteria, while the last two terms are neutral with respect to the mechanism of the process. The term disappearance will be used here because it is related to the mass-balance approach which will be recommended in this article.

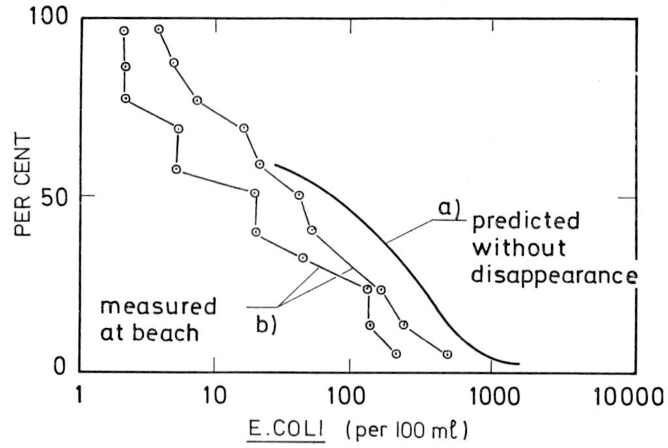

FIG. 1. Frequency distributions of *Escherichia coli* counts at a beach[1] (*a*) as predicted by dilution without bacterial disappearance, assuming flow of 1·6 m^3/s and initial concentration of 3×10^6 *E. coli* per 100 ml; (*b*) as measured

A Specific Statistical Approach

Another statistical approach has been tested by the author[1]. Tracer experiments gave rise to a statistical expression for the dilution of the sewage at a beach situated approximately 20 h drift time from an outfall close to where the tracer was released. A statistical presentation of expected bacterial concentration without regard to disappearance was obtained by dividing the bacterial concentration in the sewage by the predicted dilutions, Fig. 1. The discrepancy between the predicted and the observed statistical distributions is an estimate of the disappearance. By assuming an exponential disappearance it becomes possible to determine a T_{90} value, which expresses the drift time required for 90 per cent disappearance. This method is only as accurate as practical methods for prediction of pollution, because it is similar to such methods. The approach is thus acceptable from a practical point of view, but it is primarily of local significance because the results can be related to specified micro-environmental parameters in rather broad terms only.

The Experimental Approach

With an experimental approach it is attempted to determine the disappearance under specified conditions. There are three ways of doing this.

Laboratory experiments

Numerous experiments with sea-water samples in bottles or beakers exposed to various conditions have been made. Such experiments are indispensable for identifying the significance of specific micro-environmental factors, as in the British experiments on the effect of radiation[9]. However, laboratory experiments cannot be expected to reproduce field conditions so long as the effect of the various factors is as vaguely known as at present. The better the phenomena of disappearance become known the greater becomes the expectation of simulating field conditions in a laboratory experiment.

Confined field experiments

Laboratory experiments have been transplanted to the field in an attempt to approach field conditions. Bottles[6] or polyethylene bags[10] may be immersed in the sea in order to take account of fluctuations in radiation and temperature, and dialysis membranes have been used[11] to reach a better imitation of field conditions, but none of these approaches can be expected to come much closer to actual field conditions than do the laboratory experiments and they should be regarded as no better than an extension of them. The mere confinement is sufficient cause for being sceptical with respect to the degree to which field conditions are simulated by such experiments.

In situ experiments

As a means of determining rates of bacterial disappearance in the sea, *in situ* experiments are the only practical approach with the present knowledge of the phenomena involved. Such tests require accurate measurements of dilution in existing sewage fields in order to distinguish between the effects of dispersion and disappearance. Such results may be applied to localities similar to and in the vicinity of that of the experiment. If a sufficient number of such experiments are performed and micro-environmental factors have been properly identified and recorded, then a more general relationship and greater applicability *may* be reached.

Concentration-Ratio Approach

Continuous sewage and tracer release

The most widely used approach to measurements of bacterial disappearance *in situ* is to determine the dilution in an existing sewage field, by continuous release of a tracer, and the ratio between the actually measured bacterial concentration and the concentration obtained by dilution alone. Obtaining data corresponding to different residence times, preferably in the centreline of the sewage field, gives a relationship between disappearance ratio and residence time, which is approximated to by an exponential function for determination of a T_{90} value or of other functions.

The most comprehensive investigation of this kind has been performed by Foxworthy and Kneeling[7] who used Rhodamine-B dye as tracer with an initial concentration in the undiluted sewage of 3 mg/l. The dilution at any given point of the sewage field was determined by dividing the dye concentration by the concentration in the sewage boil. Figure 2, which is reproduced from their report[7], shows the concentration ratios for both dye and coliform bacteria. Residence time, t, is determined by dividing distance by observed current speed. The figure shows scatter ranging up to one order of magnitude, but the volume of data reduces the uncertainty in the mean curve. The disappearance data were found to fit the general expression

$$ae^{-K_1 t} + (1 - a)e^{-K_2 t},$$

the values of the 'constants' involved having the following ranges: a, 0·115–90·2; K_1, 0·0013–0·0324 min^{-1}; K_2, 0·0154–0·0728 min^{-1}. Fitting the data to a single exponential function (Chick's law) gave T_{90} values ranging from 23 to 328 min, but the single exponential approximation gave a much poorer fit. The most dominating environmental factors appeared to be water temperature and season.

Abrupt sewage and tracer release

Gameson and Gould[12] have used a similar approach in a series of 20 experiments, but since the discharge of sewage and tracer generally lasted for only 5–6 min it is more appropriate to consider it as

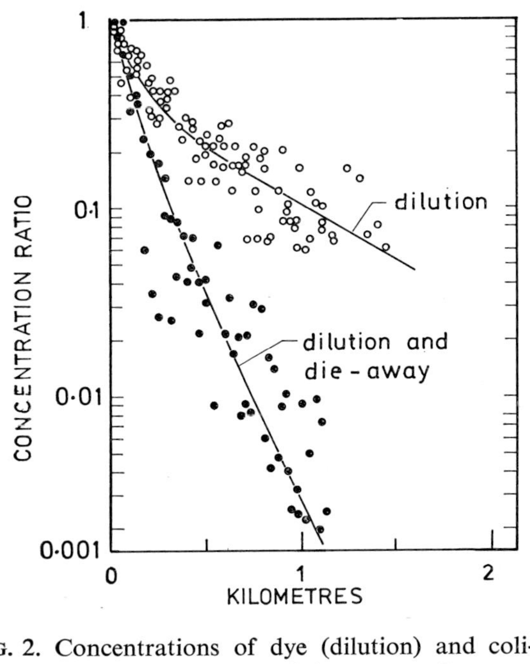

FIG. 2. Concentrations of dye (dilution) and coliforms (dilution and die-away) in centre of sewage field relative to concentrations in sewage boil[7]
Ventura outfall, U.S.A., 11 November 1966

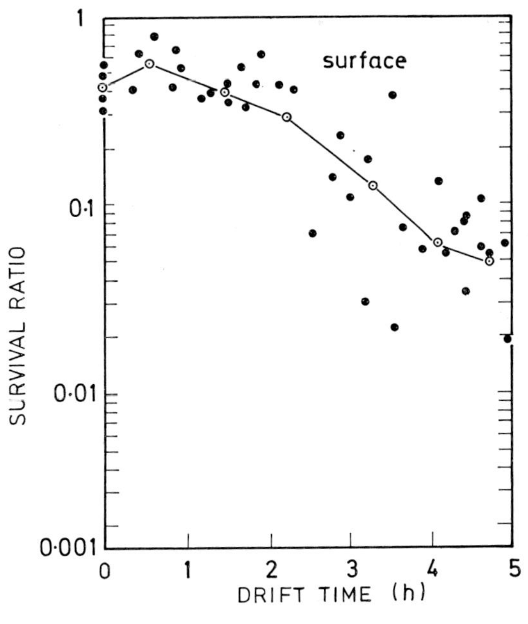

FIG. 3. Coliform survival ratios for surface samples following release of labelled sewage discharge
Sidmouth, England, 21 August 1969

an abrupt rather than as a continuous discharge. A radioactive tracer (Br-82) is added to the sewage. Pumping is continued until the labelled sewage is about to enter the sea and then stopped to allow unlabelled sewage to be carried away. The labelled sewage is then discharged and pumping is again stopped in order to prevent unlabelled sewage from interfering at the tail of the labelled sewage field. Quite a few practical problems seem to be connected with this approach, but the authors consider that these difficulties are not likely to influence substantially the results obtained if sufficient care is taken. Samples are taken in the labelled field and analysed for radioactivity and coliforms.

Figure 3 shows the coliform survival ratio plotted against drift time in the earliest of these experiments. The concentration of tracer and bacteria in the raw sewage rather than in the sewage boil is used as reference. The scatter is similar to that experienced by Foxworthy. The encircled points correspond to grouped and averaged data.

The experiments convincingly showed that radiation is an important environmental factor—to the extent that the disappearance may be expressed relative to radiation more consistently than to drift time.

Mass-Balance Approach

Continuous sewage and tracer release

Details of this approach and the following are described elsewhere [8]. A tracer is continuously released to the sewage discharging through an outfall. The sewage field is monitored by measuring tracer and bacterial concentrations across the sewage field at different distances from the outfall. The discharge of tracer and bacteria per unit time through the outfall can be calculated. By integration of the concentration distribution across the field and multiplying by the current speed the surface transport of tracer and bacteria can also be calculated. Some fraction of the tracer released per unit time will pass through a particular strip of the surface layers normal to the direction of flow. A smaller fraction of the released bacteria will pass through the same strip. The ratio between the two is the disappearance ratio. Each strip at different distances from the outfall provides a point on a plot similar to Figs. 2 and 3. The advantage of this approach is that rather than comparing many point measurements hopefully taken at or close to the centreline of the field and then averaging the data to reduce the effects of scatter, the points of measurement are distributed systematically and the grouping and averaging procedure is made into a meaningful integration according to a mass-balance concept. An additional advantage is obtained because the reasonableness of the data can be checked by mass balance. This will be illustrated by a concrete comparison. The same approach has been used in Israel [13].

Continuous sewage discharge, abrupt tracer release

Further advantage can be gained by using an abrupt tracer release instead of a continuous release. By criss-crossing the patch of tracer it is possible to determine the fraction of the injected tracer which is present at the surface after a certain drift time. The fraction of the abruptly released tracer which

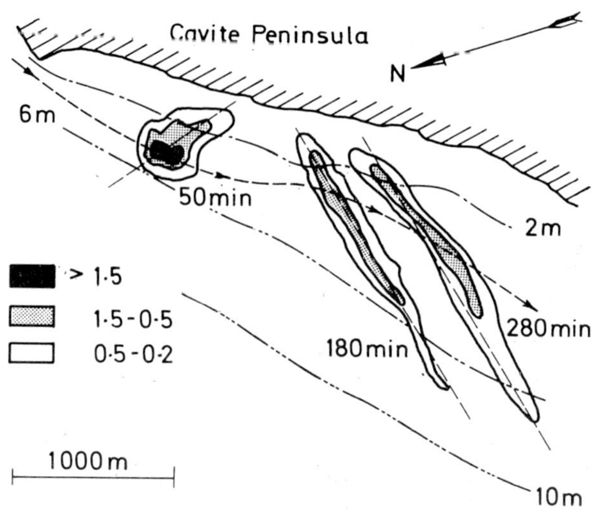

FIG. 4. Movement of tracer patch over a sloping sea bed [8]
Shading shows surface tracer concentrations in $\mu Ci/m^3$

remains at the surface is equal to the fraction for the continuously discharged sewage. The fraction of bacteria passing through a particular cross-section of the sewage field is determined as mentioned for continuous release. The disappearance ratio is determined as the ratio between the two fractions. The advantage is that the drift time is determined with much greater accuracy. The importance of this is apparent from the plot of disappearance ratio against drift time, where a 10 per cent error in the determination of drift time is equal to a 25 per cent error in the determination of disappearance ratio. This indicates that determination of drift time by measurement of current velocity may introduce considerable inaccuracy in addition to that inherent in the bacterial counting methods. Figure 4 is an example of tracer movement and dispersion[8]. Such slanting of a tracer patch is not unusual, especially along a sloping sea bed. The importance of this for the determination of drift time is apparent. These advantages are gained at the expense of the ease of the practical execution of the experiment. However, in a not-too-fast current the practical problems of sailing and sampling can be overcome[8].

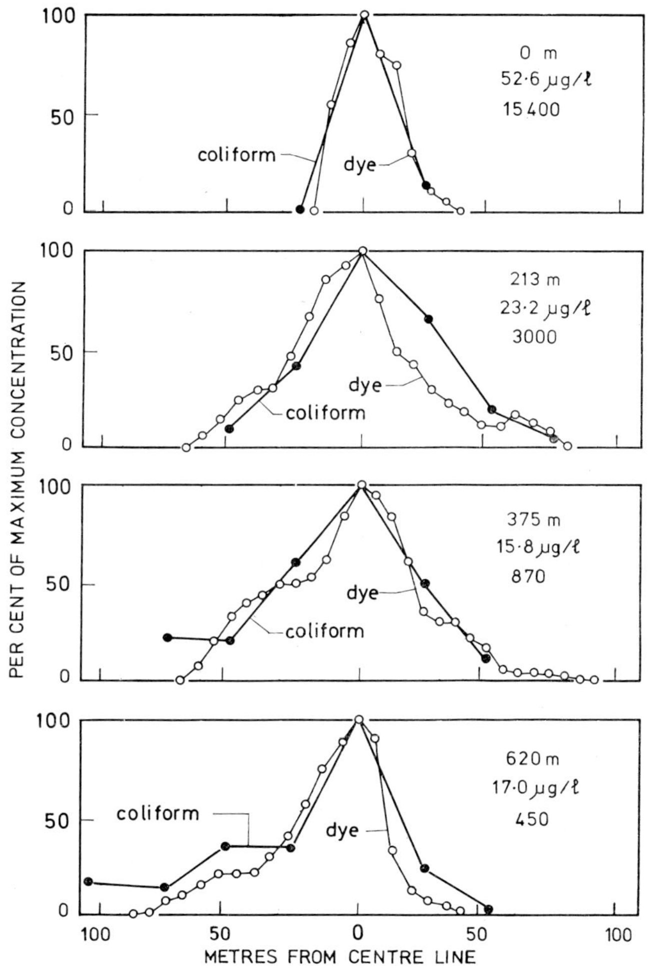

FIG. 5. Cross-sectional distributions of dye and coliform concentrations at different distances from Ventura outfall on 16 August 1966[7]

Figures in each section show distance from outfall, maximum dye concentration, and maximum coliform count (per ml)

Continuous Sewage Discharge and Total-Mass-Balance Approach

The final refinement of the mass-balance approach is to measure the vertical distribution of concentration in each cross-section of the sewage field and thus to obtain the total transport of bacteria through each cross-section. In this way the disappearance can be found without use of a tracer—in principle but not in practice. The use of a tracer is a good check, because the mass balance should be satisfied for the tracer—the lack of tracer balance indicates the errors and inaccuracies encountered in any method for *in situ* determination of bacterial disappearance. This will be illustrated by the following example.

Comparison with concentration-ratio approach

The two most comprehensive investigations[7,12] have been based on the concentration ratio approach. Figures 5 and 6 show the offshore data for a particular experiment at Ventura, and Table 1 lists other relevant data taken from the report[7]. Table 2 contains the calculated data, based on integration of the curves in Figs. 5 and 6 and a comparison of the amounts discharged and those being transported with the sewage field.

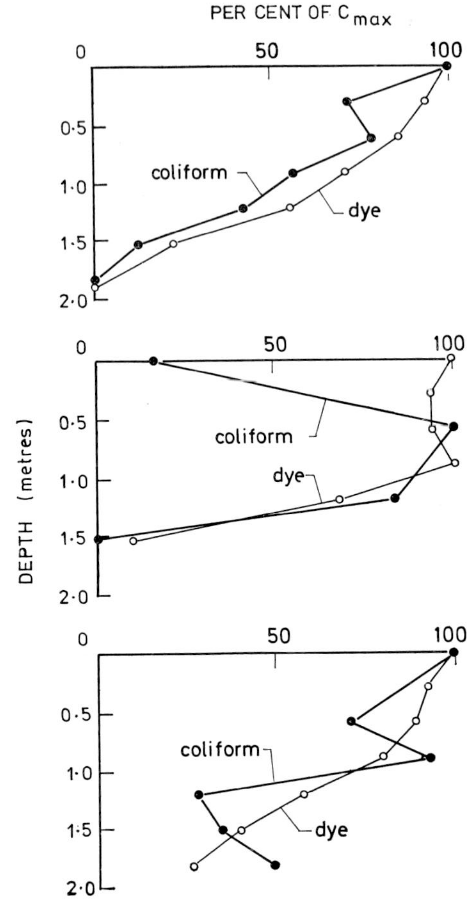

FIG. 6. Vertical distributions of dye and coliform concentrations at different distances from Ventura outfall on 16 August 1966[7]

TABLE 1. Main data for a field experiment[7] at Ventura (coast of California) on 16 August 1966

Outfall length:	1100 m
Outfall diameter:	0·89 m
Sea depth:	10·5 m
Current velocity:	0·21 m/s[a]
Discharge:	0·153 m^3/s
Average concentration of coliforms at plant:	$1{\cdot}6 \times 10^6$ MFC[b]/ml[c]
Dye concentration at plant:	3·00 mg/l.
Plus four cross-sections of the sewage field for dye and coliforms (Fig. 4) and three depth profiles (Fig. 5)	
Initial flow of coliforms:	25×10^{10} MFC/s
Initial flow of dye:	0·46 g/s

[a] Correspondence with Prof. Foxworthy has revealed that the published value[7] of 0·18 kt (0·09 m/s) was incorrect and that the correct value should be 0·21 m/s. The latter value has been used in all subsequent calculations in the present paper.

[b] MFC, Membrane Filter Count.

[c] No data available for 16 August, but data from 11 other experiments at Ventura gave a geometric mean value as shown above.

TABLE 2. Data for cross-sections of sewage field[7]

Distance from outfall (m)	0	213	375	620
Drift time (min)	0	18	30	49
Area under dye conc. curve (g/m²)	1·46	1·26	0·91	0·75
Area under coliform curve (10^{10}/m²)	39·6	18·0	5·3	2·6
Surface dye flow (g/s m)	0·30	0·27	0·191	0·156
Surface coliform (10^{10}/s m)	8·4	3·8	1·10	0·54
Surface dye ratio (m^{-1})	0·65	0·58	0·42	0·34
Surface coliform ratio (m^{-1})	0·34	0·15	0·044	0·022
Surface disappearance ratio	0·52	0·27	0·105	0·063
Area of dye conc.–depth curve (m)	1·15	1·26	—	1·40[a]
Area of coliform–depth curve (m)	0·95	1·04	—	1·28[a]
Total dye flow (g/s)	0·35	0·34	—	0·22
Total coliform flow (10^{10}/s)	8·0	3·9	—	0·69
Total dye ratio	0·76	0·74	—	0·48
Total coliform ratio	0·32	0·160	—	0·028
Total disappearance ratio	0·42	0·23	—	0·060

[a] Extrapolated below 2 m depth

Figure 7 shows the amounts transported through a surface strip across the sewage field and the total transport with the sewage field divided by the amount discharged from the pumping station, for both dye and coliforms. Figure 8 shows the resulting disappearance curve, which fits an exponential disappearance at a rate T_{90} = 53 min. The data reveal the following information about the experiment. The amount of dye in the sewage is reduced by 24 per cent during 80-min transport from the injection point at the pumping station to the outlet, corresponding to a disappearance rate with T_{90} equal to 700 min. There is reason to believe that this loss of dye is due to adsorption on particles. However, the loss continues, though at a different rate: 'T_{90}' for dye equal to 210 min. The reason for this may be continued adsorption or chemical degradation. Whatever the mechanism, it does affect the result, because the dye is considered conservative and this is apparently not the case. This is a systematic source of error which, when taken into account, reduces T_{90} to 46 min. The greatest advantage reached by the mass-balance approach is the check on the reliability of the data. This is obtained at little extra cost—just another sampling procedure. The coliforms disappear in the pipeline too—at a rate of approximately T_{90} = 160 min.

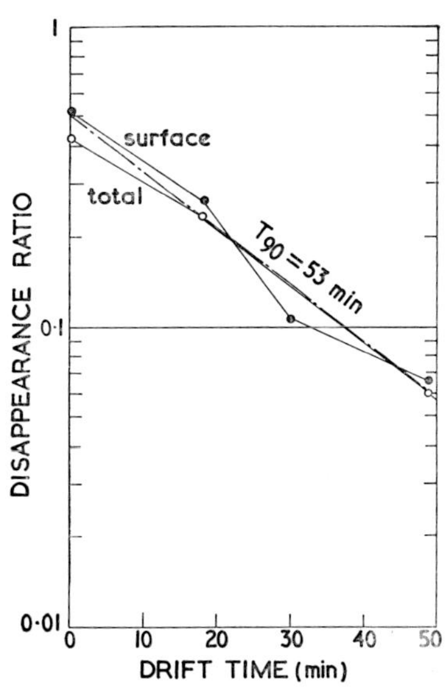

FIG. 7. Mass-balance ratios, through surface strip and through depth of sewage field, and discharge through outfall for dye and coliforms
From data of Foxworthy and Kneeling[7]

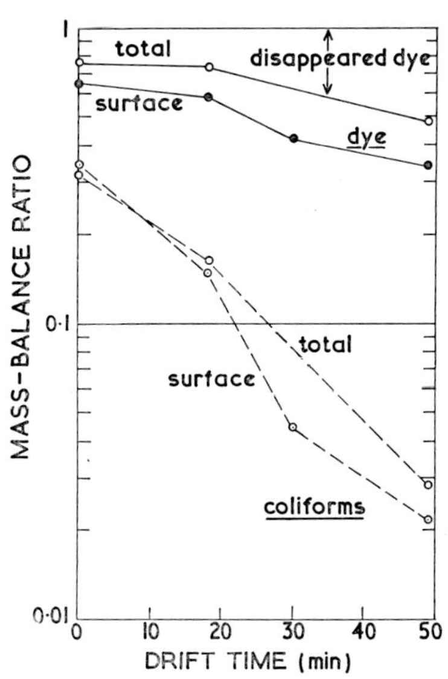

FIG. 8. Disappearance ratio by mass balance (from Figs. 5 and 6) as function of drift time

Figure 9 is a plot of the maximum concentration ratio in each cross-section against drift time, and indicates the difficulty of obtaining data at the centreline of the sewage field. This explains why the data on Figs. 2 and 3 show such a large scatter. It is noteworthy that this inconsistency of the data disappears in the data on surface and total transport (Fig. 7) in spite of the fact that the maximum concentration in the cross-section carries greatest weight in contribution to the integrated value. The other data in the cross-section provide a significant increase in the value of the data. Still greater accuracy can be reached by taking more samples near the maximum in a cross-section. There is greater value in these data than in random sampling 'along the centreline' with the same total number of samples. The inconsistency of the data in Fig. 9 is removed in the disappearance curve in Fig. 10 which gives a T_{90} value of 45 min.

The information contained in Fig. 8 is much greater than in Fig. 10 and makes $T_{90} = 53$ min a much more likely value. None of the figures accounts for the disappearance of dye, and $T_{90} = 53$ min should therefore be considered an upper value and $T_{90} = 46$ min a likely lower value.

Data corresponding to Fig. 2 for the experiment are not available in the report[7], but the following results have been recalculated, based on a velocity of 0·21 m/s (see Table 1): $a = 0{\cdot}387$, $K_1 = 0{\cdot}0136$ min^{-1}, $K_2 = 0{\cdot}085$ min^{-1}, and $T_{90} = 100$ min. It turns out that this T_{90} value is not determined by a best fitting straight line but is the drift time corresponding to the intercept of the best fitting curve with the 90 per cent disappearance line. The curve corresponding to the first three constants is shown in Fig. 10; this line of best fit deviates significantly from the straight lines of both the concentration-ratio and mass-balance approaches after 25 min drift. There seems to be some conflict between the information contained in the cross-sectional data and that contained in the centreline data which are the basis for the best fit.

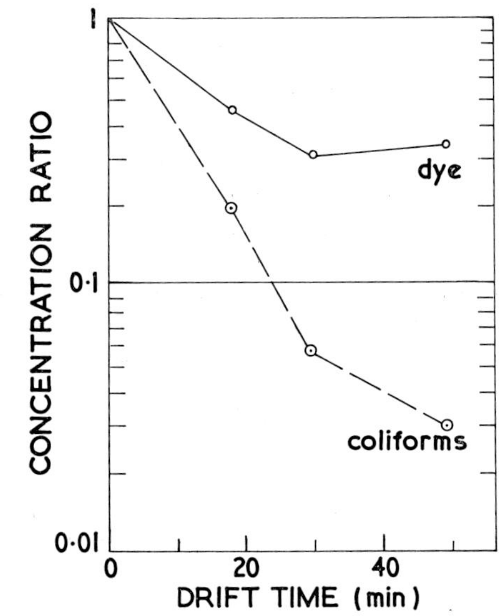

FIG. 9. Concentration ratios for dye and coliforms (from Figs. 5 and 6) as functions of drift time

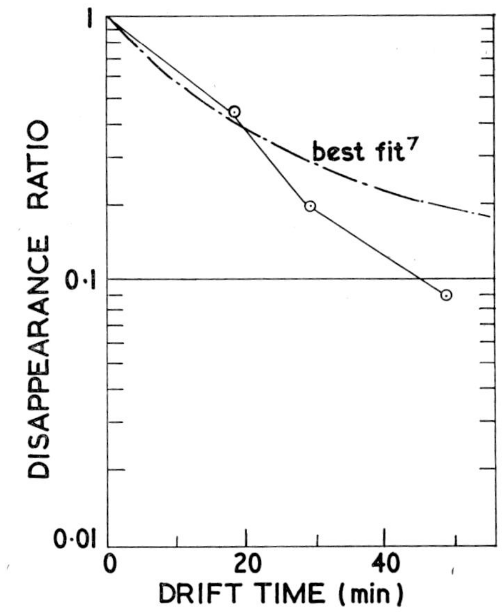

FIG. 10. Disappearance ratios (from Fig. 9) as functions of drift time. Continuous curve and points represent data from maximum concentrations in cross-sectional runs ($T_{90} = 45$ min). Further data (also based on concentration ratios) from sewage field give curve of best fit, and an intercept with the 90 per cent disappearance line after 100 min

Conclusion

In general the procedures for measuring bacteria in water are considered to be the greatest source of error and inaccuracy in field experiments for determination of bacterial disappearance in sea water. However, significant sources of error and inaccuracy are encountered by the procedures for eliminating physical dilution from the phenomena. This is shown by recalculation of data from one of the most comprehensive and well performed investigations of the kind. It might be postulated that the variability of data on T_{90} from investigations all over the world be due to errors and inaccuracies rather than due to actual differences. In order to account for such errors and to reduce inaccuracy it is recommended that a mass-balance approach for data collection and treatment be used instead of the usual concentration-ratio approach.

REFERENCES

1. Harremoës, P. Prediction of pollution from planned wastewater outfalls. *J. Wat. Pollut. Control Fed.*, 1966, **38**, 1323–1333.
2. Bonde, G. J. Badevandsundersøgelser i Øresund 1965–69. Øresundsvandkomiteens undersøgelser 1965–1970, Lund, 1971.
3. Bonde, G. J. T_{-90}, criterium or chimaera? *Revue int. Océanogr. méd.*, 1972, **25**, 5–15.
4. Bonde, G. J., and Thomsen, A. M. Computer analysis of a bacteriological monitoring system for pollution control. *Proc. 5th int. Colloquium med. Oceanogr., Messina*, 1973, pp. 73–79.
5. Pearson, E. A. An investigation of the efficacy of submarine outfall disposal of sewage and sludge. California State Water Pollution Control Board, Publ. No. 14, Sacramento, 1956.
6. Gameson, A. L. H., and Saxon, J. R. Field studies on effect of daylight on mortality of coliform bacteria. *Wat. Res.*, 1967, **1**, 279–295.
7. Foxworthy, J. E., and Kneeling, H. R. Eddy diffusion and bacterial reduction in waste fields in the ocean. University of Southern California, Los Angeles, 1969.
8. Harremoës, P. Field determination of bacterial disappearance in seawater. *Wat. Res.*, 1970, **4**, 737–749.
9. Gameson, A. L. H., Pike, E. B., and Davey, K. W. Mortality of coliform bacteria in sea water samples in daylight. *Proc. 5th int. Colloquium med. Oceanogr., Messina*, 1973, p. 353–362.
10. Stewart, R. E., Putman, H. D., Jones, R. H., and Lee, T. N. Diffusion of sewage effluent from ocean outfall. *J. sanit. Engng Div. Am. Soc. civ. Engrs*, 1971, **97**(SA4), 485–503.
11. Head, P. C., and James, A. The influence of the River Tyne on the bacteriological and chemical quality of inshore waters of the north-east coast of England. *Revue int. Océanogr. méd.*, 1970, **17**, 147–158.
12. Gameson, A. L. H., and Gould, D. J. Effects of solar radiation on the mortality of some terrestrial bacteria in sea water. This volume, pp. 209–219.
13. Gilath, C., Clit, S., and Melzer, M. Dispersion of sewage and disappearance rate of enteric organisms from a marine outfall. Report to the Oceanographic and Limnological Research Co., Haifa, Israel Atomic Energy Commission.

In his opening remarks, Prof. Harremoës said that the aims of the paper were twofold. First, to present and exemplify an approach different from that used traditionally. A mass-balance approach gave a constant check on the reliability of the information used—the benefit of this has been unintentionally demonstrated by inconsistencies revealed by treating data[7] in this way, leading to a change in the velocity data in Table 1 and to recalculation of the rest of the data. Secondly, to demonstrate the sources of inaccuracy dominating the traditional approach to die-away studies; for instance, the relative inaccuracy of T_{90} was no less than the inaccuracy of velocity measurements and it could be very inaccurate. In view of these uncertainties, he wished to encourage caution with respect to interpretation of T_{90} data from the literature, such as the lists of values presented in Papers 25 and 41.

Finally, he wanted to look at field measurements in the light of the discussions on standards. It had to be realized that the health risk related to swimming in polluted sea water appeared to be relatively small—at least no one had yet found any epidemiological evidence of a health risk. It was thus fair to say that the standards used for bacterial pollution of beaches were uncertain by orders of magnitude. In view of this, there had to be a limit to the degree of refinement at present needed for putting into predictive models—including T_{90} values. Consequently, he had not devoted any effort of his laboratory to further pursuance of methods of *in situ* determination of T_{90} values beyond the stage published[8] in 1970. For the purpose of the Symposium he had therefore turned to Prof. Foxworthy's data[7] in order to exemplify the mass-balance approach. For those who wanted to pursue the disappearance of bacteria in sea water further, he recommended the mass-balance approach for field studies.

DISCUSSION

Prof. J. E. Foxworthy (U.S.A.) agreed with Prof. Harremoës that *in situ* experiments were the only practical approach for determining bacterial disappearance in the sea, and that the mass-balance method provided an important means of checking the reasonableness of the data collected. Prof. Foxworthy had used this approach in analysing a series of experiments[14] in Southern California. The results for twenty-four experiments within surface waste fields showed mass balances of 64–294 per cent (mean, 140), and those for six experiments in background areas 37–78 per cent (mean, 51). These results appeared to substantiate earlier laboratory studies[14] on the absorptive properties of the Rhodamine-B used in the field experiments, and to be in accord with the findings of Feuerstein and Selleck[15]. It was interesting to note that Okubo[16], after evaluating the results of many similar. tracer studies, had considered mass balances between 50 and 150 per cent to be permissible in such experiments. The relatively wide range of mass balances Prof. Foxworthy had reported was attributed more to errors in field determinations of current speeds, dye concentration distributions, rates of flow, etc., than to adsorption or other phenomena[14]. A case in point was the data Prof. Harremoës had used in Tables 1 and 2. As discussed in the original report[7], it had been difficult to measure precisely the parameters required for the calculations; for instance, there had been incorrect interpretation of the data for current speed, the sewage flow was not constant, no coliform measurement was made for the effluent on the day in question, and measurements of cross-plume and vertical dye distributions gave consistently low values of maximum concentration as compared with values obtained by approximately centreline runs. Detailed consideration of these factors could materially alter Prof. Harremoës's calculations and conclusions.

It had been partly because of the above difficulties that the concentration-ratio method was selected to assess the distribution of dye and coliforms in the experiments of Foxworthy and Kneeling[7], it being concluded that data obtained by longitudinal sampling along the line of apparent maximum concentration yielded more definitive information than did cross-plume or vertical sampling techniques.

Prof. Foxworthy had several reservations concerning Prof. Harremoës's interpretation of Foxworthy and Kneeling's data[7] and the conclusions drawn therefrom. However, since these reservations were based on the original version of the paper (before the Author had substantially revised the latter part—following the amendment of Table 1), the Editor feels that to state them at this stage would only confuse the issue.

Mr. C. G. Gunnerson (Canada), on the other hand, said that the Author's conclusions were valid, but did not go far enough. Sampling artifacts ranged from seasick technicians to lack of analytical quality control within and between laboratories. Mass-balance approaches were fine, but had to include the sedimentation factor. To argue that T_{90} was unaffected by the degree of treatment was to argue that sewage treatment accomplished nothing in the way of solids removal, though it was known that shellfish beds were polluted by raw sewage with large amounts of solids deposited on the sea floor and that sewage treatment removed these solids.

Mr. R. G. Ludwig (Brazil) said that all T_{90} data had been determined in surface fields where the action of sunlight and other forms of radiation played a role. He asked if there were any data on the probable relationship between T_{90} values for sub-surface and surface fields, since this factor could be of importance where designs were based on predictions for submerged fields. On this point, Mr. G. C. van Dam (Netherlands) mentioned that the sewage from the Hague outfall was submerged during an initial stage of varying duration, but he could not say whether this was the main cause of the high T_{90} values found[17]. He added that Prof. Harremoës had said that the methods described for determining T_{90} were inapplicable in the case of turning tidal currents. This was not true when T_{90} was small compared with the duration of the periods with constant current direction. For the Hague outfall, with its extremely long T_{90} values and its tidal current turning 4 times a day, it had been necessary to use a different method[17].

REPLY TO DISCUSSION

Prof. Harremoës said he knew only too well the difficulties to which Prof. Foxworthy had referred and that he greatly appreciated his comments which, in essence, provided additional support to his main conclusion: that T_{90} values in the literature were very uncertain—not only due to the uncertainties inherent in bacterial enumeration methods but also due to the difficulties related to the measurements of the physical conditions—even in an investigation as thorough and comprehensive as that performed by Prof. Foxworthy.

He maintained that cross-plume measurements could be made in such a way that the peak concentration value would be picked up in the process. This required continuous monitoring for tracer (dye or radioactive) measurements and increased sampling for bacterial determinations when near the peak, and was sufficient for the surface-ratio approach; he thought this did not give rise to any technical problems. On the other hand, he agreed that it was very difficult to obtain reliable depth profiles without using advanced gear for submerged tracing and sampling. He would continue to recommend a mass-balance approach for theoretical as well as practical reasons.

He had reservations with respect to Mr. Gunnerson's suggestion that sedimentation was a dominant factor for T_{90}—the data were suggestive but inconclusive. The statement that coliforms were found in large numbers in sediments near a sea outfall, while perfectly true, did not prove sedimentation to be a significant factor regarding T_{90}.

In reply to Mr. Ludwig, he said he was not aware of any data on T_{90} in submerged sewage fields and that he hated to think of the complexity involved in performing an investigation at a depth of 10–40 m. [See also the second paragraph on p. 216—*Ed.*]

ADDITIONAL REFERENCES

14. ALLAN HANCOCK FOUNDATION and UNIVERSITY OF SOUTHERN CALIFORNIA. An investigation on the fate of organic and inorganic wastes discharged into the marine environment and their effects on biological productivity. California State Department of Water Resources, Sacramento, Rep. 29, 1965.

15. FEUERSTEIN, D. L., and SELLECK, R. E. Tracers for diffusion measurements in surface waters. University of California, Sanitary Engineering Research Laboratory Rep. 63-1, 1963.

16. OKUBO, A. A new set of oceanic diffusion diagrams. Chesapeake Bay Inst., Johns Hopkins Univ., Tech. Rep. No. 30, 1968

17. DAM, G. C. v. The Hague outfall. This volume, pp. 393–401.

Paper No. **20**

Effects of Antibiosis in a Marine Environment

M. Aubert, D. Pesando, and M. J. Gauthier
C.E.R.B.O.M. (I.N.S.E.R.M.), Nice, France

Before discussing the phenomena of antibiosis in a marine environment, we consider it necessary to recall what is meant by the term 'antibiotic'. "Antibiotics are chemical substances of microbic origin that are effective in inhibiting the growth of, or even destroying, micro-organisms." This general definition can be found in Encyclopaedia Universalis. Such a definition restricts the study of antagonism to the field of microbiology, i.e. it excludes the antibacterial action of such macro-organisms as animals and upper algae.

A systematic investigation of antibiotic substances was not undertaken until recently, although they have been known to exist for several centuries, and their effects duly reported. Out of hundreds of antibiotics, only a few are used at present in agriculture and in the treatment of diseases.

A characteristic of antibiotics is the selectivity of their range of activity: some affect mainly Gram-positive bacteria, others Gram-negative types, while a few are effective towards both groups. In a marine environment, certain micro-organisms have an antibacterial action. The purpose of this study is to show what their role is and how they work.

Bacterial Antibiosis

Antagonisms between specific marine bacteria and exogenous polluting bacteria

Microbial antagonisms of this sort were already suspected back in 1889 by De Giaxa[1], but were investigated exhaustively only much later. The phenomenon was definitively substantiated by the discovery and isolation of antibiotic-producing marine microbial strains. In 1947, Rosenfeld and ZoBell[2] showed that, out of 58 strains of marine bacteria investigated, nine, mostly belonging to the genera *Bacillus* and *Micrococcus*, proved antibiotic towards the Gram-positive organisms in the faecal flora.

Since then, a number of studies have shown similar antagonistic activity in many other marine bacteria and made it possible to predict their action *in situ*. Thus, Grein and Meyers[3] isolated bacterial strains which produced antibiotic substances, starting from marine sediments. More recently, Krasil'nikova[4] described the antibacterial properties of typically marine micro-organisms isolated at various depths: out of the 326 species investigated, 27 (i.e. 9 per cent) proved active against terrestrial bacteria, the highest intensity being reported in the sporogenous types. Mention may also be made of studies carried out by Buck, Meyers, *et al.*[5–8] which demonstrated significant antagonistic activity in a number of marine bacteria belonging to the genera *Pseudomonas*, *Aeromonas*, *Vibrio*, *Flavobacterium*, *Alcaligenes*, and *Caulobacter* towards yeasts present in the sea. In view of the fact that many other micro-organisms (yeasts, actinomycetes) in this environment also produce active substances, and that many authors have witnessed these phenomena in several oceans and in miscellaneous biotopes, the least we can say is that they are decidedly ubiquitous. The studies of the Danish school, as reported by Bonde[9], also emphasized the importance of the subject. In recent years, Baam, Gandhi, and Freitas (Xavier's College, Bombay)[10] showed that, out of 60 strains of marine bacteria (including 45 sporogenous strains) isolated from 41 samples of sea water taken in various areas, 51 were highly antibiotic towards *Salmonella typhosa* and moderately antibiotic towards *Staphylococcus aureus*.

Although the studies carried out in this field seem to establish beyond doubt the existence *in situ* of biochemical interactions between marine micro-organisms and terrestrial bacteria, some authors have reached different conclusions. For instance, Carlucci and Pramer[11] observed no antagonistic activity, although they investigated over 200 marine bacterial strains versus *Escherichia coli* and one *Bacillus*. In contrast, Burkholder *et al.*[12] succeeded in isolating a low molecular weight antibiotic from a marine *Pseudomonas*, and identifying it as a pyrrolic compound. To our knowledge, these are the only studies in this field that have led to the chemical identification of an active substance of bacterial origin.

Our own contribution to this general study has been especially directed towards the determination of the main bacterial groups responsible for such an activity in the Mediterranean Sea, and the definition of the activity spectrum of the strains that produce antagonistic substances[13].

They belong to several genera, in particular *Pseudomonas*, *Achromobacter*, *Flavobacterium*, *Chromobacterium*, and *Vibrio*, which make up the bulk of the marine bacterial population. Some of these species are very common and we found them several times during isolations. In this connection, and based on relevant data in the literature, it seems that the role of species belonging to genus *Chromobacterium* is fairly significant as regards phenomena of antagonism at the inter-microbial level, owing to their marked activity towards many Gram-positive and Gram-negative bacteria and also to their ecological characteristics. The studies carried out on these species led to the conclusion that they are widely distributed geographically, but likely to be limited to the well-aerated water of the euphotic level since they are strictly aerobes[14–17].

Nature of antibiotic substances produced by some marine bacteria

The chemical nature of these substances is still hardly known. Only Burkholder *et al.*[12] were able to isolate and identify an antibiotic compound with a pyrrolic structure from a marine *Pseudomonas*. In this field, we shall also mention a few tentative studies achieved by Buck and Meyers[8] on certain properties of an inhibiting compound also extracted from a marine *Pseudomonas*, close to *Pseudomonas piscicida*, whose micro-molecularity only was evidenced. It should be noted, however, that the active products had been isolated from 'pulverates' of microbial cells and thus do not correspond to the products secreted into the environment by the producing micro-organisms, although they can be freed by autolysis of the cells.

We investigated in detail the synthesis of antibiotic substances by micro-organisms belonging to the genera *Pseudomonas* (Strain No. 18 in our collection) and *Chromobacterium* (Strains Nos. 50, 101, and 113–118) which had proved active in a very similar way during our preliminary study.

Their activity, which is quite similar against several terrestrial bacteria, is due to polysaccharidic macro-molecules released in a nutrient medium. These polysaccharides are non-dialysable, thermostable, and can be extracted and purified through conventional biochemical procedures. They are active at very low concentrations in the nascent state. However, they are inhibited by certain bacterial enzymes like catalase and peroxidase.

There is a possibility that the production of such polysaccharides is fairly extensive. Our study revealed it in two marine species, one of which at least, *Chromobacterium marinum*, seems to be quite common in all well-aerated biotopes. Actually, the fact that an antibiotic activity was observed on each and every strain of *Chromobacterium* during the period of test suggests that such an activity is a characteristic of the genus.

In a marine environment, especially at the benthic sedimentary level where the concentration of bacteria resembles that of an artificial culture, such products may play a significant role in controlling the fate of indigenous as well as exogenous bacterial populations.

Planktonic Antibiosis

How antibiotic properties were demonstrated

As early as 1961, the antibacterial activity of phytoplankton was observed by M. Aubert[14,18,19]. At about the same time, reports were issued both in the United States and in Denmark on similar subjects. We shall summarize the latter studies before we describe at length the experiments made by C.E.R.B.O.M.

Early research in this field related to green algae of the genus *Chlorella* (*Chlorella vulgaris* and *Chlorella pyrenoidosa*) from which Pratt[20] isolated an antimicrobial agent which he called chlorellin. His studies were verified and continued by Spoehr *et al.*[21] in 1949, Blaauw-Jansen[22] in 1954, Steemann Nielsen[23] in 1955, and Jørgensen and Steemann Nielsen[24] in 1961–62. Chlorelline is highly anti-staphylococcic. Unicellular marine algae were investigated much later for these properties.

Let us also mention the observations of Sieburth and Pratt[25] on *Skeletonema costatum* as an inhibitor of the growth of *Escherichia coli*; by studying cultures of that diatom, it was shown that the anticoliform activity first appears after a week then gradually increases. It may therefore be assumed that active substances are not freed into the environment until after the death of the cells when the cultures are senescing (Sieburth[26]). In addition, Jørgensen[24], starting from cultures of *Nitzschia*, obtained cell extracts that are active against *Staphylococcus aureus*.

In recent years, the number of publications dealing with antibiosis by phytoplankton has increased. The activity towards important micro-organisms has been described for various groups: diatoms, chrysophyceae, chlorophyceae, dinoflagellates, etc. Thus, only recently, Duff, Bruce, and Antia[27] showed the antimicrobial action of several planktonic micro-algae in an axenic monospecific culture.

With a view to determining the role and action of each of the species, M. Aubert and M. J. Gauthier[28] carried out a series of experiments with over 70 planktonic strains in monospecies cultures, with or without contaminating bacteria. The following determinations were made: (1) systematic screening of antibacterial strains; (2) full testing of these strains, after they were made axenic, in solid and liquid media in order to define their type of activity and their antibacterial spectrum; (3) extraction of antibiotic substances in the cell extract and the culture, using mass cultures.

The tests showed that over 25 per cent of the organisms investigated are more or less antibiotic towards the test organisms mentioned. The results arrived at by C.E.R.B.O.M. are for the most part unprecedented as far as diatoms are concerned, since, to our knowledge, the major part of those species were not known to have an antimicrobial action.

As regards the full study of the activity of these micro-organisms, some results have already been obtained for a number of diatoms:

Asterionella japonica

This diatom proved active towards *Sarcina lutea*, *Neisseria flava*, *Staphylococcus aureus*, and *Staphylococcus epidermidis*.

The nature of the antibiotic activity is light-dependent. We noted that the activity observed initially during experiments in the dark was of the bacteriostatic type, i.e. the bacteria would start growing again after several days' incubation, whereas the cultures placed in test tubes exposed to light resulted in a bactericidal type of activity, with no re-growth of bacteria. This is suggestive of a phenomenon of photo-oxidation, as applied to certain long-chain fatty acids: after they have been exposed to light, they split into shorter chains that are bactericidal (Spoehr *et al.*[21]).

Also, thanks to the courtesy of Prof. Prévot, we have been able to use his laboratory of Institut Pasteur to ascertain the activity of the diatom we had obtained in an axenic culture towards a number of anaerobes. The activity was demonstrated on the following bacteria: *Welchia perfringens*, *Clostridium septicum*, *Clostridium chauvoei*, *Clostridium sordellii*, *Clostridium botulinum*, *Clostridium histolyticum*, *Clostridium oedematiens*, and *Plectridium tetani* (J. Aubert, D. Pesando, and H. Thouvenot[29]).

Asterionella notata

The inhibitory activity of this diatom was demonstrated towards 24 species of bacteria, most of which were enterobacteria (strains supplied by Institut Pasteur)[30].

The tests were carried out either in agar media at various levels of pH (using antibiograms or auxograms), or in fluid media—either rich (peptone water with phenol red) or poor (Lwoff and Quirido synthetic medium).

The tests in a fluid medium showed an inhibitory action on the following bacteria: *Staphylococcus aureus*, *Neisseria*, *Pseudomonas aeruginosa*, coliforms, *Salmonella*, *Shigella*, *Klebsiella*, *Proteus*, *Providencia*, *Aerobacter*, *Hafnia*, and *Serratia*.

In addition we determined broadly, and only on a qualitative basis, that the biosecretions of *Asterionella notata* are very inhibitory towards: (a) several anaerobes—*Welchia perfringens*, *Clostridium septicum*, *Clostridium chauvoei*, *Clostridium histolyticum*, *Clostridium sordellii*, *Micrococcus prevoti*, *Micrococcus niger*, *Ristella pseudoinsolita*, *Sphaerophorus varius*, *Fusiformis fusiformis*, and *Plectridium tetani*; (b) several terrestrial fungi and yeasts—*Candida albicans*, *Saccharomyces cerevisiae*, and various species of *Penicillium*, *Aspergillus*, *Mucor*, etc.; (c) marine bacteria; (d) other phytoplanktonic organisms—*Nannochloris oculata*, *Tetraselmis striata*, *Tetraselmis maculata*, *Tetraselmis tetrathele*, *Phaeodactylum tricornutum*, *Asterionella japonica*, *Nitzschia ascicularis*, and *Coccochloris elebens*.

We have also demonstrated that the order of magnitude of active substance minimal concentrations in the extract, as opposed to the bacteria tested, is very low: from 1 to 0·005 mg/l. or less, which represents an activity at least as high as that of the most active conventional purified antibiotics. In view of the low level of active concentrations of the factors present in a marine environment that enhance the antibiosis attributable to this substance (or these substances), very high luminosity, the highly oxidizing character of sea water, bacterial concentration always lower than in the laboratory tests, and, above all, the possible concentration of the active elements owing to their long life span, these experiments bear out the fact that the bactericidal capabilities of sea water can be accounted for by the mere presence of phytoplanktonic organisms.

Chaetoceros teres

This diatom was submitted to tests similar to those we have just described. Its range of activity seemed to us much more extensive than that of *Asterionella japonica*, since we obtained positive antibiograms on *Staphylococcus aureus* (*209 P*), *Sarcina lutea*, *Neisseria flava*, *Bacillus subtilis*, *Proteus vulgaris*, *Proteus rettgeri*, *Proteus morgani*, and *Shigella dysenteriae*.

Research of the same type carried out on anaerobes showed the activity of *Chaetoceros teres* towards *Clostridium septicum*, *Clostridium histolyticum*, *Clostridium sordellii*, *Welchia perfringens*, *Plectridium tetani*, *Micrococcus niger*, *Ristella pseudoinsolita*, *Clostridium chauvoei*, and *Fusiformis fusiformis* (J. Aubert and H. Thouvenot[29]).

Other diatoms

Skeletonema costatum proved active towards *Sarcina lutea*, *Escherichia coli*, *Welchia perfringens*, *Clostridium sordellii*, *Plectridium tetani*, *Sphaerophorus fusiformis*, *Ramibacterium pleuriticum*, and mildly active towards *Clostridium septicum*, *Clostridium histolyticum*, *Ristella pseudoinsolita*, and *Sphaerophorus varius*.

Lithodesmium undulatum proved active towards *Clostridium septicum* and *Micrococcus prevoti*.

Nitzschia ascicularis proved active towards *Clostridium septicum*.

Biochemical study of antibacterial substances of phytoplanktonic origin

The majority of the tests described were carried out on ground, filtered samples—that is, on cell extracts.

Some authors have attempted to achieve an accurate chemical determination of the substances responsible for the overall activity of these extracts. Here again, it would be no use to try to describe the large number of studies devoted to that subject. We shall mention only the most striking facts.

It is the use of cultures that made it possible to demonstrate *in vitro* and to determine the chemical nature of the substances secreted, excreted, or however released by some phytoplanktonic organisms, whether they be diatoms, dinoflagellates, cryptophyceae, or chrysophyceae. Thus, Sieburth[31] established that the antibacterial substance found in extracts from *Phaeocystis pouchetii* is acrylic acid, a fatty acid. When freed enzymatically, this acid is especially active in the nascent state, because it readily polymerizes to inactive derivatives.

Certain dinoflagellates, in addition to their toxic properties *in situ*, give off substances that are antibiotic; thus, *Gonyaulax*, investigated by Burkholder[32], decidedly inhibits the growth of *Staphylococcus aureus* and *Candida albicans*. This is not unlike the properties of zooxanthellae, which are dinoflagellates that live inside corals and can free themselves of the coralline tissues under certain conditions: those unicellular algae secrete antibacterial substances whose chemical composition is close to that of the terpenoid and hydrocarbonaceous compounds similar to those found in coralline tissues (Burkholder[32]).

In addition, there are many studies in which the chemical determination of active components has not been achieved. For instance, in a recent work[27], Duff, Bruce, and Antia reported the extraction of antimicrobial intracellular substances from cultures of phytoplanktonic organisms, using several methods of solvent extraction, but they did not mention the characteristics of the components responsible for the antibiotic activity.

The only studies in this field have been made on diatoms at C.E.R.B.O.M.—M. Aubert and M. J. Gauthier[33], M. Aubert, J. Aubert, M. J. Gauthier, D. Pesando, and S. Daniel[34]. They were mainly based on cell extracts from *Asterionella japonica* and *Chaetoceros teres*.

Using chromatography, spectrography, and specific reactions, we demonstrated that the antibacterial action of *Asterionella japonica* is due to two types of substance: a stable one (probably nucleosidic) and one of the fatty acid type which is photo-oxidizable. The chemical structure of the latter was determined; it is a C_{20} poly-unsaturated fatty acid with 5 malonic type double bonds (D. Pesando[35]) which develops bactericidal properties after photo-oxidation. Also, a polysaccharide with antibiotic properties was isolated from extracts of *Chaetoceros teres*.

Role of Antibiosis in Marine Ecology

Antibiosis in benthic environment

Many organisms belonging to the more representative genera of the marine bacterial microflora release substances that are antagonistic to the bacteria and fungi of the terrestrial flora. From observations made *in situ* as well as *in vitro*, the activity of these substances in their natural sea environment seems to be well established and confirms the role played by these biosecretions in the benthic micro-ecology.

Within a bacterial population, interspecific competitions probably exist over all immersed surfaces. In fact, the phenomena of bacterial antibiosis towards terrestrial bacteria represent only a portion of the micro-ecological picture. They should be considered as placed in a wide system of interactions, antagonistic or synergistic, which control the actual purifying properties of the benthic and coastal waters towards the terrestrial microflora [13].

Antibiosis in pelagic environment

The antibacterial activity of several phytoplanktonic species towards some polluting faecal germs was demonstrated *in situ* through a comparative study of the bacterial and planktonic populations in selected marine biotopes—Steeman Nielsen [23], Sieburth [31], C.E.R.B.O.M. [18,19].

There is a clear relationship between the bactericidal power of sea water and its phytoplankton content. This implies that a number of antibiotic-producing species are present (M. Aubert [14]).

As in the case of marine bacteria, the antibiosis phenomenon is only an element in the sea life, which can be achieved through chemical substances (chemical telemediators) acting upon interspecific relationships [36,37].

Conclusion

The phenomenon of self-purification, so important for the equilibrium of sea life, is partly dependent on the specificity of a number of mechanisms that are adapted to each oceanic environment.

In the marine environment taken as a whole, pollution by terrestrial bacteria seems to be continually controlled through a number of mechanisms based on alternative antagonisms between species.

REFERENCES

1. De Giaxa, V. Ueber das Verhalten einiger pathogener Mikroorganismen im Meerwasser. *Z. Hyg. Infect-Krankh.*, 1889, **6**, 162–225.
2. Rosenfeld, W. D., and ZoBell, C. E. Antibiotic production by marine micro-organisms. *J. Bact.*, 1947, **54**, 393–398.
3. Grein, A., and Meyers, S. P. Growth characteristics and antibiotic production of actinomycetes isolated from littoral sediments and materials suspended in sea-water. *J. Bact.*, 1958, **76**, 457–463.
4. Krasil'nikova, E. N. Antibiotic properties of microorganisms isolated from various depths of world oceans. *Microbiology*, 1962, **30**, 545–550.
5. Buck, J. D., Meyers, S. P., and Kamp, K. M. Marine bacteria with anti-yeast activity. *Science, N.Y.*, 1962, **138**, 1339–1340.
6. Buck, J. D., Ahearn, D. G., Roth, F. J., and Meyers, S. P. Inhibition of yeasts by a marine bacterium. *J. Bact.*, 1963, **85**, 1131–1135.
7. Buck, J. D., and Meyers, S. P. Anti-yeast activity in the marine environment. I—Ecological considerations. *Limnol. Oceanogr.*, 1965, **10**, 385–391.
8. Buck, J. D., and Meyers, S. P. 'In vitro' inhibition of *Rhodotorula minuta* by a variant of the marine bacterium *Pseudomonas piscicida*. *Helgoländer wiss. Meeresunters.*, 1966, **13**, 171–180.
9. Bonde, G. J. Studies on the dispersion and disappearance phenomena of enteric bacteria in the marine environment. *Revue int. Océanogr. méd.*, 1968, **9**, 17–44.
10. Baam, B. R., Gandhi, N. M., and Freitas, Y. M. Antibiotic activity of marine micro-organisms. *Helgoländer wiss. Meeresunters.*, 1966, **13**, 181–185.
11. Carlucci, A. F., and Pramer, D. An evaluation of factors affecting the survival of *Escherichia coli* in sea water. III—Antibiotics. *Appl. Microbiol.*, 1960, **8**, 251–254.
12. Burkholder, P. R., Pfister, R. M., and Leitz, F. H. Production of a pyrrole antibiotic by a marine bacterium. *Appl. Microbiol.*, 1966, **14**, 649–653.
13. Gauthier, M. J. Propriétés antibactériennes de micro-organismes marins. Thesis, Université Nice, 1970, 113 pp.
14. Aubert, M. Le comportement des bactéries terrigènes en mer. Relations avec le phytoplancton. Thesis, Université Marseille, 1971, 285 pp.
15. Aubert, M. Théorie générale de l'auto-épuration de la mer. *Revue int. Océanogr. méd.*, 1971, **24**, 61–125.
16. Gauthier, M. J. Substances antibactériennes produites par les bactéries marines. I—Etude systématique de l'activité antagoniste de souche bactériennes marines vis-à-vis de germes telluriques aérobies. *Revue int. Océanogr. méd.*, 1969, **15–16**, 41–59.
17. Halbert, S. P., and Magnusson, H. J. Studies with antibiotic-producing strains of *Escherichia coli* *J. Immun.*, 1948, **58**, 397–415.

18. Aubert, M., Vaissiere, R., Aubert, J., Pechon, J. C. L., and Vellard, M. P. Antibiogrammes obtenus par des extraits planctoniques sur des germes d'origine entérique. *Bull. Acad. natn. Méd.*, 1964, **140**, 507–513.

19. Aubert, M., Lebout, H., and Aubert, J. Rôle du plancton dans le pouvoir antibiotique du milieu marin. *Annls Inst. Pasteur, Paris*, 1964, **106**, 147–150.

20. Pratt, R., Daniels, T. C., Eiler, J. J., Gunnison, J. B., Kumler, W. D., Oneto, J. F., and Strait, L. A. Chlorellin, an antibacterial substance from *Chlorella. Science, N.Y.*, 1944, **99**, 351–352.

21. Spoehr, H. A., Smith, J. H. P., Strain, H. H., Milner, H. W., and Hardin, G. J. Fatty acid antibacterial from plants. *Publs Carnegie Instn*, 1949, **586**, 67.

22. Blaauw-Jansen, G. On the light induced transformation of chlorophyllide in a growth inhibiting substance. *Proc. K. med. Akad. Wet.*, C, 1954, **57**, 408–506.

23. Nielsen, E. S. The production of antibiotics by plankton algae and its effect upon bacterial activities in the sea. *Deep Sea Res.*, 1955, **3** (Suppl.), 281–286.

24. Jørgensen, E. G., and Nielsen, E. S. Effect of filtrates from cultures of unicellular algae on the growth of *Staphylococcus aureus. Physiol. Plant.*, 1961, **14**, 896–908.

25. Sieburth, J. M., and Pratt, D. M. Anticoliform activity of sea-water associated with the termination of *Skeletonema costatum* blooms. *Trans. N.Y. Acad. Sci.*, II, 1962, **24**, 498–501.

26. Sieburth, J. M. Role of algae in controlling bacterial populations in estuarine waters. *Symp. Commn int. Explor. scient. Mer Médit., Monaco, 1964*, pp. 217–233.

27. Duff, D. C. B., Bruce, D. L., and Antia, N. J. The antibacterial activity of marine planktonic algae. *Can. J. Microbiol*, 1966, **12**, 877–884.

28. Aubert, M., and Gauthier, M. J. Origine et nature des substances antibiotiques présentes dans le milieu marin. VIII- Etude systématique de l'action antibactérienne d'espèces phytoplanctoniques vis-à-vis de certains germes telluriques aérobies. *Revue int. Océanogr. méd.*, 1967, **5**, 63–71.

29. Aubert, J., Pesando, D., and Thouvenot, H. Action antibiotique d'extraits planctoniques vis-à-vis de germes anaérobies. *Revue int. Océanogr. méd.*, 1968, **10**, 259–265.

30. Gauthier, M. J. Activité antibactérienne d'une diatomée marine: *Asterionella notata* (Grun). *Revue int. Océanogr. méd.*, 1969, **15–16**, 103–171.

31. Sieburth, J. M. Antibiotic properties of acrylic acid, a factor in the gastro-intestinal antibiosis of polar marine animals. *J. Bact.*, 1961, **82**, 72–79.

32. Burkholder, P. R., Burkholder, L. M., and Almodóvar, L. R. Antibiotic activity of some marine algae of Puerto Rico. *Botanica mar.*, 1960, **2**, 149–156.

33. Aubert, M., and Gauthier, M. J. Origine et nature des substances antibiotiques présentes dans le milieu marin. V—Mise en évidence des fractions antibiotiques d'un extrait phytoplanctonique par séparation sur Sephadex. *Revue int. Océanogr. méd.*, 1966, **2**, 63–66.

34. Aubert, M., Aubert, J., Gauthier, M. J., Pesando, D., and Daniel, S. Origine et nature des substances antibiotiques présentes dans le milieu marin. VI—Etude biochimique des substances antibacteriénnes extraites d'*Asterionella japonica* (Cleve). *Revue int. Océanogr. méd.*, 1966, **4**, 23–32.

35. Pesando, D. Etude chimique et structurale d'une substance lipidique antibiotique produite par une Diatomée marine: *Asterionella japonica. Revue int. Océanogr. méd.*, 1972, **25**, 49–69.

36. Aubert, M. Télémédiateurs chimiques et équilibre biologique océanique. I—Théorie générale. *Revue int. Océanogr. méd.*, 1971, **21**, 5–16.

37. Aubert, M., and Pesando, D. Télémédiateurs chimiques et équilibre biologique océanique. II—Nature chimique de l'inhibiteur de la synthèse d'un antibiotique produit par une diatomée. *Revue int. Océanogr. méd.*, 1971, **21**, 17–22.

Introducing the paper, Dr. Gauthier said that the phenomenon of self-purification was dependent on the specificity of a number of mechanisms that were adapted to each part of the marine environment. An important part of the biological activity responsible for the destruction of non-marine bacteria in the sea was due to antibiotic substances produced by phytoplankton or by other specific marine micro-organisms—bacteria and yeasts. Such activity had been observed *in vitro* and *in situ* for more than 30 years.

In the pelagic zones, there were some clear relationships between the bactericidal power of sea water and its phytoplanktonic content. This implied the presence of a number of antibiotic-producing species. In fact, those species were quite common and had been found in many oceans and seas. Similarly, numerous antibiotic-producing strains of marine bacteria had been isolated from different biotopes, and they seemed to be ubiquitous. At the benthic level, or on any submerged surface, it had to be considered in terms of microbial ecology that the antagonistic activity of marine microflora against non-marine micro-organisms was a secondary consequence of endogenous competitions, since the terrestrial bacteria were known to be far less competitive than the marine ones in natural conditions.

The antibiotic phenomena, of both planktonic and bacterial origin, were therefore only an element in the sea life, which could be achieved through chemical substances, or telemediators, acting upon interspecific relationships.

DISCUSSION

Prof. H. I. Shuval (Israel) noted that in the paper, as in most of the previous papers and reports of the Authors' group, little information had been given on the kinetics of inactivation of terrestrial bacteria by the various bacteria and plankton reported to possess antibiotic capabilities. He asked whether they had data on the T_{90} attributable to the specific antibiotic factors studied, and whether any of those studies cast light on T_{90} rates of 60–90 min found in field experiments in the sea. He also enquired if there was any evidence that their findings—which were of great scientific interest—related to actual ecological conditions.

Prof. R. Mitchell (U.S.A.) remarked that large numbers of antibiotic-producing micro-organisms were present in soil, yet antibiosis was not a significant factor in soil ecology. The wide discrepancy between observed T_{90} values in the field and those observed using antibiotic-producing bacteria suggested that antibiosis was not a significant factor in this context.

Dr. M. Waldichuk (Canada) said that the results given in the paper had important practical possibilities. However, he wondered why investigators like Carlucci and Pramer[11] had observed no antagonistic activity although they had studied over 200 strains, and he asked Dr. Gauthier to explain this discrepancy in findings. He also referred to the culturing of particularly active strains of algae or other micro-organisms for treatment of sewage, as a replacement for chemical treatment with, for instance, chlorine. He thought that, some 10 years previously, ZoBell had presented a paper in which he noted that such organism selection and culturing had exciting possibilities in sewage treatment, and he asked Dr. Gauthier if he considered that progress had been made in this field.

Prof. R. R. Colwell (U.S.A.) pointed out that effects of antibiosis might be negligible in the die-off of *E. coli*, but were highly significant in colonization of substrates and surfaces, communication amongst members of populations, and in other aspects of microbial ecology. Nevertheless, with the excessive environmental abuse caused by sewage input, these rather delicate interactive effects might well be overwhelmed.

Mr. D. J. Gould (U.K.) said that the Authors had mentioned the light dependency of the bactericidal activity of *Asterionella japonica*, and had suggested a photo-oxidation process of liberated materials to form bactericidal short chains. He asked what results had been obtained when control tubes were incubated in the absence of the diatom, and if it was not possible that this activity was due, at least in part, to the direct effects of light on the bacterial cells.

REPLY TO DISCUSSION

Dr. Gauthier, replying to Prof. Shuval, said that their data indicated that about 90 per cent of non-marine test bacteria were inactivated after at least 12–24 h, both by planktonic and bacterial antibiotics at natural concentrations. Although this effect might be too slow to explain the quite short T_{90} values found near outfalls, it did not follow that antibiosis was not significant in the sea—it indicated only that the conditions for life and growth of antibiotic-producing marine species were not present in such a modified environment. In surrounding zones, where eutrophication effects of sewage often led to planktonic blooms, this antibiosis became an important factor for destruction of the remaining terrestrial bacteria; the observations of Sieburth[25,26] were highly significant in this field.

To Prof. Mitchell he said that it was invalid to extrapolate the conclusion that antibiosis was not a significant factor in soil ecology to conditions in the sea. There were several observations which had demonstrated that effective antibiotic phenomena occurred in the natural marine environment, but such activity had never been found in soil. He thought it impossible to compare soil and sea water in the field of antibiosis: in sea water, every antibiotic substance could diffuse, and then form an 'inhibitory sphere' around the antibiotic-producing cells, or be diluted in the whole medium where it could accumulate.

In reply to Dr. Waldichuk's first question, Dr. Gauthier pointed out that in all marine areas investigated there were large fluctuations in the antibiotic-producing populations. The spatial distribution of such species was also heterogeneous, depending on such local ecological factors as temperature, salinity, and nature of substrates, so that it was sometimes difficult to find the active strains. This might explain the failure of Carlucci and Pramer and others in this field, but it was quite rare: the isolation of antibiotic-producing marine macro-organisms was generally easy and had been carried out in various parts of the world. Concerning the use of algal cultures for sewage treatment, he reminded Dr. Waldichuk that fresh-water algae had been used successfully in Belgium, and that some investigators, such as Stirn in Yugoslavia, proposed the use of micro-algae for sewage treatment and biomass production from sewage nutrients.

Replying to Mr. Gould, Dr. Gauthier said that they had demonstrated the photo-activation of the fatty acid extracted from *A. japonica* by irradiation in separate tubes before carrying out the biological tests which were then made in total darkness. They had observed that this fatty acid was inactive so long as it was not exposed to visible light. The test bacteria had never been directly irradiated, either in control or in test tubes, so that the experimental method excluded any direct effect of light on the mortality of the bacteria.

Heavy Metal Ions as the Principal Bactericidal Agent in Caribbean Sea Water

Galen E. Jones and Andre B. Cobet*

Jackson Estuarine Laboratory and Department of Microbiology
University of New Hampshire, Durham, New Hampshire 03824

Bacteria alien to the sea enter during run-off from land, from rivers and estuaries, and from sewage outfalls. The ultimate fate of these alien microbes is death[1–5], although the death rate may vary[6] and bacterial increases may even occur temporarily when enriched with sewage[7,8]. The increasing volumes of sewage discharged into the ocean accentuates the need to understand the survival of enteric pathogenic bacteria in sea water. The widening use of the coastal zone for recreational activities and commercial shellfish makes the public health significance of the purification process of sea water evident. *Escherichia coli*, the common enteric organism, has been the most extensively studied bacterium as an indicator of more serious enteric pollution.

Factors responsible for the rapid decline of fresh-water and sewage bacteria in the sea include physical, chemical, and biological effects. The influence of seasonal variation in temperate regions[9–11], temperature of incubation[12,13], the destructive action of sunlight[14,15], the adsorption and sedimentation of contaminating microorganisms with particles in suspension[16,2,17], osmotic shock[18], the lack of sufficient organic energy for maintenance[19,12,18,20–22], heavy metal ion toxicity[23–26,13], the production of antibiotic substances by marine organisms[27,28,5], predators of all types such as protozoa, filter-feeding animals, and bacterial parasites such as *Bdellovibrio*[29,30,5], and the presence of destructive bacteriophage[31,32] are all contributing factors to the disappearance of non-marine bacteria from sea water. Any one of the factors considered above may be effective at a particular location in the marine environment[4].

In the current investigation, sea water was collected aseptically, filter-sterilized, and frozen at stations across the Caribbean Ocean for subsequent bactericidal evaluation in the laboratory. The purpose was to determine the basic physiological effects of sea-water samples from different depths and locations in the Caribbean Ocean on several non-marine and marine bacteria. An evaluation of the toxicity of Caribbean sea water, energy concentrations, and the concentration of chelating agents for bacteria was undertaken.

Materials and Methods

Sea-water samples

During December 1964, at stations selected in the Caribbean Ocean, sea water was collected aseptically at various depths using a sterile Cobet water sampler[33]. These samples were filtered immediately after collection to effect sterilization and remove particulate matter through membrane filters (Millipore Corporation, Bedford, Mass.) of a porosity of 0·22 μm which had been pre-sterilized by autoclaving at 69 kN/m^2 pressure for 8 min. The filtered sea-water samples were entered aseptically into sterile plastic bags (Whirl-pak of Scientific Products, McGraw Park, Illinois) and frozen aboard ship at -20°C.

Bacterial cultures

Four species of fresh-water bacteria plus two additional strains of *Escherichia coli* and two species of marine bacteria were tested for their survival in sea water (Table 1).

Testing procedure

The bacterial cultures were inoculated over the entire surface of a slant of Nutrient Agar (Difco; Detroit, Michigan) or 2216(E) Marine Agar (Difco) for the fresh-water and marine organisms, respectively, in 16 × 125 mm screw-capped test tubes. These inoculated slants were incubated for 24 h at

* Present address: Naval Biomedical Research Laboratory, Naval Supply Center, University of California (Berkeley), Oakland, California 94625

TABLE 1. Bacterial cultures tested for survival in sea water

Culture	ATCC[a] or Code No	Description
Enterobacter aerogenes	ATCC 12658	Noted for polysaccharide production
Escherichia coli	ATCC 4157	Original *E. coli* strain
Escherichia coli	ATCC 9637	Strain W of Waksman
Escherichia coli	ATCC 11229	Germicide testing
Pseudomonas aeruginosa	ATCC 10145	Antibiotic testing
Flavobacterium sp.	449	A colony isolated from Atlantis II Station 449 (8° 52′ N, 51° 34′ W) from 500-m depth in Caribbean Ocean on 5 November 1964
Pseudomonas sp.	A-3	Isolated from Antarctic sea water (77° 21′ S, 44° 30′ W) from 290 m on 20 January 1957

[a] ATCC = American Type Culture Collection

32°C. The resultant growth was diluted and suspended by adding 2–3 ml of sterile saline (0·85 per cent) to each slant and gently teasing the bacterial growth off the agar. The 2–3 ml were then returned to the 99-ml sterile saline dilution bottle from which it originated and vigorously shaken to disrupt bacterial clumps. Each suspension was diluted by adding 1 ml to a second 99-ml sterile saline bottle, shaking and subsequently to a third saline dilution blank; 1 ml of the third dilution was used to inoculate 30 ml of sea-water samples in 125-ml Erlenmeyer flasks. In this procedure, a dilution of 3×10^7 times consistently produced a bacterial suspension in the test flask of 200–600 cells/ml at inoculation time.

The sea-water samples were defrosted just before testing and 30 ml pipetted to sterile 125-ml Erlenmeyer flasks. Artificial sea water (ASW)[34] was always tested as a control. Special precautions were taken to assure that the glassware had been cleaned carefully with Haemo-sol (Scientific Products) and rinsed copiously with tap water and finally three times with distilled water.

The inoculated sea-water samples were incubated statically at 32°C. After 24, 48, and 72 h the contents of the flasks were shaken, plated, and enumerated.

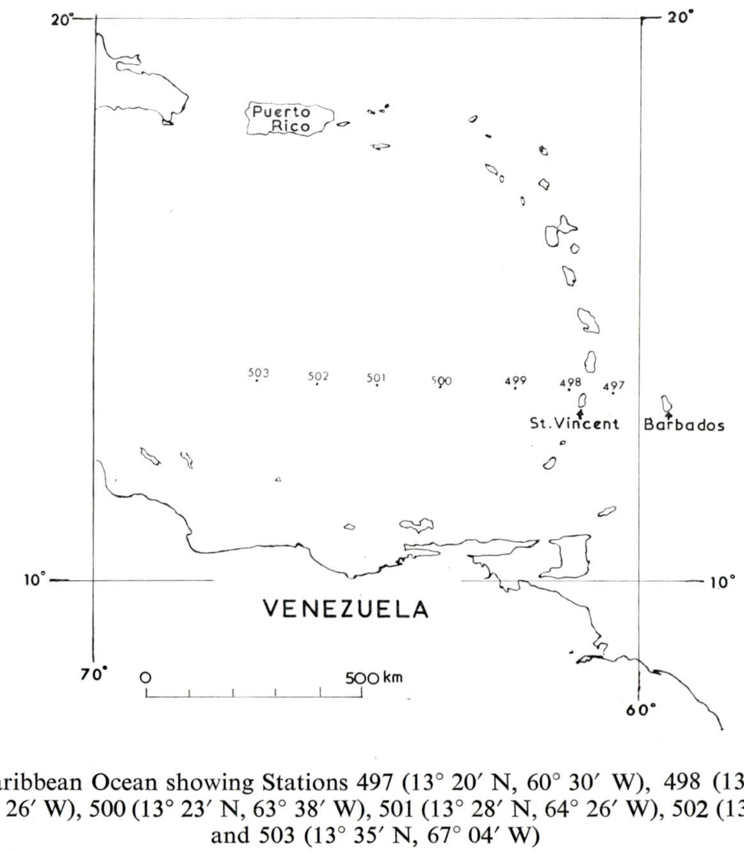

FIG. 1. Map of Caribbean Ocean showing Stations 497 (13° 20′ N, 60° 30′ W), 498 (13° 26′ N, 61° 17′ W), 499 (13° 27′ N, 62° 26′ W), 500 (13° 23′ N, 63° 38′ W), 501 (13° 28′ N, 64° 26′ W), 502 (13° 36′ N, 65° 50′ W), and 503 (13° 35′ N, 67° 04′ W)

Immediately after entry and mixing of bacterial suspensions in each sea-water treatment, duplicate 1-ml aliquots were pipetted into sterile plastic petri dishes. Pour plates for fresh-water and marine bacteria were prepared by entering sterile 42°C molten trypticase glucose extract agar (Baltimore Biological Laboratory, Baltimore, Maryland) and 2216E Marine Agar (Difco), respectively, into the petri dishes in 10–15 ml quantities. When solidified, the plates were inverted and incubated for 24 h at 32°C before enumeration. Plates harboring *Ps. aeruginosa* and the marine bacteria were incubated for 48 h at 32°C before counting.

Solutions of 8-hydroxyquinoline sulfate, thioglycolic acid, peptone, and glucose were prepared in double-distilled water and sterilized by passage through sterile 0·22 μm membrane filters (Millipore Corporation). The solutions were further diluted with double-distilled water so that a 100-fold dilution gave the desired final concentration of the additives in the sea water.

Results

Sea-water samples were obtained at station locations in the Caribbean Ocean (Fig. 1).

The average percentage survival of *E. aerogenes* (ATCC 12658), *E. coli* (ATCC 4157, 9637, and 11229), *Ps. aeruginosa* (ATCC 10145), *Flavobacterium* (449), and *Pseudomonas* (A-3) in the thawed Caribbean sea water from Stations 497–502 after 24, 48, and 72 h incubation at 32°C is shown in Table 2.

TABLE 2. Average percentage survival of marine and non-marine bacteria exposed to Caribbean sea-water samples from 6 stations (497–502) at various depths after 24, 48, and 72 h of static incubation at 32°C
(Bacterial species as in Table 1)

Sea-water samples	Time (h)	ATCC 12658	ATCC 4157	ATCC 9637	ATCC 11229	ATCC 10145	449	A-3
ASW	24	3	0	0·1	7	0	160	157
	48	0·3	0	0	7	0·1	9	2
	72	0·1	0	0	17	0·1	3	1
1 m	24	16	0·3	10	3	3	92	46
	48	3	0·5	2	1·5	45	59	915
	72	2	—	3	8	64	32	8 000
25 m	24	7	2	7	17	2	80	45
	48	0·7	0	1·5	7	4	56	774
	72	0·3	0	1·5	7	80	39	3 400
50 m	24	5	3	11	12	3	77	50
	48	0·5	0	2	2	6	55	1 423
	72	0·1	0	2	2	23	33	4 450
100 m	24	4	2	4	8	1	88	45
	48	0·1	0	1	1	8	55	820
	72	0·1	0	0·5	2	30	41	7 320
250 m	24	3	2	3	7	2	83	61
	48	0·7	0	0	7	12	64	942
	72	0	0	0	17	130	39	6 120
500 m	24	2	0·1	1	3	1	87	29
	48	0·2	0	0·2	2	8	71	333
	72	0·1	0	0	1	77	61	984
800 m	24	1	4	0	0	0	89	51
	48	0·1	0	0	0	0	83	1 850
	72	0	0	0	0	0	71	10 684
1900 m	24	7	5	9	1·5	0·1	80	43
	48	5	0	5	0·5	1	54	1 500
	72	7	0	8	1·0	7	42	17 500

Several features of these data should be mentioned which are not readily apparent from the table. Firstly, the survival of all of the non-marine bacterial populations was greatest in sea water from Station 498, followed by sea water from Station 497, and then Stations 499, 500, 501, and 502. This suggested a beneficial effect of proximity to land for the survival of non-marine bacteria in sea water. The marine bacterial cultures, *Flavobacterium* (449) and *Pseudomonas* (A-3), survived or developed better in sea water from the open ocean stations (499–502) than the near-shore stations. Secondly, the average percentage survival of bacterial cultures may represent one or two high or low values on particular stations which skew the data for an average percentage survival. For example, no survival of *Pseudomonas* (A-3) occurred at any depth on Station 499 for some unknown reason, although growth occurred at almost all depths in water from Stations 498, 500, 501, and 502. Survival of A-3 was apparent in sea water from Station 497.

It may be seen from Table 2, that of the non-marine bacteria, *E. coli* (ATCC 4157) demonstrated the least survival in Caribbean sea water, the greatest average survival of the culture in 24 h at 32°C being 5 per cent; *E. aerogenes* (ATCC 12658) and *E. coli* (ATCC 9637) were almost equally susceptible to the Caribbean sea-water samples but not as sensitive as *E. coli* (ATCC 4157); *E. coli* (ATCC 11229) was the most resistant strain to sea water although it also died off.

TABLE 3. Percentage survival of *E. coli* (ATCC 11229) after 4, 24, 48, and 72 h static incubation at 32°C following exposure to Caribbean upper (1–100 m) sea water from Station 503 and treatment with various energy sources, chelating agents, and combinations

Treatment	Time (h) 4	24	48	72	pH
None	73	1	0	0	8·12
		8-hydroxyquinoline SO_4			
10^{-5} M	112	104	100	63	7·39
10^{-6} M	99	77	73	185	8·14
10^{-7} M	72	1	0	0	8·18
10^{-8} M	32	0	0	0	8·18
		thioglycolic acid			
10^{-3} M	100	231	590	5 900	7·89
10^{-4} M	93	98	177	1 770	8·15
10^{-5} M	88	68	158	1 221	8·17
10^{-6} M	70	63	129	1 120	8·18
		peptone			
10 mg/l.	94	121	5 000	—	8·18
1 mg/l.	30	3	0	—	8·18
0·1 mg/l.	29	6	0	—	8·18
		glucose			
10 mg/l.	36	5	0	0	8·19
1 mg/l.	36	3	0	0	8·19
0·1 mg/l.	8	0	0	0	8·18
		10^{-6} M 8-hydroxyquinoline SO_4 + glucose			
10 mg/l.	—	185	750	100 000	7·89
1 mg/l.	—	108	828	50 000	8·11
0·1 mg/l.	—	79	222	1 240	8·10
		10^{-7} M 8-hydroxyquinoline SO_4 + glucose			
10 mg/l.		26	83	4 500	8·08
1 mg/l.		11	1	90	8·11
		10^{-8} M 8-hydroxyquinoline SO_4 + glucose			
10 mg/l.		10	800	16 000	8·10
1 mg/l.		5	75	600	8·13

The most resistant non-marine bacterium was *Ps. aeruginosa* (ATCC 10145) which was reduced to a few per cent survival within 24 h incubation at 32°C, followed by increasing populations during the next 48 h—except for ASW and 800-m sea water.

Survival of the marine *Flavobacterium* (449) in ASW occurred on four out of six occasions. In all the Caribbean sea-water samples, gradual attrition of this marine bacterium took place over the 72-h incubation period. Appreciable populations of *Flavobacterium* (449) remained in most samples (Table 2). Growth of this bacterium did not occur in any sea-water sample.

Pseudomonas (A-3) perished in ASW in four out of six cases within 48 h. On the three stations further at sea, growth of this bacterium was demonstrated at all depths—except 1900 m at Station 501 and 500 m at Station 502.

Survival of all these cultures was poor in ASW. The sea water from above 100 m was not as toxic as that below 100 m.

The evident die-off of non-marine bacteria in the sea stimulated an investigation to determine how to reverse the toxicity. Since two major factors, heavy metal toxicity and low nutrient concentrations, have been considered the major physiological factors in sea water[4], their influence on the survival of *E. coli* (ATCC 11229) in Caribbean sea water from Station 503 was determined. Sea-water samples from

TABLE 4. Percentage survival of *E. coli* (ATCC 11229) after 4, 24, 48, and 72 h static incubation at 32°C following exposure to Caribbean middle (250–1000 m) sea water from Station 503 and treatment with various energy sources, chelating agents, and combinations

Treatment	Time (h) 4	24	48	72	pH
None	0	0	0	0	8·16
		8-hydroxyquinoline SO_4			
10^{-5} M	94	91	77	65	7·43
10^{-6} M	95	51	47	31	—
10^{-7} M	60	0	0	0	—
10^{-8} M	68	0	0	0	—
		thioglycolic acid			
10^{-3} M	82	62	143	1053	7·98
10^{-4} M	96	57	44	32	—
10^{-5} M	80	52	24	15	—
10^{-6} M	71	15	20	8	—
		peptone			
10 mg/l.	79	26	24	43	8·22
1 mg/l.	8	0	0	0	—
0·1 mg/l.	0·1	0	0	0	—
		glucose			
10 mg/l.	25	0·1	130	5000	8·22
1 mg/l.	0	0	0	0	—
0·1 mg/l.	2	0	0	0	—
		10^{-6} M 8-hydroxyquinoline SO_4 + glucose			
10 mg/l.	—	83	115	182	8·16
1 mg/l.	—	95	221	500	8·16
0·1 mg/l.	—	86	195	610	8·16
		10^{-7} M 8-hydroxyquinoline SO_4 + glucose			
10 mg/l.	—	2	7	112	8·15
1 mg/l.	—	1	3	19	8·16
		10^{-8} M 8-hydroxyquinoline SO_4 + glucose			
10 mg/l.	—	1	0	0	8·16
1 mg/l.	—	0	0	0	8·17

(a) 1, 25, 50, and 100 m, (b) 250, 500, and 1000 m, and (c) 2000, 3000, and 4000 m, were thawed, pooled, and labelled upper, middle, and deep water, respectively. When 200–600 *E. coli* (ATCC 11229) cells/ml were added to filter-sterilized, untreated samples of this pooled sea water, no surviving cells remained after 48 h in the upper sea water (Table 3), or after 4 h in the middle (Table 4) or deep (Table 5) sea-water samples from Station 503. When chelating agents, such as 8-hydroxyquinoline sulfate in concentrations of 10^{-5}–10^{-8} M and thioglycolic acid in concentrations of 10^{-3}–10^{-6} M, were added to these samples, die-off was retarded or reversed at appropriate concentrations (Tables 3–5).

In the upper sea water, survival of *E. coli* (ATCC 11229) was enhanced by 10^{-5} and 10^{-6} but not 10^{-7} and 10^{-8} M 8-hydroxyquinoline sulfate (Table 3). Growth occurred in all concentrations of thioglycolic acid; the most rapid growth and the largest number of cells were associated with the highest concentrations of thioglycolic acid. Peptone is known to serve as a chelating agent as well as an energy source[25]. At 10 mg/l. the effect of peptone was to cause growth of *E. coli* whereas die-off at 1 and 0·1 mg/l. peptone was similar to the untreated upper sea water (Table 3). Glucose, another energy source but with poor chelating ability, at 10, 1, and 0·1 mg/l. did not spare the bactericidal action of the upper sea water (Table 3). However, when low concentrations of 8-hydroxyquinoline sulfate and glucose were

TABLE 5. Percentage survival of *E. coli* (ATCC 11229) after 4, 24, 48, and 72 h static incubation at 32°C following exposure to Caribbean deep (2000–4000 m) sea water from Station 503 and treatment with various energy sources, chelating agents, and combinations

Treatment	Time (h) 4	24	48	72	pH
None	0	0	0	0	8·15
		8-hydroxyquinoline SO_4			
10^{-5} M	76	74	69	64	7·50
10^{-6} M	99	79	104	261	—
10^{-7} M	65	7	0·1	1	—
10^{-8} M	47	0	0	0	—
		thioglycolic acid			
10^{-3} M	91	119	335	3355	7·94
10^{-4} M	93	53	55	66	—
10^{-5} M	94	76	64	86	—
10^{-6} M	84	29	61	67	—
		peptone			
10 mg/l.	92	772	7720	—	8·22
1 mg/l.	51	1	12	107	—
0·1 mg/l.	3	0	0	0	—
		glucose			
10 mg/l.	31	2	0·1	0	8·15
1 mg/l.	8	0	0	0	—
0·1 mg/l.	0·1	0	0	0	—
		10^{-6} M 8-hydroxyquinoline SO_4 + glucose			
10 mg/l.		119	380	1030	—
1 mg/l.		125	73	820	—
0·1 mg/l.		66	75	290	—
		10^{-7} M 8-hydroxyquinoline SO_4 + glucose			
10 mg/l.		18	170	7500	—
1 mg/l.		20	170	7500	—
		10^{-8} M 8-hydroxyquinoline SO_4 + glucose			
10 mg/l.		0	11	150	—
1 mg/l.		0·1	11	100	—

added to the upper sea water in combinations, the toxicity of sea water was overcome most effectively at the highest concentration of 8-hydroxyquinoline sulfate, 10^{-6} M, and glucose, 10 mg/l. (Table 3).

In the middle sea water from Station 503, results consistent with the upper sea water were obtained from somewhat more toxic sea water (Table 4). Aside from more chelating agents and energy sources required in the middle sea water to produce retarded death or growth compared to upper sea water, an anomaly was the less effectiveness of 10 mg/l. peptone than 10 mg/l. glucose. When glucose and 10^{-6} M 8-hydroxyquinoline sulfate were combined, the highest populations of *E. coli* developed at 1 and 0·1 mg/l. glucose rather than 10 mg/l. (Table 4). Generally, the results supported the importance of metal ion-binding and supplemental energy sources for coliform survival in sea water.

Deep sea water from Station 503 (Table 5) was slightly less toxic than middle sea water (Table 4) to *E. coli* populations. The non-metabolizable chelating agent, 8-hydroxyquinoline sulfate, was more effective at reversing toxicity in deep sea water at 10^{-6} than 10^{-5} M concentration (Table 5). The metabolizable chelating agent, thioglycolic acid, was more effective at reversing the toxicity of deep sea water, the higher its concentration (Table 5). Peptone was more effective than glucose in supporting *E. coli* in deep sea water. Glucose in combination with 8-hydroxyquinoline sulfate again reversed the toxicity of sea water, although the glucose functioned rather similarly at all concentrations tested whereas the 8-hydroxyquinoline sulfate was more effective the higher the concentration (Table 5).

Discussion

The disappearance of viable non-marine bacterial cells in sea water is a complex phenomenon. Enteric bacteria live in the high-energy environment of the intestinal tract. *Escherichia coli* grows competitively in sea water under continuous culture conditions when the energy levels are higher than 100 mg/l.[22] Below this concentration of organic matter, the autochthonous marine bacterial population was able to compete successfully with *E. coli*[22]. The concentrations of dissolved organic carbon in the open ocean range between 350 and 700 μg/l.[35] Soluble organic nutrients in this concentration range limit the multiplication of marine bacteria which may explain why organic substances in these concentrations persist in the ocean[21]. The nature of this organic matter is complex, variable, and, much of it, biologically refractory[36–39].

The precise interaction between organic matter in the sea such as specific amino-acids and inorganic ions such as heavy metal is difficult to calculate since both are available in transitory microquantities[40]. How much of the metals are complexed and how much are free ions in the water is unknown but varies from place to place and time to time. Ratios of organic matter to nickel of 125:1 in the ocean have been postulated[41].

The rapid death of non-marine bacteria in the Caribbean Ocean sea water in this investigation is attributed primarily to heavy metal ion toxicity. Where chelating agents were added in concentrations calculated to react with concentrations of heavy metal ions in the sea (Tables 3–5), the survival of *E. coli* (ATCC 11229) was improved or cell multiplication took place. The 8-hydroxyquinoline sulfate which served as a non-metabolizable chelating agent with high stability constants for metals was most effective at 10^{-6} M in upper sea water from Station 503 (Table 3), at 10^{-5} M in middle sea water (Table 4), and 10^{-6} M in deep sea water (Table 5), suggesting a greater toxicity for the middle sea water.

Where thioglycolic acid, a metabolizable chelating agent with lower stability constants, was entered into the sea-water samples, growth occurred at all concentrations within 48 h at 32°C in upper sea water (Table 3), only at a concentration of 10^{-3} M in middle sea water (Table 4), and deep sea water (Table 5). Survival was better in the deep sea water than in middle sea water in the presence of similar concentrations of thioglycolic acid, indicating maximum heavy metal ions in middle sea water.

When glucose and 8-hydroxyquinoline sulfate were combined in low concentrations, *E. coli* (ATCC 11229) was able to grow eventually in most cases (Tables 3–5). Only the lowest concentrations (0·1 mg/l. glucose and 10^{-8} M 8-hydroxyquinoline) failed to effect survival of *E. coli* in middle sea water from Station 503 (Table 4). The toxicities of the three pooled sea-water samples was middle > deep > upper. The bactericidal action of sea water in these samples is attributed primarily to heavy metal ion toxicity since 10^{-6} M 8-hydroxyquinoline sulfate allowed *E. coli* to grow at all concentrations of glucose added for all depths (Tables 3–5). Glucose served as an energy source for development after toxicity had been overcome.

An evaluation of temperature, salinity, dissolved organic carbon, trace metals, and diatom metabolites on the growth and death of *Salmonella typhimurium* in artificial and natural sea water "supported the hypothesis that the trace metals were a major influence"[13]. The importance of the oligodynamic effect of heavy metals for the survival of *S. typhimurium* in sea water was recognized as was the value of organic matter as an energy source and chelating agent[13].

When the test organisms were entered into the Caribbean sea-water samples, the non-marine bacteria perished rapidly and the marine bacteria showed prolonged survival or growth (Table 2). The effect of the depth from which the sea-water samples were obtained upon bacterial survival was not marked although upper sea water was less bactericidal than deep sea water. The viability of the three strains of *E. coli* was: ATCC 11229 surviving better than ATCC 9637 surviving better than ATCC 4157 which barely survived (Table 2). *Enterobacter aerogenes*, likewise, demonstrated rapid die-off in sea water (Table 2). *Pseudomonas aeruginosa* survived better than other non-marine bacteria, occasionally even multiplying in 72 h at various depths from Stations 497–499. As stations were taken further and further from land, the bactericidal effect for non-marine bacteria became increasingly greater.

The better survival of the marine bacteria (Table 2) reflects their adaptation to sea water with low concentrations of dissolved organic carbon and ability to tolerate heavy metal ions[42].

ACKNOWLEDGMENT

The authors would like to thank the Woods Hole Oceanographic Institution for providing space on the *R.V. Atlantis II* during November–December 1964. Some of the experimental work was performed in the Department of Biology, Boston University, Boston, Massachusetts, while both authors were associated with that university. This work was supported by a research grant, WP-00650-03, Federal Water Pollution Control Agency, Department of the Interior, Washington, D.C.

REFERENCES

1. Greenberg, A. E. Survival of enteric organisms in sea water. *Publ. Hlth Rep., Wash.*, 1956, **71**, 77.
2. Orlob, G. T. Viability of sewage bacteria in sea water. *Sewage ind. Wastes*, 1956, **28**, 1147.
3. Carlucci, A. F., and Pramer, D. Factors affecting the survival of bacteria in sea water. *Appl. Microbiol.*, 1959, **7**, 388.
4. Jones, G. E. The fate of freshwater bacteria in the sea. *Devs ind. Microbiol.*, 1971, **12**, 141.
5. Aubert, M., and Aubert, J. Pollutions marines et aménagement des rivages. *Revue int. Océanogr. méd.*, 1973, Suppl., p. 20.
6. Pearson, E. A. Viability of sewage bacteria in sea water. In 'An investigation of efficacy of submarine outfall disposal of sewage and sludge'. California State Water Pollution Control Board, Publ. No. 14, Sacramento, 1956, p. 31.
7. Slanetz, L. W., and Bartley, C. H. Survival of fecal streptococci in seawater. *Hlth Lab. Sci.*, 1965, **2**, 142.
8. Savage, H. P., and Hanes, N. B. Toxicity of seawater to coliform bacteria. *J. Wat. Pollut. Control Fed.*, 1971, **43**, 854.
9. Ketchum, B. H., Carey, C. L., and Briggs, M. P. Preliminary studies on the viability and dispersal of coliform bacteria in the sea. In 'Limnological aspects of water supply and waste disposal' (ed. F. R. Moulton and F. Hitzel). American Association for Advancement of Science, Washington, D.C., 1949, p. 64.
10. Vaccaro, R. F., Briggs, M. P., Carey, C. L., and Ketchum, B. H. Viability of *Escherichia coli* in sea water. *Am. J. publ. Hlth*, 1950, **40**, 1257.
11. Moebus, K. Seasonal changes in antibacterial activity of North Sea water. *Mar. Biol.*, 1972, **13**, 1.
12. Carlucci, A. F., and Pramer, D. An evaluation of factors affecting the survival of *Escherichia coli* in sea water. I—Experimental procedures. *Appl. Microbiol.*, 1960, **8**, 243.
13. Graham, J. J., and Sieburth, J. M. Survival of *Salmonella typhimurium* in artificial and coastal sea water. *Revue int. Océanogr. méd.*, 1973, **29**, 5.
14. Gaarder, T., and Spärck, R. Biochemical and biological investigations of the variations in the productivity of the West Norwegian oyster pools. *Rapp. P.-v. Réun. Cons. perm. int. Explor. Mer*, 1931, **75**, 47.
15. Gameson, A. L. H., and Saxon, J. R. Field studies on effect of daylight on mortality of coliform bacteria. *Wat. Res.*, 1967, **1**, 279.
16. Rubentschik, L., Roisin, M. B., and Bieljansky, F. M. Adsorption of bacteria in salt lakes. *J. Bact.*, 1936, **32**, 11.
17. Rittenberg, S. C., Mittwer, T., and Ivler, D. Coliform bacteria in sediments around three marine sewage outfalls. *Limnol. Oceanogr.*, 1958, **3**, 101.
18. Carlucci, A. F., and Pramer, D. An evaluation of factors affecting the survival of *Escherichia coli* in sea water. II—Salinity, pH, and nutrients. *Appl. Microbiol.*, 1960, **8**, 247.
19. Korinek, J. Ein Beitrag zur Mikrobiologie des Meeres. *Zentbl. Bakt. ParasitKde.*, II, 1927, **71**, 73.
20. Jannasch, H. W. Continuous culture in microbial ecology. *Lab. Pract.*, 1965, **14**, 1162.

21. Jannasch, H. W. Growth of marine bacteria at limiting concentrations of organic carbon in seawater. *Limnol. Oceanogr.*, 1967, **12**, 264.
22. Jannasch, H. W. Competitive elimination of Enterobacteriaceae from seawater. *Appl. Microbiol.*, 1968, **16**, 1616.
23. Jones, G. E. Suppression of bacterial growth by seawater. *Bact. Proc.*, 1961, p. 61.
24. Scarpino, P. V., and Pramer, D. Evaluation of factors affecting the survival of *Escherichia coli* in sea water. VI—Cysteine. *Appl. Microbiol.*, 1962, **10**, 436.
25. Jones, G. E. Effect of chelating agents on the growth of *Escherichia coli* in seawater. *J. Bact.*, 1964, **87**, 483.
26. Jones, G. E. Growth of *Escherichia coli* in heat- and copper-treated synthetic seawater. *Limnol. Oceanogr.*, 1967, **12**, 167.
27. Carlucci, A. F., and Pramer, D. An evaluation of factors affecting the survival of *Escherichia coli* in sea water. III—Antibiotics. *Appl. Microbiol.*, 1960, **8**, 251.
28. Sieburth, J. M. The influence of algal antibiosis on the ecology of marine microorganisms. In 'Advances in microbiology of the Sea', Vol 1 (ed. M. R. Droop and E. J. F. Wood). Academic Press, New York, 1968, p. 63.
29. Waksman, S. A., and Hotchkiss, M. Viability of bacteria in sea water. *J. Bact.*, 1937, **33**, 389.
30. Mitchell, R. Ecological control of microbial imbalances. In 'Water pollution microbiology' (ed. R. Mitchell). John Wiley, New York, 1972, p. 273.
31. Arloing, F., Sempe, and Chavanne. Propriétés antimicrobiennes de diverses eaux fluviales ou marines. Pouvoir bacteriophagique. *Bull. Acad. Méd.*, 1925, **93**, 184.
32. Carlucci, A. F., and Pramer, D. An evaluation of factors affecting the survival of *Escherichia coli* in sea water. IV—Bacteriophages. *Appl. Microbiol.*, 1960, **8**, 254.
33. Jones, G. E. In 'Marine biology. IV—Unresolved problems in marine microbiology' (ed. C. H. Oppenheimer). New York Academy of Sciences Interdisciplinary Communications Program, New York, 1968, p. 77.
34. Lyman, J., and Fleming, R. H. Composition of sea water. *J. mar. Res.*, 1940, **3**, 134.
35. Menzel, D. W., and Ryther, J. H. Distribution and cycling of organic matter in the oceans. In 'Symposium on organic matter in natural waters' (ed. D. W. Hood). University of Alaska, College, Alaska, 1970, p. 31.
36. Vallentyne, J. R. The molecular nature of organic matter in lakes and oceans, with lesser reference to sewage and terrestrial soils. *J. Fish. Res. Bd Can.*, 1957, **14**, 33.
37. Duursma, E. K. The dissolved organic constituents of seawater. In 'Chemical oceanography', Vol 1 (ed. J. P. Riley and G. Skirrow). Academic Press, New York, 1965, p. 433.
38. Barber, R. T. Dissolved organic carbon from deep sea waters resists microbial oxidation. *Nature, Lond.*, 1968, **220**, 274.
39. Blumer, M. Dissolved organic compounds in sea water: saturated and olefinic hydrocarbons and singly branched fatty acids. In 'Symposium on organic matter in natural waters' (ed. D. W. Hood). University of Alaska, College, Alaska, 1970, p. 153.
40. Duursma, E. K. Organic chelation of ^{60}Co and ^{65}Zn by leucine in relation to sorption by sediments. In 'Symposium on organic matter in natural waters' (ed. D. W. Hood). University of Alaska, College, Alaska, 1970, p. 387.
41. Gonye, E. R., and Jones, G. E. An ecological survey of open ocean and estuarine microbial populations. II—The oligodynamic effect of nickel on marine bacteria. In 'Estuarine microbial ecology' (ed. L. H. Stevenson and R. R. Colwell). University of South Carolina Press, Columbia, S.C., 1973, p. 243.
42. Jones, G. E. An ecological survey of open ocean and estuarine microbial populations. I—The importance of trace metal ions to microorganisms in the sea. In 'Estuarine microbial ecology' (ed. L. H. Stevenson and R. R. Colwell). University of South Carolina Press, Columbia, S.C., 1973, p. 233.

Prof. Jones said, when introducing the paper, that the sea had been called 'the world's largest septic tank' due to its ever-increasing use as a receptacle for human waste; yet the ocean was finite. Non-marine bacteria usually disappeared from the marine environment within a period of hours, and knowledge of the mechanism of microbial death in the sea might permit evaluation of the long-term capabilities of the sea as a septic tank.

The primary cause of death of non-marine bacteria in the Caribbean Ocean was attributed to heavy metal ion toxicity, the basis for this conclusion being the rapid die-off of such bacteria in samples from at least seven depths and six sea stations. By contrast, two species of marine bacteria, *Flavobacterium* and *Pseudomonas*, survived or grew within 72 h at 32°C in the same samples. Addition of ecologically relevant amounts of chelating agents, or energy sources with metal-complexing capabilities, retarded or reversed the toxicity. Glucose, an energy source with minor chelating potential, was generally unsuccessful in promoting growth of non-marine bacteria.

He presented evidence showing that the die-off rate increased markedly with increasing temperature, due to the regulation of metabolic activity.

Escherichica coli, when added directly to Caribbean sea-water samples, became electron-dense under the electron microscope, whereas addition of 10^{-6} M 8-hydroxyquinoline to the same samples produced cells with normal structural detail.

He concluded that, at a given place or time in the ocean, any one of a number of factors might contribute to the death of non-marine bacteria, but from a fundamental physiological viewpoint the principal bactericidal phenomena were heavy metal ion toxicity, the availability of organic energy sources, and their mutual interaction.

DISCUSSION

Dr. V. J. Cabelli (U.S.A.) said that since he believed most of the bacteria from sewage were not 'free' in the sense of a pure culture of *E. coli* added to sea water, or to a suspension of sewage in sea water, he questioned the reality of data derived from studies with pure cultures—particularly with regard to comparison of coliform and virus T_{90} values. He and his colleagues had recently completed survival studies with *E. coli*, *S. faecalis*, *Klebsiella*, *Enterobacter*, *A. hydrophila*, and *P. aeruginosa* in sea water with or without added sewage, the organisms being added as (a) single species and (b) mixed and 'natural' (sewage) suspensions. The decay rates were usually higher for organisms in mixed suspensions. Although field studies (or laboratory studies approximating to field conditions) maximized reality, they did so at the expense of control and hence of understanding. On another topic, he asked Prof. Jones if it were possible that, in his studies, thioglycollate was effective because of its antioxidant rather than its nutritive properties, and if this could also be true of 8-hydroxyquinoline.

Prof. P. V. Scarpino (U.S.A.) asked whether Prof. Jones had any information on the clump-size distribution (aggregation) of the pure cultures he had used, since these patterns could be important in understanding the survival of bacteria and viruses. He noted that the pure cultures had not been washed before being added to the sea water, and asked if the results had been affected in any way by the carry over of organic matter from the slants.

Dr. H. O. Bouveng (Sweden) remarked that in Sweden they had reached the same conclusions as Prof. Jones concerning bacterial survival; the complexing and speciation of the complex were very important. He added that the ability of a sewage effluent to interact with inorganic ions in sea water varied because the microbes produced different products, and asked if Prof. Jones had looked into this problem.

REPLY TO DISCUSSION

Prof. Jones, in reply to Dr. Cabelli, said he knew of no studies in which die-off had been compared in aerobic and anaerobic conditions, but Pramer, Carlucci, and Scarpino[43] had examined the effect of redox potential on bacterial survival in sea water. It appeared from their studies with ascorbic acid, and other agents causing minor alterations in redox potential, that (with the concentrations employed) the metal complexing ability of the organic additions outweighed the redox alternatives. In his opinion, the redox change in the medium would have to be far greater than that observed if it were to be more important than the interaction of organic compounds and trace metals effected by the addition of minute amounts of a good chelating agent. The 8-hydroxyquinoline should have no nutritional value and should give very high stability constants for most metal ions. He had not measured changes in the redox potential of the sea water in his experiments.

Replying to Prof. Scarpino he said that the bacterial colonies had been carefully teased from the surface of the nutrient agar slants. The incidence of clumping was minimal after shaking these colonies vigorously in sterile saline blanks through three 1/99 dilutions. This had been controlled by zero-hour plate counting of all samples which showed very good agreement between different samples. In the dilution of 3×10^7 cells in sterile saline, and their careful removal from the agar slants, organic matter from the original slant would be diluted proportionately to the cells which were washed as they were mixed and diluted.

Commenting on the first part of Dr. Bouveng's contribution, Prof. Jones said that toxicity of sea water for coliform bacteria was apparent when free toxic ions were available in the milieu after the organic and inorganic speciation had been exceeded; some forms of organometallic species might also have deleterious effects on the cells. In reply to Dr. Bouveng's question he said that the organometallo-complexes in sterile complex media had not been studied to any extent in the presence of growing cells. He agreed that in the presence of living organisms such complexes would probably be formed and utilized—due to modification of the media by micro-organisms. This would be the next stage of his investigations.

ADDITIONAL REFERENCE

43. PRAMER, D., CARLUCCI, A. F., and SCARPINO, P. V. The bactericidal action of sea water. *Bact. Proc.*, 1961, p. 45.

Paper No. **22**

Effects of Solar Radiation on the Mortality of Some Terrestrial Bacteria in Sea Water

A. L. H. Gameson and D. J. Gould
Water Research Centre—Stevenage Laboratory, England

The rate of die-off of coliform bacteria in samples of sea water exposed to daylight has been found to be very much greater than in the dark. Consequently, the possibility that sunlight has an important effect on bacterial mortality in the sea merits consideration.

The results and implications of shore-based studies will be discussed before examining the findings of *in situ* experiments. Most of the work to be reported is in terms of coliform bacteria, enumerated by a membrane-filtration method using a recovery medium consisting of a lactose–peptone broth containing 3 per cent Teepol 610 with phenol red as indicator[1] and with incubation for 18 h at 37°C. Some work with other organisms is discussed later.

Shore-Based Experiments

In the dark

Since 1965 more than 400 experiments on the mortality of coliform bacteria in the dark have been performed in connection with the Laboratory's coastal pollution research studies; a summary of the findings has already been published[2]. The major part of the work was done on the coast at Bridport, Dorset, in 1967–68 using natural waters containing coliform bacteria, already present or introduced by addition of domestic sewage, but about 100 experiments involved the study of *Escherichia coli* NCTC 9002 in artificial sea water under laboratory conditions.

The mortality curves for coliform bacteria in natural waters, with samples normally examined daily for 5 days, were approximately log-linear, following first-order kinetics, so that the number of organisms, N, remaining after time t is given by

$$N = N_0\, e^{-kt} \tag{1}$$

where N_0 is the initial number present and k the mortality rate constant. It is customary to denote the time for disappearance of 90 per cent of organisms by T_{90}, but the term T_D will be used to denote the value of T_{90} in the dark. It follows that $T_D = (\ln 10)/k$. The laboratory experiments with *E. coli* 9002, however, were less satisfactory in that they frequently displayed either a lag phase or a steadily increasing mortality rate.

The effect of temperature was examined in a series of 188 experiments, generally using Bridport sea water with coliform bacteria already present but sometimes fortified by addition of domestic sewage. The range of temperature examined was from 2° to 40°C and it was found that T_D decreased with increasing temperature, the regression equation being

$$\log T_D = 2{\cdot}292 - 0{\cdot}0295\theta \tag{2}$$

where T_D is in hours and θ is the temperature in °C. Thus a 10°C rise is accompanied by a reduction in T_D by a factor of 1·97, and at 0°, 20°, and 40°C the values of T_D are 196, 50, and 13 h respectively.

Although T_D had been found to increase with season from May to September 1966 at Sidmouth, Devon[3], the more extensive studies at Bridport showed no significant seasonal variations in a series of 67 experiments at about 20°C.

In daylight

Some 300 bacterial mortality experiments have been performed with samples of sea water (with or without added sewage) exposed to daylight, generally in 600-ml clear glass beakers or in 14-cm polystyrene petri dishes. A summary of most of this work has been published elsewhere[4].

In bright summer sunshine values of T_{90} as short as 20 min are typical, the rate of mortality thus being of the order of 100 times as great as in the dark. The rate has been found to be approximately proportional to the total radiation intensity as measured with a Plessey integrating solarimeter which consists essentially of a silicon photo-voltaic cell, the voltage output being converted to a frequency and fed to an electro-magnetic counter.

It has seemed reasonable to assume that the lethal effect of the radiation is additional to and independent of the effect of the inherent toxicity of sea water. Thus the term S_{90} will be used to denote the radiation received while the coliform count falls by 90 per cent, and S_L to denote the value of S_{90} after allowing for the mortality that would occur in the dark. The rate constant for the light-mortality term is K cm²/cal which is equal to $(\ln 10)/S_L$ when S_L is in cal/cm².

From this long series of experiments it appears that K is normally about 0·1 cm²/cal (or S_L is 23 cal/cm²), but that there are times when the rate is appreciably higher. This may be due to bactericidal substances released by indigenous marine organisms. Nevertheless, comparative studies using raw, filtered, and artificial sea water suggest that the effect of substances released during exposure to sunlight is small. There is no significant difference between the values of K for bright sunshine and overcast sky, and no effect of temperature on K has been found in experiments where the range of average temperatures has been from 2° to 29°C.

Wavelength of lethal radiation

In one series of experiments an attempt was made to determine the wavelengths of radiation responsible for the lethal effect on coliform bacteria. Samples were exposed in the 14-cm petri dishes in pairs, one dish being exposed to daylight and the other placed under a wooden frame in which one or more sheets of Perspex, 30 cm × 30 cm, were mounted. In five such experiments three different thicknesses were used[4]. The experiments were repeated with five different gelatin photographic filters placed between two 3-mm Perspex sheets. Portions of the Perspex sheets, with and without the filters, were subsequently examined in a spectrophotometer which gave curves for the transmittance plotted against wavelength. Each filter transmitted light into the infra-red region but had a cut-off in the visible or ultra-violet region.

The temperature was generally measured at the time of taking each sample. The average temperature in each dish was used for calculating T_D by means of Equation 2—except for occasions when simultaneous experiments were made in the dark and the observed values for T_D used instead. The values of T_{90} were then adjusted to allow for the die-off that would have been expected in the absence of radiation.

Although most of these experiments were made in conjunction with a solarimeter inferior to the Plessey instrument, it is preferable to take account of differences in radiation intensity during the concurrent pairs of experiments as the duration of the one with the filter generally exceeded that of the one without. However, it was found that the ratios of S_L values were usually virtually identical with those of the adjusted T_{90} values.

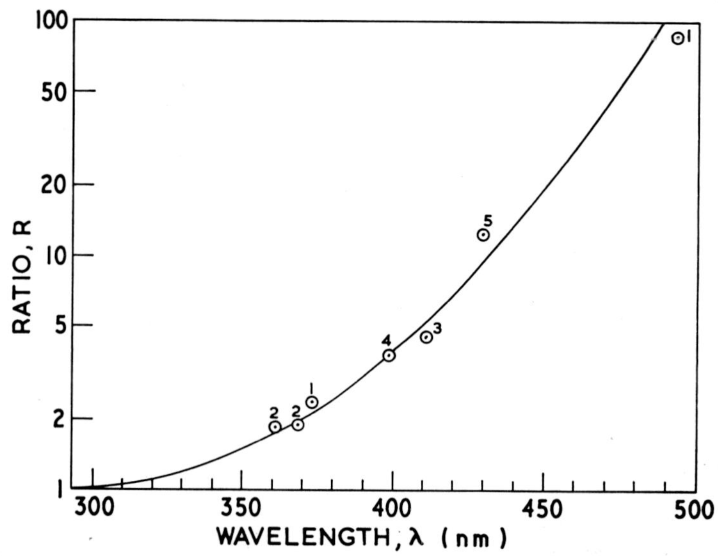

FIG. 1. Ratio of S_L for daylight spectrum of wavelength exceeding λ to that of whole spectrum

The ratio (R) of S_L with filter to that without was then plotted against the wavelength of 50 per cent transmittance and R was assumed to be unity at 293 nm, this being the shortest wavelength normally found at the earth's surface[5]. A curve drawn through the plotted points does not show the true effect of wavelength because the transmittance of each filter does not pass abruptly from 0 to 100 per cent, but by using this curve and the data for the transmission characteristics of each filter it is possible to make adequate allowance for these characteristics. The corrected data are plotted in Fig. 1 where each point represents the geometric mean of the number of experiments indicated. (This diagram supersedes fig. 4 of the earlier paper[4] which did not include the adjustment for transmission characteristics.) It is concluded that about half the lethal effect of solar radiation is attributable to wavelengths below 370 nm, a quarter to the near-visible ultra-violet (370–400 nm), and a quarter to the blue–green region of the visible spectrum (400–500 nm); the effect of wavelengths greater than 500 nm is negligible.

Mechanism of Inactivation

Ultra-violet inactivation is now recognized[6a] as resulting principally from highly localized DNA lesions produced by deletion of hydrogen bonds and consequent dimerization of pyrimidines. The wavelength of maximum sensitivity of *E. coli* is generally stated to be in the range 250–270 nm, which is below the limit of the daylight spectrum. As the wavelength is increased, the probability of inactivation decreases; it has been calculated from the results of Fig. 1 and the distribution of energies in the daylight spectrum, and comparison with the work of Gates[7], that the energy required for inactivation of coliform bacteria at 370 nm is probably about 1000 times as great as at 260 nm.

Hollaender[8] demonstrated the lethal effect on *E. coli* of wavelengths of 350–490 nm and concluded that the incident energy required to give 50 per cent kill in this wave-band was 10^4–10^5 times as much as at 265 nm. More recently, Eisenstark[9] has confirmed the lethal effects of the same wave band; he suggests that the chromophore is not DNA but some other site.

Most of the work done with bacteria has employed pure cultures and controlled light sources, and the results show the sensitivity to ultra-violet radiation to vary greatly—depending on the presence or absence of genetically coded repair mechanisms in a particular strain. It is thus perhaps remarkable that in the Laboratory's work, where the heterogeneous coliform flora of sewage is being studied, there is no consistent evidence of heterogeneity of inactivation. For instance, if 1 per cent of the bacteria were markedly more resistant than the remainder, the curve of the logarithm of the coliform count plotted against radiation received would flatten out within the second order of magnitude reduction in count, whereas the curves are virtually linear through at least five orders of magnitude in some experiments.

Photoreactivation

It was observed by Kelner[10] that the survival of bacteria inactivated by ultra-violet radiation was increased by subsequent exposure to visible light. The Laboratory's experiments[4], however, have failed to demonstrate this photoreactivation process. It would therefore appear that damage produced by daylight is not photoreversible to any significant extent; this observation is in accordance with other work reported by Eisenstark[9].

Dark recovery

Mechanisms are known to exist for dark repair of DNA lesions[6b]. In several of the Laboratory's experiments a sample has been exposed to sunlight for long enough to produce one or two orders of magnitude reduction in coliform count, and then kept in the dark for times ranging between 30 min and 2 h before being again exposed, this procedure being repeated several times. In other experiments, changes in coliform count in the dark after exposure to daylight (but without subsequent re-exposure) have been examined. From 38 relevant experiments it was concluded that on removing the sample from daylight the coliform mortality is not inconsistent with the hypothesis that the normal dark mortality rate then applies.

Submerged-Bottle Experiments

The relevance of the shore-based light mortality experiments to what happens in the sea depends on the extent to which lethal radiation penetrates. In oceanic waters the path length for 90 per cent attenuation (which will be denoted by L_{90}) may exceed 10 m for radiation of 310 nm, and the path length increases with increasing wavelength[11]. The turbidity of inshore waters, however, may result in much lower values of L_{90}.

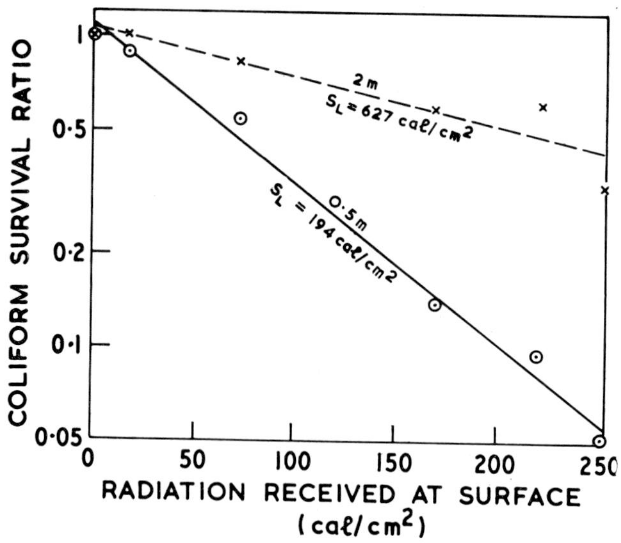

FIG. 2. Coliform survival ratios for submerged-bottle experiment at Sidmouth on 21 August 1969

In 1966 and 1969 some coliform mortality experiments were performed with mixtures of sea water and sewage in bottles hung at different depths in the sea at Sidmouth. Mortality in the dark was examined concurrently both in samples in similar submerged bottles wrapped in aluminium foil and also in samples kept ashore in a water bath at approximately sea temperature.

The six experiments in 1966 showed that even at 4 m the mortality rate was substantially greater than in the dark. The value of S_L (in terms of radiation received at the sea surface) increased, on average, by a factor of 1·67 for each metre in depth[3] which is equivalent to an L_{90} of 4·5 m. Light-attenuation measurements made with a submersible photocell with a filter giving a peak transmittance at 440 nm, and 10 per cent of the peak value at 350 and 480 nm, gave a reduction factor equivalent to an L_{90} value of 4·2 m.

Three experiments in 1969 gave reduction factors of 1·64, 2·19, and 1·72 m^{-1} for bottles at depths of 0·5 to 2 m. The results for the second experiment are plotted in Fig. 2 where each point represents the average proportion of coliforms surviving (after allowing for dark mortality) in pairs of bottles removed at the same time. In the third experiment, samples at 5 m showed a mortality barely greater than in the dark. The submersible photocell was not available for these experiments, but measurements of L_{90} were made in a vertical tube containing the water sample, through which was passed an optically filtered collimated beam from a medium-pressure mercury vapour lamp, the transmitted radiation being mainly of wavelength about 360 nm; addition of various depths of sample, with measurement of the transmitted radiation, provided the data for evaluating L_{90}. The samples examined were taken near the sea surface and the method of measurement will undoubtedly underestimate the penetration of radiation as it takes account of direct transmittance only and thus omits the contribution due to scattering. The average value of L_{90} found in these three experiments was 1·85 m, which is in poor agreement with the average of 3·7 m deduced from the results of the mortality experiments.

Extrapolating the S_L values in the first two experiments back to zero depth gives values of 93 and 131 cal/cm², which may be compared with those of 18 and 20 cal/cm² obtained from concurrent beaker experiments ashore. Mortality was thus far more rapid in the beakers than in the submerged bottles—a result which cannot adequately be explained in terms of reflection of radiation at the sea surface and the differences in the materials of the containing vessels.

In Situ Experiments

Method

The mortality of coliform bacteria in the sea off Sidmouth and Bridport has been investigated *in situ* on 20 occasions. The principles of the method adopted are as follows. A volume of comminuted sewage passing through the pumping station is labelled with radiotracer (bromine-82) and pumped to near the end of the outfall pipe. Sewage is then stored at the pumping station and in the main sewer to allow dispersion of recently discharged sewage. The labelled sewage is then pumped through the outfall (generally at slack water) and the sewage flow again stopped for as long as practicable. The dispersing patch of labelled sewage is followed by means of a γ-scintillation counter in the bilge of the tracking vessel, and samples are taken (usually at three depths) at intervals for several hours. The samples are examined ashore for content of γ-activity and coliform bacteria. The physical dilution in relation to a bulk sample taken at the outfall 'boil' is then calculated from the activity data, due allowance being made for the background activity level and radioactive decay. The coliform counts are similarly related to the value at the boil, and the results are expressed in terms of the proportion surviving after allowing for the physical dilution and are plotted both against the time of travel from the boil and against the solar radiation received during this time.

Some details of the experiments are shown in Table 1. Outfall C is at Sidmouth (discharging 430 m beyond the low-water mark); D4 is the new outfall at Bridport (1370 m), while D3 and D2 are outlets at 680 and 430 m respectively on the same outfall pipe which was constructed with these alternative discharge points for the purpose of investigations by the Laboratory. The question marks in Expts. 2–4 indicate the absence of records. Experiment 5 was the initial discharge from the new outfall; sewage previously discharged only about 20 m beyond the low-water mark. In the subsequent experiments at Bridport the volume of sewage labelled was normally about 100 m^3, the labelled sewage being discharged by use of all four pumps at the pumping station where a single pump is sufficient to cope with the dry-weather flow.

TABLE 1. Some details of *in situ* mortality experiments

No.	Year	Month	Outfall	Duration of discharge (min)	Previous storage (h)	Subsequent storage (h)	Duration of tracking (h)	Activity added (Ci)
1	1969	Aug.	C	30	0·9	2	5·2	0·39
2	1969	Aug.	C	10	1·0	?	4·1	0·45
3	1969	Aug.	C	10	0·9	3	3·7	?
4	1969	Aug.	C	10	0·8	?	4·3	?
5	1970	May	D4	157	*See text*	2	8·7	20
6	1971	July	D3	5	4·6	1·4	4·1	2·8
7	1971	July	D3	5	2·8	1·4	4·0	2·6
8	1971	Sept.	D2	5½	0·8	2·0	3·8	0·9
9	1971	Sept.	D2	6	1·2	1·3	6·7	1·0
10	1971	Sept.	D2	5	0·6	1·9	5·0	1·6
11	1971	Sept.	D2	6	2·0	2·3	7·6	1·0
12	1972	Apr.	D2	5½	1·0	2·4	7·0	3
13	1972	May	D2	5½	1·1	3·1	6·8	3
14	1972	May	D2	8	2·0	1·6	11·6	3
15	1972	May	D2	6	1·3	1·4	1·5	4
16	1973	May	D4	6	1·4	1·4	3·6	11·2
17	1973	May	D4	6	0·9	0·8	7·2	13·1
18	1973	May	D4	6	1·1	1·8	4·0	12·4
19	1973	May	D4	6	2·9	3·4	5·4	12·7
20	1973	May	D4	6	1·2	1·7	6·2	7

When examining mortality rates it is essential that the delay between taking and examining samples is both as short and as constant as possible. In most of these experiments a delay of 90 min was found to be the shortest which could be maintained in view of the time taken to transport samples from the tracking vessel to the shore.

Experiments including periods of darkness

Individual data for Expt. 11, when the labelled sewage was discharged at dawn, are plotted against time in Fig. 3(a) which shows no die-off during the first 3 h but a rapid die-off thereafter. Data for samples taken at 5–10 m depth are concordant with this diagram during the first 5 h, after which the activity levels were generally not significantly above background. The regression line shown is for the combined data for 0 and 2 m from 2·9 h onwards, the values of T_{90} for the two depths separately being 1·49 and 1·83 h respectively.

The same data are plotted against the radiation received in Fig. 3(b) which shows no great departure from linearity. The values of S_{90} for 0 and 2 m separately were 41 and 50 cal/cm^2 respectively. Figure 3 thus strongly suggests that *in situ* mortality is markedly affected by solar radiation.

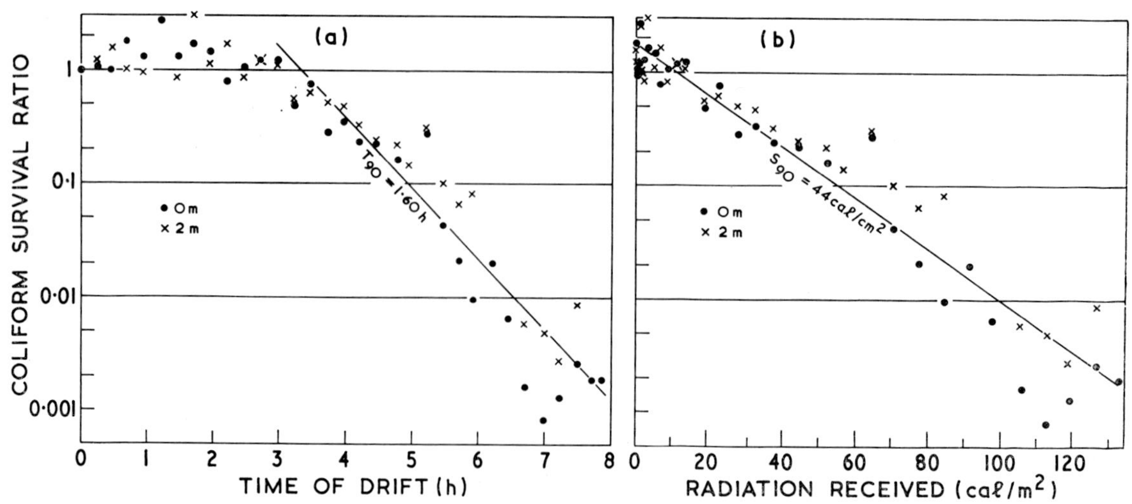

FIG. 3. Coliform survival ratios for *in situ* Expt. 11 plotted against (*a*) time and (*b*) radiation

Five other experiments include periods of darkness or of negligible radiation before sunrise or after sunset. Experiment 2 was conducted entirely at night (2330–0335 h) during which time there was no measurable die-off. Experiment 5 was exceptional in two respects, firstly in that the labelled discharge continued for 2·6 h (thus making it difficult to assign the origin of the time scale in plotting the results) and secondly in that after the first 4¾ h since the mid-time of discharge the coliform survival ratio started to rise even though radiation was still being received. Immediately after the end of the labelling with activity, spores of *Bacillus subtilis* var. *niger* were added to the sewage to give a concentration comparable with that of the coliforms so that all the sewage discharged during the remainder of the experiment was labelled with *B. subtilis* spores. Most of the samples taken from the sea were examined for the spores and the results indicate that the rise in coliform survival ratio cannot be ascribed to the effects of contamination of the sea water by sewage not labelled with activity.

Experiment 12 was started late in the afternoon and continued until after midnight. During daylight the mortality rate was low (T_{90} = 8·2 h, S_{90} = 101 cal/cm^2); after dusk the scatter of the data masks any change in survival ratio that may have occurred, but the average value suggests that mortality had virtually ceased.

Experiment 13 finished at dusk. Plotting survival ratio against time shows a decreasing mortality rate towards the end of the experiment, but plotting against radiation shows no significant departure from linearity. The same applies, though to a lesser extent, to Expt. 14 which lasted for 10·6 h and finished after dusk. Finally, Expt. 19 (which started soon after dawn) showed no significant mortality during the first 2 h.

Summary of results

From these experiments it would seem difficult to refute the hypothesis that solar radiation greatly enhances the rate of mortality of coliform bacteria in the sea. Nevertheless, further examination of the results shows a confused picture. The results of the 20 experiments are summarized in Table 2. The scale of sea state goes from glassy calm (0) to estimated wave height exceeding 1·2 m (5). The values of S_L have been derived from those of S_{90} after allowing for the dark mortality rate by means of Equation 2.

Unresolved topics

Comparison of the results for Expts. 1, 3, 4, and 5 shows that whereas the value of T_{90} was always about 3 h, that of S_{90} varied between 30 and 84 cal/cm^2, thus suggesting that T_{90} was not dependent on radiation intensity. It may be noted that Foxworthy and Kneeling[12], in a series of 34 *in situ* experiments made during 1966–69, failed to find any relation between T_{90} and solar radiation.

In Fig. 4 are shown logarithmic probability plots of the distribution of the 18 experimental values for T_{90} and S_{90}, the left-hand scale expressing the results as ratios to their medians. The values of S_{90} are seen to be approximately lognormally distributed and to be more constant than those of T_{90}.

There is a tendency for S_{90} (and S_L) to increase with decreasing temperature, with decreasing L_{90}, and with increasing sea state, but only for L_{90} is the relation just significant at the 95 per cent confidence

TABLE 2. Further details of *in situ* coliform mortality experiments

No.	Temp. (°C)	Sea state	Average radiation intensity (cal/cm² h)	L_{90} (m)	T_{90} (h)	S_{90} (cal/cm²)	S_L (cal/cm² h)
1	16	2	25	1·6	3·2	84	88
2	(16)[a]	1	0	—[b]	—[c]	—[d]	—[d]
3	16	2	19	1·8	3·0	63	66
4	(16)	—	18	—	2·9	60	63
5[e]	11	1–4	9·4	—	3·2	30	31
6	17	—	33	2·8	—[f]	—[f]	—[f]
7	17	1–3	38	2·6	0·57	22	22
8	17	1	10·3	1·5	3·9	40	43
9	(17)	0–1	6·1	1·7	9·0	53	62
10	(17)	(1–2)	27	1·3	1·4	37	38
11	(17)	(1)	17	1·5	1·6[g]	44	45
12[h]	9	(2)	14·5	0·8	8·2	101	109
13	10	(2–3)	18	0·9	5·3	91	96
14	10	0–3	4·8	0·5	8·8	46	51
15	(9)	4–5	21	0·3	4·0	84	87
16	11	4–5	40	2·0	3·0	117	120
17[i]	11	2–3	26	2·3	0·92	25	25
18	11	1–3	35	3·7	0·73	28	28
19	10	3–5	19	2·3	4·0	76	79
20	12	2–3	15	1·4	3·9	57	60

[a] Figures in parentheses are probable values in absence of records
[b] Dashes without other notes indicate absence of records
[c] Indeterminate, no die-off found in 4 h
[d] Indeterminate, no radiation
[e] First 4¾ h of experiment, thereafter some regrowth
[f] Indeterminate, data too scattered
[g] After first 2·9 h (average radiation intensity 25 cal/cm² h)
[h] For first 1¾ h
[i] For first 2 h

level. The L_{90} measurements are by no means satisfactory as the samples were perforce examined at least a day after the experiment—when radioactivity counts and salinity determinations had been completed; in addition the lamp, filter, and photocell of the instrument were all changed during the period of the investigation and it is not certain that the characteristics of the replacements were identical with those of the original parts.

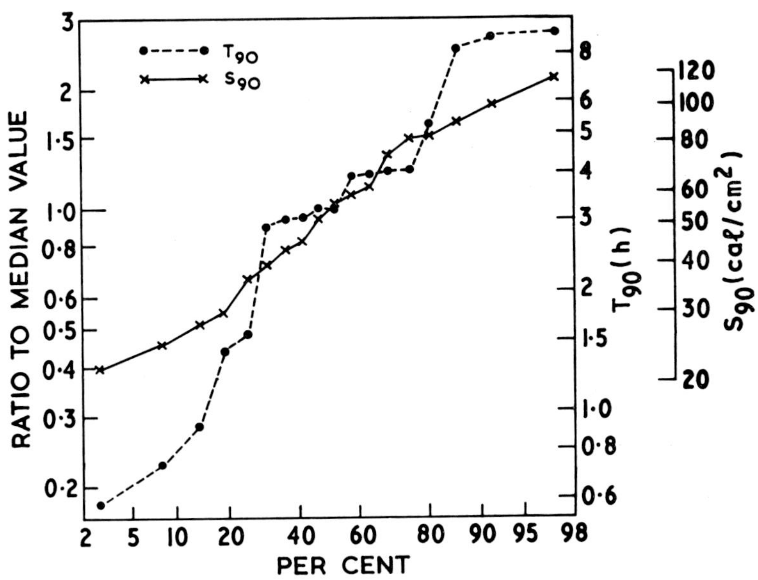

FIG. 4. Frequency distributions ot T_{90} and S_{90} in 18 *in situ* studies of coliform mortality

Large changes in S_{90} are found over short periods of time, thus Expts. 16 and 17 were only 2 days apart and gave, respectively, the highest and next-to-lowest values of S_{90} for the whole series of experiments. The cause of this variability remains unresolved.

In most of the *in situ* experiments samples were taken at depths of 0, 2, and 5 or 10 m, the deepest samples being within a few metres of the bed during the early stages of the experiments. In general, the results show comparable mortality rates at all depths. Since little lethal radiation reaches the greater depths, this implies either that the die-off of coliforms is little affected by radiation, or that vertical mixing results in the bacteria spending roughly equal times at each depth. If the results of beaker experiments made in conjunction with the *in situ* studies are considered to be relevant to the sea surface, then from the values obtained for L_{90} it is possible to calculate the depth of complete mixing required to give the observed S_{90} of each *in situ* experiment. In the last five *in situ* experiments this calculated depth is found to have an average value of only about 3 m, whereas the data for the 10-m samples showed mortality rates at least half as great as at the surface. Although the experimental values of L_{90} might well be too low (for reasons stated earlier), the errors involved are unlikely to account for these discordant results which imply that mortality in the sea is more rapid than in a beaker for the same intensity of radiation reaching the bacteria.

On only two occasions have submerged-bottle experiments been carried out in conjunction with *in situ* studies. The bottle experiment to which Fig. 2 refers gave S_L values of 194 and 627 cal/cm^2 at depths of 0·5 and 2 m respectively, the value for the corresponding *in situ* experiment (No. 1 of Tables 1 and 2) being 88 cal/cm^2. Four days later the submerged bottles gave S_L values of 95, 215, and roughly 6000 cal/cm^2 at depths of 0·5, 2, and 5 m, whereas for the *in situ* experiment (No. 3) the value was 66 cal/cm^2. In both *in situ* experiments the mortality rates at the two sampling depths, 0 and 2 m, were similar. Since the materials used for the submerged bottles and their holders will have resulted in only a small absorption of incident radiation it is evident that die-off in the sea was substantially more rapid than in the bottles.

The application of the *in situ* mortality rates to prediction of distributions of coliform bacteria in the sea is considered in a later paper at this Symposium[13].

Organisms Other than Coliform Bacteria

Occasional determinations have been made of the mortality rates of other organisms, both in shore-based and *in situ* studies. Membrane-filtration techniques have been used throughout.

Escherichia coli

The recovery medium used for enumeration of *E. coli* was a 'membrane enriched Teepol broth' containing 0·4 per cent Teepol 610[14], with incubation for 4 h at 30°C followed by 14 h at 44°C.

The *in situ* Expts. 16 and 18–20 included examinations for *E. coli*. The S_{90} ratio of *E. coli* to coliforms varied between 1·25 and 1·50 with an average value of 1·33. Four beaker experiments (in three of which the samples were taken from the outfall boil at the start of *in situ* experiments) gave similar ratios, with a mean value of 1·36. Thus the die-off of *E. coli* was appreciably slower than that of total coliforms. Indeed, in the later stages of the experiments it was found that the *E. coli* count exceeded the coliform count—an unacceptable situation. The reason for this result is considered to be that the conditions of the test for *E. coli* (a smaller concentration of Teepol, a more nutritious medium, and the use of a resuscitation period) are more favourable for the recovery of damaged bacteria than are those of the coliform test.

Faecal streptococci

A broth formulation of the Slanetz and Bartley medium[15] was used for the recovery of faecal streptococci, with incubation for 4 h at 30°C followed by 44 h at 44°C.

Three beaker experiments gave S_{90} values for faecal streptococci that were 10–39 times as great as for coliforms. *In situ* Expts. 16, 18, and 19 gave indeterminate values, no die-off being detectable during the sampling period; Expt. 20, however, gave an S_{90} value of 308 cal/cm^2 or only 5·4 times that for coliforms.

Serratia marcescens

Use has been made of an antibiotic-resistant strain of *Serratia marcescens* as a labile tracer at three sites. The recovery medium used is a yeastral-peptone broth with added antibiotics and fungicide[16]; incubation is for 36–44 h at 30°C.

Fifteen beaker experiments have been carried out with *Serratia* added to samples containing sewage coliforms. In general, the mortality rates of the two organisms were found to be similar, but on a number of occasions the mortality rate for *Serratia* increased throughout the experiment. In 13 experiments where values of S_{90} were calculated, the *Serratia* died more rapidly than coliforms in 8 and more slowly in 5.

Serratia cells were added to the labelled sewage during *in situ* Expts. 7–10; the results obtained gave die-off curves of form similar to those for coliforms, but always with a slower rate, the S_{90} ratios for *Serratia* being 1·4, 3·8, 2·6, and 2·1 respectively.

Bacillus subtilis var. *niger*

Spores of *B. subtilis* have been used as a conservative tracer[16]. Samples are heated for 30 min at 63°C before filtration, the recovery medium being tryptone (Difco) with added glucose and mannitol[16] with incubation for 18 h at 30°C.

A single beaker experiment gave an S_{90} value some 200 times that for coliform bacteria.

Summary

Shore-based experiments have amply demonstrated the large effect of solar radiation on the mortality of coliform bacteria in sea water. *In situ* studies, although showing that mortality rates are much higher during the day than at night, have failed to demonstrate a consistent quantitative relation between mortality rate and incident solar radiation.

REFERENCES

1. TAYLOR, E. W., and BURMAN, N. P. The application of membrane filtration techniques to the bacteriological examination of water. *J. appl. Bact.*, 1964, **27**, 294–303.
2. PIKE, E. B., GAMESON, A. L. H., and GOULD, D. J. Mortality of coliform bacteria in sea water samples in the dark. *Revue int. Océanogr. méd.*, 1970, **18–19**, 97–107.
3. GAMESON, A. L. H., and SAXON, J. R. Field studies on effect of daylight on mortality of coliform bacteria. *Wat. Res.*, 1967, **1**, 279–295.
4. GAMESON, A. L. H., PIKE, E. B., and DAVEY, K. W. Mortality of coliform bacteria in sea water samples in daylight. *Proc. 5th int. Colloquium med. Oceanogr., Messina*, 1973, pp. 353–362.
5. HENDERSON, S. T. Daylight and its spectrum. Adam Hilger, London, 1970, p. 30.
6. SMITH, K. C., and HANAWALT, P. C. Molecular photobiology: inactivation and recovery. Academic Press, New York, 1969: (*a*) Chap. 1, (*b*) Chap. 7.
7. GATES, F. L. Study of bactericidal action of ultra-violet light. III—The absorption of ultra-violet light by bacteria. *J. gen. Physiol.*, 1930, **14**, 31–42.
8. HOLLAENDER, A. Effect of long ultra-violet and short visible radiation (3500–4900 Å) on *Escherichia coli*. *J. Bact.*, 1943, **46**, 531–541.
9. EISENSTARK, A. Mutagenic and lethal effects of visible and near-ultraviolet light on bacterial cells. *Adv. Genet.*, 1971, **16**, 167–198.
10. KELNER, A. Photoreactivation of ultraviolet-irradiated *Escherichia coli* with special reference to the dose-reduction principle and to ultraviolet-induced mutation. *J. Bact.*, 1949, **58**, 511–522.
11. JERLOV, N. G. Optical oceanography. Elsevier, Amsterdam, 1968, p. 120.
12. FOXWORTHY, J. E., and KNEELING, H. R. Eddy diffusion and bacterial reduction in waste fields in the ocean. University of Southern California, Los Angeles, 1969, p. 144.
13. MUNRO, D. Observed and predicted coliform distributions near a sea outfall. This volume, pp. 353–362.
14. DEPARTMENT OF HEALTH AND SOCIAL SECURITY, WELSH OFFICE, and MINISTRY OF HOUSING AND LOCAL GOVERNMENT. Report 71—The bacteriological examination of water supplies. H.M. Stationery Office, London, 1969, p. 42.
15. SLANETZ, L. W., and BARTLEY, C. H. Numbers of enterococci in water, sewage and feces determined by the membrane filter technique with an improved medium. *J. Bact.*, 1957, **74**, 591–595.
16. PIKE, E. B., BUFTON, A. W. J., and GOULD, D. J. The use of *Serratia indica* and *Bacillus subtilis* var. *niger* spores for tracing sewage dispersion in the sea. *J. appl. Bact.*, 1969, **32**, 206–216.

In presenting the paper, Mr. Gameson emphasized the almost total irrelevance of shore-based experiments to mortality rates in the sea. Although such experiments had been found invaluable in studying the effects of certain controllable factors, if knowledge was required of the magnitude of T_{90} in the sea, the sea was the only place where it could be measured.

He then outlined the five stages of the Authors' work on coliform mortality, in order of increasing relevance to the marine environment. (1) Shore-based experiments in the dark, which showed a remarkable approximation to first-order kinetics and a pronounced effect of temperature. (2) Similar experiments with exposure to daylight which resulted in greatly enhanced mortality; for instance, in bright summer sunshine T_{90} was about 20 min, whereas the corresponding value in the dark was at least 1 day. Information had been obtained on the wavelength of the lethal radiation. No effect of temperature could be found, nor could photoreactivation be demonstrated. (3) Submerged bottle experiments which had shown that lethal radiation could penetrate to a depth of at least 4 m in the sea. (4) *In situ* experiments which had confirmed the virtual cessation of coliform mortality in the dark. (5) Application of results—knowledge of T_{90} was academic unless used in predictive models, and such models needed validating.

He concluded by saying that: they did not maintain that solar radiation was the only cause of the mortality of terrestrial bacteria in the sea, nor that it was always an important one; they knew that other participants believed mortality to be the wrong term and that bacterial disappearance was due mainly to sedimentation (though it would be difficult to explain why sedimentation occurred only during the hours of daylight); they were all too well aware of the difficulties inherent in the *in situ* measurement of mortality, and they were disturbed and disappointed by the discrepancies between observed and predicted coliform distributions in the sea when using the experimental data reported in the paper. Nevertheless, they pleaded with all who made future *in situ* studies not to dismiss the relevance of solar radiation before conducting an experiment at night.

DISCUSSION

Prof. J. E. Foxworthy (U.S.A.) said that while it was true that he and his colleagues had been unable to establish any relationship between coliform disappearance and total solar radiation input for surface waste fields[12], it should be noted that they had measured the effects of all environmental factors on T_{90}. Thus the effects of solar radiation were in all probability masked by those of other factors. In his opinion, solar radiation was likely to be an important factor in the disappearance of coliform bacteria in surface waste fields in the sea.

Mr. C. G. Gunnerson (Canada) said that no attempt had been made in either the Santa Monica Bay[17] or Bosphorus[18] studies to evaluate the bactericidal effect of sunlight. However, re-examination of the limited night-time data from a 22-h experiment, and comparison of the data for generally cloudy winter days with those for generally brighter summer ones, failed to disclose any differences in T_{90}. This suggested that effects of sunlight were limited to a thin surface layer and that most of the coliforms throughout a thicker sewage field were unaffected.

Mr. H. Schrøder (Denmark) reported that T_{90} values had been determined in the course of an investigation of a Danish estuary by the Danish Hydraulic Institute. The determinations were based on the application of a one-dimensional (box-type) estuary model which was calibrated by means of field observations of salinity variations at a number of positions in the estuary. Five longitudinal *E. coli* concentration profiles were available for direct calculation of T_{90}, the values of which were found to lie in the range 1–30 h. The model and the applied data in this case were not of a nature to provide accurate T_{90} values, but examination of meteorological data for the days of the *E. coli* surveys revealed that T_{90} was strongly correlated with solar radiation intensity. There was no doubt in his mind that this investigation supported the Authors' findings, and that the approach used was an important alternative to the *in situ* measurement of T_{90}.

Prof. H. I. Shuval (Israel) said that he and his colleagues[19] had found, in sea water under laboratory conditions, coliform T_{90} values of about 1 h in sunlight but more than 1 day in the dark; viruses persisted for similar times in both light and dark with T_{90} values of 1–2 days. This raised the question as to the real value of a coliform T_{90} as a predictor of pathogen disappearance. An indicator organism was satisfactory only if it could be used for predicting the behaviour of pathogens. Since it appeared that coliform bacteria died in the sea much more rapidly than viruses, he had to ask whether there was really any value in coliform T_{90} calculations. For outfall design purposes he would prefer to assume no die-away of pathogens. This might lead to a standard based on dilution alone, say about 1:1000, and would meet both the aesthetic criteria and the potential health risk. Prof. Y. Kott (Israel), however, pointed out that since solar radiation attacked nucleic acids in the micro-organisms, the viruses would probably be inactivated whereas the bacteria could utilize repair mechanisms and thus survive longer, to an extent depending on the solar energy and the organic matter in the sea. For instance, an experiment at the Technion, with tomatoes immersed in sewage and subsequently exposed to sunlight, showed that the viruses disappeared within a few hours, while there was little change in the numbers of coliform bacteria.

Dr. G. E. B. Kullenberg (Denmark) asked the Authors if, on the basis of their findings, they thought that surface discharges were generally preferable to discharges from submerged outfalls where the sewage was trapped at a sub-surface level. He also commented that the influence of solar radiation on the bacteria must depend greatly on the condition of the water (such as the presence of slicks) and on the atmospheric optical conditions.

Dr. M. J. Suess (W.H.O.) pointed out that solar radiation was known to have an effect on photosensitive substances which were present in both sewage and sea water, and asked if it were possible that such substances might be at least partially responsible for changes in bacterial concentrations. Dr. L. J. Guarraia (U.S.A.) added that there might be quite important effects of sunlight on chemical compounds, and secondary effects of the radicals produced on bacteria; for example, ferricyanide was converted to free cyanide. He asked if the Authors had examined the effects on bacteria of chemicals exposed to sunlight.

REPLY TO DISCUSSION

Mr. Gameson said, in reply, that the first three speakers had reported conflicting findings. Thus, whereas Mr. Schrøder had found a strong correlation between T_{90} and solar radiation, Prof. Foxworthy had been unable to demonstrate such a correlation (but did not consider he had demonstrated the absence of an effect), and Mr. Gunnerson maintained that in his experiments solar radiation was a factor of no importance. It was not possible to resolve this conflict without further information—particularly on the penetration of the lethal wavelengths in each of the receiving waters—and these results suggested (as Mr. Gameson had implied in his opening remarks) that solar radiation might not always be an important cause of bacterial mortality.

Profs. Shuval and Kott also reported conflicting results, on the relative susceptibilities of coliforms and viruses. The Authors felt it wrong to dismiss the value of coliform mortality studies if the die-off of viruses was markedly different, since their work had been part of wider investigations on the dispersion and degradation of sewage in the sea; coliforms had been used as indicators of sewage rather than of pathogens.

In reply to Dr. Kullenberg, Mr. Gameson said there was no simple answer to his question. The depth of water at the outfall, the offshore distance, the turbidity, and the extent of vertical mixing would all be relevant factors. However, in a situation where there was no possibility of trapping the sewage at a sub-surface level, and where there was complete vertical mixing before the sewage approached the shore, it might well be preferable to allow the sewage to reach the surface over the outfall, though account would have to be taken of any resulting adverse effects such as the formation of slicks.

Mr. Gould agreed with Dr. Suess that photodynamic action resulting from irradiation of photosensitizers present in sewage or sea water was quite likely to have been a factor in the experiments reported. He had examined the oxygen dependence of the inactivation process in a few experiments, the results of which could be interpreted as adding support to this view. However, he pointed out that he had some reservations in this interpretation, since photodynamic responses produced by visible light on endogenous photosensitizers could be an alternative interpretation. The only relevant work by the Authors had been some shore-based experiments with fluorescein (since this was used as a visual marker during *in situ* studies), and these had failed to demonstrate any photodynamic effect.

Replying to Dr. Guarraia, Mr. Gould said that apart from their work with fluorescein the Authors had not studied the effects of individual chemicals on bacteria. Nevertheless, the results of a substantial number of experiments comparing the mortalities of pure-culture and sewage coliforms, in natural and artificial sea waters, suggested that the effect of any such chemicals present in domestic sewage or natural sea water was not very great.

ADDITIONAL REFERENCES

17. GUNNERSON, C. G., SUNGUR, O. E., BILAL, E., and ÖZTURGUT, E. Sewage disposal in the Turkish Straits. *Wat. Res.*, 1972, **6**, 763–774.

18. GUNNERSON, C. G. Sewage disposal in Santa Monica Bay. *Trans. Am. Soc. civ. Engrs*, 1959, **124**, 823–851.

19. SHUVAL, H. I. Detection and control of enteroviruses in the water environment. In 'Developments in water quality research' (ed. H. I. Shuval). Ann Arbor–Humphrey Science Press, Ann Arbor, Mich., 1970, pp. 47–71.

Paper No. **23**

Other Factors Determining Life Expectancy of Microorganisms in the Marine Environment

J. D. Walker

Department of Microbiology, University of Maryland
College Park, Maryland 20742

and

L. J. Guarraia

Water Quality Protection Branch, Environmental Protection Agency
Washington, D.C. 20460

Many factors are responsible for the decrease of freshwater bacteria in the marine environment, including viruses, antibiotics, and heavy metals. The biological, physical, and chemical factors affecting survival of freshwater bacteria in the marine environment are discussed in this paper.

Physical and Chemical Factors

Several physical and chemical parameters may affect the fate of microorganisms in the marine environment (Table 1). Several of these factors may act synergistically to limit the existence of microorganisms, e.g. pH and temperature[2], salinity and temperature[3], or pressure and temperature[4]. The effects of salinity, temperature, and pressure on marine microorganisms are discussed below.

Salinity

The salt content of estuaries and oceans varies from 0·001 to 30 per cent (Table 1) and is influenced by dilution and evaporation of water. Inorganic salts become potentially more toxic as their concentration increases[5]. Salt may exert its influence by osmotic effect or requirements for a specific ion. The osmotic effects of salt on marine bacteria have been reviewed by Jones[6], and the specific ion requirements that affect metabolism have been reported for marine bacteria[7] and fungi[8,9].

TABLE 1. Physical and chemical extremes for oceans/estuaries and microbial growth
Modified from ZoBell[1]

Parameter	Oceans/estuaries[a]		Microbial growth	
	Min.	Max.	Min.	Max.
Salt content (per cent)	<0·001	30[b]	<0·001	30
Temperature (°C)	−18[c]	45	−18	100
Pressure (atm)	1	1000[d]	1	1250
pH	*ca* 3	9	0·1	11·5
Eh (mV)	−450	750	−550	950
Oxygen tension (mg/l.)	0	14	0	70
Surface tension (dynes/cm)	55	75	25	75

[a] Including bottom deposits
[b] Saturated NaCl solution
[c] Freezing point of saturated NaCl solution
[d] Hydrostatic pressure at 10 000 m

Temperature

Effects of temperature on marine microorganisms may be considered as extremes of heat or cold for obligate thermophiles or psychrophiles, respectively, or on a periodic basis, e.g. seasonally. Extremes of heat in the marine environment are usually associated with thermal pollution. Changes in both the bacterial and fungal flora due to thermal pollution have been found in estuaries[10,11]. Although these microbial ecological alterations may be significant, a more severe effect is probably created by chlorination which often accompanies thermal discharge[12]. Extremes of cold in the marine environment are usually associated with the polar regions or the deep sea, of which 90 per cent of the water is in the temperature range of 2–3°C. Although a majority of the microorganisms from these regions probably are psychrophilic, hence survive and grow, it should be noted that large numbers of obligate thermophiles have been isolated from the abyssal regions of the Pacific Ocean[13]. A majority of these isolates resembled *Bacillus stearothermophilus*, a terrestinal spore former, which may be dormant in the deep ocean. The natures of thermophilic and psychrophilic microorganisms have been reviewed by Brock[14] and Stokes[15] respectively.

Effects of temperature on seasonal distribution of heterotrophic marine bacteria have been noted[16–18]. Similar studies have been conducted for marine sulfur bacteria[19], estuarine pathogenic bacteria[20], and mercury-resistant and petroleum-degrading bacteria[21–23]. These studies are significant in that they have provided data showing a correlation of temperature stress with ecological functioning of microbes, an interesting area of microbial ecology which merits further study.

Pressure

Many microorganisms common to surface soil and shallow water are killed in logarithmic phase of growth in 48 h at 30°C by pressures of only 500–600 atm. Multiplication of some of these microbes is inhibited by 300 atm[24]. Hydrostatic pressure affects multiplication, protein synthesis, and nutrient uptake in marine bacteria[25–29,4]. The data which are available demonstrate that inhibition of growth of marine bacteria caused by hydrostatic pressure is due primarily to inhibition of protein synthesis at the ribosomal level. In addition, nutrient uptake in the deep sea may be 10 to 100 times slower than in shallow water. The effect of hydrostatic pressure on hydrocarbon degradation by deep-sea bacteria has been examined by Schwartz, Walker, and Colwell[30]. Experiments demonstrated that significant hydrocarbon degradation could take place at 22°C and *in situ* pressure. The relatively slow rate of nutrient uptake by deep-sea bacteria and the low levels of non-refractory material in deep-sea sediment indicate that other agents, such as the intestinal flora of deep-sea invertebrates, may play an important role in mineralization occurring in the deep ocean.

Organic Pollutants

Contamination of the marine environment by organic pollutants represents a potentially serious threat to human health. The addition of organic pollutants include: (1) selective concentration of toxic materials, (2) effect of contamination on indigenous organisms, (3) rates of mixing, transport, distribu-

TABLE 2. Estimated masses (10^9 U.S. tons*) of organic matter in the oceans
From Hood[31]

Dissolved	3000
Particulate detrital	300
Living	7
Hydrocarbons	400

TABLE 3. Estimated rates of entry (10^5 U.S. tons*/year) of petroleum to the marine environment
From National Academy of Sciences[32]

Offshore oil production	1
Tanker operations	5
Other ship operations	5
Accidental spills	2
Deliberate dumping	5
Refinery operations	3·5
Industrial and automotive wastes	4·5
Atmospheric input	900
Natural seepage	<1

* 1 U.S. ton = 0.91 t

tion and degradation, (4) synergism, and (5) food-chain effects. A significant proportion of organic matter in the marine environment is derived from hydrocarbons of natural or petroleum origin (Table 2). The atmospheric input from vaporization of petroleum products is about 35 times that derived from man's other activities which, in turn, is some 30 times that derived from natural seepage (Table 3). The effects of two organic pollutants, petroleum and pesticides, on microorganisms in the marine environment are discussed below.

Petroleum

Many of the ecological effects of petroleum on marine macroorganisms have been summarized[33] by Cowell. However, only a very small amount of such information is available for marine microorganisms. The effects of petroleum depend on: (1) the type and amount of petroleum released into the environment, (2) the persistence or removal of the petroleum, (3) the organic content of the sediment at the site of the petroleum input, (4) the season in which the petroleum is released, and (5) the partitioning of toxic materials from the environment into the petroleum and vice versa. Obviously, these points are related, e.g. a more toxic petroleum that contains refractory components would tend to persist[34] longer than that comprised of non-toxic components which are easily biodegraded[35]. High organic matter in sediments reduces the incorporation of oil into sediment[36]. This incorporation may affect biodegradability and toxicity. Seasonal effects of petroleum on microorganisms may be expressed in terms of the toxicity of petroleum for the autochthonous microbial population or the susceptibility to biodegradation regulated by temperatures. Autochthonous microbial populations cultured in estuarine and ocean water may be suppressed by petroleum hydrocarbons[37,38]. Seasonal temperature also affects hydrocarbon solubility and, hence, a significant amount of microbial degradation of petroleum hydrocarbons occurs at low temperature, when toxic hydrocarbons are less soluble in sea water[22]. As noted above, certain materials may be preferentially partitioned or extracted into oil or from oil because of solubility properties. This has been reported for pesticides[39] and, more recently, for mercury, where concentrations of mercury in benzene-extractable material were 4000 and 300 000 times those detected in sediment and water respectively[40]. The water-soluble components of petroleum[41] may interfere with chemotaxis of marine bacteria[42,43], thereby affecting predator–prey relationships among microorganisms in the marine environment[44].

Effects of petroleum on microorganisms in the marine environment may also be complicated by the use of emulsifiers, such as BP 1002 which is toxic to some marine algae and lichens[45,46] or dispersants, such as Corexit 7664 which is non-toxic to beach microflora[47].

Pesticides

As stated in a recent review by Bourquin[48], little is known about the degradation of organochlorine pesticides in the marine environment. In fact, the effects of pesticides on microorganisms in the marine environment are poorly understood, except, perhaps, in the case of those studies on algae showing reduced photosynthesis and species composition alterations[49,50]. Some work has been done with the hydrocarbon-utilizing fungus, *Cladosporium resinae*[51]. In the presence of high concentrations of pesticides (such as may be found in oil slicks), the following results were observed: (1) aldrin, dieldrin, and malathion enhanced growth on hexadecane, (2) aldrin, dieldrin, DDT, and malathion enhanced growth on glucose, and (3) none of the pesticides served as sole carbon source, inhibited growth, or affected oxygen consumption on glucose or *n*-hexadecane. These results suggest that additional *in vitro* and *in situ* studies on the fate and effects of pesticides in the marine environment would be useful.

Predation

It is now well known that many marine protozoa, grazers, and filter feeders ingest and digest bacteria[52]. A number of reports have appeared in the literature which demonstrate that mussels[53], worms[54,55], sponges[56], and crayfish[57] ingest bacteria. The predation of these marine organisms acting to maintain a microbial balance in nature is only just beginning to receive attention.

ACKNOWLEDGMENTS

Support from Contract No. N00014-69-A-0220-0006 between the Office of Naval Research and the University of Maryland is acknowledged. J. D. Walker derived partial support from an Environmental Conservation Postdoctoral Fellowship awarded by the National Wildlife Federation and American Petroleum Institute.

We express appreciation to Dr. R. R. Colwell for her suggestions and criticisms during the preparation of this manuscript.

REFERENCES

1. ZoBell, C. E. Microbial and environmental transitions in estuaries. In 'Estuarine and marine ecology' (ed. L. H. Stevenson and R. R. Colwell). University of South Carolina Press, Columbia, S.C., 1973, pp. 9–34.
2. Brock, T. D., and Darland, G. K. Limits of microbial existence: temperature and pH. *Science, N.Y.*, 1970, **169**, 1316–1318.
3. Cooper, M. F., and Morita, R. Y. Interaction of salinity and temperature on net protein synthesis and viability of *Vibrio marinus*. *Limnol. Oceanogr.* (*in press*).
4. Jannasch, H. W., and Wirsen, C. O. Deep-sea microorganisms: *in situ* response to nutrient enrichment. *Science, N.Y.*, 1973, **180**, 641–643.
5. Carlucci, A. F., and Pramer, D. Factors affecting the survival of bacteria in sea water. *Appl. Microbiol.*, 1959, **7**, 388–392.
6. Jones, G. E. The fate of freshwater bacteria in the sea. *Devs ind. Microbiol.*, 1971, **12**, 141–151.
7. MacLeod, R. A. On the role of inorganic ions in the physiology of marine bacteria. In 'Advances in microbiology of the sea' (ed. M. R. Droop and E. J. F. Wood). Academic Press, New York, 1968, pp. 95–126.
8. Siegenthaler, P. A., Belsky, M. M., and Goldstein, S. Phosphate uptake in an obligately marine fungus: a specific requirement for sodium. *Science, N.Y.*, 1967, **155**, 93–94.
9. Belsky, M. M., Goldstein, S., and Menna, M. Factors affecting phosphate uptake in the marine fungus *Dermocystidium* sp. *J. gen. Microbiol.*, 1970, **62**, 399–402.
10. Buck, J. D. Effect of thermal effluent on the microbiology and chemistry of the Connecticut River. In 'Estuarine microbial ecology' (ed. L. H. Stevenson and R. R. Colwell). University of South Carolina Press, Columbia, S.C., 1973, pp. 283–292.
11. Shearer, C. A. A study of the ecology, taxonomy and physiology of some wood-inhabiting Ascomycetes and fungi imperfecti from the Patauxent River. Ph.D. dissertation, University of Maryland, College Park, Md, 1971, 104 pp.
12. Hamilton, D. H., Flemer, D. A., Keefe, C. W., and Mihursky, J. A. Power plants: effects of chlorination on estuarine primary production. *Science, N.Y.*, 1970, **169**, 197–198.
13. Bartholomew, J. W., and Paik, G. Isolation and identification of obligate thermophilic sporeforming bacilli from ocean basin cores. *J. Bact.*, 1966, **92**, 635–638.
14. Brock, T. D. High temperature systems. *Ann. Rev. Ecol. Syst.*, 1970, **1**, 191–220.
15. Stokes, J. L. Nature of psychrophilic microorganisms. In 'Low temperature biology of foodstuffs' (ed. J. Hawthorne). Pergamon Press, New York, 1968, pp. 221–233.
16. Sieburth, J. M. Seasonal selection of estuarine bacteria by water temperature. *J. expl. mar. Biol. Ecol.*, 1967, **1**, 98–121.
17. Murchelano, R. A., and Brown, C. Heterotrophic bacteria in Long Island Sound. *Mar. Biol.*, 1970, **7**, 1–6.
18. Nedwell, D. B., and Floodgate, G. D. The seasonal selection by temperature of heterotrophic bacteria in an intertidal sediment. *Mar. Biol.*, 1971, **11**, 306–310.
19. Nedwell, D. B., and Floodgate, G. D. The effect of microbial activity upon the sedimentary sulfur cycle. *Mar. Biol.*, 1972, **16**, 192–200.
20. Kaneko, T., and Colwell, R. R. Ecology of *Vibrio parahaemolyticus* in Chesapeake Bay. *J. Bact.*, 1973, **113**, 24–32.
21. Colwell, R. R., Nelson, J. D., and Walker, J. D. Microbial mobilization of nutrients in estuaries. *Proc. 2nd int. estuar. Res. Conf., Myrtle Beach, South Carolina, 1974* (*in press*).
22. Walker, J. D., and Colwell, R. R. Microbial degradation of model petroleum at low temperatures. *J. microb. Ecol.*, 1974, **1**, 59–91.
23. Walker, J. D., and Colwell, R. R. Microbial degradation of model petroleum at warm temperatures. *J. microb. Ecol.* (*in preparation*).
24. ZoBell, C. E., and Johnson, F. H. The influence of hydrostatic pressure on the growth and viability of terrestrial and marine bacteria. *J. Bact.*, 1949, **57**, 179–189.
25. ZoBell, C. E., and Oppenheimer, C. H. Some effects of hydrostatic pressure on the multiplication and morphology of marine bacteria. *J. Bact.*, 1950, **60**, 771–781.
26. Paul, K. L., and Morita, R. Y. Effects of hydrostatic pressure and temperature on the uptake and respiration of amino acids by a facultatively psychrophilic marine bacterium. *J. Bact.*, 1971, **108**, 835–843.
27. Schwarz, J. R., and Landau, J. V. Hydrostatic pressure effects on *Escherichia coli*: site of inhibition of protein synthesis. *J. Bact.*, 1972, **109**, 945–948.
28. Schwarz, J. R., and Landau, J. V. Inhibition of cell-free protein synthesis by hydrostatic pressure. *J.Bact.*, 1972, **112**, 1222–1227.

29. Jannasch, H. W., Eimhjellen, K., Wirsen, C. O., and Farmanfarmaian, A. Microbial degradation of organic matter in the deep sea. *Science, N.Y.*, 1971, **171**, 672–675.

30. Schwarz, J. R., Walker, J. D., and Colwell, R. R. Growth of deep-sea bacteria on hydrocarbons at ambient and *in situ* pressure. *Devs ind. Microbiol.*, 1974, **15** (*in press*).

31. Hood, D. W. Pollution of the world's oceans. *Topics Ocean Engng*, 1972, **4** (*in press*).

32. National Academy of Sciences. Marine environmental quality. Ocean Science Committee; National Academy of Sciences and National Research Council, Washington, D.C., 1971.

33. Cowell, E. B. The ecological effects of oil pollution on littoral communities. Elsevier, London, 1971, 250 pp.

34. Blumer, M., and Sass, J. Oil pollution: persistence and degradation of spilled fuel oil. *Science, N.Y.*, 1972, **176**, 1120–1122.

35. Kinney, P. J., Button, D. K., and Schell, D. M. Kinetics of dissipation and biodegradation of crude oil in Alaska's Cook Inlet. In 'Proc. API/FWPCA Conf. on prevention and control of oil spills'. American Petroleum Institute, Washington, D.C., 1969.

36. Meyers, P. A., and Quinn, J. G. Association of hydrocarbons and mineral particles in saline solution. *Nature, Lond.*, 1973, **244**, 23–24.

37. Walker, J. D., and Colwell, R. R. Degradation of hydrocarbons and model petroleum by microorganisms from Chesapeake Bay. *Prog. Wat. Technol.*, 1975, **7** (*in press*).

38. Walker, J. D., and Colwell, R. R. Degradation of petroleum hydrocarbons by bacteria isolated from the ocean environment. Submitted to *Limnol. Oceanogr.*

39. Seba, D. B., and Corcoran, E. F. Surface slicks as concentrators of pesticides in the marine environment. *Pestic. Monit. J.*, 1969, **3**, 190–193.

40. Walker, J. D., and Colwell, R. R. Mercury-resistant bacteria and petroleum degradation. *Appl. Microbiol.*, 1974, **27**, 285–287.

41. Boylan, D. B., and Tripp, B. W. Determination of hydrocarbons in sea-water extracts of crude oil and crude oil fractions. *Nature, Lond.*, 1971, **230**, 44–47.

42. Mitchell, R., Fogel, S., and Chet, I. Bacterial chemoreception: an important phenomenon inhibited by hydrocarbons. *Wat. Res.*, 1972, **6**, 1137–1140.

43. Young, L. Y., and Mitchell, R. Negative chemotaxis of marine bacteria to toxic chemicals. *Appl. Microbiol.*, 1973, **25**, 972–975.

44. Chet, I., Fogel, S., and Mitchell, R. Chemical detection of microbial prey by bacterial predators. *J. Bact.*, 1971, **106**, 863–867.

45. Boney, A. D. Toxicity studies with an oil-spill emulsifier and the green alga *Prasinocladus marinus*. *J. mar. biol. Ass. U.K.*, 1970, **50**, 461–473.

46. Brown, D. H. The effect of crude Kuwait oil and a solvent emulsifier on the metabolism of the marine lichen *Lichena pygmaea*. *Mar. Biol.*, 1972, **12**, 309–315.

47. Bloom, S. A. An oil dispersant's effect on the microflora of beach sand. *J. mar. biol. Ass. U.K.*, 1970, **50**, 919–923.

48. Bourquin, A. W. Estuarine microbes and organochlorine pesticides (a brief review). In 'The microbial degradation of oil pollutants' (ed. D. G. Ahearn and S. P. Meyers). Louisiana State University, Publ. No. LSU-SG-73-01, 1973, pp. 237–243.

49. Wurster, C. F. DDT reduces photosynthesis by marine phytoplankton. *Science, N.Y.*, 1968, **159**, 1474–1475.

50. Mosser, J. L., Fisher, N. S., and Wurster, C. F. Polychlorinated biphenyls and DDT alter species composition in mixed cultures of algae. *Science, N.Y.*, 1972, **176**, 533–535.

51. Walker, J. D., Cofone, L., and Cooney, J. J. Microbial petroleum degradation: The role of *Cladosporium resinae*. In 'Proc. API/EPA/USCG Conf. on prevention and control of oil spills'. American Petroleum Institute, Washington, D.C., 1973, pp. 821–825.

52. ZoBell, C. E. The occurrence of bacteria in the deep sea and their significance for animal life. *Int. Union biol. Sci.*, 1954, Ser. B, No. 16, 20–26.

53. ZoBell, C. E., and Landon, W. A. Bacterial nutrition of the California mussel. *Proc. Soc. expl. Biol. Med.*, 1937, **36**, 607–609.

54. ZoBell, C. E., and Feltham, C. B. Bacteria as food for certain marine invertebrates. *J. Mar. Res.*, 1938, **7**, 312–324.

55. Wavre, M., and Brinkhurst, R. O. Interactions between some tubificid oligochaetes and bacteria found in the sediments of Toronto Harbour, Ontario. *J. Fish. Res. Bd Can.*, 1971, **28**, 335–341.

56. Pourbaiz, N. Nota soba la nutricon bacteria de las Esponza. *Not. as. Resum.*, 1932, **64**, 1–4.

57. Cviic, V. Contribution à la connaissance du rôle des bactéries dans l'alimentation des larves de langouste. *Rapp. P.-v. Réun. Commn int. Explor. scient. Mer Méditerr.*, 1960, **25**, 45–47.

Dr. Guarraia, introducing the paper, said that microbial survival in marine environments was dependent upon a host of factors, as previously discussed in other papers at the Symposium. Particular interest centred upon those organisms considered as indicators of faecal pollution, and potential disease-causing microbes. The factors to be considered when a fresh-water bacterium was placed into the ocean environment were many. Physical, chemical, and biological factors within the new environment would exert an effect on the longevity of the organism in question, and the particular organism under consideration was also of importance.

Physical factors were the first of the 'other factors' affecting the microbe when placed in the marine or estuarine environment. Although the stresses placed on it by any individual factor might be negligible, the combined effects of several factors could be lethal. For example, pH–temperature, salinity–temperature, or pressure–salinity–temperature might (in these or other combinations) result in an untoward reaction by the microbe.

DISCUSSION

Dr. G. E. B. Kullenberg (Denmark) asked Dr. Guarraia if he would say that thermal pollution was of any significance in the vicinity of coastal outfalls. Dr. Guarraia replied that the macrobiological effects were likely to be of more significance than the microbiological ones.

Paper No. **24**

Mortality of Enteric Viruses in Marine and Other Waters

Elmer W. Akin, William F. Hill, Jr., and Norman A. Clarke
Water Supply Research Laboratory, National Environmental Research Center
U.S. Environmental Protection Agency, Cincinnati, Ohio 45268

Historically, a high endemic level or the occurrence of frequent epidemics of a fatal or crippling infectious disease served as a mandate for public health workers to elucidate and break the cycle of transmission. Resultant investigations often uncovered unique and complex routes of transmission of infectious agents and generally led to control measures which dramatically reduced the level of disease. In the case for water transmission of viral agents, we believe a most unusual situation exists. During the last decade a world-wide increase in concern has developed over the possibility, and indeed the likelihood, that significant levels of viruses are being transmitted through recreational and potable water. This concern has heightened in spite of the fact that no hard evidence exists to document the transmission of enteric viral disease through this route. The only exceptions have been rare outbreaks of infectious hepatitis where standard sanitary practice or water treatment had obviously broken down. Even the suspicion that poliomyelitis, which last peaked in the United States in 1952 at approximately 58 000 cases, was transmitted through swimming pools and other water routes was not proven to the satisfaction of most epidemiologists.

Several observations would seem to justify this concern: (1) laboratory data have shown most enteric viruses to be more resistant than indicator bacteria to inactivation by water disinfectants, (2) infectivity tests have shown infection could be caused by one poliovirus unit, (3) a consistently high endemic level of infectious hepatitis (IH) has occurred (in the U.S.) with the concomitant knowledge that the IH agent is relatively resistant to inactivation in the aquatic environment, (4) sporadic outbreaks of non-bacterial gastroenteritis suspected of being waterborne have occurred coupled with a most likely high endemic level of the disease, and (5) surface-water domestic pollution has increased to the point that direct recycling of waste water has all but become a reality for some water systems.

These concerns have led public health workers to re-examine the likelihood of water transmission of enteric viral disease. Research efforts have been expended in several areas: (1) development of sensitive methods for recovering viruses from water, (2) monitoring of water and waste water for enteric viruses, (3) evaluation of the virus removal efficiency of treatment procedures for water and waste water, and (4) the study of survival times of these viruses in different types of water. This paper is concerned with this last area of research.

Observations on the Inactivation of Enteric Viruses in Water

It has been known for some 30 years that poliovirus could survive for extended periods of time outside an animal host. Paul and Trask[1] reported in 1941 the isolation of poliovirus from sewage. Since that time, virus recovery methods have improved, and investigators from all parts of the world have reported the isolation of numerous members of the enteric virus group in waste waters and surface waters receiving domestic waste. However, only in the last few years have studies been conducted which were designed to quantify the survival rate of these viruses in water. Literature reports concerning the survival of enteric viruses in various types of water have recently been summarized[2]. The vast majority of the viruses tested lost 3 logs within 3–21 days at 20–25°C, but persisted considerably longer at 4–6°C.

Of course, our main concern in this Symposium is the survival of these viruses in marine waters (the term marine waters will be used in this report to denote water which contains any concentration of sea salts). Obviously, enteric viruses can survive environmental conditions long enough to reach the sea through direct domestic-waste discharge and the inflow of polluted rivers and streams. The first critical studies on the survival of viruses in marine water *per se* were reported in 1965 by a Swedish group[3,4]. Of 24 water samples they collected from the Baltic and North Seas, 19 possessed a virus inactivating capacity (VIC) for poliovirus 3. The investigators defined VIC as a loss of infectivity of 2 logs or more in 8 days at 23°C as measured by the $TCID_{50}$ method. One of these sea-water samples was tested and found to possess VIC for five other enteric virus serotypes. River and lake water samples showed no

VIC for poliovirus. The VIC was eliminated by: (1) heating of sea water to 45°C for 1 h, (2) filter sterilization of the sea water, (3) the addition of 10 mg/ml of glycine, or (4) shaking the sea water with 5 per cent ether for 1 h. The VIC was not affected by ultrasonication (15 min at 10 kHz). Based on other experiments reported in these papers, the authors stated that the VIC of sea water could not be explained as an effect of salt concentration, oxidation, or a photodynamic action upon the virus. A marine bacterium isolated from sea water, grown in pure culture and suspended in heat-treated sea water, was found to impart a VIC to the water[5]. This observation was consistent with the data showing that the Seitz filtration of the sea water removed the VIC. However, the picture was confused by the discovery that stored sea water having no detectable viable organisms maintained its virucidal reactivity.

Shuval *et al.*[6] have reported similar findings with poliovirus suspended in waters of the eastern Mediterranean and the Red Sea. Three logs of virus infectivity were lost in 9 days in normal sea water at 15°C. Heating, ether treatment or filtration through a 0·22-μm porosity membrane filter, eliminated the marine antiviral activity (MAVA). Eight of 58 marine bacteria which were isolated from the sea water and grown in pure culture imparted a MAVA to heat-treated sea water. These bacteria lost their antiviral capacity on laboratory passage, suggesting that the MAVA may have been a metabolic by-product of these organisms which was not produced under artificial cultivation conditions. However, MAVA could not be demonstrated in any water samples free of viable marine bacteria nor could it penetrate a dialysis bag. It seemed that the MAVA required the presence of the bacteria in high numbers or was associated with a very short-lived agent produced by the bacteria. Mitchell[7] has also observed that the presence of indigenous marine microorganisms enhanced the inactivation of bacteriophage (Φx-174) in sea water collected off the New England (U.S.A.) coast. He reported that the chemical component, presumably the heavy metals, of sea water also exerted a detrimental effect on the phage.

Matossian and Garabedian[8] have attempted to identify the factors in natural Mediterranean sea water which inactivate poliovirus 1. These investigators used water freshly collected from the sea near Beirut (Lebanon) in their experimentation. Their findings were significantly different in one respect from the previous work reviewed; i.e. the virucidal factor of the water readily passed through a Seitz filter. Three logs of virus infectivity were lost in 6–7 days in both untreated and filtered sea water at 25°C. Water boiled for 5 min or stored at room temperature for 2 weeks prior to the addition of virus, required about twice as long to achieve a 3-log reduction when compared to fresh untreated sea water. Sterile artificial sea water constituted to simulate natural sea water in its major salt content produced an inactivation pattern similar to that of boiled natural sea water. These workers therefore concluded that the virucidal property of sea water was due to the combined effect of its chemical composition and to an unknown heat-labile inhibitory substance.

Metcalf and Stiles[9] have observed the inactivation patterns of coxsackievirus B-3, echovirus 6, and poliovirus 1 in sea water collected off the coast of New Hampshire (U.S.A.). They found the virus inactivation rates in autoclaved and ultraviolet-irradiated sea water to be the same as the inactivation rates in natural sea water when studied under laboratory conditions. These studies which were conducted at 5°, 24°, and 37°C indicated that temperature was the only decisive factor influencing the inactivation curves of these viruses under laboratory conditions. Field studies with these viruses revealed that factors other than temperature influenced virus survival in the natural environment. Viruses placed in cellulose dialysis bags and immersed in estuarine water in the natural environment were inactivated in a shorter time than under laboratory conditions at 5°C, even though the water temperature hovered near zero during the field experimental period.

Ruschi and Bendinelli[10] reported that the survival time of five enteric viruses was not significantly enhanced by the Seitz filtration of sea water from the Ligurian Sea collected off the coast of Italy when compared to raw sea water. Interestingly, the presence of domestic waste in the sea water decreased the viral survival times. Filtering or autoclaving the mixture of sea water and sewage returned the survival time to that of the pure sea water.

Leogrande and Iandolo[11] compared the survival of ten strains of poliovirus 1 suspended in Hanks balance salt solution, raw sea water, and autoclaved sea water; the sea water was collected 6–8 km off the Italian coast in the Adriatic Sea. The authors concluded that there was no difference in the survival times among the three suspending media. They further concluded that no VIC was manifest in these raw sea-water samples.

Akin *et al.*[12] have conducted survival studies with poliovirus 1 in marine water from the Gulf of Mexico collected off the coast of Alabama (U.S.A.). Our initial studies showed that these waters also possess a VIC. However, in contrast to findings with marine waters from other bodies of water, we found that autoclaving or filtering the water did not prevent the viral loss (Fig. 1). This study was conducted with the LSc2ab strain of poliovirus 1 in freshly collected estuarine water having a salinity

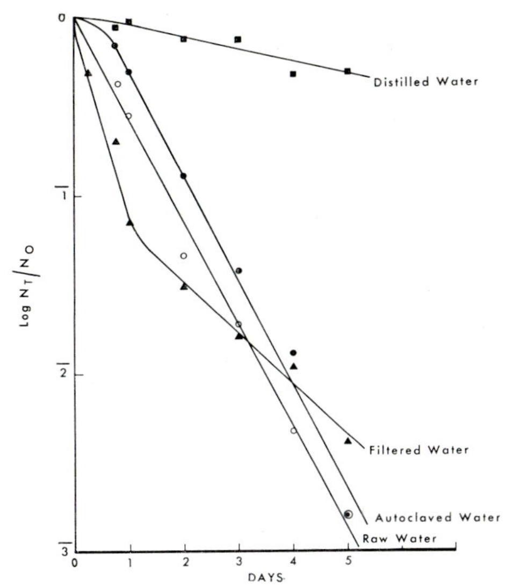

FIG. 1. Loss of poliovirus 1 infectivity in raw and treated marine water and distilled water[12]

of 15 g/kg; 500 ml of water were placed in sterile 1-l. bottles, dosed with virus then placed in the dark on a reciprocal shaker at 24°C. It is obvious from the figure that the autoclave temperature (121°C) and 0·45-μm filtration did very little to eliminate the VIC of the marine water sample. The inactivation rate change which occurred after 1–2 days in the filtered water remains unexplained. However, throughout the course of our survival studies in marine water, we often observed a two-component or retardant die-away curve.

We were concerned that these laboratory-based studies conducted in a closed system with small volumes of water may not represent the true virus inactivation that occurred in nature. It has been shown that the natural balance of open waters is upset when small volumes of water are placed in a closed system[13]. For this reason a large volume flow-through system was designed to conduct survival experiments in estuarine water under more natural conditions[14]. The system is diagramically shown in Fig. 2. The 3785-l. reservoir was filled with fresh estuarine water prior to a survival study. Poliovirus was then added to the reservoir and mixed well. To begin the experiment a constant volume of 'fresh' estuarine water was added to the reservoir via the distribution manifold. The reservoir contents were mixed well before the excess volume was discharged through the central standpipe. Composite samples were taken at 0 h and periodically throughout the experimental period. The virus remaining at each sampling period was corrected for loss by dilution, which then gave the number of infectious units of virus not inactivated. Table 1 shows the results of six experiments conducted over an 8-month period. The hydrographic data reflect the seasonal changes typical of the Mobile Bay (Alabama) Estuary. Time

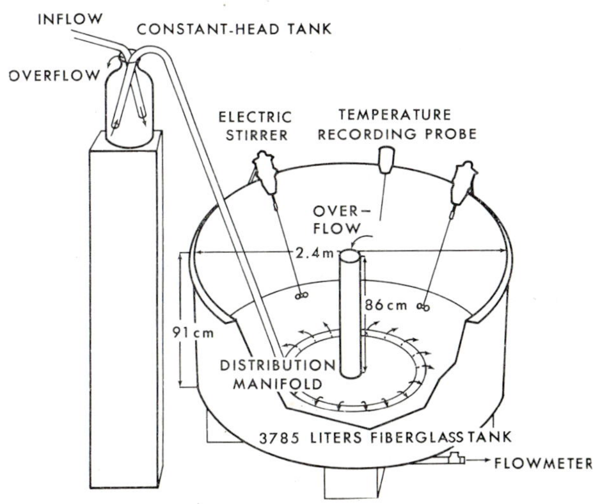

FIG. 2. Flow-through system used for conducting virus-inactivation studies in marine water under field conditions[14]

TABLE 1. Experimental conditions and percentage reduction times of poliovirus 1 in flowing estuarine water

Adapted from Akin and Presnell[14]

Expt. No.	Month	Temperature range (°C)	Salinity range (g/kg)	Turbidity range (JTU)	Time (h) for inactivation percentage of 99	99·9
1	Aug.	30–34	27	26–39	4	5
2	Sep.	28–32	21–22	26–36	4	5
3	Oct.	22–25	18–25	36–64	14	31
4	Oct.	20–23	21–23	24–39	6	20
5	Jan.	3–11	13–21	16–31	36	68
6	Mar.	10–17	2–5	54–123	38	63

for 3 logs (99·9 per cent) of viral inactivation ranged from 5 h at 28–34°C to 68 h at 3–11°C. Clearly, VIC as originally defined was manifested by the estuarine water during each experimental period. Water temperature was singularly the most important factor in determining reduction time. The result of Expt. 6 suggested that salinity may also play a role, i.e. less inactivation at a lower salinity.

To determine more clearly the influence of salinity on poliovirus inactivation, a laboratory study was conducted with artificial sea water prepared after Kester[15]. This chemically-defined sea water was prepared with analytical grade chemicals to contain the eleven basic chemical elements in the concentration and proportion that they are found in 35-g/kg salinity natural sea water. Salinity was lowered by diluting with glass-distilled water. Figure 3(a) shows the inactivation curves of poliovirus 1 in water at three salinity levels at 24°C. The experiment was conducted in the dark on a reciprocal shaker (120 cycles/min). From 3·0 to 4·5 logs of inactivation occurred within 11 days depending upon salinity. Greater inactivation occurred at a salinity of 10 g/kg than at 1 or 20 g/kg—suggesting an optimal salinity for poliovirus inactivation of approximately 10 g/kg. This pattern of inactivation was not observed, however, in a subsequent study with natural sea water of 32 g/kg, collected from the Gulf of Mexico, filtered prior to dosing with virus, and diluted with sterile glass-distilled water. The result of this experiment is shown in Fig. 3(b). The data show that under the same experimental conditions as the study with artificial sea water, inactivation of poliovirus 1 was directly proportioned to salinity, thus as salinity was reduced from 32 to 14 g/kg and then to 1 g/kg, inactivation was delayed. The factors responsible for these differences in inactivation patterns remain unexplained.

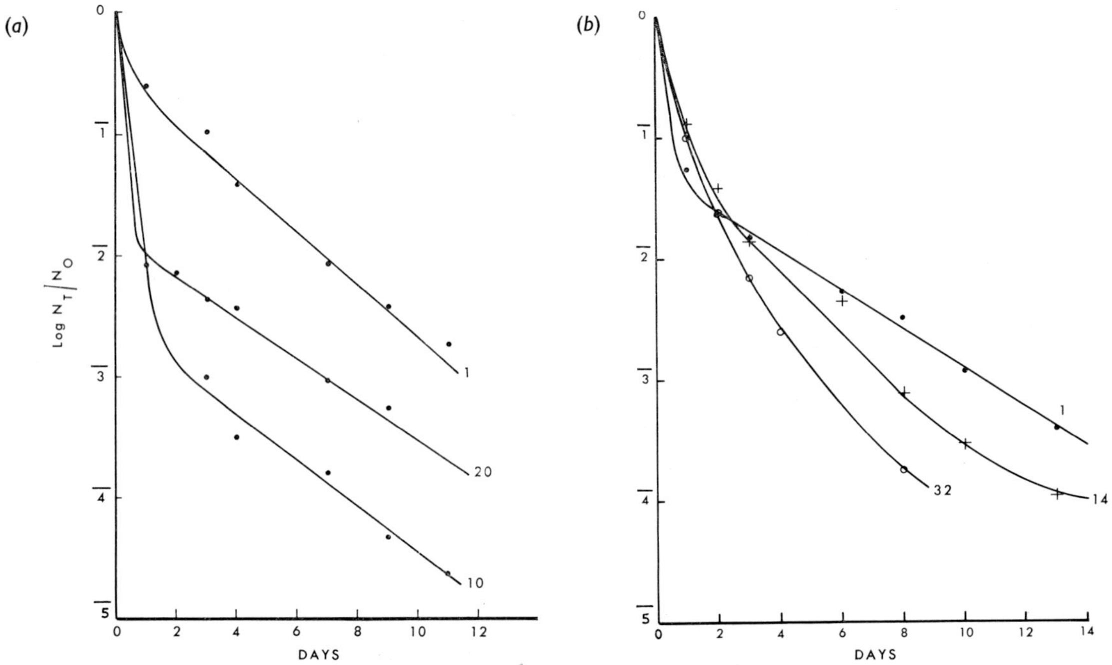

FIG. 3. Loss of poliovirus 1 infectivity in sea water at salinity levels (g/kg) indicated[12]
(*a*) Artificial sea water; (*b*) filtered Gulf sea water diluted with distilled water to give lower salinities

Evaluation of Factors which may Influence Loss of Viral Infectivity in Marine Water

Loss of virus by adsorption

Poliovirus adsorption to and elution from containers made of ten different materials have recently been studied at our laboratory[12]. From 3 to 5 logs of virus were lost in 8 days when suspended in marine water and placed in these containers. Elution studies revealed that no significant loss could be attributed to container adsorption. All laboratory survival experiments prior to this study were conducted in flint-glass or pyrex-glass containers.

Occurrence of resistant virus to explain persistent fraction

It has been previously stated that many inactivation studies with poliovirus suspended in marine water produced two-component curves. Sharp rate changes often occurred during the first 3 logs of inactivation. In one typical experiment with poliovirus 1 suspended in natural estuarine water of 14·5-g/kg salinity, we observed a dramatic rate change on the 6th day after approximately 2·5 logs of infectivity had been lost. A very gradual loss occurred from the 6th day through the 13th day at which time 44 PFU/ml of the original 18 000 PFU/ml remained. In an effort to determine if the virus remaining on the 13th day represented a more resistant strain within the stock virus, the following study was conducted. The 13th-day virus sample was passaged three times in HEp-2 cells to produce a high titer of virus. The virus suspension was then filtered through a 50-nm porosity treated membrane filter to yield monodispersed virus in the filtrate. The filtrate-virus was then diluted 100-fold in distilled water and 1 ml added to 100 ml of natural estuarine water having a salinity of 14·5 g/kg. The rates of the loss of infectivity of these viruses are shown in Fig. 4. The original and the selected (13-day sample) virus both lost infectivity at statistically identical rates through 6 days ($P > 0{\cdot}05$) and at similar rates from the 6th through the 13th day. These data indicated that a genetically more resistant population of virus was not responsible for the second component of the poliovirus inactivation curve obtained when the virus was suspended in natural estuarine water.

Influence of viral aggregation on loss of infectivity of poliovirus

Chang[16] has reported that viral clumps are responsible for many of the aberrations observed in viral inactivation curves with chemical virucides. We have shown that two-component curves occurred with stock virus that had been filtered to remove viral clumps. This observation suggested that the second component of the curve did not represent the inactivation rate of viral clumps. However, the possibility of individual virions clumping together during the agitation of experimental samples had not been ruled out. If clumping was occurring, it would reduce the number of infectious foci and would be interpreted as viral inactivation by reducing the number of plaque units remaining. This phenomenon could explain the rapid rate of loss of the first component. The fact that certain hydrophilic organic macromolecules tend to aggregate in the presence of cations has been known for sometime. The following experiment was conducted to determine whether or not this phenomenon was occurring with poliovirus 1 in marine water.

An equal volume of 50-nm filtered virus was placed in two 300-ml volumes of filtered marine water (15 g/kg salinity). One container remained static and the other was shaken for 20 h on the reciprocal shaker to enhance virion collision frequency. After 20 h the virus in the shaken container had been reduced by 68 per cent of its original infectivity while the stationary sample was reduced by only 20 per cent. Rate-zonal centrifugation studies were conducted on these two 20-h samples; 20-ml aliquots of the samples were layered on sucrose density gradients, and 46 fractions of the sucrose gradient were collected after a 90 min centrifugation run and were then assayed for viral infectivity.

Figure 5 shows the distribution of the viral infectivity through the gradient. The major infectivity peaks in both distributions are practically superimposed at a sedimentation coefficient of 150. Computer analysis indicated both peaks represented virion monomers (single particles). Over 97 and 95 per cent of the initial infectious virus remained as monomers in the static and shaken sample, respectively. The few remaining per cent of virus penetrated the gradient further, most likely in the form of aggregates. Had viral aggregation been responsible for the significant loss of infectious foci in the shaken sample, the infectivity peak should have moved further down the gradient during the centrifugation period. The failure to do so indicated that significant viral clumping had not taken place within the 20-h shaking period even though the plaque count had dropped substantially.

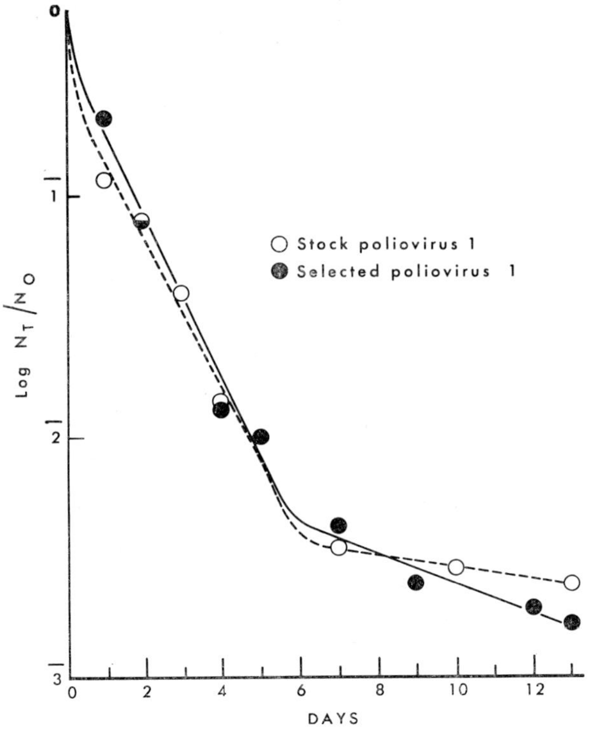

FIG. 4. Loss of poliovirus 1 infectivity in marine water of stock virus and selected virus which had replicated from the stock virus remaining on 13th day[12]

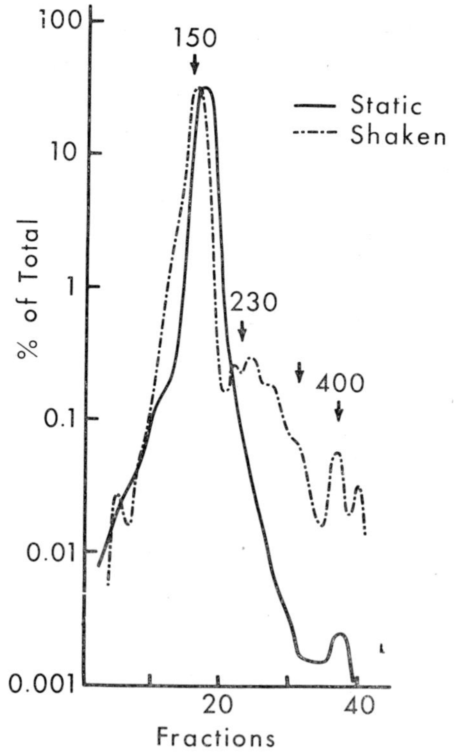

FIG. 5. Distribution of poliovirus 1 infectivity through a sucrose density gradient after rate-zonal centrifuga- tion, following 20 h static or shaken[12]

The Nature of Virion Inactivation in Water

It has been known for some time that the inactivation of viruses was directly related to temperature. Empirically, this increased viral inactivation rate at increasing temperatures was thought to be due to an accelerated oxidation of the virion. The first insight into the nature of thermal inactivation of viruses was provided in 1962 by Pohjanpelto[17] who concluded that the addition of cystine to a poliovirus 1 suspension increased the thermal stability of the virus by preventing oxidation. More importantly, he concluded from his experiments that oxidation was a major factor in the thermal inactivation of poliovirus at 50° but not at 32°C. Younger[18] had earlier suggested that various strains of poliovirus 1, 2, and 3 were inactivated by different mechanisms at 36·5° and 50°C. A short time later, Lund[19] came to the same conclusion with all three poliovirus types. Halsted *et al.*[20] added sufficient weight to this hypothesis with their studies on viral inactivation at 45°C in the presence of a sulfhydryl-reducing substance. They found that by adding the reducing agent dithiothreitol, thermal stability was imparted to poliovirus 1 and 2, coxsackievirus B-5, and echovirus 7. All three investigators found approximately 37°C to be the critical temperature at which the type of inactivation was determined. At temperatures below this level, inactivation was of the ill defined 'spontaneous' type. Recently, Lund[21] has concluded from her own experimentation that oxidation is not the mechanism of inactivation of the small RNA-viruses in water at 20°C or below. At temperatures between 37° and 50°C, oxidation appeared to be the most important factor.

Dimmock[22] offered convincing evidence that the primary mechanism of inactivation at higher temperatures (above 44°C) was due to damage of the viral protein coat whereas inactivation at lower temperatures was due to damage to the nucleic acid. The protective effect observed with molar $MgCl_2$ was thought to be through the stabilization of the protein coat.

In response to reports that showed antiviral activity of some sea waters required the presence of microorganisms, Cliver and Herrmann[23] attempted to determine the antiviral nature of bacteria-derived proteolytic enzymes. Their results were inconclusive but showed that some enteroviruses, notably coxsackievirus A-9, were inactivated by proteolytic bacteria. Poliovirus 1 was shown to be resistant to enzymatic inactivation by the organisms tested.

These reports have occurred in the literature during the past 12 years and represent our limited knowledge of the viral inactivation process in nature. Obviously, the viral inactivation processes working at the near neutral pH of most waters at moderate temperatures are basically unknown. Even nature's original disinfectant, the ultra-violet range of the electromagnetic spectrum from the sun, may be ineffective as a virucide in the aquatic environment. These rays are known to destroy the integrity of viral nucleic acid. However, their penetration of the upper centimeters of most untreated surface waters are negligible due to suspended particulate matter.

General Discussion and Conclusions

Known virucides, such as chlorine and ultra-violet radiation, produce a reproducible rate of inactivation from one experiment to another. This is generally true when experimental conditions are not varied and when the virucide is present in such a concentration that its loss during the experimental period is insignificant. If this condition is required for a substance to be considered a virucide, then marine water cannot be so classified. Our observations after many inactivation studies with poliovirus 1 showed that virus inactivation in marine water is unpredictable. Marine water collected from the same site on different days with the same salinity has shown wide variations in inactivation patterns. The wide variation in viral inactivation patterns suggests that natural die-off occurs randomly depending on virus type, the physiochemical differences within virus of the same type, and the physical, biological, and chemical composition of the marine water. The divergence of opinion reported in the literature lends support to such a conclusion. Undoubtedly, marine organisms play a major role in virus inactivation under certain conditions. However, the studies done with sterile marine water suggested that a filterable heat-resistant component of marine water may play the dominant role in the inactivation of viruses. Obviously, marine water cannot be considered as a single entity when evaluating its effect on enteric virus inactivation. It is equally clear that all enteric viruses do not follow a similar inactivation rate in a given water sample. Liu *et al.*[24] have shown that a wide variation exists among the enteric viruses to a single inactivating agent (chlorine). One would assume that a similar variability would occur with 'natural' inactivation. Consequently, data collected concerning the inactivation pattern of one virus serotype cannot necessarily be extended to other virus serotypes.

The only finding that seems to be consistent is the direct relationship between virus inactivation and temperature. Under otherwise identical conditions, enteric viruses will be inactivated in a shorter time period with increasing temperature.

The public health safety of human contact with virus-contaminated water cannot rely on current knowledge of virus survival-time. Presently, these decisions must be made without definitive guidelines on virus inactivation patterns in marine water. In what position then are we who are interested in the safety from a virus standpoint of these waters? Obviously, the classical statement "more work needs to be done" is appropriate. Conflicting data which show that viruses are rapidly inactivated in sterile artificial and natural sea water from one location and yet remain in full titer in sterile sea water from another sea are without adequate explanation. The explicit component(s) of sterile sea water responsible for virus inactivation needs to be exposed. Much of the available virus survival data are of a quantal nature rather than rate-reduction data. Clearly, greater sophistication in experimental design and data evaluation and presentation is needed. Studies must be made with a wider spectrum of the 100 known enteric virus types so that the most resistant virus can be determined. Laboratories interested in this area of research should be equipped to conduct inactivation studies with infectious hepatitis virus when a culture system is found. Greater care should be given to the chemical and biological characterization of water samples being used in the viral inactivation studies.

Even with all the above factors considered, the inactivation of viruses in the complex and dynamic aquatic environment may be so multifactoral that an *a priori* prediction would be futile. Such a conclusion before sufficient research is conducted, however, is premature and should not be used to thwart the appropriate research to accumulate the data.

REFERENCES

1. PAUL, J. R., and TRASK, J. D. The virus of poliomyelitis in stools and sewage. *J. Am. med. Assn.*, 1941, **116**, 493.
2. AKIN, E. W., BENTON, W. H., and HILL, W. F., JR. Enteric viruses in ground and surface waters: a review of their occurrence and survival. *Proc. 13th Wat. Qual. Conf.*, University of Illinois, Urbana, 1971, p. 59.
3. LYCKE, E., MAGNUSSON, S., and LUND, E. Studies on the nature of the virus inactivating capacity of sea water. *Arch. ges. Virusforsch.*, 1965, **17**, 409.
4. MAGNUSSON, S., HEDSTROM, C. E., and LYCKE, E. The virus inactivating capacity of sea water. *Acta path. microbiol. scand.*, 1966, **66**, 551.
5. MAGNUSSON, S., GUNDERSEN, K., BRANDBERG, A., and LYCKE, E. Marine bacteria and their possible relation to the virus inactivation capacity of sea water. *Acta path. microbiol. scand.*, 1967, **71**, 274.
6. SHUVAL, H. I., THOMPSON, A., FATTAL, B., CYMBALISTA, S., and WIENER, Y. Natural virus inactivation processes in sea water. *J. sanit. Engng Div. Am. Soc. civ. Engrs*, 1971, **97**(SA5), 587.
7. MITCHELL, R. Destruction of bacteria and viruses in sea water. *J. sanit. Engng Div. Am. Soc. civ. Engrs*, 1971, **97**(SA4), 425.
8. MATOSSIAN, A. M., and GARABEDIAN, G. A. Virucidal action of sea water. *Am. J. Epidemiol.*, 1967, **85**, 1.
9. METCALF, T. G., and STILES, W. C. Survival of enteric viruses in estuary waters and shellfish. In 'Transmission of viruses by the water route' (ed. G. Berg). Wiley-Interscience, New York, 1967, p. 439.
10. RUSCHI, A., and BENDELLI, M. Richerche sulla inattivazione di enterovirus nell'acqua di mare. *Boll. Ist. sieroter. milan*, 1966, **45**, 525.
11. LEOGRANDE, G., and IANDOLO, E. Effect of seawater on survival of virulent and attenuated poliovirus. *Ig. mod.*, 1969, **62**, 705.
12. AKIN, E. W., HILL, W. F., JR., CLINE, G. B., and BENTON, W. H. The loss of poliovirus 1 in marine waters (*to be published*).
13. JANNASCH, H. W. Growth of marine bacteria at limiting concentrations of organic carbon in sea water. *Limnol. Oceanogr.*, 1967, **12**, 264.
14. AKIN, E. W., and PRESNELL, M. W. Poliovirus 1 and coliform bacteria in estuarine water: survival in a large volume flow-through system (*to be published*).
15. KESTER, D. R., DUEDALL, I. W., CONNORS, D. N., and PYTKOWICZ, R. M. Preparation of artificial seawater. *Limnol. Oceanogr.*, 1967, **12**, 176.
16. CHANG, S. L. Statistics of the infective units of animal viruses. In 'Transmission of viruses by the water route' (ed. G. Berg). Wiley-Interscience, New York, 1967, p. 219.
17. POHJANPELTO, P. Oxidation in thermoinactivation of poliovirus. *Virology*, 1962, **16**, 92.
18. YOUNGER, J. S. Thermal inactivation studies with different strains of poliovirus. *J. Immun.*, 1957, **78**, 282.
19. LUND, E. Oxidative inactivation of poliovirus at different temperatures. *Arch. Virusforsch.*, 1963, **13**, 375.

20. HALSTED, C. C., SETO, D. S., SIMKINS, J., and CARVER, D. H. Protection of enteroviruses against heat inactivation by sulphydryl-reducing substances. *Virology*, 1970, **40**, 751.
21. LUND, E. Inactivation of viruses. In 'Progress in water technology. Vol 3—Water quality: management and pollution control problems' (ed. S. H. Jenkins). Pergamon Press, Oxford, 1973, p. 95.
22. DIMMOCK, N. J. Difference between the thermal inactivation of picornoviruses at 'high' and 'low' temperatures. *Virology*, 1967, **13**, 338.
23. CLIVER, D. O., and HERRMANN, J. E. Proteolytic and microbial inactivation of enteroviruses. *Wat. Res.*, 1972, **6**, 797.
24. LIU, O. C., SERAICHEKAS, H. R., AKIN, E. W., BRASHEAR, D. A., KATZ, E. L., and HILL, W. F., JR. Relative resistance of twenty human enteric viruses to free chlorine in Potomac waters. *Proc. 13th Wat. Qual. Conf.*, University of Illinois, Urbana, 1971, p. 171.

Introducing the paper Dr. Akin said that since enteric viruses had been shown to survive significantly longer in distilled, tap, and fresh surface waters than in marine waters, a virucidal property had been claimed for the latter. Various factors—including marine micro-organisms, trace heavy metals, and sea salts—had been credited with imparting this property. However, no virucidal agent common to all marine waters had been isolated, and the literature was inconclusive and seemingly contradictory on this point.

Experiments at the Authors' laboratory with marine water from the Gulf of Mexico had failed to reveal any phenomenon other than true virion inactivation to explain the loss of infectivity of viruses suspended in this medium, but the lack of reproducibility and the variability of the deleterious effects seemed to preclude the classification of marine water as a virucide. The fact that investigators had isolated enteric viruses from coastal waters and shellfish up to several kilometres from the nearest pollution source would lend support to the view that marine water *per se* was not virucidal. Indeed, the deleterious effect on viruses had been shown to be significantly reduced when domestic sewage was added to the marine water in conjunction with the virus. It appeared that natural die-off could not be relied on adequately to eliminate the virus risk arising from the domestic use of polluted marine waters.

DISCUSSION

Prof. R. Mitchell (U.S.A.) said that recent work in his laboratory supported Dr. Akin's observations on the protective action of marine silt on viruses; low concentrations of pure clays were shown to provide a high level of protection for viruses. However, he disagreed with the conclusion that the viruses were carried into the sediments—they could be attached to either biological or abiotic colloids, remain in suspension, and ultimately find their way to shellfish beds. Prof. G. Berg (U.S.A.) remarked that for several years he had been trying to recover viruses from natural fresh and saline waters, and had consequently found more viruses in the suspended solids than in the waters themselves. He believed that the adsorbed viruses acquired some degree of protection from the environment. There was clearly protection on the side where the virus was adsorbed to the solid matter, and absorption to larger particles might result in viruses being engulfed by small organisms. In view of this adsorption, Mr. J. B. Kirby (U.K.) asked whether one would not expect viruses to be present in coastal storm-sewage discharges which tended to contain sediments scoured from the sewers.

Prof. H. I. Shuval (Israel) agreed with Dr. Akin that further study of marine viral inactivation should be made—but for reasons of scientific curiosity rather than of relevance to outfall design. The inactivation in the sea was much too slow to be of engineering consequence. Nevertheless, a specific marine bacterium, isolated in Prof. Shuval's laboratory, had been shown to produce anti-viral effects; they had not yet been able to isolate the chemical or enzymic constituent that was effective in the absence of the living bacteria, but recent data suggested that the viruses might adsorb to the living cell and not appear as infective virus in tissue culture.

Prof. R. R. Colwell (U.S.A.) said that the synergistic effects of many of the factors affecting micro-organisms in the marine environment needed to be emphasized. For example, data were accumulating which suggested a high incidence of tumours of fish and shellfish near sewage outfalls. Experiments to investigate co-carcinogenicity—employing as co-carcinogens the polyaromatic hydrocarbons resulting from microbial action on crude oil and a known virus system—had been started at her laboratory. Carcinogenicity of chemical compounds of sewage, and the presence of human oncogenic viruses in sewage discharge from sea outfalls, were matters of great human health significance; the persistence of the chemicals and the survival of the viruses were unknown, and it might well be that examination of these problems was of higher priority than some of the others discussed at the Symposium.

REPLY TO DISCUSSION

In response to Prof. Mitchell, Dr. Akin said that his statement on the removal of viruses from marine waters by sedimentation in the presence of turbidity had not been a general conclusion, but referred only to the experimental results he had shown in one of his slides where over 99 per cent of the poliovirus had been removed from suspension with centrifugation at 7000 *g* in the presence of a turbidity of 150 JTU. If the virus were adsorbed to particulate matter in the colloidal size range, settling out would obviously not be a significant factor in the removal of the virus from the water. In reply to Dr. G. E. B. Kullenberg (Denmark), who here interjected a question of the relation between JTU and solids concentrations, he regretted he could not give the corresponding figure for concentration. The Jackson turbidity units had been determined by a transmitted-light method, whereas the recommended method[25] was now a scattered-light one; standardization of the measurements of this parameter was badly needed.

In reply to Prof. Berg he said that the viruses did attach to many types of particles and surfaces, their adsorption being greatly enhanced by the presence of cations—especially multivalent ones such as Mg^{2+}, Ca^{2+}, and Al^{3+}. Sea water had a sufficient concentration of these cations to assure a high adsorption level unless interfered with by proteinaceous material in polluted water. He did not know if there would be enough residual virus in sewer deposits to pollute storm sewage significantly, as Mr. Kirby had suggested. Comparative tests on the viral content of dry-weather and storm sewage would be needed to determine whether a potential virus hazard existed.

Replying to Prof. Shuval, he said that the intriguing and puzzling question concerning the fate of viruses in marine water was the diversity of findings reported in the literature. The Authors had found that heating and filtering water from the Gulf of Mexico did not remove its deleterious effect on the viruses, which was contrary to Prof. Shuval's findings[6] with waters from the Mediterranean and the Red Sea. He could not explain these differences, and considered that this area of research should be pursued because there was clearly some ingredient of sea water that had a deleterious effect on viruses.

He was pleased to learn from Prof. Colwell that someone was looking into the synergistic aspects of the effects of water pollution. Knowledge in that area was needed to assess the health risk to man.

ADDITIONAL REFERENCE

25. A.P.H.A., A.W.W.A., and W.P.C.F. Standard methods for the examination of water and wastewater. American Public Health Association, Washington, 13th Edn, 1971, p. 350.

Paper No. **25**

Factors Influencing the Survival of Enteric Microorganisms in the Sea: An Overview

R. Mitchell and C. Chamberlin
Harvard University, Cambridge, Massachusetts

Within the last 25 years, researchers from a variety of disciplines have attempted to unravel the complex problem presented by the rapid decline of enteric microbial populations in the sea. Investigation of this problem both in the laboratory and in the field led this diverse group to consider sedimentation, solar radiation, predation, bacteriophages, nutrient deficiencies, algal toxins, bacterial toxins, and physico-chemical factors as possible causes of the decline or 'die-off'. However, in the flood of research generated by this question, the rationale for the effort has often been lost in the background. As a result, laboratory studies on single factors have frequently been conducted in isolation from the field studies providing definition of the field phenomena.

Such conceptual separation between laboratory and field work would certainly be avoided if estimates of the significance of each factor were based on the factor's relative contribution to the die-off of enteric bacteria in the marine environment. This is not to suggest that many questions that have arisen in the course of these studies are not intriguing or lacking in scientific interest on their own merits. However, the central issue of the bulk of these studies remains: which factors are responsible for the observed rapid die-off of enteric bacteria in the sea?

In attempting to answer this question, a mathematical and conceptual framework for interpreting and comparing laboratory and field results would serve as a valuable tool. The framework utilized in this brief paper will be a multi-factor model of die-off and dispersion of enteric bacteria in the sea. An outline of the development of the model will be presented partly as a brief survey and summary of relevant literature, and partly as a functional analysis and subsequent formulation of the model itself. In the final section, the model will be used to evaluate quantitatively the relative contribution of the various factors in producing the observed die-off and to draw some tentative conclusions. Other applications of the model will also be discussed briefly.

Consideration will also be given to the problems of viral inactivation in the sea. Since this field is less well developed only a summary of the current literature will be attempted.

Development of Model

Discussion of modelling

No argument will be made that the model presented here has any special nobility or that it is more 'real' than any of the innumerable other models that could be proposed on the basis of the same literature. The specifics of the model to be presented reflect its intended use as a tool in interpreting and comparing laboratory and field studies and in describing and, hopefully, predicting the phenomena observed in the natural system. In Levins's terminology[1], precision has been sacrificed to generality and realism in this model.

Before proceeding, there are a few frequently confusing points worth considering. First, the literature contains references to *Escherichia coli*, coliforms, and fecal coliforms. Although it is recognized that these terms are not interchangeable, the discussion to follow will not strive to distinguish between these groups since the factors seem to produce comparable effects on each. Use of the term enteric bacteria is intended to include all bacteria of interest, pathogens and indicators alike. Secondly, the coliform population decline observed in the field and in the laboratory is often called die-off. It should be recognized that die-off may encompass processes other than actual loss of viability, e.g. sedimentation and aggregate formation (reducing the apparent number of cells). Finally, it should be emphasized that it is the die-off of enteric bacteria in the mixture of sea water and sewage (in contrast to unpolluted sea water) that is of general concern.

With those points in mind, let us begin by considering each suggested factor individually. The following discussion will be restricted to three specific issues:

(1) Has it been demonstrated by laboratory study that this factor influences die-off rates?
(2) Do the results of relevant field studies indicate that this factor is probably of general significance?
(3) What mathematical expression of the influence of this factor seems most appropriate?

Factors affecting viral inactivation

Surprisingly little can be stated with certainty about the inactivation of enteric viruses in sea water. In the review by Akin *et al.*[2] the inconsistencies in the capacity of various treated and untreated natural and artificial sea waters to inactivate enteric viruses are very evident. Clearly, the effect depends on viral type, temperature, and chemical and biological composition of the sea water. Effects resulting from particle aggregation, adsorption to walls, and resistant population fractions seem negligible. Increased temperature consistently produces an accelerated rate of inactivation although the mechanism of the inactivation is completely unknown. Marine microorganisms appear to be important in inactivating viruses in some instances[3] and not in others[2]. Suspended organic material has been shown to provide a protection for viruses[3] that soluble organic material not only does not provide[4] but in some cases has been shown to increase the inactivation rate[2].

No *in situ* investigations have been reported which suggests an obvious and very necessary extension of these laboratory studies. The significance of sedimentation and solar radiation do not seem to have received appropriate attention.

Given the complex and ill-defined nature of viral inactivation in sea water, it seems premature to propose either a conceptual or mathematical model. Such an effort must await further research.

Factors affecting bacterial die-off

Sedimentation

Speculation that sedimentation was involved in the rapid decline of coliforms in the surface waters of the wastewater plume has been a recurrent but minor theme of reviews on die-off for the past 35 years[5], but surprisingly little research has been forthcoming. Since removal efficiencies of conventional primary sewage treatment processes indicate that generally 50 and sometimes up to 75 per cent of coliforms in sewage are associated with particles with sizable settling velocities ($>0{\cdot}05$ cm/s), it seems reasonable to suggest that with the added effects of flocculation by the sea water and adsorption by marine sediments and estuarine silts, sedimentation might be responsible for a significant fraction of the observed removal. These tendencies would be opposed by vertical mixing primarily due to wind, and by bottom scour and consequent resuspension of settled bacteria.

In the relatively limited investigation of sedimentation, several researchers[6–8] have demonstrated that clays have a higher adsorptive capacity for bacteria than silts or sands. Weiss[8] went on to show that the degree of adsorption is a function of particle size and composition and that the general range of turbidity found in natural waters was high enough to produce significant removal of *E. coli*. However, sea water reduced the adsorptive capacity of silts and may have caused desorption of some bacteria. Orlob[9] attempted to evaluate directly the effect of sedimentation by adding settled sewage to two sea-water samples (1:1000 dilution) and periodically determining the number of surviving units forming coliform colonies. One sample was vigorously agitated prior to testing while the other was left undisturbed. An abrupt decrease in bacterial numbers occurred in the undisturbed sample in contrast to an initial increase in the agitated sample. At all times the agitated sample had higher bacterial concentrations, suggesting that sedimentation might be a significant factor. A series of sedimentation tests has been reported by Gunnerson[10] in which various sea-water and sewage dilutions were centrifuged for 30 min at 1000 *g*. Primary effluent mixtures showed greater reductions than secondary effluent with greater reductions at higher dilutions.

Supplementing these laboratory investigations, the results of several field studies also indicate that sedimentation may be a significant factor. Both Nusbaum and Garver[11] and Rittenberg *et al.*[12] argue that the occurrence and areal distribution of coliform fields in the sediments surrounding marine sewage outfalls demonstrates the significance of sedimentation as a major factor in coliform reductions. More directly, the considerable differences between rates of coliform reduction for primary and secondary effluents observed in the Hyperion Outfall field studies[13,14] strongly suggested the significance of sedimentation. Assuming that the sedimentation characteristics were the only influences on the differences between the observed removal rates, it was possible to derive the exponential removal rate-constants due to sedimentation for primary effluent and secondary effluent: 0·43 and 0·11 per h respectively.

It seems evident that adsorption and sedimentation generally are significant factors in the removal of coliforms from surface waters but that their relative importance will depend on the degree of treatment and the general nature of the sewage, and the magnitude of vertical dispersion processes. A first effort in modelling this process might be directed toward inclusion of a sedimentation rate and a vertical dispersion coefficient in the physical modelling of the sewage plume itself. Since the physical skeleton of the model is presented below, further discussion of modelling of this factor will be deferred until then.

Solar radiation

Sunlight has long been known to have a lethal effect on coliform and other bacteria[15] but most of the information gathered has concerned ultraviolet light and not the complete spectrum. Although Gaarder and Spärck[16] considered sunlight to contribute most significantly to the destruction of bacteria in sea water in their study on oyster pools, later investigators attributed minimal importance to sunlight in view of the high attenuation coefficients in sea water of ultraviolet light especially and visible light in general[9,17–19].

Consequently, few investigations on the effect of visible radiation on the survival of enteric bacteria in sea water have been reported, in spite of the fact that Hollaender[20] demonstrated that near-visible and visible light was lethal to *E. coli* and that the effect was enhanced by suspension in physiological saline. Subsequent studies, recently reviewed by Harrison[21] and Eisenstark[22], confirmed that such light was lethal to a great many bacteria, and also revealed that the lethal effect was greatly enhanced by the addition of photosensitizers such as fluorescent dyes and exposure to oxygen. Absence of light, photosensitizer, or oxygen greatly reduces the rate of kill. Sistrom, Griffiths, and Stanier[23] and Dworkin[24] have further shown that the photosensitizer may be endogenously produced and that the primary function of carotenoid pigments in bacteria may be to prevent photodynamic sensitization by endogenous compounds such as bacterial chlorophylls. The prevalence of pigmented strains among airborne and marine bacteria in surface waters would tend to support this. Bitton *et al.*[25] have also shown that adsorption to clays and humic acids may protect bacteria against ultraviolet radiation.

With the single exception of the work by Reynolds[26] that strongly suggested the influence of light on the survival of *E. coli* in sea water, serious consideration of this factor was initiated by Gameson and Saxon[27] in a series of submerged light and dark bottle experiments that demonstrated the dependence of die-off rate on light intensity. These and subsequent studies have already been discussed at this Symposium[28]. It appears clear from the submerged bottle experiments and the *in situ* experiments that solar radiation is a very significant factor. Experiments carried out totally or in part in darkness very convincingly suggest the marked influence of solar radiation on *in situ* mortality.

Several relationships observed in these experiments[28] are involved in the mathematical expression of the influence of solar radiation on *E. coli* die-off: the die-off rate is related to depth with an effective attenuation coefficient of about 0·22 m^{-1}; the rate is proportional to light intensity; and the die-off is essentially first order with respect to coliform concentration. This relationship could be formulated as:

$$\frac{dC}{dt} = -kI_0 e^{-\alpha z} C \qquad (1)$$

where C is the concentration of coliform bacteria at time t and depth z, k a proportionality coefficient, I_0 the light intensity just below water surface (generally a function of time and latitude), and α the effective attenuation coefficient.

Predation

Biological factors of some sort have been known to influence the survival of enteric bacteria in sea water since de Giaxia's early investigations[29]. Predators, bacteriophage, competition for nutrients, algal toxins, and bacterial toxins have all been proposed as the causal agent. Each factor will be considered individually since more than one may be significant.

Orlob[9] presented data indicating that the bactericidal agent or property associated with sea water was diminished by boiling, autoclaving, filtration, pasteurization, and chlorination. Increased temperature increases the rate of decline and shortens the lag phase. Carlucci *et al.*[30] showed that *E. coli* survived longer in filtered than in raw sea water. However, the effect was inconsistent. Using double-layer-plate predation assays, marine bacteria capable of killing *E. coli* in artificial sea water have been isolated by Mitchell and Nevo[31]. Mitchell, Yankofsky, and Jannasch[32] found that the rate of *E. coli* kill in natural sea water was proportional to the size of the marine microflora. Three groups of microorganisms were associated with the kill: (1) cell-wall lytic marine bacteria, (2) marine amoebae primarily of the genus *Vexillifera*, and (3) marine bacterial parasites similar to *Bdellovibrio bacteriovorus*. This last group has been repeatedly isolated from soil, sewage, and sea water; the organisms are capable of passing through a 0·45-μ pore size filter, and this might explain the inconsistent results of filtration on observed rate of decline. Repeated inoculation of *E. coli* in sea water has been shown by Mitchell[33], Mitchell and Morris[34], and Moebus[35] to increase the rate of decline. Moebus has further shown that periodic re-inoculation of *E. coli* produces an increase in marine bacteria by a factor of ~ 100 and maintains this level as long as the inoculations are continued.

In a series of plume field studies, Foxworthy and Kneeling[36] tried to assess the influence of biological processes on die-off rates. It was suggested that if the die-off were strongly influenced by naturally occurring biological agents then there should be a seasonal variation in the initial rate of die-off at a given site and variation in initial die-off rates should be related to initial dilution rate of wastewater with background sea water. Marked seasonal variation in initial rate of change was clearly demonstrated, but the die-off rate was not shown to be related to initial dilution rate.

Although it seems probable that predation is a significant factor in laboratory studies, its significance in the field has been evaluated only indirectly with generally inconclusive results. At this stage, it would be useful to include the factor in the model for purposes of evaluation.

Empirical and theoretical aspects of predator–prey models are discussed in detail by May[37] and Canale[38]. Theoretical considerations confirmed by several laboratory studies[39–41] suggest a saturation predation model of the form:

$$\frac{dC}{dt} = \frac{-\lambda_m C}{(K_C + C)\,Y_C} P \qquad (2)$$

where P is the predator concentration, λ_m the maximum predation rate, K_C the half-saturation constant (value of C producing half-maximum predation rate), and Y_C the yield of predators per unit prey.

Bacteriophages

Suggestions that bacteriophages contribute to enteric bacterial kill in sea water have been around for a long time[42]—as have counter-arguments that they occur too sporadically to be considered important factors[43]. The results on sterilization by Orlob[9] would also support this. Carlucci and Pramer[44] made the only serious attempt to investigate this factor. They found that bacteriophages active against *E. coli* and other bacteria could be isolated without difficulty from natural sea water by enrichment procedures. However, coliphages were found to be rapidly inactivated by natural sea water but not by either heat- or filter-sterilized sea water. In addition, coliphages were only effective in killing *E. coli* when sufficient nutrients were present in the sea water to support growth and multiplication of host cells. In the absence of any reported determinations of unenriched coliphage levels in sea water even the kill observed may not be relevant to the field phenomenon due to the very high inocula of coliphage ($\sim 10^5$ units/ml) used in these experiments.

From the sparse data available, the indications are that bacteriophages are of minor significance in influencing survival of enteric bacteria in sea water.

Nutrient deficiencies

The mixture of sea water and sewage making up the plume from the outfall may be considered a dilute culture medium in which some of the most important nutrients (specifically carbon, nitrogen, and phosphorus) are present in very low concentration. In many cases, additions of nutrient supplements to sea water have decreased the rate of die-off of enteric bacteria and/or stimulated bacterial growth. Vaccaro *et al.*[45], Orlob[9], Carlucci and Pramer[46], Won and Ross[4], and Moebus[47] showed that inorganic nutrients additions such as $(NH_4)_2SO_4$ and $(NH_4)_2PO_4$ and organic supplements such as peptone, lactose broth, sewage volatile solids, and others produced population growth or reduced die-off rates.

However, the conclusions of several of these studies are called into question by Jones's findings[48] that *E. coli* would not grow at 37°C either in natural sea water supplemented with glucose, NH_4Cl, and K_2HPO_4 or in synthetic sea water that had been filter-sterilized. This inhibition could be reversed by autoclaving, by addition of 0·01 per cent or less organic matter (e.g. casein hydrosylate, peptone, or yeast extract), by addition of 10^{-5} M cysteine, by addition of organic chelating agents (e.g. histidine, glycine, methionine, and EDTA), or by addition of metal complexing agents (e.g. $Na_2S_2O_3$, Na_2S, NaCN). Jones concluded that the toxicity was caused by heavy metals, and demonstrated that very low levels of heavy metals could produce the effect. Scarpino and Pramer[49] reached the same conclusion in their more restricted study on the effect of cysteine on *E. coli* survival in sea water. It had also been shown earlier by Carlucci and Pramer[46] that glucose alone had no significant effect on the survival of *E. coli* in sea water. But these and later studies[4,47,50] on the effect of nutrient supplements either to natural or to filter-sterilized sea water have repeatedly shown reduced rates of die-off and sometimes population increases. The fact that several of the studies showed that the addition of low amounts of sewage volatile solids or fecal material reduced die-off rates or caused population increases suggests that if chelation of toxic heavy metals is involved it is almost totally achieved at relatively low levels of organic pollution[50]: 1–10 mg/l. BOD. Carlucci and Pramer[46] also showed that $(NH_4)_2SO_4$ or $(NH_4)_2HPO_4$ had a similar effect.

Several additional complications arise here. First, the marine microflora compete for the same nutrients as the enteric bacteria. Chemostat studies by Jannasch[51] have demonstrated that *E. coli* can compete successfully with marine bacteria for nutrients in sea water only at high growth rates and high substrate levels. Restated, *E. coli* generally has a lower affinity than marine bacteria for organic substrates in sea water. Second, McGrew and Mallette[52] and Marr *et al.*[53] have shown that bacteria consume carbon and energy sources for purposes of maintenance in addition to growth. Although this specific maintenance requirement is relatively small, it is not a function of growth rate, so that at low growth rates it can form a major fraction of carbon and energy requirements. The results of Savage and Hanes[50] suggest that this concept is relevant to the growth and death of enteric microorganisms in the sea. Finally, whatever predators are present in the medium might preferentially utilize the supplemental nutrients instead of the enteric bacteria and thus relieve one pressure on the bacterial population.

With regard to applicable field studies, Foxworthy and Kneeling[36] found that in roughly half the field experiments the *E. coli* concentrations corrected for dilution rose initially before beginning to die-off. The tendency for initial growth was correlated strongly with season and water temperature.

These results coupled with the laboratory findings tend to suggest a significant role for nutrients in determining survival of enteric bacteria in sea water. The relationship between nutrient concentrations, growth rates, and bacterial concentration can be expressed by the formulation developed by Pirt[54]:

$$\frac{dC}{dt} = \left(\frac{\mu_m S}{K_S + S} - a\right)C \tag{3}$$

where μ_m is the maximum growth rate, S the substrate level, K_S the half-saturation constant, and a is a specific maintenance constant. Certain points about this expression may require further explanation. A Monod substrate-uptake model has been used which assumes saturation growth kinetics with regard to the given substrate. A single substrate variable has been assumed which in practice might be total organic carbon, chemical oxygen demand, biochemical oxygen demand, or some other complex function.

Algal toxins

Anti-bacterial toxins produced by planktonic algae have attracted increasing attention since Sieburth[55,56] demonstrated the antibacterial capabilities of Antarctic phytoplankton. Subsequent research has revealed numerous and widespread planktonic algal species which produce antibacterial substances under experimental conditions[57–61]. Efforts by Jørgensen[62], Sieburth[63], Saz *et al.*[64], and others[60] to isolate and characterize the compounds involved have revealed a wide variety of materials, generally large heat-labile molecules. One was a chlorophyllide that was active only during illumination.

Since laboratory results do not necessarily imply actual ecological importance, the observed correlations between the life cycles of phytoplankton communities and seasonal changes in antibacterial activity of sea water reported by Sieburth and Pratt[65] and Moebus[66] assume special importance. In general both studies found that the most pronounced antibacterial activity of untreated sea water was associated with breakdown of phytoplankton blooms. The diatom *Skeletonema costatum* was implicated in both studies. However, Moebus[67] showed that filter-sterilized sea water was less effective than raw sea water and did not vary significantly with season, and also that marine bacteria showed the greatest increases when the antibacterial activity of the sea water was greatest. From these results, Moebus concludes that the antibacterial capacity of sea water depends considerably on the availability of nutrients (primarily organic matter) which in turn is dependent on phytoplankton development. Another hypothesis consistent with the data at hand suggests itself: as the phytoplankton reach the peak of the bloom, microbial predators also approach the peak of their bloom in response to the algal growth and peak during the rapid reduction in algal numbers following the bloom, with the predators serving as the antibacterial agent. This is in keeping with work by Mitchell[68] and Canter and Lund[69].

Before proceeding, it should be noted that in both studies, the incubation of samples was in the dark. If, as suggested by Jørgensen[62], the chemicals released by the algae serve as photoactivators in photodynamic kill as in the discussion above, the experimental technique would not reveal their presence.

In field studies, Foxworthy and Kneeling[36] were unable to determine any relationship between phytoplankton activity and coliform survival.

Based on the studies available, the significance of algal toxins in influencing the survival of enteric microorganisms in sea water is unclear and will not be included in the model.

Bacterial toxins

Since deGiaxa[29] demonstrated that some marine bacteria inhibited the growth of *Vibrio comma*, it has been speculated that the rapid death of bacteria in the sea was to some extent due to an indigenous microflora of antibiotic producing organisms. Rosenfeld and ZoBell[70] tested 58 species of marine bacteria and found 9 that were antagonistic to colony formation by various non-marine bacteria on agar media. When grown in sea-water nutrient broth, however, little or no antibiotic was produced by these same organisms. In a series of similar experiments, Krasil'nikova[71] also found that despite the large number of antibiotic-producing organisms isolated from sea water, little if any antimicrobial activity was present in natural sea-water cultures of these organisms. Carlucci and Pramer[72] were unable to demonstrate any antibiotic production effective against either *E. coli* or *Bacillus subtilis* in tests of 200 marine bacteria isolates.

On the basis of these studies, there appears to be no evidence that antibiotics are produced under natural conditions by marine bacteria, or that they contribute to death of enteric bacteria in the sea.

Physicochemical factors

Physicochemical factors such as osmotic effects, pH effects, and specific ion toxicity (e.g. NaCl, iodate, perchlorate, and heavy metals) have been proposed as possibly significant factors in the die-off of enteric bacteria in sea water. Several studies[11,73–75] revealed no significant antagonism to enteric bacteria due to normal saline constituents of sea water. In contradiction to these conclusions, Carlucci and Pramer[46] found that the survival of *E. coli* was comparable in natural sea water and in NaCl solutions of equal salinity and generally varied inversely with salt concentration, and also that survival was favored by more acid conditions (pH < 8) in both solutions. In a later study[30] artificial sea water was found to be at least as bactericidal as natural sea water.

Heavy metals were implicated as important factors by Jones[48] who demonstrated that *E. coli* growth was inhibited in natural, artificial, or filter-sterilized sea water and that the inhibition could be reversed by autoclaving or by addition of chelating agents. He also presented data indicating that iodate was probably not the inhibitory factor and that the toxicity probably arose from trace impurities in reagent-grade chemicals, especially in NaCl (which reflects on Carlucci and Pramer's study[46]). This and later work[76] clearly indicate inhibition of *E. coli* growth on either natural, artificial, or filter-sterilized sea water supplemented with glucose, NH_4Cl, and K_2HPO_4. However, this growth inhibition is not necessarily applicable to the killing of *E. coli* in the mixture of sea water and sewage of the wastewater plume (see discussion above on nutrient deficiencies). Cells unable to grow in a given medium may still remain viable for extended periods. Since *E. coli* is generally more rapidly killed in natural than in filter-sterilized sea water, the presence of an agent more actively antagonistic than heavy metals seems probable. Seasonal variations also seem to argue against these factors.

Although physicochemical factors in sea water do not appear to be of primary significance, osmotic pressure, pH, and heavy-metals concentrations do not favor the survival of *E. coli* in sea water and will contribute to some extent to their rapid death. If this factor is assumed to be uniformly distributed in sea water and to be unconsumed in its action, the influence may be described by a simple Chick's Law relationship:

$$\frac{dC}{dt} = -rC \tag{4}$$

where r is the die-off rate due to physicochemical factors.

Formulation of model

Various models of enteric bacterial die-off and dispersion have been developed over the last three decades, ranging from deterministic models[77,78] that attempt mechanistically to relate specific physical and biological factors to the observed concentration distributions, to stochastic or statistical models[79,80] that are concerned with developing hopefully precise descriptions of the observed concentration distributions as functions of the relevant physical and biological parameters with little attention given to the mechanisms involved. Since the model developed here is intended as a mathematical and conceptual framework for relating laboratory and field results, a model explicitly including the factors considered to significantly influence the dispersion and population variations of the enteric bacteria (i.e. a deterministic model) seems most appropriate.

Summarized from the conclusions reached in the discussion above, the processes and factors generally and significantly influencing the concentration distribution of enteric bacteria in the far-field of a steady-state wastewater plume are horizontal advection, sedimentation, lateral dispersion, vertical dispersion, solar radiation, predation, nutrient deficiencies, and physicochemical factors. The model developed includes the explicit formulations developed above for these individual factors and makes several simplifying physical assumptions:

(1) Diffusion is much less significant than dispersion and is neglected.
(2) The plume is in steady state.
(3) The current has a constant mean velocity.
(4) Longitudinal dispersion is much less significant than lateral or vertical dispersion and is neglected.
(5) The lateral dispersion coefficient is constant.
(6) The vertical dispersion coefficient is constant.

These assumptions are quite similar to those made in other models[78] and have led to reasonable conclusions in those models. With these assumptions the resulting partial-differential equation governing the system becomes:

$$U\frac{\partial C}{\partial x} - V_s\frac{\partial C}{\partial z} = \epsilon_y\frac{\partial^2 C}{\partial y^2} + \epsilon_z\frac{\partial^2 C}{\partial z^2} + \left(\frac{\mu_m S}{K_S + S} - a\right)C - \left(\frac{\lambda_m P}{(K_C + C)Y_C}\right)C - (kI(t)e^{-\alpha z})C - rC \quad (5)$$

where C is the concentration of bacteria in the x, y, z, t spatial/temporal coordinate system, U the mean current velocity and V_s the sedimentation velocity of sewage particles, ϵ_y, ϵ_x are the dispersion coefficients in the indicated directions, μ_m is the maximum coliform growth rate (a function of temperature), K_S the half-saturation constant for the concentration, S, of nutrient, a the endogenous respiration rate, λ_m the maximum predation rate (a function of temperature), K_C the half-saturation constant for C, P the concentration of predators and Y_C the yield of predators on bacteria, k the die-off rate due to light, $I(t)$ the solar radiation intensity (a function of time), α the attenuation coefficient of light in the plume, and r the die-off rate due to physicochemical factors.

As the equation stands, no analytical solution exists. However, solutions are available for certain restricted cases. Thus, if the last four terms are omitted, the solution will represent the isolated effect of sedimentation. Using the solution developed by van der Hoven[81] in his consideration of the deposition of particles from an atmospheric plume, and the ratio of the deposition of particles in surface ($z = 30$ cm) concentration between a sedimentary material (e.g. sewage particles) and a neutrally buoyant material (e.g. dye tracer), an expression for effective die-off rate due to sedimentation can be derived:

$$\frac{dC}{dt} = -\left(\frac{V_s^2}{4\epsilon_z}\right)C. \quad (6)$$

This expression should not be confused with those developed for other factors since it is derived from a restricted solution of the model and is not derived from specific laboratory or field observations.

Application of Model

Interpretation of laboratory results

Let us begin applying this model by examining with its framework the relative significance of the various factors discussed above. It will first be necessary to review the results of *in situ* plume experiments so that the nature and magnitude of the field phenomena can be brought to bear in interpreting laboratory and restricted field studies, since it is our basic assumption that description and prediction of the field phenomena is the central issue.

In situ plume studies are conducted by introduction of a conservative tracer (e.g. radioactive isotope or fluorescent dye) into the sewage prior to discharge into the sea, and by subsequent sampling and analysis in the plume both for tracer material and enteric bacteria. Tracer concentrations are then used to correct bacterial concentrations for effects due solely to dilution. The precise procedure followed varies from study to study but two general approaches are most often used. In the first, the tracer–sewage mixture is continuously pumped into the sea for a sufficient period for an approximately steady-state plume to have become established. Concentrations of tracer and bacteria in the area where the sewage first reaches the surface (generally called the boil) are taken as the initial conditions with concentrations at various down-current positions in the plume translated from distance to time coordinates assuming constant current velocity. In the second approach, the tracer–sewage mixture is released over a relatively brief period so that samples taken from the dispersing 'cloud' have a straightforward time coordinate. Both cases can be mathematically described by the techniques employed in atmospheric plume modelling.

By a variety of techniques, the results of these studies are used to estimate die-off rates, reported generally in terms of T_{90} (the time required for the dilution-corrected enteric bacterial level to decrease to 10 per cent of its initial value) or k_e (the exponential die-off rate constant). T_{90} and k_e are related by the simple function:

$$k_e = \frac{\ln 10}{T_{90}}. \tag{7}$$

The data analysis techniques generally employed have been criticized by Harremoës[82]; by using a mass-balance approach and re-evaluating data from Foxworthy and Kneeling[36], he shows that the conservative tracer was in fact disappearing at a significant rate: 'T_{90}' for the dye of 500 min. Because of this source of error, Harremoës suggests that the procedures for measuring physical dilution and also bacterial concentrations might explain the bulk of variability in die-off data.

In Table 1, results from 69 separate *in situ* plume studies are summarized; T_{90} as observed in the field is on the order of 2 h and ranges from about 30 min to 8 h. Likewise, k_e was found to be approximately lognormally distributed with a geometric mean of 1·0 h^{-1} and a total range from 0·3 h^{-1} (in an experiment conducted partially during darkness by Gameson and Gould[28]) to 5 h^{-1}. Thus it would seem reasonable to doubt the primary significance of a factor unable to produce die-off rates within this range. This will provide a simple yardstick to evaluate the significance of the various factors under discussion.

TABLE 1. Summary of field study results for bacterial die-off rates
μ is mean value for n determinations with standard deviation σ

Location	Date	Previous sewage treatment	T_{90} (h)	k_e (h^{-1})
Santa Monica, California[14]	1954–56	Secondary	6·5	0·354
		Primary + chlorine	3·8	0·606
		Primary	4·0	0·576
		Primary	3·4	0·677
		Primary	1·5	1·53
Titahi Bay, New Zealand[83]	1959–60	None	0·65	3·54
Rio de Janeiro, Brazil[84]	1963	None	< 1·0	> 2·3
Tema, Ghana[85]	Aug. 1964	None	1·33	1·73
Dawlish, England[86]	1965	None	0·78	2·95
			3·5	0·66
Nice, France[87]	—	—	1·50	1·54
Denmark[88]	1968	—	2·00	1·15
Gentofte, Denmark[89]	—	None	1·16	1·98
Manila Bay, Philippines[89]	May 1969	—	2·16	1·06
			2·84	0·81
Santa Barbara, California[36]	1966–67	Primary	μ = 2·35 σ = 1·1 n = 37	0·98
Sidmouth and Bridport, England[28]	1969–73	None	μ = 3·7 σ = 2·5 n = 18	0·62

Table 2 summarizes the factors included in the model, the mathematical formulations, general influences on their magnitude, and an estimate of the maximum die-off rates under circumstances in which the factor would be most significant. For example, sedimentation would be most significant during conditions of minor turbulence (small ϵ_z) and minimal treatment (high mean V_s). The actual values of the rates are drawn from the literature discussed above.

It should be emphasized that these rates make no allowance for factor interaction. In fact, the interaction of certain factors might be expected to be very strong: sedimentation would clearly reduce the effect of solar radiation; polysaccharides produced by nutrient utilization might enhance sedimentation and predation; physicochemical effects might increase the susceptibility of bacteria to predation. How-

TABLE 2. Summary of factors in model

Factor	Mathematical representation	Sensitivity	Estimated maximum k_e (h^{-1})
Sedimentation[14]	$-\dfrac{V_s^2}{4\epsilon_z}$	Degree of treatment Turbulence	0·6
Solar radiation[28]	$-k\,I(t)\,e^{-\alpha z}$	Season Latitude Turbidity	4·0
Predation[33]	$\dfrac{-\lambda_m P}{Y_C(K_C + C)}$	Temperature	0·3
Nutrient utilization[51]	$\dfrac{\mu_m S}{K_S + S} - a$	Temperature Organic pollutants Degree of treatment	−0·6
Physicochemical effects[46]	$-r$	Temperature	0·15

ever, the imprecision introduced by neglecting these interactions does not significantly interfere with the crude evaluation intended. In addition, these rates are not intended to characterize the general relative significance of the factors. Certainly, the relative significance might depend on latitude, season, time of day, prior treatment, condition of the receiving water, and other factors. Nor is consideration given to influences due to petrochemical pollution, pesticide contamination, or pressure effects as discussed by Walker and Guarraia[90].

This crude analysis does provide some valuable generalizations. Solar radiation is probably primarily responsible for the extremely rapid die-off observed in some *in situ* studies. Physicochemical effects and predation are probably only rarely primary factors in bacterial die-off. Their main effect is to enhance the effect of other factors. Sedimentation and nutrient utilization seem to be significant factors varying with prior treatment and other modifying influences.

Other applications

Ranking and evaluating factors is one profitable application of the multifactor model proposed. However, the proposed multifactor model also provides the skeleton of a valuable predictive tool in evaluating sites for submarine outfall disposal of sewage. It would certainly be necessary to refine the formulation and validate the model via *in situ* plume studies but it would extend the sensitivity of site selection and permit a more rational site evaluation approach.

In addition, a modified model of this form might provide the theoretical foundation for evaluating coliforms as biological tracers for the organisms of true concern: enteric pathogens. If it is assumed that the same or similar general factors are active in reducing both coliform and enteric pathogen concentrations in the sea, but that the significance of each factor might vary, it would be possible to determine the system parameters for the specific pathogens of interest. Parallel models might be developed for the organisms, thus taking one more step in the direction of reality.

Conclusions

Based on the extensive laboratory and field studies that have been carried out over the past several decades, it has been possible to develop a mathematical and conceptual framework for interpreting and comparing experimental results on the die-off of bacteria in the sea. The work presented in this Symposium and elsewhere attests both to the scope and to the detail of the multi-disciplinary effort in this field that ranges from concern with viral inactivation to hydrodynamic modelling. Drawing upon these diverse areas, it has seemed possible to deduce the potential significance of various factors that have been investigated. It is proposed that solar radiation, sedimentation, and nutrient-related effects account for the bulk of the observed die-off. It is further proposed that this or similar models might prove useful in outfall site evaluation and more refined consideration of enteric pathogen die-off in the sea.

ACKNOWLEDGEMENT

This work was supported in part by Grant No. R800433 from the U.S. Environmental Protection Agency.

REFERENCES

1. Levins, R. Strategy of model-building in population biology. *Am. Scient.*, 1966, **54**, 421–431.
2. Akin, E. W., Hill, W. F., and Clarke, N. A. Mortality of enteric viruses in marine and other waters. This volume, pp. 227–236.
3. Mitchell, R., and Jannasch, H. W. Process controlling virus inactivation in seawater. *Envir. Sci. Technol.*, 1969, **3**, 941–943.
4. Won, W. D., and Ross, H. Persistence of virus and bacteria in seawater. *J. envir. Engng Div. Am. Soc. civ. Engrs*, 1973, **99** (EE3), 205–211.
5. Waksman, S. A., and Hotchkiss, M. Viability of bacteria in sea water. *J. Bact.*, 1937, **33**, 389–400.
6. Rubentschik, L., Roisin, M. B., and Bieljansky, F. M. Adsorption of bacteria in salt lakes. *J. Bact.*, 1936, **32**, 11–31.
7. Waksman, S. A., and Vartiovaara, U. The adsorption of bacteria by marine bottoms. *Biol. Bull.* 1938, **74**, 56–63.
8. Weiss, C. M. Adsorption of *E. coli* on river and estuarine silts. *Sewage ind. Wastes*, 1951, **23**, 227–237.
9. Orlob, G. T. Viability of sewage bacteria in sea water. *Sewage ind. Wastes*, 1956, **28**, 1147–1167.
10. Gunnerson, C. G. Discussion of 'The settling properties of suspensions'. *J. Hydraul. Div. Am. Soc. civ. Engrs*, 1960, **86** (HY7), 29–32.
11. Nusbaum, I., and Garver, R. M. Survival of coliform organisms in Pacific Ocean coastal waters. *Sewage ind. Wastes*, 1955, **27**, 1383–1390.
12. Rittenberg, S. C., Mittwer, T., and Ivler, D. Coliform bacteria in sediments around three marine sewage outfalls. *Limnol. Oceanogr.*, 1958, **3**, 101–108.
13. Hyperion Engineers. Ocean outfall design. Los Angeles, Calif., 15 Oct. 1957.
14. Gunnerson, C. G. Sewage disposal in Santa Monica Bay, California. *J. sanit. Engng Div. Am. Soc. civ. Engrs*, 1958, **84** (SA1), Paper 1534, 28pp.
15. Ward, H. M. The action of light on bacteria. *Proc. R. Soc.*, 1893, **52**, 393–400; **53**, 23–44.
16. Gaarder, T., and Spärck, R. Biochemical and biological investigations of the variations in the productivity of the West Norwegian oyster pools. *Rapp. P.-v. Réun. Cons. perm. int. Explor. Mer*, 1931, **75**, 47–58.
17. Pearson, E. A. An investigation of the efficacy of submarine outfall disposal of sewage and sludge. California State Water Pollution Control Board, Publ. No. 14, Sacramento, 1956.
18. Carlucci, A. F., and Pramer, D. Factors affecting the survival of bacteria in sea water. *Appl. Microbiol.*, 1959, **7**, 388–392.
19. Greenberg, A. E. Survival of enteric organisms in sea water. *Publ. Hlth Rep., Wash.*, 1956, **71**, 77–86.
20. Hollaender, A. Effect of long ultraviolet and short visible radiation (3500 to 4900 Å) on *Escherichia coli*. *J. Bact.*, 1943, **46**, 531–541.
21. Harrison, A. P. Harmful effects of light, with some comparisons with other adverse physical agents. *A. Rev. Microbiol.*, 1967, **21**, 143–156.
22. Eisenstark, A. Mutagenic and lethal effects of visible and near ultraviolet light on bacterial cells. *Adv. Genet.*, 1971, **16**, 167–198.
23. Sistrom, W. R., Griffiths, M., and Stanier, R. Y. Carotenoid-less mutant of *Rhodopseudomonas spheroides*. *J. cell. comp. Physiol.*, 1956, **48**, 473–515.
24. Dworkin, M. The effect of dark, aerobic growth on the photosensitivity of carotenoid-less *Rhodopseudomonas spheroides*. *Biochim. biophys. Acta*, 1960, **39**, 88–92.
25. Bitton, G., Henis, Y., and Lahau, N. Effect of several clay minerals and humic acid on the survival of *Klebsiella aerogenes* exposed to ultraviolet irradiation. *Appl. Microbiol.*, 1972, **23**, 870–874.
26. Reynolds, N. The effect of light on the mortality of *E. coli* in sea water. *Symp. Commn int. Explor. scient. Mer Médit., Monaco, 1964*, pp. 241–249.
27. Gameson, A. L. H., and Saxon, J. R. Field studies on effect of daylight on mortality of coliform bacteria. *Wat. Res.*, 1967, **1**, 279–295.
28. Gameson, A. L. H., and Gould, D. J. Effects of solar radiation on the mortality of some terrestrial bacteria in sea water. This volume, pp. 209–219.
29. de Giaxa, V. Ueber das Verhalten einiger pathogener Mikroorganismen in Meerwasser. *Z. Hyg. InfektKrankh.*, 1889, **6**, 162–224.
30. Carlucci, A. F., Scarpino, P. V., and Pramer, D. Evaluation of factors affecting the survival of *Escherichia coli* in sea water. V—Studies with heat- and filter-sterilized sea water. *Appl. Microbiol.*, 1961, **9**, 400–404.
31. Mitchell, R., and Nevo, Z. Decomposition of structural polysaccharides of bacteria by marine microorganisms. *Nature, Lond.*, 1965, **205**, 1007–1008.
32. Mitchell, R., Yankofsky, S., and Jannasch, H. W. Lysis of *Escherichia coli* by marine microorganisms. *Nature, Lond.*, 1967, **215**, 891–893.
33. Mitchell, R. Destruction of bacteria and viruses in sea-water. *J. sanit. Engng Div. Am. Soc. civ. Engrs*, 1971, **97** (SA4), 425–432.

34. MITCHELL, R., and MORRIS, J. C. The fate of intestinal bacteria in the sea. *Proc. 4th int. Conf. Wat. Pollut. Res., Prague, 1969*, pp. 811–821.
35. MOEBUS, K. Factors affecting survival of test bacteria in sea water: marine bacteria, test bacteria, and solid surfaces. *Helgoländer wiss. Meeresunters.*, 1972, **23**, 271–285.
36. FOXWORTHY, J. E., and KNEELING, H. R. Eddy diffusion and bacterial reduction in waste fields in the ocean. University of Southern California, Los Angeles, 1969.
37. MAY, R. M. Stability and complexity in model ecosystems. Princeton University Press, Princeton, N.J., 1973.
38. CANALE, R. P. An analysis of models describing predator–prey interactions. *Biotechnol. Bioengng*, 1970, **12**, 353–378.
39. PROPER, G., and GARVER, J. C. Mass culture of the protozoa *Colpada steinii*. *Biotechnol. Bioengng*, 1966, **8**, 287–296.
40. CURDS, C. R., and COCKBURN, A. Continuous monoxenic culture of *Tetrahymena pyriformis*. *J. gen. Microbiol.*, 1971, **66**, 95–108.
41. TSUCHIYA, H. M., DRABE, J. F., and FREDERICKSON, A. G. Predator–prey interactions of *Dictyostelium discoideum* and *Escherichia coli* in continuous culture. *J. Bact.*, 1972, **110**, 1147–1153.
42. D'HERELLE, F. The bacteriophage and its behavior. Williams and Wilkins, Baltimore, Md, 1926.
43. GILDEMEISTER, E., and WATANABE, H. Untersuchungen über das Vorkommen von Bakteriophagen in Oberflächenwasser. *Zentbl. Bakt. ParasitKde*, I, Orig., 1931, **22**, 556–575.
44. CARLUCCI, A. F., and PRAMER, D. An evaluation of factors affecting the survival of *Escherichia coli* in sea water. IV—Bacteriophages. *Appl. Microbiol.*, 1960, **8**, 254–256.
45. VACCARO, R. F., BRIGGS, M. P., CAREY, C. L., and KETCHUM, B. H. Viability of *Escherichia coli* in sea water. *Am. J. publ. Hlth*, 1950, **40**, 1257.
46. CARLUCCI, A. F., and PRAMER, D. An evaluation of factors affecting the survival of *Escherichia coli* in sea water. II—Salinity, pH, and nutrients. *Appl. Microbiol.*, 1960, **8**, 247–250.
47. MOEBUS, K. Bactericidal properties of natural and synthetic sea water as influenced by addition of low amounts of organic matter. *Mar. Biol.*, 1972, **15**, 81–88.
48. JONES, G. E. Effect of chelating agents on the growth of *Escherichia coli* in seawater. *J. Bact.*, 1964, **87**, 483–499.
49. SCARPINO, P. V., and PRAMER, D. Evaluation of factors affecting the survival of *Escherichia coli* in sea water. VI—Cysteine. *Appl. Microbiol.*, 1962, **10**, 436–440.
50. SAVAGE, H. P., and HANES, N. B. Toxicity of seawater to coliform bacteria. *J. Wat. Pollut. Control Fed.*, 1971, **43**, 854–861.
51. JANNASCH, H. W. Competitive elimination of Enterobacteriaceae from seawater. *Appl. Microbiol.*, 1968, **16**, 1616–1618.
52. MCGREW, S. B., and MALLETTE, M. F. Maintenance of *E. coli* and the assimilation of glucose. *Nature, Lond.*, 1965, **208**, 1096.
53. MARR, A. G., NILSON, E. H., and CLARK, D. J. The maintenance requirement of *Escherichia coli*. *Ann. N.Y. Acad. Sci.*, 1963, **102**, No. 3, 536–548.
54. PIRT, S. J. Maintenance energy of bacteria in growing cultures. *Proc. R. Soc.*, B, 1965–66, **163**, 224–231.
55. SIEBURTH, J. M. Gastrointestinal microflora of Antarctic birds. *J. Bact.*, 1959, **77**, 521–531.
56. SIEBURTH, J. M. Antibacterial activity of Antarctic marine phytoplankton. *Limnol. Oceanogr.*, 1959, **4**, 419–424.
57. AUBERT, M., PESANDO, D., and GAUTHIER, M. J. Effects of antibiosis in a marine environment. This volume, pp. 191–197.
58. AUBERT, M., AUBERT, J., and GAUTHIER, M. J. Pouvoir auto-épurateur de l'eau de mer et substances antibiotiques produits par les organismes marins. *Revue int. Océanogr. méd.*, 1968, **10**, 137–207.
59. DUFF, D. C. B., BRUCE, D. L., and ANTIA, N. J. The antibacterial activity of marine planktonic algae. *Can. J. Microbiol.*, 1966, **12**, 877–884.
60. SIEBURTH, J. M. The influence of algal antibiosis on the ecology of marine microorganisms. In 'Advances in microbiology of the Sea', Vol 1 (ed. M. R. Droop and E. J. F. Wood). Academic Press, New York, 1968, pp. 63–94.
61. BURKHOLDER, P. R., BURKHOLDER, L. M., and ALMODÓVAR, L. R. Antibiotic activity of some marine algae of Puerto Rico. *Botanica mar.*, 1960, **2**, 149–156.
62. JØRGENSEN, E. G. Antibiotic substances from cells and culture solutions of unicellular algae with special reference to some chlorophyll derivatives. *Physiologia Pl.*, 1962, **15**, 530–545.
63. SIEBURTH, J. M. Acrylic acid, an 'antibiotic' principle in *Phaeocystis* bloom in Antarctic waters. *Science, N.Y.*, 1960, **132**, 676–677.
64. SAZ, A. K., WATSON, S., BROWN, S. R., and LOWERY, D. L. Antimicrobial activity of marine waters. I—Macromolecular nature of antistaphylococcal factors. *Limnol. Oceanogr.*, 1963, **8**, 63–67.
65. SIEBURTH, J. M., and PRATT, D. M. Anticoliform activity of sea water associated with the termination of *Skeletonema costatum* blooms. *Trans. N.Y. Acad. Sci.*, II, 1962, **24**, 498–501.
66. MOEBUS, K. Seasonal changes in antibacterial activity of North Sea water. *Mar. Biol.*, 1972, **13**, 1–13.

67. Moebus, K. Studies on the influence of plankton on antibacterial activity of sea water. *Helgoländer wiss. Meeresunters.*, 1972, **23**, 127–140.
68. Mitchell, R. Role of predators in the reversal of imbalances in microbial ecosystems. *Nature, Lond.*, 1971, **230**, 257–258.
69. Canter, H. M., and Lund, J. W. C. Studies on phytoplankton parasites. *New Phytol.*, 1948, **47**, 238.
70. Rosenfeld, W. D., and ZoBell, C. E. Antibiotic production by marine microorganisms. *J. Bact.*, 1947, **54**, 393–398.
71. Krasil'nikova, E. N. Antibiotic properties of microorganisms isolated from various depths of world oceans. *Microbiology*, 1962, **30**, 545–550.
72. Carlucci, A. F., and Pramer, D. An evaluation of factors affecting the survival of *Escherichia coli* in sea water. III—Antibiotics. *Appl. Microbiol.*, 1960, **8**, 251–254.
73. Korinek, J. Über Susswasser-Bakterien im Meere. *Zentbl. Bakt. ParasitKde*, Abt. II, 1926, **66**, 500–505.
74. Burke, V., and Baird, L. A. Fate of fresh water bacteria in the sea. *J. Bact.*, 1931, **21**, 287–298.
75. ZoBell, C. E., and Feltham, C. B. Are there specific marine bacteria? *Proc. 5th Pacif. Sci. Congr.*, 1933, Vol 3, pp. 2097–2100.
76. Jones, G. E., and Cobet, A. B. Heavy metal ions as the principal bactericidal agent in Caribbean sea water. This volume, pp. 199–208.
77. Ketchum, B. H., Ayers, J. C., and Vaccaro, R. F. Processes contributing to the decrease of coliform bacteria in a tidal estuary. *Ecology*, 1952, **33**, 247–258.
78. Brooks, N. H. Diffusion of sewage effluent in an ocean current. *Proc. 1st int. Conf. Waste Disposal mar. Envir., Univ. Calif., 1959*. Pergamon Press, New York, pp. 246–267.
79. Gameson, A. L. H., Bufton, A. W. J., and Gould, D. J. Studies of the coastal distribution of coliform bacteria in the vicinity of a sea outfall. *Wat. Pollut. Control*, 1967, **66**, 501–523.
80. Lee, T. N., McGuire, J. B. An analysis of marine waste disposal in southeast Florida's coastal waters. *Proc. 6th int. Conf. Wat. Pollut. Res., Jerusalem, 1972*, pp. 865–880.
81. Hoven, I. v. d. A diffusion-deposition model for particulate effluents from ground-tested nuclear engines. *Int. J. Air Wat. Pollut.*, 1963, **7**, 1023–1032.
82. Harremoës, P. *In situ* methods for determination of microbial disappearance in sea water. This volume, pp. 181–190.
83. Thomas, R. H. Marine dilution and inactivation of sewage. *Proc. 1st int. Conf. Wat. Pollut. Res., London, 1962*, Vol 3, pp. 57–83.
84. DaSilva, Moreira, *et al.* Pesquisas para o lancamento submarino dos esgotos na costa maritima. Sursan Department de Esgotos Sanitarias, Estado de Guanabara, Brazil, 1963.
85. Annan, C. K., and Wright, A. M. Marine waste disposal in Ghana—water quality evaluation. *Proc. 5th int. Conf. Wat. Pollut. Res., San Francisco and Hawaii, 1970*, Vol 2, Paper III-13.
86. Wheatland, A. B., Agg, A. R., and Bruce, A. M. Some observations on the dispersion of sewage from sea outfalls. *J. Proc. Inst. Sew. Purif.*, 1965, 291–314.
87. Aubert, M., Guizerix, J., Aubert, J., Molinari, J., and Daniel, S. Etude de la dispersion bactérienne d'eaux résiduaires en mer au moyen de traceurs radioactifs. *Revue int. Océanogr. méd.*, 1966, **1**, 56–91.
88. Bonde, G. J. Studies on the dispersion and disappearance phenomena of enteric bacteria in the marine environment. *Revue int. Océanogr. méd.*, 1968, **9**, 17–44.
89. Harremoës, P. Field determination of bacterial disappearance in seawater. *Wat. Res.*, 1970, **4**, 737–749.
90. Walker, J. D., and Guarraia, L. J. Other factors determining life expectancy of microorganisms in the marine environment. This volume, pp. 221–226.

Prof. Mitchell, introducing the paper, said that he and Mr. Chamberlin had attempted to develop a mathematical and conceptual model to describe and interpret laboratory and field data relating to the decline of enteric bacteria in the sea. Five factors were associated with coliform die-off in the sea: solar radiation, sedimentation, predation, nutrient utilization, and physico-chemical effects. These factors had been shown to be significant on the basis of analysis of a large quantity of field data. The paper described the general sensitivity of each factor to local environmental conditions, and presented a preliminary model to determine the significance of each factor in coliform die-off. This type of model could be used as a starting point in the development of a more sophisticated equation to predict the die-off in the sea, and ultimately to aid the design of more efficient outfall systems.

DISCUSSION

Prof. A. G. Occhipinti (Brazil) took the opportunity of presenting the results of more than fifty recent *in situ* experiments on bacterial disappearance performed in Santos coastal waters, São Paulo, Brazil. Raw sewage discharged at 1 m^3/s over a weir, and spread over the water to give less than a tenfold initial dilution. The sewage field thickness was never greater than 50 cm, so that the experiments were conducted with a surface spreading flow. The sewage dilution was determined from measurements of added bromine-82 and of the phosphate originating in the sewage; there was generally good agreement between the results of the two methods.

Rhodamine-B and drift cards were also used—as visual auxiliary tracers. The experiments were of the type categorized by Prof. Harremoës[82] as: concentration-ratio approach, continuous sewage and tracer release.

In twenty-nine determinations of sewage coliform MPN's the values (in millions per 100 ml) ranged between 50 and 950 with a median of 250. The coliform disappearance curves were approximately log-linear with respect to time.

The thirty-six measured T_{90} values were found to be almost lognormally distributed [see Fig. 1], ranging from 35 to 220 min with a median of 84 min. About 75 per cent of the values were between 1 and 2 h, corresponding to exceedences of 87 and 18 per cent respectively. Prof. Occhipinti considered the values outside this interval to arise from inaccuracies inherent in the experimental procedures. No correlation was found between T_{90} and either season or temperature, although the water temperature varied between 18° and 28°C. Details of the methods used and the results obtained had been published elsewhere[91, 92].

Since other workers had used dialysis membranes in microbial mortality studies, this technique was used (on more than thirty occasions) simultaneously with, and at the same site as, the *in situ* experiments. Correlation between the two sets of results was very poor, and in all but two cases the coliforms within the membranes disappeared more slowly than those in the sea.

Prof. Occhipinti concluded by saying that the many factors and processes affecting the survival of enteric micro-organisms in the sea were dependent on the random meteorological and oceanographic conditions which prevailed; consequently, mathematical models representing these processes should be considered as stochastic, and T_{90} itself should be treated as a stochastic parameter or a stochastic design variable. Finally, he claimed that the *in situ* investigations he had described ought to be included among the significant sets of data on the subject.

Editorial comment. The final point of Prof. Occhipinti's contribution is well taken; it resulted from remarks by speakers in earlier discussions to the effect that there had been only two extensive series of *in situ* studies reported —those of Foxworthy and Kneeling[36] and of Gameson and Gould[28]. It appeared that none of the speakers had been acquainted with Prof. Occhipinti's work. In Fig. 1 are shown these three sets of data on logarithmic–probability plots with fitted regression lines. Prof. Occhipinti's results for thirty-six experiments show a remarkably close approximation to a lognormal distribution. The line shown in his report[92] has a slope only 70 per cent of that in Fig. 1 since his line refers only to the data for which T_{90} was between 1 and 2 h. The results for the 37 experiments reported by Foxworthy and Kneeling[36] show a rather poorer fit to a lognormal distribution, a wider variability (greater slope), and generally higher values of T_{90}. The data plotted from Gameson and Gould's paper[28] are the eighteen values of T_{90} relating almost entirely to daylight conditions, and show further departure from lognormality, increased variability, and generally slower mortality. The median values of T_{90} in these three series of experiments (in the order of discussion) are 1·4, 2·1, and 3·1 h.

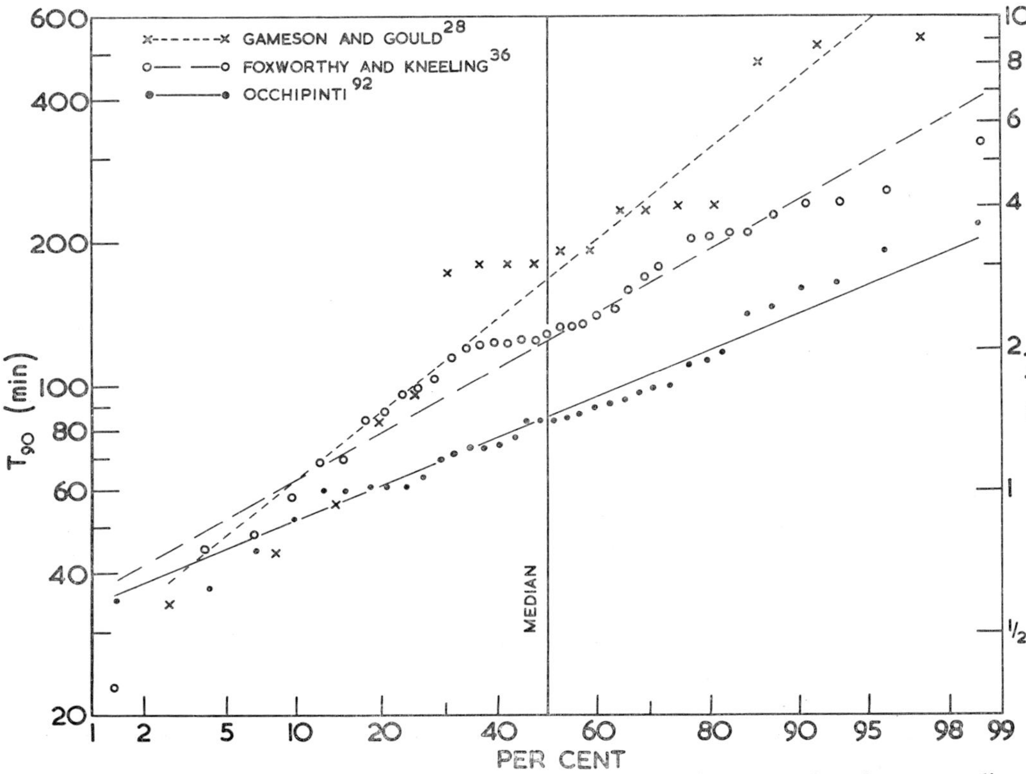

FIG. 1. Logarithmic–probability plots of T_{90} values obtained in three series of *in situ* studies

Mr. R. M. Hedgland (New Zealand) said that it had been evident from earlier discussions that, for the general engineering design of a sea outfall, it was pointless to carry out expensive and time-consuming surveys to evaluate T_{90}, and that an average of world studies to date should be used for design purposes. Results of individual T_{90} studies were often highly unreliable for reasons such as absence of steady-state velocities, variations in mixing with weather conditions, differences in die-off rates between day and night, and (in his experience) difficulties arising from variations in background coliform counts with tidal movement. Although he did not imply that no further research should be carried out on this subject, he felt that local determinations of T_{90} should not be made in conjunction with each outfall design. Prof. P. Harremoës (Denmark) said that he would certainly use T_{90} values for engineering design, but that (in view of the inaccuracy of the ultimate epidemiological grounds for judgement) he agreed that there had to be a limit to the efforts in determining local T_{90} values and their statistical distribution. Nevertheless, he would find it most interesting that the processes governing bacterial die-away be studied from a scientific aspect.

Dr. G. E. B. Kullenberg (Denmark) stated that all investigations on bacterial disappearance had been made in very calm conditions, and asked how representative were the results with regard to average conditions. He added that the natural fluctuations in the environment were extremely important, and felt that they ought somehow to be included in any model which aimed at covering anything more than very short time-scales.

Dr. R. J. Callaway (U.S.A.) pointed out that, for completeness, Equation 5 should be accompanied by similar equations for predation and substrate, the resulting coupled equations then being solved simultaneously by numerical methods. He asked if the Authors had conducted sensitivity analyses on the partial differential equations.

Dr. J. J. Barnard (South Africa) said that no mention had been made in any of the Symposium papers of the effect of dissolved oxygen on the die-off rate of organisms in the marine environment. Human enteric micro-organisms were exposed to a foreign environment on entering water, and then to a less favourable one in the sea. In thus passing from anaerobic to aerobic conditions they would tend to die-off naturally, and the die-off would be greatly assisted if aerobic conditions were maintained.

Prof. R. R. Colwell (U.S.A.) suggested that, with the increasing use of long outfalls discharging into deep water, the combined effects of lower temperature and higher pressure would tend to inhibit microbial action.

Prof. Y. Kott (Israel) asked the Authors how important they thought *Bdellovibrio* was in sea outfalls or the marine environment.

Prof. T. G. Metcalf (U.S.A.) referred to factors involving the adsorption, survival, and inactivation of viruses. The basic mechanism of adsorption was related to electrostatic attraction. Adsorption to a membrane filter was maximized by the presence of hydrogen ions and an enhancing salt, whereas reversal of adsorption was maximized by adjusting the pH value to the 9–11 range. Data obtained over a long period by many workers had demonstrated a direct relationship between the silt content of a gauze pad and the likelihood of virus isolation. Studies of virus adsorption to particles had shown deposition of such particles from tributary rivers when exposed to saline estuarial water, thus causing sedimentation of viruses. This phenomenon might be considered inducive to virus carriage by shellfish through flocculation of particles over shellfish beds. Virus adsorption could be inhibited by organic compounds occupying surfaces to which viruses would ordinarily adsorb. Thus, adsorption of viruses to the surfaces of particulates could account for both survival and transport, as well as the conditions favourable to accumulation of virus by shellfish. Prof. Jones's studies[76] on bacterial inactivation by heavy metals might provide some clues to virus inactivation in sea water. For example, it was recognized that virus survival was prolonged in the presence of domestic sewage; this had been attributed to an ill-defined sparing or protective effect exhibited by the organic matter—an effect which might be a function of chelating action on the part of the organic matter in which heavy metals normally virucidal would be sequestered and rendered incapable of action against virus. Finally, many faulty interpretations of laboratory findings had resulted from failure to recognize the occurrence of an inhibition of virus adsorption to monolayers caused by inhibitory components found in virus inocula.

Prof. H. I. Shuval (Israel) said that all the coliform die-away studies had been intended to serve as predictors of pathogen die-away, so that there was a need to carry out similar field studies with viruses—rigorously controlled and with good tracer techniques. Such studies, he considered, would help to answer one of the key questions facing outfall designers.

Mr. A. L. H. Gameson (U.K.) did not speak during the discussion but made a written contribution. The two-fold range in median values for T_{90}, given at the end of the editorial comment on Prof. Occhipinti's contribution suggested that Mr. Hedgland's recommendation for adopting an average of world studies might be inadvisable. It was of interest to note that the latitudes of the experimental areas of the studies in Brazil, California, and England were roughly 24°, 34°, and 51° respectively, so that both the median values and the variability increased with latitude; this was consistent with the hypothesis that solar radiation was an important factor, since the radiation was likely both to decrease in average intensity and to become more variable with increasing latitude. Dr. Kullenberg's statement that all investigations on bacterial disappearance had been made in very calm conditions was erroneous: the average of the sea state recorded in Table 2 of Gameson and Gould's paper[28] was, in fact, slightly higher than the average of some 3500 observations made during 1967–69 at Bridport[93]. In reply to Prof. Shuval, these coliform mortality studies had never been intended to serve as predictors of pathogen die-away; coliform bacteria had been studied because they were extremely sensitive indicators of recent sewage pollution, and because they provided some measure of the purification of sewage in the sea. Mr. D. J. Gould (U.K.) added with regard to Dr. Barnard's comment that Eisenstark[22] reported inactivation by visible radiation to be inhibited by anaerobic conditions, and that he himself had made a series of five experiments which gave limited support to the contention that lowered oxygen concentration was accompanied by a lessening in the effect of radiation on coliform mortality.

REPLY TO DISCUSSION

Mr. Chamberlin agreed with Mr. Hedgland that carrying out expensive and time-consuming *in situ* determinations for every proposed outfall site would not be very productive. The limited extent of such studies could not provide adequate estimation of seasonal, diurnal, and other variability. Use of some average value taken from other studies, however, would also be inadequate owing to the site-to-site variability to which Mr. Gameson had referred. He admitted that current designs, for lack of any better alternative, had to use a conservative estimate of T_{90} or even to assume no die-off. It was the overall objective of the current work by Prof. Mitchell and himself to develop estimates of the variability of die-off rates attributable to specific factors and thus to refine the design parameter.

He completely agreed with Dr. Callaway that conservation equations for predators and substrate were needed to describe the system fully. The Authors had developed preliminary forms of these equations, but had felt them to be insufficiently advanced for inclusion in their paper. In addition, their primary object had been to evaluate, qualitatively, the maximum potential influence of the various factors contributing to die-off, and this did not require the added conservation equations. To conduct the sensitivity analysis would, however, require these added equations. The Authors felt that, due to the primitive nature of these equations and the uncertainty of the various model parameters (particularly K_S, K_C, and k), they should defer carrying out a sensitivity analysis until a later stage.

Mr. Chamberlin referred Dr. Barnard to work by one of his countrymen, Dr. Marais, who had studied the effect of dissolved oxygen levels on coliform die-off rates in maturation ponds[94] and found that under anaerobic conditions the rate was less than half that when the oxygen concentration was 1–2 mg/l. However, he suggested that if the marine receiving waters were anaerobic this fact would itself deserve consideration.

Prof. Mitchell, replying to Prof. Kott, said that during several years' work they had found *Bdellovibrio* to be very common close to ocean outfalls, and also to be present in large numbers in coastal waters some distance from outfalls—suggesting that these organisms were entering with the sewage and adapting to the marine environment. *Bdellovibrio* appeared to be a very important predator of enteric bacteria, not only in the sea but also in the fresh-water and soil environments.

Prof. Metcalf's comment emphasized the complex phenomena underlying the die-off of enteric viruses in the sea. Several speakers had expressed the opinion that further refinement of the coliform die-off model was unnecessary because of the difficulty in the refinement, but Prof. Mitchell submitted that the reverse was true: more detailed laboratory and field studies would lead to better understanding of the factors controlling coliform die-off, and should yield more accurate models. Prof. Metcalf's contribution suggested that the factors controlling virus inactivation might be quite different; it would therefore be useful to study viruses in more detail and find out if a model for the die-off of viruses was the same as that for coliforms. After all, the survival of viruses was of the greater interest since they presented the primary human health hazard from sewage in the sea because of their concentration in shellfish.

ADDITIONAL REFERENCES

91. Occhipinti, A. G. Estudos para o sistema de disposição oceânica dos esgotos de Santos e São Vicente. *Proc. XIII Congresso Interamericano de Engenheria Sanitária, Ascunsion, Paraguay, 1972.*

92. Occhipinti, A. G. Pesquisas de decaimento bacteriano e virulógico nas águas do litoral santista. *Proc· VII Congresso Brasileiro de Engenharia Sanitária, Salvador, Bahia, Brazil, 1973.*

93. Gameson, A. L. H., Munro, D., and Pike, E. B. Effects of certain parameters on bacterial pollution at a coastal site. In 'Water pollution control in coastal areas'. *Institute of Water Pollution Control, London,* 1970, pp. 34–54.

94. Marais, G. v. R. Faecal bacterial kinetics in stabilization ponds. *J. envir. Engng Div. Am. Soc. civ. Engrs*, 1974, **100** (EE1), 119–139.

Initial Dilution of Effluent Discharging into Stagnant Sea Water

Taizo Hayashi
Professor Dr. Eng., Head, Hydraulic Laboratory
Chuo University, Tokyo

and

Masaki Ito
Postgraduate Student
Department of Civil Engineering
Chuo University, Tokyo

In the design of submarine outfalls, the engineer is faced with the problem of determining the dilution necessary to meet disposal objectives. A fundamental problem of ocean disposal is the dilution of effluent in the vicinity of the outfall. Normally, variable ocean currents exist. Therefore, it is when there is no current that a limiting condition with minimum dilution is supposed to occur.

The problem to be discussed is the determination of the concentration and velocity fields of sewage discharging vertically from an orifice into a calm sea. Attention will be confined to a circular orifice. Initial diffusion of a vertical buoyant plume in quiescent environment has been studied previously by several investigators[1-12], each of them adopting a different approach. Priestley and Ball[12], in their study on continuous convection from an isolated source of heat, showed that the assumption of the same profile shape for the two quantities, temperature and vertical velocity, of a plume facilitates the mathematical analysis of initial diffusion of the plume, no matter whether the ambient fluid is homogeneous or has a density gradient.

The Authors have studied forced plumes issuing into heavier water[13-15], based on Priestley and Ball's theory. The present paper presents a theory made by the extension and remoulding of Priestley and Ball's work, for the determination of the concentration and velocity fields of forced plumes of sewage in a calm sea.

Mathematical Formulation

As shown in Fig. 1 the origin of the coordinates is taken at the virtual point source which is fixed by extending the widths of spread of the effluent discharged from an orifice. The location of this virtual point source will be discussed later. The position in a circular-symmetric column of effluent will be defined by the distance, r, from the axis, and by either the height, x, above the origin or the height, z, above the outlet. Consideration of the physical processes controlling the condition of the plume of effluent in the steady state yields the following equations:

$$\frac{\partial}{\partial x}(rw\rho) + \frac{\partial}{\partial r}(ru\rho) = 0, \tag{1}$$

$$\frac{\partial}{\partial x}(rw^2\rho) + \frac{\partial}{\partial r}(rwu\rho) = r\rho' g + \frac{\partial}{\partial r}(r\tau), \tag{2}$$

$$\frac{\partial}{\partial x}[rw(\rho_{e1} - \rho)] + \frac{\partial}{\partial r}[ru(\rho_{e1} - \rho)] = -\frac{\partial}{\partial r}(rG), \tag{3}$$

where $\rho = \rho(x, r)$ is the density in the plume, $u = u(x, r)$ the radial velocity, $w = w(x, r)$ the vertical velocity, $\tau = \tau(x, r)$ the vertical turbulent shearing stress, ρ_{e1} the density of the receiving water at the level of the outlet (taken as the reference of density for the system), and $G = G(x, r)$ the radial turbulent diffusion flux of density deficit.

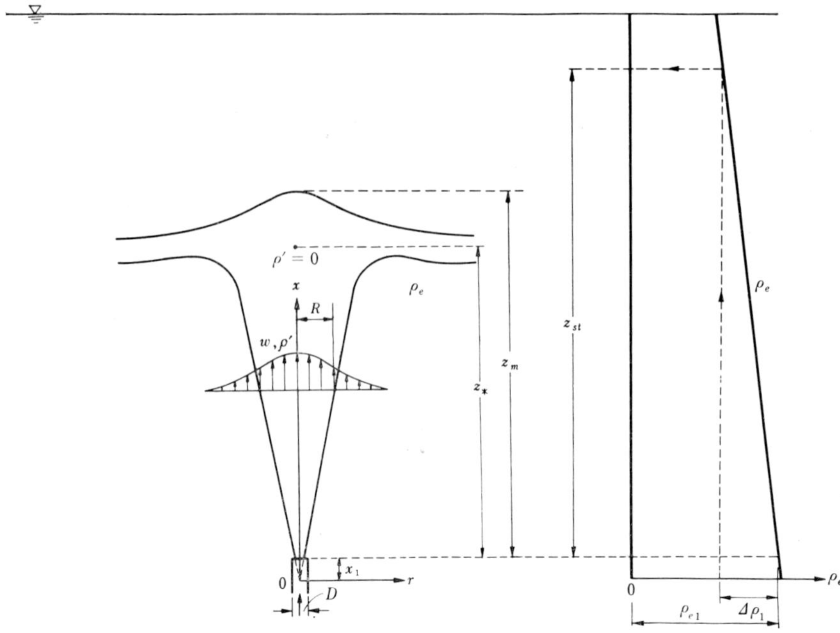

FIG. 1. Plume and notations

By use of Equation 1, Equations 2 and 3 may be rewritten respectively as

$$\frac{\partial}{\partial x}(\tfrac{1}{2}rw^3\rho) + \frac{\partial}{\partial r}(\tfrac{1}{2}rw^2u\rho) = rw\rho'g + w\frac{\partial}{\partial r}(r\tau), \tag{4}$$

$$\frac{\partial}{\partial x}(rw\rho') + \frac{\partial}{\partial r}(ru\rho') = -\frac{\partial}{\partial r}(rG) + rw\frac{d\rho_e}{dx}, \tag{5}$$

where $\rho_e = \rho_e(x)$ is the density of the receiving water, and $\rho' = \rho'(x, r) = \rho_e(x) - \rho(x, r)$ the deficiency of density at a point in the plume relative to the density of environmental water at the same level.

The assumptions are made that the lateral profiles of w and ρ' are of similar shape at different distances from the origin and that the shearing stress is a quadratic function of the relative velocity. This similarity assumption is expressed by the following equations:

$$\frac{w}{w_m} = f\left(\frac{r}{R}\right), \qquad \frac{\rho'}{\rho'_m} = h\left(\frac{r}{R}\right), \qquad \frac{\tau}{\frac{1}{2}\rho w_m^2} = j\left(\frac{r}{R}\right), \tag{6}$$

where m is a suffix pertaining to the values on the axis of the plume, R is some linear characteristic of the lateral extent of the plume, and f, h, and j are functions to be given on the basis of experiment.

Further assumptions are made that the profiles of w and ρ' are Gaussian and that they are the same for these two quantities. Thus, they are taken as

$$f = h = \exp(-r^2/2R^2). \tag{7}$$

Since w, ρ', rG and $r\tau$ are all considered to tend to zero as r tends to infinity, integration of Equations 2, 4, and 5 from $r = 0$ to $r = \infty$, in consideration of Equations 6 and 7, yields respectively:

$$d(R^2w^2)/dx = 2R^2(\rho'/\rho_e)g, \tag{8}$$

$$d(R^2w^3)/dx = 3R^2w(\rho'/\rho_e)g - \beta Rw^3, \tag{9}$$

$$d(R^2w\rho')/dx = 2R^2w(d\rho_e/dx), \tag{10}$$

where the suffix m has been dropped for the sake of brevity. Equations 8–10 are those derived and used by Priestley and Ball in their work[12].

Boundary conditions at outlet

As shown in Fig. 2, let $\bar{w}$ represent the velocity of discharge of effluent at the outlet; also let $\Delta\rho_1$ and $\bar{c}$ denote the density deficiency and the concentration, respectively, of the waste water issuing from the outlet, the section of the outlet being at $x = x_1$. The real conditions of the effluent at the outlet section are described by the rectangular profiles

$$\bar{w} = Q/(\pi/4)D^2 \quad \text{and} \quad \overline{\rho'} = \Delta\rho_1, \tag{11}$$

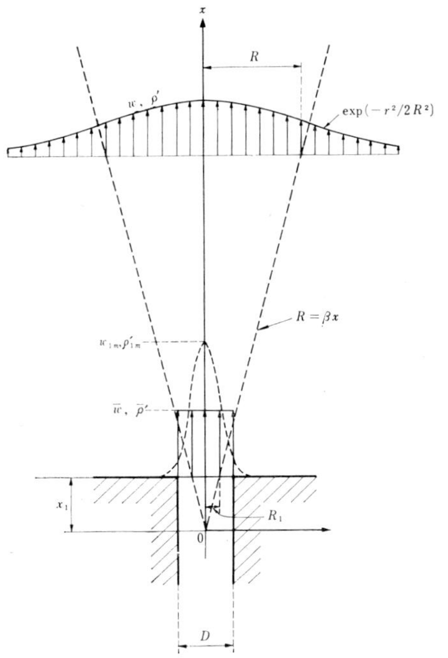

FIG. 2. Equivalent Gaussian profiles at an outlet

where Q is the rate of discharge of effluent, and D the effective diameter of the orifice. However, since we have made the assumption that the profiles of w and ρ' are Gaussian, we need also to approximate the profiles of Equation 11 to 'equivalent' Gaussian forms. For this purpose let us take the Gaussian profiles, $w_1(r) = w_{1m} \exp(-r^2/2R_1^2)$ and $\rho_1'(r) = \rho_{1m}' \exp(-r^2/2R_1^2)$, which are equivalent in such a way as to make (i) the rate of discharge of effluent, (ii) the momentum of effluent, and (iii) the rate of discharge of density deficiency, equal to those corresponding quantities given by Equation 11. This equality yields the following relations:

$$w_{1m} = 2\bar{w}, \qquad \rho_{1m}' = 2\Delta\rho_1, \quad \text{and} \quad R_1 = D/4, \tag{12}$$

which are the values to be imposed on w_m, ρ_m', and R, or w, ρ', and R, as the boundary conditions at $x = x_1$.

Continuous Discharge Into Uniform Receiving Water

Solution of equations

In this case when $d\rho_e/dx = 0$ the solutions of Equations 8–10, on the boundary conditions expressed by Equation 12, are obtained as follows:

$$R = \beta x, \tag{13}$$

$$\frac{w}{\bar{w}} = \left(\frac{16\beta F^2 - 3}{128\beta^4 F^2}\right)^{1/3}\left(\frac{z}{D} + \frac{1}{4\beta}\right)^{-1}\left\{1 + \left(\frac{48\beta^2}{16\beta F^2 - 3}\right)\left(\frac{z}{D} + \frac{1}{4\beta}\right)^2\right\}^{1/3}, \tag{14}$$

$$\rho' = \tfrac{1}{4}D^2\bar{w}\Delta\rho_1/R^2 w; \tag{15}$$

in Equation 14

$$F = \bar{w}/\sqrt{(\Delta\rho_1/\rho_{e1})gD}, \tag{16}$$

this being the internal Froude number of the effluent at the outlet. Also in Equation 14 the following relation has been used:

$$x = z + x_1 = z + D/4\beta. \tag{17}$$

In the meantime, the concentration of effluent, c, can be expressed by

$$c/\bar{c} = \rho'/\Delta\rho_1. \tag{18}$$

Therefore, from Equations 13–15 and 18, c is given as follows:

$$\frac{c}{\bar{c}} = \frac{1}{\beta}\left(\frac{2\beta F^2}{16\beta F^2 - 3}\right)^{1/3}\left(\frac{z}{D} + \frac{1}{4\beta}\right)^{-1}\left\{1 + \left(\frac{48\beta^2}{16\beta F^2 - 3}\right)\left(\frac{z}{D} + \frac{1}{4\beta}\right)^2\right\}^{-1/3}. \tag{19}$$

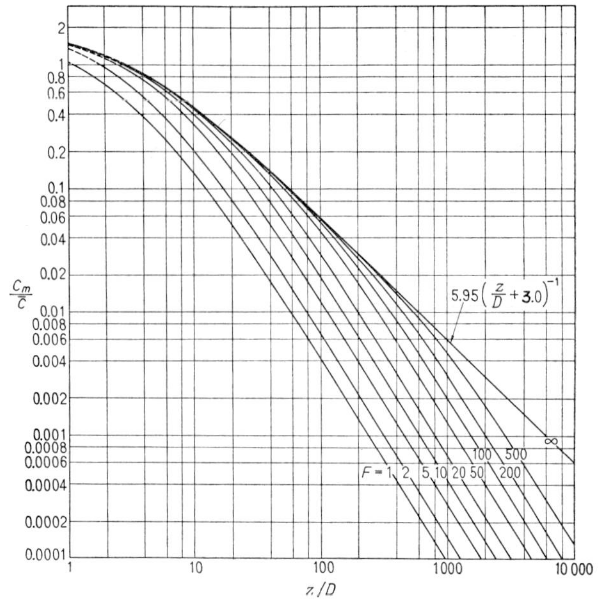

FIG. 3. Axial concentration as given by Equation 19

Figures 3 and 4 illustrate the relations (19) and (14) respectively. For the calculation of these figures it has been assumed that $\beta = 0{\cdot}084$ on the basis of the experiments of Rouse, Yih, and Humphreys[2] conducted on the gravitational convection of air from a point heat source. They found that $\rho'(x, r) = X(x) \exp(-71r^2/x^2)$. The value of β has been assumed to be $\sqrt{1/2 \times 71} = 0{\cdot}084$ in view of the relations, $\exp(-71r^2/x^2) = \exp(-r^2/2R^2) = \exp(-r^2/2\beta^2x^2)$.

Experiments

The experiments were conducted in a tank 2·5 m × 2·1 m in plan and 1·8 m high. The experiments were performed by injecting diluted ethyl alcohol vertically upwards through a 0·25-cm diameter nozzle into fresh water (Fig. 5). A small amount of table salt of high purity was mixed as a tracer in the diluted ethyl alcohol to be injected into the receiving water, and the variation of the concentration of salinity along the plume axis was measured by means of a calibrated conductivity probe. The electrodes were 0·50 cm square platinum plates placed 0·50 cm apart.

Figure 6 shows comparison between the experimental results and the theory expressed by Equation 19. The coefficient of contraction, C_d, has been assumed as 0·85 in these experiments. In spite of the wide ranges of F and z/D the agreement between the theory and experiments is good as a whole.

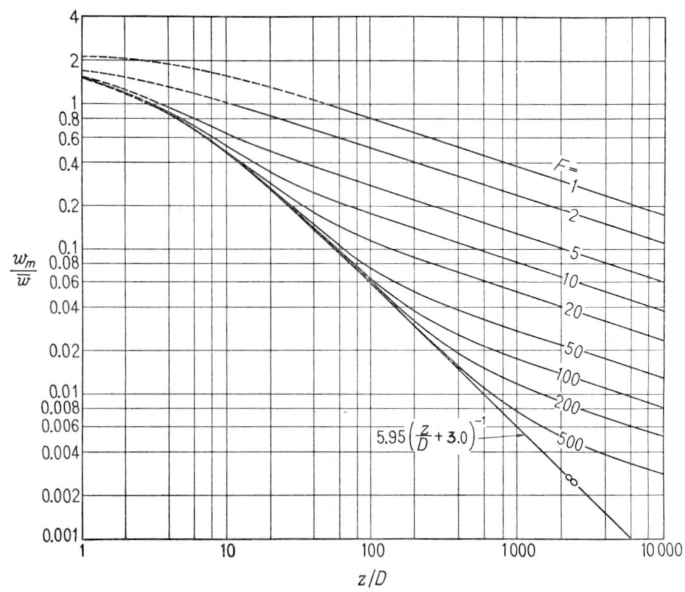

FIG. 4. Axial velocity as given by Equation 14

FIG. 5. Forced plume entering homogeneous water

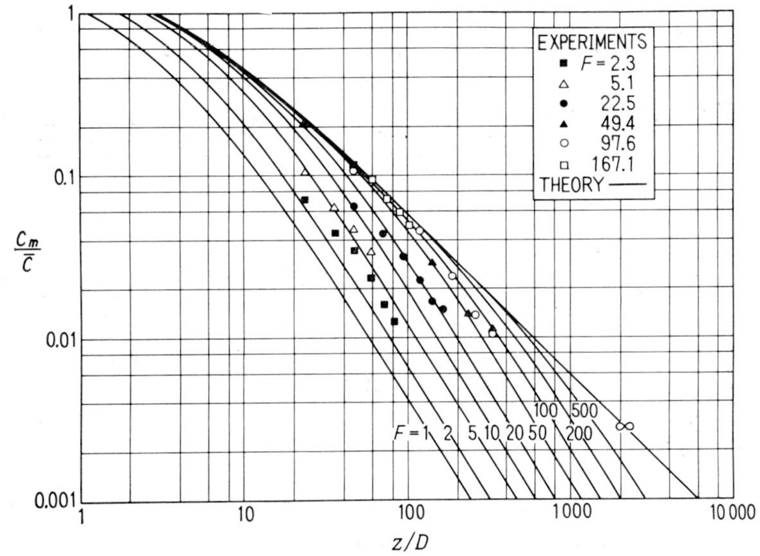

FIG. 6. Comparison between theoretical and experimental values of axial concentration

Comparison with previous theories

Two of Abraham's formulae which are widely used in the case of small values of F are[5]

$$\frac{c_m}{\bar{c}} = 9{\cdot}7F^{2/3}\left(\frac{z}{D} + 2\right)^{-5/3}, \qquad \frac{w_m}{\bar{w}} = 3{\cdot}65F^{-2/3}\left(\frac{z}{D} + 2\right)^{-1/3}. \tag{20}$$

For the values of F and z/D in the range where the relations $16\beta F^2 \ll 3$ and $16\beta^2 (z/D + 1/4\beta)^2 \gg 1$ are satisfied, Equations 19 and 14 reduce respectively to

$$\frac{c_m}{\bar{c}} = 9{\cdot}4F^{2/3}\left(\frac{z}{D} + 3{\cdot}0\right)^{-5/3}, \qquad \frac{w_m}{\bar{w}} = 3{\cdot}76F^{-2/3}\left(\frac{z}{D} + 3{\cdot}0\right)^{-1/3}, \tag{21}$$

which are close to Abraham's formulae cited above.

On the other hand, when $16\beta F^2 \gg 3$, and for such a range of z/D that $(3\beta/F^2)(z/D + 1/4\beta)^2 \ll 1$, Equations 19 and 14 reduce respectively to

$$\frac{c_m}{\bar{c}} = 5{\cdot}95\left(\frac{z}{D} + 3{\cdot}0\right)^{-1}, \qquad \frac{w_m}{\bar{w}} = 5{\cdot}95\left(\frac{z}{D} + 3{\cdot}0\right)^{-1}, \tag{22}$$

Here, the coefficient 5·95 comes from the term $1/2\beta$, and these equations agree with Abraham's formulae[7] for the case when $F = \infty$.

Continuous Discharge Into Uniformly Stratified Receiving Water

Solutions of equations

In this case when $d\rho_e/dx$ is constant, the solutions of Equations 8–10, on the boundary conditions of Equation 12, are obtained as follows:

$$R = \beta x, \tag{23}$$

$$\int_{\Phi_1}^{\Phi} \sin^{1/2}\Phi\, d\Phi = \frac{1}{4\beta^2F^2} D_r(1 + D_rF^2)^{1/2}\gamma^3(x_r^2 - x_{r1}^2), \tag{24}$$

$$\rho' = \frac{1}{R^2w}\left(A^2 + \frac{\rho_{e_1}}{g}\frac{d\rho_e}{dx}R^4w^4\right)^{1/2}, \tag{25}$$

where

$$\Phi = \sin^{-1}(\gamma x_r w_r)^2, \tag{26}$$

$$\left.\begin{aligned} x_r &= x/\{\Delta\rho_1/(-d\rho_e/dx)\}, \qquad D_r = D/\{\Delta\rho_1/(-d\rho_e/dx)\},\\ w_r &= w/\overline{w}, \qquad \Phi_1 = \sin^{-1}(\gamma D_r/2\beta)^2,\\ \gamma^4 &= 16\beta^4F^2/D_r^3(1 + D_rF^2), \end{aligned}\right\} \tag{27}$$

and

$$A = \frac{D^2}{4}\,\overline{w}\,\Delta\rho_1\left\{1 - \frac{\rho_{e1}}{g(\Delta\rho_1)^2}\frac{d\rho_e}{dx}\,\overline{w}^2\right\}^{1/2}. \tag{28}$$

It is illustrated in Fig. 1 that $\Delta\rho_1/(-d\rho_e/dx) \equiv x_{st}$ represents the height to which the plume would rise if there were no dilution in the course of its rising. Therefore we call x_{st} the 'static height of rise' of a plume.

Height of rise of a plume

Figure 7 shows a typical example of the behaviour of a plume in a uniformly stratified environment. Of special interest is the height of rise of the plume, found by setting the vertical velocity, w, equal to zero, and hence $w_r = 0$. It is understood from Equation 26 that $\Phi = \pi$ corresponds to $w_r = 0$. Therefore from Equation 24 we obtain the following equation:

$$z_{rm} = \left\{\frac{D_r^2}{16\beta^2} + \frac{1}{2\beta}F^{1/2}D_r^{5/4}(1 + D_rF^2)^{1/4}\left(2{\cdot}40 - \int_0^{\Phi_1}\sin^{1/2}\Phi\, d\Phi\right)\right\}^{1/2} - \frac{D_r}{4\beta} \tag{29}$$

where z_{rm} is the dimensionless height of rise of plume. Equation 29 is illustrated in Fig. 8.

When $D_rF^2 \ll 1$ and $\Phi_1 \ll \pi/2$, and for $\beta = 0{\cdot}084$, the above equation becomes

$$z_m - \frac{D}{4\beta} \simeq \frac{5{\cdot}34}{\pi^{1/4}}\left(Qg\frac{\Delta\rho_1}{\rho_{e1}}\right)^{1/4}\left(\frac{\rho_{e1}}{g|d\rho_e/dx|}\right)^{3/8} \tag{30}$$

where z_m is the height of rise of plume and $Q = \frac{1}{4}\pi D^2\overline{w}$. Comparison of Equation 30 with the experimental results of Crawford and Leonard[16] shows excellent agreement.

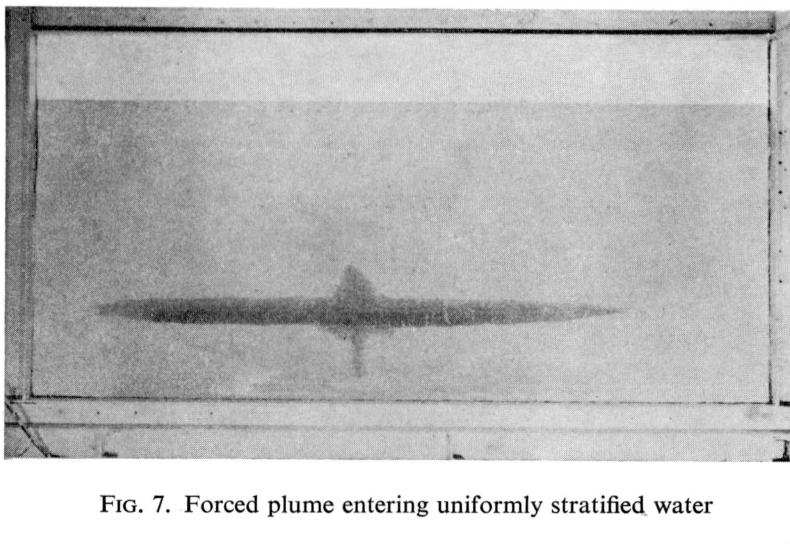

FIG. 7. Forced plume entering uniformly stratified water

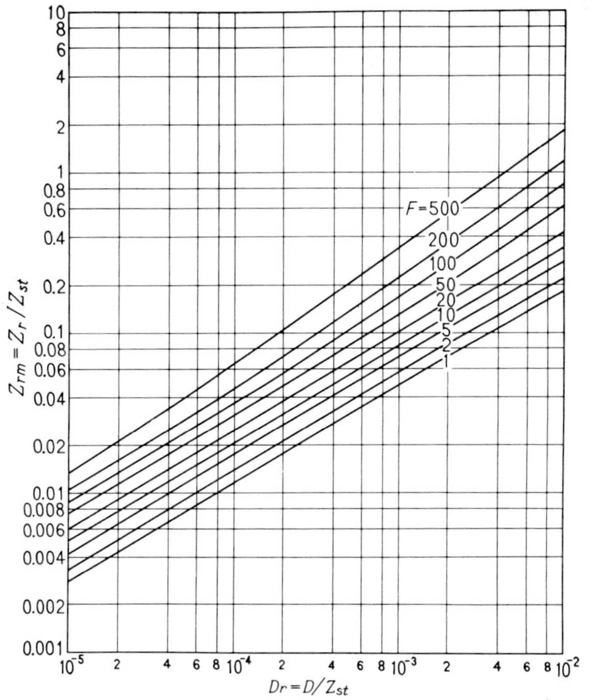

FIG. 8. Relative ceiling height of plume as given by Equation 29

Height of neutral position of a plume

After a plume reaches its maximum height it sinks back to a lower position where its density is in balance with that of the ambient fluid. In this position the plume spreads laterally. The height of this position can be found by finding the height of the position where ρ' becomes zero. In this position it follows from Equation 25 that

$$A^2 + (\rho_{e1}/g)(d\rho_e/dx)\beta^4 x^4 w^4 = 0. \tag{31}$$

In the case of a plume in a uniformly stratified fluid the relative axial concentration can be expressed by

$$c_r = c/\bar{c} = \{\rho' - x(d\rho_e/dx)\}/\Delta\rho_1. \tag{32}$$

It follows, therefore, that

$$c_{r*} = -x_*(d\rho_e/dx)/\Delta\rho_1 = x_{r*} \tag{33}$$

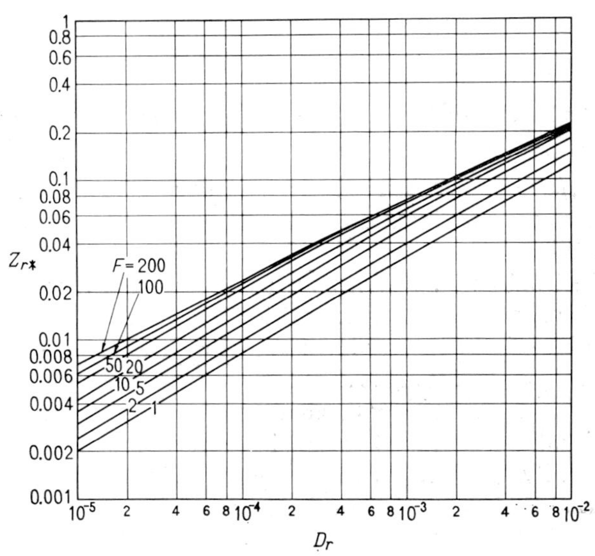

FIG. 9. Relative height of neutral position of plume

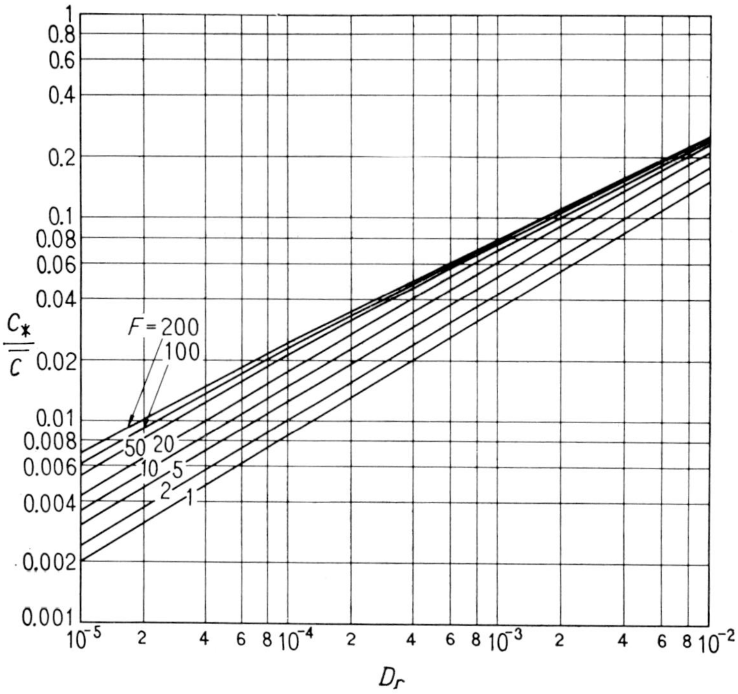

FIG. 10. Relative axial concentration of plume in neutral position

where c_{r*} is the relative axial concentration in the neutral position of a plume, and x_{r*} the relative height of the neutral position. Substitution of Equations 27, 28, and 33 into Equation 31 yields

$$\gamma^4 x_{r*}^4 w_{r*}^4 = 1, \quad \text{or} \quad x_{r*} = 1/\gamma w_{r*}. \tag{34}$$

Substituting Equation 34 into Equation 26, it follows that the value of Φ which makes ρ' equal to zero is given by

$$\sin \Phi = 1, \quad \text{or} \quad \Phi = \pi/2. \tag{35}$$

With the value of Φ thus determined x_{r*} and c_{r*} are obtained from Equations 24 and 33 as

$$x_{r*} = z_{r*} + \frac{D_r}{4\beta} = c_{r*} = \left\{ \frac{4\beta^2 F^2}{D_r(1 + D_r F^2)^{1/2}\gamma^3} \left(1{\cdot}20 - \int_0^{\Phi_1} \sin^{1/2} \Phi \, d\Phi \right) + x_{r1}^2 \right\}^{1/2}, \tag{36}$$

Figures 9 and 10 respectively show the relative height of, and relative axial concentration in, the neutral position of a plume; it has been assumed that $\beta = 0{\cdot}084$ as in the case of homogeneous environment.

Spread of a Plume in a Uniformly Stratified Surface Layer

Height and axial concentration for neutral position of plume

The density-depth relationship in environmental waters can often be approximated by $d\rho_e/dx = 0$ for the lower layer and $d\rho_e/dx$ = constant for the upper, or surface, layer. As shown in Fig. 11, the suffix 2 relates to the values of quantities at the level where the density gradient of the environmental water changes abruptly.

It is required to find the height of the position where a plume is theoretically supposed to spread laterally and also the axial concentration in that position. As has been shown in the previous sections, the linear spreading of a plume is a property independent of whether the value of $d\rho_e/dx$ is zero or a function of x. Therefore, a linear spreading is assumed throughout this analysis.

In order to make possible the application of the analyses made in the foregoing sections, an imaginary outlet with a diameter D_2 is taken in Section 2 and an imaginary rectangular profile of flow of effluent is assumed. Let $\Delta\rho_2$ be the density deficiency of the flow discharging from the imaginary outlet placed in Section 2, and F_2 the internal Froude number of the flow discharging from the same imaginary outlet. Then, from the similar consideration that was made for the derivation of Equation 12, it follows that

$$\bar{w}_2 = w_2/2, \qquad \Delta\rho_2 = \rho'/2 = (c_2/\bar{c})\Delta\rho_1/2, \quad \text{and} \quad D_2 = 4R_2 \tag{37}$$

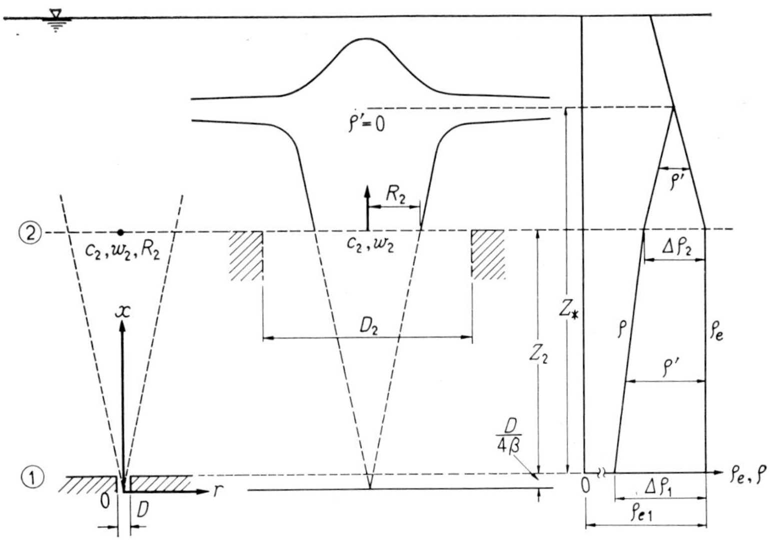

FIG. 11. Uniformly stratified surface layer

where $R_2 = \beta x_2$, and $\bar{c}$ and $\Delta\rho_1$ are the concentration and density deficiency of the effluent at the real outlet at $x = x_1$ as before.

By the use of Equation 37, $F_2 = \bar{w}_2(gD_2\,\Delta\rho_2/\rho_{e1})^{-\frac{1}{2}}$ is rewritten as

$$F_2 = \frac{1}{2\sqrt{2\beta}}\frac{w_2/\bar{w}}{\sqrt{c_2/\bar{c}}}\sqrt{\frac{D}{x_2}}\,F_1. \tag{38}$$

Here, F_1 is the internal Froude number of the flow of effluent at the real outlet, it being given by Equation 16.

It follows from Equation 36 that the height of the neutral position of a plume is given by

$$x_{r*} = \left\{\frac{1}{2\beta}F_2^{1/2}(4\beta x_{2r})^{5/4}(1 + 4\beta F_2^2 x_{2r})^{1/4}\int_{\Phi_2}^{\pi/2}\sin^{1/2}\Phi\,d\Phi + x_{2r}^2\right\}^{1/2} \tag{39}$$

where

$$x_{2r} = x_2\bigg/\frac{\Delta\rho_2}{-d\rho_e/dx} = \left(z_2 + \frac{D}{4\beta}\right)\bigg/\frac{(c_2/\bar{c})\Delta\rho_1}{-2d\rho_e/dx}, \tag{40}$$

$$\Phi_2 = \sin^{-1}(\gamma_2 x_{2r} w_{2r})^2, \quad \text{and} \quad \gamma_2^4 = \frac{16\beta^4 F_2^2}{D_{2r}^3(1 + D_{2r}F_2^2)}. \tag{41}$$

Therefore, the relative height of the neutral position above Section 2 is given by

$$x_{r*} - x_{2r} = z_{r*} - z_{2r} \cong \frac{\beta F_2^2}{2\gamma_2^3(1 + 4\beta x_{2r}F_2^2)^{1/2}x_{2r}^2}\int_{\Phi_2}^{\pi/2}\sin^{1/2}\Phi\,d\Phi, \tag{42}$$

provided that $x_{r*} - x_{2r} \ll x_{2r}$.

The definite integral in Equation 42 may be approximated in this case by

$$\int_{\Phi_2}^{\pi/2}\sin^{1/2}\Phi\,d\Phi = \left(\frac{\pi}{2} - \Phi_2\right) - \frac{1}{12}\left(\frac{\pi}{2} - \Phi_2\right)^3 + \cdots. \tag{43}$$

When $\beta = 0{\cdot}084$, Equation 42 becomes

$$x_{r*} - x_{2r} \cong 0{\cdot}76F_2^{1/2}x_{2r}^{1/4}(1 + 0{\cdot}336x_{2r}F^2)^{1/4}\int_{\Phi_2}^{\pi/2}\sin^{1/2}\Phi\,d\Phi, \tag{44}$$

with

$$\Phi_2 = \sin^{-1}[0{\cdot}58x_{2r}^{1/2}F_2/(1 + 0{\cdot}336x_{2r}F_2^2)^{1/2}], \tag{45}$$

and

$$x_{2r} = 2(z_2 + 3{\cdot}0D)(-d\rho_e/dx)/\{(c_2/\bar{c})\Delta\rho_1\}. \tag{46}$$

On the other hand, the axial concentration of the plume in its neutral position, c_*, can be calculated by

$$c_*/\bar{c} = -(x_* - x_2)(d\rho_e/dx)/\Delta\rho_1, \tag{47}$$

whence we obtain

$$c_*/\bar{c} = (c_2/2\bar{c})(x_{r*} - x_{2r}). \tag{48}$$

Numerical example

Let us consider a discharge from an orifice at a depth of 18·5 m, the effective diameter of the orifice being $D = 20$ cm. There is a density gradient in the upper layer. The densities and the values of F are as follows:

Sea water, surface ($z = 18{\cdot}5$ m)	ρ_{es}	$= 1{\cdot}0249$ g/cm³
Sea water, Section 2 ($z = 17{\cdot}0$ m)	ρ_{e2}	$= 1{\cdot}0257$ g/cm³
Sea water, bottom ($z = 0$)	ρ_{e1}	$= 1{\cdot}0257$ g/cm³
Effluent	ρ_1	$= 0{\cdot}9995$ g/cm³
$F = 10$ and 20.		

The difference of the densities, ρ_{es} and ρ_{e2} is equivalent to temperature difference $\Delta T \sim 4$°C.

It is required to determine the height of, and axial concentration in, the neutral position of the plume.

The procedure of numerical calculation is as follows. (1) Read the values for $c_2/\bar{c}$ and $w_2/\bar{w}$ in Figs. 3 and 4, respectively. (2) Determine F_2 from Equation 38. (3) Calculate x_{2r} from Equation 46 with the values $d\rho_e/dx = (1{\cdot}0257 - 1{\cdot}0249)/1{\cdot}50$ (g/cm³)/m and $\Delta\rho_1 = 1{\cdot}0257 - 0{\cdot}9995$ g/cm³. (4) Find the value of Φ_2 from Equation 45. (5) Calculate $x_{r*} - x_{2r}$ from Equation 44. (6) Calculate the value $z_* - z_2$ by the relation $z_* - z_2 = (x_{r*} - x_{2r})(c_2/\bar{c})\Delta\rho_1/(-2d\rho_e/dx)$. (7) Determine $c_*/\bar{c}$ from Equation 48. The values thus obtained are listed in Table 1.

TABLE 1. Numerical values in worked example

	F = 10	F = 20
$c_2/\bar{c}$	0·025	0·038
$w_2/\bar{w}$	0·185	0·125
F_2	1·52	1·67
x_{2r}	28·7	18·9
Φ_2	78° 12′ = 1·36 rad	76° 47′ = 1·34 rad
$x_{r*} - x_{2r}$	1·00	0·98
$z_* - z_2$ (m)	0·61	0·92
z_* (m)	17·61	17·92
$c_*/\bar{c}$	0·013	0·019

Conclusions

A theoretical study has been made of the initial dilution of effluent issuing vertically from a circular orifice into a calm sea. It leads to the following conclusions:

1. When the environment is homogeneous the axial concentration and vertical velocity of a plume of sewage are given by Equations 19 and 14 respectively. They are illustrated in Figs. 3 and 4.
2. In the case when the environment has a density gradient, the ceiling height of a plume is given by Equation 29 which is illustrated in Fig. 8. In the same case, the height of, and the axial concentration in, the plume are given by Equations 36 which are illustrated in Figs. 9 and 10.
3. When a depth of homogeneous water is covered by a uniformly stratified layer, the relative height of, and the axial concentration in, the plume are given by Equations 44 and 48.

REFERENCES

1. SCHMIDT, W. Turbulente Ausbreitung eines Stromes erhitzer Luft. *Z. angew. Math. Mech.*, 1941, **21**, 265.
2. ROUSE, H., YIH, C. R., and HUMPHREYS, H. W. Gravitational convection from a boundary source. *Tellus*, 1952, **4**, 201.
3. MORTON, B. R., TAYLOR, G. I., and TURNER, J. S. Turbulent gravitational convection from maintained and instantaneous sources. *Proc. R. Soc.* A, 1956, **234**, 1.
4. MORTON, B. R. Forced plumes. *J. Fluid Mech.*, 1959, **5**, 151.
5. ABRAHAM, G. Jet diffusion in liquid of greater density. *J. Hydraul. Div. Am. Soc. civ. Engrs*, 1960, **86**(HY6), 1.

6. HART, W. E. Jet discharge into a fluid with a density gradient. *J. Hydraul. Div. Am. Soc. civ. Engrs*, 1961, **87**(HY6), 171.
7. ABRAHAM, G. Jet diffusion in stagnant ambient fluid. Delft Hydraulics Laboratory, Publ. No. 29, 1963.
8. FRANKEL, R. J., and CUMMING, J. D. Turbulent mixing phenomena of ocean outfalls. *J. sanit. Engng Div. Am. Soc. civ. Engrs.*, 1965, **91**(SA2), 33.
9. FAN, L. N. Turbulent buoyant jets into stratified or flowing ambient fluids. California Institute of Technology, Rep. No. KH-R-15, 1967.
10. FAN, L. N., and BROOKS, N. H. Numerical solutions of turbulent buoyant jet problems. California Institute of Technology, Rep. No. KH-R-18, 1969.
11. ABRAHAM, G., and EYSINK, W. D. Jets issuing into fluid with a density gradient. *J. Hydraul. Res.*, 1969, **7**, 145.
12. PRIESTLEY, C. H. B., and BALL, F. K. Continuous convection from an isolated source of heat. *Quart. Jl R. met. Soc.*, 1955, **81**, 144.
13. HAYASHI, T., TSUCHIYA, T., and ITO, M. Diffusion of concentration of a buoyant plume of effluent. *Proc. 19th Conf. Coastal Engng, Jap. Soc. civ. Engrs, 1972*, p. 253 (in Japanese).
14. HAYASHI, T., and ITO, M. Diffusion of a forced plume in a uniformly stratified layer. *Proc. 17th Conf. Hydraul., Jap. Soc. civ. Engrs, 1973*, p. 6 (in Japanese).
15. HAYASHI, T., and ITO, M. Further study on the diffusion of forced plumes of effluent. *Proc. 20th Conf. Coastal Engng., Jap. Soc. civ. Engrs, 1973*, p. 189 (in Japanese).
16. CRAWFORD, T. V., and LEONARD, A. S. Observations of buoyant plumes in calm stably stratified air. *J. appl. Met.*, 1962, **1**, 251.

When presenting the paper, Prof. Hayashi said that hydraulic knowledge of initial dilution played an essential role in the prediction of water quality in the immediate proximity of a sewage outfall. The dispersion of sewage from the top of the rising column discharging from a port could be computed by various models, but it was more difficult to calculate the concentration at the top of the column. A limiting condition over the outfall was established with minimum dilution when there was no current in the surrounding water. Despite many excellent earlier works (including Dr. Abraham's pioneering studies[5,7,11]) on the determination of this limiting condition, Prof. Hayashi thought that further accumulation of knowledge of the fluid-mechanics aspects of initial dilution was desirable to match the technological progress in methods of discharging sewage.

The aim of the paper was to develop the analytical formulae and to provide diagrams for determining the limiting conditions over the outfall for various cases of vertical density gradient in the surrounding water, attention being confined to a circular orifice.

DISCUSSION

Mr. P. Ackers (U.K.) said that as Prof. Hayashi was President of the International Association for Hydraulic Research it was not surprising that the hydraulic theory given in the paper was elegant and convincing; the experimental results confirmed again that the present methods of calculating initial dilution in still water were reliable—they were perhaps the least controversial aspect of the performance of submarine outfalls. However, he invited Prof. Hayashi's comments on the small discrepancy in Fig. 6 at low densimetric Froude numbers, the results at $F = 2{\cdot}3$ consistently showing a dilution factor about 20 per cent below the calculated value. As this was a condition commonly found at outfalls, it would be unfortunate if the theory were less perfect in this range than at high Froude numbers.

He then questioned the Authors' philosophy that minimum dilution occurred when there was no current. In coastal regions with strong tidal currents this state was transient or non-existent, because the currents often had a rotary component and were not reciprocal—they reversed by revolving to a new direction rather than by going through zero. The performance of an outfall should be judged in relation to the full range of tidal velocities; each case had to be considered individually, but designing for a brief transient condition of still water could, in many situations, be unnecessarily conservative. If indeed there were a long period of still water the outfall design would be dominated by other considerations, such as the ultimate mechanism (density spread, degradation, or turbulent dispersion) by which the pollutants were removed. He thought it likely that if the water was virtually still the buoyant plume would be trapped below a stratified layer which would be self-generated if it did not arise from other circumstances. He asked if the method used on the last page of the paper was applicable in this case, and if so whether Prof. Hayashi could indicate how to calculate the effect of the build-up of a surface layer of diluted effluent in a permanently still water situation. The Chairman, Dr. G. Abraham (Netherlands), gave a reference to a paper[17] dealing with Mr. Ackers's final point.

Commenting on the subject rather than the paper itself, Dr. Abraham said that jet-diffusion theory had developed sufficiently far for the presentation of theoretical solutions for single round jets issuing into an infinite flowing fluid. Slawson and Csanady[18,19] had solved the problem of buoyant jets with negligible initial momentum entering a cross-flow of fluid with linear density gradient; they specified the length of the zone where the turbulence generated by the jet was large compared with that generated by the ambient fluid. Theoretical solutions for buoyant jets issuing with initial momentum into a cross-flowing homogeneous ambient fluid had been presented by Abraham[20], Ooms[21], Hoult and Weil[22], and Hirst[23]. The methods by which these solutions had been derived seemed applicable to jets in flowing ambient fluid where density varied with height[21].

REPLY TO DISCUSSION

Prof. Hayashi thanked Mr. Ackers for his kind comments and for his contribution. The small discrepancy at low Froude numbers seemed to have resulted from an error in estimating the coefficient of contraction, C, used in the experiments. In the Authors' theory, D should be interpreted as the effective diameter of the orifice, which required accurate estimation of C. They had assumed that C was 0·85 for the whole range of F, but it was well known that C depended on the Reynolds number of a jet, and that C approached unity when the Reynolds number became small. In their experiments, when F was small the Reynolds number was also small, hence when F was 2·3 the value of C should be almost unity. It could be seen that increasing C from 0·85 to 1 gave almost perfect agreement between the observed data and the theory.

With regard to Mr. Ackers's comment on their philosophy, he replied that their aim had been to study the limiting condition itself, rather than to examine how conservative it might be. However, he agreed that the effect of currents ought eventually to be considered, and that the build-up of spreading layers in shallow still water was pertinent.

ADDITIONAL REFERENCES

17. JIRKA, G. H., and HARLEMAN, D. R. F. The mechanics of submerged multiport diffusers for buoyant discharges in shallow water. Ralph M. Parsons Laboratory for Water Resources and Hydrodynamics, Rep. No. 169, Massachusetts Institute of Technology, 1973.

18. SLAWSON, P. R., and CSANADY, G. T. On the mean path of buoyant, bent-over chimney plumes. *J. Fluid Mech.*, 1967, **28**, 311–322.

19. SLAWSON, P. R., and CSANADY, G. T. The effect of atmospheric conditions on plume rise. *J. Fluid Mech.*, 1971, **47**, 33–49.

20. ABRAHAM, G. The flow of round buoyant jets issuing vertically into ambient fluid flowing in a horizontal direction. *Proc. 5th int. Conf. Wat. Pollut. Res., San Francisco and Hawaii, 1970*, Vol. 2, Paper III-15.

21. OOMS, G. A new method for the calculation of the plume path of gases emitted by a stack. *Atmos. Envir.*, 1972, **6**, 899–909.

22. HOULT, D. P., AND WEIL, J. C. Turbulent plume in a laminar cross flow. *Atmos. Envir.*, 1972, **6**, 513–531.

23. HIRST, E. A. Buoyant jets with three-dimensional trajectories. *J. Hydraul. Div. Am. Soc. civ. Engrs*, 1972 **98** (HY 11), 1999–2014.

Outlet Design

W. R. White
Hydraulics Research Station, Wallingford, England
and
A. R. Agg
Water Research Centre—Stevenage Laboratory, England

The design of an outfall for the discharge of liquid wastes to the sea is of fundamental importance. After discharge the effluent will be dispersed, diluted, and degraded under the influence of environmental forces, both long and short term, which vary substantially, and whose cumulative effect in a particular situation is difficult to predict[1]. The siting of the point of discharge affects the rate at which dispersion will take place and the minimum time required for any effluent to reach the shoreline. The method of discharge, however, and specification of the number, size, and shape of the outlets can be chosen so that a prescribed level of initial dilution can be achieved at the water surface subject to certain limits of the hydraulic parameters. The designer thus has considerable choice in the mode of discharge to be selected according to the range of conditions which will occur, and the dilution conditions prescribed for the effluent.

Whatever criteria may ultimately be proposed for particular effluents or considered appropriate in areas of prescribed amenity value, the technology of outfall design should be available to allow the method of discharge to be decided which will meet the stipulated requirements at minimum cost.

Review of Current Practice in the U.K.

Information on discharges of domestic sewage and associated trade effluents to coastal waters, outside the limits of 'controlled waters' as defined by the Clean Rivers (Estuaries and Tidal Waters) Act 1960, is taken from a survey[2] based on data for early 1972. Details relating to outfall length, depth of discharge, and type of discharge are summarized in Table 1. Although the majority of sea outfalls do not extend beyond the low-water mark of ordinary spring tides (LWMOST) they discharge only 38 per cent of the total dry-weather flow (DWF) and their number includes many relatively minor pipes serving small coastal communities.

TABLE 1. Data relating to U.K. sewage outfalls discharging at different distances offshore (at early 1972)

Distance of outfall from LWMOST (m)	Percentage of outfalls	Percentage of total DWF	Percentage of outfalls discharging at given depths (m) at LWMOST					Type of discharge (per cent of number at each distance)	
			0[a]	0–1·5	1·5–4·6	4·6–9·2	>9·2	Continuous	Intermittent
0[a]	61	38	61	—	—	—	—	80	20
0–90	24	27	—	15	8	0·6	0	78	22
90–270	5	7	—	1·8	1·8	1·2	0·3	35	65
270–1600	9	27	—	0·9	2·7	4·5	0·9	53	47
>1600	1	1	—	0	0	0·3	0·3	100	0

[a] Includes discharges landward of LWMOST

The use of outfall length as a comparative parameter is complicated by variations in the separation between high and low water marks on different parts of the coast. This distance may be only a few metres on steep rock or shingle beaches with a small tidal range but can approach 1 km on some beaches where the tidal range exceeds 6 m. Thus a discharge at LWMOST where there is a large tidal excursion across the beach may be considered more satisfactory for much of the tidal cycle than an

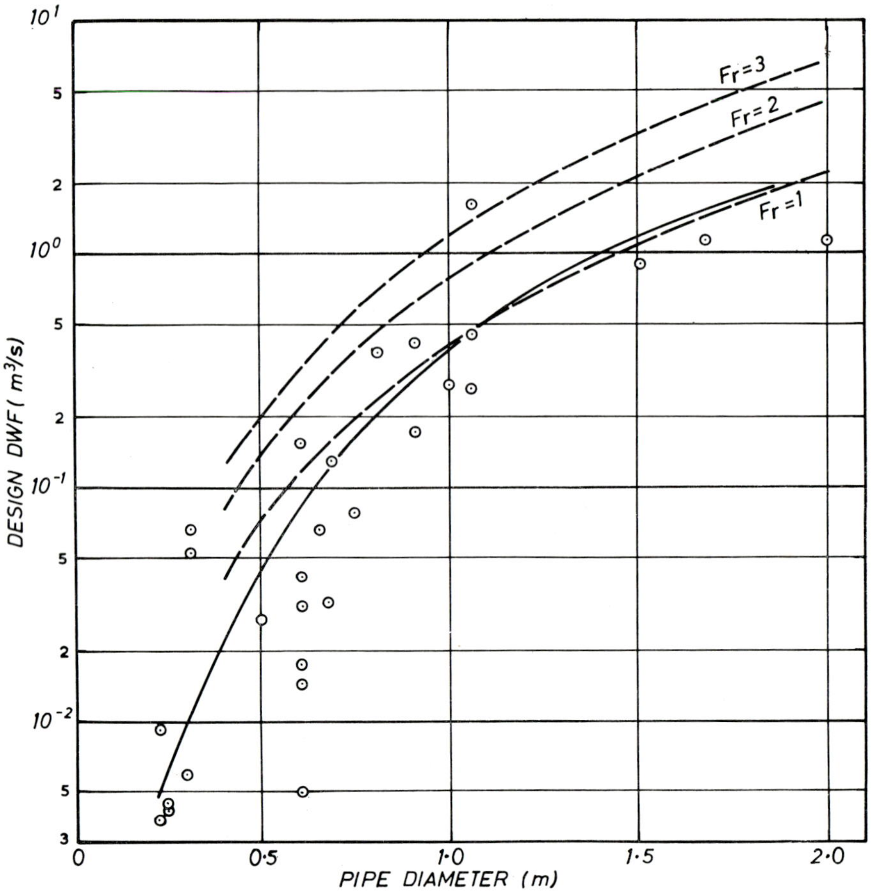

FIG. 1. Relation between design DWF and pipe diameter for 28 U.K. sewage outfalls *Fr* indicates densimetric Froude number

outfall extending a short way beyond LWMOST but where the discharge point remains relatively close to the waterline. In the former case normal discharge may be restricted to the period around high water by providing storage for DWF at other times and automatically controlling the periods of discharge. Excess flows normally pass to the outfall at any state of the tide. Some indication of the use of tidal storage, whether by design or by tide-locking of the outfall by high tides, can be seen from the figures for type of discharge in Table 1.

From information made available by the Directorate General of Water Engineering of the Department of the Environment and the Welsh Office it appears that there have been at least 45 schemes for new outfall installations since 1960. Data for design DWF for 28 of these outfalls were available and are plotted in Fig. 1 against pipe diameter, *D*. The broken lines represent the distributions of constant densimetric Froude number, *Fr*, defined by

$$Fr = u_j(g'D)^{-1/2} \tag{1}$$

where u_j is the velocity of the jet at the outfall and g' the effective acceleration due to gravity—defined by $g(\rho_a - \rho_0)/\rho_0$ where ρ_a is the density of the receiving water and ρ_0 the density of the effluent. Although the diurnal pattern of sewage flow has not been taken into account, nor the fact that sewage may be pumped intermittently in some cases rather than flowing continuously by gravity at a variable rate, it can be seen that under dry-weather conditions the densimetric Froude number of the jet will probably be fairly low.

To allow for diurnal variations in flow and for infiltration, provision should be made for at least three, and if possible six, multiples of the DWF to be discharged via the main sea outfall with excess storm sewage being diverted to a shorter overflow outfall. Where land levels allow sewage to flow by gravity to the outfall this is preferred, thus avoiding the costs of pumping. Taking the foregoing factors into account the working range of Froude number for sewage outfalls is probably 1–10.

Tidal velocities in the U.K. are usually in the range 0–1 m/s, but may exceed 3 m/s in certain localities. In general the direction of flow is parallel to the coast and reverses direction twice in each tidal cycle.

Design Methods

Design criteria

The basic criteria applicable to discharges of domestic sewage and industrial effluents to the sea in Great Britain[3] are that solids, identifiable as of sewage origin, should not reach areas used by the public for recreation; that the discharge should either be sufficiently diffuse or so remote that visual offence is not caused to the public; and that the effects on local fishing interests should be minimized. It is considered that such measures are adequate for the protection of public health.

The point of maximum concentration of effluent is in the immediate vicinity of the outfall, and the parameters which control initial dilution are sufficiently well understood to permit dilution factors to be calculated for various outfall arrangements. Dispersion models describing physical dilution can be constructed provided sufficient information is obtained to assess the rate of three-dimensional mixing under an appropriate range of conditions. When the waste is degraded in the sea some rate constant must be used to allow for this self-purification factor under different conditions. Should this correspond, however, to worst, median, or some other statistical average condition? Provided it can be established that there is sufficient residual tidal movement to prevent build-up of effluent in the area it would be appropriate to design for a given level of initial dilution, at least in those situations such as occur in the U.K. where discharges of sewage are no more than the present maximum of 2 m^3/s. Such a concept requires the definition of acceptable dilution factors at the water surface related to distance of the point of outfall from the shoreline.

Design philosophy

It is suggested, as a basis for discussion, that an outfall positioned close to the shoreline would, in general, be required to produce an initial dilution higher than one further out to sea, and at some position further offshore initial dilution would become unimportant. Superimposed on this general pattern would be some criteria for minimum tolerable dilution levels related to the formation of visible slicks. It is known that slicks will occur with initial dilution levels below about 100, that they will occur only infrequently with initial dilution levels in excess of about 400, and that in between these limits the grease content of the effluent will influence slick formation[4]. There is some evidence from observations off the Dutch coast[5] that subsequent vertical mixing of the sewage field with underlying sea water will not be restricted if the initial dilution factor exceeds 50.

The design requirements in terms of initial dilution are not listed in any design code and hence hypothetical criteria are proposed for discussion. Curves of suitable shape are given by an equation of the form

$$\frac{C_0}{C_m} - 1 = \frac{\alpha}{x^2} \tag{2}$$

where C_0/C_m is the initial dilution ratio at the surface, α is a constant related to the requirements of individual sites, and x is the distance of the outfall from the shoreline in metres. Four curves are plotted in Fig. 2 for different values of α.

Curve A ($\alpha = 10^8$ m^2) might be applicable to a site of high amenity value where a high degree of initial dilution is required while Curve B ($\alpha = 5 \times 10^6$ m^2) might be applicable to a remote site where initial dilution is relatively unimportant. The intermediate curves, A′ and B′, are for coefficients α of 5×10^7 and 10^7 m^2 respectively.

These four curves in Fig. 2 represent design requirements in the form of initial dilution levels at various distances from the shoreline for different degrees of 'sensitivity' related to local situations. It is interesting to compare the initial-dilution levels which the designer can, in practice, achieve. There is no unique answer because the engineer can vary some of the relevant parameters in his design procedure. For example, he can consider outfalls of various diameters, multiple outlets, capped outlets, or pre-dilution devices, and these options are considered later. However, two principal factors which influence initial dilution are the depth of the outfall beneath the surface and the ambient current velocity; these can be altered only by changing the position rather than the form of the outfall.

As examples of the initial dilutions which can be achieved using a simple horizontal pipe in still and moving water, two 'attainable' curves are plotted in Fig. 2 (Curves C and D) for a pipe diameter of 0·75 m, an effluent flow of 0·3 m^3/s, and an offshore sea-bed slope, S_0, of 1 in 100, all typical U.K. figures. The densimetric Froude number of the effluent discharge is thus 1·55. Curve C is for still (slack) water situations while Curve D assumes an ambient current of 0·25 m/s, again a typical U.K. figure.

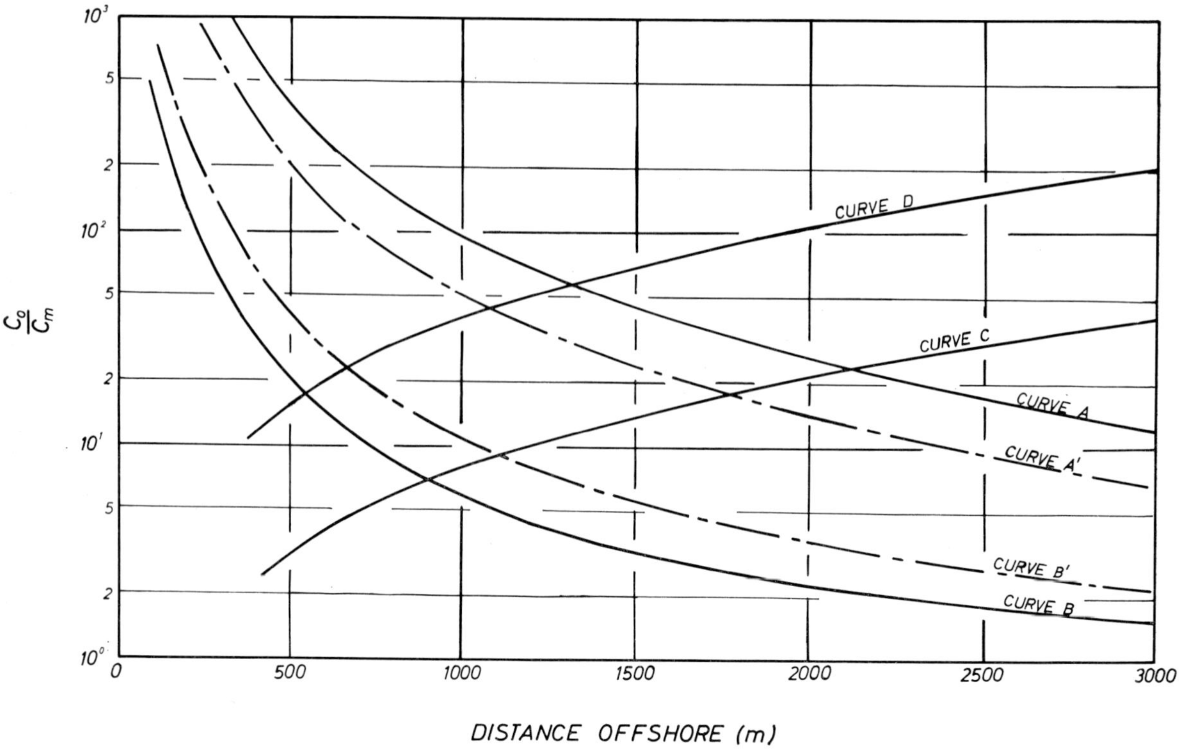

FIG. 2. Schematic diagram relating initial dilution (C_0/C_m) to distance offshore
Curves A, A′, B′, B show hypothetical design criteria for values of α in Equation 2 equal to 100, 50, 10, and 5×10^6 m^2. Curves C and D show attainable dilution for offshore slope of 1 in 100, outfall diameter 0·75 m, jet velocity 0·68 m/s, and ambient velocities of zero and 0·25 m/s respectively

Clearly, the design length for the outfall is where the required and attainable dilution curves coincide. At the 'difficult' site, where high initial dilutions are required and the ambient velocities are low, the two curves will cross furthest from the shoreline. Conversely at sites with no requirement for high initial dilutions and where ambient currents are strong, the intersection will be much closer to the shoreline. This approach is suggested as a logical method of determining outfall length. It is later elaborated further in the design example where methods of increasing initial dilution values are considered. There is obviously pressure on the engineer, particularly when faced with a difficult site, to find the most economic solution to the problem and in the context of this design method the problem resolves into considerations of the ways, means, and costs of modifying the curves for the attainable initial dilutions so that the stipulated requirements are met.

Performance data

Initial-dilution levels for still water can be calculated from established performance data. These data are readily available for plain circular outlets, the jet issuing horizontally, vertically[6], or at some intermediate angle[7]. Additional dilution associated with an ambient current can also be estimated. The performance of outlets other than the plain circular type are not documented to the same extent.

The performance data for circular outlets operating in still water have been presented by a number of workers in the form

$$\frac{C_0}{C_m} = \phi_1\left(\frac{y}{D}, Fr\right) \tag{3}$$

where C_0 is the initial concentration of the effluent, C_m the concentration at height y above the outfall, and D the diameter of the outfall.

When the receiving water is moving with a velocity u_a, the increase in dilution will be related to the velocity ratio u_a/u_j so that Equation 3 may be modified for the moving water case to

$$\frac{C_0}{C_m} = \phi_2\left(\frac{y}{D}, Fr, \frac{u_a}{u_j}\right). \tag{4}$$

Agg and White[8] showed that the above non-dimensional parameters, while being a comprehensive set capable of describing the phenomenon and suited to the requirements of the research worker, were

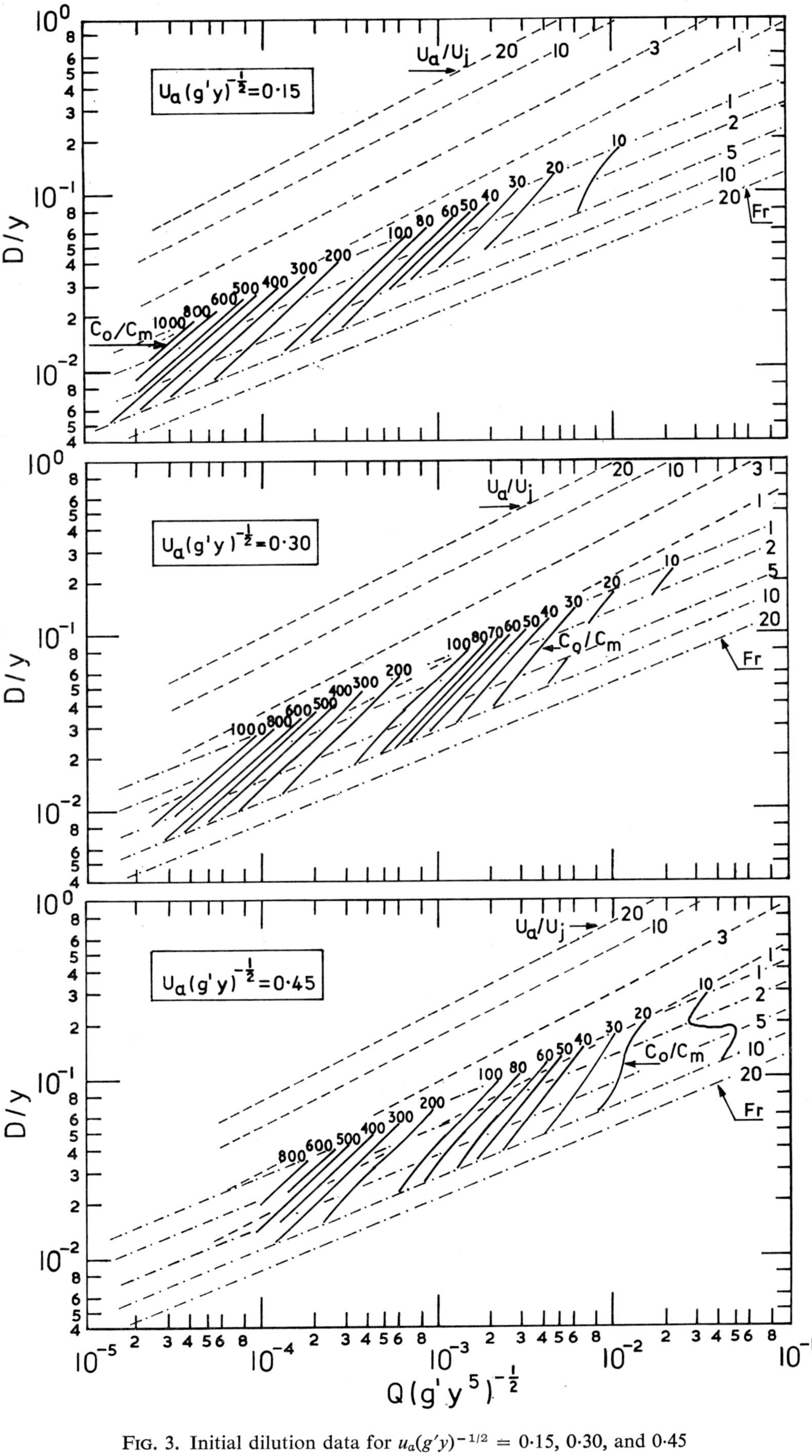

FIG. 3. Initial dilution data for $u_a(g'y)^{-1/2} = 0{\cdot}15$, $0{\cdot}30$, and $0{\cdot}45$

somewhat inconvenient for the designer since the pipe diameter, D, is implicit in all three quantities on the right-hand side of Equation 4. They suggested a re-arrangement of the basic variables to give the following new groupings

$$\frac{C_0}{C_m} = \phi_3\left\{\frac{D}{y}, \frac{u_a}{(g'y)^{1/2}}, \frac{Q}{(g'y^5)^{1/2}}\right\} \tag{5}$$

where Q is the effluent discharge.

Existing performance data were given in this new form of presentation and three graphs corresponding to $u_a(g'y)^{-1/2} = 0{\cdot}15$, 0·30, and 0·45 are given in Fig. 3. The positions of the C_0/C_m contours are based on the data of Abraham[6] for slack water and the correction factor for the moving water case of Agg and Wakeford[9]. It will be noted that the parameter $u_a(g'y)^{-1/2}$ is dependent upon the location of the outfall; it is independent of the effluent flow and details of the geometry of the outfall. Thus the engineer can choose the appropriate design chart before giving thought to the details of the outfall itself.

One interesting conclusion from these charts is that, for a given set of ambient conditions, and a given sewage flow and depth of water, values of C_0/C_m increase with increasing values of D/y, i.e. increasing values of the diameter, D. Thus the most efficient outfall in terms of initial dilution at the surface is one with the largest practicable diameter, subject to the densimetric Froude number being greater than unity if entrainment of sea water is to be avoided. In general, initial dilutions increase as the densimetric Froude number falls in the range $1 < Fr < 10$. There is a marginal reversal of this trend at high Froude numbers ($Fr > 10$) but this is not of practical significance to the design engineer who seeks to minimize head losses at the outfall.

Types of Outfall

The outfall system consists of a pipe or pipes designed to transmit the effluent from the shoreline to the point of discharge and the outfall itself. The outfall may consist of a simple orifice, or a device designed either to spread the effluent over a wider area or to mix the effluent with sea water thus achieving a degree of pre-dilution before emission.

The diameter of the delivery pipe is chosen to minimize head losses while maintaining self-scouring velocities within the pipe. The commonly accepted self-cleansing velocity for raw sewage[10] is 0·76 m/s but for sea outfalls, where primary settlement of gross solids or comminution is frequently carried out, a value of 0·5 m/s seems appropriate. Maximum velocities of 3 m/s are occasionally reached but head losses are significant and pumping is normally required in such cases.

Outfalls with no pre-dilution

The most usual outfall is a simple circular orifice which emits the effluent in a near-horizontal direction. Abraham[6] has shown that a horizontal jet gives higher dilution than a vertical jet although the difference is not great when the kinetic energy of the emergent jet, and hence the densimetric Froude number, is low. Frankel and Cumming[7] have provided information of the performance of jets at other angles.

Rawn, Bowerman, and Brooks[11] obtained increased initial dilutions by spreading the effluent over a wider area from a series of outlets, or a diffuser. The outlets are aligned at right angles to the ambient current where this has a fairly well defined and unique direction, but where the tidal currents vary in direction a forked pipe system is preferable such that the outlets constitute an area rather than a line source. One criterion for such a design is that the spacing of the individual outlets should be sufficient to prevent interference between adjacent jets.

Hansen and Schrøder[12] attempted to increase initial dilution by placing obstructions in front of a single outlet to divide the flow into two or more components. The best device tested doubled the level of dilution but such an improvement could have been achieved by discharging through two separate outlets. Spiral inserts in the end of the pipe showed little improvement and increased head losses within the system.

Cover plates form another means of achieving additional initial dilution. The outfall is arranged such that the buoyant effluent is intercepted by a horizontal disc or plate with the result that the emergent jet is spread over a wider plan area and initially travels in a horizontal or downward direction. Both factors help to increase dilution at the surface.

Circular outlets, while having the advantage of ease of construction, are not the most efficient in terms of obtaining high dilution factors. Dilution must depend to some extent on the ratio of the cross-sectional area of the emergent jet to its perimeter; a larger perimeter for a given cross-sectional area should produce greater dilution. Hence rectangular or triangular orifices are probably more efficient than circular ones. A long slot has the highest perimeter/area ratio and Anwar[13] has reported favourably on the performance of such a device.

Outfalls with pre-dilution

One way of increasing initial dilutions is to mix the effluent with ambient fluid before emission from the outfall. This method, in effect, concertinas the lower section of the plume within the outfall device since the mixed effluent emerges with a concentration which would otherwise have occurred well above the outfall.

Nece, Goldstern, and Black[14] obtained entrainment of sea water into the effluent using a venturi section in a model outfall pipe. Pressures within the venturi section were lower than the surrounding ambient pressure and sea water was drawn in through a separate pipe. An additional dilution factor of two at the outfall was achieved but this device involved a substantial increase in head losses in the outfall system.

Research on water jet pumps has been reported by Silvester[15] and further work has since been carried out by Patarapanich[16]. The water jet pump operates by fluid being discharged from a driving nozzle into a larger-diameter concentric mixing tube where entrainment of the ambient fluid takes place. Dilutions at the outfall of more than two were obtained by Silvester using relatively high velocity jets and with the receiving water apparently at rest. Patarapanich examined various ratios of driving and mixing tubes and predicted an improvement in initial dilution somewhat less than two at the surface.

Agg and White[8] investigated various types of mixing devices and concluded that, of those tested, the most efficient type was of the form shown in Fig. 4. The flow of effluent created a weak vortex within a conical shaped device which has a circular opening at the apex of the cone and an annular opening around the base. The density difference between the effluent and the receiving fluid created a spiral column of mixed effluent which entrained the ambient fluid at the base of the cone and emitted the mixed effluent through the upper orifice. The amount of pre-dilution achieved by the device depended on both the effluent and ambient velocities and dilution factors varied from 4 to 18. Initial dilutions at the water surface were increased by a factor of about two for typical U.K. tidal situations.

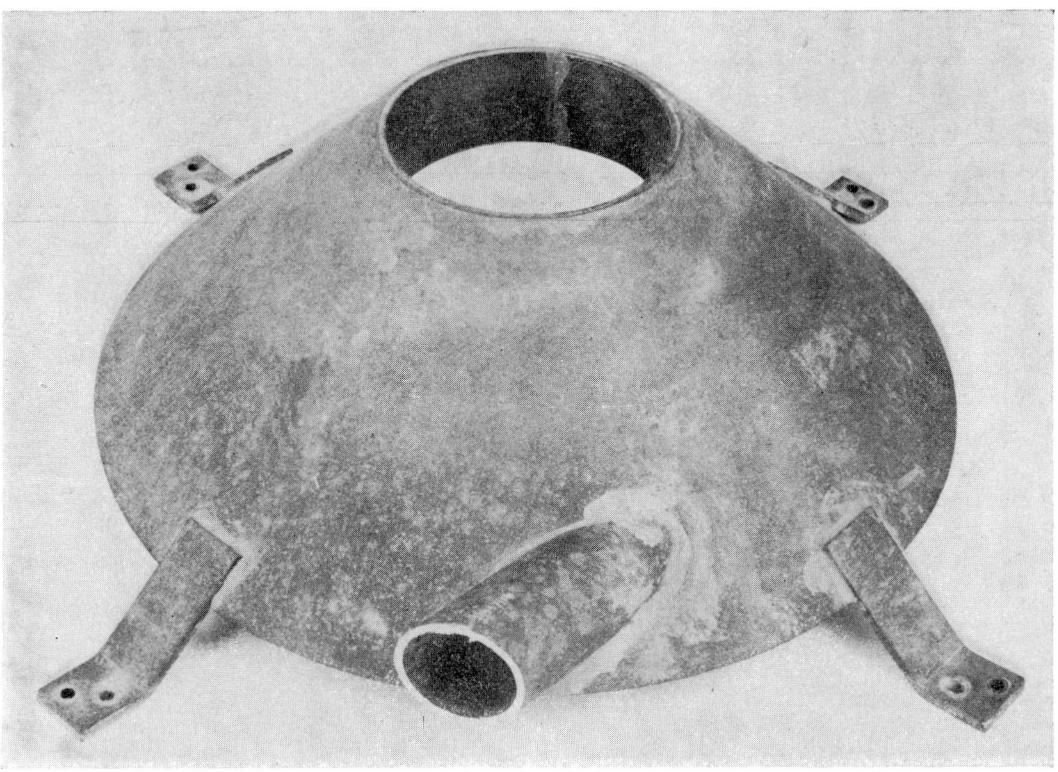

FIG. 4. Pre-dilution device, Agg and White[8]

Design Example

An outfall is to be designed which will satisfy the requirements of demand curve A, Fig. 2. Ambient flows vary from $u_a = 0{\cdot}05$ to $0{\cdot}25$ m/s and the offshore slope of the sea bed is 1 in 100. The design effluent flow is $0{\cdot}3$ m^3/s with $(\rho_a - \rho_0)/\rho_0 = 0{\cdot}026$.

Consideration is to be given to the relative merits of four specific designs: a single plain pipe outlet, a single plain pipe outlet with pre-dilution device, a multiple-port system (with say 10 ports), and a similar system in which each port is fitted with a pre-dilution device. It is required to calculate the relevant design parameters such that each type of outlet will meet the initial dilution requirements under the most adverse ambient conditions ($u_a = 0{\cdot}05$ m/s) and to calculate the additional dilution when $u_a = 0{\cdot}25$ m/s. Limiting conditions relevant to subsequent mixing and slick formation are also to be considered.

1. Single plain pipe outlet

Agg and White[8] have shown that outfalls operating at or near a densimetric Froude number of unity produce optimum initial dilutions at the water surface. This criterion gives an outfall diameter of 0·9 m and a jet velocity of 0·47 m/s. Using Abraham's data[6] for still water conditions together with Agg and Wakeford's correction[9] for moving water (see Fig. 3) the following dilution figures are obtained at various depths when $Fr = 1$ and the ambient velocity is at a minimum of 0·05 m/s.

Depth (m)	5	10	20	30	50
C_0/C_m	4	11	31	59	140

This curve is plotted in terms of distance offshore ($S_0 = 1/100$) as Curve 1 in Fig. 5 and the design requirements of Curve A are met 1900 m offshore where a dilution of 28 is achieved. The initial dilution when $u_a = 0{\cdot}25$ m/s is then 130. Initial dilution factors of 50 and 100 are obtained at 2700 and 4100 m offshore respectively for $u_a = 0{\cdot}05$ m/s; these would be the appropriate lengths if mixing or slick formation were limiting factors.

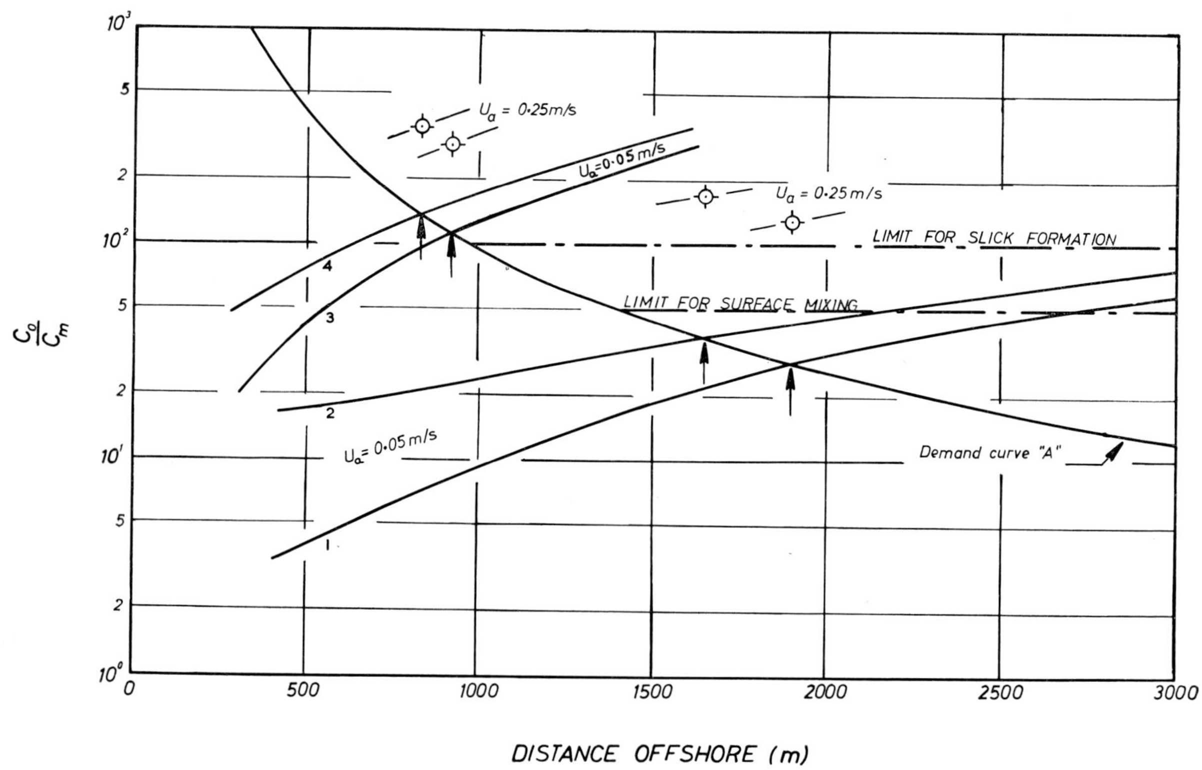

FIG. 5. Design example, initial dilution factors
Curve A, hypothetical demand curve from Fig. 2
Curves 1–4 for ambient velocity of 0·05 m/s: 1 and 2 with single orifice, 3 and 4 with 10-port system; 1 and 3 with plain outlets, 2 and 4 with pre-dilution devices

2. Single plain pipe outlet with pre-dilution device

The device shown in Fig. 4 produces a dilution of around 10 within the cone when the inlet densimetric Froude number is unity. The outlet diameter is $0{\cdot}9\sqrt{10} = 2{\cdot}84$ m and the emission of the mixed effluent is at a densimetric Froude number of $1 \times 10^{1/4} = 1{\cdot}78$. Again using the data of Abraham, and of Agg and Wakeford (see Fig. 3) to determine the dilution occurring within the rising jet and then multiplying these figures by the pre-dilution of ten gives Curve 2 of Fig. 5 for obtainable initial dilution values when $u_a = 0{\cdot}05$ m/s. In this case the design requirements of Curve A are met 1650 m offshore where dilution of 37 is achieved. The initial dilution when $u_a = 0{\cdot}25$ m/s is 170. Initial dilution factors of 50 and 100 are obtained at 2150 and 3150 m offshore respectively when $u_a = 0{\cdot}05$ m/s.

3. Multiple-outlet system (10 ports)

The design flow per port is 0·03 m^3/s and this indicates a diameter of 0·35 m and jet velocity of 0·31 m/s to produce a densimetric Froude number of unity. Assuming that the ports are spaced such that interaction of the ascending plumes is avoided the attainable surface initial dilutions are as shown in Curve 3 of Fig. 5. The outlets have to be 925 m offshore where the initial dilutions are 110 and 290 for the minimum and maximum ambient velocities respectively.

Brooks[17] studied the geometry of plumes in still water and, for low densimetric Froude numbers, his graphs indicate an approximate expression for the width, w, of the plume just below the surface:

$$w = 0{\cdot}3y. \tag{6}$$

There seems to be little data on the geometry of plumes in moving water, but if it is assumed that the approximate width of the plume rises with the cube root of the additional dilution caused by the moving water, then, using Agg and Wakeford's expression for this additional dilution, a valid equation would be

$$w = y\left(0{\cdot}3 + 0{\cdot}4\sqrt{\frac{u_a}{u_j}}\right). \tag{7}$$

Using these criteria the spacing of the outlets would have to be 6 m to meet the requirement when $u_a = 0{\cdot}25$ m/s.

4. Multiple-outlet system (10 ports), each port with a pre-dilution device

The inlet densimetric Froude number to the pre-dilution devices is again unity which indicates a pre-dilution of around 10. The outlet diameter is 1·11 m with a mixed effluent velocity of 0·31 m/s. Using the method outlined in 2 above the obtainable initial dilutions are as shown by Curve 4 in Fig. 5. The outlet system has to be 830 m offshore and the initial dilution factors are 140 and 350 for the minimum and maximum ambient velocities respectively. A small increase in the spacing of the ports would be desirable.

The example shows the marked advantages of a multiple outlet system. The ten-port system reduced pipe lengths by approximately 50 per cent. The pre-dilution devices further reduced pipe lengths by 12 per cent but this figure would be much higher in conditions where the required initial dilutions are less severely prescribed or where the offshore slope is shallower than 1 in 100. The pre-dilution devices are also useful where the existence of concentrated effluent near the sea bed is undesirable.

ACKNOWLEDGEMENTS

This paper is based on work carried out at the Hydraulics Research Station and the Water Pollution Research Laboratory, and is published with the approval of the respective Directors, Mr. R. C. H. Russell and Mr. G. E. Eden.

REFERENCES

1. MUNRO, D. Observed and predicted coliform distributions. This volume, pp. 353–362.
2. DEPARTMENT OF THE ENVIRONMENT and WELSH OFFICE. Report of a survey of the discharges of foul sewage to the coastal waters of England and Wales. H.M. Stationery Office, London, 1973.
3. AGG, A. R. Criteria for marine waste disposal in Great Britain. In 'Marine pollution and marine waste disposal' (ed. E. A. Pearson and E. d. F. Frangipane) Pergamon Press, 1975 (*in press*).
4. NEWTON, J. R. Factors affecting slick formation at marine, sewage outfalls. In 'Pollution criteria for estuaries' (ed. P. R. Helliwell and J. Bosanyi). Techno Press, London, 1965, Chap. 12 (*in press*).

5. Abraham, G., and Brolsma, A. A. Diffusers for disposal of sewage in shallow tidal water. International Association for Hydraulic Research Congress, Leningrad, 1965, Vol 2, p. 1.
6. Abraham, G. Jet diffusion in stagnant ambient fluid. Delft Hydraulics Laboratory, Publ. No. 29, 1963.
7. Frankel, R. J., and Cumming, J. D. Turbulent mixing phenomena of ocean outfalls. *J. sanit. Engng Div. Am. Soc. civ. Engrs*, 1965, **91**(SA2), 33.
8. Agg, A. R., and White, W. R. Devices for the pre-dilution of sewage at submerged outfalls. *Proc. Instn civ. Engrs*, Part 2, 1974, **57**, 1.
9. Agg, A. R., and Wakeford, A. C. Field studies of jet dilution of sewage at sea outfalls. *Instn publ. Hlth Engrs J.*, 1972, **71**, 126.
10. British Standards Institution. B.S. Code of Practice CP2005: 1968. Sewerage.
11. Rawn, A. M., Bowerman, F. R., and Brooks, N. H. Diffusers for disposal of sewage in sea water. *J. sanit. Engng Div. Am. Soc. civ. Engrs*, 1960, **86**(SA2), 65.
12. Hansen, J., and Schrøder, H. Horizontal jet dilution studies by use of radioactive isotopes. *Acta polytech. scand.*, Ser. (b), 1968, No. 49.
13. Anwar, H. O. Two-dimensional buoyant jet in a current. *J. engng Math.*, 1973, **7**, 297.
14. Nece, R. E., Goldstern, P. P., and Black, J. L. Single-port suction manifolds. *J. Hydraul. Div. Am. Soc. civ. Engrs*, 1966, **92**(HY1), 43.
15. Silvester, R. Jet mixers in sewage outfalls. *J. Instn Engrs, Aus.*, 1967, **39** (March), 33.
16. Patarapanich, M. M.Eng. thesis. Asian Institute of Technology, Bangkok, 1971.
17. Brooks, N. H. Conceptual design of submarine outfalls. In 'Pollution of coastal and estuarine waters', University of California, Wat. Resour. Engng Educational Ser., Progm VII, Berkeley, Calif., January 1970.

Introducing the first part of the paper, Mr. Agg amplified details relating to the current practice of sewage disposal in the United Kingdom where about 8 per cent of the total dry-weather flow of 12 million m^3/day was discharged to sea through over 300 separate outfalls with contributory populations ranging from a few hundred to 100 000 (plus one of 300 000). Commenting on the operating characteristics of these outfalls, he said that in the studies of the effect of ambient flow on initial dilution at five outfalls the densimetric Froude number had always been less than 5. He emphasized that hydrographic conditions around the U.K. were particularly favourable for dispersion, and outlined the philosophy behind the design procedure given in the paper.

Dr. White then gave details of the derivation of the design charts shown in Fig. 3, and explained the differences between performance in still and moving receiving waters. He concluded by describing the tests of possible pre-dilution devices which had led to the design shown in Fig. 4.

DISCUSSION

Mr. P. Ackers (U.K.) referred to Equation 4, which extended the usual Abraham type of relationship of dilution to densimetric Froude number and relative depth—multiplying it by an ambient current factor which (following Agg and Wakeford[9]) depended solely on the ratio of current to outlet velocity. This was a case where a function satisfying dimensional analysis might not satisfy arguments based on physical similarity. If an outfall operated at low densimetric Froude numbers, the momentum of the outflow was unimportant compared with its buoyancy efflux. He asked how, in those conditions, the efflux velocity could be significant in determining the additional dilution due to an ambient current. Taking buoyancy flux ($Q\Delta\rho g$), ambient velocity (u_a), density (ρ), and water depth (y) as the variables, the group that emerged was $u_a^3 y/Qg'$. A more generally acceptable added dilution factor might have emerged if Agg and Wakeford had related their field observations for their low-momentum outfalls to that parameter rather than to the velocity ratio. An argument could be developed from this, suggesting that over most of the range of Fig. 3 initial dilution in the presence of a current should depend on $u_a/\sqrt{(g'y)}$ and $Q/\sqrt{(g'y^5)}$ but not on D/y. It was the flow that counted, not the size of the hole from which it emerged. It might thus be unwise to rely on Fig. 3 outside the range of Agg and Wakeford's results. He asked the Authors to indicate the range of conditions covered by the field data on which Fig. 3 had been based.

Mr. Ackers concluded by citing results of calculations of the performance of alternative forms of diffuser in an actual case in British waters. Whereas, under particular conditions, twelve 15-cm outlets would give an initial dilution of 200, twelve pre-dilution devices (as in Fig. 4) would give one of 350, and a line-source of 240 outlets of 2·5 cm would give 500.

The Chairman, Dr. G. Abraham (Netherlands), said that the Authors had stated that the combined effects of dilution and degradation occurring after discharge of an effluent were difficult to predict. This observation seemed correct[1, 18] and resulted partly from the fact that shoreward transport of effluent was due solely to diffusion only when currents were parallel to the shore. Inaccuracies in the description of the diffusion might also lead to errors in the estimates of the time taken by the effluent to reach the shore. This implied that predictions of the concentrations of non-conservative effluents at shore were particularly affected by these inaccuracies[19]. Because of this difficulty the Authors had suggested defining acceptable initial dilution factors at the water surface related to the offshore distance of the outfall, these factors being applied at the point of discharge—provided it could be established that there was sufficient tidal movement to prevent build-up of effluent in the discharge area. However, it seemed that the cumulative effect of dilution and of the degradation taking place after discharge had to be determined before deciding on the potential build-up. The Authors suggested Equation 2 as a hypothetical design criterion, but Dr. Abraham maintained that the relationship between the concentration of a conservative effluent reaching the shore and the offshore distance (x) of the outfall was far more complicated than a simple power function[20]; for non-conservative substances the concentration at the shore clearly decreased more rapidly with x.

The data of Fig. 3 showed that, for a given set of conditions, initial dilution increased with increasing jet diameter (D), but Dr. Abraham said that this did not agree with jet-diffusion theory for round jets issuing horizontally into stagnant ambient fluid. For such jets the initial dilution did not vary with D for values of Fr less than about $50y/D$, while for larger values it had been shown[21] to increase with decreasing values of D. Finally, he asked for which types of outlet Fig. 3 had been established, how the data had been obtained, and how the initial dilution represented by Fig. 3 compared with jet-diffusion theory in stagnant and flowing receiving fluid.

Dr. J. A. French (U.S.A.) also asked for further experimental details, and commented that the effectiveness of the pre-dilution devices described must necessarily be limited to small dilutions. To obtain dilutions of about 100, one had to involve fluxes of ambient fluid about 100 times the port discharge, and this meant involving a large volume of slow-moving ambient fluid which, in practice, could be entrained only by a conventional plume rising through many metres of water. A solid structure to accomplish this would be impractically large, or the energy used would be excessive.

Mr. B. Sterregaard (Denmark) said he believed that a rational approach to outfall design was to determine the length on the basis of some standards or criteria for pollution or dilution, such as at a beach or in a shellfish area, and observations of current, dispersion, and other characteristics, in the receiving waters—even if predictions were uncertain. At the point of discharge thus determined, a diffuser should be designed to obtain a required initial dilution, taking into account factors mentioned in Paper 29. He asked if the demand curves in Fig. 2 had been arbitrarily laid down or how the values of α had been obtained.

Prof. P. Harremoës (Denmark) said he was strongly in favour of a conceptual approach in which the objectives were identified, even if those objectives were difficult to express quantitatively; otherwise the problem was obscured. He considered that the design approach shown in Fig. 2 was evading the real issues.

Mr. J. A. Wakefield (U.K.) remarked that one was tempted to accept the Authors' design philosophy because of its simplicity; however, since many beaches were situated in bays where the currents impinged on the shore, it was necessary to employ more sophisticated methods for siting sewage outfalls than was suggested in the paper.

Mr. B. E. Waters (U.K.), whose firm had been involved with long sea outfalls for a number of years, showed some slides illustrating the great practical importance of achieving adequate initial dilution. With a 380-m outfall discharging into 3 m of water the effluent could be seen along some 6 km of coast. Even with an outfall in deep water the absence of a diffuser could result in a surface boil—as he showed for a 1-km outfall (1 m in diameter) discharging horizontally at high slack tide about 18 m below the surface. He then gave an example of the application of research: in a scheme currently being constructed in tunnel, the diffuser design had been simplified as a direct result of information supplied by Agg and Wakeford[9], with a resulting saving of £200 000 on the tendered cost of construction. Looking at dilution devices, the slot was an elegant solution, but any practical application considered by his firm would have required an unacceptably narrow slot.

Dr. R. J. Taylor (U.K.) pointed out that reference had been made in the paper to the spacing of outlets to prevent interference between adjacent jets, but that apart from some calculations made by Brooks[17] there appeared to be very little information on this subject; he asked if the Authors could give any guidance. He also asked if there had been any developments to prevent the deposition of solids which might present long-term problems near outfalls, and if any recommendations could be made on diffuser arrangement where there was clear evidence that the tidal flow was parallel to the shore at the outfall.

REPLY TO DISCUSSION

Whilst agreeing with Mr. Ackers's cautionary remarks, the Authors pointed out that the data presented in Fig. 3 had deliberately been restricted to the range covered by the field experiments[9]. At very low densimetric Froude numbers the velocity ratio u_a/u_j might diminish, but at what Froude number and by how much was unknown.

Commenting on Dr. Abraham's discussion, the Authors emphasized that their design philosophy was related primarily to discharges of domestic sewage with low concentrations of industrial effluents. Although the performance data were applicable to effluents containing conservative pollutants, there might be other constraints on their discharge in particular locations. Drs. Abraham and French had asked for details of the original data from which Fig. 3 was derived; these had been given in an earlier paper[9]. The experiments had been carried out with round outlets, discharging horizontally, using either salinity or a radioactive tracer for dilution determinations. No effect of jet alignment to the direction of the ambient flow on dilution at the surface was detected, but this was perhaps not surprising in view of the low densimetric Froude numbers of the jets. The Authors recognized the variability in the field data, and had advocated that confirmatory laboratory experiments should be carried out. The experiments with pre-dilution devices had been designed to obtain tenfold dilution with sea water within each device; in practice, dilution factors varied between 4 and 18. There was no suggestion, as Dr. French implied, that 100-fold dilution could be obtained within such a device. The relative benefits of multiple outlets and pre-dilution devices were illustrated by the design example.

Mr. Sterregaard and Prof. Harremoës had expressed preferences for standards or criteria, although admitting the inaccuracies of methods of prediction. The Authors had defined qualitatively the objectives applicable to U.K. conditions, and believed they could be expressed quantitatively in terms of the initial dilution which could be predicted with some confidence. However, any attempt at producing a generalized procedure was open to criticisms on the grounds of particular site characteristics such as those cited by Mr. Wakefield. The Authors agreed that each site must be assessed individually. The application of the design procedure required information on local current velocities, the determination of which would indicate particular problems associated with local tidal movement or unusual topography. Nevertheless, such indications should not invalidate the method, although they might suggest that alternative lines for the outfall should be considered or that an additional length be added to provide further dilution during transit to the shore.

Replying further to Mr. Sterregaard, the Authors agreed that the form of Equation 2 was arbitrary. However, Fig. 5 included limits for surface mixing and slick formation based on the admittedly limited observational data available. Experience gained from existing outfalls in the U.K. suggested that the demand curves were realistic—particularly when the effects of other sources of pollution were recognized.

In reply to Dr. Taylor, the Authors agreed that information on the interference between jets from multiple outlets was fragmentary and empirical, especially in the practical case where the ambient water was moving past the outfall. The width of an individual jet was related to the depth of water, but would be substantially affected by the behaviour of the effluent on reaching the water surface. The problem was thus further complicated by the residual momentum of the jet and the density difference between the diluted effluent and the sea water. Diffuser sections should be arranged to intercept the maximum flow of water normal to the rising jets. There were recommended minimum velocities which should be maintained in pipelines to ensure self-scouring conditions[8].

ADDITIONAL REFERENCES

18. Dam, G. C. v. The Hague outfall. This volume, pp. 393–401.

19. Abraham, G., and Dam, G. C. v. On the predictability of waste concentrations. In 'Marine pollution and sea life' (ed. M. Ruivo). Fishing News (Books) Ltd, West Byfleet, England, 1972, pp. 135–160.

20. Brooks, N. H. Diffusion of sewage effluent in an ocean current. *Proc. 1st int. Conf. Waste Disposal mar. Envir.*, Univ. Calif., Pergamon Press, New York, 1960, pp. 246–267.

21. Abraham, G. Horizontal jets in fluid of other density. *J. Hydraul. Div. Am. Soc. civ. Engrs*, 1965, **91** (HY 4), 139–154.

Paper No. **28**

Surface Spreading

Hans Hydén

Assistant Professor, Royal Institute of Technology, Stockholm

and

Ian Larsen

Head of Submarine Piping Section and Off-shore Systems, VBB-SWECO, Stockholm

The surface spreading phase starts when the vertical motion of a diluted waste-water plume ceases, and ends when the difference in density between the diluted waste water and the surrounding waters is no longer a significant factor for the pattern of motion.

The object of this paper is to describe what may be expected to happen when diluted waste water at the rate of Q arrives at the water surface—or is trapped at an intermediate level—with almost the same horizontal velocity, U_0, as the main current. Due to the relative difference in density, Δ, the diluted waste water will try to spread in all directions.

Buoyancy Spread without Vertical Turbulent Diffusion

A description of the buoyancy spread has already been published[1]. It was found that the spreading of waste water due to density differences was chiefly determined by the buoyancy flux and to a large extent was unaffected by dilution and diffusion. The theory was inexact in many details, but was expected to give results of the correct order of magnitude. Calculations for outfalls from larger cities showed that places normally considered sheltered from the outfall zone by a region of considerable turbulent diffusion could in some cases be polluted by poorly diluted waste water and that the influence of wind could add significantly to this effect.

The sheet of waste water formed on top of the main current is supposed to have a uniform thickness, h, in the direction perpendicular to the main current. A density front will advance with a velocity $v = k\sqrt{\Delta gh}$ relative to the main current, g being the gravitational acceleration and k a constant between 1 and 1·4. With the notation of Fig. 1 the front condition states that

$$U_0 \sin \theta = k\sqrt{\Delta gh}. \qquad (1)$$

When turbulent diffusion is neglected, the continuity equation is

$$2BhU_0 = Q. \qquad (2)$$

Integration yields

$$\frac{B}{B_0} = 1 + \left(\frac{3x}{2B_0}\right)^{2/3} \qquad (3)$$

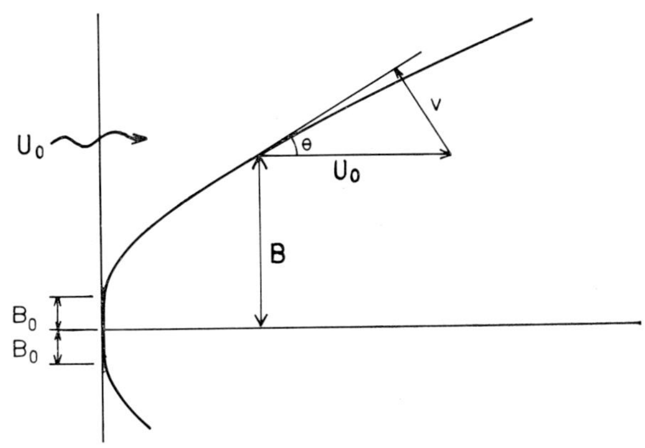

FIG. 1. Plan view of plume boundary in steady current of velocity U_0

where x is the distance from the upper edge of the plume, and B_0 is the initial half-width of the sewage field, chosen so that the front velocity at this section equals the velocity of the main current:

$$B_0 = \frac{k^2 \Delta g Q}{2 U_0{}^3}. \tag{4}$$

Only the product of Δ and Q, being equivalent to the buoyancy flux, appears in Equations 3 and 4. For a homogeneous recipient the spread is evidently not dependent on the initial dilution. Normally B_0 is larger than the diameter of the rising jet of waste water or the width of the outlet of cooling-water. If not, the equation of spreading (Equation 3) will have to be modified, but the dependence of the buoyancy flux will remain.

Turbulent Diffusion

Vertical exchange was then considered, both the case of one-sided vertical diffusion from the top layer to the underflowing recipient, and that of symmetrical turbulent exchange between these layers.

It was found that the variation of B with x was little affected by turbulent diffusion so long as the surface layer existed. The width, B, of the surface field may therefore be calculated according to Equation 3. Further, the transport of density deficit, T, was found to depend mainly on the level of turbulence in the receiving water. In the case of symmetrical exchange the dilution, S, was determined by

$$S^{-1} = 1 - \frac{Mx}{B_0}\left\{0{\cdot}78\left(\frac{x}{B_0}\right)^{2/3} + 1\right\} \tag{5}$$

where M is the dimensionless exchange rate given by

$$M = \frac{T \Delta g B_0}{U_0^3}. \tag{6}$$

An evaluation of the distance over which the waste water is able to spread due to density effects requires the knowledge of the turbulent exchange mechanism.

Turbulent diffusion between two layers of slightly different density has been studied theoretically and experimentally by the Authors, and some preliminary results have been published[2]. It has been shown[3] that in homogeneous water the diffusion is mainly determined by the larger eddies and that a density gradient will act as an eddy filter reflecting these eddies. Consequently in the presence of a density difference between two layers the turbulent mixing between them will be strongly damped.

By dimensional analysis it was found that the turbulent diffusion across an interface, as expressed by the velocity of exchange, u_e, can be described by the functional relationship

$$u_e(\nu w)^{-1/4} = F(N_1, N_2, N_3) \tag{7}$$

where ν = kinematic viscosity, w = energy dissipation rate per unit mass and N_1, N_2, N_3 are dimensionless parameters defined by

$$N_1 = L(\nu^3/w)^{-1/4}, \tag{8}$$

$$N_2 = t(w/\nu)^{1/2}, \tag{9}$$

$$N_3 = \frac{\rho w}{\Delta \rho g}(\nu w)^{-1/4}, \tag{10}$$

where L = eddy size characterizing the energy supply, t = time, ρ = density, and $\Delta\rho$ = density difference.

It was further argued that if L is larger than the largest eddy size that can contribute to the turbulent diffusion, and if the process may be considered quasi-stationary, then Equation 7 can be simplified to

$$u_e(\nu w)^{-1/4} = F_1(N_3). \tag{11}$$

In order to support the theoretical results, laboratory experiments were carried out using different types of agitators[2]. The results can be summarized by the equation

$$u_e(\nu w)^{-1/4} = N_3^{4/3}/G \tag{12}$$

where G is a constant depending on the type and the positions of the agitators. Thus, for each type of agitator the experimental result (Equation 12) agrees with the theoretical result (Equation 11).

The variation of the constant G in Equation 12 with the type of agitator can be explained by the fact that some part of the energy supplied to the system was dissipated in the immediate vicinity of the agitators due to their presence. This part should be omitted when calculating the energy supply to the turbulence causing the diffusion in the interface. This fact must be taken into account, for example when calculating the energy supply from wind-shear stresses in a natural water body. In this case some of the energy is dissipated by breaking waves, etc., which is an effect similar to the energy dissipation around the agitators.

In the experiments it was also found that the diffusion in the interface was dependent on the ratio between the distance from the energy source to the interface and the maximum eddy size. This dependency can be described by the formula

$$G = 0{\cdot}3 \exp (3{\cdot}3\, Z/L_{max}) \tag{13}$$

where Z is the distance from energy source to interface and L_{max} the maximum eddy size.

Equations 12 and 13 were derived from a wide range of laboratory and field data and, considering the uncertainty of these data, the agreement is satisfactory. Thus, by using these equations, it is possible to evaluate the diffusion across the interface for any situation provided that the position of the energy source, the energy supply rate, and the eddy structure of the water body are known. One common situation is when the energy is supplied by shear stresses in the interface. In this case the factor $Z/L_{max} = 0$, and Equation 13 gives $G = 0{\cdot}3$. The energy supply is

$$w = \frac{\tau_i u_d}{\rho H} \tag{14}$$

where τ_i = shear stress in the interface, u_d = velocity difference across the interface, ρ = density of water, and H = layer thickness over which the energy is distributed (this can be estimated from the velocity distribution). Another common situation is when the energy is supplied by wind-shear stresses on the water surface. In this case the factor $Z/L_{max} \approx 1$, and Equation 13 gives $G = 8$. The energy supply is

$$w = \frac{\tau_w u}{\rho H} \tag{15}$$

where τ_w is the wind-shear stress, u the wind-induced velocity in the water surface, and H the thickness of surface layer. (The assumption $Z/L_{max} = 1$ should be investigated from case to case.)

The results given above can be included in the general mathematical model describing surface spreading which is presented below.

General Mathematical Model of Surface Spreading

Consider an element of a surface layer thickness $h(x, y, t)$ in an x-y coordinate system, Fig. 2. The mean velocities in the x and y directions are denoted $u(x, y, t)$ and $v(x, y, t)$ respectively. If the velocities in the z direction can be neglected the pressure can be considered hydrostatic and the equations of motion in the x and y directions are

$$\frac{\partial u}{\partial t} + u\frac{\partial u}{\partial x} + v\frac{\partial u}{\partial y} + g\left(\frac{\partial h}{\partial x} + \frac{\partial h_0}{\partial x}\right) = \frac{\tau_{ix}}{\rho h} + \frac{\tau_{wx}}{\rho h} + \phi v, \tag{16}$$

$$\frac{\partial v}{\partial t} + u\frac{\partial v}{\partial x} + v\frac{\partial v}{\partial y} + g\left(\frac{\partial h}{\partial y} + \frac{\partial h_0}{\partial y}\right) = \frac{\tau_{iy}}{\rho h} + \frac{\tau_{wy}}{\rho h} - \phi u, \tag{17}$$

where h_0 is the thickness of the bottom layer, ρ the density of surface water, τ_{ix}, τ_{wx}, τ_{iy}, τ_{wy} are shear stresses in the interface and in the water surface in the x and y directions, ϕ is the Coriolis parameter, $2\omega \sin \psi$, where ω is the angular velocity of the earth and ψ is the north latitude.

When the surface layer is thin compared to the bottom layer, h/h_0, du_0/dt, dv_0/dt are all small (u_0 and v_0 are the velocities in the bottom layer) and the equations of motion for the lower layer then simplify to

$$g\left(\lambda\frac{\partial h}{\partial x} + \frac{\partial h_0}{\partial x}\right) = \frac{\tau_{0ix}}{\rho_0 h_0} + \phi v_0, \tag{18}$$

$$g\left(\lambda\frac{\partial h}{\partial y} + \frac{\partial h_0}{\partial y}\right) = \frac{\tau_{0iy}}{\rho_0 h_0} - \phi u_0, \tag{19}$$

where ρ_0 is the density of the bottom layer, $\lambda = \rho/\rho_0$, and τ_{0ix}, τ_{0iy} are shear stresses in the interface.

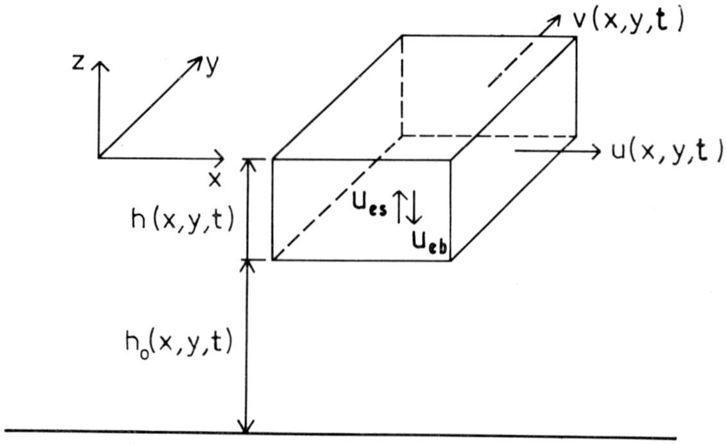

FIG. 2. Nomenclature for element of surface layer

By using Equations 18 and 19, h_0 can be eliminated from Equations 16 and 17 which now can be written

$$\frac{\partial u}{\partial t} + u\frac{\partial u}{\partial x} + v\frac{\partial u}{\partial y} + g(1 - \lambda)\frac{\partial h}{\partial x} = \frac{\tau_{ix} + \tau_{wx}}{\rho h} + \phi(v - v_0), \tag{20}$$

$$\frac{\partial v}{\partial t} + u\frac{\partial v}{\partial x} + v\frac{\partial v}{\partial y} + g(1 - \lambda)\frac{\partial h}{\partial y} = \frac{\tau_{iy} + \tau_{wy}}{\rho h} - \phi(u - u_0). \tag{21}$$

The consequence of the simplifications made above is an uncoupling of the equations of motion for the surface and bottom layers[4], which will make it much easier to solve the equations.

Two more equations governing the state of the surface layer have to be formulated, namely an equation of continuity of mass and a transport-dispersion equation:

$$\frac{\partial h}{\partial t} + u\frac{\partial h}{\partial x} + v\frac{\partial h}{\partial y} + h\left(\frac{\partial u}{\partial x} + \frac{\partial v}{\partial y}\right) - \left(\frac{u_{es}}{\lambda} - u_{eb}\right) = 0, \tag{22}$$

$$\frac{\partial \rho}{\partial t} + u\frac{\partial \rho}{\partial x} + v\frac{\partial \rho}{\partial y} - \frac{\rho}{\rho_w}\left(D_x\frac{\partial^2 \rho}{\partial x^2} + D_y\frac{\partial^2 \rho}{\partial y^2}\right) + \frac{u_{es}}{h}(\rho - \rho_0) = 0, \tag{23}$$

where u_{es}, u_{eb} = velocity of entrainment of the surface and bottom layers, D_x, D_y = diffusion coefficients in the x and y directions, and ρ_w = reference density.

Equations 20–23 are the system of equations describing the motion of a surface layer over a bottom layer of large depth. The solution of the system must be made numerically, as in similar problems tackled elsewhere[5].

Boundary conditions

The boundary conditions of the problem are illustrated in Fig. 3. One set of boundary conditions is formed at Boundary I where the vertical motion of the waste-water plume ceases and the horizontal spreading starts. A reasonable estimation of the conditions within this area can be made with the help of data reported in the literature. Another set of boundary conditions is formed along the outer boundary

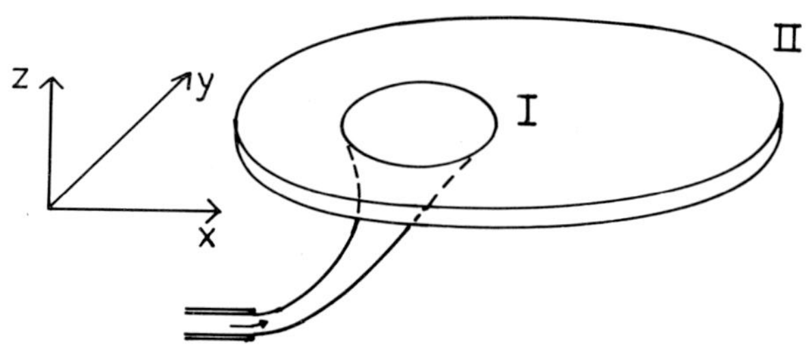

FIG. 3. Boundaries between which surface spreading occurs

(II in Fig. 3) of the surface layer. The absolute velocity of this boundary is the vector sum of the velocity of the recipient and of the surface-layer front relative to the recipient, the latter being of magnitude $k\sqrt{gh}$ directed perpendicularly to the front. The computations must cover the whole area over which the density difference is a significant factor for the surface spreading. The criteria for this condition remain to be stated.

Conclusions

With a mathematical model like the one outlined above it will be possible to determine the area over which a discharged sewage effluent will spread as a surface layer during non-stationary conditions. The effects of recipient currents, wind-shear stresses, and vertical turbulent diffusion are taken into account.

REFERENCES

1. LARSEN, I., and SÖRENSEN, T. Buoyancy spread of waste water in coastal regions. *Proc. coastal Engng Conf. London, 1968*, p. 1397.
2. HYDÉN, H., and LARSEN, I. Turbulent diffusion in the presence of a density gradient. *Proc. int. Ass. hydraul. Res. Congr., Istanbul, 1973*, Vol 2, p. 17.
3. LARSEN, I. On turbulent diffusion in the presence of a density gradient. Royal Institute of Technology, Stockholm, IVFA Publ. 68:3, 1968.
4. ABBOTT, M. B. An introduction to the method of characteristics. Elsevier, New York, 1966, p. 112.
5. ABBOTT, M. B., and GRUBERT, J. Towards a design system for stratified nearly horizontal flows. *Int. Ass. hydraul. Res. Symp. on Stratified Flow, Novosibirsk, U.S.S.R., 1972*, Paper 1.

Prof. Hydén said, by way of introduction, that the particular part of the establishment of a sewage field considered in the paper was the spreading outside the area where the buoyant plume reached the water surface, so that conditions in the 'boil' were considered to be known. Significant factors for surface spreading were gravitational effects, currents in the receiving water, and wind-shear stresses on the water surface.

Gravitational spreading occurred as long as the density difference between the sewage field and the receiving water exceeded some particular value. Dilution of the sewage field was caused by turbulent mixing across the transition zone between the surface layer and the receiving water. Theoretical and experimental investigations had been carried out to determine this turbulent mixing, the results being summarized in Equations 12 and 13 of the paper. Comparison with results from different investigations reported in the literature showed fair agreement, and the results could be used in a mathematical model of surface spreading. Such a model was being developed from the basic equations given in the paper. The work was incomplete and, so far, only some special cases could be treated. The model had not yet been applied in any practical cases but could already be used for studying the relative effects of such factors as wind forces and currents in the receiving water.

Figure 4 shows results from one of the studied cases. A sewage plume meets the surface 125 m from the shore in a water area 300 m wide where there is a uniform current of 0·1 m/s—from left to right in the diagrams. The sewage plume is assumed to spread equally in all directions around a circle 50 m in diameter, which thus forms the inner boundary of the computations. The diluted waste-water flow is 12 m^3/s, and the initial density difference between it and the receiving water is 10 kg/m^3. In the opposite direction to the water current there is a wind current of 5 m/s. Calculation of the position of the outer boundary of the surface layer is included in the model. The density, thickness, and velocity distributions in the surface layer are shown in Fig. 4 (a), (b), and (c) respectively.

With a more developed mathematical model it should be possible to study the gravitational spread of a sewage field; the effect of different factors could easily be studied and the size of the area affected by the sewage could be estimated.

DISCUSSION

Dr. M. Waldichuk (Canada) said there were situations where the effluent jet from a submarine outfall brought to the surface a mixture of effluent and sea water that was denser than the surrounding surface water. Consequently, the effluent field at the surface was rather confined, the mixture sinking beyond the periphery of the boil to a depth where its density was equal to that of the ambient water. He asked if the Authors' model took account of such a condition.

Prof. P. Harremoës (Denmark) drew attention to the fact that the same measure of surface spreading could be used in calculating a boundary condition for the sewage field. There was a subcritical range where the sewage field could spread upstream, thus making the boil into a line source much wider than the diameter of the jet at the surface. In a higher current velocity the sewage boil would be in a supercritical stage where the buoyancy

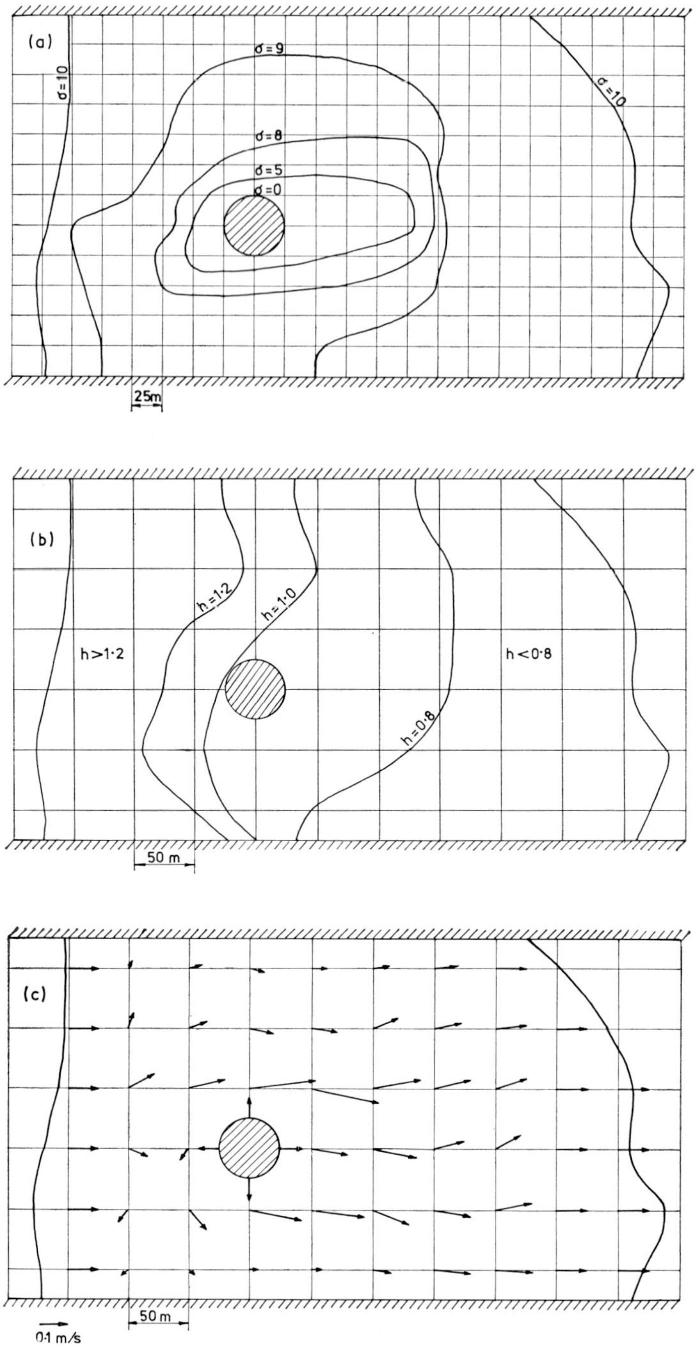

FIG. 4. Results from mathematical model of surface spreading: (*a*) Density distribution (*density* = *1* + *0·001σ g/cm*^{3}); (*b*) Surface-layer thickness (*in metres*); (*c*) Velocity distribution

discharge was too small in relation to the current velocity to cause spreading upstream. This phenomenon had not been dealt with in the paper, but Prof. Harremoës had considered it himself[6].

Dr. J. A. Charlton (U.K.) pointed out that the horizontal buoyant spread of sewage preceded (and might even be part of) any subsequent dispersion of the sewage plume. This initial surface spread would also be influenced by tidal flow and wind, and would determine the initial concentration and width of the ensuing plume. This stage of dilution might not always be so simple, and the surface buoyant spread might (under some conditions) be restrained by a form of inverted hydraulic jump. The comment that a mathematical model was often an aid to thought, rather than a design tool, then became more valid. Prof. H. B. Fischer (U.S.A.) asked if the model allowed for the possible formation of an internal hydraulic jump.

Dr. G. E. B. Kullenberg (Denmark) asked Prof. Hydén to elaborate on the Authors' use of the inertial sub-range eddies for the mixing mechanism, since he thought this was not quite correct. He also thought that the dependence on the distance from the energy source would be affected by conditions in the ocean, since persistent winds would result in transference of the shear stress to a considerable depth.

The Chairman, Dr. G. Abraham (Netherlands), pointed out that the Authors had used the same functional relationship to describe mixing induced both by agitators and by shear. Long[7] had summarized the results of a large number of experiments and concluded that different relationships existed, that for agitators being[8,9]

$$u_e \propto \Delta\rho^{-3/2}\, w^{4/3} \tag{24}$$

and that for shear being[10–12]

$$u_e \propto \Delta\rho^{-1}\, w, \tag{25}$$

whereas the Authors' relationship for both was

$$u_e \propto \Delta\rho^{-4/3}\, w^{5/4}. \tag{26}$$

Equation 25 coincided with one proposed by Larsen and Sörensen[1], and had been confirmed by wind-mixing experiments[13] with a continuous supply of energy from outside the body of water where the mixing took place. Wu's experiments[12] were also made for wind-induced mixing. It remained to be seen whether Equation 25 held for flows induced by density differences within the water where the main energy transition was from potential into kinetic energy.

REPLY TO DISCUSSION

Prof. Hydén, replying to Dr. Waldichuk, said that one of the assumptions when formulating the basic equations for the model had been that the spreading occurred at the surface. It that were not the case, the basic equations had to be reformulated to some extent and somewhat different front conditions applied. Basically, only nearly horizontal flow was considered; if that formulation was not valid the mathematical formulation of the hydrodynamic equations would be much more complicated.

He agreed with Prof. Harremoës and Dr. Charlton that different conditions might occur inside and outside the boil area, but these phenomena could be included only as boundary conditions, mainly because the model considered nearly horizontal flow only. If the boil area were to be included in the model, it had to be considered separately since the mathematical formulation would be quite different from that for the subsequent spreading. He admitted that discussion of the different types of conditions which might occur (as pointed out by Prof. Harremoës) ought to have been included in the paper.

In reply to Drs. Kullenberg and Abraham he said that the theoretical basis for Equations 7–11 has been discussed more fully elsewhere[3]. Within the range of variation in the significant parameters considered, Equations 12 and 13 seemed to hold fairly well. The conditions referred mainly to coastal waters where the depth was relatively shallow compared with that in the oceans. The results could not, of course, be extrapolated to conditions in deep waters without further investigations.

ADDITIONAL REFERENCES

6. Harremoës, P. Diffuser design for discharge to a stratified water. *Wat. Res.*, 1968, **2**, 737–752.
7. Long, R. R. The influence of shear on mixing across density interfaces. Johns Hopkins University, Department of Mechanics and Material Science and Earth and Planetary Sciences, Tech. Rep. No. 4 (Series C), 1974.
8. Rouse, H., and Dodu, J. Turbulent diffusion across a density discontinuity. *La Houille Blanche*, 1955, **10**, 405–410.
9. Turner, J. S. The influence of molecular diffusivity on turbulent entrainment across a density interface. *J. Fluid Mech.*, 1968, **33**, 639–656.
10. Kato, H., and Phillips, O. M. On the penetration of a turbulent layer into a stratified fluid. *J. Fluid Mech.*, 1969, **37**, 643–655.
11. Moore, M. J., and Long, R. R. An experimental investigation of turbulent stratified shearing flow. *J. Fluid Mech.*, 1971, **49**, 635–655.
12. Wu, J. Wind induced tubulent entrainment across a stable density interface. *J. Fluid Mech.*, 1973, **61**, 275–287.
13. Delft Hydraulics Laboratory. Wind induced mixing. Rep. M 1235, 1974 (in Dutch).

The Relevance of Initial Dispersion

B. Sterregaard
Danish Isotope Centre, Copenhagen

The design of the diffuser or other discharge structure is an important part of the total design of a sea outfall for sewage. Models are available for calculation of jet dilution, enabling the designer to meet stipulated requirements for initial dispersion (dilution) at the point of discharge, when information about the sewage and the receiving water is given. To specify requirements for the initial dilution, it is naturally necessary to know the relevance of this parameter to water quality. This paper discusses the relevance of the initial dilution of sewage to water quality in coastal receiving waters, including estuaries.

Most of the paper considers discharge of 'ordinary' sewage which is lighter than sea water. In a short paragraph, discharge of heavy (industrial) liquid waste is touched on. Primary treatment is often considered sufficient when discharging sewage to the sea, and the points raised are mainly relevant for discharge of primary effluents. The further the treatment is carried, the less important is the initial (and the subsequent) dispersion.

Surface or Sub-Surface Trapping

When vertical density gradients occur in the receiving water, it is possible to trap the sewage below the surface; see, for instance, papers by Hayashi and Ito[1] and Hansen[2].

Dilutions in the sewage jet and trapping levels depend on the density of the ambient water, strength of the density gradient, water depth, current, and conditions of discharge, i.e. outlet diameter, discharge velocity, and sewage density. In shallow water, trapping below the surface may be impractical even if significant density gradients occur. In deep water a weak gradient may still permit trapping below the surface. In a homogeneous water the sewage will reach the surface regardless of the conditions of discharge. The two situations—surface and sub-surface trapping—are demonstrated in Fig. 1.

Variations in the vertical density gradients with time are generally to be expected. In order to predict the performance of a particular diffuser, statistics of the density profiles at the point of discharge should be established. Figure 2 shows an example of trapping statistics, computed at the Danish Isotope Centre for a discharge at a Danish locality (the Sound). In many places the variations in tidal level are very significant in comparison with the total depth at the point of discharge. Such variations may greatly influence the trapping statistics and should be taken into account.

Once density (and water level) statistics have been established, decision can be made on the frequency of trapping at different levels which the diffuser design should aim at. Some considerations which are relevant when making this decision are discussed in the following paragraphs. Economics—mainly expenses for pumping—are often very relevant but will not be dealt with here.

Hygiene and aesthetics

For hygienic and aesthetic reasons it is generally preferable for the sewage to be trapped below the surface at the outfall. The observations[3] at Whites Point, Los Angeles, before application of diffusers are illustrative of the objectionable qualities of a (poorly diluted) sewage field at the surface over an outfall: surface layers of poorly diluted sewage sometimes travelled with wind-driven currents towards the shore; sewage fields were quite stable and maintained their identity for extended periods of time and long distance of travel; the hygienic water quality at the shoreline at times was unsatisfactory; the sewage field was visible from the shore; odour problems were created in and near the sewage field.

If the sewage is trapped below the surface, the aesthetic problems—i.e. visible signs of sewage and odour—are eliminated or reduced very significantly. Sewage which eventually penetrates to the surface through mixing will most often be so diluted as not to create aesthetic nuisances. Floatables—faecal particles, grease, oil, etc.—must be efficiently removed before the sewage is discharged, since they will come to the surface regardless of the trapping of the liquid. Screening or comminuting will generally not be sufficient to avoid problems with floatables[3].

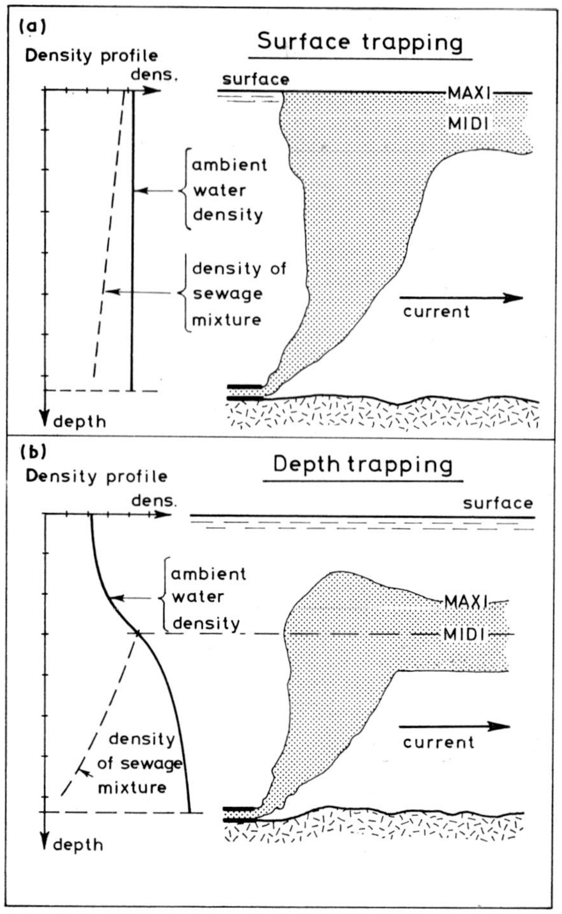

FIG. 1. Trapping of sewage in (*a*) homogeneous and (*b*) stratified water. MIDI and MAXI are levels where sewage of mean and minimum (jet-axis) densities respectively are trapped

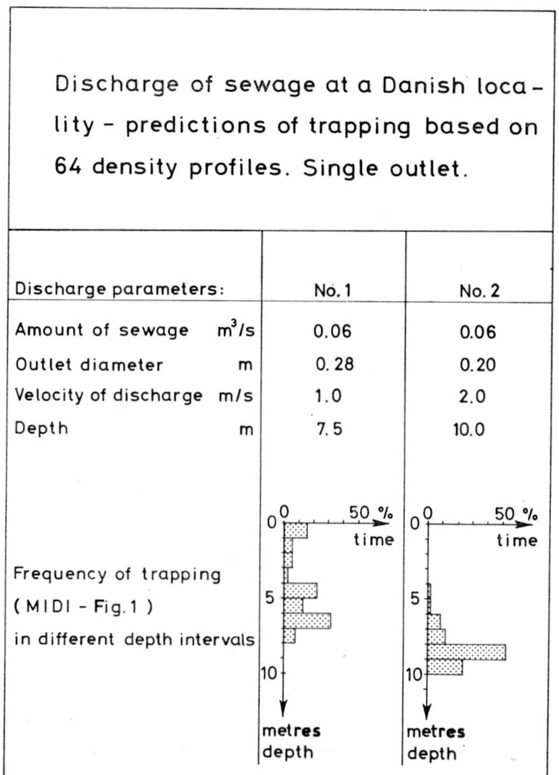

Discharge of sewage at a Danish locality - predictions of trapping based on 64 density profiles. Single outlet.

Discharge parameters:		No. 1	No. 2
Amount of sewage	m^3/s	0.06	0.06
Outlet diameter	m	0.28	0.20
Velocity of discharge	m/s	1.0	2.0
Depth	m	7.5	10.0

FIG. 2. Trapping statistics for sewage mixtures of mean density (MIDI)

When the sewage is trapped below the surface, the hygienic risks created by the wind carrying a surface field of poorly diluted sewage on to the shore are eliminated. While the submerged field is carried away by the current, sewage may penetrate to the surface by mixing, but most often in high dilutions only. Current velocities are generally less at depth than at the surface, so that the time of travel to the shore is increased. This means a longer time for reduction of micro-organisms by natural die-away. Sedimentation plays an important role in the process of die-away, and sedimentation is enhanced by the lower current velocities at depth and the smaller distance to the bottom. Also currents at depth often have a tendency to run along the coast rather than towards it, even if there is a wind-driven surface component towards the coast. Upwelling at the coast is an important exception (see 'Local circulation' below).

It has been shown[4] that sunlight has a lethal effect on coliform and other bacteria. Seen from this angle it would be expedient to let the sewage rise to the surface to expose it as strongly as possible to the sunlight. The Author, however, believes that the above mentioned advantages of trapping below the surface will generally outweigh the advantage of exposing the bacteria to daylight. The sewage itself may also shield the bacteria from the sunlight, so that its effect is only significant at the very surface. Further research into this matter seems to be desirable.

Biology

Low oxygen contents may often be found in the deepest layers of a natural water, caused by oxidation of natural organic matter near the bottom and in the sediments and a low rate of oxygen renewal from the atmosphere. Directly adding a substantial BOD load from sewage to these layers may result in serious oxygen depletion.

The low current velocities in the deep layers may lead to faster and more concentrated sedimentation of particulate material from sewage trapped there than would have been the case if trapping had been in the upper layers. A high load of material from the sewage in the area close to the outfall may damage the benthic life, whereas spreading the load over a greater area might mean that no significant damage was caused in any place.

The above considerations lead to the conclusion that for the best protection of fauna and flora it is generally not desirable to trap the sewage in the deepest layers of the receiving water. Specific toxic materials in the sewage may necessitate demands for a high initial dilution regardless of trapping levels.

Mixing

A submerged sewage field has two interfaces for vertical mixing compared with the single interface of a field at the surface—see Fig. 1. Also the density differences between the submerged field and the ambient water are generally smaller than the corresponding density differences at a surface field. These facts make mixing of the sewage with sea water more rapid when the field is submerged than when it is at the surface.

If, on the other hand, the sewage is trapped below a relatively strong pycnocline (density jump), the turbulence in the trapping layer and consequently the mixing may be much weaker than in the surface layer.

Local circulation

Vertical water circulation in the area where sewage is discharged should be taken into account when the most desirable trapping levels are considered. In a stratified estuary there will be an upstream flow of saline water in the deep layers, while the fresh water moves towards the ocean at the surface. In principle, trapping of sewage can only take place above the interface between the two water masses. Vertical density gradients may, however, occur in the saline water moving upstream, making trapping possible also in this layer. In such a case it should be considered whether upstream transport of sewage in the deep layer is desirable or not; in many cases it will not be.

When a steady wind blows away from a coast, it carries water with it. This results in a reversed (compensating) current in the deeper layers replacing the removed water by upwelling at the coast. In such a case, trapping the sewage in the deep layers would result in pollution of the beach zone, while surface trapping would lead to the pollutants being carried away from the beach—see Fig. 3. Onshore winds reverse the situation—sewage at the surface is carried towards the beach and sewage at depth away from the beach. If both situations occur frequently, it may be profitable to try to trap the sewage in the shear zone between the reversed currents.

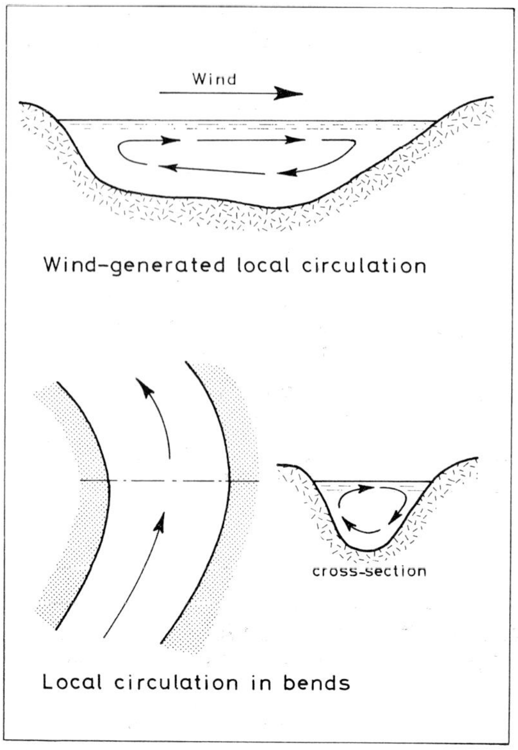

FIG. 3. Vertical circulation generated by wind and by current in narrow bends

In bends in narrow straits or sounds similar upwelling and sinking phenomena occur—see Fig. 3. On the inside of the bend upwelling takes place, and on the outside the opposite situation is found.

Sewage at the Surface

Under some conditions the trapping level of the sewage must be at the surface. This may be because there are reasons (such as those already mentioned) why it is not desirable to trap the sewage at depth. It may also be because the vertical density gradients and/or depth at the point of discharge do not permit trapping at depth at all, or only permit it part of the time.

The initial dilution of the sewage at the surface influences the water quality of the receiving water in a number of ways which should be considered when designing the outfall and diffuser.

Stable sewage field

A stable sewage field exists when the density gradient across the interface between the diluted sewage and the ambient sea water is so great that it significantly hampers mixing across the interface. A stable sewage field at the surface is undesirable for a number of reasons:

1. By buoyant surface spreading the field may reach the shore fast and with high concentrations of sewage resulting in serious pollution.
2. Onshore winds may carry the sewage field to the beach without much further dilution, resulting in serious pollution.
3. Further mixing of the sewage with the ambient sea water while the field is travelling with the current is hampered, so that sewage may be found in high concentrations at a considerable distance from the outfall.
4. Prediction of pollution is often carried out by tracer measurements. Significant density gradients do not occur between the tracer field and the ambient water, and spread of the tracer is not hampered in the same way as spread of a stable sewage field. Thus the predictions may become inaccurate—i.e. optimistic. The same is true for predictions based on mathematical models. The dispersion parameters used (which may have been determined by tracer measurements) do not take into account a density gradient between the sewage field and the ambient water.

When the jet from the outfall reaches the surface, a density difference between the diluted sewage and the sea water generally remains, the magnitude of this difference depending on the initial dilution. The density difference creates a horizontal buoyancy spreading of the diluted sewage. Such spreading is discussed in the paper by Hydén and Larsen[5]: buoyant spreading of the sewage field is significant so long as the density differences are significant for the pattern of motion of the field, but the criteria for this condition still remain to be stated. Where significant buoyancy spreading occurs, mixing across the interface between the field and the ambient water is hampered, and the field is what above is termed 'a stable sewage field'[5,6].

It has been observed[6] that the initial (jet) dilution should be so strong that the stability of the sewage field formed over the discharge is sufficiently weak. The stability is sufficiently weak, when the turbulence set up by transport of the field, for instance towards the shore, can destroy the interface between the sewage field and the ambient sea water. It is understood that the interface is broken down early enough for the further mixing to ensure that quality standards on the beach are kept.

It is assumed[6] that the density difference should remain below a critical value of 0·5 kg/m^3. This value was indicated by Prof. J. C. Schönfeld, who considered specific conditions at an outfall on the coast of the Netherlands. A given distance from the diffuser to the beach, and an assumed difference in velocity between the sewage field and the underlying water, led to the above figure. For other conditions another number could have been found. For a sea-water density of 1025 kg/m^3 and a sewage density of 1000 kg/m^3 the above figure (0·5 kg/m^3) for the density difference corresponds to a required dilution at the surface of about $S_{req} = 50$:

$$S = \frac{\text{parts sewage} + \text{parts sea water}}{\text{parts sewage}}.$$

In the paper by White and Agg[7] this value of S_{req} is adopted as a possible design criterion.

Dilution in the sewage jet, Fig. 1, varies over a cross-section of the jet—the concentration distribution is approximately Gaussian. Consequently the dilutions and the densities in the sewage field above the discharge cannot be assumed to be equal in a vertical cross-section—see MIDI and MAXI in Fig. 1. This gives rise to a question of which density in the field should be considered in relation to a specified upper limit for the density difference. Should it be the minimum density, the mean density, or some figure in between?

The above mentioned upper limit of the density difference (0·5 kg/m^3) was determined for specific conditions, and may be regarded as a guidance only in other cases. Further research into the matter of this upper limit for the density difference, and into the question of which densities to consider in the sewage field, seems very relevant.

Size of the surface field

The mixing efficiency of the eddies in the receiving water grows with the size of the eddies. See, for instance, the paper by Kullenberg[8]. Also the efficiency of mixing of a sewage field drifting in the current depends on the relative size of the field to the eddies—eddies of the same size as the field are most efficient.

It has been found that the relation between the effective dispersion coefficient and the scale is of the form

$$k = aL^n \tag{1}$$

where k is the dispersion coefficient, L the scale of the field (Fig. 4), and a and n are constants.

Experimental values of a and n for different L are given by Kullenberg[8]. For a range of conditions $n \approx 4/3$ has been found to be valid.

It is thus evident that the dispersion of the drifting field is promoted if the initial dilution of the sewage results in a large field at the surface. In this context not only the initial dilution in a single sewage jet is important, but also the dispersion obtained with a diffuser by discharging through a number of ports along a line. The diffuser should, of course, be in a direction normal to the predominant or the critical current to serve this purpose.

Mathematical models describing the dispersion of sewage discharged from a point source are often used for prediction of pollution. Thus the initial size of the sewage field is neglected. If the scale of the initial field is small compared with the distance to shore this approximation will not result in any significant error, Fig. 4(a). In any case the error will give concentrations on the safe side—i.e. too large. If a long diffuser is used, a model describing the dispersion from a line source may be applied—as discussed, for instance, by Brooks[9].

To obtain the best approximation of dispersion in the sea the dispersion coefficient in the mathematical models should be used as a variable. The variation can for instance be in accordance with Equation 1.

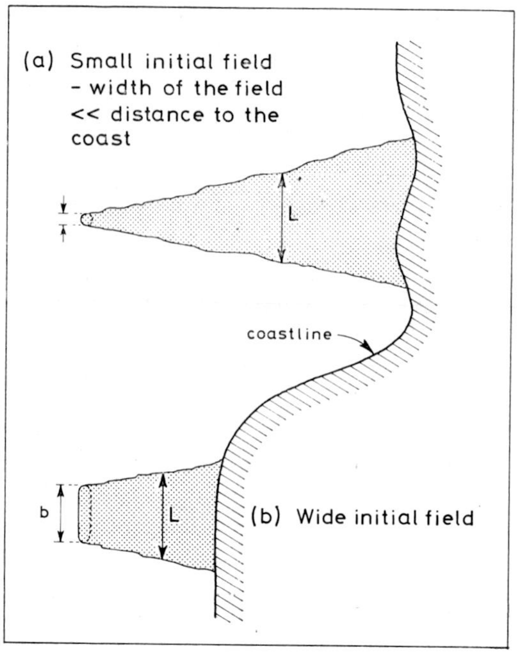

FIG. 4. Size of initial field at point of discharge relative to distance to coast

In this case the initial size of the sewage field determines the starting value of k. Brooks[9] has suggested models for dispersion from a line source applying variable k according to

$$\frac{k}{k_0} = \frac{L}{b} \tag{2}$$

or

$$\frac{k}{k_0} = \left(\frac{L}{b}\right)^{4/3} \tag{3}$$

where k_0 is the initial value of k, and b the initial width of the sewage field—Fig. 4(b).

In many cases a constant k is used in the models for reasons of simplicity. The value of k is then chosen as one which is thought to represent mean conditions in the drifting sewage field. This will lead to prediction of concentrations which are too small close to the outfall and too large after a long distance of travel. The actual size of the initial sewage field is decisive for how much the real concentrations close to the point of discharge are larger than the ones predicted.

Slick formation—floatables

A surface slick may be formed over the discharge by the contents of greasy and oily material in the sewage. This material forms a very thin film on the surface, down to a thickness of 1 molecule[10]. The film dampens waves and ripples on the surface of the receiving water and is clearly visible. The slick has been found to have a high concentration of bacteria, up to 10 times the concentrations in the diluted sewage below the surface layer[11]. Newton[10] reported the most important factors governing slick formation to be sea state, initial dilution, and contents of greasy material in the sewage. The frequency of occurrence of slicks decreased when wave heights increased above 0·3 m. At 0·6 m wave heights and above, slicks were seldom observed. Newton also found that slicks were usually formed when the initial dilution of the sewage field was less than about 100, but that they formed only infrequently when it was greater than 400. The sewage discharged by the outfall, where these figures were observed, was untreated. Primary treatment removes much of the greasy material in the sewage. For a primary effluent the dilutions necessary to avoid formation of surface slicks are probably much lower than the above mentioned figures.

Floatable particles in the sewage are likely to be a nuisance regardless of the magnitude of initial dilution. Such particles should be removed before discharge.

Odour

The smell of sewage from a surface field decreases as the initial dilution increases. At two outfalls at Whites Point, Los Angeles, a distinct smell of sewage was noticeable in the vicinity, while the initial dilution was about 40, i.e. before diffusers were applied[3]. Later, when initial dilutions at the surface were increased to about 170 and 300 respectively, the odour problems ceased almost completely; only on calm days could a faint odour be detected immediately over the discharges. The sewage at Whites Point receives primary treatment. Newton[10] reported that a faint smell of sewage was detectable immediately over an outfall at Bridport, England, when dilution was about 1000. When dilutions were in the range 50–200 a much more distinct smell was reported. This sewage is untreated.

Discoloration and turbidity

Problems of discoloration and turbidity also decrease as initial dilution increases. At the above mentioned two outfalls at Los Angeles a discoloured sewage field was visible from the shore before diffusers were constructed[3]. Since then the sewage field has become almost indiscernible. Newton[10] concluded that visible discoloration of sea water is not likely to occur if the initial dilution is greater than about 50. Natural turbidity of the receiving water has of course a considerable influence in this matter.

Distance from coast

The distance of the discharge from the coast is relevant to consider together with the requirements for initial dilution. If the point of discharge is very far from the coast, the initial dilution of the surface field may become unimportant, at least if only conditions at the beaches are considered.

White and Agg[7] suggest a criterion of the form

$$S_{req} = \frac{A}{x^2} \tag{4}$$

where x is the distance from the coast and A is a constant related to the requirements of the individual sites. A suitable value of the parameter A may be hard to lay down. It should in principle include all the factors considered in this paper.

In many cases it is questionable if the conditions in the vicinity of the point of discharge can be ignored, even if it is far out to sea.

Sewage Trapped below the Surface

As already mentioned, sewage should preferably not be trapped in the deepest parts of the water, where oxygen contents and oxygen renewal may be low for natural reasons. Low oxygen contents are often found in stratified waters in the layer between the bottom and the deepest pycnocline. In such a case it is worth noting that the sewage will always be trapped higher than the deepest pycnocline, and consequently not in the layer where oxygen conditions are worst. Trapping the sewage just above the deepest pycnocline should not be attempted, but may be unavoidable in some cases, if conditions at the surface are not to be sacrificed.

A sharp pycnocline in the receiving water may cause the sewage to be trapped in high concentrations in a rather thin layer just above the pycnocline. This condition is similar to the formation of a stable sewage field at the surface. Furthermore, currents and turbulence are often weak at depth, and so further mixing may be slow, even despite the fact that two horizontal boundaries are available at the submerged field (Fig. 1). When designing the diffuser it should be attempted not to trap the sewage in a thin layer like the one mentioned.

Heavy Effluents

Waste waters from some industries have higher densities than sea water. Suspensions of gypsum in water discharged from industries producing phosphoric acid may be mentioned as an example.

When discharging heavy effluents the considerations regarding the initial phase are similar to the ones for sewage lighter than sea water, only the buoyancy forces are reversed. When the heavy liquid or suspension is discharged, it sinks towards the bottom being progressively diluted by entrainment of sea water on the way. If the dilution is insufficient, a stable field of waste water may be formed at the bottom. The further mixing of this stable field is hampered due to the density differences, just as was the case with a stable surface field. It is, however, a further disadvantage that current and turbulent mixing is often weak close to the bottom.

The stable field of heavy waste water at the bottom may be avoided by securing a high initial dilution. This may be achieved by directing the discharge upwards at a suitable velocity or discharging at a level high over the bottom. At the Danish Isotope Centre a computer programme has been developed for prediction of dilution and trapping of a heavy jet in sea water[12].

In a stratified receiving water it will generally be desirable to design for an initial dilution so great that the waste water is trapped above the bottom. If trapping in the bottom layer is inevitable, the density differences between the field and the ambient water should be small. How small depends upon the local current and mixing conditions. No general criteria have been arrived at so far.

Summary

In many cases it is possible to control the trapping depth of sewage at a sea outfall through the initial (jet) dilution. When deciding about trapping levels it is important to take into account that:

(a) for the hygienic and aesthetic water quality it is generally preferable to trap the sewage below the surface,
(b) for best protection of fauna and flora the sewage should in general not be trapped in the deepest layers of the receiving water, and
(c) the vertical water circulation in the area of discharge is important for predictions of the directions in which the sewage will travel after trapping.

When sewage is trapped at the surface, the initial dilution of the sewage field is important for a number of reasons:

1. The initial dilution is decisive for whether a stable sewage field is formed or not. In a stable sewage field, pollutants may be carried far away in high concentrations by density spread or directly to the beach by the wind. The subsequent dilution is hampered in the case of a stable field, and predictions of pollution by tracer measurements or model calculations may be optimistic.
2. The efficiency of mixing of the surface field as it travels with the current depends upon the initial size of the field. The bigger the field, the more efficient is the mixing.
3. The formation of surface slicks depends on the initial dilution. For untreated sewage it has been found that slicks usually occur when the dilution is less than 100. For a primary effluent the limit is probably considerably lower.
4. Problems of odour, discoloration, and turbidity are reduced as initial dilution increases.

When conditions permit trapping of sewage below the surface, the possibility of formation of a stable sewage field with high concentrations at depth should be kept in mind and avoided.

When discharging a heavy effluent, formation of a stable field of waste water at the bottom should be avoided. This may be achieved by directing the jet upwards at a suitable velocity thus obtaining a high initial dilution.

If specific toxic materials are present in the sewage, a high initial dilution may be necessary to protect fauna and flora regardless of trapping level.

REFERENCES

1. Hayashi, T., and Ito, M. Initial dilution of effluent discharged into stagnant sea water. This volume, pp. 253–263.
2. Hansen, J. A. Predetermining trapping of sewage in stratified receiving waters. In 'Nuclear techniques in environmental pollution'. International Atomic Energy Agency, STI/PUB/268, Vienna, 1971.
3. Rawn, A. M., Bowerman, F. R., and Brooks, N. H. Diffusers for disposal of sewage in sea water. *Trans. Am. Soc. civ. Engrs*, 1960, No. 3179, p. 344.
4. Gameson, A. L. H., and Saxon, J. R. Field studies on effect of daylight on mortality of coliform bacteria. *Wat. Res.*, 1967, **1**, 279.
5. Hydén, H., and Larsen, I. Surface spreading. This volume, pp. 277–281.
6. Abraham, G., and Brolsma, A. A. Diffusers for disposal of sewage in shallow tidal water. Delft Hydraulics Laboratory, Publ. No. 37, 1965.
7. White, W. R., and Agg, A. R. Outlet design. This volume, pp. 265–274.
8. Kullenberg, G. Effects of environmental factors, Part 1. This volume, pp. 333–342.

9. Brooks, N. H. Diffusion of sewage effluent in an ocean current. *Proc. 1st int. Conf. Waste Disposal mar. Envir.*, Univ. Calif. Pergamon Press, New York, 1960, p. 246.
10. Newton, J. R. Factors affecting slick formation at marine sewage outfalls. In 'Pollution criteria for estuaries' (ed. P. R. Helliwell and J. Bosanyi), Techno Press, London, 1975, Paper 12 (*in press*).
11. Wheatland, A. B., Agg, A. R., and Bruce, A. M. Some observations on the dispersion of sewage from sea outfalls. *J. Proc. Inst. Sew. Purif.*, 1965, 29.
12. Hansen, J. A. Marine disposal of heavy suspensions. Scandinavian Congress of Chemical Engineering, Copenhagen, 1971.

Mr. Sterregaard, introducing his paper, said that to specify requirements for initial dispersion it was necessary to know the relevance of this parameter to water quality. When density gradients occurred in the receiving water the sewage was trapped below the surface at a depth depending on the density gradient and the magnitude of initial dispersion. Once density statistics had been established, decision could be made on the trapping levels to be aimed at.

For hygienic and aesthetic reasons it was generally preferable to trap the sewage at depth, but upwelling of sub-surface water at the coast (as during offshore wind conditions) was an important exception. For biological reasons the sewage should not be trapped in the deepest layers, and specific toxic materials might necessitate a high initial dilution regardless of trapping levels.

Trapping in the surface layer was often unavoidable; the initial dilution should then be great enough to avoid a stable field which was characterized by great density differences between the field and the ambient water. These density differences hampered mixing, and the field might drift onto the shore without much further dilution.

The rate of mixing of a sewage field in a turbulent current increased with growing size of the field; thus mixing was promoted by high initial dilution. Aesthetic nuisances of a surface field—such as slicks, odour, discoloration, and turbidity—were reduced with increasing initial dilution.

If the discharged effluent was denser than the sea water, the considerations regarding initial dilution were similar, but with the buoyancy forces reversed. A stable sewage field in the bottom layers should be avoided by securing high initial dilution.

DISCUSSION

Prof. J. E. Foxworthy (U.S.A.) related some of Mr. Sterregaard's conclusions to the results of field studies off the coast of Southern California[13–15], where the flow from several treatment plants had been labelled with Rhodamine-B and the resulting fields sampled for dye and coliforms; most of the outfalls included multiport diffuser sections. In a slide (fig. 40 of Ref. 13) he showed the labelled waste field from the Orange County outfall before construction of a diffuser section. The average minimum dilution (as found from dye measurements) was 7·5, and the existence of a stable field was shown by the clearly defined edge effect at the plume origin and for a considerable distance downstream. Vertical profiles indicated that the field depth increased from about 0·6 m at the origin to 3·4 m at a distance of 1·6 km downstream. The centreline dye concentration fell rapidly at first and reached 10 per cent of the initial value at a distance of 2·3 km. Similar results were obtained in five other experiments on the Orange County and Ventura outfalls[14,15] without diffusers, the range of initial dilutions in the six experiments being 7·5–34. Since subsequent dilution was found to be significant, the results appeared to cast doubt on Mr. Sterregaard's statements that subsequent dilution of such fields was suppressed as they drifted with the current and that high waste concentrations might be found at considerable distances from the outfall.

The experiments on outfalls with multiport diffusers showed that for minimum initial dilutions of 21–115 there was considerable transverse spreading at the field origin, and that this sometimes persisted well downstream. The initial spread, expressed as the ratio of the plume width to the length of diffuser section normal to the current, showed a tendency to decrease with increasing initial dilution, but to be more closely related to the mean surface current. This appeared to support the theory presented by Harremoës[16] and others, corresponding to so-called subcritical current conditions. According to Harremoës, for a given flow rate the density spread of a series of jets merging over an outfall should be independent of the initial dilution and inversely related to the surface current speed. The initial depth of the plumes was found to be inversely related to initial dilution and independent of current speed, thus conflicting with Harremoës's theoretical predictions which indicated that it should be related to the initial dilution divided by the square of the current.

Many investigators had shown a scale dependence of the transverse dispersion coefficient similar to that described by Equation 1. The results of the Californian work[13–15] indicated that the variance of the transverse dye concentration distributions was approximately proportional to the drift time, so that the dispersion coefficient was nearly constant. However, this coefficient varied with the initial size of the plume, the relationship following Equation 1 with n approximately equal to 4/3.

Finally, Prof. Foxworthy showed a diagram (fig. 37 of Ref. 14) illustrating the relation between the estimated decrease in maximum waste concentration and distance for two surface fields resulting from identical discharge rates from two diffusers of different lengths. It was apparent that both the maximum concentration and the concentration ratio decreased more rapidly for the smaller plume, thus indicating more efficient subsequent dilution. However, the maximum concentration at any point along the plume was much less for the larger diffuser because of the higher initial dilution—not because of a higher rate of subsequent dispersion which seemed to be suggested by Mr. Sterregaard. Similar results had been described by Hyperion Engineers[17].

The Chairman, Dr. G. Abraham (Netherlands), said that when density variations in the receiving water were strong, sub-surface trapping of sewage could not be avoided unless poor initial mixing were to be accepted. For instance, if the density of an upper layer was 1024 kg/m^3, and 1 kg/m^3 lower than that of the water at greater depth where sewage of density 1000 kg/m^3 was discharged, then the initial mixing should be below 25 to avoid sub-surface trapping. The choice was then between a stable sewage field at the surface and a better mixed field some distance below. Selecting the latter alternative it was necessary to determine whether upwelling might occur and to evaluate the biological disadvantages.

No mathematical models were yet available for a three-dimensional description of the flow of the receiving water and the sewage discharged into it. For sewage fields at the surface, two solutions seemed available. First, by accepting a wind-induced transport of a stable sewage field towards the shore as a design condition and applying theoretical solutions for this flow field[18]; this would produce conservative results, and estimating the thickness of the sewage field (both initially and on arrival at the shore) was one of the crucial points in this procedure. Secondly, by making the jet mixing strong enough to weaken the stability of the sewage field. Then, after the sewage had travelled some distance towards the shore by a wind-induced current, part of it might enter the compensating current at great depth (see Fig. 3). The problem was then shifted to determining (a) the degree of initial mixing required to obtain this situation, and (b) the combined effect of the wind-induced and compensatory currents on the concentration at the shore. No solution of these problems seemed available yet[19, 20], though there were approximate solutions[21] which showed that increasing values of the vertical diffusion coefficient (and hence decreasing stability of the sewage field) affected the concentration at the shore. These considerations applied especially for the situation of discharge of sewage into alternating tidal currents parallel to the shore with a small net residual: the peak concentrations at the shore might then be due to onshore wind currents. The design condition of a wind-induced transport of a stable field towards the shore seemed to Dr. Abraham to be a more realistic one than that represented by Equation 4.

Prof. A. G. Occhipinti (Brazil) said that it was obviously desirable to predict statistically the sewage-field trapping levels. However, prediction was dependent not only on the stratification of the receiving waters, but also on the discharge rate (which would be subject to daily, seasonal, and yearly fluctuations) and on the tidal variations. He asked if Mr. Sterregaard's model took account of these factors.

Dr. R. J. Taylor (U.K.) asked if the Author was aware of any recent developments in the study of wind and wave action on the dispersion of sewage fields.

REPLY TO DISCUSSION

Mr. Sterregaard thanked Prof. Foxworthy for his valuable comments, but said that (on the basis of his own experience with tracer measurements in Danish waters) he did not consider a 90 per cent reduction in concentration over a distance of 2·3 km to be very substantial, so he did not feel it cast much doubt on his statement that the density difference between the sewage field and the ambient sea water should be small to enhance further dilution. However, he stressed that the density difference of 0·5 kg/m^3, and the corresponding dilution of 50 mentioned in the paper, had been calculated for a specific situation on the Dutch coast. Other conditions might yield other values of the required initial dilution, so that the value of 50 should be used for guidance only and with great reservation.

A scale dependence of the dispersion coefficient k had been pointed out by many authors, such as Kullenberg[8]. A number of experiments performed by the Danish Isotope Centre generally showed an increase in k while the radioactive patch was small, and a tendency towards a more constant value at greater scales, whereas Prof. Foxworthy had found k to be constant in a particular plume, regardless of the growth in size, and to vary with the initial size of the sewage field. It was not clear why k had varied with the initial scale but not with the scale in a particular field, and further investigation into this matter seemed desirable. If the initial size of the sewage field was small, the transverse concentration distribution in the plume would be approximately Gaussian from the start of plume development. Maximum concentration was thus found at only one point in a cross-section, and plume dilution would cause a decrease in maximum concentration immediately from the position of the outfall along the plume axis. If the initial field was wide, the concentration would be constant over much of the cross-section near the outfall, and the maximum concentration would thus start to decrease only after some distance of travel—when the transverse concentration distribution had become approximately Gaussian. Consequently, Mr. Sterregaard agreed that the decrease in maximum concentration was most rapid when the initial size of the field was small (though it started from a higher value). The overall dispersion of the plume, however, was greater in the wide field owing to the effect of scale on the dispersion coefficient, or was at least as great as the dispersion in a narrow field.

He agreed with the observations made by Dr. Abraham. If the demands of initial dilution and/or trapping could not be satisfied at a particular place, either the outfall had to be moved or the sewage given further treatment before discharge.

In reply to Prof. Occhipinti, he said that the factors he had mentioned had not been taken into account in the model. So far, the calculation had been made with characteristic values of discharge and water level, and with relevant extreme values. He was unable to give Dr. Taylor any new information.

ADDITIONAL REFERENCES

13. Allan Hancock Foundation and University of Southern California. An investigation on the fate of organic and inorganic wastes discharged into the marine environment and their effects on biological productivity. California State Department of Water Resources, Sacramento, Rep. 29, 1965.
14. Foxworthy, J. E., and Kneeling, H. R. Eddy diffusion and bacterial reduction in waste fields in the ocean. University of Southern California, Los Angeles, Rep. 69-1, 1969.
15. Foxworthy, J. E. Eddy diffusion and the four-thirds law in near-shore coastal waters. University of Southern California, Rep. 68-1, 1968.
16. Harremoës, P. On the significance of density effects to marine disposal. Technical University of Denmark, Copenhagen, 1967.
17. Hyperion Engineers. Ocean outfall design. Hyperion Engineers Inc., Los Angeles, 1957.
18. Fischer, H. B., and Brooks, N. H. Technical aspects of waste disposal in the sea through submarine outfalls. In 'Marine pollution and sea life' (ed. M. Ruivo). Fishing News (Books) Ltd, West Byfleet, England, 1972, Paper R-4.
19. Munro, D. Observed and predicted coliform distributions near a sea outfall. This volume, pp. 353–362.
20. Dam, G. C. v. The Hague outfall. This volume, pp. 393–401.
21. Munro, D., and Mollowney, B. M. The effect of a bottom counter current and vertical diffusion on coliform concentrations at the shore due mainly to wind-induced onshore currents. *Rapp. P.-v. Réun. Cons. int. Explor. Mer*, 1975, **167**, 37.

Subsurface Horizontal Dispersion of Pollutants in Open Coastal Waters

R. J. Callaway

Coastal Pollution Branch, Pacific Northwest Environmental Research Laboratory
U.S. Environmental Protection Agency, Corvallis, Oregon 97330

One method of assessing the efficacy of planned or existing ocean outfalls is through utilization of mathematical models. Such models usually treat the near and far fields separately and differ mainly in that they include buoyancy and momentum-flux terms in the near field while providing for dispersion or eddy-diffusion terms in the far field. In this paper the far field is treated and attention is directed to one aspect of the use of such models—the determination of horizontal diffusion coefficients at depth.

Generally, attention is directed to the surface layer only. A complete treatment requires the determination of the spread of pollutants at all depths, for instance for material trapped below the thermocline. Additionally, increasing use of multi-layer numerical models of offshore areas[1] requires a determination of the relative magnitude of dispersion in each layer.

The Taylor hypothesis[2] for expressing diffusivity in terms of Lagrangian statistics is used as the basis for this study. Briefly, Taylor showed that the diffusivity or dispersion (terms which are used synonymously in this paper although dispersion is usually meant to include diffusion *and* advection effects), ϵ, of substances could be expressed as the product of the square of the turbulent eddy velocity, v', and the Lagrangian integral time scale for large diffusion times:

$$\epsilon = v'^2 \int_0^\infty R_L(\tau) d\tau \tag{1}$$

where τ is a dummy lag variable and $R_L(\tau)$ is the Lagrangian correlation coefficient of an assumed stationary time series of velocity.

In the derivation given below, it is shown that, with suitable approximations, the eddy diffusivity can be expressed in the easier-to-measure Eulerian frame:

$$\epsilon = v'\overline{U}\beta \int_0^\infty R_E(\tau)\, d\tau \tag{2}$$

where $\overline{U}$ is the mean velocity, $R_E(\tau)$ is the Eulerian correlation coefficient and β is the ratio of the Lagrangian to Eulerian integral time scales. Hay and Pasquill[3] have shown that $R_L(\eta) = R_E(\tau)$, where $\eta = \beta\tau$ and the magnitude of β depends on the scale of turbulence. Making a further assumption about the magnitude of β, oceanographic data collected during the Coastal Upwelling Experiment (CUE-1) off the Oregon coast of the United States are utilized to compute ϵ at depths of 20, 40, 60, 70, 80, and 120 m.

Eddy-Transport Relations

The basis for the computations to be described assumes a homogeneous flow field and averages taken over a large number of particles. It can be shown[4] that the eddy diffusion of a scalar quantity can be expressed as

$$\epsilon(\alpha) = v'^2 \int_0^\infty f(\alpha, \tau) R_L(\tau)\, d\tau \tag{3}$$

where $f(\alpha, \tau)$ signifies the effect of exchange during the path of the particles. The Lagrangian turbulence velocity $v' = \sqrt{\overline{v^2}}$ and the Lagrangian time correlation is given by

$$R_L(\tau) = \frac{\overline{v(t)v(t+\tau)}}{v'^2}. \tag{4}$$

For fluid particles, $f(\alpha, \tau) = 1$, and the eddy diffusivity is given by

$$\epsilon = v' \Lambda_L \tag{5}$$

where $\Lambda_L = v' \int_0^\infty R_L(\tau)\, d\tau$ is the spatial integral scale.

Averaging with respect to time for the velocity u at a fixed point, the Eulerian correlation coefficient is given by

$$R_E(\tau) = \frac{\overline{u(t)u(t+\tau)}}{u'^2}. \tag{6}$$

The integral scale is given by $\int_0^\infty R_E(\tau)\, d\tau$, which is related to the longitudinal integral scale, Λ_f, by $\Lambda_f = \overline{U} \int_0^\infty R_E(\tau)\, d\tau$, when the uniform mean velocity is greater than the turbulence component of the Eulerian velocity: $\overline{U} \gg u$.

Lagrangian and Eulerian Scales

Several theoretical, field, and laboratory studies have been conducted to determine the ratio of the Lagrangian to Eulerian integral scales. Krasnoff and Peskin[5] summarized some of these and found values ranging from 1·5 to 11·3 for experimental conditions of laboratory grid turbulence and the atmospheric boundary layer, respectively. All of the values reported were >1; however, Kao[6] computed $\beta = 0{\cdot}33$ from free-drifting meteorological balloon flights at the 300-mbar level. He also related β to the particular scale involved and the angular wave number. For the high Reynolds numbers of oceanic flow it is assumed here that $\beta = 1$. Whatever value of β exists, the conclusions drawn in this paper will remain unchanged only if it is not a strong function of depth. To confirm the value of β, rather difficult Eulerian–Lagrangian field measurements are required. To the Author's knowledge, these experiments have not been undertaken.

Data

The first Coastal Upwelling Experiment was conducted off the Oregon coast during the summer of 1972. The purpose of CUE-1 was to study the dynamics of upwelling and it was primarily a physical oceanographic endeavor. The CUE-1 setting is shown in Fig. 1. Periodically, hydrographic stations were taken within the area; some of the standard oceanographic data collected are used here to illustrate the nature of the oceanographic setting of the area. Of particular interest are the current-meter stations; data from these are extensively employed in this study and form the basis for it. The locations of these stations are also shown in Fig. 1.

Aanderaa current meters were used by the CUE group. The meters continuously monitor rotor turns and output the data at, usually, 10-min intervals. The data are recorded within the meter on

TABLE 1. Current-meter station information, August 1972

Station	Longitude (W)	Latitude (N)	Depth (m)	Start time (Date)	Start time (GMT)	Stop time (Date)	Stop time (GMT)
NH3	124°08·0′	44°39·1′	35	10th	0601	31st	1301
NH6	124°12·0′	44°39·1′	20	5th	0156	31st	1056
			40	5th	0157	31st	1057
			60	5th	0158	31st	0958
NH10	124°17·6′	44°39·1′	20	2nd	2333	31st	0933
			60	2nd	2330	31st	0930
POL	124°44·8′	44°44·7′	40	14th	2311	29th	1111
			60	14th	2302	29th	1102
			80	14th	2308	29th	1108
DB7	124°12·7′	44°51·5′	20	2nd	0002	29th	1102
			40	1st	2359	29th	1459
			60	2nd	0000	20th	1800
			80	1st	2359	29th	1459
DB13	124°19·0′	44°54·0′	20	10th	0218	29th	1418
			40	10th	0217	29th	1417
			70	10th	0229	29th	1229
			120	10th	0231	29th	1231

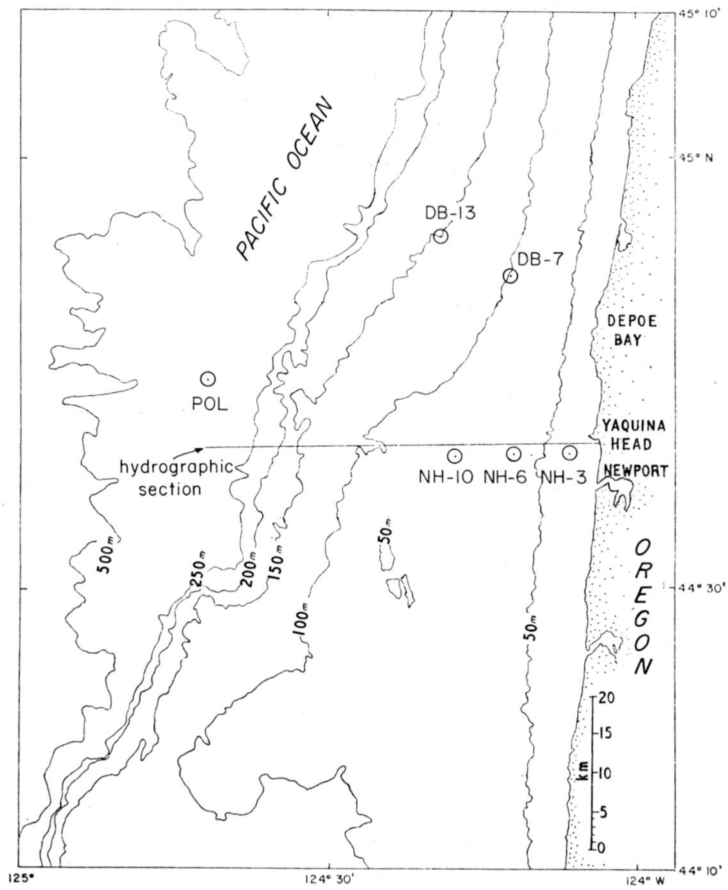

FIG. 1. CUE area, and current-meter stations and hydrographic section discussed in text

magnetic tape which is then used in the data-processing and numerical-filtering procedures as described by Pillsbury *et al.*[7] The filter is designed to preserve as much information as a time series with a data interval of one hour can carry without aliasing. The Nyquist frequency is $\frac{1}{2}$ cycle/h; the half power of the filter lies at that point and the frequency response has a fairly sharp cut-off. The maximum possible amount of information is passed without aliasing from frequencies above the Nyquist frequency. Pillsbury *et al.*[7] feel that the output of this filter is preferable to the simple hourly average usually employed.

The current meters were taut-line moored and were retrieved by triggering sonic release mechanisms. Table 1 shows station names, locations, meter depths, and GMT starting and stopping times available for the periods considered. All of the data at DB13 were used in the analysis. Those data where a comparable period (15–20 August) was common to the remainder of the meters were also used.

Computations

The basic computations required to solve Equation 2 are for the evaluation of v', $\overline{U}$, and $\int_0^t R_E(\tau)\, d\tau$. Each CUE-1 data record consisted of filtered 10-min velocities averaged over 1 h as has been described. The north-south (NS) and east-west (EW) components of each velocity vector are treated as separate records.

A computer program presented by Ahn and Smith[8] was modified for use in this study. The turbulent velocity was computed by breaking the velocity record into segments, fitting a second-order polynomial to these segments, and taking the residual from the fit as v'. The mean velocity was computed from the sum of the absolute values of the velocities for each hour, divided by the number of observations. The integral time scale was computed by calculating the autocorrelation coefficient and integrating to the point of the first zero crossing as the limit. When $R_E(\tau) = 0$, it is assumed that the 'memory' of the diffusion event has lapsed. The program lists the input data, takes data averages (if any), and evaluates covariance, correlation coefficient, integral to the zero point of the autocorrelation, $\overline{U}$, v', ϵ, $v'/\overline{U}$, and Λ_f.

The Eulerian autocorrelation coefficient was computed from the NS and EW velocity components according to the relation

$$R_E(\tau) = \frac{\sum_{t=0}^{T-\tau} u(t)u(t+\tau)}{\left\{\sum_{t=0}^{T-\tau} u^2(t) \sum_{t=0}^{T-\tau} u^2(t+\tau)\right\}^{1/2}} \tag{7}$$

where T is the record length. The scale of turbulence involved is dependent on the record length and sampling interval. The time scale of v' is a function of the record length; the scale dependency is thus the same for all computed cases. Bowden and Fairbairn[9] used a 3·5-min record length and a sampling interval of 0·6 s in a study of turbulent fluctuations in the Mersey estuary. The first zero crossing was, on the average, about 40 s for the longitudinal velocity component. Although they did not compute diffusivity, their data provides a value of $\epsilon = 1 \cdot 12 \times 10^3$ cm²/s. Kao[6], in a study of planetary waves, showed a zero crossing of about 48 h for the transverse Eulerian coefficient.

Some Oceanographic Features during CUE-1

Smith[10] has described variations in currents, wind, and sea level for CUE-1 during July–August 1972. For current-frequency variations of 0·6 cycle/day, the mean longshore current at 20 m was 18 cm/s southward while at 80 m it was 6 cm/s northward. Spectral analysis showed a very high mutual coherence between wind, sea level, and clockwise currents only in a narrow band at 0·15 cycle/day. Huyer and Patullo[11], in a study of wind and weather in the summer of 1969 in the same general area, demonstrated that wind and current are related at periods longer than two days. Currents were roughly parallel to bottom contours and behaved similarly to the wind when the wind varied slowly.

Currents

Figure 2 is a progressive vector diagram (PVD) of DB13 currents at all depths for 10–28 August 1972. The 20 and 40 m PVD's show a clockwise rotation ending in a constant southward motion. The 70-m PVD shows a northward motion then a counterclockwise rotation towards the south. The net displacement of the vector is of the order of 25 km from the beginning point, the least of those shown. At 120 m

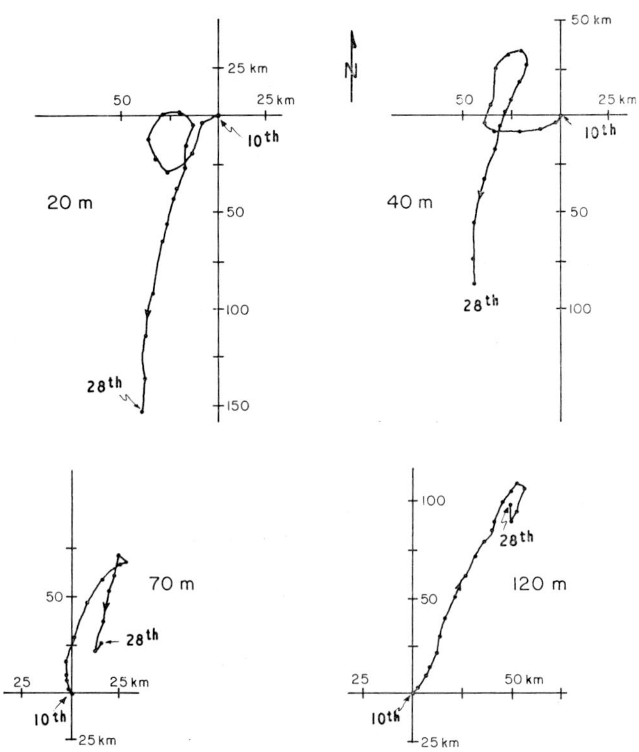

FIG. 2. Progressive current-vector diagrams for Station DB13, 10–28 August 1972
Starting and finishing dates shown

the PVD shows that the current is counter to that at 20 and 40 m most of the time; there appears to be little coupling between the upper layers and that at 120 m.

Winds

For the period 15–18 August 1972, a typical $1\frac{1}{2}$-day wind event indicates cyclonic rotation through the north, then decelerating winds followed by an increase in speed to 8 m/s turning through the south. On 18 August, calm periods prevailed. This wind picture is not atypical of the variability to be found in the upwelling area during certain low wind speed periods. The winds during 10–28 August averaged about 3·6 m/s; from 10 to 22 August the direction was to the north and thereafter to the south.

The effect of winds on dispersion has been discussed by Wu[12]. His argument mainly concerns the surface layer where suspended materials may be directly acted upon by wind stress; at depth the effect of wind is to generate dispersion via shear and to maintain turbulence. It is expected that the direct effect on diffusivity can be approximated by computing coherence between wind, current measurements, and the Eulerian time scale although this has not been attempted here.

Temperature, salinity, and density

Temperature, salinity, and density sections for 22 August, during a period of calm winds, have been presented by Huyer[13] along 44°40′N. Exhibited were temperature anomalies and an upward slope of the salinity and density contours toward the shore, features usually associated with an upwelling situation. Time series of temperature and salinity at 44°40′N, 124°18′W, illustrate the occurrence of warm anomalies at about 15 and 45 m. There is no strong evidence of a regular semidiurnal periodicity in these data although the predominant fluctuations occur at that scale

The temperature, salinity, current, and wind data discussed here are only a minute portion of the extensive CUE-1 data records which are still undergoing analysis. They do indicate, however, the temporal and spatial variations one may expect in this coastal environment.

In the time-mean case, we might expect large-scale lateral mixing to occur along isentropic surfaces. In the limited scale associated with most coastal pollution problems, more attention must be devoted to short-time variations in density in order to prescribe satisfactorily the dispersion of effluents likely to occur.

Deductions from the Diffusion Equation

As shown by Okubo[14], solutions of the radially symmetric diffusion equation

$$\frac{\partial S_I}{\partial t} = \frac{1}{r}\frac{\partial}{\partial r}\left\{af(t)\cdot r^{m+1}\frac{\partial S_I}{\partial r}\right\}, \tag{8}$$

with generalized diffusion coefficient $ar^m f(t)$, can be expressed in terms of $\epsilon \sim l$ or $\epsilon \sim l^{4/3}$ where l is the scale of turbulence; S_I is the mean concentration for a diffusing patch from an instantaneous point source, r is patch radius, $f(t)$ an arbitrary function of time t, and a, m are arbitrary constants.

The mean dispersion rate of change, $d\langle r^2\rangle/dt$, is shown to be proportional to $\langle r^2\rangle^{1/2}$ in solutions to Equation 8 by Joseph and Sendner[15] and Okubo and Pritchard[16], while Ozmidov[17], Obukhov[18], and Okubo[14] have it as $\langle r^2\rangle^{2/3}$. The last three solutions employ an 'energy dissipation' parameter (dimensions $[L^{2/3}T^{-1}]$) consistent with a 4/3 hypothesis, while the former two use a 'diffusion velocity' term $[LT^{-1}]$ consistent with a mixing length concept.

While solutions of Equation 8 have been extensively studied only for surface distributions, there is no obvious reason in the absence of boundary effects why they should not apply to subsurface far-field dispersion situations such as may occur around deep coastal outfalls. An examination of Equation 8 suggests that difficulties will exist relative to the measurement of the dispersion term in a Lagrangian frame since the coordinate r is taken at the moving center of the diffusing patch.

Direct measurements of dispersion, for instance by fluorometric tracing of dye releases, are notoriously difficult to obtain, especially in open coastal regions where navigation, among other things, is a very real problem. Tracing a subsurface release is at least an order of magnitude more difficult than tracing a surface release; fixed-point velocity measurements, however, are considerably easier to obtain although still expensive and arduous undertakings.

Results

Representative values of the autocorrelation coefficient for this study are shown in Fig. 3 for the EW and NS components at four depths for Station DB13. In general, the first crossing occurred at 1–15 and 1–16 h for the EW and NS components, respectively.

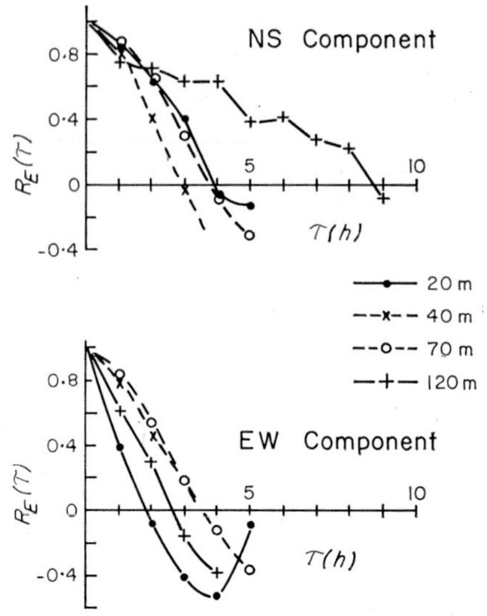

FIG. 3. Representative Eulerian autocorrelation coefficients, Station DB13, 18 August 1972

In Fig. 4, daily values of ϵ are shown for Station DB13 for both components for 11–18 August 1972. Values at 20, 40, 70, and 120 m are given. The values range from 4×10^4 to 10^6 cm^2/s with the average for all values being about 10^5 cm^2/s. The 120-m diffusivities are somewhat less than the others most of the time, but there are periods when all the ϵ's are of about the same magnitude. The EW component at 120 m is consistently lower than those at the shallower depths. The average values are somewhat less than for the NS component.

The components for the remaining stations for 15–20 August are shown in Fig. 5. There are uniform fluctuations with depth in the NS data, most noticeable for Station NH6. The average values of all the data are again about 10^5 cm^2/s. The average values are noticeably lower, about 8×10^4 cm^2/s for Station NH6. The average values at depth are given in Table 2 for each station. The averaging period

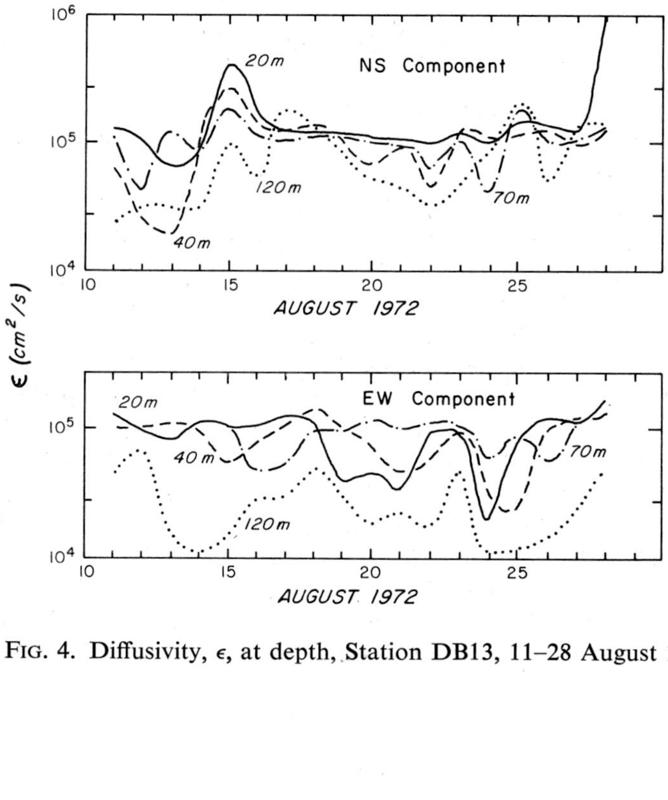

FIG. 4. Diffusivity, ϵ, at depth, Station DB13, 11–28 August 1972

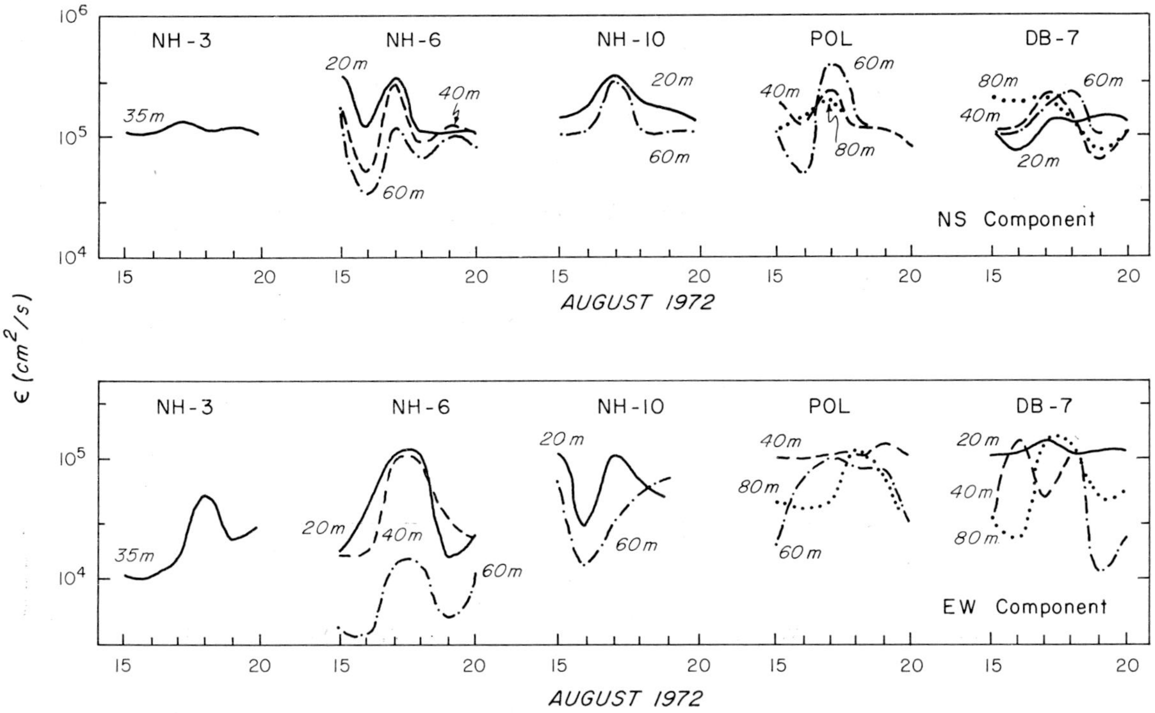

FIG. 5. Diffusivity, ε, at depth, all stations except DB13, 15–20 August 1972

TABLE 2. Average values of eddy diffusivity (10^4 cm²/s)

Station	Depth (m)	Component East	Component North	Station	Depth (m)	Component East	Component North
POL	40	9·29	22·6	NH3	35	3·59	17·4
	60	11·6	20·9	DB7	20	16·9	18·3
	80	8·68	19·6		40	11·2	20·8
NH10	20	8·85	23·3		60	11·6	22·4
	60	12·8	45·2		80	6·29	26·6
				DB13	20	12·6	26·3
NH6	20	6·47	27·2		40	11·9	18·1
	40	5·91	21·5		70	12·6	16·4
	60	1·33	13·8		120	4·71	12·8

is 15–20 August for all stations except DB13 where the period was 11–28 August. There is a general decrease in ϵ with depth at most stations although at DB7 the trend is reversed for the NS component.

Integral length scale

In Fig. 6, diffusivity is plotted against the Eulerian integral length for both components at all depths. Also shown are the slopes for an integral scale to the 4/3 and to the unit (3/3) powers. It can be seen that, although regression analysis was not performed, the unit power provides a reasonable fit.

In Table 3, the average integral lengths are given for Station DB13. The NS component length decreases with depth whereas the EW component increases to 70 m, the 120-m value being considerably less than any of the others. If the integral lengths are given a physical significance, then the longitudinal and transverse components can be visualized as being the major and minor axes of an ellipse. The area of the ellipse is then equivalent to

$$A_E = \frac{\pi}{4}\left\{\overline{U}\overline{V}\beta^2 \int_0^t R_{Eu}(\tau)\, d\tau \int_0^t R_{Ev}(\tau)\, d\tau\right\} \qquad (9)$$

where $\overline{V}$ is the transverse equivalent of $\overline{U}$, and R_{Eu}, R_{Ev} represent the longitudinal and transverse Eulerian autocorrelation coefficients, respectively. These areas are also shown in Table 3 where it can be seen that they decrease with depth.

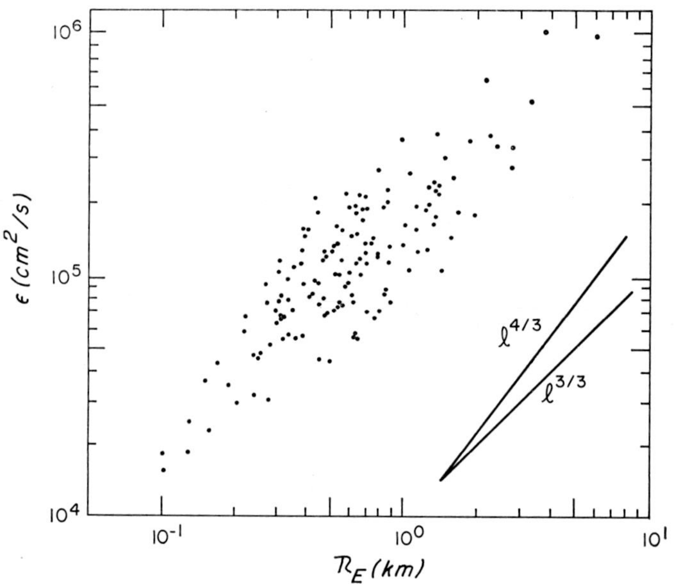

FIG. 6. Diffusivity versus Eulerian integral scale, all data, both components
Inset: slopes of $l^{3/3}$ and $l^{4/3}$

TABLE 3. Integral length averages and Eulerian areas
Station DB13, 11–18 August 1972

Depth (m)	Component		Area (km²)
	EW (km)	NS (km)	
20	0·41	1·27	0·66
40	0·51	1·06	0·62
70	0·56	1·00	0·62
120	0·30	0·97	0·50

Approximations for a 4/3 hypothesis

As shown above, diffusivity can be expressed as

$$\epsilon = v'\overline{U} \int_0^t R_E(\tau)\, d\tau. \tag{10}$$

The Eulerian integral length is

$$\Lambda_E = \overline{U} \int_0^t R_E(\tau)\, d\tau, \tag{11}$$

so that

$$\epsilon = v'\Lambda_E. \tag{12}$$

The energy-dissipation relation for diffusion invoking a 4/3 law is

$$\epsilon = \rho\varepsilon^{1/3} l^{4/3} \tag{13}$$

where ρ is an arbitrary constant, ε is the energy dissipation term, and l is the scale of turbulence.

Equating Equations 12 and 13, and solving for the Eulerian scale gives

$$\Lambda_E = \frac{\rho\varepsilon^{1/3} l^{4/3}}{v'} = \gamma l^{4/3}. \tag{14}$$

Some typical values of the parameters in this equation are: $\rho \simeq 0{\cdot}02$, $\varepsilon \simeq 5 \times 10^{-4}$ cm^2/s^3, $v' \simeq$ 2 cm/s; hence $\gamma \simeq 8 \times 10^{-4}$ cm$^{-1/3}$ and $l \simeq 1{\cdot}25 \times 10^3 \Lambda_E^{3/4}$.

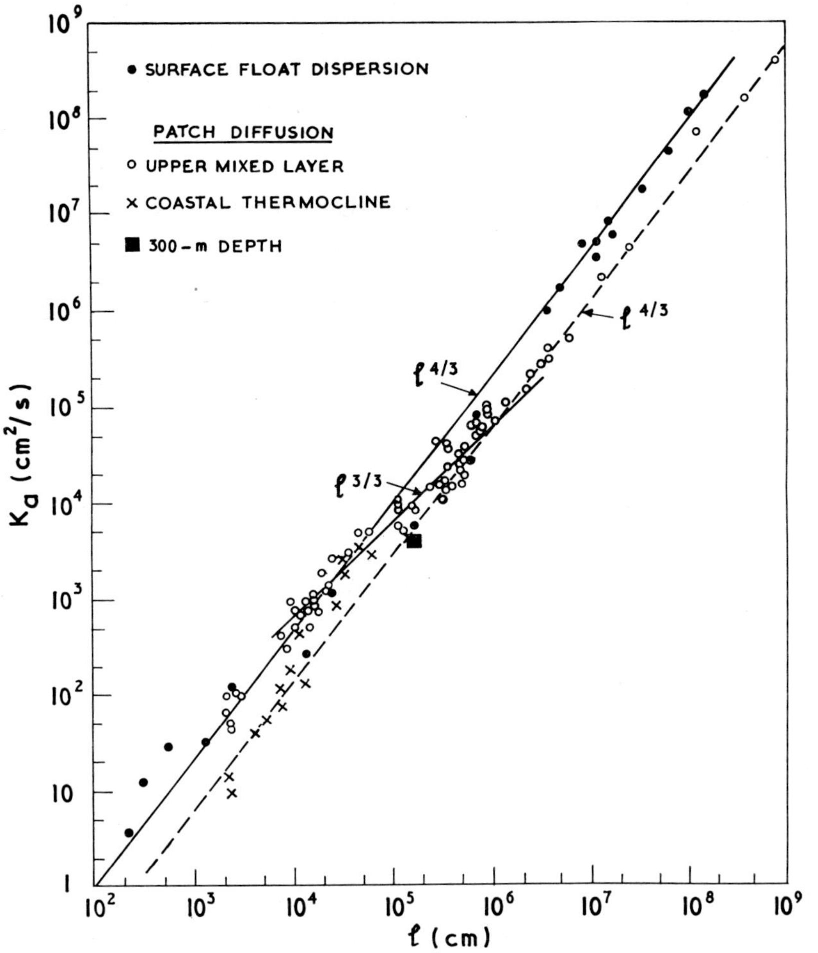

FIG. 7. Okubo's unified diagram of patch diffusion and float dispersion[19]
Line added: slope of $l^{3/3}$

It should be noted that the evidence for a 4/3 diffusion law is not always convincing. For instance, in Fig. 7 in the region $l = 10^4$–10^6 cm, a linear scale has been superimposed on Okubo's unified diagram[19] of dispersion versus scale length. Although a regression analysis was not run on the data, the $l^{3/3}$ fit is more obvious than $l^{4/3}$. Thus, the linearity expressed by Equation 12 lends itself more easily to a mixing-length hypothesis than to a 4/3 law.

Discussion

There is no obvious relation between distance offshore and the computed eddy diffusivities. The NS diffusivity component always exceeded the EW component; since the prevailing direction of the mean current was NS there is evidence for an elongation of a diffusing patch in the absence of vertical shear. There is a decline in the diffusivities with depth, the value at 120 m being about half that at 20 m for the NS component and about 37 per cent for the EW component. Ahn and Smith[8] computed diffusivities for current meters set at a depth of about 4 m at four stations having water depths of 10–25 m. They found values of ϵ averaging 3×10^5 and 6×10^6 cm²/s for the EW and NS components respectively. These values are about an order of magnitude greater than those computed here at the 20-m depth. The most obvious explanation for this difference is a combination of the water-column depths involved, and the fact that the CUE-1 meters were below the wind-mixed layer depth.

There are indications that the 4/3 hypothesis is not necessarily a good functional descriptor of the scale of turbulence studied here, shown both by the data here described and by an examination of Okubo's 1972 data[19]. A mixing-length concept is suggested.

The framework exists for carrying out simultaneous Eulerian-Lagrangian field experiments. The progressive vector representation of currents could be used to indicate the motion of certain layers of water for short time periods and the computed diffusivities used to determine the pseudo-Lagrangian spread of materials. These results should be compared with true Lagrangian measurements. One benefit of conducting pollution studies in nearshore regions on a Eulerian basis would be the reduction in field work required. Since field surveys in the marine environment are usually the largest part of a research or engineering budget, the potential for Eulerian application should be exploited.

ACKNOWLEDGEMENTS

I would especially like to thank Dr. R. L. Smith, Oregon State University (OSU), for making available CUE current-meter records. The Coastal Upwelling Experiment is a program sponsored by the International Decade of Oceanography under National Science Foundation Grant GX28746. Dr. Dale Pillsbury and Messrs J. S. Bottero and D. Root, also of OSU were very helpful. Without their cooperation, it would not have been possible to undertake this work.

Dr. C. S. Ahn provided helpful comments on aspects of the computer program; his report provided the initial impetus for this work.

Drs. A. Okubo and M. Shirazi also provided constructive criticism of the manuscript.

Mrs. Judy Maher modified the computer program and carried out the computations with her usual efficiency.

REFERENCES

1. LAEVASTU, T., and CALLAWAY, R. J. Currents, tides and dispersal of pollutants in New York Bight. Part 4—The large grid size model in single and two layers for New York Bight from Block Island to Atlantic City. Environmental Prediction Research Facility, U.S. Navy Postgraduate School, Tech. Note No. 4–74, 1974.
2. TAYLOR, G. I. Diffusion by continuous movements. *Proc. Lond. math. Soc.*, 1921, Ser. 2, **20**, 196–202.
3. HAY J. S., and PASQUILL, F. Diffusion from a continuous source in relation to the spectrum and scale of turbulence. *Adv. Geophys.*, 1959, **6**, 345–364.
4. HINZE, J. O. Turbulence—an introduction to its mechanism and theory. McGraw Hill, New York, 1959.
5. KRASNOFF, E., and PESKIN, R. L. The Langevin model for turbulent diffusion. *Geophys. Fluid Dynamics*, 1971, **2**, 123–146.
6. KAO, S. K. The large-scale Lagrangian turbulence in the atmosphere. In 'Random functions and turbulence' (by S. Panchev). Pergamon Press, London, 1971, Appendix.
7. PILLSBURY, R. D., BOTTERO, J. S., STILL, R. E., and GILBERT, W. E. A compilation of observations from moored current meters, Oregon continental shelf, April–October 1972. Oregon State Univ., Sch. Oceanogr., Ref. 74–1, Vol. 6, Data Rep. No. 57, 1974.
8. AHN, C. S., and SMITH, P. F. The first report of a study to forecast nuclear power plant effects on coastal zones. U.S. Govt. Printing Office, Washington, D.C., 1972.
9. BOWDEN, K. F., and FAIRBAIRN, L. A. Measurement of turbulent fluctuations and Reynolds stresses in tidal currents. *Proc. R. Soc.* A, 1956, **237**, 422–438.
10. SMITH, R. L. A description of current, wind, and sea level variations during coastal upwelling off the Oregon coast, July–August 1972. *J. geophys. Res.*, 1974, **79**, 435–443.
11. HUYER, A., and PATULLO, J. G. A comparison between wind and current observations over the continental shelf off Oregon, summer, 1969. *J. geophys. Res.*, 1972, **77**, 3215–3220.
12. WU, J. An estimation of wind effects on dispersion in wide channels. *Wat. Resour. Res.*, 1969, **5**, 1097–1104.
13. HUYER, A. Vertical distribution of temperature, salinity, and sigma-t from observations from R/V *Yaquina* during coastal upwelling experiment, 1972. Oregon State Univ., Sch. Oceanogr., Ref. 73–6, 1973.
14. OKUBO, A. Horizontal diffusion from an instantaneous point source due to oceanic turbulence. Johns Hopkins Univ., Chesapeake Bay Inst., Tech. Rep. 32, 1962.
15. JOSEPH, J., and SENDNER, H. Über die horizontale Diffusion im Meere. *Dt. hydrogr. Z.*, 1958, **11**, 49–77.
16. OKUBO, A., and PRITCHARD, D. W. Unpublished note. (See PRITCHARD, D. W. The application of existing oceanographic knowledge to the problems of radioactive waste disposal into the sea. In 'Disposal of radioactive wastes' Vol 2. International Atomic Energy Agency, Vienna, 1960, pp. 229–253.)
17. OZMIDOV, R. V. On the calculation of horizontal turbulent diffusion of the pollutant patches in the sea. *Dokl. Akad. Nauk. SSSR*, 1958, **120**, 761–763.
18. OBUKHOV, A. M. Description of turbulence in terms of Lagrangian variables. *Adv. Geophys.*, 1957, **6**, 113–116.
19. OKUBO, A. Some speculations on oceanic diffusion diagrams. *Rapp. P.-v. Réun. Cons. int. Explor. Mer*, 1975, **167**, 77–85.

In presenting his paper Dr. Callaway took the opportunity to summarize ongoing work not reported in his paper. In efforts to model coastal circulation and dispersion, he had been engaged in applying the two-dimensional models of Leendertse[20] and Hansen[21]. The Leendertse circulation model was being applied to the Oregon coast; the Hansen multi-layer dispersion model had been used in New York Bight[1] and was currently being applied to the south-east coast of Florida, Los Angeles Bight, and Valdez Harbor in Prince William Sound, Alaska. This work was being performed in co-operation with the U.S. Navy at Monterey, California, Dr. T. Laevastu being the principal investigator. Other work related to an investigation of sludge dispersion from barges operating from New York Harbor; series of vertical profiles of temperature, salinity, velocity, and light transmission were obtained, and water samples collected, after a barge release, the samples being examined for suspended sediments and particle counts. Two field investigations had been made so far, and a data report was being prepared.

DISCUSSION

Dr. G. E. B. Kullenberg (Denmark) pointed out that, since the factor β depended on the turbulence intensity, it would probably vary with depth; this would explain the depth variation reported.

Mr. J. W. Talbot (U.K.) asked the Author to comment on the magnitude of the diffusion coefficients shown in Fig. 4, since they appeared to be rather large—ranging up to 10^6 cm^2/s which (according to Fig. 7) corresponded to a scale of more than 10 km.

Prof. H. B. Fischer (U.S.A.) said that Equation 2 could be substituted for Equation 1 only if $\overline{U}$ was much greater than v'. Pasquill[22] had quoted evidence that the 'Taylor hypothesis' was valid for values of $v'/\overline{U}$ as large as 0·26, and Lumley and Panofsky[23] had mentioned experimental evidence that the hypothesis was usable for values of $v'/\overline{U}$ less than $\frac{1}{3}$. Taylor had intended his hypothesis to apply to velocity in one direction only, but Dr. Callaway had applied the hypothesis by components; Prof. Fischer therefore asked if he knew of any theoretical basis for applying it to the components of a directionally varying velocity vector.

REPLY TO DISCUSSION

Dr. Callaway referred Dr. Kullenberg to p. 298 where it was stated that β was assumed to be unity and where comments were made on the depth variation. If further field experiments of the RHENO type were to be made in the future, efforts should be expended to ensure that Eulerian–Lagrangian measurements were made.

In reply to Mr. Talbot, he said that the absolute values of the diffusion coefficients were, of course, debatable without a real knowledge of β (which, however, would be not less than unity). Nevertheless, the main purpose of the paper was to give some insight into the vertical variation of horizontal diffusion coefficients, namely the relative magnitudes at different depths. In any event, he did not think the values were too large, considering the sampling interval used, and he suggested that Mr. Talbot—as one of the more able and experienced dye (Lagrangian) experimenters around—should consider the possibility of simplifying his work by using (Eulerian) current meters in future work.

Replying to Prof. Fischer, Dr. Callaway said that, for Station DB13, the values of the turbulence intensity ($v'/\bar{U}$) at 20, 40, 70, and 120 m had been respectively 0·43, 0·83, 0·88, and 0·68 for the EW component, and 0·19, 0·37, 0·59, and 0·36 for the NS component. Thus, for the data and averaging period used in the paper, the fluctuations were of the same order of magnitude as (but consistently less than) the mean velocities. This suggested that the Stokes velocity might be important, since the Eulerian and Lagrangian mean velocities would be equal only if the mean was very much greater than the fluctuations. With regard to the application of the Taylor hypothesis, he pointed out that the individual daily experiments were carried out on a uni-directional basis, the direction shifting from day to day. However, he noted that the hypothesis itself was especially subject to doubt in shear flow. The main point was not the absolute values of the diffusion coefficients, but rather the relative values with respect to depth. Aspects of the theoretical development had been discussed by Kao[6].

ADDITIONAL REFERENCES

20. Leendertse, J. J. Aspects of a computational model for long-period water-wave propagation. Rand Corp., Santa Monica, Memo. RM-5294-PR, 1967.

21. Hansen, W. The reproduction of motion in the sea by means of hydrodynamical-numerical methods. *Mitt. Inst. Meeresk., Hamburg*, 1966, **5**, 57pp.

22. Pasquill, F. Atmospheric diffusion. Van Nostrand, 1962, p. 39.

23. Lumley, J. L., and Panofsky, H. A. The structure of atmospheric turbulence. Interscience, New York, 1964, p. 57.

Subsequent Dispersion—Methods of Measurement

K. Cederwall, C-G. Göransson, and T. Svensson
Hydraulics Division, Chalmers University of Technology
Göteborg, Sweden

For the study of dispersion in a receiving water, two properties are often of special interest: dilution, and residence time of a discharged pollutant. Residence time should be understood as the time after the discharge of the waste-water element in question. These two properties are of great interest with reference to later consideration of, for example, ecological and hygienic consequences of a planned discharge of waste water.

The objective of a field study is either to describe—qualitatively or quantitatively—the actual pattern of spreading of the sewage from an existing outfall or, in case of a planned outfall installation, to predict concentrations and residence times of pollutants in the receiving waters considered. In both cases a thorough planning of the field survey is required, particularly for the mapping by *in situ* measurements of tracer concentrations in order to provide adequate statistical information about the distribution of pollutants in the receiving waters. The framework for these tests is a thorough knowledge of the meteorological and hydrographic conditions.

Planning

There are different strategies for the prediction of the dispersion pattern from a waste-water outfall:

1. *In situ* simulation of the transport and mixing of the waste effluent by tracer methodology in the field. The purpose of these studies is to get a quantitative or qualitative picture of the spreading of the waste water.
2. A deterministic mathematical model of the dynamics of the disposal area including a dispersion model for the waste-water discharge. This theoretical work must be supplemented by field tests in order to verify the models (calibration) and also for feeding the models with adequate values of the governing parameters.
3. Predictions based on a stochastic model as an alternative to a deterministic model—Case 2 above. A substantial amount of field work is required.
4. Scale test by means of a physical model. Great emphasis must again be put on the calibration of the model using field data.

The choice of strategy depends on experience from other studies in similar water bodies and preliminary surveys in the disposal area. This is certainly also a decision that comprises economics. Field experiments are quite expensive and the 'optimal strategy' is likely to be a combination of two or several of the approaches listed above. However, for receiving waters exhibiting more complicated current structures, tracer tests will provide information that otherwise would be very difficult to obtain. This emphasizes the value of *in situ* simulation of the waste spreading which should be supplemented by statistical information from recording current meters in selected positions in the disposal area.

Hence, the strategy for prediction of waste-water dispersion in case of a more regular current structure (a straight coastline) is more towards mathematical modelling supplemented by field measurements, whereas for receiving waters having a complex circulation (the archipelago type of water environment) tracer tests are probably the most adequate alternative to a necessarily rather complicated mathematical model.

A field study should be based on thorough knowledge of the dynamic factors governing the dispersive mechanisms. The systematic planning of a series of tracer tests should then cover the full range of these parameters.

For the planning of the dispersion tests there are several questions that have to be cleared up. Alternative disposal sites should be specified. The initial mixing of the waste water in the immediate proximity of the outfall as related to outfall (for instance multiple-port diffuser) operation must be analyzed. In a stratified receiving environment there is a possibility of submerged sewage fields forming above a submarine outfall. Information on this point will provide the starting condition for the dispersion tests.

Tracer Technique

Most tracer experiments carried out in recent years for dispersion studies have used the fluorescent dye Rhodamine-B as the tracer and the Turner Model III Fluorometer for the assay. This technique will be described as well as the constraints with which the fluorometer operates. Radioactive tracers have also been used extensively to study pollution problems.

It is necessary to distinguish between conservative and non-conservative tracers. When the tracer concentration decreases with time—assuming no dilution effect—the tracer is non-conservative or decaying. Very often, rapidly decaying tracers are preferred to more conservative ones. For instance, the presence of the tracer may be wanted only for a short period of time. Use of a slowly decaying tracer then could give rise to undesirable effects of accumulation.

There are two major approaches for tracer release: continuous injection during a certain period of time, and repeated instantaneous injections covering the full range of environmental conditions. The most adequate simulation of a continuous waste discharge is certainly attained by applying a continuous tracer-release technique. Assume that the waste product which is to be traced may be characterized by a decay parameter, k, corresponding to the most critical or most conservative component of the waste. Then a continuous injection of a tracer material with a decay parameter equal to or close to k is the most straightforward and reliable approach. This situation is, however, very seldom at hand. In order to simulate the continuous discharge of a conservative contaminant by means of a decaying tracer, the injection must be steadily reduced at a rate equal to the tracer decay compensating for the loss of tracer material in the recipient. This severely restricts the length of the period of injection, especially as radioactive isotopes normally have to be used (the method works for tracers of constant decay rates only) due to the fact that both the initial and final rates of tracer injection, for safety and practical reasons have to be neither too high nor too low. With continuous release of one tracer, residence time must be estimated from hydrographic observations. It may be almost impossible to get reliable information by such a procedure if the receiving water has a complicated flow pattern. In that case a dual tracer technique may be the best alternative technically as well as economically.

To be used for a continuous, parallel injection two decaying tracers are used which fulfil the following conditions: the decay parameters must be well defined, and the tracer substances must differ in at least one property, detectable in low concentrations of the substance.

Assume that the two tracers, with decay parameters k_1 and k_2, are simultaneously and continuously released at a constant rate. A time parameter, τ, is now defined by:

$$\begin{cases} C_1 = C_1' e^{-k_1\tau} & (1) \\ C_2 = C_2' e^{-k_2\tau} & (2) \end{cases}$$

where C_1 and C_2 are the tracer concentrations at the measuring site, and C_1', C_2' corresponding concentrations of the tracers assumed conservative. For commonly used tracers and normal dispersive properties of the receiving water τ is very close to the mean residence time, see Cederwall[1]. Consequently, a τ value may be determined for each site in the receiving water where tracers are found in measurable concentrations. Thus a dual tracer technique makes possible the simultaneous registration of concentration distributions and residence time in a receiving water area. An investigation using the radioactive tracer bromine-82 and the fluorescent tracer Pontacyl Brilliant Pink has been reported by Cederwall and Hansen[2].

Instead of using a continuous-release method, spread tests can be carried out by means of instantaneous injections of the tracer. Provided that the pattern of circulation is steady or repeated regularly (tidal oscillations), the dispersion from a continuous source can be calculated by integration of the results obtained from a single injection. The theoretical background to the instantaneous injection technique as well as case reports have been described by Harremoës[3,4].

Choice of tracer

When studying transport phenomena in coastal waters any detectable property of the water that has gradients in the system can be used as a tracer. Such natural tracers include salinity, temperature, silica, dissolved oxygen, and phosphorus. Their main disadvantage is that the boundary conditions are difficult to control and some of them undergo chemical alterations in the water. Added tracers are therefore more useful and in many cases they offer the only possibility of studying water movements and turbulent spreading. Radioactive isotopes with suitable decay times and fluorescent dyes are most frequently used, both with considerable success. These tracers are detectable at very low concentrations. Some radioactive isotopes—for instance Br-82—have the advantage of not being significantly adsorbed by suspended particles. They have a constant decay velocity and are somewhat easier to detect *in situ*. However, due to the risks and restrictions in connection with the use of radioactive isotopes, fluorescent tracers have so far been more widely used. Though exhibiting a rather high adsorption on particles, Rhodamine-B has proved to be a suitable fluorescent tracer in most waters for reasons of lightfastness, low price, and high detectability.

Rhodamine-B

Those properties of Rhodamine-B which are of concern for the tracer experimentalist are thoroughly examined by Feuerstein and Selleck[5] and Carter[6]. In natural waters the most important parameters are:

Temperature. The fluorescence decreases with increasing temperature

$$F_t = F_{t_0} e^{-0{\cdot}027(t-t_0)} \tag{3}$$

where F_t and F_{t_0} are fluorescence at the temperature (°C) of calibration, t_0, and actual temperature, t, respectively.

Suspended solids. Organic or inorganic particles account for both the background reading and measuring errors due to physical adsorption. The latter effect is strongly dependent on the salinity as shown in Fig. 1 where the measured Rhodamine-B concentration, expressed as a percentage of the absolute value, is plotted against the content of suspended solids.

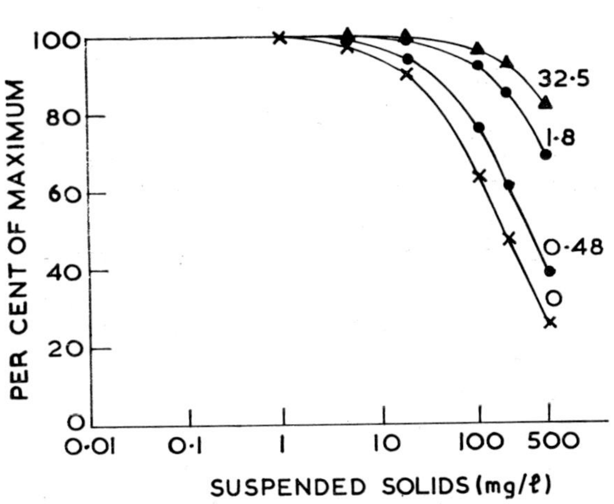

FIG. 1. Effect of suspended solids on Rhodamine-B fluorescence at four salinities (g/kg)

Data from Refs.[5,6]

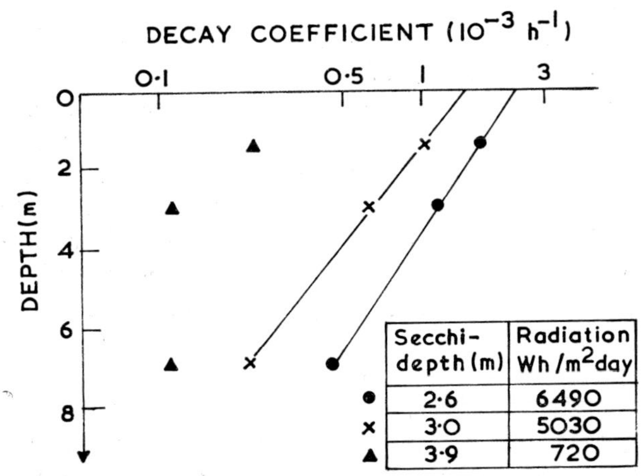

FIG. 2. Photochemical decay coefficients of Rhodamine-B

Byfjorden, Sweden, 1972

Photochemical decay. Violet or ultra-violet radiation causes decomposition of the tracer. Mechanical stirring may accelerate this process. In natural waters the light conditions and hence the decay rate vary considerably, geographically as well as with water depth.

Simple decay tests with glass bottles containing Rhodamine-B solutions at different depths were carried out by the Authors. The solution strength was measured regularly and exponential decay coefficients calculated as shown in Fig. 2. Light conditions are indicated by Secchi-disc readings and measured total radiation on a horizontal surface above the water table.

Chlorine and *pH* could be of some concern but their effects are usually neglected.

The fluorometer

The Turner Model III Fluorometer, though originally designed as a laboratory instrument, has proved to be well suited for field measurements. Increased accuracy and ease of handling can, however, be obtained by modifying the instrument[6]. The lower limit of detection is approximately 5×10^{-11} or 10^{-10} (volume concentration of a 44 per cent ICI manufactured solution) for continuous flow cell and discrete samples respectively, varying to some extent with the specific instrument used, the lamp and filter combination, and also with the actual water quality and background reading.

The main instrumental problems experienced by the Authors concern calibration changes due to suspended particles sticking to the inside of the flow cell (regular cleansing is necessary), the heating of the instrument which takes a long time especially in cold weather (check the stability with a standard), and frequency instability of the power source.

Tracer injection and field mapping

There are several possible procedures for performing tracer tests with Rhodamine-B. The practical arrangements in connection with tracer injection and subsequent mapping of the tracer field may vary from case to case depending primarily on the size and degree of exposure to, for example, wave activity of the water area. Sometimes it is adequate just to pour out a quantity of the dye and visually or photographically observe its course and spreading. A crude measure used in experimental work is then to assume that the observed width of the tracer 'cloud' is equal to 4 times the standard deviation of the horizontal concentration distribution[7]. On the other hand, measurements in the ocean normally require complicated and expensive equipment and large boats with accurate navigational aids. Relatively simple equipment will be described that has been used by Chalmers University of Technology with a high degree of efficiency.

A specially equipped 8-m motor boat is used for the surveying. Power is supplied by a 24-V generator driven by the diesel engine of the boat. The fluorometer is inside the cabin in relative darkness but is easy to reach from the steering position for range selection or direct reading. Fluorometer outputs are fed to a Minigor potentiometric recorder which can be operated at different speeds. Temperature is read from a thermometer. The pump and thermometer set-up is placed downstream of the fluorometer to avoid air bubbles in the flow cell. A paravane helps to keep the perforated suction pipe at constant depth during tracking at a speed of 3–6 knots depending on the depth of tracing. A second suction pipe is used for vertical profiling. Measurements are easily carried out by two men including the helmsman.

An 'instantaneous tracer injection' means in practice that a tracer patch is produced by mixing in a pre-defined volume of water. Initial concentrations within the cloud are set to make measurements possible without risk of contaminating the instrument and piping. A small boat with outboard motor carries the dosing equipment, and dilution water for the 100-l. injection tank is taken from the depth of tracer injection to eliminate density differences. A circular patch with approximately constant concentration is obtained by steaming on a spiral course from a central drogue.

For continuous tracer release a piston-type dosage pump has been used. When studying the waste-water dispersion from existing outfalls the tracer injection can usually be made at the treatment plant. For a future ocean outfall the tracer can be injected either directly at the predicted level of establishment of the sewage field or sometimes by a pumping procedure simulating initial mixing.

Monitoring the tracer field can be made either relative to the centre of the cloud or in absolute coordinates, and navigating is made accordingly. One or more free-floating drogues or parachute-buoys dropped in the cloud may serve as 'fix point', and surveys can be made radially from this point. The wind drift of the buoy relative to the cloud can be significant and hence a rapid mapping of the tracer field is required. Especially for vertical profiling, the lag-time in the fluorometer piping has to be compensated for. Absolute navigation is made with the aid of landmarks, anchored buoys, radar, radio-navigation systems (Decca), or other methods. Even in this case a drogue could be of great help localizing the centre of the cloud. It may be necessary to compensate for the movements of the clouds and diffusion during the time of monitoring, which can be done from data obtained by starting and finishing each zig-zag mapping of the cloud with a crossing of the centre. For tracing the dispersion field following a continuous tracer injection the navigation is less complicated. In a steady flow field the navigation can usually be done relative to anchored buoys and the mapping of the tracer field is perpendicular crossings of the plume axis.

Check list for fluorescent tracer experiments

Before starting the field tests a detailed plan for the practical arrangements should be made. This is especially important for instantaneous tracer experiments when time usually is short. The planning can be as follows:

1. Calculate an adequate amount of tracer for each test (or the injection rate in case of a continuous tracer release) from estimations of spreading. Concentrations within the experimental area should not be too high—$C_{max} < 100\ \mu$g/l.—to avoid contamination and public offence but high enough to allow accurate measurements.

2. Make a survey of relevant hydrographic conditions: salinity, temperature, currents, tidal conditions, wind, etc. and make a plan for the measurements of relevant environmental factors during the experiment.
3. Make usable maps and forms, and establish routines for accurate recording and data handling of all parameters.
4. Check the equipment (including boats) thoroughly and plan for maintenance, service, and exchange of faulty parts.
5. Measure background concentrations, estimate adsorption and photochemical decay and, if necessary, schedule further measurements during the experiment.
6. Prepare possible routes for criss-crossing the cloud or plume and the procedure for vertical tracer profiling. Provide for necessary navigation aids.
7. Make a detailed time schedule for the field tests.

Isotope investigations

Procedures for radioactive isotope investigations follow very much the same pattern as outlined for fluorescent tracers[8]. The radioactive isotope Br-82 is frequently used as a standard method by, for instance, the Danish Isotope Centre.

Current Measurements

The evaluation of the general circulation pattern must be one of the main objectives of a field study and, due to the complex nature of ocean currents, this means a rather extensive programme of measurements. Various methods can be used to determine the circulation. Tracer tests reveal significant information about not only the dispersive properties of the disposal area but also the magnitude and the direction of the currents. To get the overall picture of the circulation, however, tracer tests have to be supplemented by current measurements. Methods for these fall into two categories: measurements of net motion over a period of time by means of drogues or drift cards, and measurements of velocities and directions at a fixed point using current meters. Both these types of study yield useful information.

The objective of a drogue study is often restricted to a qualitative description of the water movements in the receiving water during different environmental conditions. In embayments and archipelago areas where both bottom topography and the configuration of the coastline may cause complex patterns of circulation this path tracking method is of special importance. The use of drogues for tracking coastal currents has the advantage of a simple and straightforward detecting procedure. On the other hand, for comprehensive surveying covering large areas it is a laborious method which therefore may turn out to be rather expensive.

For determining the drift of surface currents, drift cards have proved to be useful, especially for beach pollution studies[9]. These cards are released, usually in large quantities, in the disposal area and then recovered later on at the beach. This means that they provide no information on the path from the point of release to the point of recovery.

Fixed-point measurements by means of self-contained or transmitting current meters provide the natural basis for a statistical analysis of the current structure. These current meters can provide data for periods of a month or more without service, and the data handling of the modern current meters is directly adapted to computer calculations. Current meters are often used in combination with (short-term) tracer simulations or drogue measurements in order to relate these observations to the prevailing environmental conditions.

An extensive summary of methods and instruments for current measurements used prior to 1959 is given by Wiegel and Johnson[10]. Since then there has been a rapid development of self-contained current meters and subsequent data-handling procedures.

Drogue studies

A drogue consists of a submerged drag device, which is connected through a cord to a surface float carrying the indicator unit, see Fig. 3. A path associated with the movements of a certain volume of water at the depth of the drag device is then recorded. However, as deflecting forces are exerted on the cord and surface parts of the drogue the trajectory of the drogue is merely an approximation of the Lagrangian movement of that particular water mass. The main principle for the design of the drogue is to minimize wind drag and other deflective forces so that the movement of the drogue as closely as possible represents the path of the 'tagged' water volume at the desired depth. Nevertheless, if not properly considered, the deflection may—in cases with generally weak currents in the observational area and/or large spatial current variations—cause direct misleading results. Therefore an adequate estimate of these deflections must be made and subtracted from the recorded path for each space interval separately.

Forces exerted on a drogue

The forces exerted by the flow field on the drogue having a deviating speed emerge from drag and inertia. The effect of inertia is difficult to estimate and is usually neglected although this is obviously a crude approximation in case of, for instance, heavy wave action on the surface float.

The drag force on a submerged body in a stationary flow field can be expressed as

$$F = \tfrac{1}{2} C_D \rho A u|u| \tag{4}$$

or

$$F = K u|u| \tag{5}$$

where F is the drag force, C_D the drag coefficient, ρ the density of the fluid, A the projected area of the body, and u the velocity of the fluid relative to the body; K is a lumped coefficient including the drag coefficient and the area of projection.

The equilibrium (vector) equation for the drag forces exerted on the drogue can be written

$$K_1(u_1 - u_0)|u_1 - u_0| + K_2(u_2 - u_0)|u_2 - u_0| + K_3(u_3 - u_0)|u_3 - u_0| + K_4(u_4 - u_0)|u_4 - u_0| = 0 \tag{6}$$

where u_0 is the drogue velocity, u_1, u_2, u_3 are the current velocities at the depth of the drag device, at an intermediate depth, and at the water surface respectively, and u_4 is the wind velocity. Provided that the coefficients of drag have been determined for the different parts of the drogue, corrections of the recorded paths of the drogues can be made once these velocities u_0, u_2, u_3, u_4 are measured.

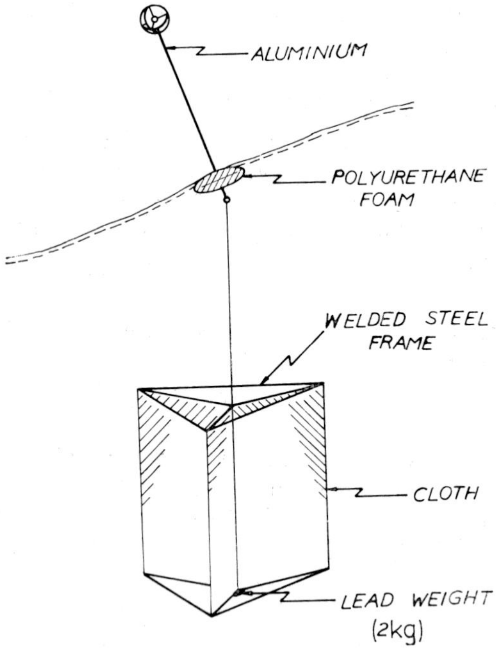

	Dimensions (m)	Lumped drag coefficient, K (kg/m)
Radar reflector	ϕ 0·15	$5{\cdot}3 \times 10^{-3}$
Stick	0·8 × 0·008	$3{\cdot}0 \times 10^{-3}$
Surface float	0·30 × 0·15	5·1
Sub-surface stick	0·15 × 0·008	0·8
Cord	ϕ 0·0028	2·2 per m
Drogue	1·0 × 1·0 (totally) 1·0 × 0·57 (wing)	550

FIG. 3. Drogue (equipped with radar target) of size convenient for more sheltered water areas

Drogue design

To reduce the deflections the submerged drag device should be large relative to the surface units and the signal device must be properly designed to permit good tracking possibilities. Finally, the drogues must be easy to handle. Obviously, the ultimate design of the drogue has to be a compromise of these three requirements. Two different types of sub-surface drag devices are frequently used: winged types with projected areas of 2–3 m² or less and, especially for tracing currents over large coastal areas, parachutes with projected areas larger than 5 m².

The Authors have conducted laboratory and field investigations on the proper design of the winged-type drogue. The drag forces on the different parts of the drogue were determined in a hydraulic flume and in a wind tunnel. *In situ* studies of drogue behaviour were carried out under various field conditions.

As a result of this study a drogue design was suggested as illustrated in Fig. 3. The lumped drag coefficients are listed in the figure.

The drag force of the drag device is measured in the direction corresponding to the smallest area of projection. As the drogue is rotating in the water due to turbulence and shear, the drag force exerted on it may vary. The 3-winged drogue type had a somewhat smaller drag variation when rotated in the flow field compared to the 4-winged drogue. The suspending cord from the surface float to the drag device should be as thin as practically possible (2·8 mm for the drogue shown in Fig. 3). This gives a coefficient of drag which for low current velocities (low values of Reynolds number) varies with the speed. For practical calculations, however, a constant value $C_D = 1{\cdot}6$ can be used.

If the surface float is shaped like an ellipsoidal disc, the drag is reduced to about 50 per cent relative to that of an equivalent spherical float. The float should be properly balanced so as to produce only a small excess buoyancy. In that case the float will cut through steep waves reducing the vertical motion of the drogue. A stick attached to the suspension cord provides for the necessary floating stability.

The largest deflective force is usually caused by the wind drag on the signal device. For the drogue in Fig. 3 the wind correction amounts to roughly 0·4 per cent of the wind speed. Hence, to reduce this effect the design of the observation target must be carefully considered. For radar and laser detection, spherical reflectors of aluminium preferably with a centre hole to reduce wind drag can be used. Visual or optical tracking requires small lamps, stiff streamers (better than flags), ping-pong balls, or other similar targets. For aerial surveying plastic sheets of 1–2 m^2 floating in the water surface are convenient.

Methods for drogue tracking

Drift studies with drogues over long periods of time usually cover large areas. A surveying boat equipped with radar is then the best method, as the mapping of the migration of the floats will be less dependent on weather conditions. If the height of the radar is 10–15 m, and the reflector target of the drogue is not less than 1 m above sea level, a good tracking efficiency can be expected. The position of the drogue is determined either by searching up the drogue—if the drogues cannot be distinguished from each other—or the position can be determined by radar at a distance. The surveying boat must have adequate navigational aids. In confined water areas, bays or archipelagos, drogue tracking is simpler. Land-based types of tracking equipment can be used: radar, laser (ranging up to 5 km), optical distance meters (range 1 km), radio direction-finding instruments, theodolites, and so on. Tracking from small boats can also be performed by means of sextant or with special compass equipment.

Evaluation of results

Correction for the deflection of the drogues is a laborious calculation which is preferably carried out on a computer. Such a computer program has been worked out at Chalmers University of Technology. The computational procedure is as follows: measured drogue positions are evaluated and converted to a given coordinate system, and the trajectories of the drogues are plotted (see Fig. 4); additional data on wind and current velocities are used for calculation of corrected current vectors which are plotted in the graph; the corrected drogue trajectory is calculated separately for each drogue.

If spatial variations of current velocities in the direction of drogue deflections are found to be relatively small the corrected drogue trajectories can be plotted as representative for true Lagrangian movements. Otherwise this graphical illustration can be quite misleading and a systematic multiple-drogue tracking technique is required in order to provide adequate data for estimates of current pattern. Usually, however, the plotted graphs of Fig. 4 are sufficient for a qualitative description of the circulation.

Current meters

In the past, almost every measurement with a current meter was carried out with the instrument suspended from the boat. During the last decades, however, current meters anchored in the sea have been widely used. Among commercially available self-contained meters are the Norwegian Aanderaa, the American Geodyne, and the English Plessey.

Capacity and performance

The Aanderaa and Geodyne current meters operate with a so-called Savonius rotor for measuring the current velocity. The rotating element is mounted on a vertical axis, hence the absolute velocity is always recorded, whereas the current meters employing a propeller (such as the Plessey meter) measure the velocity in the direction of the propeller axis. Although the rotor-type current meters have a tendency of over-registration due to turbulence, long waves, and motion induced by the mooring system, these

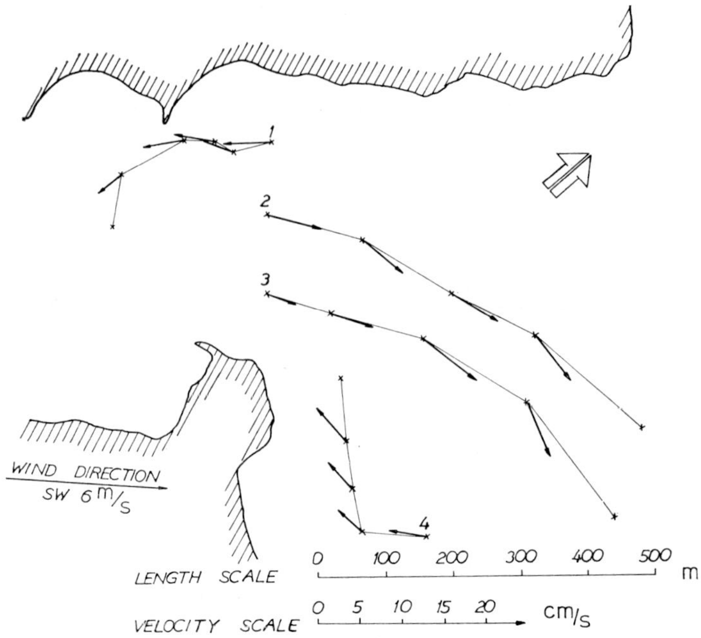

FIG. 4. Plotted drogue trajectories and evaluated local current vectors

current meters are nowadays widely used. The propeller-type meter shows less sensitivity to these disturbances. Pendulum, ultrasonic, and electromagnetic current meters are examples of instruments based on other principles for velocity measurements.

Measuring range, sampling intervals, and time of operation as well as recording system, procedures, and costs for the subsequent evaluation of the measured data vary considerably between different long-term operating current meters.

The Aanderaa current-meter records on magnetic tape easily convertible to computer use, current speed range from 3 to 200 cm/s, sampling interval 5–60 min. Time of operation is $1\frac{1}{2}$–2 months for 10 minutes' sampling. A mooring system using a submerged buoy for the suspension of the current meters is advantageous as it reduces wave disturbances substantially. One to three anchoring points are used depending on water depth and measuring levels, wave exposure, etc.

With a proper handling of the current meter as well as of the mooring system, accurate registrations can be expected during at least 75 per cent of the period of operation. Errors and failures are caused by fouling, collapse of the rotor unit caused by sea plants or animals, corrosion, leakage, and complete losses.

Placing of the instruments

The number of current meters that have to be used and the choice of positions depend on several factors—but primarily on the degree of complexity of the circulation pattern as indicated by, for instance, preliminary studies. For such studies very simple gelatin pendulums can be used[11]. By means of a rapid and easy procedure for release and recovery, a synoptic mapping of instantaneous current velocities covering a large area can be achieved with just one boat.

In open coastal waters a network of sparsely placed current meters covering the water area of interest may be adequate. For a more irregular coastline, current meters are usually placed in straits and in other strategic positions to provide maximum information on water exchange and the overall water circulation; 2–4 current meters are frequently used at different depths using the same mooring system.

Evaluation procedures

The recording and processing of data pertaining to current-meter measurements can be highly systematized. Due to the large amount of data involved, these calculations have to be carried out on a computer. For checking the recordings, graphs of for instance N–S and E–W current components for the measuring period should be plotted. A progressive vector summation is useful for correlation with the results of drogue or tracer studies. For statistical analysis, polar-system histograms of current direction and volume flux as well as histograms of current velocities are frequently used. The current recordings are analyzed for the tidal components. In areas with little tidal activity, where other long-period oscillations may be significant, it is appropriate to start with a spectral analysis followed by a harmonic analysis for found oscillations. After subtraction of the tidal components, correlations with meteorological parameters as well as wave characteristics can be made. Finally, it is possible to obtain correlations between current recordings for different stations and at different depths.

Summary

In this paper we have described experimental procedures for field studies of the spreading of discharged waste water in the coastal environment. Tracer technique and drogue measurements are discussed and methods for carrying out (supplementary) current measurements by means of automatic current meters are outlined. The importance of a thorough planning of the field investigation has been stressed. Our experience stems largely from dispersion studies in water areas exhibiting a rather complex circulation pattern. In the open sea, the main circulation and dispersive characteristics are easier to define whereas the experimental conditions, as far as navigation and wave activity are concerned, are generally much more severe.

There is a fundamental distinction between the research worker performing experiments in the sea and the engineer conducting field studies on a purely commercial basis. It is important not to underestimate the role of economics. Highly sophisticated instruments and tracing procedures are not competitive if a simpler but otherwise adequate engineering technique is available.

REFERENCES

1. Cederwall, K. Hydraulics of marine waste water disposal. Chalmers University of Technology, Göteborg, Hydraulics Division, Rep. No. 42, 1968.
2. Cederwall, K., and Hansen, J. Tracer studies on dilution and residence time distribution in receiving waters. *Wat. Res.*, 1968, **2**, 297.
3. Harremoës, P. Prediction of pollution from planned sewage outfalls. *Teknill. Kem. Aikak.*, 1964, **21**, No. 6, 251.
4. Harremoës, P. Tracer studies on jet diffusion and stratified dispersion. *Proc. 3rd int. Conf. Wat. Pollut. Res., Munich, 1966*, Vol 3, p. 65.
5. Feuerstein, D. L., and Selleck, R. E. Fluorescent tracers for dispersion measurements. *J. sanit. Engng Div. Am. Soc. civ. Engrs*, 1963, **89**(SA4), 1.
6. Carter, H. H. The measurement of rhodamine tracers in natural systems by fluorescence. *Rapp. P.-v. Réun. Cons. int. Explor. Mer*, 1975, **167**, 193.
7. Csanady, G. T. Dispersal of effluents in the Great Lakes. *Wat. Res.*, 1970, **4**, 79.
8. Hansen, J. Tracer engineering in coastal pollution control. International Atomic Energy Agency, SM-142/37, 1970.
9. Tibby, R. B. Inshore circulation patterns and the oceanic disposal of waste. *Proc. 1st int. Conf. Waste Disposal mar. Envir.* University of California, Berkeley, 1960, p. 296.
10. Wiegel, R. L., and Johnson, J. W. Ocean currents, measurements of currents, and analysis of data. *Proc. 1st int. Conf. Waste Disposal mar. Envir.* University of California, Berkeley, 1960, p. 175.
11. Haamer, J. Current measurements with gelatin pendulums. Chalmers University of Technology, Göteborg, Geology Division, Rep. A4, 1973.

In his opening remarks Dr. Cederwall said that the Authors had emphasized the need for thorough planning of environmental studies on marine waste disposal problems. One had to know the overall objectives and the basic features of the problem; something should also be known about the dynamics of the receiving waters and the economic constraints on the field studies. Reliable predictions could usually be made from relevant tracer experiments. Fluorescent dyes and radioactive isotopes were commonly used, drogues or driftcards could be used (alternatively or additionally) in determining large-scale horizontal circulation, and fixed current meters provided the data essential for statistical analysis of the current structure. He stressed that the Authors' experience of such field surveying was mainly in sheltered waters which had circulation characteristics rather different from those of open coastal waters. The principles of the methods available for reproducing dispersion phenomena were (a) a theoretical model leading to mathematical formulation and a numerical solution, (b) a physical model as an alternative to (a) or a means of testing it, and (c) *in situ* observations to give direct simulation and estimates of diffusivities; the approach could be in the order (a)–(c) or the reverse. The objectives were to predict the environmental effects of different disposal schemes or measures for pollution control, and to establish the dilution capacity of the receiving waters.

Dilution and transport of pollutants could be categorized in terms of characteristic scales. For the open coast with a continental shelf there was first the initial dilution and plume mixing, then the currents and natural diffusion in the outfall area, and finally the flushing to deep water off the shelf. The time scales for these three stages progressed from minutes to months, and the length scales from tens of metres to tens of kilometres. In sheltered waters (isolated from main coastal currents) wind and topography were important factors, and fresh-water run-off could lead to stratification.

Dr. Cederwall concluded by saying that field studies were always needed, but that they differed greatly from case to case.

DISCUSSION

Mr. P. Ackers (U.K.) said he found the paper of particular interest because his firm had recently undertaken studies in a U.K. coastal region using most of the methods described, namely dye tests with Rhodamine-B and a Turner fluorometer, drogue studies, and moored recording current-meter measurements. Judging from his experience, the Authors had given an excellent appraisal of the problems, and of the advantages and the scope of these methods. All field exercises were expensive, and British waters were notoriously affected by almost unpredictable weather, but outfall performance was also affected by the same factors, so it was important to relate environmental conditions during an experimental period to the range of conditions that might occur. Because dispersion depended on weather conditions, which affected both drift and the mixing by waves, a single tracer exercise might not catch worst, typical, or best conditions. He asked what advice the Authors could give on this aspect of planning a survey campaign.

He strongly supported the Authors in recommending the use of a current meter with a computer-compatible record. His firm had used: an analogue plot of currents, a statistical analysis giving the percentage of time that a series of current strengths was exceeded, a polar scatter plot of velocity vectors showing whether currents were rotary or reciprocal and their dominant strengths and directions, a progressive vector plot showing the variations in current direction during a tidal cycle and the residual drift day by day over the spring-neap cycle, and a tabulation of residual drifts over 25-h periods as a basis for correlation with tidal range, wind velocity, etc.

Although drogue tests were a necessary part of a comprehensive outfall survey, a large number was needed to cover the essential variables. Thus a recent exercise had required over 100 h of drogue time, six release points, release at 2-hourly intervals, tracking over 13 h, and at least two weather situations. The interpretation of this amount of data was not easy, but any less would have given a very incomplete picture. Costs of analysis and interpretation were rather daunting too. In this connection the Authors' analysis of the wind effect and the method of correcting for it were of particular interest. Computer methods of converting raw field data into current vectors in hourly sections of the tidal period were most desirable. He enquired how far the Authors' programs met this need, and if these had been published so that others might use them.

Dr. J. A. Charlton (U.K.) reported that the Tay Estuary Research Centre was currently making site investigations using methods very similar to those of the Authors. Tracking of multiple float-drogues was combined with fluorometric tracking of a dye plume. The current vectors produced by the drogues were corrected for wind drift (the float–drogue combination having been calibrated), and the plume travel amended accordingly. The normalized plume track could then be used in a synthesis of dispersion under adverse wind conditions as part of a design exercise. It was often found that a critical coastal pollution condition existed at the turn of the tide, and that this state needed careful observation. A mobile land-based radar unit was used for following the floats and the dye-tracking boat, and very close control could be maintained of the survey as it developed.

Mr. R. M. Hedgland (New Zealand) reported that in New Zealand they had recently been tracking surface-water movements by placing 250-g bags of fluorescein powder from an aeroplane on a grid pattern over a wide area, and following the resulting movement by vertical aerial photography at 10-min intervals. The data could be plotted simply and accurately by superimposing the photographic slides on scale plans and then sketching the dye patches. This method did not have the inherent errors of surface floats due to wind effects, and could be used over a much wider range of weather conditions.

Dr. G. E. B. Kullenberg (Denmark) asked the Authors what time-scales had been covered in their instantaneous release experiments and to what depth they had managed to do the tracing. He made a plea for using *in situ* fluorometers in studies relating to submerged coastal discharges, since stratification in the sea was usually very complex; it varied over small scales in the vertical, and relatively thin layers of contaminants could be expected to be trapped at density interfaces. Such features could not be detected with the Authors' technique, and he thought it important to consider the problem in connection with submerged sewage outfalls.

REPLY TO DISCUSSION

In reply to Mr. Ackers's question on survey planning, the Authors said that an adequate statistical description of pollution from a future outfall had to be based on study of a sufficient number of situations to represent those parameter combinations which were significant to the spreading pattern. This was sometimes difficult to achieve, and would thus favour a long-term test with continuous release of a single tracer or (alternatively) a dual-tracer technique as described in the paper. For efficient planning of a tracer study—irrespective of the methods adopted for tracer release—it was essential to analyse available field data describing the circulation characteristics of the water body exposed to waste discharge. In this context, long-term measurements from fixed current meters could be used for determining the most significant parameters governing the mechanisms for water movements in the area, and from a practical point of view it was advisable to start the tracing on a day when weather was likely to be favourable.

With regard to Mr. Ackers's second question, the answer was that the Authors' computer program did not convert the raw drogue field data into hourly sections of the tidal cycle; however, the program could easily be supplemented by such or similar procedures—as had been done for calculating the tide-induced water transport through a fjord entrance. In both cases it was important that drogue-tracking observations were made with a time interval much shorter than to the tidal period, as interpolation between longer intervals might give misleading results. The tracking procedure therefore needed to be rapid, and the difficulties increased with the size of the area to be covered. Radar techniques were especially suitable for such synoptic mapping. More information on the Authors' program would shortly be presented in the bulletins of their Institute and hence be available to those interested.

The Authors thought the technique described by Mr. Hedgland was very attractive. They had once used a similar aerial photographic method for studying horizontal diffusion by means of several drogues released in a grid pattern.

Replying to Dr. Kullenberg, they said their tracing procedures usually covered several days, or even a week in sheltered waters when studying transport and mixing of a sewage field established at or near the surface. For releases in confined bodies of water—such as the bottom water of a sill fjord—the time scales were much longer. For the Authors' simple equipment, a depth of 20 m was a practical limit for continuous horizontal tracking at about 1·5 m/s. For greater depths, vertical profiling was more efficient, provided the weather was good or the study was in a sheltered water, but with larger boats and paravanes (as used by the Water Pollution Research Laboratory) continuous tracking could be made at significantly greater depths. *In situ* fluorometers were very suitable for studying submerged tracer fields; however, the Authors did not think that from an engineering viewpoint it was necessary to measure concentrations over such small vertical scales that the drawbacks of a pumping procedure became significant. Sewage usually formed a relatively thick layer (5 m or more) at the trapping level due to different degrees of dilution in the centre and the outer part of the rising plume.

Interpretation of Diffusion Data

J. W. Talbot

Ministry of Agriculture, Fisheries and Food, Fisheries Laboratory, Lowestoft, England

The importance of turbulent diffusion as a process of transport in the sea is now well established. As early as 1909 Gehrke[1] recognized the part played by turbulence in vertical heat transfer and by 1925, when Prandtl[2] proposed the 'mixing length' concept, G. I. Taylor[3] and others had laid the foundations of the classical theory of turbulence. Progress during the next 25 years included von Karman's application of the similarity principle[4], the formulation of Kolmogoroff's ideas on turbulence spectra[5] and Richardson and Stommel's observations on particle diffusion[6]. In 1950 Inoue[7] summarized evidence for the numerical value of the diffusion coefficient in terms of scale, using the 4/3 power law, and the same year MacEwan[8] proposed his statistical model of turbulence which was to be given a more mathematical presentation by Joseph and Sendner[9] eight years later. Up to this time work on diffusion depended heavily on statistical formulations, and experimental data of diffusion in the sea were very sparse.

Possibly the major advances since 1950 have been the recognition by Taylor[10] and Novikov[11] of the importance of the combined action of velocity gradients and turbulence in producing dispersal—the shear-diffusion effect—and the advances that have been made in experimental techniques. In the latter the introduction by Pritchard and Carpenter[12] of the Rhodamine-B technique has been of special importance. In spite of much work in the last 15 years and the accumulation of much data it remains true that we have only a limited ability to predict diffusion rates in the sea and that our understanding of the state of turbulence, on which diffusion largely depends, is incomplete. On the other hand, important practical problems frequently arise in which it is necessary to take account of diffusion. The methods of solution adopted inevitably involve a degree of empiricism and simplification of the conditions, dependent on the financial resources and time available.

Horizontal Diffusion—Variance-Time Relationship

Although the importance of vertical mixing rates has been increasingly recognized, the rate of horizontal diffusion is usually of first importance. Many workers have commented on the complex shapes of diffusing distributions and, although by no means all situations exhibit these complexities, there are many cases where any attempt to describe the diffusion process in detail is clearly hopeless. The usual solution to this difficulty is to consider the distribution in terms of the area, $A(S)$, within particular isolines, where S is concentration, and hence define an equivalent radius $r(S) = \{A(S)/\pi\}^{1/2}$. When such an approach is adopted distributions are often described by the variance σ_r^2, where

$$\sigma_r^2 = \left\{ \int_0^\infty r^2 S(r) 2\pi r \, dr \right\} \Big/ \left\{ \int_0^\infty 2\pi r S(r) \, dr \right\}. \tag{1}$$

An even simpler alternative[13,14] is to use the maximum concentration $S_{max}(t)$, but Okubo[15] has pointed out that variance is usually a more satisfactory measure of dispersion, because S_{max} may be difficult to measure reliably and is very sensitive to the presence of boundaries. However, the variance, calculated from Equation 1, will always be less than that of a measured distribution, since survey results in practice never yield circular distributions. Also, because the numerator in Equation 1 contains the cube of the radius it is often impossible to determine the variance with accuracy. For example, the results of the ICES experiment RHENO[16] were assessed by a number of scientists independently, and for the last survey three independent assessments of the radius for the unit concentration contour were 27, 31·4, and >37 km. These values take into account differences of interpretation only, but from this cause alone, uncertainty in the variance by a factor of about 3 is possible. Okubo[17] considers that mass balances of between 50 and 150 per cent are permissible in tracer experiments, so that, taking into consideration all factors, it seems that uncertainty in the variance could approach one order of magnitude for many practical results.

In spite of the difficulties mentioned above, results of horizontal diffusion studies are often interpreted in terms of variance and from this the diffusion coefficient has been defined as $\sigma_r^2/4t$ or $\frac{1}{4}\,\partial(\sigma_r^2)/\partial t$, expressions which are equivalent only if the diffusion coefficient is not time-dependent.

Much consideration has been given to the possibility that there may be some fairly general relationship between variance and time of diffusion of the form

$$\sigma_r^2 = At^n. \tag{2}$$

Although, in view of the widely different conditions under which diffusion in the sea may take place, no precisely predictable time dependence is to be expected, an impressive amount of evidence pointing towards some general relationship has emerged and much of this was summarized by Okubo[15] in the equation

$$\sigma_r^2 = 0{\cdot}0108t^{2{\cdot}34}. \tag{3}$$

Such a relationship is of more than empirical interest and is related to the well known idea of the 4/3 power-law dependence of the diffusion coefficient on scale of the diffusing system, which would lead to $n = 3$ in Equation 2. This arose in Richardson's early work[18] on atmospheric diffusion and was later applied to results for the separation of particles in the sea[6]. Further support for the idea was provided by Kolmogoroff's suggestion[5] of inertial subranges in the spectrum of turbulence and by Ozmidov's proposition[19] that energy enters the sea mainly through relatively narrow wavelength bands, between which conditions could approximate to an inertial subrange. Such idcas have been applied in practice by Brooks[20] and Gunnerson[21] in connection with the design of the Los Angeles sewage outfall, and a number of other workers, including Ozmidov[22], have produced results consistent with a 4/3 power law for the diffusion coefficient. On the other hand, Foxworthy, Tibby, and Barsom[23], Csanady[24], and Talbot and Talbot[25], working respectively in Californian coastal waters, the Canadian Great Lakes, and the southern North Sea, have reported results which deviate appreciably from the law.

The strongest evidence for inertial subrange conditions in the sea has been provided at very short wavelengths. In a tidal stream[26] and under waves[27], spectral energy distributions have been found that

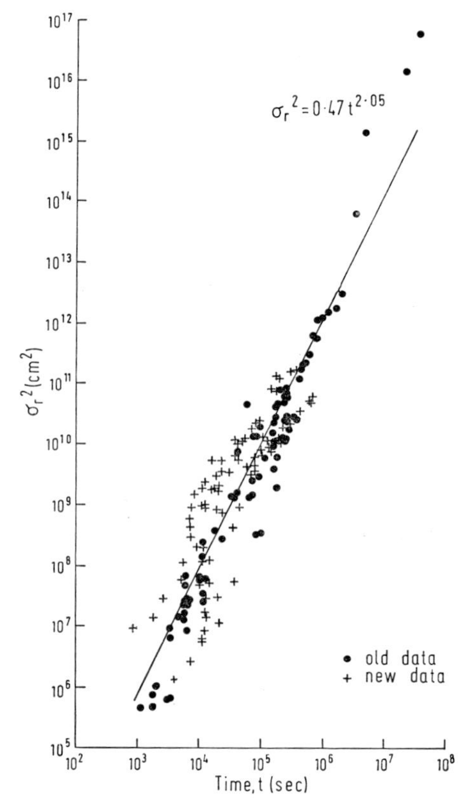

FIG. 1. Variation of the horizontal variance, σ_r^2, of Equation 1 with time

conform well with the subrange hypothesis. However, practical problems of waste dispersal in the sea are usually concerned with systems having horizontal distributions in the range 10–10^4 m and with diffusing times in the range 10^3–10^6 s. Data of variance plotted against diffusion time, summarized by Okubo[28], are shown in Fig. 1, together with similar data from other sources. Most of the additional points included on Fig. 1 are for results reported by Talbot and Talbot[25] and Kullenberg[29] and apply to English coastal waters, the southern North Sea, and the Baltic. Most of the data shown are for diffusion times in the range 10^3–10^6 s. It is perhaps not surprising that these data show a considerable spread. For a diffusion time between 10^4 and 10^5 s, values of σ_r^2 cover a range of more than two orders of magnitude. Equation 3 was based on a visual fit to the data[15] and the slope chosen depends heavily on the small number of points available for times greater than 10^6 s. In fact, the three values for the longest diffusion times are based on surveys of radioactivity in the Pacific following nuclear tests in the Bikini-Eniwetok area[30]. The next point applies to similar data reported by Folsom and Vine[31]. Thus for these four points on the σ_r^2–t graph the original sources of activity in the sea must have been of appreciable size and the practical difficulties of determining the subsequent distributions must have been considerable. Below these values the next are from the RHENO experiment, the largest experiment yet conducted using a specially released tracer. Some of the difficulties of interpretation of the RHENO results have already been mentioned. Considering the vertical distribution of dye at the last RHENO survey, van Dam[16] found that up to 50 per cent more dye than originally released was accounted for, so that the value derived for σ_r^2 is probably too large. In fact, the RHENO data plotted by Okubo[17] were based on a preliminary evaluation of the results and some of the values finally derived, which are included in Fig. 1, were appreciably lower.

Thus, examination of evidence pointing towards a general σ_r^2–t relationship reveals wide variations. From the practical point of view it seems unlikely that any refinement of the 4/3 power law can provide a useful basis for the general prediction of diffusion rates, although a restricted application in particular areas may be of value. It is important to recognize the wide differences in conditions found in various areas where information on diffusion rates is needed. Many outfalls are sited in the area of the

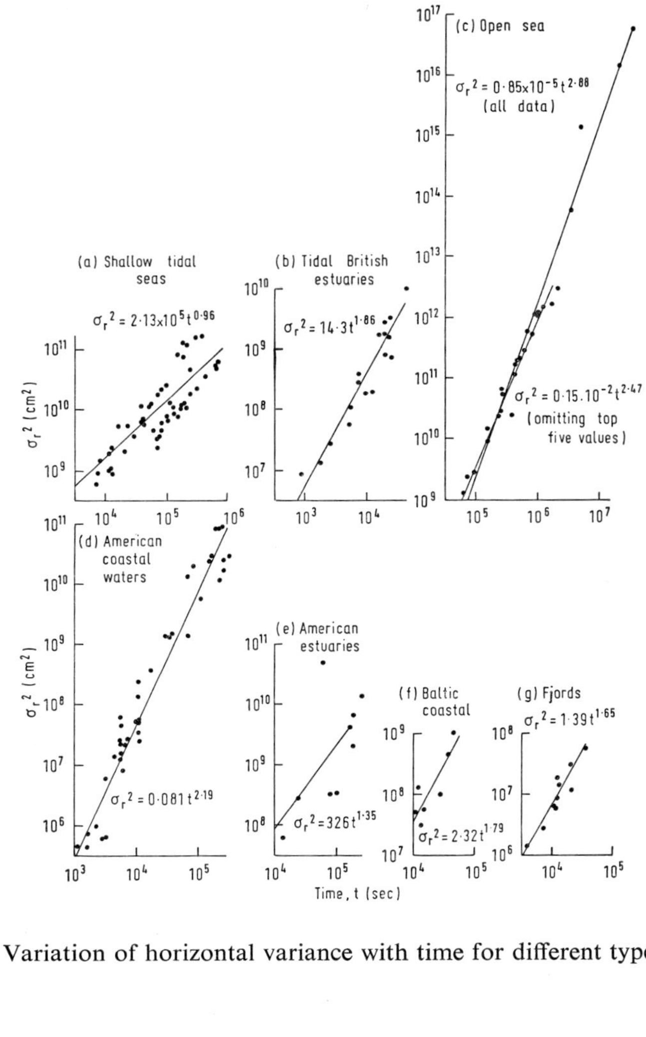

FIG. 2. Variation of horizontal variance with time for different types of area

continental shelf, generally taken to include water up to 100 or 200 m in depth. Although shallow in comparison with the open ocean, even this is large in terms of the early stages of vertical dispersal from many outfalls. The shelf has an average width of 50 km, but in some areas, for example off California, is much narrower than this. In other areas, like the North Sea, the depth at outfalls is often no more than 15 m and deep water is remote. Thus variation in depth alone is considerable. An equally important factor is the tidal velocity, which is negligible in areas like the Canadian Lakes, the Black Sea, or the Baltic, but which reaches 100 cm/s or so in many shallow shelf areas and appreciably more where the flow is converged through a narrow passage. Although Fig. 1 cannot be used as the basis of a single generally applicable relationship between variance and time, it may often be applied where diffusion data are needed in an area for which no diffusion measurements are available. In many cases it will be possible to select from the figure data for a similar area, in terms of depth, tidal velocity, and other features. To facilitate this type of comparison, the data representative of particular types of area have been plotted separately in Fig. 2. When the data are subdivided in this way some significant differences emerge; for example in the case of shallow tidal seas n in Equation 2 is 0·96, corresponding to almost constant values of the diffusion coefficient, whereas the open-sea data give n equal to 2·88, which is close to the value of 3 predicted by the Kolmogoroff inertial subrange theory[5]. However, the slope of the open-sea data depends rather heavily on the values for large diffusion times, whose accuracy is discussed above. If the five highest values are omitted the fitted slope drops to 2·47, and for practical applications it would often be more appropriate to use this value.

Irregularity of Diffusing Distributions

The fact that diffusing distributions are often complex in shape has been mentioned above, but this is not always the case, and in their analysis of shallow-sea diffusion data Talbot and Talbot[25] adopted a simple–complex classification. Such a classification reflects differences in the basic mechanism of diffusion and the relative importance of moderate-scale turbulence. Of various areas around England, the Fal and Blackwater Estuaries, the St. Austell coastal region, and the northern North Sea are complex, whereas parts of the eastern Irish and southern North Seas show simple behaviour. Other areas classified as simple are the Solent, a number of small estuaries in south-east England, and the coastal areas off Sizewell, Lowestoft, and Whitby. Even in these supposedly simple cases some irregularities are observed (for example, the outflow of the Yare at Yarmouth has been found to affect dispersal a few kilometres north of Lowestoft), but in all of these areas field measurements have yielded results showing clearly that diffusion is an essentially regular process. This behaviour was explained in terms of the major importance of vertical shear diffusion in comparison with the effects of horizontal turbulence in these areas, a matter which is discussed further below.

The simple–complex classification was based on the shapes of diffusing distributions following point releases of tracer. As would be expected, the contrast is also found in individual sections of the distributions; the smooth variation of concentration measured across a southern North Sea distribution contrasts with the irregularity shown by a northern North Sea section (Fig. 3). A similar effect is evident at smaller scale in the results of continuous-release experiments. For example, results for the

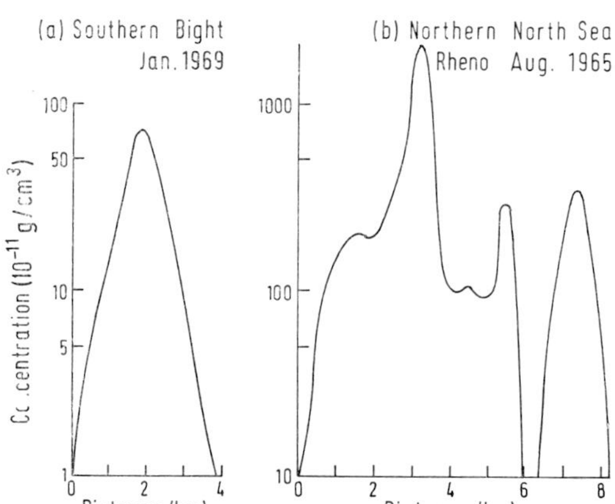

FIG. 3. Sections across Rhodamine distributions in (*a*) southern and (*b*) northern North Sea, 75 h after release

Great Lakes[25] and the Black Sea[32] exhibit a marked irregularity of individual sections and their analysis was based on the mean of a number of sections. In other results even this small-scale irregularity is absent. Off the Norfolk coast and in the Thames Estuary continuous releases of tracer have produced plumes showing smooth sections, and working off southern California, Foxworthy, Tibby, and Barsom[23] found that plumes produced from a small-source continuous release were smooth and Gaussian in form.

If the release is steady, irregularities in plume cross-sections must represent irregularities in the flow and it is to be expected that these would produce some increase in the transverse diffusion coefficient as a plume increases in width. Csanady[24] reported such an increase but only up to a plume width of about 50 m. For plumes wider than this he found the diffusion coefficient to be constant.

Transverse Diffusion

Values of the transverse diffusion coefficient may be derived for continuous-release experiments and also for point releases in cases where the distributions are of the simple type. Such information is summarized in Table 1 and presented in a more complete form elsewhere[37]. All individual values from these experiments are plotted in Fig. 4 as σ_y^2 against time from release t, where

$$\sigma_y^2 = \left\{\int_{-\infty}^{+\infty} y^2 S(y)\, dy\right\} \Big/ \left\{\int_{-\infty}^{+\infty} S(y)\, dy\right\} \tag{4}$$

and y is in the direction normal to the main flow.

The transverse diffusion coefficient K_y given in Table 1 is related to the variance by the equation

$$K_y = \sigma_y^2/2t. \tag{5}$$

The results quoted by Murthy[34] for Lake Huron show an increase in K_y with depth. Similarly a greater transverse diffusion rate has been found at 7 m depth than nearer the surface in the Thames

TABLE 1. Transverse diffusion coefficients, K_y (10^3 cm^2/s)

Place, reference, and date	Diffusion coefficient Min.	Mean	Max.	Number of values	Diffusion times (s) covered by data	Sea conditions
(a) From continuous-release experiments						
Lake Huron[33,34], 1963, 1967, 1968	0·067	0·34	1·5	38	$1{\cdot}4 \times 10^4$	
Lake Ontario[34], 1969	0·2	1·87	4·3	9	6×10^4	
Lake Erie[33], 1963	0·7	1·87	4·4	8		
Black Sea[32], 1964		0·75		1		
Holy Loch[35], 1969		1·3				
Red Wharf Bay[35], 1969		4·3				
Cumberland coast[35], 1969		20·7				
Norfolk coast (Mundesley), 1973	1·28	3·74	7·27	17	7×10^3	
Thames Estuary, 1972	0·99	4·08	15·6	43	4×10^3	Calm
Thames Estuary, 1973	0·39	1·17	4·79	19	$8{\cdot}5 \times 10^3$	Choppy
(b) From point-release experiments						
Southern Bight, Jan. 1969	2·7	6·2	7·9	11	$7{\cdot}5 \times 10^5$	Moderate to rough
Southern Bight, Oct. 1969	1·4	3·1	6·0	8	2×10^5	Calm
Yorkshire coast (Whitby), Feb. 1970	4·2	18·7	29·4	3	10^5	
Eastern Irish Sea, 1965, 1970	1·6	30·8	73·4	14	4×10^5	
Cornish coast (St. Austell), 1970	0·6	4·3	6·7	8		
Breydon Water, May 1970	0·3	0·4	0·5	3	3×10^3	
Blackwater Estuary, 1966, 1968	2·7	8·1	12·5	3	4×10^4	
River Colne, 1968		7·4		1	2×10^4	
River Crouch, 1968, 1970	1·2	1·4	1·7	2	2×10^4	
River Roach, 1968, 1969	1·4	3·8	6·3	2	2×10^4	
River Fal, 1967	1·4	2·0	2·9	3	$2{\cdot}5 \times 10^4$	
Solent, 1971	0·2	1·5	4·0	6	2×10^4	
Suffolk coast (Sizewell), 1969	6·8	14·0	22·8	3	4×10^4	
Suffolk coast (Lowestoft), 1968	2·1	4·1	6·0	8	$2{\cdot}4 \times 10^4$	
North Sea[36], 1961, 1962	1·5	7·8	18·4	9	4×10^5	

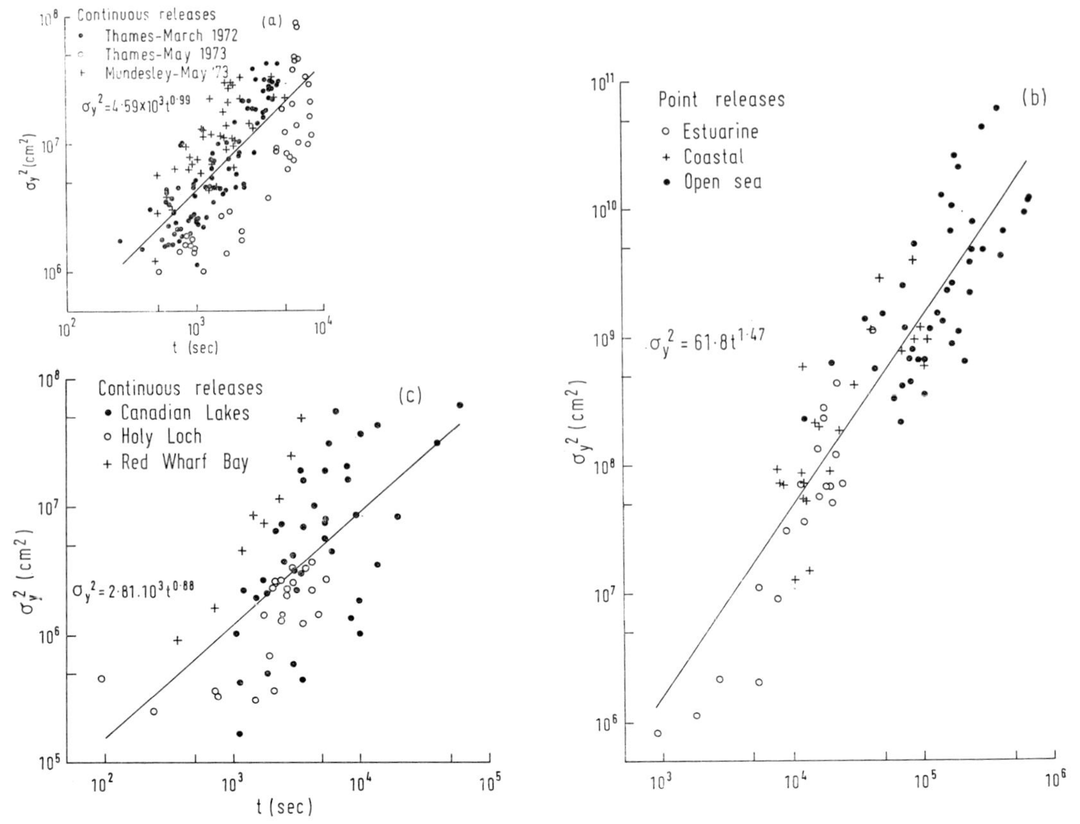

FIG. 4. Variation of the transverse variance, σ_y^2, of Equation 4 with time for (*a*) Thames Estuary and East Anglian coast, (*b*) various estuarine, coastal, and open-sea areas, and (*c*) the Great Lakes, Holy Loch, and Red Wharf Bay

Estuary[37]. An increase in diffusion rate with increasing distance from the air-sea boundary could be explained by the Prandtl mixing-length hypothesis. Alternatively in the shallow water of the Thames Estuary the increase in the vertical gradient of velocity must provide a source of turbulent energy increasing with depth. In contrast to these results Bowden and Lewis[35] found that, under certain conditions in the Holy Loch, diffusion at 2 m was greater than at 3 m depth.

The general agreement of all the results from different areas given in Table 1 and Fig. 4 is quite high, particularly in comparison with the variation often found between individual determinations at a given position. Values for the transverse coefficient cover a somewhat lower range in the case of the Great Lakes and Black Sea than is the case off California and the English east coast. In some areas, for example the southern North Sea, the results cover diffusion over several tidal cycles, and the lateral spreading observed must have resulted to an appreciable degree from vertical shear diffusion associated with transverse tidal motion. This is discussed further below, but it is noteworthy that even in these cases the overall transverse diffusion coefficients are not much greater than those observed for much shorter diffusion times and much narrower distributions. This suggests that in areas like the southern North Sea the lateral eddy-diffusion coefficient cannot be much greater than that found closer inshore and that there is very little increase in the coefficient with time of diffusion or size of the distribution.

Vertical Diffusion

Following Gehrke's observations[1], some of the earliest work on diffusion was concerned with vertical transport. In 1913 Jacobsen[38] derived some of the first values for the vertical diffusion coefficient by considering the equilibrium between horizontal advection and vertical diffusion of salt in the Baltic; this method was to be used by many other workers. Recently, improved experimental techniques have enabled workers, including Kullenberg[29] and Csanady[24], to investigate vertical mixing using tracer materials such as Rhodamine. However, in spite of the greatly improved equipment available today the measurement of vertical profiles of comparatively small patches of tracer remains a matter of considerable difficulty. Data of vertical diffusion are therefore less plentiful than for the horizontal

TABLE 2. Vertical diffusion coefficients, K_z (cm^2/s)

Place, reference, and date	Diffusion coefficient Min.	Mean	Max.	Number of values	Stability quoted[a]
Feuerschiff (Schultz's Ground)[38], 1910, 1911	0·04	0·50	2·1	8[b]	
Lake Erie[33,24], June & Aug. 1963	1	9·0	30	7	
Lake Huron[33,40], July 1963	2	6·2	15	5	
The Sound[41,29], 1965, 1967, 1968	0·04	0·46	1·2	5	$N^2 > 3 \times 10^{-3}\ s^{-2}$
	0·2	23·0	60	5	$N^2 < 4 \times 10^{-4}\ s^{-2}$
Fjord[41,29], 1965, 1968	0·04	0·09	0·12	4	$N^2 = (2{\cdot}4 - 9{\cdot}1)10^{-3}\ s^{-2}$
Coastal[42], 1966, 1967, 1968	0·05	0·43	1·5	7	$N^2 = (1{\cdot}4 - 11{\cdot}0)10^{-3}\ s^{-2}$
Baltic[42], 1969, 1970, 1971	0·02	0·25	1·5	9	$N^2 = (0{\cdot}5 - 3{\cdot}0)10^{-3}\ s^{-2}$
Mersey Estuary[43], 1959, 1960	3	21·6	71	20	
Liverpool Bay[44], 1962, 1963	0·6	12·0	39	15	$Ri < 0{\cdot}87$
California[23], 1963, 1964	0·25	5·2	22	16	
Black Sea[32], 1964		42		1	
Holy Loch[35], 1969	2	6·1	17·8	6	$< 2{\cdot}3 \times 10^{-7}\ cm^{-1}$
Red Wharf Bay[35], 1969		9		1	$3{\cdot}6 \times 10^{-7}\ cm^{-1}$
Cumberland coast[35], 1969		6		1	$2 \times 10^{-7}\ cm^{-1}$
Southern Bight, 1969	19	> 34	> 55	6	Negligible
Suffolk coast (Lowestoft), 1968	> 50	> 75	> 125	3	Negligible
Suffolk coast (Sizewell), 1969		> 117		1	Negligible
Yorkshire coast (Whitby), 1970		> 100		1	
Eastern Irish Sea, 1965, 1970	3·1	14·8	26	13	
Solent, 1971	10	97	244	5	
Thames Estuary, 1972	119	220	320	2	Negligible

[a] See relevant references for explanation of terms
[b] Measured at depths from 2·5 to 20 m

process. Defant[39] summarized data available up to 1938; more recent data, but including some results of Jacobsen[38], are summarized in Table 2 and presented in a more complete form elsewhere[37].

Shear Diffusion

The above discussion has been much concerned with the direct effect of turbulence in producing dispersal, especially in the horizontal plane, but there are many situations where this is in fact of only secondary importance. In recent years the greatest advance in our understanding of diffusion has been the recognition of the importance of velocity shear combined with turbulent diffusion across the shear[10,11,45,46]. Carter and Okubo[47] have analysed diffusion data entirely in terms of shear diffusion, Talbot and Talbot[25] have shown that in English coastal waters and in the southern North Sea the contribution to horizontal diffusion from horizontal turbulence is often of secondary importance, and Kullenberg[29] has reached a similar conclusion in the Baltic. Thus it is not surprising that data from these areas yield a slope on the σ_r^2–t diagram significantly different from that predicted by turbulence theory.

Shear diffusion is predicted by appropriate solutions of the general diffusion equation

$$\frac{\partial S}{\partial t} + u\frac{\partial S}{\partial x} + v\frac{\partial S}{\partial y} = \frac{\partial}{\partial x}\left(K_x\frac{\partial S}{\partial x}\right) + \frac{\partial}{\partial y}\left(K_y\frac{\partial S}{\partial y}\right) + \frac{\partial}{\partial z}\left(K_z\frac{\partial S}{\partial z}\right) \quad (6)$$

where u and v, the velocities along the x and y coordinate axes, are functions of z, the vertical coordinate; velocity in the z direction is zero; K_x, K_y, K_z are the eddy-diffusion coefficients in the directions of the three axes; S is concentration and t is time.

No general solution of this equation is available but solutions have been obtained for particular cases. Okubo and Pritchard[48] assumed that velocity in the y direction was zero and that the remaining velocity gradients (Ω_y and Ω_z) were constant. With the further assumption that the diffusion coefficients were constant with respect to x, y, and z the equation becomes

$$\frac{\partial S}{\partial t} + (\Omega_y y + \Omega_z z)\frac{\partial S}{\partial x} = K_x \frac{\partial^2 S}{\partial x^2} + K_y \frac{\partial^2 S}{\partial y^2} + K_z \frac{\partial^2 S}{\partial z^2}. \tag{7}$$

Okubo and Pritchard[48] gave a solution of Equation 7 for an unbounded medium, a condition which is often not fulfilled in practice since many problems occur in surface waters where the air-sea interface is a very effective boundary to vertical diffusion. The assumption of velocity gradients that are constant in space will also fail in many situations in shallow shelf seas, where Bowden[45] has shown that the vertical velocity profile may be of great significance. In these areas the function of the sea floor as a boundary to vertical diffusion is also important. The velocity gradients are often oscillatory, being produced by tidal flow, and it is useful to define a vertical mixing time, T_H:

$$T_H = H^2/K_z. \tag{8}$$

Bowden[45] considered the case where the vertical mixing time is short compared with the tidal period and Okubo[46] considered both short and long mixing times. Both authors ignored variation in the y direction. Bowden's analysis[45] came closer than Okubo's[17] to field conditions in one important respect: he considered various profiles for velocity and for the vertical mixing coefficient and concluded that for an oscillatory horizontal motion the horizontal diffusion coefficient was given by

$$K_x = -[U^2H^2\overline{f(\zeta)F(\zeta)}]/(2K_3) \tag{9}$$

where $f(\zeta)$ and $F(\zeta)$ are functions depending on the vertical profiles of velocity and K_z and their mean product is taken with respect to depth; K_3 is the maximum value of K_z in the vertical profile. Bowden[45] assumed that the vertical diffusion coefficient was equal to the vertical eddy viscosity. Talbot[49] has pointed out that measured diffusion rates are often some seven times larger than those given by the above analysis and suggested that this was due to an overestimate of the vertical mixing rate. Talbot and Talbot[25] applied an expression derived empirically by Pritchard[50] and, assuming that the friction velocity U_* is related to the depth mean velocity $\bar{u}$ by the relation

$$U_*^2 = 0{\cdot}002\bar{u}^2, \tag{10}$$

found that the horizontal diffusion coefficient was given by

$$K_x = 1{\cdot}52\bar{u}H \tag{11}$$

or

$$K_x = 4{\cdot}76\bar{u}H \tag{12}$$

for logarithmic and van Veen velocity profiles respectively. These expressions agree well with the relation

$$K_x = 3{\cdot}6\bar{u}H \tag{13}$$

derived earlier[51] on empirical grounds.

Bowden's analysis[45] did not consider the case of long vertical mixing time, but Okubo[46] found that in this case the diffusion coefficient would be directly and not inversely proportional to K_3, as in Equation 9.

Unfortunately, from the point of view of the simplified treatment of shear diffusion, the vertical mixing time is often of the order of the tidal period. Thus, in well-mixed coastal water, vertical mixing coefficients of the order 100 cm²/s may be expected, so that putting T_H in Equation 8 equal to one half of the tidal period gives a critical depth of the order of 15 m, a common value in many coastal situations.

Solutions of Equation 6 have been obtained for such cases using numerical methods[52]. Horizontal velocity components have been assumed to be of the form

$$u(z, t) = (z/h)^N\{u_0 + u_1 \sin(2\pi t/T)\}. \tag{14}$$

This is equivalent to assuming a van Veen vertical velocity profile where N is put equal to 1/5·2. Figure 5 shows a typical output obtained in this way. The method is now being used to simulate various field investigations, particular regard being paid to the value of the vertical mixing coefficient.

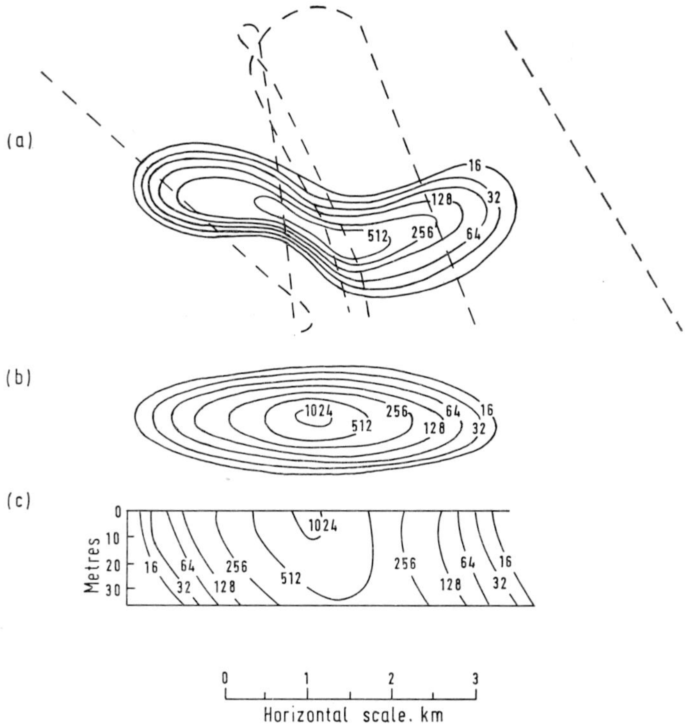

FIG. 5. Comparison of a distribution of Rhodamine-B measured $20\frac{1}{2}$ h after a point release in southern North Sea in January 1969 (*a*) with a numerical solution of shear diffusion using $K_z = 100\ \mathrm{cm}^2/\mathrm{s}$ (*b*) and (*c*)

REFERENCES

1. GEHRKE, J. Beitrag zur Hydrographie des Finnischen Meerbusens. *Finnl. hydrogr.-biol. Unters.*, 1909, **3**.
2. PRANDTL, L. Bericht über Untersuchungen zur ausgebildeten Turbulenz. *Z. angew. Math. Mech.*, 1925, **5**, 136.
3. TAYLOR, G. I. Diffusion by continuous movements. *Proc. Lond. math. Soc.*, 1921, Ser. 2, **20**, 196.
4. KARMAN, T. v. Mechanische Ähnlichkeit und Turbulenz. *Nachr. Ges. Wiss. Göttingen, Mat.-phys. Kl.*, 1930, 58.
5. KOLMOGOROFF, A. N. The local structure of turbulence in incompressible viscous fluid for very large, Reynolds numbers. *Dokl. Akad. Nauk SSSR*, 1941, **30**, 301.
6. RICHARDSON, L. F., and STOMMEL, H. Notes on eddy diffusion in the sea. *J. Met.*, 1948, **5**, 238.
7. INOUE, E. The application of turbulence theory to oceanography. *J. met. Soc. Japan*, 1950, **28**, 420.
8. MACEWAN, G. F. A statistical model of instantaneous point and disc sources with application to oceanographic observations. *Trans. Am. geophys. Un.*, 1950, **31**, 35.
9. JOSEPH, J., and SENDNER, H. Über die horizontale Diffusion im Meere. *Dt. hydrogr. Z.*, 1958, **11**, 49.
10. TAYLOR, G. I. The dispersion of matter in turbulent flow through a pipe. *Proc. R. Soc.* A, 1954, **223**, 446.
11. NOVIKOV, E. A. Concerning a turbulent diffusion in a stream with a transverse gradient of velocity. *PMM*, 1958, **22**, 576.
12. PRITCHARD, D. W., and CARPENTER, J. H. Measurements of turbulent diffusion in estuarine and inshore waters. *Bull. int. Ass. scient. Hydrol.*, 1960, **20**, 37.
13. DAM, G. C. v. Dispersie van opgeloste en zwevende stoffen in zee gebracht ter hoogte van Wijk aan Zee op 3 km uit de kust. Rijkswaterstaat, Directie Waterhuishouding en Waterbeweging, MFA 6812, 1968 (English summary).
14. CARTER, H. H., and OKUBO, A. Longitudinal dispersion in non-uniform flow. Chesapeake Bay Inst., Johns Hopkins Univ., Tech. Rep. No. 68, 1970.

15. Okubo, A. Oceanic diffusion diagrams. *Deep Sea Res.*, 1971, **18**, 789.
16. Weidemann, H. The ICES diffusion experiment, RHENO 1965. *Rapp. P.-v. Réun. Cons. int. Explor. Mer*, 1973, **163**, 1.
17. Okubo, A. A new set of oceanic diffusion diagrams. Chesapeake Bay Inst., Johns Hopkins Univ., Tech. Rep. No. 38, 1968.
18. Richardson, L. F. Atmospheric diffusion shown on a distance-neighbour graph. *Proc. R. Soc.* A, 1926, **110**, 709.
19. Ozmidov, R. V. Energy distribution between oceanic motions of different scales. *Izv. Akad. Nauk SSSR*, Fiz. Atmos. Okeana, 1965, **1**, 439.
20. Brooks, N. H. Diffusion of sewage effluent in an ocean-current. *Proc. 1st. int. Conf. Waste Disposal mar. Envir.*, Univ. Calif., Pergamon Press, New York, 1960, p. 246.
21. Gunnerson, C. G. Sewage disposal in Santa Monica Bay. *Trans. Am. Soc. civ. Engrs*, 1959, **124**, 823.
22. Ozmidov, R. V. Experimental investigation of the horizontal turbulent diffusion in the sea and in a shallow artificial water reservoir. *Izv. Akad. Nauk SSSR*, Geofiz., 1957, No. 6, p. 756.
23. Foxworthy, J. E., Tibby, R. B., and Barsom, G. M. Dispersion of a surface waste field in the sea. *J. Wat. Pollut. Control Fed.*, 1966, **38**, 1170.
24. Csanady, G. T. Dispersal of effluents in the Great Lakes. *Wat. Res.*, 1970, **4**, 79.
25. Talbot, J. W., and Talbot, G. A. Diffusion in shallow seas and in English coastal and estuarine waters. *Rapp. P.-v. Réun. Cons. int. Explor. Mer*, 1975, **167**, 93–110.
26. Grant, H. L., Stewart, R. W., and Moilliet, A. Turbulence spectra from a tidal channel. *J. Fluid Mech.*, 1962, **12**, 241.
27. Stewart, R. W., and Grant, H. L. Determination of the rate of dissipation of turbulent energy near the sea surface in the presence of waves. *J. geophys. Res.*, 1962, **67**, 3177.
28. Okubo, A. Some speculations on oceanic diffusion diagrams. *Rapp. P.-v. Réun. Cons. int. Explor. Mer*, 1975, **167**, 77–85.
29. Kullenberg, G. E. B. Results of diffusion experiments in the upper region of the sea. *Rep. Inst. phys. Oceanogr. Univ. Copenh.*, 1971, No. 12.
30. Mijaki, Y., and Saruhashi, K. Distribution of man-made radioactivity in the North Pacific through summer 1955. *J. mar. Res.*, 1958, **17**, 383.
31. Folsom, T. R., and Vine, A. C. The tagging of water masses for the study of physical processes in the oceans. *Coll. Repr. Woods Hole oceanogr. Instn*, 1957, Contr. 929.
32. Karabashev, G. S., and Ozmidov, R. V. A study of turbulent diffusion in the sea with the aid of fluorescent tracers. *Izv. Akad. Nauk SSSR*, Fiz. Atmos. Okeana, 1965, **1**, 1178.
33. Csanady, G. T. Turbulence and diffusion in the Great Lakes. *Publs gt Lakes Res. Div.*, 1964, No. 11, p. 326.
34. Murthy, C. R. Complex diffusion processes in coastal currents of a lake. *J. phys. Oceanogr.*, 1972, **2** (1), 80.
35. Bowden, K. F., and Lewis, R. E. Dispersion in flow from a continuous source at sea. *Wat. Res.*, 1973, **7**, 1705.
36. Joseph, J., Sendner, H., and Weidemann, H. Untersuchungen über die horizontale Diffusion in der Nordsee. *Dt. hydrogr. Z.*, 1964, **17**, 57.
37. Talbot, J. W. Diffusion data. (*In preparation.*)
38. Jacobsen, J. P. Beitrag zur Hydrographie der dänischen Gewässer. *Meddr Kommn Havunders.*, Hydrografi, 1913, **2** (2).
39. Defant, A. Physical oceanography. Vol. 1. Pergamon Press, Oxford, 1961, p. 104.
40. Murthy, C. R., and Csanady, G. T. Experimental studies of relative diffusion in Lake Huron. *J. phys. Oceanogr.*, 1971, **1** (1), 17.
41. Kullenberg, G. E. B. Measurements of horizontal and vertical diffusion in coastal waters. *Rep. Inst. phys. Oceanogr. Univ. Copenh.*, 1968, No. 3.
42. Kullenberg, G. E. B. Investigations on dispersion in stratified vertical shear flow. *Rapp. P.-v. Réun. Cons. int. Explor. Mer*, 1975, **167**, 86–92.
43. Bowden, K. F. The mixing processes in a tidal estuary. *Int. J. Air Wat. Pollut.*, 1963, **7**, 343.
44. Bowden, K. F., and Sharaf el Din, S. H. Circulation and mixing processes in the Liverpool Bay area of the Irish Sea. Geophys. *J. R. astr. Soc.*, 1966, **11**, 279.
45. Bowden, K. F. Horizontal mixing in the sea due to a shearing current. *J. Fluid Mech.*, 1965, **21**, 83.
46. Okubo, A. The effect of shear in an oscillatory current on horizontal diffusion from an instantaneous source. *Int. J. Oceanol. Limnol.*, 1967, **1**, 194.
47. Carter, H. H., and Okubo, A. A study of the physical processes of movement and dispersion in the Cape Kennedy area. Chesapeake Bay Inst., Johns Hopkins Univ., U.S. Atom. Energy Commn Rep. No. NYO-2973/1, 1965.
48. Okubo, A., and Pritchard, D. W. Summary of our present knowledge of the physical processes of mixing in the ocean and coastal waters, and a set of practical guidelines for the application of existing diffusion equations in the preparation of nuclear safety evaluations of the use of nuclear

power sources in the sea. Chesapeake Bay Inst., Johns Hopkins Univ., U.S. Atom. Energy Commn Rep. No. NYO-3109-40, 1969.

49. TALBOT, J. W. The influences of tides, waves and other factors on diffusion rates in marine and coastal situations. In 'Marine pollution and sea life' (ed. M. Ruivo). Fishing News (Books) Ltd, West Byfleet, England, 1972, p. 122.
50. PRITCHARD, D. W. The movement and mixing of contaminants in tidal estuaries. *Proc. 1st int. Conf. Waste Disposal mar. Envir.*, Univ. Calif., Pergamon Press, New York, 1960, p. 512.
51. TALBOT, J. W. Measurement of dispersion. Department of the Environment, Wat. Pollut. Res. Tech. Paper No. 13, H.M. Stationary Office, London, 1972, p. 62.
52. TALBOT, J. W. Investigation of the dispersal of plaice eggs and larvae in the Southern Bight of the North Sea. (*In preparation*)

Introducing the paper Mr. Talbot commented that it had been written largely as a summary of diffusion data. However, it was appropriate to emphasize the importance of interpretation in the sense of evaluation. A number of possible sources of error in tracer experiments were mentioned and it was pointed out that these should be borne in mind when applying the data obtained. After summarizing the main points in the paper, he suggested that, despite what had been implied by some earlier speakers, it could not be assumed that an area like the southern North Sea (where tides were strong) would necessarily have high dispersal rates; on the contrary it had been found that in this area, at moderate or large scale, diffusion rates (as illustrated in Table 1) were comparatively low.

[*The discussion of this paper starts on p. 350*]

Paper No. **33**

Effects of Environmental Factors, Part 1

Gunnar E. B. Kullenberg
Institute for Physical Oceanography, University of Copenhagen, Denmark

The dispersion of material discharged to the sea is influenced by such environmental factors as: wind, surface waves; radiation (incoming and outgoing), convection, evaporation and precipitation; salinity, temperature, density (or buoyancy) distributions; currents and current distributions, including tides and other rotational effects; turbulence (which in this paper always means three-dimensional turbulent motion), internal waves.

In addition to these usually highly interrelated physical factors there are chemical and biological factors and their interactions. In this paper some of the effects of the physical factors on the dispersion will be discussed. The flocculation process and other chiefly chemical effects of salinity and temperature will not be discussed. Suffice it to mention that laboratory experiments have demonstrated that flocculation considerably influences the gravitational settling velocity of the solid material contained in sewage sludge[1]. A settling velocity of the order 1 m/h seems to be representative for the flocculated state. In the surface layer of the sea the turbulent transport will effect a vertical spread at least one order of magnitude larger than this. In deeper layers and in stratified fjords and enclosed seas, the vertical transport due to mixing can often be of the same order as or less than that due to gravitational settling.

The topographic effects and the influence of the shape of the coastline on the dispersion are not discussed. The aim is to discuss various factors in the sea in relation to the very complex problem of dispersion. Despite considerable progress in the last decades on the problem, we can only parametrize and predict the dispersion in a few simplified situations.

General Considerations

Partition of the ocean environment

Most parts of the ocean, including the shelf areas, are characterized by a relatively stable density distribution. As a consequence the turbulence in the ocean is much weaker than in the atmosphere, and wave motion plays a very important role. Considering these two characteristics of the marine environment, it is pertinent to divide the water column into four different regions or layers:

(1) the surface boundary layer where atmospheric effects dominate,
(2) the pycnocline (thermocline and/or halocline) layer where the buoyancy effects dominate,
(3) the deep-water layer where internal waves are very important and turbulence occurs intermittently, and
(4) the bottom layer where bottom friction dominates.

Furthermore it is often necessary to give special consideration to regions such as upwelling areas, estuaries, fjords, and enclosed seas, where local features can determine the conditions.

Scale effect

The horizontal motion in the sea covers scales ranging from those of oceanic circulation to molecular ones. Long-term series of current measurements show that the tidal and inertial motions have high energy levels. The peak corresponding to the inertial period is often conspicuous in the power spectrum. In general it has not been possible to resolve longer-period oscillations. The possible existence of such oscillations is of great importance for the dispersion efficiency of an area.

As for the short-period fluctuations the energy density decreases rapidly with increasing frequency; Webster[2] found an energy decrease conforming with the $-5/3$ law for periods less than 6 h.

The functional dependence of the effective horizontal diffusion coefficient K_h on the scale l_h has been found on observational grounds to be

$$K_h = cl_h^n. \qquad (1)$$

Recently reported values of n are in the range 1·1–1·3 whereas c varies more and depends on the depth as well as on other conditions.

The similarity laws of turbulence predict

$$K_h = \text{constant } \epsilon^{1/3} l_h^{4/3} \tag{2}$$

where ϵ is the rate of energy dissipation per unit mass. It cannot be concluded that the assumptions behind the similarity laws are fulfilled in the case of ocean turbulence simply because the empirical value of n is close to 4/3. On the contrary, these assumptions cannot be expected to hold true in the stratified sea over the large range of scales under consideration.

In many coastal and shallow-water areas tidal oscillations together with local winds are the important energy sources. The range of scales covered by these motions is limited, furthermore the coast will prevent the formation of large-scale fluctuating motions. This implies that the contaminated area of a dispersing cloud will reach a critical size after a relatively short time, beyond which no significant increase in the effective diffusion should be expected. This situation is illustrated by large-scale dye experiments described by Talbot[3] in the North Sea. Kullenberg[4] compiled several sets of data from shelf seas and found that for scales larger than a few kilometres the value of n decreased noticeably. It should be noted that this applies to coastal waters and shelf seas.

Inertial currents, i.e. horizontal oscillations with a period of $\pi/\omega_0 \sin \varphi$ (ω_0 is the angular velocity of the rotation of the earth and φ the latitude), are very prominent features in many stratified waters some distance from the coast. They are generated by meteorological forces[5], most effectively by winds with periods less than the inertial period. The inertial motions are built up in the surface layer and penetrate gradually to deeper layers. They do not change phase with depth.

In stably stratified waters, which is the most common condition, the vertical motion is very small and the vertical convection is usually neglected, and only vertical diffusion is considered.

Very often there is vertical shear in the contaminated layer, and then the vertical shear effect will influence the dispersion. The shear effect is the combined action of vertical diffusion and vertical shear, generating an apparent horizontal diffusion. This mechanism is very important in many areas, both coastal waters and open seas. Oscillating horizontal motions with shear will affect considerably the spreading, even when the scales of the motions are larger than the dispersing cloud[4].

The separation between advection and diffusion must be considered from case to case. It is quite clear that the oscillations of the motion are of central importance for the dispersion, and the average, or residual, current is not necessarily a good measure of the dispersion.

The layers defined above will now be discussed separately.

The Surface Boundary Layer

This layer is influenced by wind and wind-generated waves, radiation, convection, precipitation, and evaporation. Its thickness is determined in most areas by a balance between the mechanical mixing induced by the wind action and the sources of negative buoyancy operating at the surface. The rotation of the earth is in general important, implying that the wind current changes both magnitude and direction with depth.

Wind and waves

The wind transfers momentum to the sea by turbulent air motion. Phillips[6] demonstrated that the momentum flux to long waves is a small fraction of the total transfer. The molecular viscosity and drag induced by short waves seem to dominate the transfer process. The wind stress is usually given in the form

$$\tau_0 = \rho_a c_d W^2 \tag{3}$$

where ρ_a is the air density, c_d the drag coefficient, and W the wind speed at 10 m. Recent observations suggest that $c_d \simeq (1{\cdot}3 \pm 0{\cdot}3) \times 10^{-3}$ for 5 m/s $\leqslant W \leqslant$ 20 m/s, and that the height of observation of the wind speed is not critical. Since the momentum transfer to short waves dominates, the drag is practically independent of the mean-square height of the waves and the fetch. The energy flux to long waves, on the other hand, is considerable. It can be of the same magnitude as or larger than the energy flux to the wind-induced current[6]. It should be pointed out that the processes active in the transfer are not yet well understood.

The transfer of momentum from the air results in the generation of turbulence and drift currents. In general the turbulence will penetrate down to the first density interface, the depth of which depends upon the strength and persistence of the wind, as well as upon the buoyancy flux at the surface. The penetration to the topmost interface (pycnocline) is usually rapid, a matter of hours for winds above 4–5 m/s. The deepening of the mixed layer, however, is a much slower process.

A considerable amount of turbulence is generated by the breaking of waves, and an intricate problem is the relative importance of the wave-induced and the directly wind-induced turbulence. This question has not yet been resolved. It seems clear that the wave-generated turbulence is concentrated near the surface[7]. Investigations by means of dye and floats in the surface layer suggest that the dispersion is stronger in the top metres than in deeper parts of the surface layer[8].

Several aspects of the surface wave dynamics are of importance for the dispersion. They are discussed by Bretschneider[9] in this Symposium.

Vertical transfer coefficients

In the surface layer the vertical turbulent eddy viscosity A can be related to the wind force as done by Ekman[10], who found that for wind speeds exceeding 5 m/s

$$A = 7{\cdot}3 \times 10^{-5} W^2. \tag{4}$$

(The numerical coefficient, measured in seconds, has been recalculated by the present author using the new values of the drag coefficient cited above.) The energy input from the wind can be expressed as $V_0 \tau_0$ where V_0 is the wind-induced current velocity at the surface. This is very nearly 2 per cent of the wind speed.

The vertical turbulent diffusivity, K_z, has also been related to the wind force, but the scatter in the observations is great. A representative range of K_z in the wind-mixed layer in shelf seas is 5–50 cm²/s. Only in homogeneous conditions can it be expected that $K_z \simeq A$. In stratified conditions $K_z < A$.

Since considerable scatter has been found in the relation between A and W it has been suggested[11] that the eddy viscosity is related more to the wave spectrum than to the wind itself. This question has yet to be resolved. The semi-empirical relations between A and W can in any case be used to estimate A.

Vertical shear effect

The development of the wind-induced current involves time scales of the order of days. The wind current causes dispersion both by advection and mixing. The shear in the wind-current system is considerable, and the vertical shear effect is an important dispersion mechanism in many areas. The shear results from wind currents, slope currents, tidal currents, and wave effects. The vertical shear is an extremely important dispersion parameter. The form of the current spectrum is also of great importance, and the characteristic periods of the currents in an area should always be investigated.

There are two different regimes of the shear effect:

(1) The case when the contaminant is occupying an internal unbounded layer; then the effective horizontal diffusion is proportional to the vertical diffusion coefficient K_z.

(2) The case of a layer bounded by two impermeable horizontal boundaries; then the effective diffusion is inversely proportional to the vertical diffusion coefficient.

Bounded case

Bowden[12] and Okubo[13], among others, treated the bounded case, considering a current field consisting of a uni-directional steady current with a superimposed oscillating component in the same direction. The amplitudes were functions of the depth z. They predicted that in a stable density gradient when K_z is decreased, the horizontal dispersion is increased since $K_x \propto 1/K_z$. Bowden considered the shear effect to be important for scales up to about 10 km. For larger scales the purely horizontal turbulent mixing would be the dominant factor.

Unbounded case

The unbounded case has been studied for instance by Carter and Okubo[14], Okubo[13], and Kullenberg[15,4]. Carter and Okubo included steady vertical and horizontal shears in a uni-directional current field U. They incorporated the small-scale motion by means of effective horizontal diffusion coefficients as well as vertical diffusion.

Kullenberg[15,4] considered a model with a steady depth-dependent current and a superimposed oscillation with a longitudinal and lateral component. The amplitude was variable with depth but the phase and period were constant. The mixing due to the small-scale horizontal turbulence was neglected (an approximation based on observational evidence), and K_z as well as the vertical shears were treated as constants in the layer under consideration. The following forms for the horizontal variances (longitudinal σ_x^2, lateral σ_y^2) were obtained for a finite size, instantaneous source with the thickness H_0:

$$\sigma_x^2 = \tfrac{2}{3}K_z\left(\frac{dU_0}{dz}\right)^2 t^3 + \frac{K_z t}{\omega^2}\left(\frac{du}{dz}\right)^2 + \tfrac{1}{12}H_0^2\left(\frac{dU_0}{dz}\right)^2 t^2, \tag{5}$$

$$\sigma_y^2 = \frac{9K_z t}{8\omega^2}\left(\frac{dv}{dz}\right)^2 + \frac{K_z t}{\omega^2}\frac{(dv/dz)^2}{(du/dz)^2}\left(2\frac{dU_0}{dz} - \frac{du}{dz}\right)\frac{dU_0}{dz}, \tag{6}$$

where U_0 is the steady current, u and v are the amplitudes of the oscillation in the x and y directions respectively, and ω is the frequency of the oscillation. Trigonometric terms have been omitted here; they are in general negligible. These forms were found to give a reasonable prediction of the horizontal diffusion. Data from experiments by the Author[15] and from the large-scale experiment RHENO[16] in the North Sea were used covering scales up to about a week. The results demonstrate the dependence of the dispersion on the vertical current shear and the oscillating current components. When two oscillating components are included in the model it is found that the oscillation with the largest period is most important, provided the gradients of the amplitudes are the same. (The validity of the term following the + sign in Equation 6 is debatable, but in the Author's experiments this term was not very significant.)

Usually the horizontal concentration distribution in a shear flow is asymmetrical and $\sigma_x^2 > \sigma_y^2$. This implies that the effective horizontal diffusion coefficient is not symmetrical. In general, rotationally symmetrical models should be avoided for the interpretation and prediction of diffusion in coastal waters.

The importance of the vertical shear effect has also been demonstrated by Talbot[3] for waters around Great Britain. It is more difficult to ascertain experimentally the influence of the horizontal shear effect, largely because it is extremely difficult to measure the horizontal current gradients.

Buoyancy generation

A number of processes operating at the surface will generate a net flux of buoyancy across the surface. This buoyancy flux is in most cases negative, and will tend to create a stable stratification which absorbs the mechanical turbulent energy generated by the wind action. The thickness of the wind-mixed layer as a function of the wind force was deduced by Kitaigorodsky[17] for open sea regions. He found that the thickness would increase from 10 m to about 30 m for a wind increase from 5 m/s to 10 m/s.

The thermocline (pycnocline) depth is determined by a balance between the locally wind-induced mixing and the heat input. The wind stirring does not penetrate beyond the topmost density interface. Since both the heat input and the wind force can vary over the warm season, several density steps can be generated and preserved in the water column until the cooling season. It should be pointed out that the density interface is not destroyed by the erosion, but is gradually sharpened and forced downwards. After a storm the mixed layer will be deep and the density interface sharp.

Vertical mixing in stratified conditions

Often a shallow mixed layer, 5–10 m deep, is followed by a gradual increase of density towards the pycnocline. In stably stratified waters the vertical component of the turbulent motion is suppressed. This means that the turbulence is weakened since the production of turbulence is coupled to the existence of a vertical component[18]. In such conditions the vertical mixing is strongly decreased.

The Richardson number and the ratio K_z/A

The momentum transport is not suppressed as much, implying that $K_z < A$. The ratio K_z/A is then of great interest and has been subject to many theoretical studies and experimental investigations in the laboratory. It is generally found to be a function of the gradient Richardson number Ri,

$$Ri = -\frac{g}{\rho}\frac{d\rho}{dz}\Big/\left(\frac{dq}{dz}\right)^2 = N^2\Big/\left(\frac{dq}{dz}\right)^2, \quad z \text{ positive upwards}, \tag{7}$$

where ρ is the density and q the horizontal current vector. The symbol N represents the stability frequency or the Brunt-Väisälä frequency. Munk and Anderson[19] found the relation

$$K_z/A = (1 + 10Ri)^{1/2}/(1 + 3{\cdot}33Ri)^{3/2} \tag{8}$$

which has been much used. It was found to compare reasonably well with recent observations in the Mersey Estuary[20].

The flux Richardson number Rf is defined as the ratio between the energy absorbed by the buoyancy forces and the total available turbulent energy; thus

$$Rf = Ri(K_z/A). \tag{9}$$

In the stable conditions encountered in many areas of the sea the ratio K_z/A is well approximated by a relation of the form

$$K_z/A = Rf_c/Ri, \tag{10}$$

where $0{\cdot}15 \geqslant Rf_c \geqslant 0{\cdot}05$. The critical value Rf_c of the flux Richardson number is the value reached by Rf when the turbulence is fully suppressed.

Ellison[21] investigated the conditions in the atmospheric boundary layer and deduced the relation

$$\frac{K_z}{A} = \frac{(K_z/A)_0(1 - Rf/Rf_c)}{(1 - Rf)^2}, \tag{11}$$

where $(K_z/A)_0 \simeq 1{\cdot}1$ is the ratio in neutral conditions. This relation follows the trend found in laboratory observations and in the few existing field observations quite well.

The gradient Richardson number Ri is a very important parameter. Usually a meaningful overall type of Ri-number $(\overline{Ri})$ can be determined from observations with a vertical resolution of 1 to 5 m; commonly $\overline{Ri}$ is in the range 0·1–10.

Vertical mixing related to environmental conditions

Several investigators have tried to relate K_z to the environmental conditions, but considerable work remains to be done on this intriguing problem. Kullenberg[22], using sub-surface tracer experiments in the depth range 5–30 m in coastal waters and open-sea regions, proposed a relation (for $W \geqslant 5$ m/s) of the form

$$K_z = a\frac{\overline{W^2}}{\overline{N^2}}\left|\overline{\frac{dq}{dz}}\right|, \tag{12}$$

where $a = \rho_a c_a Rf/\bar{\rho}$ and is in the range $(6\text{–}25) \times 10^{-8}$.

For shear-generated turbulence the following relation holds[13]:

$$K_z\overline{N^2} = Rf\frac{\tau}{\bar{\rho}}\overline{\frac{dq}{dz}}. \tag{13}$$

The stress τ is given by $\tau = -\bar{\rho}(\overline{u'w'} + \overline{v'w'})$, and we may assume that $\overline{u'w'} \propto \overline{u'^2}$ and $\overline{v'w'} \propto \overline{v'^2}$, which assumption has been used by many authors[e.g.2]. It implies that

$$K_z\overline{N^2} \propto \overline{q'^2}\,\overline{\frac{dq}{dz}} \quad \text{and} \quad \overline{q'^2} = \overline{u'^2} + \overline{v'^2}. \tag{14}$$

In order to verify this relation the fluctuating kinetic energy must be determined. This is difficult, and requires long-term series current measurements. Equation 13 is equivalent to

$$K_z = Rf\epsilon N^{-2} \tag{15}$$

using the expression

$$\epsilon = \frac{\tau}{\rho}\left(\frac{dq}{dz}\right). \tag{16}$$

Summary

Our knowledge about the dispersion and its dependence upon the environmental factors is most advanced by far for the surface layer. It can be summarized that wind, current shear, and stratification are the most important parameters, that the motion is predominantly turbulent, and that the dispersion is a function of the horizontal scale. The thickness of the wind-mixed or wind-influenced layer depends upon the latitude, and varies considerably over the year. In many areas it is shallow, much less than 50 m, being shallowest during the warm season.

The Pycnocline Layer

Permanent pycnoclines occur in many coastal areas, and in temperate latitudes seasonal thermoclines are developed during the warm season. The formation of these is linked to the influence of the local wind, as described above. The wind-induced erosion of pycnoclines is essentially a one-way transport whereby water from the pycnocline layer is entrained in the surface layer.

Buoyancy effects

In the pycnocline layer the buoyancy forces dominate the motion, suppressing the vertical exchange through the layer very markedly. Since only a small fraction of the total turbulent energy seems in general to be contained in the vertical fluctuations, the value of Rf_c is small. A representative value seems[21,18,4] to be 0·1, even though higher values have been suggested in some cases. For instance the Munk and Anderson relation[19] gives the value 0·5. The value of the gradient Richardson number Ri is mostly used as a criterion for the possible occurrence of turbulence. The value $\frac{1}{4}$ has been determined theoretically as the critical value of Ri. When $Ri < \frac{1}{4}$ in the whole water column turbulence can develop. Experimental investigations[e.g.23] suggest that the critical value of Ri falls in the range 0·1–0·3. In the pycnocline layer the overall Ri is usually greater than one; a common range is 1–20, implying a marked decrease of the vertical exchange. Vertical mixing coefficients in such layers in fjords, coastal waters, and open-shelf seas tend to fall within the range 0·5–0·01 cm^2/s.[24,25,4]

Despite this low vertical exchange the vertical shear effect is also important in pycnocline layers as a dispersion mechanism. This is the case because the current shear is often considerably enhanced in these layers. This compensates for the decrease of K_z since the shear is squared in expressions for the horizontal variances.

Small-scale structure

The pycnocline surface is irregular due to the processes active in the layer. The internal wave motion is important and turbulent motion tends to occur intermittently[26,27]. Besides the wind-induced erosion, important turbulence-generating mechanisms are shear instability and breaking internal waves. These processes combined with advection and the very weak overall vertical mixing tend to give the pycnocline layer a step-wise structure. Relatively thin sheets, of the order of decimetres in size with marked gradients, are separated by thicker, almost homogeneous, regions of the order of metres in size. Refined observational techniques are able to resolve these features of the salinity and temperature distribution, but the corresponding small-scale current structure has not yet been satisfactorily resolved. This is an important task, and until it has been solved no clear understanding of the mixing processes can be obtained. It is quite clear that the type of structure described influences the dispersion very much, since the distribution of a passive contaminant will tend to follow the natural layered distributions of salinity and temperature[4].

The Deep Water Layer

Deep water is here regarded simply as the water beneath the pycnocline layer. In many shelf areas the mixing induced by tidal currents and winds (e.g. most of the Irish Sea, the southern North Sea) is so strong that the whole water column is nearly homogeneous all the year round. In other areas (e.g. the Baltic, the central North Sea) a deep-water zone is present either for the whole year, or during the warm season.

Vertical structure

Except in certain special regions the deep water is stably stratified. The overall Richardson number is quite large. In many oceanic regions the water beneath the main thermocline is characterized by a structure similar to the one found in the pycnocline layer. The Richardson numbers defined by the small-scale structure can be considerably less than the overall number, and thus small-scale instabilities may occur.

Dispersion in deep water

Not much is known about dispersion in deep water. The influence of the stable buoyancy field is decisive and the vertical mixing is weak. Relations like Equation 15 may be used to estimate K_z, although the applicability of this relation has not been definitely established. Internal waves are a dominant feature of the motion, and possible mixing mechanisms are primarily connected with wave instabilities of various kinds[27–29]. A representative range of vertical mixing coefficients cannot be given; values in the range 0·1–0·01 cm^2/s are indicated.

Moored recording instruments are now providing us with long-term records of currents also in deep waters. Besides the tidal periods the inertial period is important[e.g.2]. Inertial motions have generation and decay times of the order of a few days. It is important from the point of view of dispersion that significant long-period oscillations are usually not resolved; more observations are needed. Current measurements of this kind can be used to estimate the mixing by means of the Taylor hypothesis[30] (see also Callaway[31]).

Mean current velocities, as well as the tidal velocities, in general decrease with depth in the oceanic regions. The energy density of the time-dependent motion and the rate of energy dissipation seem to decrease nearly proportionally to the depth[e.g.2,24]. Dispersion can be expected to have a similar depth dependence, but the experimental investigations are as yet too few to allow any definite conclusions as to the deep-water dispersion.

The Bottom Boundary Layer

Thickness

The thickness of the bottom boundary layer is much less than the surface boundary layer although the same physics are involved. In some oceanic regions the thickness is about 1 m, and in most areas the bottom friction velocities seem to be less than the critical friction velocity for resuspension of non-cohesive sediments.

In shallow waters where tidal currents are strong and/or the wind-induced motion reaches the bottom, the boundary layer is of great importance. Observations[e.g.32] suggest a drag coefficient of about $(2\text{–}4) \times 10^{-3}$. The bottom friction velocity is $u_* = (\tau_b/\rho)^{1/2}$ and in the layer where the stress τ_b is constant, which can be of the order of metres thick, the eddy viscosity $A_b = \kappa u_* z$ where z is the height above the bottom and κ the von Kármán constant. In the same layer the rate of energy dissipation per unit mass can be estimated as $\epsilon = u_*^3/\kappa z$.

Transport along the bottom

In many shallow-water areas high velocities at the bottom lead to resuspension of deposited material. Stratification of the water column, tidal currents, and wind action are decisive physical factors. In addition, the bottom conditions and the type of sediments are important. Thus, resuspension will occur in some areas all the year round during the strong tidal currents, but in other areas where the tides are weaker, resuspension will only occur on a seasonal basis when stratification is weak and the wind-induced currents can penetrate to the bottom. It should be noted that available observations suggest that the turbulence generated by bottom friction only influences near-bottom layers.

When bottom erosion occurs the concentration profile of suspended matter in the constant-stress layer is expected to be of the form

$$C - C_1 = \frac{FPr}{\kappa u_*} \ln \frac{z}{z_1} \qquad (17)$$

where C_1 is the concentration at the level z_1, F is the flux of particles, and $Pr = K_z/A_b$ is the turbulent Prandtl number. When the net flux is zero the following profile is expected:

$$\ln \frac{C}{C_1} = -\frac{wPr}{\kappa u_*} \ln \frac{z}{z_1} \qquad (18)$$

where w is the settling velocity. Thus, by observing the near-bottom profiles of suspended matter, information about the conditions can be obtained.

Finally, the feature of compensating bottom currents in shallow waters should be mentioned. Their ability to transport material deposited at a distance from the coast back to the coast has been amply demonstrated in estuaries, river mouths, and along relatively shallow water coastlines where wind-induced upwelling is a regular feature.

The Lateral Boundary Layer

In addition to these vertically defined regions of the sea the horizontally defined lateral boundary layer between the land and the sea is of primary interest. In this layer, with an overall width of the order of 10 km, a number of transport and mixing processes occur which involve interactions among environmental factors.

Large-scale events, where the rotation of the earth plays a role, are generated by atmospheric forcing. Strong jet-like coastal currents of baroclinic and essentially transient nature are generated and are important for the renewal of the coastal water mass[e.g.25]. In stratified waters internal Kelvin waves, i.e. waves with no transverse velocity, of tidal origin, and often trapped along the coastline, are notable features[e.g.33]. Coastal upwelling, generated by a persistent wind forcing the surface water from the coast gives rise to a compensating inflow at sub-surface levels.

Special transport phenomena are generated by the adjustment of a stratified field, with quasi-horizontal isolines in the interior of the sea, to sloping lateral walls[34]. The importance of this transport has not yet been clarified. Investigation of the lateral boundary layer in enclosed seas like the Baltic where the internal vertical mixing is extremely weak due to the permanent pycnocline would provide valuable information. The effects of the processes in the lateral boundary layer must be recognized in connection with coastal waste disposal.

Observations of Environmental Factors

With the above discussion in mind it is pertinent to consider briefly the type of observations which should be carried out before a site is chosen for sewage disposal.

The vertical distributions of salinity, temperature, and current should be observed over periods long enough to permit the determination of characteristic oscillations and variations, i.e. the spectral distributions. It is desirable to obtain some knowledge of the residual or drift currents in the area, but it is at least equally important to determine the long-period fluctuations. These are not only semi-diurnal and diurnal tides, and inertial motions, but also other meteorologically induced motions with longer time scales. Furthermore, this should be done on a seasonal basis since winter and summer conditions are often very different.

The thickness of the more or less well-mixed surface layer and its variations over the year should be observed. Clearly this is a relevant parameter which, together with information about the current, directly gives a guideline to the capacity of the region.

In stratified waters the vertical resolution of the observations should preferably be such that a multi-layered structure can be resolved. It is difficult to investigate the relative roles of internal wave motion and turbulent motion. In an internal wave field the fluctuating kinetic energy can be expected to be proportional to the Brunt-Väisälä frequency[6,2]. The fluctuating kinetic energy can be estimated by means of long-term series current measurements.

In connection with current measurements the problem of the relation between Eulerian and Lagrangian fields should be noted. Longuet-Higgins[35] showed that there is a difference between the mean (Eulerian) velocity in a point and the mean mass transport past the point in a time-dependent current field. This difference, which is due to the Stokes velocity, can be appreciable in areas where the residual velocity is low and the currents mainly have a tidal or inertial character. In order to resolve this from the observations, tripartite current-meter stations should be used.

The meteorological conditions in the area must also be observed, primarily the wind spectrum. The wave conditions are also of great importance, but more difficult to observe on a long-term basis.

REFERENCES

1. Crickmore, M. J. Initial behavior of sludge. In 'Out of sight out of mind. Report of Working Party on Sludge Disposal in Liverpool Bay', Vol 2. H.M. Stationery Office, London, 1972, pp. 45–54.
2. Webster, F. Turbulence spectra in the ocean. *Deep Sea Res.*, 1969, **16**, Suppl., 357–368.
3. Talbot, J. W., and Talbot, G. A. Diffusion in shallow seas and in English coastal and estuarine waters. *Rapp. P.-v. Réun. Cons. int. Explor. Mer*, 1975, **167**, 93–110.
4. Kullenberg, G. E. B. An experimental and theoretical investigation of the turbulent diffusion in the upper regions of the sea. *Rep. Inst. phys. Oceanogr.*, 1974, **25**, 250 pp.
5. Pollard, R. T., and Millard, R. C. Comparison between observed and simulated wind-generated inertial oscillations. *Deep Sea Res.*, 1970, **17**, 813.
6. Phillips, O. M. The dynamics of the upper ocean. Cambridge University Press, 1966, 261 pp.

7. Stewart, R. W., and Grant, H. L. Determination of the rate of dissipation of turbulent energy near the sea surface in the presence of waves. *J. geophys. Res.*, 1962, **67**, 3177–3180.
8. Okubo, A. Some speculations on oceanic diffusion diagrams. *Rapp. P.-v. Réun. Cons. int. Explor. Mer*, 1975, **167**, 77–85.
9. Bretschneider, C. L. Effects of environmental factors, Part 2. This volume, pp. 343–350.
10. Ekman, V. W. On the influence of the earth's rotation on ocean currents. *Ark. Mat. Astr. Fys.*, 1905, **2**, No. 11, 52 pp.
11. Wiegel, R. L. Some engineering aspects of wave spectra. In 'Ocean wave spectra'. National Academy of Sciences, 1963, p. 323.
12. Bowden, K. F. Horizontal mixing in the sea due to a shearing current. *J. Fluid Mech.*, 1965, **21**, 83–95.
13. Okubo, A. The effect of shear in an oscillatory current on horizontal diffusion from an instantaneous source. *Int. J. Oceanol. Limnol.*, 1967, **1**, 194–204.
14. Carter, H. H., and Okubo, A. A study of the physical processes of movement and dispersion in the Cape Kennedy area. Chesapeake Bay Institute, Johns Hopkins University, Final Report, 1965, 150 + 12 pp.
15. Kullenberg, G. E. B. Apparent horizontal diffusion in stratified vertical shear flow. *Tellus*, 1972, **24**, 17–28.
16. Weidemann, H. (ed.) The ICES diffusion experiment RHENO 1965. *Rapp. P.-v. Réun. Cons. perm. int. Explor. Mer*, 1973, **163**, 111 pp.
17. Kitaigorodsky, A. On the theory of turbulent mixing in the sea. The calculations of the thickness of the upper isothermal layer. In 'Problems in dynamical oceanography' (ed. A. J. Fel'zenbaum). *Acad. Sc. USSR Trans. Int. Oceanol.*, Vol. 5, 1961.
18. Stewart, R. W. The problems of diffusion in a stratified fluid. *Adv. Geophys.*, 1959, **6**, 303–311.
19. Munk, W. H., and Anderson, E. R. Notes on a theory of the thermocline. *J. mar. Res.*, 1948, **7**, 276.
20. Bowden, K. F., and Gilligan, R. M. Characteristic features of estuarine circulation as represented in the Mersey Estuary. *Limnol. Oceanogr.*, 1971, **16**, 490–502.
21. Ellison, T. H. Turbulent transport of heat and momentum from an infinite rough plane. *J. Fluid Mech.*, 1957, **2**, 456–466.
22. Kullenberg, G. E. B. Vertical diffusion in shallow waters. *Tellus*, 1971, **23**, 129–135.
23. Turner, J. S. Buoyancy effects in fluids. Cambridge University Press, 1973, 366 pp.
24. Okubo, A. Oceanic mixing. Chesapeake Bay Institute, Johns Hopkins University, Tech. Rep. 62, 1970, 117 + 20 pp.
25. Csanady, G. T. Mass exchange episodes in the coastal boundary layer, associated with current reversals. *Rapp. P.-v. Réun. Cons. int. Explor. Mer*, 1975, **167**, 41–45.
26. Grant, H. L., Moilliet, A., and Vogel, W. M. Some observations of the occurrence of turbulence in and above the thermocline. *J. Fluid Mech.*, 1968, **34**, 443–448.
27. Nasmyth, P. W. Some observations on turbulence in the upper layers of the ocean. *Rapp. P.-v. Réun. Cons. int. Explor. Mer*, 1972, **162**, 19.
28. Orlanski, I., and Bryan, K. Formation of the thermocline step structure by large-amplitude internal gravity waves. *J. geophys. Res.*, 1969, **74**, 6975–6983.
29. Woods, J. D., and Wiley, R. L. Billow turbulence and ocean microstructure. *Deep Sea Res.*, 1972, **19**, 87–121.
30. Taylor, G. I. Diffusion by continuous movements. *Proc. Lond. math. Soc.*, 1921, Ser. 2, **20**, 196.
31. Callaway, R. J. Subsurface horizontal dispersion of pollutants in open coastal waters. This volume, pp. 297–307.
32. Bowden, K. F. Measurements of turbulence near the sea bed in a tidal current. *J. geophys. Res.*, 1962, **67**, 3181–3186.
33. Mortimer, C. H. Advances in lake hydrodynamics—a selective review. Center for Great Lakes Studies, Milwaukee, Wis., 1972.
34. Phillips, O. M. On flows induced by diffusion in stably stratified fluid. *Deep Sea Res.*, 1970, **17**, 435–443.
35. Longuet-Higgins, M. S. On the transport of mass by time-varying ocean currents. *Deep Sea Res.*, 1969, **16**, 431–447.

In presenting his paper Dr. Kullenberg said it gave a summary of present knowledge on the dependence of turbulent diffusion on sea conditions, and that in his opening remarks he would consider a few aspects of the marine environment which had profound influence on dispersion characteristics. It was primarily by means of a proper dispersion, including turbulent diffusion and transport by currents, that possible basic criteria regarding the fate of pollutants could be fulfilled. The most important feature of the sea in this connection was its stable stratification, vertical processes being confined to much smaller scales than horizontal processes. Internal layers on the decimetre scale with distinguishable features could exist for days, and internal waves were also characteristic of the motion resulting from stratification.

Much of the mixing in the sea was generated at the boundaries. The depth penetration of the direct wind-generated mixing depended on the balance between the energy input and the buoyancy forces; the erosion of a thermocline was a very slow process. The wind-driven current penetrated to relatively greater depths and also had a large influence on the dispersion, and the transient motion generated by strong winds was of particular significance.

Tidal motion was another important source of energy for the mixing. In many shelf seas the tidal mixing prevented the generation of stable stratification; in other areas the tides generated internal waves capable of transferring energy and causing local internal mixing.

Due to the time-dependence and fluctuations of the driving forces, the motion and the density field in the sea were highly variable, and this variability had to be kept in mind in connection with sewage disposal in the sea. Both the time and space variability were extremely important features for the mixing characteristics. In particular, the significance of the vertical shear should be emphasized.

[*The discussion of this paper starts on p. 350*]

Paper No. **34**

Effects of Environmental Factors, Part 2

Charles L. Bretschneider

Professor of Ocean Engineering of the University of Hawaii

This paper is limited to the problem of winds and the corresponding actions on the water surface. Winds acting on the water surface create currents and waves, and the waves themselves produce mass transport. It is of fundamental importance to have a clear definition of wind speed since it is correlated with current, storm tide, and waves through the so-called drag, pressure, or friction coefficients.

Two major mechanisms for wave generation are (a) a scalar quantity called pressure and (b) a vector quantity called wind stress. The mechanisms which limit wave generation are primarily sheltering effects and breaking waves in deep water, and breaking waves and bottom friction in shallow water. Other factors of local interest are also important, such as streams of water like the Gulf Stream, river flows, and reverse winds.

Changes in wind speed are reflected in pressure changes, and being scalar will generate circular waves. For moderate to high winds the variability in wind speed is between ± 35 and ± 45 per cent of the 10-min average speed.

The vector wind is also variable for all wind speeds. There seems to be less variability in direction for strong than for light winds. For moderate to strong winds it appears that the directional variability is about $\pm 45°$ from the primary direction, but in some cases it may be larger. The vector wind then generates waves in all these directions, and perhaps also in other directions.

Both scalar pressure and vector wind increase with increasing mean wind speed, and the scalar circular waves and vector directional waves will always combine to form a directional wave spectrum.

The directional spectrum of the sea surface is relevant to the directional spreading of sewage discharged from sea outfalls, particularly since such directional aspects are not only at the sea surface but also extend to some meters below.

The present paper is primarily concerned with the 'probabilistic' wave theory, wherein are included various forms of the non-dimensional period, frequency, and symmetrical directional spectra.

The U.S. Army Corps of Engineers and the U.S. Weather Service made certain definitions of wind speed[1] in order to correlate the responses of the water of Lake Okeechobee, Florida, with hurricane winds. The winds were 10-min averages at the 10-m anemometer level.

Action of Wind on the Water Surface

The action of wind on the water surface is to create shear stresses and pressure forces. There is transport of momentum across the boundary layer from the air to the water immediately below. Figure 1 illustrates the wind profile and the resulting water-current profile.

It is important to recognize the boundary layer. The turbulent boundary layer is given by the von Karman logarithmic profile for the air. The viscous boundary layer moves forward with the speed of the water surface current. This is a quasi-steady-state problem. The steady-state problem can be solved by use of the Rayleigh principle, or the moving coordinate system which must move forward with the same velocity as the mean surface current. The surface current consists of three parts: (a) that due to the wind shear stresses in the absence of waves, and averaging 3–5 per cent of the mean wind speed at the standard anemometer level, (b) that due to mass transport of the waves, which results from wind shear stresses and the pressure forces, and (c) that due to large-scale circulation and tidal currents—factors not dealt with in this paper.

Logarithmic velocity distributions

The logarithmic distribution of velocity U_z relative to the surface water current, for either air or water, at distance z from the surface is given by

$$\frac{U_z}{u^*} = \frac{1}{k} \ln \frac{z + z_0}{z_0} \tag{1}$$

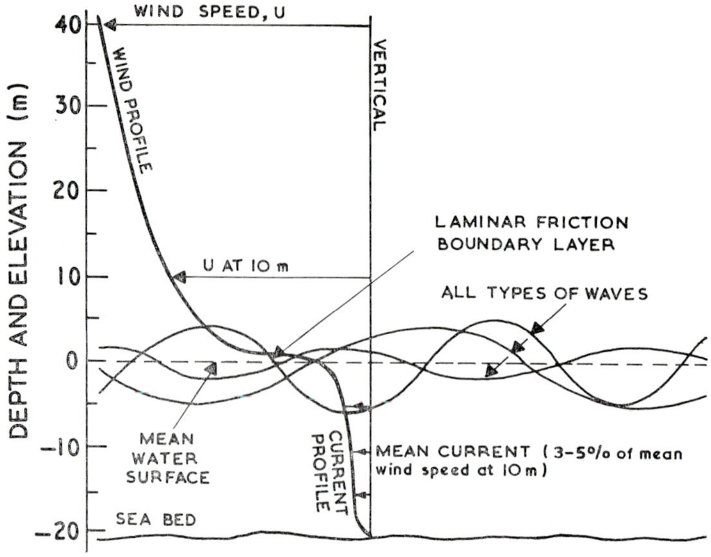

FIG. 1. Schematic wind and current profiles

where u^* is the so-called friction velocity (the square root of the ratio of shear stress to density), k is the universal von Karman constant (0·4), and z_0 the parametric friction length for which typical values are between 0·5 and 1·0 cm. This equation does not apply immediately below the sea surface during wind generation, the exponential or Ekman theory then being applicable.

Power-law velocity distribution

The power law of the velocity distribution (which has many engineering applications) is

$$\frac{U_z}{U_r} = \left(\frac{z}{z_r}\right)^{1/n} \qquad (2)$$

where U_r is the mean wind speed at the reference height z_r which is 10 m according to U.S. Weather Service practices, although it has been advocated[2] that 19·5 m above the mean sea surface is more appropriate because many large ships' anemometers are at about that height. The value of n normally varies between 3 and 7 (depending on wind speed and roughness of the surface) but may be as high as 12 for extreme gusts.

Wind Definitions and Wind Variability

For moderate to high strength winds Equation 1 gives[3] the ratio of the 19·5-m to the 10-m wind speed as 1·08, whereas Equation 4 gives 1·10 and 1·057 for values of n equal to 7 and 12 respectively. The 1/7 power law supposedly represents the variation in wind speed with elevation for the 10-min average.

Different definitions lead to different interpretation of data, and this is clearly of importance as wind speeds may be expressed in terms of: (a) the fastest mile, (b) the maximum gust, (c) the 1-min average, (d) the 10-min average, or (e) the average value determined from the pressure gradient and/or the geostrophic wind scale. The parameters determined must be correlated with a common wind speed at a common plane of reference. For example, there may be large differences between the 1-min average wind speed and that calculated from the pressure gradient—as has been demonstrated by Goodyear[3] whose data are reproduced in Fig. 2.

Variability in wind speed and direction leads to variability in the height, period, and direction of waves which comprise the directional wave spectrum.

Deterministic Wave Theories

There is quite a difference between the concepts of deterministic and probabilistic wave theories; the former will be mentioned only in passing.

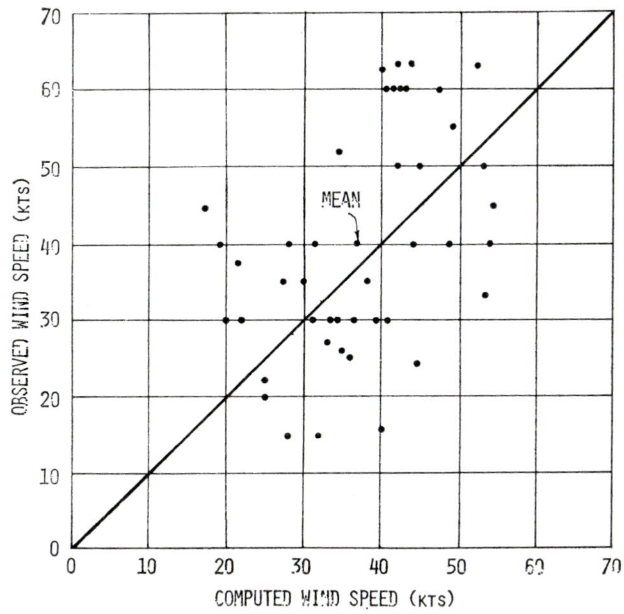

FIG. 2. Observed and computed surface wind speeds for 43 randomly selected points
After Goodyear[3]

In general, the deterministic wave theories all represent symmetrical repetitive waves, whether linear or higher-order Stokes or solitary waves, etc. The deterministic wave theory has been used[4] in the study of the influence of waves and currents on dispersion over the continental shelf.

Probabilistic Wave Theory

By probabilistic waves it is inferred that the wave heights and periods are variable in magnitude as well as in direction. The sea is said to be Gaussian. However, before presenting the problem of wave statistics and variability, it is important to refer to the Weibull distribution function[5] which has a wide application for statistical representation of random data. For example, the Rayleigh distribution for wave heights, the Bretschneider period distribution, period, and frequency spectra[6,7], and the Pierson-Moskowitz spectrum[8] are all special forms of the Weibull function. It has been demonstrated[9] that this function can also be used to describe the directional wave spectrum.

Weibull distribution function

In 1951 Weibull[5] introduced a new distribution function which he applied to random data for the yield strength of a Bofors steel, the size distribution of fly ash, and the fiber strength of Indian cotton. In each case he checked his distribution function by applying the chi-square test and found it to be in good agreement with the observed distribution. Four other sets of random data had to be treated, however, as complex distributions; after splitting each into two simple distributions there was again good agreement.

The Weibull distribution function has been used in regard to short-term and long-term ocean wave statistics by various authors, such as Bretschneider[10,11], Collins[12], Faltinsen and Pederssen[13], and Nordenstrom[14].

In terms of wave variability it is convenient to write the Weibull distribution function as

$$F(x) = 1 - e^{-Ax^m}, \tag{3}$$

and the probability density as

$$f(x) = \frac{d}{dx} F(x) = mAx^{m-1}e^{-Ax^m}. \tag{4}$$

Re-arranging the terms of Equation 3, and taking the natural logarithm twice, gives

$$\ln \ln \frac{1}{1 - F(x)} = \ln A + m \ln x \tag{5}$$

which is nothing more than the equation of a straight line. Plotting the double-logarithm term against ln x on linear scales, and fitting a straight line, then gives ln A as intercept and m as slope. Some special forms of the Weibull function are given in Table 1.

TABLE 1. Special forms of Weibull distribution function

m	$f(x)$	Type
1	Ae^{-Ax}	Exponential
2	$2Axe^{-Ax^2}$	Wave height—Rayleigh distribution
3	$3Ax^2e^{-Ax^3}$	Directional wave spreading
4	$4Ax^3e^{-Ax^4}$	Wave period[a]
−4	$4Ax^{-5}e^{-Ax^{-4}}$	Wave frequency[b]

[a] Bretschneider period distribution and wave-period spectrum[6,7]
[b] Bretschneider frequency spectrum for all seas[6,7] and Pierson-Moskowitz spectrum for fully developed seas[8]

Wave variability

A very elaborate report on wave variability was given by Bretschneider[6,7]. The important features of this report confirmed the Rayleigh distribution of wave height (H), the Rayleigh distribution of wave length (L)—or period (T) squared—and the formulation of the wave-period and wave-frequency spectra. The following expressions were obtained for the probability densities:

$$p(\eta) = 2\eta e^{-\eta^2} \tag{6}$$

$$p(\lambda) = 2\lambda e^{-\lambda^2} \tag{7}$$

$$p(\tau) = 4\tau^3 e^{-\tau^4} \tag{8}$$

where $\eta = H/H_r$, $\lambda = L/L_r$, and $\tau = T/T_s$; H_r and L_r are the root-mean-square values of H and L, and T_s is the significant wave period.

In terms of significant wave height (H_s) and period (T_s) the probability distributions are

$$p(H) = \frac{4H}{H_s^2} \exp\left\{-2\left(\frac{H}{H_s}\right)^2\right\} \tag{9}$$

and

$$p(T) = \frac{4T^3}{T_s^4} \exp\left\{-\left(\frac{T}{T_s}\right)^4\right\}. \tag{10}$$

Equations 9 and 10 are obtained[6,7] by a suitable choice of H_s and T_s. When non-dimensional wave height and period are uncorrelated, the joint probability distribution is given by

$$p(\eta, \tau) = p(\eta).p(\tau). \tag{11}$$

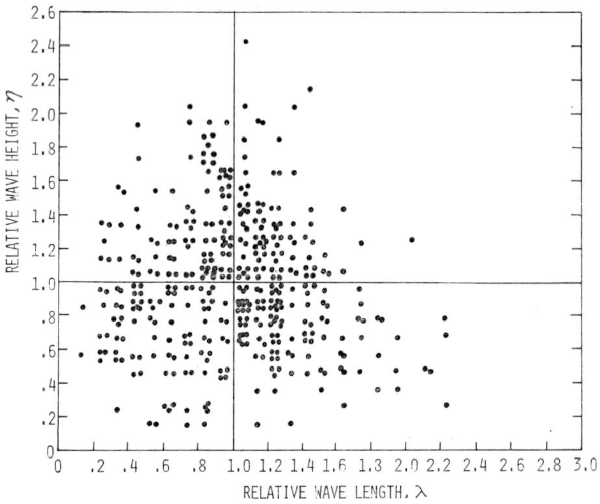

FIG. 3. Scatter diagram of η and λ for 400 consecutive waves from Gulf of Mexico

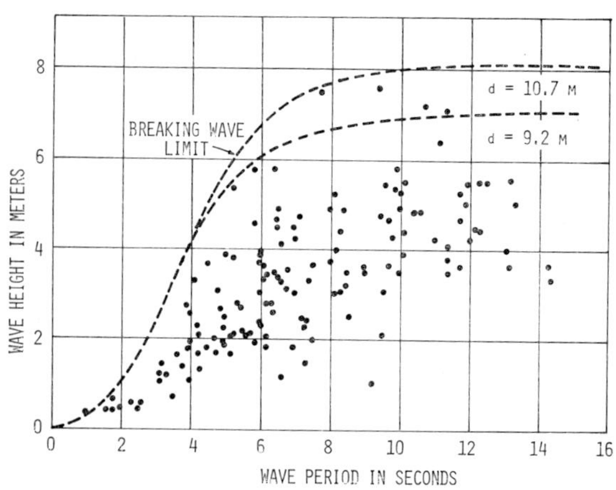

FIG. 4. Wave heights and period, Hurricane Audrey, Gulf of Mexico, 27 June 1957
d = water depth

This distribution may be extended to include wave direction. Since, however, the non-dimensional angular spreading (α) is dependent on wave period, it must be expressed as

$$p(\eta, \tau, \alpha) = p(\eta) . p(\tau) . p_\tau(\alpha). \tag{12}$$

For deep water the distribution of non-dimensional wave height η and length λ in Fig. 3 is the Rayleigh distribution based on 400 consecutive waves. However, in shallow water the conditions of limiting depth and finite steepness will alter the distributions, as seen in Fig. 4.

Wave spectrum

During the past few decades much progress has been made in defining the wave spectrum, but there has been some misunderstanding on the proper interpretation of the various forms presented. Figures 5 (a) and (b) show the generally accepted forms of the normalized period and frequency spectra respectively—though each is only one particular form. Table 2 has been especially prepared to show alternative forms of the frequency spectrum. The symbol $s(\nu)$ denotes the wave spectrum, whereas $p(\nu)$ etc. in Equations 6–12 denote probability density which, in normalized form, looks the same.

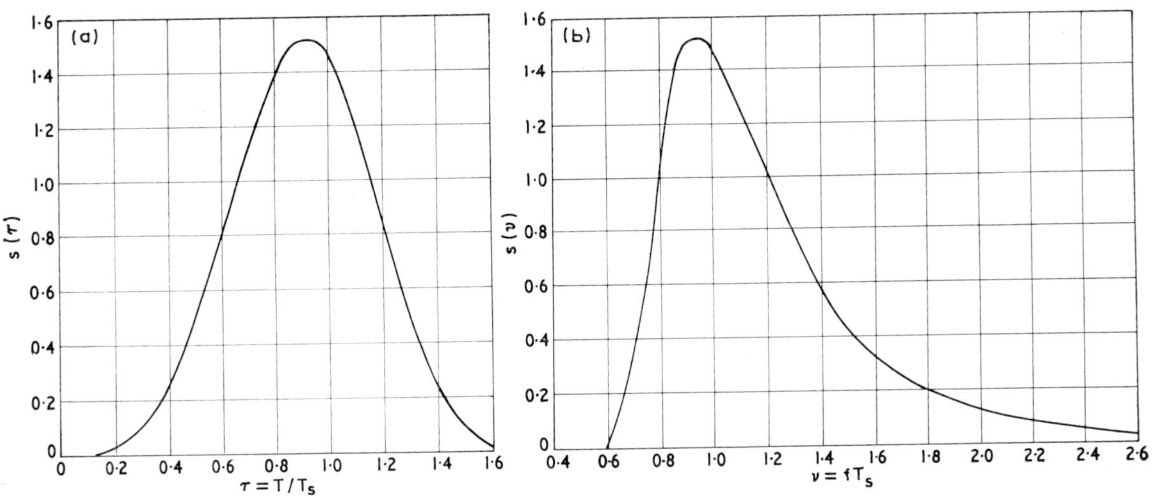

FIG. 5. Non-dimensional (*a*) period and (*b*) frequency spectra

TABLE 2. Alternative forms of frequency spectrum, for $m = -4$ and $x = \nu$ in Table 1
Dimensional frequency spectrum is $s(f) = Bs(\nu)$ where f is real frequency and $B = H_s^2\nu/f$; H_s^2 is area under spectrum

ν	A	Non-dimensional frequency spectrum equation
fT_s	1	$s(\nu) = 4\nu^{-5}e^{-\nu^{-4}}$
$f\bar{T} = 0{\cdot}907fT_s$	0·675	$s(\nu) = 2{\cdot}7\nu^{-5}e^{-0.675\nu^{-4}}$
$fT_0 = 0{\cdot}931fT_s$	$\frac{3}{4}$	$s(\nu) = 3\nu^{-5}e^{(-3/4)\nu^{-4}}$
$f\tilde{T} = 0{\cdot}751fT_s$	$\frac{1}{\pi}$	$s(\nu) = \frac{4}{\pi}\nu^{-5}e^{(-1/\pi)\nu^{-4}}$
$f/f_0 = 1{\cdot}057fT_s$	$\frac{5}{4}$	$s(\nu) = 5\nu^{-5}e^{(-5/4)\nu^{-4}}$
$f/f_{50} = 0{\cdot}912fT_s$	ln 2	$s(\nu) = (4 \ln 2)\nu^{-5}e^{-(\ln 2)\nu^{-4}}$

Note: $\dfrac{s(f)}{s(f_0)} = e^{5/4}(f/f_0)^{-5}e^{-5/4(f/f_0)^{-4}}$, where $s(f_0) = 5e^{-5/4}H_s^2/f_0$

However, not much has been done to relate the directional sea-state spectrum to the directional wind-field spectrum. Equation 12 is applicable, though it is convenient to replace the wave period τ by its reciprocal—the frequency ν. This modified form of Equation 12 must be coupled with a similar form of wind spectrum, such that

$$\{s(\eta, \nu, \alpha)\}_{waves} = \text{function}\{s(\eta, \nu, \alpha)\}_{wind} \tag{13}$$

Directional spreading

The spreading relationship is frequency dependent, higher frequencies having a larger range of spreading. The earliest work on directional spreading was published in 1960 by Cote *et al.*[16] in what is known as Project SWOP. In this study the spreading was related to a constant, plus some function of $\cos^{2b}\theta$, where b is related to the frequency and θ is the dimensional wave direction. The limits of this spreading function were $-\pi/2$ and $\pi/2$. Since then, others have made directional wave-spectrum field measurements—notably Longuet-Higgins *et al.*[17] who showed that the spreading could be completely polar to $\theta = \pm\pi$ and confirmed that it was frequency dependent. Some of their data have been analysed by Bretschneider[9]. The symmetric directional spreading function is based on the Weibull distribution function. This is only one of several ways of attacking the problem, but it does lead to an easy engineering solution.

The Weibull distribution function for wave spreading can be written as $P(\alpha) = 1 - e^{-A\alpha^m}$, where $\alpha = 1 + \theta/\pi$. By stipulating that the spreading is equally distributed about $\alpha = 1$, or that $P(1) = \frac{1}{2}$, it follows that $A = \ln 2$. Making this substitution, and differentiating the distribution function, it follows that

$$p(\alpha) = (m \ln 2)\alpha^{m-1}e^{-(\ln 2)\alpha^m}. \tag{14}$$

The problem for the symmetrical spreading function is to determine m as a function of wave frequency or period. Equation 14 is not symmetrical about $\alpha = 1$, and therefore we will define a final form of the spreading function as

$$\hat{s}(\alpha) = \tfrac{1}{2}\{p(\alpha) + p(2-\alpha)\} \tag{15}$$

where $p(2-\alpha)$ is the mirror image of $p(\alpha)$ about $\alpha = 1$. It appears from records such as shown in Table 3 that m will fall within the range from 2·25 to 3·25; values of $\hat{s}(\alpha)$ for $m = 2$ to $m = 4$ are shown

TABLE 3. Some experimental values for the m of Equation 14
Using data of Longuet-Higgins et al.[17]

Wave period, T (s)	2·24	2·62	3·93	6·28	7·85
Wave angular frequency (rad/s)	2·8	2·4	1·6	1·0	0·8
m	2·62	2·30	2·78	3·25	3·07

in Fig. 6. More records are being analysed by this method. Perhaps m depends also on the contribution of energy at each particular frequency.

The symmetrical spreading function as given by Equation 15 should be more complete than that given by $\cos^{2b}\theta$; in fact it is similar[17] to that given by $\cos^{2b}\frac{1}{2}\theta$ if m varies from 2 to 4 as $2b$ varies

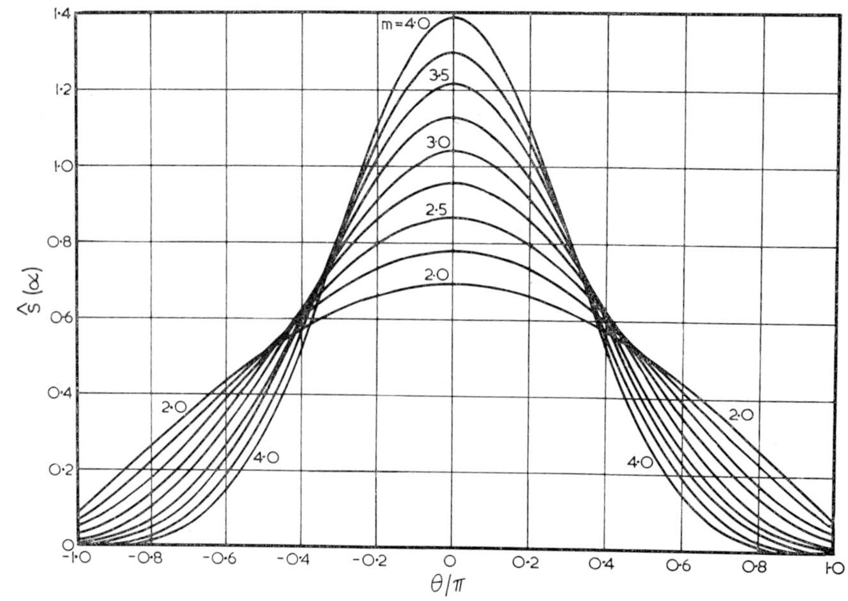

FIG. 6. Symmetrical spreading function $\hat{s}(\alpha)$, given by Equations 14 and 15, based on Weibull distributions normalized to unit area

from 1 to 4. The main difference is that Equation 15 gives waves travelling in all directions, whereas $\cos^{2b} \frac{1}{2}\theta$ does not allow waves to travel opposite to the predominant direction.

Perhaps the most sensible or reasonable first assumption to make in regard to the wave-energy spreading function is that the dimensionless m (or $2b$, as the case may be) is related to the non-dimensional period τ, or frequency ν. This would allow for the strength of a storm and the age of decay of swell to be included in the spreading function.

Asymmetrical spreading

The asymmetrical spreading function should be applicable to more-or-less straight fetches as indicated by the isobars. Asymmetrical spreading would result in skewing of the spreading function. For strong cyclonic air flow the spreading function will be skewed to the right in the northern hemisphere and to the left in the southern hemisphere. For anticyclonic air flow the reverse applies. Skewness in the spreading function can be noted in the data presented by Longuet-Higgins *et al.* More research, both theoretical and experimental, is needed on this topic. The Weibull distribution function may still apply by shifting α to $\alpha + \Delta\alpha$, where $\Delta\alpha$ is the shift from the primary wind and wave direction to the actual peak direction of the spreading function, and replacing m by m' to allow for the possibility of a change in slope. This is where more research is required. It is important because the wave-directional spreading function should be correlated with the turbulent diffusion spreading of pollutants.

REFERENCES

1. Myers, V. A. Characteristics of United States hurricanes pertinent to levee design for Lake Okeechobee, Florida. *Hydromet. Rep. U.S.*, 1944, **32**.
2. Pierson, W. J. The interpretation of wave spectrums in terms of the wind profile instead of the wind measured at a constant height. *J. geophys. Res.*, 1964, **69**, 5191.
3. Goodyear, H. V. Reconstructed wind fields for the east coast storm of March 5–8, 1962. Presented at Ann. Meet. Am. Met. Soc. U.S. Weather Bureau, New York, 1963.
4. Bretschneider, C. L. On the influence of waves and currents on dispersion over the continental shelf. *Rapp. P.-v. Réun. Cons. int. Explor. Mer*, 1975, **167**, 13–20.
5. Weibull, W. A. A statistical distribution function of wide applicability. *J. appl. Mech.*, 1951, **18**, 293.
6. Bretschneider, C. L. Wave variability and wave spectra for wind generated gravity waves. U.S. Army, Corps of Engineers, B.E.B.T.M. 118, 1959.
7. Bretschneider, C. L. A one-dimensional gravity wave spectrum. In 'Ocean wave spectra'. Prentice-Hall, Englewood Cliffs, N.J., 1963, p. 41.
8. Pierson, W. J., and Moskowitz, L. A proposed spectral form for fully developed wind seas based on the similarity theory of S. A. Kitaigorodskii. *J. geophys. Res.*, 1964, **69**, 5181.
9. Bretschneider, C. L., and Ou, S. H. On the directional wave spectrum determined from published data and the Weibull distribution function. Taiwan Hydraulic Laboratory, Taiwan (*in press*).
10. Bretschneider, C. L. Generation of waves by wind: state of the art. National Engineering Science Co., Tech. Rep. SN-134-6.
11. Bretschneider, C. L. Probability distribution of wave force. *J. WatWays Harb. Div. Am. Soc. civ. Engrs*, 1967, **93** (WW2), Paper 5217.
12. Collins, J. I. Wave statistics from hurricane Dora. *J. WatWays Harb. Div. Am. Soc. civ. Engrs*, 1967, **93** (WW2), Paper 5236.
13. Faltinsen, J. H., and Pederssen, B. Distribution of wave height in different ocean areas around the world. Det Norske Veritas, Res. Dep. Rep. No. 70-7-S, 1970.
14. Nordenstrom, N. Methods for predicting long term distribution of wave loads and probability of failure for ships. Part I—Environmental conditions and short term response. Det Norske Veritas, Res. Rep. No. 71-2-S, 1971.
15. Bretschneider, C. L. *J. WatWays Harb. Div. Am. Soc. civ. Engrs*, 1964, **90** (WW1), Paper 3524, Discussion.
16. Cote, L. J., Davis, J. O., Marks, W., McGough, R. J., Mehr, E., Pierson, W. J., Ropek, J. F., Stephenson, G., and Vetter, R. C. The directional spectrum of wind generated sea as determined from data obtained by the stereo wave observation project (SWOP). *Met. Rep. New York Univ.*, 1960, **2** (6).
17. Longuet-Higgins, M. S., Cartwright, D. E., and Smith, N. D. Observations of directional spectrum of sea waves using the motions of a floating buoy. In 'Ocean wave spectra'. Prentice-Hall, Englewood Cliffs, N.J., 1963, p. 111.

Prof. Bretschneider, introducing his paper, said he wished to discuss only ocean waves and wave spectra; he would not deal with the deterministic wave theory which he had covered in an earlier paper[4]. His present paper was primarily concerned with the 'probabilistic' wave theory, wherein were included various forms of the non-dimensional, frequency, and symmetrical directional spectra. He hoped to make a difficult problem simple for engineering purposes.

The definition of wind speed was important, the proper wind speed being correlated through pressure and drag with the corresponding significant waves and the directional wave spectrum; it had to be defined as some average measured at a common level or plane of reference, and in his paper he had defined it as the 10-min average at the normal 10-m anemometer level. The 1-min average varied considerably about the 10-min average, and this variability existed in both magnitude and direction.

Changes in wind speed were reflected in pressure changes, and pressure being scalar would generate circular waves. The vector wind speed would generate vector components of the wave spectra, and the combined effects of scalar wind pressure and vector wind stresses were coupled with the directional sea spectra. Data were available for correlating significant waves with mean wind speed, which in turn could be used to determine the unidirectional wave spectra. Data were also available for correlating directional sea-spectra frequency components according to frequency, so there were complete relationships for obtaining the directional sea spectra. Although there were other forms of directional spectra, the Weibull distribution function had been used to derive a completely polar spreading function; an account could thus be made for energy of various frequencies moving opposite to the primary wave direction. The wave-directional spreading function should be correlated with the turbulent diffusion spreading pollutants, because such wave actions extended decreasingly from the surface to many metres below it. This was where more research was required.

GENERAL DISCUSSION ON PAPERS 32–34

The Chairman, Dr. M. Waldichuk (Canada), opened the session by saying that there had been a great deal of discussion during the previous week on the cases for and against microbial standards and for and against treatment. He thought it had become abundantly clear that the debate had eventually not been so much for and against those requirements as about whether hydrographic conditions along a particular coast were such as to warrant them. The United Kingdom, having strong tidal currents and turbulent mixing along most of its coast, obviously did not require the same precautions as a partially enclosed embayment or inlet like Oslofjord. Although legislation in some countries, such as the U.S.A. and Canada, required treatment of effluents to meet certain minimum standards, he considered it would not be long before account would also be taken of environmental characteristics when formulating pollution control standards; this was both inevitable and desirable—environmentally as well as economically. In the papers to be discussed, the Authors had considered how the coastal environment and the waters further offshore were naturally equipped to deal with sewage and other effluents after initial dispersion from outfalls.

Mr. P. Ackers (U.K.) said that one of the features of Mr. Talbot's paper was the variability of the results reported from place to place—as shown in Tables 1 and 2. This suggested that diffusion could not be predicted by analogy with results from other places. He asked Mr. Talbot if he saw much hope of future progress in terms of predicting the various diffusion coefficients from knowledge of a few of the governing parameters, such as offshore distance, tidal range, and water depth. He also asked if it was correct to conclude from Fig. 1 that the area of an effluent patch from an instantaneous point source would increase with approximately the square of the time since injection. In recent dye experiments he had mentioned earlier (p. 318) the point injection that was tracked the longest showed areas increasing with about the 2·5 power of time, but injections at different tidal states followed for shorter periods gave areal increases with a lower power. This was for a 'shallow tidal sea', but it appeared that the rate of dispersion was quite variable even at the one site. He therefore enquired if this was a common observation and, if so, how it affected plots such as Figs. 1 and 2.

In reply, Mr. Talbot said that the tables in his paper summarized data available in the literature or obtained by himself. They were from areas having widely different hydrographic characteristics, and variability in the diffusion coefficients must be related to the conditions in the area at the time of measurement. He had realized that these data could be discussed only rather briefly within the context of the paper, and for this reason he was producing a further report (Ref. 37 of his paper). Some consideration of the problem of predicting diffusion rates had, however, been given on pp. 324 and 326. It would be unwise to conclude from Fig. 1 that the area of a patch resulting from a point source would increase as the square of the time, and it was necessary to define what was meant by the area of the patch—a difficulty of definition that had been avoided by many workers who had used variance in the analysis of diffusion results. The rate of diffusion in a shallow sea would depend on the water velocity and therefore on the tidal phase. Thus short-period experiments might give varying results, the value of which would be in contributing to an understanding of the diffusion process rather than in a direct application to longer-term predictions.

Prof. J. E. Foxworthy (U.S.A.) reported that he and his colleagues, in work off the coast of Southern California[18], had found results similar to those of Mr. Talbot relating the transverse variance of dye distributions to time, values of the mean transverse diffusion coefficient ($\bar{K}_y$) being more nearly constant than variable. However, they had established an empirical proportionality between $\bar{K}_y$ and the 1·29 (or essentially 4/3) power of the initial size of a patch or plume. He agreed with Dr. Kullenberg's conclusion regarding the importance of the vertical diffusion coefficient (K_z) on the overall dispersion process of dye patches and plumes. Similar results had been obtained in Southern California, the mean value ($\bar{K}_z$) being found to be a function of water column stability which, in turn, was a function of mean wind speed ($\overline{W}$). A relationship showing $\bar{K}_z$ to be proportional to $\overline{W}^{2.48}$ had been obtained; no relationship between K_z and K_y could be established for an essentially unbounded system. In reply, Dr. Kullenberg said he had found K_z to depend on wind, stability, and vertical current shear. The dependence of K_y on K_z would be subject to the presence of both vertical mixing and oscillating currents with vertical shear.

Dr. R. J. Callaway (U.S.A.) reminded participants that Richardson and Stommel's original 'four-thirds law' experiments [19] had been made with parsnips, and Stommel's later ones [20] with mimeograph paper. Both substances were constrained to remain in the surface so that there could be no loss by shear. The law might well work for this type of tracer experiment, but more variable results might be obtained where the tracer was subject to shear. Dr. Kullenberg, in reply, admitted that these experiments excluded the vertical shear effect. However, it appeared that the concepts of shear diffusion and dependence on the 4/3 power scale conformed. He believed the varying results to be due to natural variations; no general power law could be expected—as had been pointed out by Mr. Talbot.

Mr. G. C. van Dam (Netherlands) showed a sketch illustrating the variability in horizontal diffusion in areas off the Dutch coast, referring to a large number of experiments (with time scales from 0·3 to 120 h), and covering almost all weather conditions except severe storms. The range in values attributable to varying weather conditions did not cause as great differences as might have been expected when used in predicting coliform distributions at the shore near the Hague outfall—as shown by the left part of Fig. 4(b) in Paper 39.

Prof. A. G. Occhipinti (Brazil) asked if Prof. Bretschneider could report on research on mixing in the surf zone, and if he believed the breaking of waves close to the shore was a phenomenon relevant to sewage dispersion. Prof. Bretschneider replied that he knew of no research on mixing in the surf zone primarily in regard to sewage dispersion. Dye studies had been made in Hawaii, on the mainland of the United States, and in Europe, and there had been drogue and current-meter studies also, but all these related to longshore currents and sediment transport. He was familiar with one study on the dispersion of fly ash [21]. He thought it important that field research be conducted on sewage dispersion in the surf zone. It could be done in conjunction with aircraft, helicopters, or perhaps even satellite photography—satellite infra-red oceanography had been used successfully for defining the role off river discharges of sediments into the ocean.

Mr. D. W. Mackay (U.K.) said that for the past 8 years the Clyde River Purification Board had been making the kind of dye-diffusion and drifter studies described by Mr. Talbot and Dr. Kullenberg. In relation to the discharge of sewage from sea outfalls, many of which extended no more than 1 km from the shore, it had frequently been observed that the dye and floats moved towards the shore but then underwent an abrupt change in behaviour about 100 m from the shore in a water depth of about 3 m—instead of reaching it they were transported parallel to t. If this was a general phenomenon it had obvious significance in terms of the control of beach pollution. He asked if any of the Authors had noted such behaviour and if they could explain it. In reply, Mr. Talbot said he had not observed this effect, his work had mostly been in areas having moderately strong tides where movement had been generally parallel to the shore except around slack water. Dr. Kullenberg added that he had not worked so close to the shore, but he thought this phenomenon was probably due to the circulation cells very near to the shore; the results in two recent papers [22, 23] seemed to conform with Mr. Mackay's observations. Prof. Bretschneider said he had not observed a similar situation elsewhere without waves, but he believed it was possible. It might be due to local geographical effects of the long-term general circulation, such as the boundary shear, or to large vortex shedding which would be destroyed along a coastal boundary. With waves breaking at an angle to the coastline, both rip currents and longshore currents would be generated, and since the angles of breaking would be variable, such a situation could be expected, but this was a different problem from that with no waves.

Mr. A. R. Agg (U.K.) mentioned that some years ago the South Africans had reported [24] an effect of discontinuous flow inside the surf zone on a long straight coast subject to oblique wave motion. Dye added to this stream had formed a river-like flow along the beach with a marked interface on the offshore side. Prof. Bretschneider commented that there was a directional spreading of angle of breaking waves, and this would cause a directional variability in the observed dye movement.

Mr. H. M. Cotton (New Zealand) said that his country was in the Roaring Forties, and he reported that the possibility of constructing an outfall on a coast in the south-west of North Island exposed to these westerly winds was being investigated. He was concerned about storm conditions which, although causing considerable turbulence, might result in the rapid transport of pollutants and previously deposited pathogens to shellfish areas and to the shore near proposed and existing outfall sites. He invited Prof. Bretschneider's comments on assessing the importance of such storm conditions on the receiving water when it was physically impossible to carry out tests from boats. Prof. Bretschneider replied that he was familiar with the Roaring Forties and had studied swell off Southern California arriving from Southern Hemisphere storms in the Roaring Forties and Fifties. The same kind of waves were experienced at similar northern latitudes on the western coast of the U.S.A. He had also seen 6-m breakers off the north shore of Hawaii during winter storms 1000 km away, and waves breaking on the shore making direct measurements impossible. The choices were then to theorize, to watch from a vantage point, or to see what happened on a long-term basis. He could only philosophize: if the waves did not move the shellfish, the pollutants could certainly cause damage, but the extent of the damage was a problem for the marine biologist. As an engineer he could say only that the big surf would transport the pollutants shorewards; this was evident from the large quantities of debris washed ashore after a storm.

Dr. A. L. Downing (U.K.) referred to a situation on the east coast of New Zealand where the sewage from a population of roughly 40 000 was discharged into shallow water in the surf zone only about 50 m offshore. The sewage contained effluents from slaughter-houses and canneries giving a BOD load over 25 times that of the domestic sewage. It was not possible to prevent some of the blood from the slaughter-houses from entering the drains, and this resulted in a distinct red coloration of the plume in the sea. Aerial photographs clearly showed an area of contamination bounded by the shore and fanning out seawards along the direction of the tidal drift. The red colour appeared as a series of arcs transverse to the longitudinal axis of the plume, though whether this reflected fluctuations in the pattern of dilution and dispersion or some other cause (such as intermittent discharge from the factories) was unknown. A pronounced slick, with a clearly defined edge, extended over a wider area than that discoloured. Solids obviously of sewage origin could be seen stranded on the beach—though their amounts were less than might have been expected in view of the size of the discharge and the shortness of the outfall. Present conditions were recognized as unsatisfactory, and it was proposed that the domestic sewage and most of the more offensive coloured wastes would eventually be discharged through a new outfall terminating nearly 3 km offshore. It had been of considerable interest, when flying over the area, to observe that the discharge from a nearby town of similar size, released over 1 km offshore, was relatively inconspicuous. One surprising feature, however, was that the area of the boil was quite small in relation to the fairly long diffuser which he understood had been installed.

Mr. G. B. McBride (New Zealand) pointed out that the discharge from the meat works into the surf zone had a high content of salt (used on the cleaning floors), thus the effluent field would resemble a jet rather than the usual plume in its flow characteristics. Also, the Napier–Hastings outfall (like that at Gisborne) included an end port larger than the side ports, so that the dilution from this port would be lower than that from the side ports and hence this part of the discharge would be the more easily observed.

Dr. J. J. Barnard (South Africa) asked Dr. Downing if he could report on the possible increase in shark activity in the vicinity of the discoloration of the sea by blood. Similar discoloration had been noticed at a whaling station on the Durban Bluff, and this could be a factor contributing to the number of shark attacks on humans near bathing beaches in South Africa. Although the sea temperature of 21–27°C favoured shark activity, the presence of high concentrations of organic material—particularly blood—appeared to be an added attraction. Dr. Downing replied that he was not aware of any shark activity in the area of the outfall to which he had referred.

ADDITIONAL REFERENCES

18. Foxworthy, J. E., and Kneeling, H. R. Eddy diffusion and bacterial reduction in waste fields in the ocean. University of Southern California, Los Angeles, 1969.
19. Richardson, L. F., and Stommel, H. Notes on eddy diffusion in the sea. *J. Met.*, 1948, **5**, 238.
20. Stommel, H. Horizontal diffusion due to oceanic turbulence. *J. mar. Res.*, 1949, **8**, 199–225.
21. Foster, D. N., and Stone, D. M. Data provided for N.S.W. coastal engineering works. *Aus. civ. Engng Construc.*, April 1967.
22. Csanady, G. T. Mass exchange episodes in the coastal boundary layer, associated with current reversals. *Rapp. P.-v. Réun. Cons. int. Explor. Mer*, 1975, **167**, 41–45.
23. Bowen, A. J., and Inman, D. L. Nearshore mixing due to waves and wave-induced currents. *Rapp. P.-v. Réun. Cons. int. Explor. Mer*, 1975, **167**, 6–12.
24. Harris, T. F. W., Jordaan, J. M., McMurray, W. R., Verwey, C. J., and Anderson, F. P. Mixing in the surf zone. *Proc. 1st int. Conf. Wat. Pollut. Res., London, 1962*, Vol 3, pp. 177–198.

Paper No. **35**

Observed and Predicted Coliform Distributions near a Sea Outfall

D. Munro
Water Research Centre—Stevenage Laboratory, England

Bacteria added to the sea are carried around by a current system defying exact description, and approximations in modelling their movements are necessary because of the high cost of accuracy. It is important to the user of models to know the degree of inaccuracy brought about by the many simplifying assumptions. The aims of the coastal pollution research programme of the Water Pollution Research Laboratory (now the Water Research Centre—Stevenage Laboratory) have been to gain and disseminate experience on levels of pollution (particularly at the shore) attributable to sea outfalls, with a view to advising on the effects of proposed outfall sitings. A mathematical model of non-extreme dispersion conditions has been developed, it being accepted that by ignoring many of the actual transport mechanisms at least the highest and lowest concentrations are likely to be poorly estimated. In this paper some comparisons are shown between the results given by the model and the concentrations of coliform bacteria found by the Laboratory during sampling programmes at Bridport, Dorset, in 1971–73, using three alternative outlets from the marine sewage outfall, then available for experimental purposes, which allowed the sewage to be discharged at 430, 680, or 1370 m from the shore. The methods and results are discussed to show what has been achieved and in the hope of clarifying priorities in the development of models for similar situations.

Choice of Model

The essential elements of the transport of bacteria are described when the region is sub-divided into small volumes, and the exchanges between adjacent volumes are known. In a confined situation the velocity field is at its most predictable and Leendertse[1] has predicted dispersion patterns using a finite-difference model. Flows on an open coastline are, in general, much less predictable, since they are relatively unconstrained by local features and are a small dependent part of a much bigger system. The large current variations found above an ocean outfall have been reported by Lukin *et al*[2]. Knowledge of flows between adjacent volumes in this type of situation is usually rather limited, and it is then common to consider the dispersion process as diffusion superimposed on a homogeneous flow, since the result of diffusion can be given analytically for a variety of assumptions about the diffusion coefficient. Okubo[3] quotes analytic solutions for an instantaneous vertical line source in a horizontally unbounded region.

In this paper it is assumed that the effect of a straight shoreline may be simulated by an image system of sources—the reflection of the real system in the shoreline. One assumption inherent in this is that the depth of water is the same everywhere, and at Bridport this exaggerates by 60 per cent the volume of sea water inshore of the 1370-m outfall position. The error due to this assumption can be indicated by comparing the results of two-dimensional Fickian diffusion in the vertical plane perpendicular to the shoreline for a uniformly sloping sea bed and for a uniform depth right to the shore. With the decay rates used in this study the sloping-bed model gives coliform concentrations at the shoreline up to three times those given by the constant-depth model. The sea-bed profile at Bridport lies between these two types. Although a finite-difference model would cope with variations in depth, it was decided that dealing with the complications of such a model should be postponed until a fuller understanding of the velocity field had been obtained.

It is assumed in the model that inputs of a pollutant occur every hundredth part of a tidal cycle, and are uniformly spread over a cuboid about the sewage outfall, throughout the depth at the outfall, over a length equal to the distance moved during one hundredth of a tidal cycle, and over a width of 60 m (a distance chosen from concentration profiles made by the Laboratory across sewage plumes nearby at Sidmouth, Devon). At any later time the position of this cuboid is known when the current system, which is assumed to be essentially parallel to the shoreline, is specified. The pollutant is no longer entirely inside the cuboid and is assumed to have spread out horizontally according to the solution given by Joseph and Sendner[4] with modifications to allow for the finite size of the initial source and for the effect of the shoreline. The Laboratory had found[5], in a series of 20 experiments at Sidmouth, that Joseph and Sendner's solution gave a reasonable description of the areal spread of concentration contours. Although in practice the contours were elongated in approximately the wind direction and the velocity of diffusion was dependent on wind speed, the model simply uses the average value of 18 m/h, which is half that found by Joseph and Sendner to be suitable for describing oceanic phenomena.

The solution for horizontal diffusion from a point source is used to describe that from a rectangular source by considering whether having nine equal sources on a 3 × 3 grid giving nine unequal contributions is a worthwhile improvement in accuracy on a single point source giving the equivalent of nine equal contributions. If it is, then each of these nine sources is split into a further nine separate sources on a smaller grid, and so on until there is no worthwhile improvement. The criterion is that if the ratio of the contributions for the nearest and farthest sources is less than two the improvement is not worthwhile. The accuracy that this gives is best illustrated by an example. Consider a single release of coliforms when the tidal movement is 120 m in one per cent of the tidal cycle, and suppose that 4 h later 9×10^9 coliforms per metre of depth are left after decay and that the initial 120 × 60 m rectangle has moved according to the supposed translations to position ABCD (Fig. 1). Knowledge is required of the contribution that this rectangular source is making at a point X, 100 m inshore and 10 m alongshore from the centre of gravity of the rectangle. The nine sources we consider first are E–M, each one representing a source of 10^9 coliforms per metre of depth. The solution to Joseph and Sendner's equation is that the coliform concentration, C per m³, at a distance r m from a source of Q coliforms per m of depth, when the patch has been diffusing for t h, is

$$C = \frac{Q}{2\pi P^2 t^2} \exp\left(-\frac{r}{Pt}\right), \tag{1}$$

where P is the velocity of diffusion in m/h. Taking P as 18, the concentrations from the nine sources E to M are those shown in parentheses in Fig. 1. The ratio of largest to smallest contributions is 1·98 and it is therefore considered that the rectangle ABCD is adequately represented by a single point source, where distance from X is taken as the average of the distances from the farthest and nearest sources, in this example (EX + LX)/2. This contribution works out to be 63 983 coliforms/m³, approximately 2 per cent less than the more accurate value of 65 296 found by summing contributions from the nine separate sources. Furthermore, in order to shorten computations the rectangle is not subdivided when its contribution is less than 1 per cent of the concentration attributable to earlier discharges. The accuracy of this method is probably better than 10 per cent.

The bacterial decay terms used in the model are described more fully by Gameson and Gould[6] at this Symposium. Essentially the number of coliforms left in a diffusing patch is assumed to decrease ex-

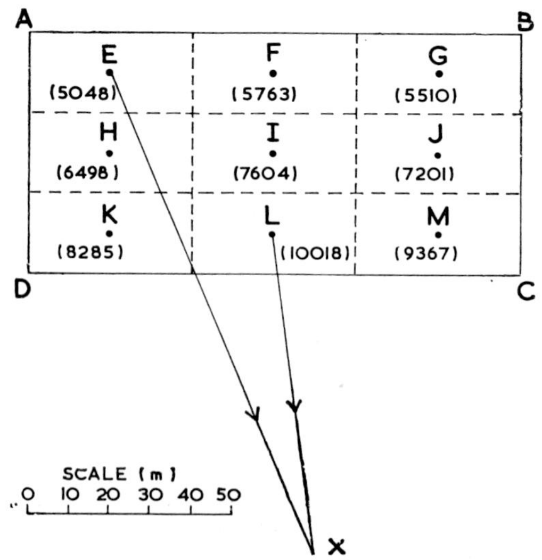

FIG. 1. Example of subdividing rectangular source ABCD into nine equal point sources to illustrate relative contributions sources make to concentration (see text)
Contributions (coliforms/m³) at X from nine sources E to M are shown in parentheses
Single source just acceptable for required accuracy

ponentially with time in the absence of solar radiation, and a further exponential decrease with the cumulative radiation received is superimposed. Rates of decay are quoted here in terms of the length of time, T, and amount of radiation, S, which separately reduce the number of coliforms by a factor of 10.

It is important to take some account of the component of velocity perpendicular to the shoreline since the model is used to predict concentrations at sampling stations at about the outfall distance offshore as well as elsewhere. There was no evidence from float tests and current-meter readings that this component was related to tidal state, but experiments to track labelled patches of water showed it to be correlated with wind direction. In the model, however, patches have been considered to move sinusoidally in the direction perpendicular to the shoreline so as to account for the rise and fall in tidal level inshore of the outfall. As with other assumptions made, this contributes to the discrepancies between observed and predicted concentrations.

Application

Although the eventual purpose of a model is to predict the future performance of an outfall system using estimated pollution loads and decay rates, more detail is included in the present development stage. Thus the model is given all available data on sewage flow and coliform concentration, and solar radiation levels throughout the sampling periods (about two weeks), together with predicted times of high water, tidal ranges, and a longshore velocity curve describing the variation through the tidal cycle found by floats and current meters at each of the outfall positions (these were similar in phase and magnitude and would move a patch about 5 km in half a tidal cycle for the average tidal range of 2·4 m). The velocities were assumed to be proportional to tidal range and to have a residual movement of about 500 m per tidal cycle, though the actual residual probably varied greatly.

During each test period, sea-water samples were taken two or three times a day at the surface and at one or two depths over a grid of sampling stations from just offshore of the outfall to the shore and extending about 1·5 km in both directions along the shore. Sampling was omitted on some days when mortality rates of coliforms in the sea were being measured directly. These rates were sometimes far from constant[6] but single values of S and T were used in predictions corresponding to each sample taken. The assumed decay rate is so much faster by day than by night that the coliform concentrations of samples taken in the morning are expected to be much greater than those for later in the day, although this effect may be partially offset by the increase in the rate of discharge of sewage coliforms in the morning. The model thus predicts large ratios between morning and afternoon concentrations; however, the observed ratios are much smaller. Thus to obtain reasonable fits to the observed data it is necessary to use decay coefficients different from those found by direct experiments, and in this paper the results shown are derived from the values of S and T giving the most consistent comparisons. The effects of using the experimentally determined decay rates are discussed in an earlier paper[7]. However,

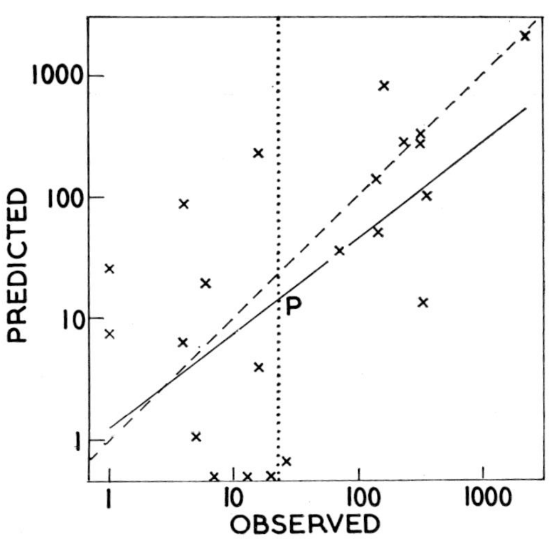

FIG. 2. Plot of predicted against observed coliform counts (per 100 ml) at a single sampling station in September 1971

Continuous line is result of regression of predicted on observed, broken line is line of perfect prediction, and dotted line passes through median of observed counts; P is point on regression line where departure from line of perfect prediction is measured

the predictions shown there should be reduced by factors of between 2 and 5, as it has since become apparent that sewage coliform concentration measurements were overestimated owing to difficulties in obtaining representative samples of sewage.

A description of goodness of fit is illustrated in Fig. 2 by an example from the first period to be discussed; predictions for a sampling station at the outfall distance offshore and at an edge of the area sampled are plotted against observed coliform counts at the surface. Logarithms of coliform concentrations are best compared, because the distributions found are closer to lognormal than normal. Concentrations less than 0·5 coliforms per 100 ml are taken as that value. In this paper the goodness of fit is specified by the correlation coefficient, by the slope of the regression line of predicted on observed, and by a parameter called the ratio, defined as the predicted value given by the regression line at the median of the observed values divided by that same median. The logarithm of this ratio is the distance in logarithmic units between the line of perfect prediction and the point P, where the dotted line (through the observed median) intersects the regression line. In the example this distance is negative, showing the prediction to be too low. The correlation coefficient in this example is 0·56 which is significant at the 95 per cent confidence level.

Some details of the sampling periods are given in Table 1.

TABLE 1. Data on sampling periods

Length of outfall (m)	430		680	1370
Mean depth of outfall (m)	8		10	16
Month	Sept. 1971	April/May 1972	July 1971	May 1973
No. of samples per station	21–23	21–31	21–25	9–15
Estimated median coliform load (10^{14}/d)	17	6	12	5
Estimates of i) time in dark[a], T(h) ii) solar radiation[b], S(cal/cm^2) causing 90 per cent coliform mortality	65 (12) 46 (138)	100 (20) 74 (276)	65 (65) 23 (64)	100 (100) 46 (64)
Average daily radiation total (cal/cm^2)	203	218	336	263
Median ratio of coliform counts at depth (5 or 10 m) to those at surface	1·4	1·3	4	2·2
Percentage of ratios between half and twice median	50	41	21	26

[a] Derived from shore-based experiments; numbers in brackets are values used in model
[b] Derived from *in situ* experiments; numbers in brackets are values used in model

430-m outfall, September 1971

It may be seen from the last two lines of Table 1 that the coliform concentrations were nearest to being homogeneous throughout the depth when the 430-m outfall was in use. It thus appears least unreasonable with this outfall to ignore the depth variations (as in the model) and to compare predictions with surface samples only. At each of 24 offshore sampling stations and 8 beach stations the medians of the observed surface coliform concentrations have been evaluated together with the 10 and 90 percentiles of the distributions. These are shown in Fig. 3(a). The River Brit, which enters the sea halfway along the length of beach sampled, may be seen to have increased the coliform concentration at the sampling station nearest its mouth at the median and 90-percentile levels. Other anomalies are obscured since the contour levels are factors of 10 apart, required by the large variations of concentration in time and space.

The same percentiles of the distributions of predicted concentrations are shown in Fig. 3(b) and bear some similarity to those observed. The goodness of fit is illustrated in Fig. 3(c) by the results of regression analyses. Apart from some stations near the shore the regression lines are within a factor of two of the

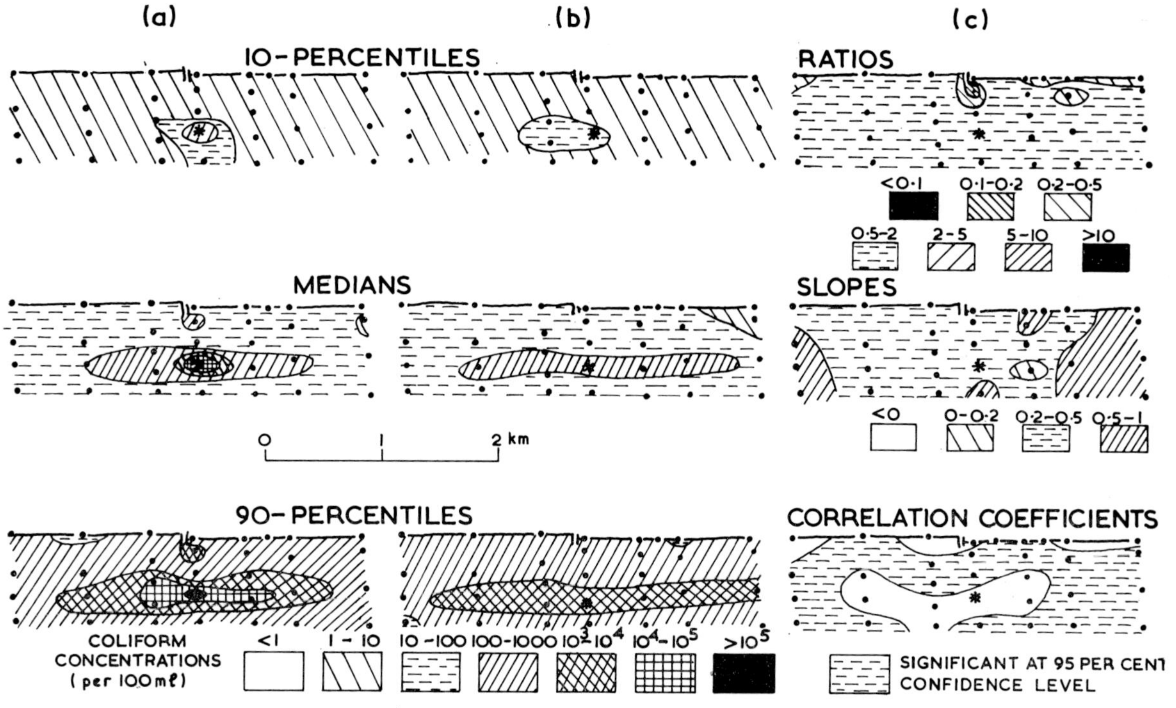

FIG. 3. (*a*) Observed and (*b*) predicted distributions of 10-, 50-, and 90-percentile coliform concentrations during sampling at Bridport, Dorset, in September 1971; (*c*) goodness of fit—defined in Fig. 2 and accompanying text

Plotted points show positions of sampling and prediction, asterisks the sewage outfall

line of perfect prediction at the median observed concentration. The slopes of the regression lines lie between 0 and 1, indicating that extreme values of observed concentrations are associated, on average, with less extreme predictions. In general, the extreme conditions appear to be best predicted at the two ends of the sampling area, presumably because there the effect of the oscillating tidal current is greatest and that effect is basic to the model. The distribution of correlation coefficients between the logarithms of predicted and observed is shown in the lowest part of Fig. 3(c), the shaded area indicating where correlation coefficients are better than those expected by chance on 19 occasions out of 20. Near the outfall it appears that linear relations between the logarithms of observed and predicted concentrations are not meaningful. It should be borne in mind that these predictions were obtained after selecting decay rates to give a reasonable fit for both morning and afternoon samples. Even if the overall fit is acceptable, the problem remains of choosing decay constants applicable to other periods and at other sites.

430-m outfall, April/May 1972

Samples were not taken at the shore during this sampling period. The 10, 50, and 90 percentiles of the surface distributions of concentration are shown in Fig. 4(a). It is more difficult here to distinguish between the dominant effects of river or outfall at several stations near the river entrance, and concentrations are mostly higher than in the previous period. Again, after choosing decay rates to give good fits to samples which could have received much radiation and to those having received little, the 10, 50, and 90 percentiles of the predicted distributions, shown in Fig. 4(b), appear fairly similar to the observed. The goodness of fit is examined more closely in Fig. 4(c). The uppermost part shows that the regression lines (of the type shown in Fig. 2) give predictions, corresponding to the median observed, within a factor of 2 of the median observed at 70 per cent of the stations. The extremes of the predictions are strongly associated with those of the observed at the ends of the sampling area, possibly owing to the strong tidal influence in those regions. The correlations are poor at more than half the sampling stations, indicating that the processes causing the large variability have not been satisfactorily modelled.

680-m outfall, July 1971

Samples were taken at 27 offshore stations and 8 beach stations. It may be seen in Table 1 that there was little similarity between concentrations at the surface and 10-m depth. A reasonable explanation of

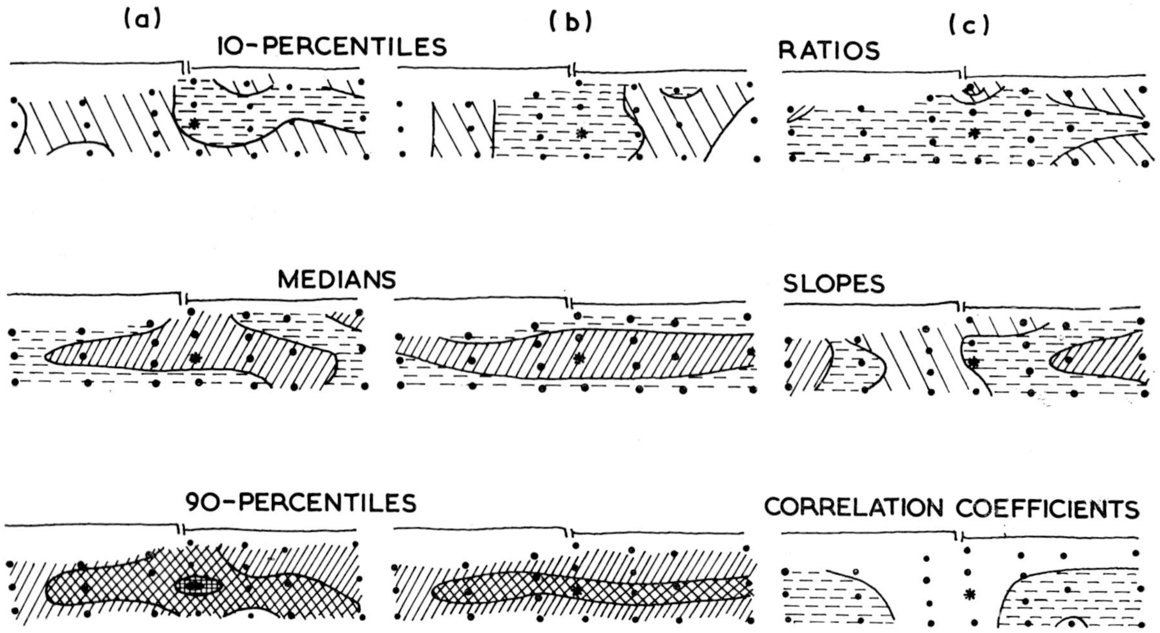

FIG. 4. (*a*) Observed and (*b*) predicted distributions of 10-, 50-, and 90-percentile coliform concentrations during sampling at Bridport, Dorset, in April/May 1972; (*c*) goodness of fit
Key as in Fig. 3

the observations therefore requires a three-dimensional model, though the two-dimensional model is applied here to try to fit the surface observations. The 10, 50, and 90 percentiles of the surface distributions of concentration are shown in Fig. 5(a). The concentration patterns at the 10 and 50 percentile levels give no indication of the position of the outfall and they are generally an order of magnitude lower than with the shorter outfall, but at the shoreline some concentrations are an order of magnitude higher than those of Fig. 3(a). Thus it is more difficult than before to distinguish between coliforms originating from the outfall and those from other sources. The 10, 50, and 90 percentiles of the predicted concentrations are shown in Fig. 5(b). At only a few stations do the distributions appear similar. The

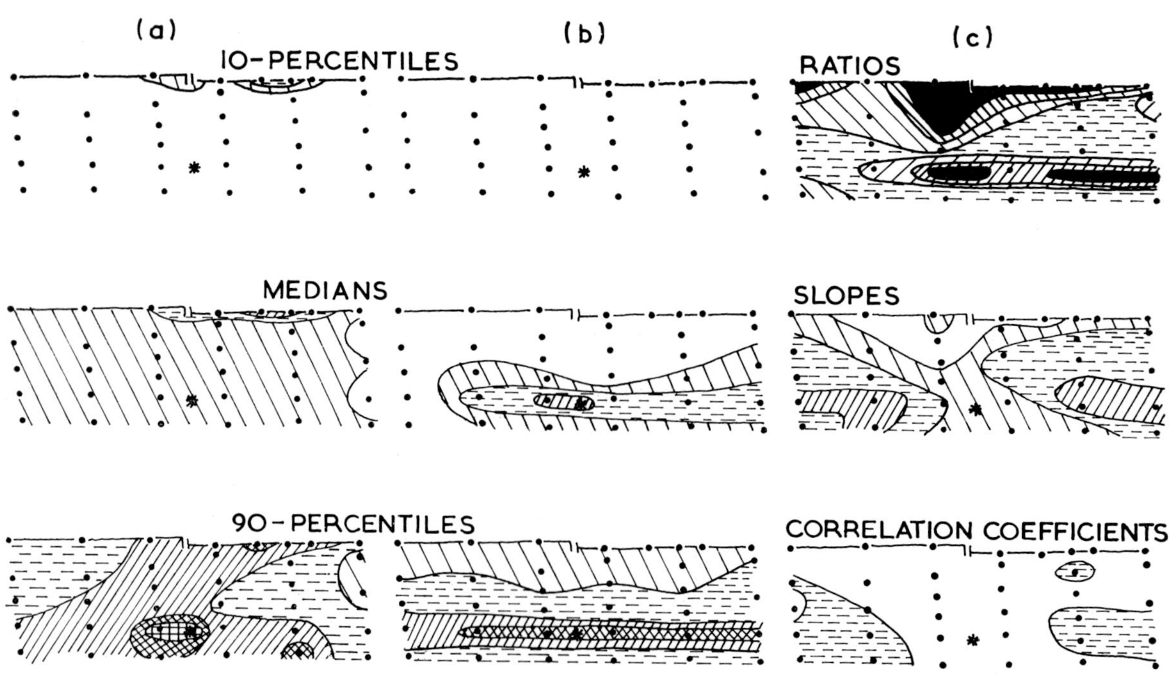

FIG. 5. (*a*) Observed and (*b*) predicted distributions of 10-, 50-, and 90-percentile coliform concentrations during sampling at Bridport, Dorset, in July 1971; (*c*) goodness of fit
Key as in Fig. 3

failure of the model in this situation is shown quite clearly in Fig. 5(c). The predicted values are about an order of magnitude too high at the outfall distance offshore and too low by an even greater factor at the shoreline. In between, the area showing agreement to within a factor of two is quite small. The high values of the slope and correlation coefficient at the ends of the sampling area again indicate that the longshore velocity probably accounts for much of the variability there. For most stations, however, the model fails to predict a significant part of the variability.

1370-m outfall, May 1973

The degree of vertical homogeneity in this period lay between the cases already discussed. The surface observed coliform concentrations are compared in Fig. 6 with the predictions. Observed concentration levels are again lower than during the two periods with the 430-m outfall, but the fields of pollution arising from the outfall and the shore appear reasonably distinct. If the poor fit well inshore of the outfall is ascribed to other sources of coliforms, then the observed and predicted distributions appear to be similar. In Fig. 6(c) it may be seen that at about half the stations influenced mainly by the outfall (a judgement made from Fig. 6(a)) the median of the observed data is predicted to within a factor of two. The slope of the regression line is generally less than in the other periods, and is even negative at two stations. Thus, although the range of concentrations has been fairly well predicted the agreement appears to be fortuitous. The small number of samples at each station may be partly responsible for the lack of significant correlations.

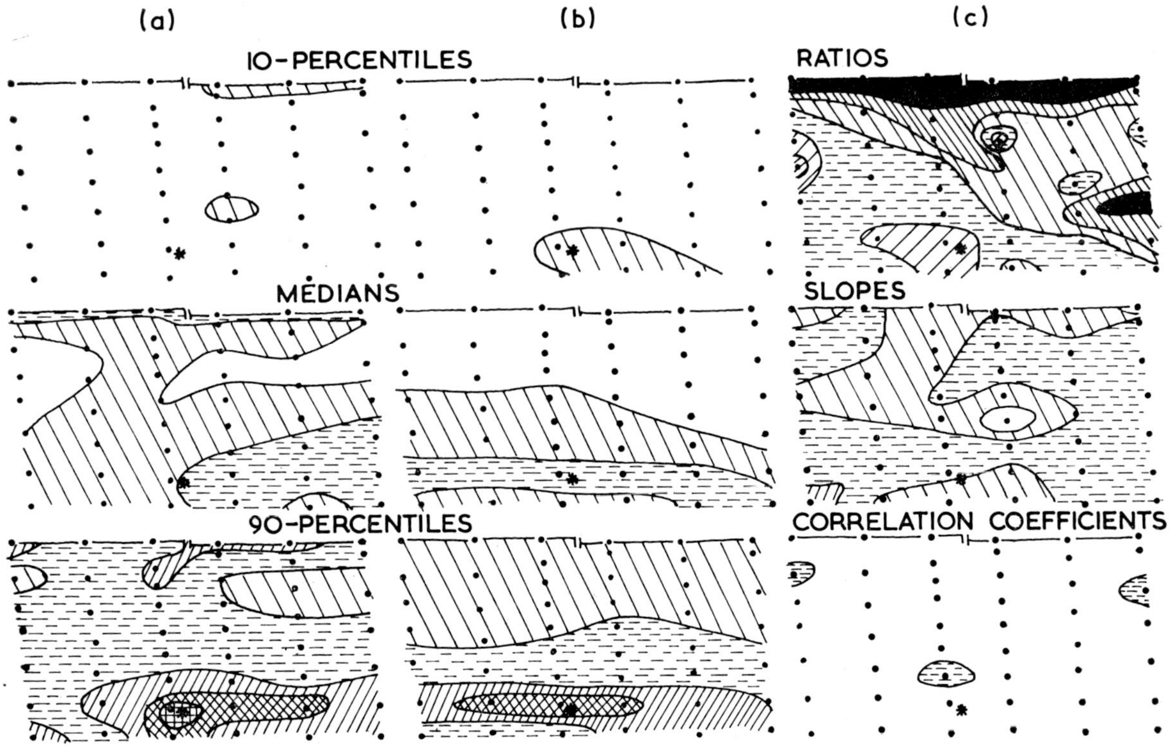

FIG. 6. (*a*) Observed and (*b*) predicted distributions of 10-, 50-, and 90-percentile coliform concentrations during sampling at Bridport, Dorset, in May 1973; (*c*) goodness of fit
Key as in Fig. 3

Discussion

It has been shown that the two-dimensional model gives a poor fit to the observed concentrations at the surface when circumstances produce large concentration ratios between the 10-m depth and the surface. Undoubtedly a three-dimensional model would describe the dispersion better if it included important mechanisms not considered in the two-dimensional model. There was a tendency for concentrations found at the surface to be lower than those under 10 m of water, the greatest effect being in the period with the 680-m outfall when the sea was calmer and the sun stronger than in other periods, both of these conditions encouraging vertical stability of the water column. The processes responsible for higher concentrations at depth are possibly sedimentation or slower decay 10 m below the surface, and the effect is augmented if vertical diffusion is slow. Another possibly important process requiring vertical inhomogeneity is explicit transport in the direction perpendicular to the shore by wind and waves, since from continuity considerations there cannot be a consistent movement either towards or away from the shore at all depths. At present the information available on these processes is at best qualitative, and the added information included in a three-dimensional model would be speculative. Some careful speculation is possible in a two-dimensional model since one effect of vertical stratification could be increased horizontal mixing[8], a conjecture which is supported by Fig. 5(a) and (b) where the observed data indicate a faster horizontal spread of coliforms than that predicted.

The difference between the decay constants established by direct experiment and those found reasonably acceptable in the model is very large. The daily decay (expressed in orders of magnitude) experienced by a patch of coliforms is $24/T +$ (solar radiation total during day)$/S$. It can be calculated from Table 1 that this ranged from 3 to 15 orders of magnitude from the experimental data for a day of average solar radiation, and from 2 to 6 for the decay constants found suitable in the model. When the periods are ranked in order of daily decay rate they show the same rankings for directly observed decay coefficients as for those used in the model but the range of observed values is considerably larger than that of those calculated.

The range of variability in concentration is given surprisingly accurately by the model; other processes might have been expected to govern the extreme concentrations. The extremes are best modelled about 1 km alongshore from the outfall where, even in the absence of variations in rates of addition and decay, large variations are to be expected owing to the alternating longshore current. Nearer the outfall, correlations between observations and predictions are weaker showing the dispersion processes in that region to be poorly modelled. Possibly a three-dimensional model considering the dispersion of coliforms through the depth could give an improvement, or perhaps a more accurate knowledge of the currents around the slack-water periods is necessary. The slopes of the regression lines were almost invariably positive, indicating that the coliforms from other sources were affected by some of the mechanisms considered in the model. This was particularly apparent in the period with the 1370-m outfall, when the fields of pollution from the outfall and other sources were reasonably distinct. The River Brit, which appears to be the chief source of pollution near the shore, is almost directly inshore from the outfall and thus variations due to the longshore current are in phase from these two sources. The rate of decay has a considerable effect on the inshore predictions, since diffusion in the model takes about a day to bring the highest concentrations to the shoreline. There is thus a suggestion that the rate of discharge of coliforms from the river to the sea could show an inverse relationship with the preceding day's solar radiation. Yet other complex links may exist—for example, much of the pollution found at the shore has been attributed to gull droppings.

In conclusion it has been shown that even when decay rates are chosen to describe morning and afternoon coliform concentrations equally well, a difficult enough task in itself, agreement is sometimes poor between observations and predictions, being poorest when the biggest variations occur between concentrations at the surface and at depth. Improved reliability in predicting for fairly stratified conditions seems to need a three-dimensional model, employing a mechanism to reduce concentrations at the surface faster than those at the sea bed and perhaps incorporating data on the transport due to wind and waves. Little can be said about the contributions of the various outfalls used at Bridport to pollution at the shore because of other sources, although the predictions are not discredited there, being generally less than the observed concentrations.

REFERENCES

1. LEENDERTSE, J. J. A water quality simulation model for well-mixed estuaries and coastal seas. Vol 1—Principles of computation. Rand Corp., Santa Monica, Report RM-6230-RC, 1970.
2. LUKIN, L. D., LEE, T. N., LACKEY, J. B., MCGUIRE, J. B., DAVIS, T. L., and BARRY, J. Limitations and effects of waste disposal on an ocean shelf. Florida Ocean Sciences Institute. U.S. Environmental Protection Agency, Washington, D.C., 1971.
3. OKUBO, A. A review of theoretical models for turbulent diffusion in the sea. *J. oceanogr. Soc. Japan*, 1962, **20**, 286.
4. JOSEPH, J., and SENDNER, H. Über die horizontal Diffusion im Meere. *Dt. hydrogr. Z.*, 1958, **11**, 49.
5. BARRETT, M. J., MUNRO, D., and AGG, A. R. Radiotracer dispersion studies in the vicinity of a sea outfall. *Proc. 4th int. Conf. Wat. Pollut. Res., Prague, 1969*, p. 863.

6. Gameson, A. L. H., and Gould, D. J. Effects of solar radiation on the mortality of some terrestrial bacteria in sea water. This volume pp. 209–219.
7. Hart, I. C., Barrett, M. J., and Gameson, A. L. H. Forecasting pollution in rivers, estuaries and the sea. In 'Computer uses in water systems'. Water Research Association, Medmenham, England, 1973, p. 129.
8. Bowden, K. F. Horizontal mixing in the sea due to a shearing current. *J. Fluid Mech.*, 1965, **21**, 83.

In presenting his paper Mr. Munro emphasized the need for feedback in design processes and hoped that engineers would examine the performance of their outfalls after commissioning them—perhaps by the methods outlined in his paper. He showed some observed surface distributions of coliform bacteria and corresponding results of a predictive model, and discussed probable reasons for the disparity between them. After the dominance of other sources at the shore, the main differences were in the total numbers of coliforms at the surface and in the longshore distribution at the outfall distance offshore. In the examples shown, the dispersion of pollutant towards the shore seemed to be satisfactorily modelled. To illustrate the overall accuracy of the predictions, the comparisons had been summarized into Figs. 3–6.

He said that the most disappointing feature of the work had been the need to use decay coefficients differing from those directly measured. The trouble had been that predictions made for samples taken in the afternoon had been generally too low, whereas those for morning samples had been in better agreement or too high. Using more suitable decay coefficients, the model was shown to be least reliable when there was evidence of the greatest vertical stability. He thought that a three-dimensional model would be required for improved predictions in this situation, but that such a model would need more physical information about water movements than was currently available. He ended by reiterating a statement by Prof. Bretschneider that simple solutions could be found when the total system was too complicated to be fully describable.

DISCUSSION

Mr. B. E. Waters (U.K.) said he would like to add a rather unscientific footnote to the paper. His firm had recently studied the dispersion of sewage from three outfalls on the north-east coast of England with a contributory population of about 50 000. Two bays formed the main recreational areas and it had been reported that visual observations and limited bacteriological tests showed the water in these areas to be relatively clean. The outfalls were so short that their outlets were uncovered at low tide; initial dilutions of 1–10 were therefore expected. After tracking floats and making current-meter measurements, sampling points were selected from which it was confidently expected that high coliform counts would be obtained. The Author's laboratory collected and analysed 136 samples during a period of little wind, calm seas, and no sunshine. Coliform counts in the raw sewage were in the range 11–30 million per 100 ml. Stations within 200 m of one of the larger outfalls gave counts of between 10 000 and $6{\cdot}5 \times 10^6$, with a median value of 50 000. Stations 100 m from the other large outfall (where the tidal currents were stronger) gave counts between 900 and 10 000, with a median of 5000. He was quite unable to postulate any model which would give dilutions greater than about 100 and thus an expected median count less than about 200 000.

Dr. J. A. French (U.S.A) asked if, by using a dye tracer or a sewage constituent such as phosphorus, the capacity of the model to predict separately the physical dilution and the coliform disappearance had been analysed. Use of a tracer with properties more predictable than those of coliforms would better indicate the capability of the model to predict physical dilution.

Mr. J. W. Talbot (U.K.) asked if the Author agreed that when considering possible pollution near the shore it was particularly important to take account of material released over the slack-water period, since it was over this period that greater concentrations and a more marked vertical density structure would develop (because of the lower velocities) and that movement towards the shore occurred and thus caused dispersal in that direction. Diffusion measurements off the East Anglian coast had demonstrated that dispersal normal to the shore could occur mainly over the slack-water period. He also suggested that in a situation where shear diffusion was important it might be necessary to relate the diffusion rate to variation of velocity during the tidal cycle.

Mr. P. Ackers (U.K.) said that the apparent disadvantage of Mr. Munro's model (which had emerged from his frank reporting of testing in a true predictive fashion) might well arise from the vertical variability of currents and from the vertical distribution of the emerging sewage. Of course, these depended greatly on the tidal state.

Dr. J. F. Macqueen (U.K.) asked how the Author was sure that the time interval between opening a new outfall port and the subsequent sampling had been long enough for coliforms from the previous discharge point to have decayed or dispersed from the sampling area.

The Chairman, Mr. M. G. Gross (U.S.A.), expressed surprise that no-one had mentioned the scarcity (or poor quality) of environmental data as a factor causing difficulties in developing or validating predictive models for sewage outfalls. In the United States there was a lack of data on coastal currents and wind fields suitable for the rigorous testing of predictive models. This was particularly true for long-term data and for data on sub-surface waters, and the situation would become more troublesome as longer outfalls were designed (and modelled) to operate in deeper water. The remedy was for groups from engineering societies to formulate their data needs and to communicate them to the appropriate government agencies. In the United States, the National Weather Service and the Environmental Data Service of the National Oceanic and Atmospheric Administration (NOAA) would be sources of environmental data and might collect the additional data needed.

Mr. A. L. H. Gameson (U.K.) said that Mr. Munro's paper was not entirely a success story, but there was far too little information published on methods of predicting the effects of the discharge of sewage from sea outfalls, accompanied by field data to demonstrate the validity (or otherwise) of the predictions. Indeed, it was a series of such papers that Mr. Gameson had hoped would provide the climax of the Symposium. It was easy to design an outfall system (with or without treatment) to give bacterial counts below a particular value at the shore, but it was by no means easy to do this at minimum cost. If the optimum length of a particular outfall was 1 km, and the outfall constructed was 2, 3, or even 5 km long, the design criteria would have been satisfied and all parties would be happy in their ignorance of the money wasted; but if it was made only 800 m long and there was occasional trouble, the designer would also be in trouble. Thus, in the absence of reliable predictive methods, there was probably a tendency to over-design, outfalls being made longer than the situation required—and incidentally reducing the economic attractiveness of a long outfall in comparison with treatment.

Since, after working for so long at Bridport, it had been found impossible to get good agreement between observed and predicted data—even when using decay coefficients differing substantially from those determined experimentally—it was inappropriate to use this model for purely predictive purposes at other sites. There were, of course, ways of improving the model, though they were difficult to incorporate. For instance, no account had been taken of wind which produced surface currents and compensatory sub-surface counter-currents. Wind was also an important source of horizontally anisotropic dispersion; Mr. Talbot (in Paper 32) had emphasized the non-circularity of diffusing patches, and the Laboratory's earlier work[5] had related the degree of elongation of patches to wind velocity. Although it had been concluded that it was no longer profitable to continue with this work, Mr. Gameson did not wish to discourage anyone from attempting to improve Mr. Munro's model or from developing new predictive models.

REPLY TO DISCUSSION

In reply to Dr. French, Mr. Munro said that radioactive tracer (bromine-82) had been added continuously to the sewage flow for 24 h on four separate occasions and its concentration measured in conjunction with the sampling for coliforms during the period of addition and during the two following days. It was found that the concentrations predicted were in reasonable agreement with those observed during the period of addition, but that for the subsequent days they were about an order of magnitude too high. Thus the model assumed that old pollutant remained longer in the area than was actually observed. However, he thought this had little relevance to the coliform predictions, since so few of the bacteria survived for 24 h.

He agreed with Mr. Talbot that sewage released at slack water could, at times, move shorewards relatively quickly with little dilution, because of its greater vertical stability. Whether or not the shoreward dispersion was increased was more difficult to answer; the Laboratory's studies at Sidmouth[5] had shown no marked change in dispersion rates at slack water. The two-dimensional model might give improved results if some account were taken of changes in diffusion rate, but the dissimilarities in the physical distribution could be dealt with satisfactorily only by a three-dimensional model using more information about the differences between depths. The two-dimensional model described in the paper had been used for predicting average rather than worst conditions; other types of model could be developed to consider extreme conditions.

Replying to Dr. Macqueen, he said that the changes in outlets had usually been made about a week before the start of the sampling programme; in fact, he considered that an interval of a day would have sufficed with the decay rates involved.

Prediction of Far-Field Exclusion Areas and Effects

Harry H. Carter
Chesapeake Bay Institute, The Johns Hopkins University
Baltimore, Maryland 21218, U.S.A.

The fate of sewage effluent discharged into the sea depends on many physical processes which, for purposes of analysis, may be categorized as either *near* or *far field*. The near-field processes are governed primarily by the characteristics of the discharge whereas the far-field processes depend on larger scale ambient conditions. In some cases, the degree of treatment and near-field processes may be sufficient to meet pertinent bacteriological water quality standards for certain water uses, such as water contact recreation. That is, a deep-water discharge with large initial inertia and buoyancy will achieve near-field dilutions of the order of 100 as a result of mechanical entrainment of ambient fluid in the jetted effluent which may be adequate to protect nearby bathing beaches. For this case, there are various models available for prediction purposes which have been verified in the laboratory. Each involves slightly different assumptions but all contain sufficient curve fitting so that from a practical standpoint there is little variation with model in the predicted centerline trajectories or dilutions.

In many other instances, however, near-field processes result in insufficient dilution to meet standards and the far field must be included in the analysis. For example, shoreline discharges or shallow-water submerged discharges in the case of water-contact recreation standards or even deep-water buoyant-jet discharges if the standards are high, such as for shellfish harvesting or public water supplies. This paper describes several approaches to the problem of predicting the far-field impact on a specific resource. The methodology is independent of the characteristics of the discharge since, except under very special conditions, the far-field concentrations are determined solely by ambient conditions.

The Model

I have chosen to treat the problem of a continuous source as the superposition of a large number of small diffusing patches each from an infinitesimal, instantaneous source and all moving with a constant velocity, U. For the diffusion relative to the center of mass for these elementary patches, I have chosen the Okubo–Pritchard (O–P) diffusion model[1]. This model is one of a class of radially symmetric solutions for horizontal diffusion from an instantaneous vertical line source characterized by a 'diffusion velocity'. The O–P model is based on the intuitive concept that the rate of change of local variance, i.e. the diffusivity, depends on a characteristic velocity, ω, and the time of diffusion. I recognize that in a turbulent environment the shape of an instantaneous release of material is seldom if ever circular and that by characterizing the velocity field as constant I cannot account for the highly variable trajectories taken by the various patches. Nevertheless, radially symmetric models have achieved considerable success in predicting the diffusion of introduced substances. In addition, the O–P solution, unlike many others, can be integrated in time and space to solve a wide variety of problems.

For horizontal diffusion from an instantaneous vertical line source, Okubo and Pritchard proposed the relationship

$$S_i(x, y, t) = \frac{m}{\pi\omega^2 D t^2} \exp\left\{\frac{-(x - Ut)^2 - y^2}{\omega^2 t^2}\right\} \tag{1}$$

where $S_i(x, y, t)$ is the concentration of material, at distance x from the source and y from the centerline, at time t since the single instantaneous release of mass m, distributed uniformly within a layer of thickness D; ω is the diffusion velocity, and U the ambient velocity—assumed to be constant and directed parallel to the x-axis.

Assuming that a continuous source can be considered as a superposition of a large number of instantaneous releases (N per unit time), we have at time t_l after initiation of discharge

$$\int_0^{t_l} S_i\, N\, dt = S_c(x, y, t_l) \tag{2}$$

where $S_c(x, y, t_l)$ is now the concentration from a continuous release of material at the rate $q = Nm$. Upon integration, Equation 2 gives the result along the plume centerline, i.e. $y = 0$, that

$$S_c(x, 0, t_l) = \frac{q}{2\sqrt{\pi \omega x D}} F(\theta) \tag{3}$$

where

$$F(\theta) \equiv 1 - \frac{2}{\sqrt{\pi}} \int_0^{\theta} \exp(-\Phi^2) d\Phi = 1 - \operatorname{erf} \theta, \quad \text{and} \quad x_l \equiv U t_l. \quad \theta \equiv \frac{U}{\omega}\left(\frac{x}{x_l} - 1\right).$$

For a given value of ω/U, as x/x_l increases, F reduces rapidly and then asymptotes to zero. Thus when $\theta = 2{\cdot}326$, $F = 0{\cdot}001$ (or essentially zero) and when $\theta = -2{\cdot}326$, F lies between 1·999 and 2. Accordingly, when $x/x_l \leqslant 1 - 2{\cdot}326\omega/U$, $F = 2$ (for all practical purposes) and when $x/x_l \geqslant 1 + 2{\cdot}326\omega/U$, F is essentially zero. The model is, therefore, capable of characterizing the advancing front of material as well as the equilibrium region, a feature which is essential to the prediction of the fate of effluent discharged into the sea.

For that portion of the plume in equilibrium, i.e. $F = 2$, the concentration varies inversely with downstream distance, x. If Equation 2 is numerically integrated for the case where the velocity U is an oscillating function of time, a plot of the maximum concentrations observed over the period of oscillation as a function of downstream distance retains the general features of Equation 3 above; that is, these concentrations vary inversely with downstream distance and show a rapid decrease for $x > \int_0^{T/2} U\, dt$. It seems acceptable, therefore, to apply the model to situations where tidal currents dominate, it being understood that Equation 3 represents the locus of the maximum values of concentration (over the tidal period T) for all x, and that the U in the model is essentially a free parameter determined by fitting the data to Equation 3 and has no familiar physical meaning.

Applications

The water quality standard for Maryland shellfish waters is that the median coliform concentration in the waters overlying the bars or beds shall not exceed 70 per 100 ml (based on a 5-tube decimal dilution MPN test), and that the 90-percentile shall not exceed 230 per 100 ml. To cover the contingency of a breakdown in chlorination at the treatment plant, an exclusion area is established around each sewage outfall within which harvesting of shellfish (*Crassostrea virginica* and *Mya arenaria*) is not permitted. The exclusion limit is defined by the 70 per 100 ml concentration line at $t = 24$ h where t is the time since chlorination ceased, it being assumed by regulatory authorities in Maryland that this time is sufficient to close the affected area to harvesting. Based on an estimated coliform concentration in raw domestic sewage and allowing for approximately 75 per cent reduction by conventional primary treatment and 30 per cent reduction by mortality[2], the exclusion limit has been defined as the 264 000 dilution line.

To delineate this dilution line, Rhodamine dye tracer is metered into the sewage effluent at a constant rate for 24 h and its surface distribution at $t_l = 24$ h measured in the receiving waters by fluorescence. Since a dilution of 264 000 is greatly in excess of what can be measured, the data are fitted to the diffusion model (Equation 3) and the position of the exclusion limit extrapolated. To illustrate the methodology, an experiment conducted at the Annapolis, Maryland, treatment plant during March 1965 is described below.

A solution of Rhodamine-WT was discharged at a rate of approximately 0·069 g/s of dry dye into a mean effluent discharge of 0·168 m^3/s between 0800 on 24 March and 0900 on 25 March. The tracer concentration in the area of interest was measured by standard fluorometric procedures[3] from 0900 to 1120 on 25 March. The horizontal distribution of tracer was plotted in the form of concentrations relative to the discharge concentration (400 μg/l.); these contours are shown in Fig. 1. In Fig. 2 are plotted the peak concentrations, $S_c(x, 0, t_l)$, as a function of distance, x.

In order to fit the data to Equation 3 it is necessary first to determine ω and U. From Equation 3 and the data, ω can be found by using values of $S_c(x, 0, t_l)$ and x for which $F = 2$, i.e. the steady-state region where $x/x_l \leqslant 1 - 2{\cdot}326\omega/U$ (care must be taken, however, not to select a concentration too close to the source since the model is a point-source model); ω so determined was 1·22 cm/s. The magnitude of U was determined indirectly from the data as follows. For a given value of F, the associated value of x/x_l depends on ω/U everywhere except for $F = 1$. At this point ($x = x_l$), the concentration is exactly half the steady-state value. From Fig. 2, $S_c(x_l, 0, t_l)$ and x_l were estimated to be 0·13 μg/l. and 4755 m respectively. From the definition of x_l, U was computed to be 4·87 cm/s, and ω/U was therefore 0·250.

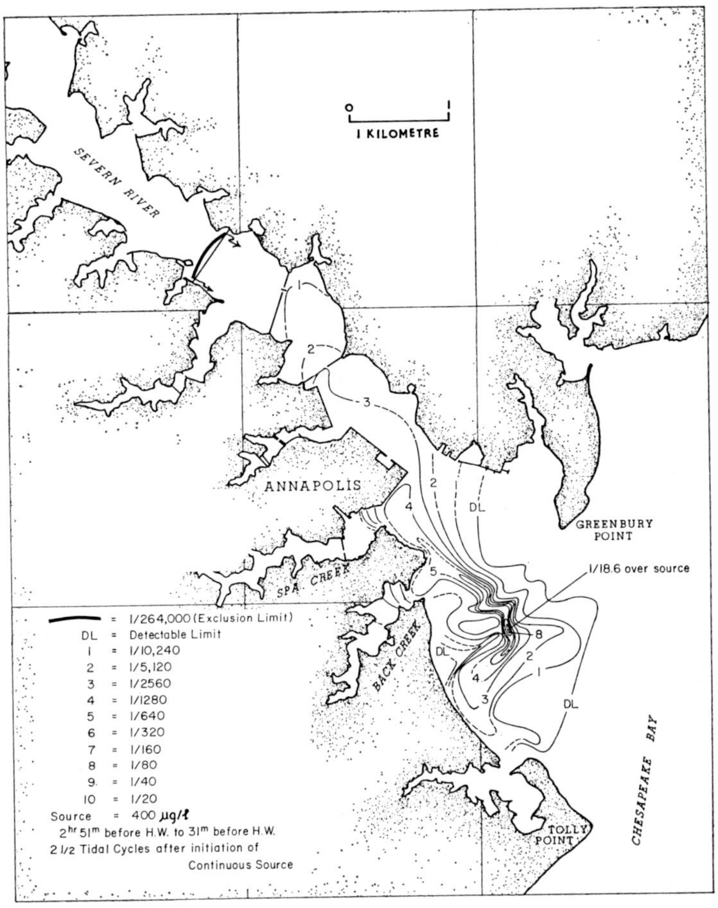

FIG. 1. Horizontal distribution of Rhodamine-WT at depth of 1 m at end of flood tide, Annapolis, Maryland, 25 March 1965

From Equation 3, setting $x = x_l$ and $F = 1$, we obtain

$$S_c(x_l, 0, t_l) = \frac{q}{2\sqrt{\pi\omega x_l D}}, \tag{4}$$

or resubstituting Equation 4 in Equation 3 we have

$$\frac{S_c(x, 0, t_l)}{S_c(x_l, 0, t_l)} = \frac{F(x/x_l, U/\omega)}{x/x_l}. \tag{5}$$

Figure 3 is a plot of $S_c(x, 0, t_l)/S_c(x_l, 0, t_l)$ as a function of x/x_l for $\omega/U = 0{\cdot}250$ together with the data points from the experiment. Since the lowest measured concentration was 0·04 μg/l., this point represents a dilution of 10 240 and $S_c(x, 0, t_l)/S_c(x_l, 0, t_l)$ is 0·30. Therefore

$$\frac{0{\cdot}30}{1/10\,240} = \frac{S_c(x, 0, t_l)/S_c(x_l, 0, t_l)}{1/264\,000},$$

or

$$\{S_c(x, 0, t_l)/S_c(x_l, 0, t_l)\}_{\text{dilution}=264\,000} = 0{\cdot}0116$$

which (from Fig. 3) occurred at $x/x_l = 1{\cdot}42$, or 6752 m from the source—as indicated by the thick curve in Fig. 1.

At times a question arises as to the effect of a *proposed* discharge on an existing resource such as a bathing beach or water intake. In this event, the problem is to estimate the probability that effluent will be found at x_i in the receiving waters together with its concentration and 'age' distribution. To do this we proceed as follows.

The mass emission rate of the source, q, may be written as

$$q = S_0 Q_c \tag{6}$$

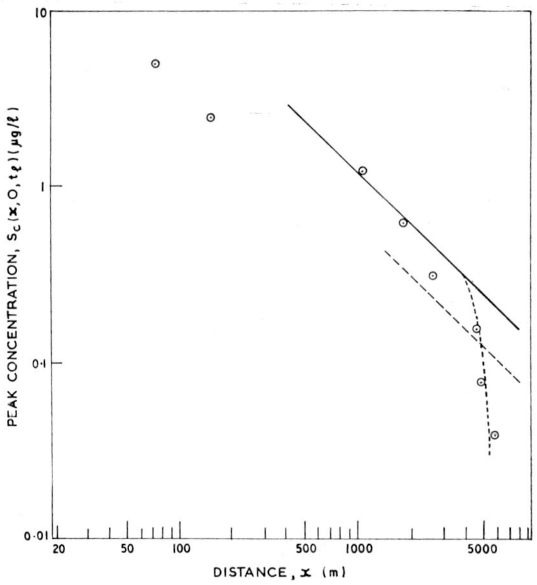

FIG. 2. Peak concentration, $S_c(x, 0, t_l)$, as function of downstream distance, x, from data of Fig. 1
Continuous line: steady-state relation
Broken line has peak concentration half that of continuous line
Dotted curve for ω/U equal to 0·250

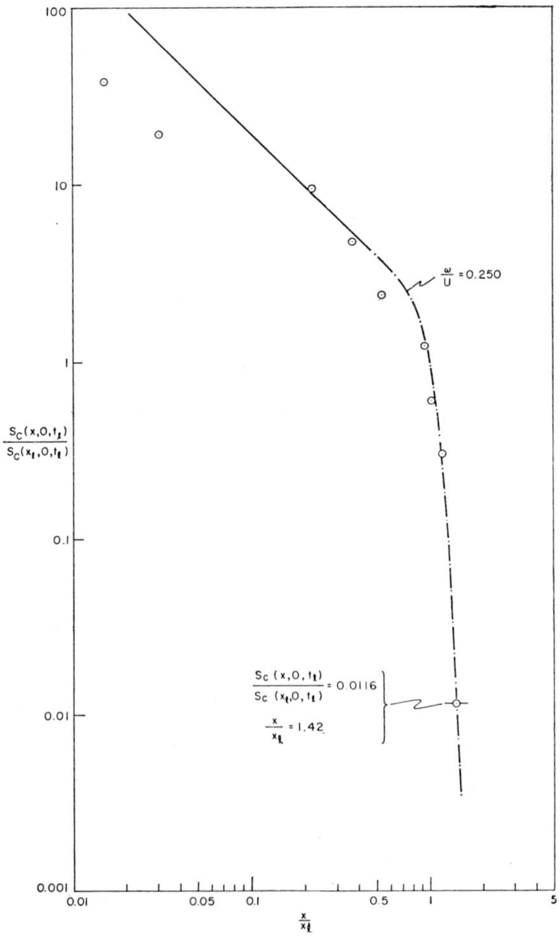

FIG. 3. Normalized peak concentration, $S_c(x, 0, t_l)/S_c(x_l, 0, t_l)$ as function of normalized downstream distance, x/x_l, from data of Fig. 2

where S_0 is the initial concentration of some constituent in the effluent such as *Escherichia coli* and Q_c is the effluent flow rate. From Equations 3 and 6 we have

$$\text{Relative concentration} \equiv S_c(x_i, 0, t^*)/S_0 = (\text{Relative concentration})_{t^*=\infty}\, \tfrac{1}{2}F(\theta) \qquad (7)$$

where Relative concentration is defined as $1/(1 + \chi)$, χ being the volume of receiving water into which unit volume of effluent is mixed, and t^* is the time since initiation of the velocity U or its duration. At x_i, and for a constant value of U/ω, it may be seen from Equation 7 that the concentration varies from essentially zero to its steady state value as θ varies from $+2{\cdot}326$ to $-2{\cdot}326$.

Rearranging the definition of θ, we have

$$\theta \equiv \frac{U}{\omega}\left(\frac{x}{x_l} - 1\right) = \frac{x_i}{\omega t^*} - \frac{U}{\omega}. \qquad (8)$$

Figure 4 is a plot of Equation 8 with θ set equal to $\pm 2{\cdot}326$; it should be interpreted as follows.

For a given U, the first curve encountered as one increases t^* defines the arrival of the plume front, i.e. the largest duration associated with zero concentration. The last curve marks the time at which steady-state dilution is achieved. The hatched area between is a zone where the concentration varies from zero to its steady-state value. To illustrate the application of the curves in Fig. 4, the diagram also includes actual current-meter data taken in Lake Michigan off Zion, Illinois, during 1973 at stations located approximately 900 m offshore in 6 m of water. The coastline is oriented in a north–south direction and x_i is assumed to be south of the source; therefore, the current-meter records were processed to give the duration and mean velocity of every period of southerly flow. It should be noted that on only three occasions did the plume reach the point x_i. In preparing Fig. 4 the highest reasonable estimate of ω has been used for the arrival curve and the lowest estimate for the steady-state curve for conservatism. In addition, the values of $1 + \chi$ shown are half those computed from Equation 7—on the assumption of complete reflection of the effluent at the shoreline.

Age is quite as important in assessing the effects of a sewage discharge as is concentration or dilution. If the effluent is very old, bacteria and viruses have been reduced by die-off and/or grazing, suspended solids have settled out, BOD has been reduced, etc. The steady-state age distribution of the effluent in unit volume of the effluent collected at x_i, however, can be estimated from the diffusion model as follows.

It is assumed that the source is started, or the flow initiated, at $t = 0$ and that observations are made at $t = t^*$. At x_i and time t^*, $S_c(x_i, 0, t^*)$ is made up of patches whose age, τ, varies from 0 to t^*; see

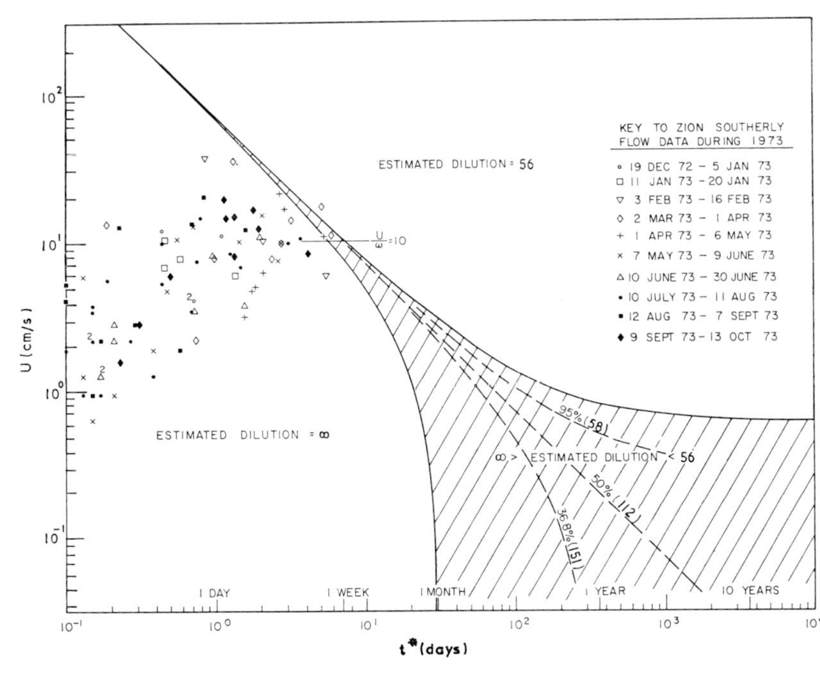

FIG. 4. Dilution, $1 + \chi$, as function of velocity, U, and duration, t^*, for $x_i = 0{\cdot}6 \times 10^5$ m, $D = 4{\cdot}57$ m, $Q_c = 10{\cdot}95$ m^3/s, and $\omega = 0{\cdot}25$–$1{\cdot}00$ cm/s

Plotted points from current-meter data in 6 m of water in Lake Michigan, about 900 m offshore of Zion, Illinois

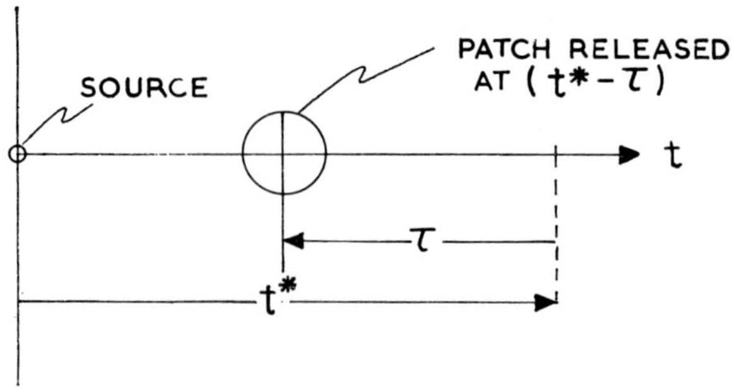

FIG. 5. Definition sketch for determining patch age

Fig. 5. At steady state, i.e. t^* large, the contribution of the patch released at $t = t^* - \tau$ to the total concentration $S_c(x_i, 0, t^*)$ measured at x_i is given by

$$\frac{\partial S_c(x_i, 0, t^*)}{\partial \tau}\,\delta\tau = S_i(x_i, 0, \tau)N\,\delta\tau, \tag{9}$$

since

$$S_c(x_i, 0, t^*) = \int_0^{t^*} S_i N\,dt = \int_0^{t^*} S_i N\,d\tau$$

from Equation 2.

The fractional contribution, f, is thus

$$f = \frac{S_i(x_i, 0, \tau)N\,\delta\tau}{S_c(x_i, 0, t^*)}\ . \tag{10}$$

This contribution may be computed as a function of τ, i.e. age, from Equation 1 (setting $y = 0$ and remembering that $q = Nm$) and from Equation 3 (setting $F(\theta) = 2$ and $x = x_i$). A sample calculation for $x_i = 0{\cdot}6 \times 10^5$ m, $\omega = 1$ cm/s, and $U = 10$ cm/s is shown in Table 1.

TABLE 1. Fractional contribution, f, as a function of age for $U/\omega = 10$

τ (days)	f	Per cent
5·45	0·000345	0·03
5·95	0·033	3·3
6·45	0·260	26·0
6·95(x_i/U)	0·410	41·0
7·45	0·230	23·0
7·95	0·060	6·0
8·45	0·010	1·0
8·95	0·002	0·2
Total	1·005	100·5

TABLE 2. Fractional contribution, f, as a function of age for $U/\omega = 0$

τ (days)	f	Per cent
27·95	0·00103	0·10
37·95	0·00948	0·95
47·95	0·021	2·1
57·95	0·028	2·8
69·51	0·035	3·5
79·51	0·029	2·9
89·51	0·027	2·7
99·51	0·024	2·4
109·51	0·022	2·2
119·51	0·0195	1·95
129·51	0·0176	1·76
139·51	0·0158	1·58
149·51	0·0141	1·41
159·51	0·0128	1·28
>164·51	0·724	72·4
Total	1·000	100·0

For the situation shown in Fig. 4, it is possible that for large t^*, $\int_0^{t^*} U\,dt \approx 0$. In this case, effluent will reach x_i by diffusive processes alone (from this point of view all advective fluxes considered to be diffusive), the leading edge arriving at x_i in approximately one month according to Fig. 4. The maximum dilution will be 112 as the minimum value of θ is 0 at $t^* = \infty$. The effluent, however, will be quite old as may be seen from Table 2, where each contribution (except the last) is for $\delta t = 10$ days, centered on the value of τ shown.

Some Additional Comments

It is not the intent of this paper to argue either for or against water-use standards but rather to present a reasonably objective methodology for the application of such standards. It should be pointed out, however, that the intent of the original federal (U.S.) regulation requiring the creation of exclusion areas was not to withdraw large areas from active harvesting but rather to require proper operation of treatment facilities through redundant systems for guaranteed post-chlorination, holding-ponds with 24-h storage capacity, and competent operators. In applying the regulation, the path of least resistance has been largely followed and exclusion areas created.

In the second application presented, a method of estimating effluent age was described. In my opinion, age together with pathogenic specificity receives insufficient attention in the establishment of water use standards. We must somewhow get away from coliform standards to organism specific standards with proper allowance made for die-away.

ACKNOWLEDGEMENTS

Contribution No. 206 from the Chesapeake Bay Institute, The Johns Hopkins University. This work was supported by the Department of Health and Mental Hygiene, State of Maryland.

REFERENCES

1. OKUBO, A., and PRITCHARD, D. W. Unpublished note. (See PRITCHARD, D. W. The application of existing oceanographic knowledge to the problem of radioactive waste disposal into the sea. In 'Disposal of radioactive wastes, Vol 2'. International Atomic Energy Agency, Vienna, 1960, pp. 229–253.)
2. CARTER, H. H. A method for delineating an exclusion area around a sewage outfall in a tidal estuary based on water quality with application to the Severn and Choptank Rivers. Chesapeake Bay Institute, Johns Hopkins University, Spec. Rep. 9, 1965.
3. CARTER, H. H. The measurement of rhodamine tracers in natural systems by fluorescence. *Rapp. P.-v. Réun. Cons. int. Explor. Mer*, 1975, **167**, 193–200.

In presenting his paper Mr. Carter made an initial comment to the effect that although there was not a consensus regarding the necessity for standards of dilution or treatment insofar as bathing beaches were concerned, there seemed to be general agreement that shellfish waters deserved special attention because of the tendency of shellfish to concentrate pathogens. He stated that the model described in the paper was created in response to water quality regulations for the co-existence of shellfish harvesting areas and sewage outfalls. He then outlined the model, illustrated both dye-tracer (Annapolis example) and current-meter (Zion example) modes of application, and described a method of estimating the steady-state age distribution at a given position from the model.

A concluding remark was made to the effect that in some shallow coastal areas, such as Maryland coastal waters, design criteria should be based on far-field effects because of (1) the possibility of intermittent coupling between the near and far fields, and (2) the process of upwelling. The former process negated the assumption of an infinite supply of diluting water inherent in most near-field models, and the latter resulted in the entire sewage field being carried to the beach as a single water mass.

DISCUSSION

Mr. B. Grover (World Bank) asked whether the Maryland water quality standard for shellfish waters defined, in a vertical profile, the sampling points for water 'overlying the bars or beds'. As the interpretation apparently related to sampling at the surface, Mr. Grover wondered if this simplification was rational—or at least scientifically rigorous—for protecting such bottom feeders; certainly, one did not sample on the bottom when considering standards for recreational waters.

Dr. R. J. Callaway (U.S.A.) asked if Mr. Carter had used numerical models, such as Leendertse's, in the area he had described, and (if not) if he thought anything would be gained by the application of such models.

Mr. P. Ackers (U.K.) pointed out that the models described in Papers 35 and 36 were examples of two radically different approaches. Mr. Munro had used a finite-difference method with discrete time steps and a box-like division of his two-dimensional space in the sort of model ideally suited to computer solution, whereas Mr. Carter's model was basically the use of an analytical formula with the unknown coefficient in it determined by one experiment using a conservative tracer. The impression had been given that the Munro model gave a very poor fit and the Carter model a good fit, but Mr. Ackers suggested that this impression might be unfair, because it appeared that in Mr. Carter's paper the test was of how well the field data from a conservative tracer experiment fitted the equation

—having fed into the equation a value for the unknown diffusion velocity that provided a reasonable fit for that particular experiment. He said he would be interested to learn how well the method predicted the three-dimensional field of a non-conservative material (such as coliform bacteria) for different tides, river flows, temperatures, winds, etc., and covering a two-dimensional area including regions as close to the outfall as Mr. Munro's model sought to represent.

Mr. R. A. Apostol (U.K.) reported that he and his colleagues had solved the diffusion–convection equation, as applied to a continuous steady-state plume directly by a finite-element method, and arrived at a reduction in concentration with distance along the plume axis approximately similar to that shown in the steady-state portions of Figs. 2 and 3. The advantage of the direct solution method was that the idealizations required to convert the instantaneous case into the continuous discharge case were avoided.

Dr. J. F. Macqueen (U.K.) asked the basis for the estimates of the parameter D associated with the line-source strength, and if it was regarded as constant.

Dr. R. E. Lewis (U.K.) noted that Mr. Carter had stated that far-field predictions should be used for outfall design. Since Prof. Foxworthy's earlier comments (p. 293) had suggested that the total dilution did not vary greatly with changes in initial dilution, he asked if this meant that the degree of initial dilution did not matter.

REPLY TO DISCUSSION

Replying to Mr. Grover, Mr. Carter said the Maryland standards (Regulation 08.05.04.03) related to 'receiving waters' and 'shellfish harvesting waters'. A reasonable interpretation of the pertinent regulation would suggest that bacteriological sampling should take place in the waters immediately over the bars or beds. In the case of the tracer sampling for his model, the sampling was three-dimensional with the horizontal tracer distribution measured at the near surface, but with sufficient vertical sampling to check that the near-surface distribution was representative of conditions adjacent to the bottom. Since the 24-h exclusion limit was always greater than a tidal excursion away from the outfall, it was well beyond the zone of significant deposition of sewage solids—at least from the source being considered.

In reply to Dr. Callaway he said a numerical model such as Leendertse's had not been used since it seemed a complicated way of solving a relatively simple problem. In addition, by tagging the sewage effluent continuously for 24 h one was assured that the absence of tracer was indicative of, at most, extremely low concentrations (of the order of 1 in 10^{11}) of effluent younger than 24 h.

As he interpreted Mr. Ackers's comment, the contrast was between Mr. Munro's and his own approaches to the same problem—namely the prediction of bacterial concentrations. Such a comparison was probably inapplicable, since his own model or approach delineated a zone outside which there was little probability of the presence of effluent which had emanated from the particular source within the past 24 h, whereas Mr. Munro had been faced with the problem of predicting the distribution of coliform bacteria from whatever source and of any age, which was, of course a much more general (and hence considerably more difficult) problem. With regard to further testing of the approach described in the paper, it had been used extensively in Maryland, some 8–10 such experiments had been performed with results comparable with those given in the paper.

In reply to Dr. Macqueen he said he had considered D—the depth of the system—to be a constant. In point of fact, it was probably more correct to state that the data provided an estimate of ωD rather than of ω, and that ω was then estimated using a representative value of D associated with the location of the tracer plume and its vertical distribution. Thus D was more properly an effective depth, or the depth of an equivalent constant-depth system.

Regarding Dr. Lewis's comment, Mr. Carter said that his statement about designing outfalls and far-field effects had been a general one, but it applied only to certain shallow coastal situations where consideration of far-field effects would be required for determining the optimum diffuser lengths. At great distances, the dilution was essentially independent of the initial value, and he considered that economics required consideration of the far field if deep water was not close at hand.

Numerical Modelling of Dispersion in Estuaries

Hugo B. Fischer
Professor of Civil Engineering, University of California
Berkeley, California 94720, U.S.A.

This paper considers the dispersion of pollutants beyond the zone of initial mixing. The emphasis is on estuaries though some of the methods described may also be applied in coastal waters. We consider wastes whose presence does not change the dynamics of the recipient flow. Where buoyancy of the waste is important, as near the source of a warm-water discharge, other techniques are required.

Methods for predicting dispersion may be grouped into three categories: the use of hydraulic models, the use of analytic models, and the use of numerical computer models. This paper will discuss only numerical models. Analytical models are omitted because their use is generally rather limited, while hydraulic models are omitted because their uses and limitations in dispersion studies have received substantial recent discussion elsewhere[1,2]. The discussion of numerical models will further emphasize those models sufficiently simple to be immediately applied.

Modelling is as yet a process of approximation. It is not yet thought possible to solve the three-dimensional equations of motion to determine completely the velocity distribution in space for even the smallest real estuary. Nor, even if an exact velocity distribution were available, would it be possible to model advective diffusion exactly. Thus all numerical models must adopt some level of approximation, both in the choice of spacing between grid points and in the number of mechanisms and dimensions to be included. Models vary widely in terms of cost and accuracy. In most cases, however, greater accuracy is achieved with increased cost. Nevertheless, certain types of estuary can be modelled with reasonable accuracy by some fairly simple models. The problem is to fit the model to the estuary. In this paper, we discuss several rather simple models which have been found to give good predictions in certain estuaries.

Dispersion Mechanisms

There are four primary dispersion mechanisms active in estuaries: circulation induced by the density difference between fresh and ocean water (gravitational circulation), dispersion through cross-sectional gradients in a relatively uniform channel (Taylor shear-flow dispersion), circulation induced by tidal flow in the estuarine bathymetry, and circulation derived from wind action. Models suited to each of these mechanisms will be discussed in turn.

Gravitational circulation

Gravitational circulation is the flow induced by the density difference between ocean water at one end of the estuary and fresh water at the other. The flow is predominantly landward at depth and seaward near the surface (the vertical gravitational circulation). The depth-averaged flow is predominantly landward in the deeper parts of the cross-section and seaward in the shallower parts. This depth-averaged movement has been called the transverse gravitational circulation and is believed to be more important than the vertical circulation in most real estuaries[3]. A sketch of the transverse gravitational circulation is shown in Fig. 1.

Whilst there have as yet been no completed laboratory studies of the effect of the transverse gravitational circulation, vertical gravitational circulation has been studied in two extensive sets of experiments[4,5], and in several analyses[6]. Figure 2, a graph of a parameter expressing the length of salinity intrusion versus a Richardson number, summarises the experimental results. In most real estuaries, however, the results shown in this figure are probably inadequate, because the original experimental designs excluded most elements of transverse circulation. In the writer's opinion, the vertical circulation will be dominant only in a highly stratified estuary such as, perhaps, Puget Sound. In less well stratified estuaries, transverse circulations are likely to be more important. Thus vertically two-dimensional models are probably not adequate. Laboratory studies of transverse gravitational circulation are now in progress at the University of California, Berkeley, under the writer's supervision. Results, however,

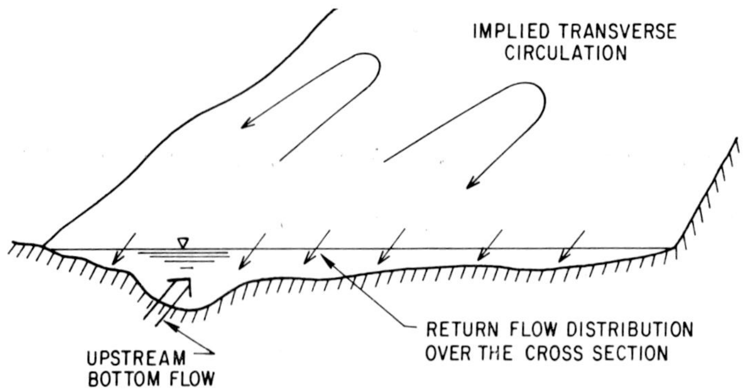

FIG. 1. Schematic illustration of transverse gravitational circulation

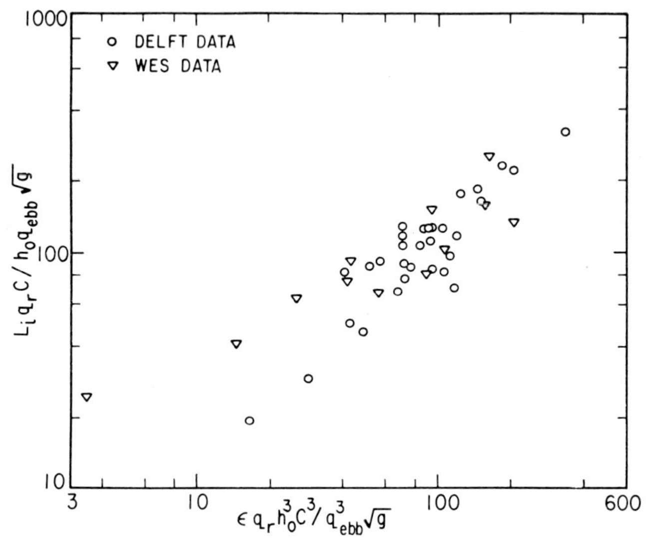

FIG. 2. Correlation of experimental observations of salinity intrusion due to vertical gravitational circulation[7], using results of Delft[5] and WES[4] experiments
L_i = length of salinity intrusion, q_r = river discharge per unit width, q_{ebb} = peak tidal discharge per unit width, C = Chezy coefficient, h_0 = mean depth, ϵ = dimensionless density difference between river and ocean water

are not yet available. Pending further research, it is probably correct to say that in most cases the effects of gravitational circulation cannot be modelled with any degree of confidence.

Shear-flow dispersion

Models of shear-flow dispersion are based on G. I. Taylor's classical analysis of longitudinal dispersion in a pipe[8]. In this analysis dispersion is produced by the cross-sectional velocity gradient, whose effect, however, is reduced by the movement of particles between stream lines of different velocities.

Taylor's analysis has been applied to steady flow in streams by Fischer[9], and to oscillating flow by Holley, Harleman, and Fischer[10]. In steady flow, as in a river, the dispersion coefficient is found to be proportional to the square of the width of the channel. In oscillating flow, however, there is an optimum width beyond which the dispersion coefficient decreases. Figure 3 shows a comparison of dispersion in steady and oscillating flows in a wide channel. Wide is defined as being of that width at which the time scale for mixing across the channel, W^2/E_z, (where W is the width and E_z is the coefficient for turbulent mixing across the channel) is much greater than the period of oscillation.

In estuaries where there is a long, reasonably uniform reach of channel, shear-flow dispersion may be the most important mechanism to be modelled. For instance, the next section of this paper describes a model of a reach of the Delaware Estuary which includes only the shear effect and turbulent mixing and which successfully described a prototype dye study.

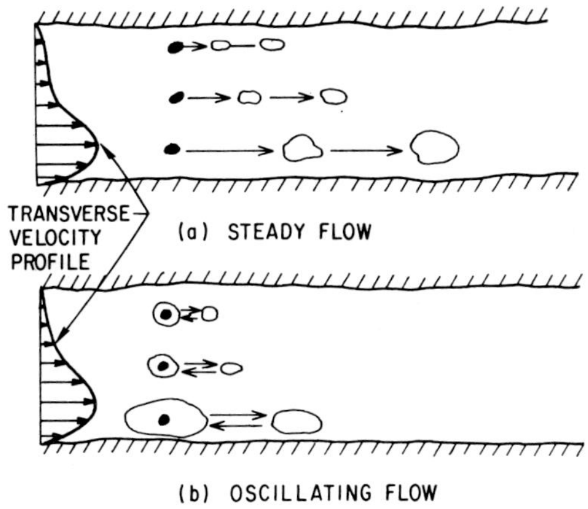

FIG. 3. Longitudinal dispersion in steady and oscillating flows

Bathymetrically induced dispersion

Much of the circulation and salinity intrusion in some estuaries comes from the interaction between the tidal movement and the boundary irregularities. We may differentiate two sub-types. Firstly, it is often observed that the tide tends to flood more in some channels and to ebb more in others. The general pattern of flood and ebb currents is not the same. Two examples of this type of flow are shown in Figs. 4 and 5. The first shows a typical flow through an opening; the inflow is as a jet whilst the outflow is as a potential flow to a sink. The second figure shows observed flood and ebb channels in the Mersey Estuary. Since the inflow is concentrated in certain channels, salinity or any other dissolved substance tends to intrude along these same channels.

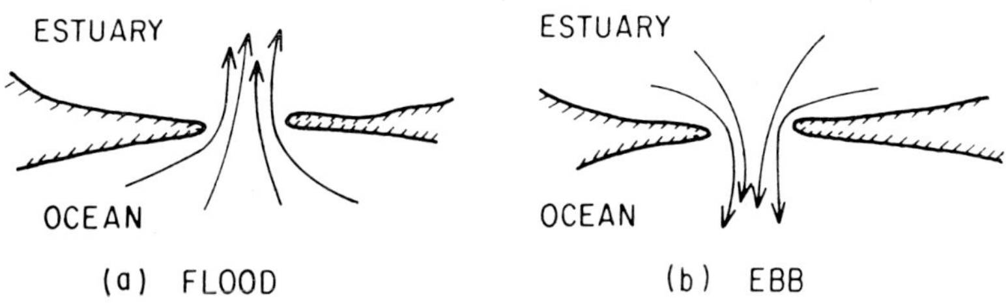

FIG. 4. Flood and ebb flow patterns for flow through a constriction[11]

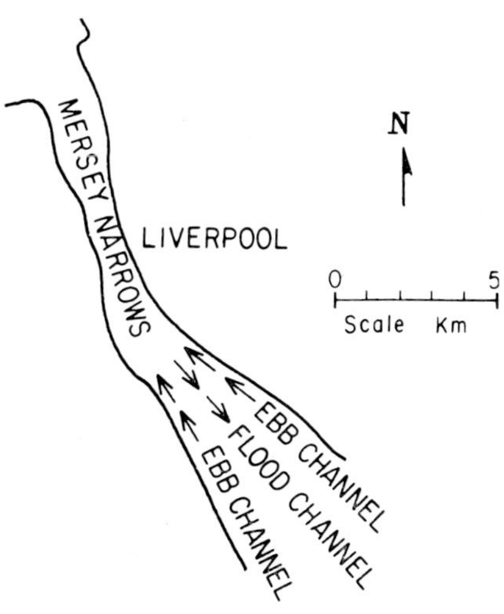

FIG. 5. Flood and ebb flow patterns in Mersey Estuary[12]

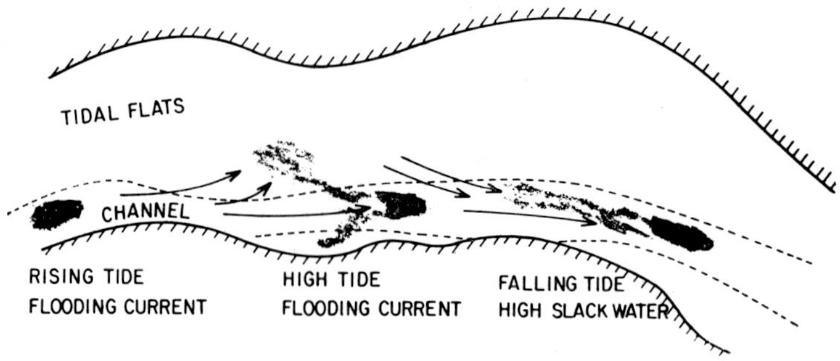

FIG. 6. Illustration of bathymetrically induced dispersion

The second sub-type of bathymetrically induced dispersion has been described by Schijf and Schönfeld[13]. In most estuaries, high tide and high slack water (or low tide and low slack water) do not occur at the same time. The result is shown in Fig. 6. Tidal flats or embayments along the sides of an estuary act as storage areas for constituents, but since they fill and empty out of phase with the flow in the main channel, they act to disperse a cloud of pollutant, or to assist in salinity intrusion. Figure 7 shows the concept, and Fig. 8 the results of a simple numerical model of a main channel and tidal flats.

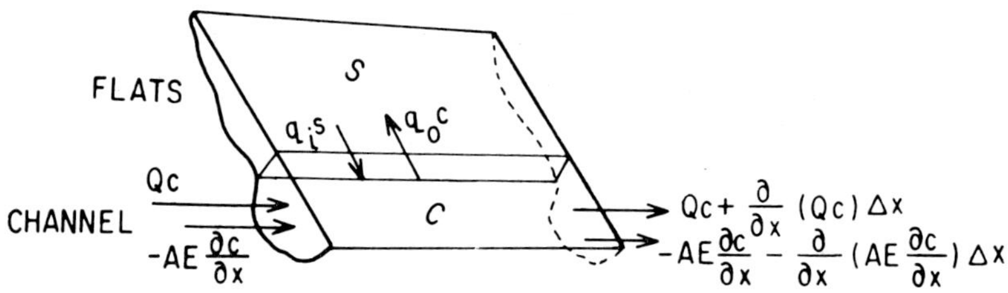

FIG. 7. Schematization of an estuary into a channel and tidal flats, for use in a simplified model

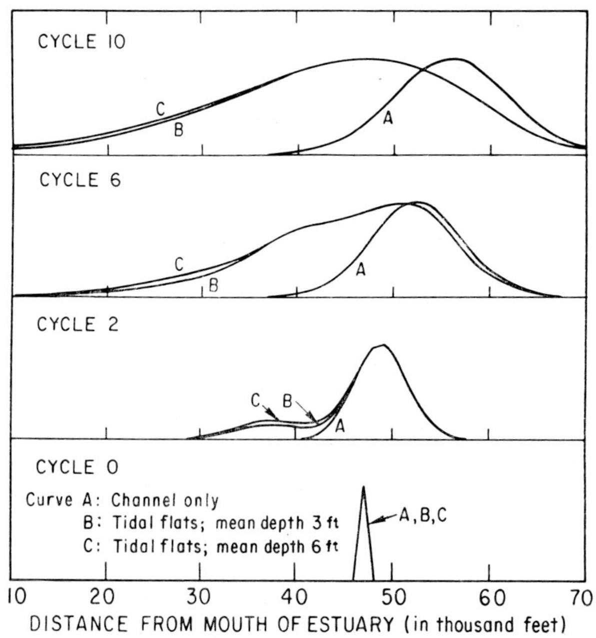

FIG. 8. Effect of tidal flats on longitudinal dispersion, as computed by a simplified model

The flats carry no longitudinal flow, but fill and empty according to the propagation of the tide in the main channel. Within the main channel and within the flats, concentration is assumed to be sectionally well mixed. In Fig. 8, Case A shows dispersion in a channel 2000 ft wide, 30 ft deep with a tidal amplitude of 3 ft and an assumed longitudinal dispersion coefficient of 40 ft^2/s. In Cases B and C, tidal flats 2000 ft wide are added; in Case B the flats are 3 ft deep and in Case C they are 6 ft deep (at mean tide). The figure makes clear that the flats greatly increase the apparent rate of longitudinal dispersion. Calculated apparent dispersion coefficients for the channel with flats are approximately 200 ft^2/s, or five times the dispersion coefficient assumed for the channel without flats.

In some bays, the channel geometry includes braided channels as well as tidal flats. The upper reaches of San Francisco Bay and the Sacramento San Joaquin Delta in California are examples. Tidal 'pumping' occurs around islands and through sloughs causing net circulations which may be a major source of salinity intrusion. This type of circulation can be modelled fairly well by existing hydrodynamic models. The effect on dispersion can be approximated by a segmented transport model such as the Lagrangian scheme described by Fischer[14] or the finite-difference scheme developed by Orlob *et al.*[15], or by a fully two-dimensional transport model such as Leendertse's model of Jamaica Bay[16].

Wind-induced circulations

Mixing induced by wind is sometimes over-rated; wind waves, unless they break, cause little mixing. Steady, large-scale circulations, however, can be induced by wind in a bay of variable depth and can be an important source of pollutant dispersion. A steady wind exerts a constant stress everywhere on the water surface. In shallows, the water has less mass per unit surface area and can accelerate faster in response to the wind. A steady circulation results from a steady wind, the direction being with the wind in the shallows and against the wind in the deeps. Figure 9 shows an example of predicted circulation around an island. The cause was a constant wind and a channel depth greater on one side of the island than on the other. Circulations of the type shown in Fig. 9 can be modelled by any of the existing two-dimensional hydrodynamic models, for example, the one developed by Leendertse[17].

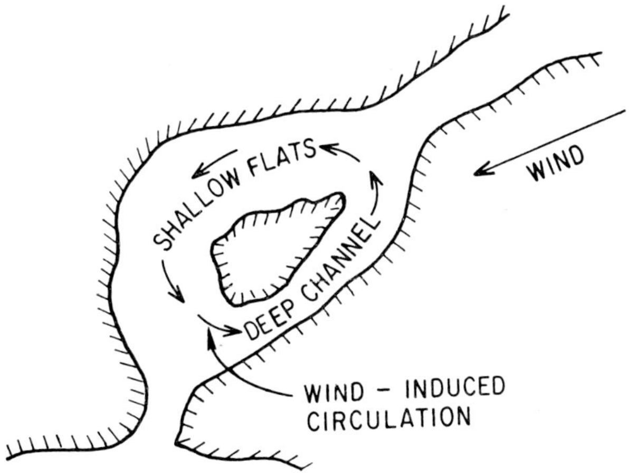

FIG. 9. Illustration of wind-induced circulation

Selection of a model

The best numerical model is one which models only the important dispersion mechanisms and omits the others. The most difficult step in the modelling process is often the definition of the important mechanisms. We can give no firm rules, but the following may help in some cases:

(a) If the estuary is quite deep and has a substantial fresh-water inflow, gravitational circulation may be dominant. One should be aware, however, that even though the gravitational circulation in a given estuary may be measurable (see, for instance, Pritchard's study of the James River[18]), other dispersion mechanisms may be substantially more important. One way to test for the importance of gravitational circulation is to use an hydraulic model, if one exists. For example, Fischer and Dudley[19], by varying the ocean salinity in an hydraulic model and observing no change in relative salinity intrusion, showed that gravitational circulation is not important in San Francisco Bay.

(b) If the estuary has a relatively uniform cross-section, with few side irregularities, embayments, or tidal flats, shear-flow dispersion may be the most important. The next section describes a simple shear-flow dispersion model which was used successfully in the Delaware River.

(c) Otherwise, channel bathymetry effects are probably the most important. A tidal-flat model such as the one used to generate the results shown in Fig. 8 can be extended to an irregular geometry of tidal flats, as long as there is only one channel. Where there is more than one channel a combination of sectional models and tidal-flat models may be possible. If wind-driven circulations are important, or if the bathymetry is too complex for a combination of simple models, one may need to use a completely two-dimensional model.

None of the present models, however, can successfully describe both an irregular boundary geometry and vertical stratification of sufficient magnitude to invalidate a two-dimensional description. At present, the only likely approach for a semi-three-dimensional model may be to combine what is known analytically of the gravitational circulation with a numerical two-dimensional treatment of the bathymetry.

A Simplified Numerical Model

This section describes a numerical model, based on the use of stream tubes, which has been found useful in a study of dispersion of waste in the Delaware estuary. The model includes the effect of shear-flow dispersion, and could be extended to include the transverse gravitational circulation (although this has not yet been done).

We begin by dividing the cross-section into stream tubes separated by vertical surfaces, as shown in Fig. 10. Within each stream tube the flow is assumed to be well mixed. The velocity of flow in the ith stream tube is given by

$$u_i = a_i \sin\left(\frac{2\pi t}{T} + k_i\right) + U_i \tag{1}$$

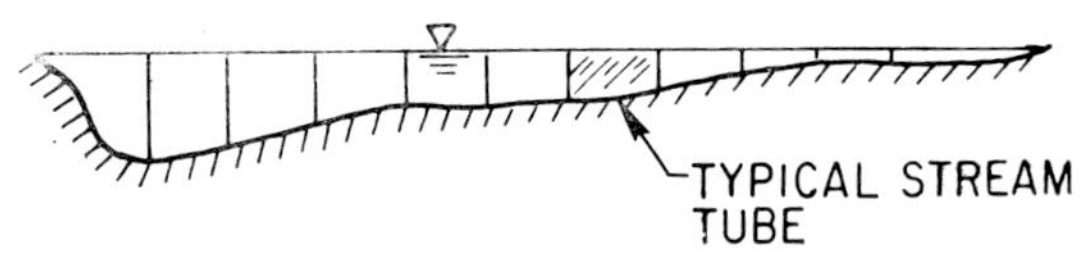

FIG. 10. Division of channel into stream tubes

in which t is time and T is the tidal period. The constants a_i, k_i, and U_i are based on field measurements; a_i is the amplitude of the tidal velocity, and of course is smaller in the shallower parts of the cross-section; k_i is a phase angle to account for the fact that the velocity reverses sooner in shallower sections than in deeper sections (in the Delaware River the velocity reverses at the banks approximately 3 h before it reverses in the center); U_i is a net downstream velocity component to account for the fresh-water discharge—U_i can also be varied across the river to model transverse gravitational circulation.

The computer program consists of two main parts, firstly a step in which concentrations are advected along the stream tubes at the velocities given by Equation 1, and secondly by a step in which transverse mixing takes place between stream tubes. Mass transport across each stream tube boundary is given by

$$M_i = \epsilon d_i \frac{\Delta c}{\Delta z} \Delta t\, \Delta x \tag{2}$$

in which ϵ is the transverse mixing coefficient, d_i is the depth at the stream tube boundary, Δx is the longitudinal grid spacing, Δz is the transverse grid spacing, Δt is the time step, and Δc is the concentration difference between adjacent stream tubes. Concentrations in each stream tube are adjusted to account for the mass transport. The details of this step have been described more fully (for steady flow, but this step is the same) elsewhere[20].

To study the dispersion of a slug of contaminant, the program is started with zero concentration at all grid points except one whose concentration represents the mass of the slug. A continuous source of contaminant is modelled by increasing the concentration at a grid point by a fixed amount at the end of every time step. Figure 11 shows the results of use of the program to model the dispersion of a continuous source of material in the Delaware River. For this application the time step was 15 min, the longitudinal distance step was 2000 ft, and the river was divided into 18 stream tubes each 250 ft wide.

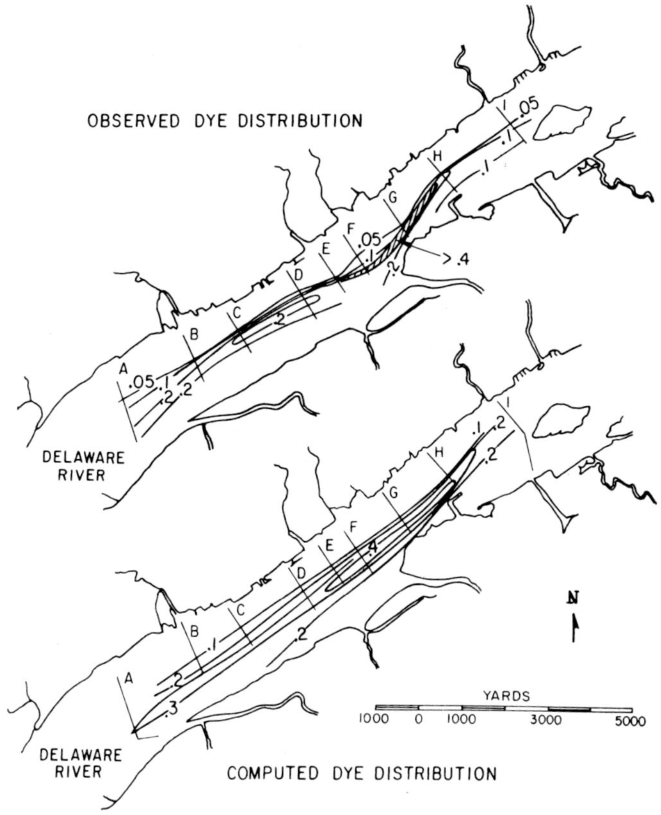

FIG. 11. Computed and actual dye distributions in Delaware River

Prototype river velocity measurements were made at two cross-sections, and the results used as a basis for the parameters in Equation 1.

Figure 11 also shows the results of a dye study in the prototype river, carried out during tidal and fresh-water flow conditions approximating those of the computer run The dye study was conducted over a period of ten days. During the first 6 h, starting at low slack tide, a high rate of dye injection was used to define the initial plume. The spread of the plume, measured carefully over six cross-sections upstream from the injection, was used to compute the transverse mixing coefficient for use in the program. The result was $\epsilon = 1{\cdot}2du^*$, where d = depth and u^* = shear velocity; this corresponds to 9·6 ft^2/s in the 40-ft deep section of the channel. The computer program was then run to simulate the ten-day dye injection. As can be seen in the figure, the agreement between computer-predicted and actual results is excellent. All the main features of the concentration distribution are reproduced, including the rate of spread across the stream, the advection downstream, and the transient plume of high concentration.

Figure 11 is typical of concentration distributions predicted at high and low slack tides for several days. We may conclude that in this reach of the Delaware River, the program described in this section works well. The reason is that the most important transport mechanisms in this reach appear to be those selected for modelling, the advection in stream tubes with a velocity differential, and the mixing between stream tubes. It seems reasonable to conclude that this model will work well in any tidal river with a reasonably uniform channel; it may also work well along a coast if the oscillating tidal currents are well defined and reasonably uniform. It would not work well, however, in a large irregular bay or along a jagged coast with deep inlets, or islands. The fundamental requirement of the model is that stream tubes can be drawn; wherever that is true, a computer model based on stream tubes is likely to be efficient.

Summary

This paper has discussed the use of simplified numerical models to predict the dispersion of pollutants in estuaries and, by extension, in coastal areas. No single model will be optimal for all problems. The first step in analysing a dispersion problem is to determine the important mechanisms, primarily to decide which may be omitted from a computer model. The second step is to select from the variety of models now available that which is most appropriate to the problem at hand. This paper has described one model which is particularly efficient for uniform reaches of tidal rivers, and along uniform coastal areas; reference has been given to other models useful for other problems. Some general, although far from complete, suggestions have been given for the selection of the most efficient model. It remains nevertheless a matter of judgement and experience to select the most efficient model for any given problem, and to be aware of its limitations.

ACKNOWLEDGEMENTS

The simplified numerical program described in this paper was developed by the writer for the firm of Dames and Moore, Consulting Engineers. The prototype dye study was conducted by the Raytheon Corp., and the prototype velocity measurements were taken by Alpine Geophysical Corp. Dr. Yuen Jen, of Dames and Moore, did much of the application of the program and obtained the results shown in Fig. 11. Permission to reproduce the results is greatly appreciated.

Figures 7 and 8, and the development of the simplified tidal flats model, are from a study by Mr. John Hamrick, a doctoral student at the University of California, Berkeley.

Funds for the preparation of this paper were provided by the Committee on Research of the Berkeley campus of the University of California.

REFERENCES

1. Fischer, H. B., and Holley, E. R. Analysis of the use of distorted hydraulic models for dispersion studies. *Wat. Resour. Res.*, 1971, **7**, 46–51.
2. Crickmore, M. J. Tracer tests of eddy diffusion in field and model. *J. Hydraul. Div. Am. Soc. civ. Engrs*, 1972, **98** (HY10), 1737–1752.
3. Fischer, H. B. Mass transport mechanisms in partially stratified estuaries. *J. Fluid Mech.*, 1972, **53**, 671.
4. Ippen, A. T., and Harleman, D. R. F. One dimensional analysis of salinity intrusion in estuaries. U.S. Corps of Engineers, Waterways Experiment Station, Vicksburg, Miss., Tech. Bull. No. 5, 1961.
5. Rigter, B. P. Minimum length of salt intrusion in estuaries. *J. Hydraul. Div. Am. Soc. civ. Engrs*, 1973, **99** (HY9), 1475–1496.
6. Hansen, D. V., and Rattray, M. Gravitational circulation in straits and estuaries. *J. mar. Res.*, 1965, **23**, 104–122.
7. Fischer, H. B. Discussion of 'Minimum length of salt intrusion in estuaries'. *J. Hydraul. Div. Am. Soc. civ. Engrs*, 1974, **100** (HY5), 708.
8. Taylor, G. I. The dispersion of matter in turbulent flow through a pipe. *Proc. R. Soc.*, 1954, **223**, 446–468.
9. Fischer, H. B. The mechanics of dispersion in natural streams. *J. Hydraul. Div. Am. Soc. civ. Engrs*, 1967, **93** (HY6), 187–216.
10. Holley, E. R., Harleman, D. R. F., and Fischer, H. B. Dispersion in homogeneous estuary flow. *J. Hydraul. Div. Am. Soc. civ. Engrs*, 1970, **96** (HY8), 1691–1709.
11. Stommel, H., and Farmer, H. G. On the nature of estuarine circulation. Woods Hole Oceanographic Institution, Woods Hole, Mass., Reports No. 52–51, 52–63, and 52–88, 1952.
12. Bowden, K. F., and Gilligan, R. M. Characteristic features of estuarine circulation as represented in the Mersey Estuary. *Limnol. Oceanogr.*, 1971, **16**, 490–502.
13. Schijf, J. B., and Schönfeld, J. C. Theoretical considerations on the motion of salt and fresh water. *Proc. Minnesota int. Hydraul. Conv., Minneapolis*, 1953, pp. 321–333.
14. Fischer, H. B. A Lagrangian method for predicting pollutant dispersion in Bolinas Lagoon, Marin County, California. U.S. Geological Survey, Prof. Paper 582-B, 1972.
15. Orlob, G. T., Selleck, R. E., Shubinski, R. P., Walsh, F., and Stann, E. Modeling of water quality behavior in an estuarial environment. *Proc. 4th int. Conf. Wat. Pollut. Res., Prague, 1969*, pp. 845–855.
16. Leendertse, J. J. A water-quality simulation model for well-mixed estuaries and coastal seas. Vol. 1—Principles of computation. Rand Corp., Report RM-6230-RC, 1970.
17. Leendertse, J. J. Aspects of a computational model for long-period water-wave propagation. Rand Corp., Santa Monica, Memo. RM-5294-PR, 1967.
18. Pritchard, D. W. A study of the salt balance in a coastal plain estuary. *J. mar. Res.*, 1954, **13**, 133–144.
19. Fischer, H. B., and Dudley, E. Salinity intrusion mechanisms in San Francisco Bay, Cailfornia. *Proc. 16th Congr. int. Ass. hydraul. Res., São Paulo, 1975* (*in press*).
20. Fischer, H. B. Methods for predicting dispersion coefficients in natural streams with applications to lower reaches of the Green and Duwamish Rivers, Washington. U.S. Geological Survey, Prof. Paper 582-A, 1968.

Prof. Fischer introduced his paper by noting that it was possible to make a formal decomposition of longitudinal mass transport into four terms[3]: seaward transport by the mean flow, landward transport by the residual currents, tidal-cycle covariance of cross-sectional averages of salinity and velocity, and cross-sectional covariance of salinity and velocity. He stressed the importance of determining which landward transport mechanisms were relevant in each modelling study, since different numerical models were successful in modelling different mechanisms, but no model could successfully model all mechanisms at once. He then reviewed the mechanisms described in his paper, with reference to how each could be modelled numerically.

DISCUSSION

Dr. J. A. Charlton and Mr. R. A. Apostol, in a joint contribution, said it was encouraging to see that the Author was attempting to divide the various dispersion mechanisms, because an understanding of them was essential in the prediction of local pollution tidal situations. In real estuaries it was probably difficult to separate the dispersion due to transverse gravitational circulation from the bathymetrically induced dispersion, as both mechanisms arose from velocity differences along the width of the estuary. Since gravitational circulation was caused by the density difference between fresh and ocean water, it was therefore seasonal, so that the dispersion coefficient due to it might be lower during periods of low fresh-water inflow[21].

The inclusion of the primary dispersion mechanisms in numerical modelling might depend on the dimension of the model employed. A two-dimensional model, for instance, took account of whether the transverse or vertical circulation was important, whereas a purely one-dimensional model did not, since the two mechanisms were combined by the diffusion coefficients used. The effect of velocity variation in a cross-section was already taken care of by the dispersion coefficient. The predicted concentration distribution depended to some extent on the magnitude of the transverse mixing coefficient which was proportional to the product of the hydraulic radius and the friction velocity. In the paper the constant of proportionality had a measured value of 1·2. However, field studies by Crickmore[2] had given a value of 0·1, while at the Tay Estuary Research Centre the speakers had found one of 0·6. Others, like Glover[22] and Yotsukura[23] gave values of 0·6–0·7. The results of field studies thus ranged over about an order of magnitude and it would therefore be interesting (by varying the transverse diffusion coefficient by this amount) to check the sensitivity of the concentration distribution and to compare it with the change due to the inclusion or exclusion of the various dispersion mechanisms mentioned in the paper. The wide variation in the measured values indicated the necessity of calibrating the numerical model in the region of particular interest—as had been stressed in a previous paper[24].

Dr. M. Waldichuk (Canada) was particularly interested in the paper because it took account of irregularities in the shoreline which might improve dispersion—even though they complicated the model. Prof. Fischer had also noted the effect of tidal pumping; there were situations on the Canadian Pacific coast where this provided a useful mechanism for disposal of wastes into coastal channels for rapid transport out of the area into one of strong tidal current and turbulence. He wondered if Prof. Fischer had considered the effect of a right-angle bend in an estuarine channel. Inlets on the Canadian Pacific coast were long, deep, and narrow, but contained sharp bends and were highly stratified due to the fresh-water inflow. Surface circulation could be greatly complicated by shoreline configuration—as had been observed by aerial photography of silt distribution and time-lapse photography of the radar screen during drogue tracking by radar. The eddies in the lee of elbow bends and other promontories actually retarded dispersion, and this had affected dispersion of pulp-mill and mine wastes discharged into Howe Sound. He therefore asked Prof. Fischer if his model could take into account the effect of bends in the channel, and particularly the eddy formation.

Dr. J. F. Macqueen (U.K.) said that the Author, among other proponents of mathematical modelling, had pleaded for improved data from large field experiments. He asked Prof. Fischer if he believed that the mathematical models developed really deserved the financial commitment this implied, and if (while awaiting results from such surveys) he saw any ways of improving the mathematical models so as to do justice to the new measurements.

Prof. A. G. Occhipinti (Brazil) asked if it was possible, in Prof. Fischer's model, to take account of meteorological tides or storm surges produced by transient meteorological disturbances, since these might be very relevant phenomena in some estuaries.

Mr. G. C. van Dam (Netherlands) observed that, although not fit for all situations, the superposition models such as those presented in Papers 35, 36, and 39 did not give the difficulties of instabilities and boundary conditions encountered with numerical models, and could often be applied successfully.

REPLY TO DISCUSSION

Prof. Fischer was in agreement with all the points raised by Dr. Charlton and Mr. Apostol. In particular, he agreed that in analysing field data it was not possible to separate residual currents caused by gravitational circulation from those induced bathymetrically, even though caused by different mechanisms. He still thought it useful to separate them conceptually, since the bathymetrically induced currents could be computed by two-dimensional numerical models, but gravitational currents could not. He added that the relation between the transverse mixing coefficient, hydraulic radius, and friction velocity might not be the correct functional dependence, since transverse mixing in estuaries was sometimes caused by large horizontal circulations whose effect should possibly be related to the mean velocity and width.

The situation described by Dr. Waldichuk would be very difficult to model because it included both vertical and lateral inhomogeneities. Prof. Fischer knew of no model presently capable of coping with a complex three-dimensional stratified water body, and even if there was no stratification the eddy at a right-angle bend was difficult to model properly since it depended on a proper inclusion of transverse frictional effects.

In reply to Dr. Macqueen, he said he did not consider himself a proponent of mathematical modelling any more than of other ways of studying pollutant dispersion. He though the current financial commitment to developing more complex numerical methods was probably out of balance with the lack of financial commitment to verifying the methods through field data. What was needed was not more field data from surveys of the type currently existing in substantial quantity, but field experiments specifically designed and executed to determine the strengths and weaknesses of the new numerical methods.

Replying to Prof. Occhipinti, he said that the existing two-dimensional models were reasonably good at predict, ing storm surges. The first use of Leendertse's model[17] had been to predict surges due to a tsunami in Tokyo Bay-and storm surge calculations would be similar.

Prof. Fischer agreed with Mr. van Dam that superposition was often a good modelling technique, and mentioned that one of his own models[25] was basically a superposition model.

ADDITIONAL REFERENCES

21. West, J. R., and Williams, D. J. A. Some basic problems of estuarine water quality control. Symposium on Statistical Aspects of Pollution Problems, Harvard Business School, Boston, Mass., 1971.
22. Glover, R. E. Dispersion of dissolved or suspended materials in flowing streams. U.S. Geological Survey, Prof. Paper 433-B, 1964.
23. Yotsukura, N., Fischer, H. B., and Sayre, W. W. Measurement of mixing characteristics of the Missouri River. U.S. Geological Survey, Water Supply Paper 1899-G, 1970.
24. Cederwall, K., Göransson, C-G., and Svensson, T. Subsequent dispersion—methods of measurement This volume pp. 309–319.
25. Fischer, H. B. A method for predicting pollutant transport in tidal waters. Water Resources Center, University of California, Contr. No. 132, 1970, 143 pp.

Interpretation of Field Data

H. Roy Oakley and Kenneth D. Staples
Partners, Messrs. J. D. & D. M. Watson, England

The Working Party on Sewage Disposal (Jeger Committee) expressed the opinion in their Report[1] that "In a well designed outfall scheme we are satisfied, on the basis of information available to us, that dangers of beach pollution can be largely, if not completely, eliminated", but "Unfortunately, it is easier to specify the need for a 'properly sited outfall' than to define one." They went on to indicate some of the variables which influence the effect of the discharge on receiving waters and said "To determine the effects of these variables it is necessary to undertake extensive tests."

Unfortunately, tests may be easier to arrange than interpret and this paper is concerned with the manner in which field data can be used to estimate water quality and the effect of polluting discharges.

It will be apparent from other contributions to the Symposium that the effort and cost of data collection can be significant. It is not always appreciated that evaluation may be no less of a problem, and that unless the method of interpretation is borne in mind, much wasted effort can be spent in the field in obtaining data which are irrelevant or unnecessary.

Interpretation can range from simple intuitive examination of a few results to sophisticated computation of a wealth of data. The Paper describes examples within the Authors' experience covering a greater part of this range, and attempts to draw some conclusions as to the relative benefit and cost of different methods of approach.

Whatever method is used, it is clearly necessary to have in mind the ultimate objectives, which broadly will be an understanding of the present situation with a view to predicting the effects of changes in future circumstances. It is, perhaps, hardly necessary to underline once again the difficulty of specifying objectives in relation to sewage and industrial waste disposal in coastal waters.

The Jeger Committee also expressed the view[1], in relation to satisfying standards, that "for public health, amenity and the protection of fisheries, coastal discharges of sewage must meet the following conditions: Particles of sewage should not be able to reach the bathing areas; The point of discharge should be far enough offshore to render the sewage slick inoffensive to people on shore or in the bathing areas; The siting of outfall should take local fisheries into consideration."

No-one would quarrel with this declaration of principle. The problems arise in assigning quantitative values to such qualitative ideas. In practice, data collection must be realised in terms of physical, chemical, bacteriological, or biological observations, while the objectives must be defined and desirable water qualities determined in similar concrete terms as a necessary precursor to any examination of the problem.

Water Quality

Precision in the determination of water quality is not justified unless consistent homogeneous conditions obtain, and this is seldom the case in coastal waters where variations of weather and tide result in constant change. Variations in water quality may be short or long term.

Longer-term variations may be ascribed to a number of causes, many of which are largely seasonal, but it is well known that biological effects may be judged over a still longer term, and periods of observation of less than five years are of limited value in assessing background quality or the effect of changing circumstances.

Such variations in time-scale dictate different methods in observing data. For short-term variations continuous methods of observation are clearly essential and, because of the ease of measurement, salinity, temperature, dissolved oxygen, and clarity are frequently the most convenient physical and chemical parameters. Such observations, however, are not directly applicable in determining background quality or the effect of polluting substances where measurements may be needed of the chemical or biochemical oxygen demand, of the concentration of both dissolved and particulate toxic chemicals, and of the concentrations of nutrients or grease. For most of these parameters it will be necessary to sample the water and determine the required parameter by analysis, and this necessarily limits both the frequency and number of observations which it is practicable to carry out in any location. Similarly, bacterial and biological observations must be undertaken by sampling and examination.

There is considerable difficulty in interpreting biological data. A typical example is the Report on Phase I of the Environmental Study of Port Philip Bay, Australia, where after two years' work costing A$1 000 000 it was concluded[2] "at present it is not possible to predict what effects these changes in chemical concentrations or other changes in the physical or chemical characteristics might have on the biological characteristics and hence on the beneficial uses of the Bay, and further studies under Phase II are necessary to improve the understanding of the Bay's physical, chemical and biological mechanisms, and thus of the levels of water quality that must be set in order to protect designated beneficial uses."

Similar difficulties have been met in assessing the effect of sewage discharges into other coastal waters. In considering biological conditions off the Tees the principal conclusions were that Tees Bay was badly polluted but that much more work would be needed if firm conclusions were to be drawn, and in considering the effects of discharging treated sewage effluent to enclosed marine waters in Hong Kong it has not been possible to obtain any quantitative assessment of critical nutrient levels, although a limited amount of biological data is available.

Water Movement

An understanding of the cause and effect of water movement is essential to any attempt to predict the behaviour of waste discharges. Because of the random nature of many of the interacting influences theoretical models are seldom of value in shallow coastal situations, and knowledge has to be gained from observations.

Measurement of current strength and direction may be ascertained from a series of fixed stations (a Eulerian system) or by observing the path of small bodies of water within a reference frame (a Lagrangian system). The former implies the use of some form of current meter, and the latter usually some form of float or drogue, and in practice combinations of both approaches are found to be the most fruitful. In either case it is necessary to cover the study area by a sufficient number of fixed points, or start points for tracking, and to make observations throughout the depth of the sea. If variations of wind, tide, and ocean currents are to be determined both long and short period observations are necessary.

Subsequent handling of the observational data is divided into two activities, analysis for interpretation and synthesis or prediction of water movement under different conditions. The latter is, of course, necessary because it is rare that a full range of conditions can be observed, and indeed, in some study areas, such as the north-east coast of Great Britain, it is impracticable to make observations under the most adverse conditions of high onshore winds.

Analysis

The main purpose of analysis of observed water movement is to separate the component parts, which normally comprise tidal currents, ocean currents, wind induced currents, and density currents.

Tidal currents normally follow a modified sinusoidal curve and can thus be broken into axes with variable velocity, direction, and phase components related to time with reference to the tidal cycle, tidal height and, where necessary, season. Tidal currents frequently predominate and are sometimes of a relatively simple pattern although many variations occur from the virtually tideless Mediterranean situation in the Gulf of Naples[3] to the abnormal situation found, for instance, in Hong Kong[4] where diurnal and semi-diurnal tidal patterns with varying amplitudes alternate with the moon's declination.

Residual movement or main ocean currents are more subtle but no less important as they frequently form one of the main mechanisms transporting accumulated pollution from the discharge area. As they may be relatively weak and vary in strength and even direction from period to period, extensive records over a long period are necessary to identify their average mode. The reduction of observed currents to remove tidal and wind effects enables an assessment to be made of the residual current local to the area or station[5].

Wind-induced currents have their greatest effect on movement of the surface layers and assessment is particularly important where winds are variable and changeable both in direction and strength, but it is in just these circumstances that the analysis of the wind-induced currents proves most difficult. Theoretical analysis serves to illustrate the relationship between the onset of wind and the development of the associated current, while regression analysis and reference to meteorological records enable relationships to be provided from observations for a particular area. Relationships between the strength and direction of the induced current relative to the wind have been developed[5]. It is interesting to note that in the case of some tropical waters, where meteorological conditions influenced by regional monsoon conditions persist, there is so little fluctuation in wind conditions that the wind-induced currents are closely related to the seasonal residual or ocean currents.

The three-dimensional nature of the problem is clearly exemplified by the wind-induced currents which vary considerably with depth, not only in strength but also in direction. This is particularly important when dealing with fresh-water discharges into sea water since density effects may lead to stratified conservative layers which can only be mixed vertically by shear turbulence associated with variations in current strength with depth.

The immediate surface layer of the sea has been shown[6] to have particular relevance in the concentrations of grease and bacteria although the identification and assessment of movement involves particular problems. Simple methods of observation used with some success include the release of drift cards, but, as these give information only of the time and location of recovery at the shore, interpretation is difficult. Nevertheless correlation of the drift card results, general water movement, and meteorological observations enable development of reasonable relationships when sufficient data are available [7].

Finally, density currents should not be neglected where massive fresh-water flows discharge into saline water. The identification of the fresh-water field derived from the efflux of the Tyne is a case in point, and clearly modifies the behaviour of any discharge in the vicinity. There are also circumstances where, in enclosed waters, the transport of pollutants may be largely effected by densimetric currents[8].

Synthesis

If the relationship between the magnitude and direction of each component of the total current can be established with the causative effect, then it is possible to predict each component for specified conditions and to combine or synthesize them to give an overall pattern of water movement. Synthesis of this type is the aim of most of the modelling techniques which have been described.

The need for synthesis, as already pointed out, arises from the difficulties, and impracticability in some instances, of obtaining observations for every combination of conditions and the consequent need to model in order to fill the gaps in the observed data. Clearly there is more confidence when the modelling is interpolative than when it is extrapolative, and in either case it is necessary to establish the validity of the model by demonstrating its ability to reproduce observed data.

Methods of synthesis fall into four main groups, the rudimentary intuitive process, graphical processes, physical modelling and mathematical modelling. Although most emphasis has been put on mathematical modelling, perhaps because of its greater complication and satisfaction to the theorist, the feasibility of alternative methods should not be overlooked since they may, in some instances, offer an equally satisfactory result at much less cost.

Intuitive interpretation

No space need be devoted to this process although it may be the only option open when the observational data are insufficient for satisfactory analysis. Its limitations, inaccuracies, and uncertainties are apparent.

Graphical interpretation

Although not often described, graphical interpretation offers advantages in some situations. For the technique it is necessary to plan carefully the siting of metering stations from preliminary information and observations so that the data from each location are representative of a sector of the study area.

The surveyed metering records for whole tidal cycles are grouped for each station and individually plotted vectorially, Fig. 1. Following correction for wind-induced current effects the major axis of tidal oscillation and the mean basic drift are determined. Separate analysis is undertaken for maximal, minimal, and intermediate tidal stream cycles and the currents resolved and plotted graphically for the major and minor axes against time relative to high water.

Comparison of a series of component curves for similar tidal ranges leads to the preparation of mean model component curves, Fig. 2. From the mean curves model vectors are plotted for each station and order of tidal range, Fig. 3, and where necessary the vectors can be adjusted by the addition of calculated wind-induced current effects.

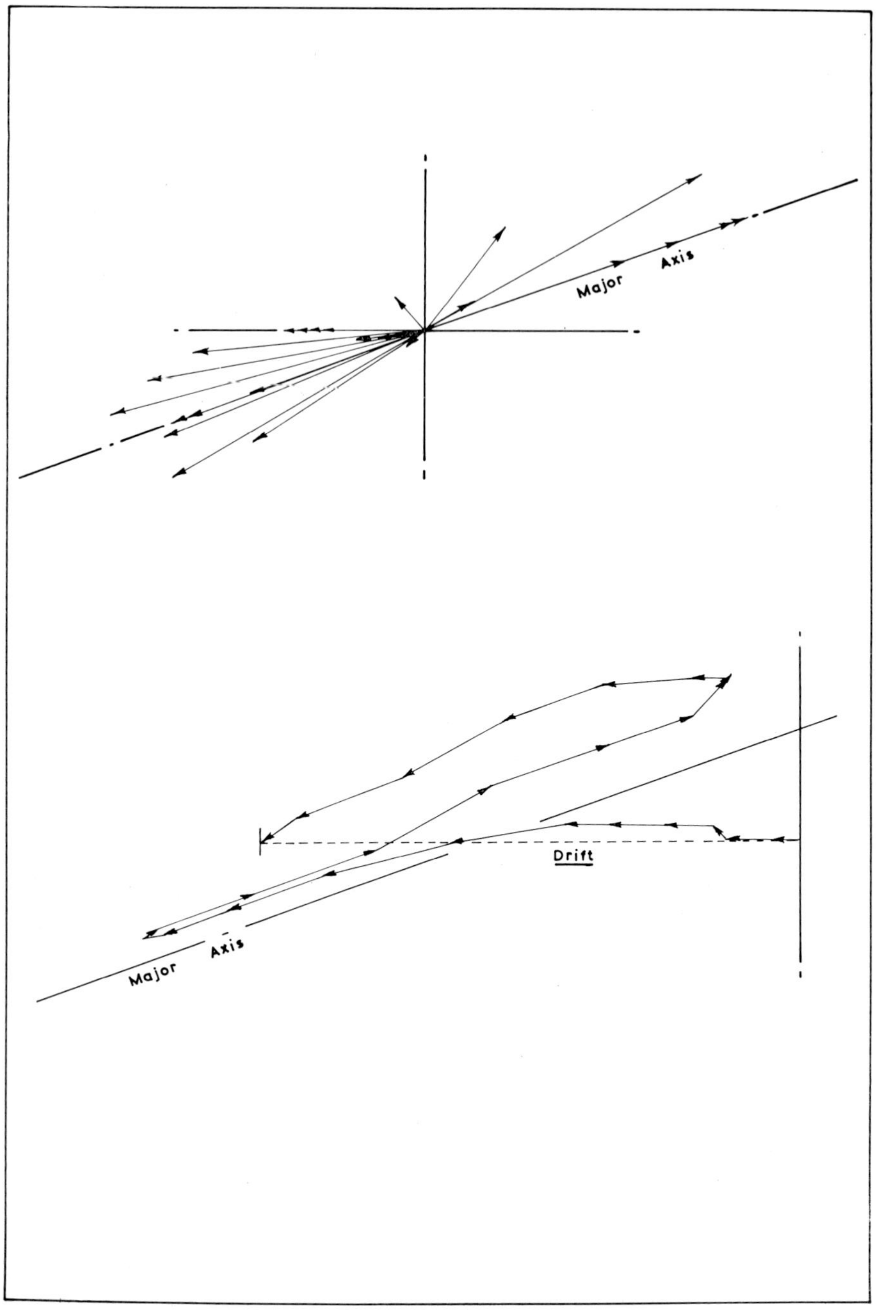

FIG. 1. Vector diagrams of observed tidal streams

The model vectors are then used for the preparation of simulated drogue tracks, successive station model vectors being applied as the track travels from one sector to the next, and it is necessary to modify intuitively the vector proportions and orientation as the track passes through the transition zone between sectors. Initially simulated tracks are produced and compared with surveyed drogue paths and adjustments can be made to the model vectors to obtain a reasonable correlation.

As with the other modelling techniques, the graphical method may be subsequently used to produce predicted tracks from the studied location at relatively short intervals of time, frequently 30 min or 1 h over the full semi-diurnal or diurnal tidal cycle. From a complete series of tracks, families of patch centres associated with any chosen time within the tidal cycle can be plotted, each centre being ascribed its age from release for the subsequent dispersion calculations, Fig. 4.

The method has been used under several different circumstances, including a survey undertaken in the Far East, and covering an area 50 km long extending some 20 km offshore, where the current pattern is complicated by the presence of numerous islands and reefs, so that the water movement is largely dictated by the coastal figuration.

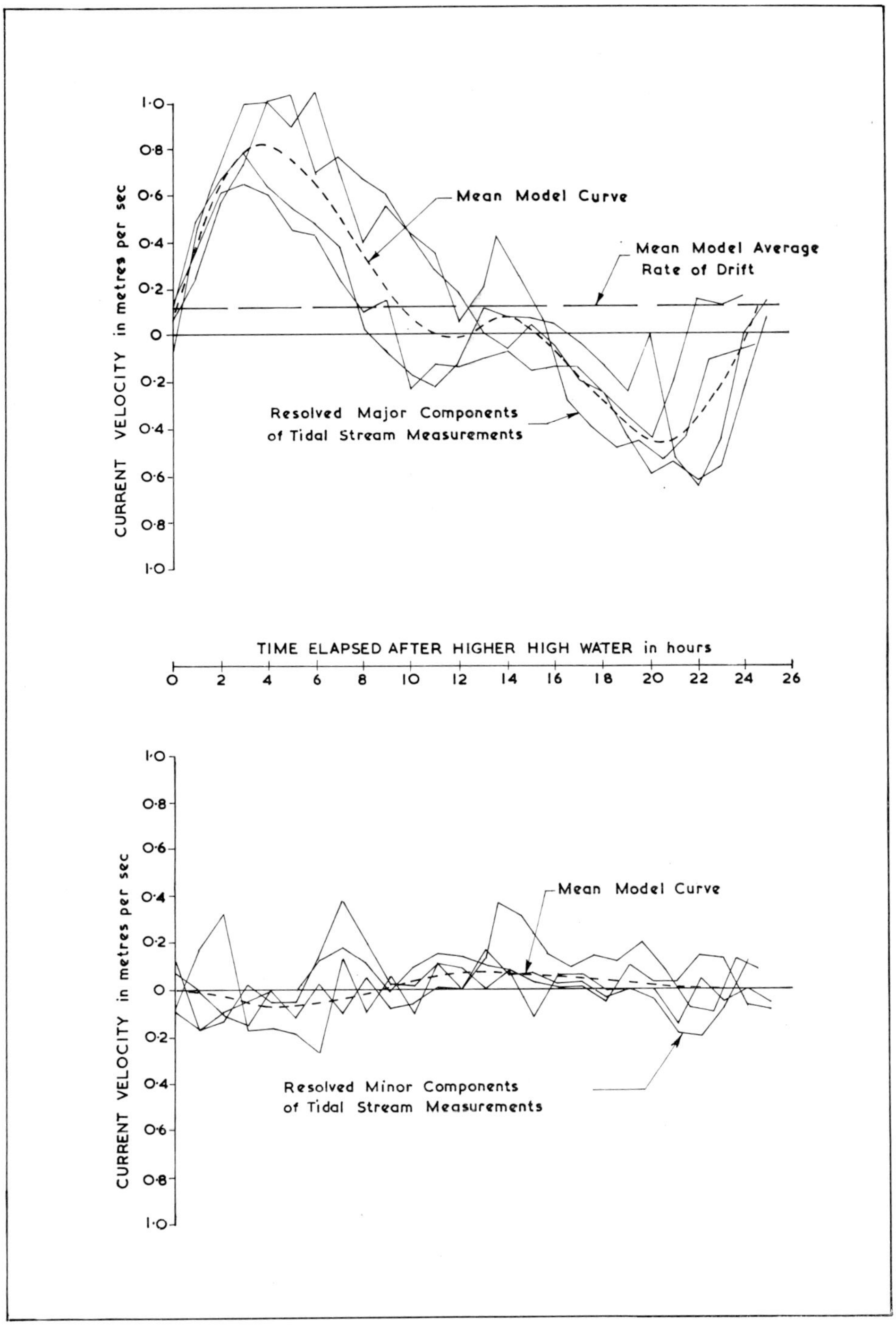

FIG. 2. Major and minor axis components of tidal stream
Spring tide

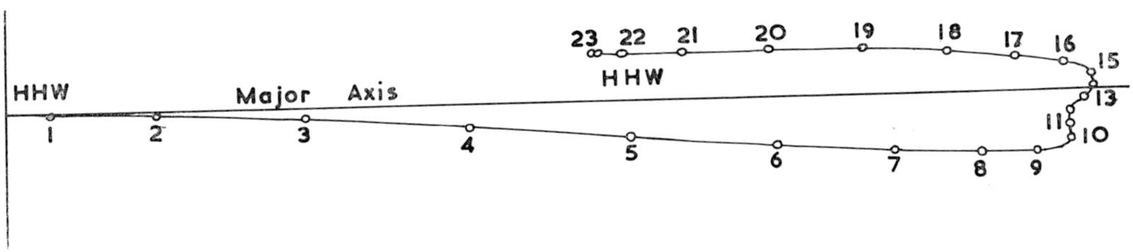

FIG. 3. Vector sum of mean model curves derived from Fig. 2

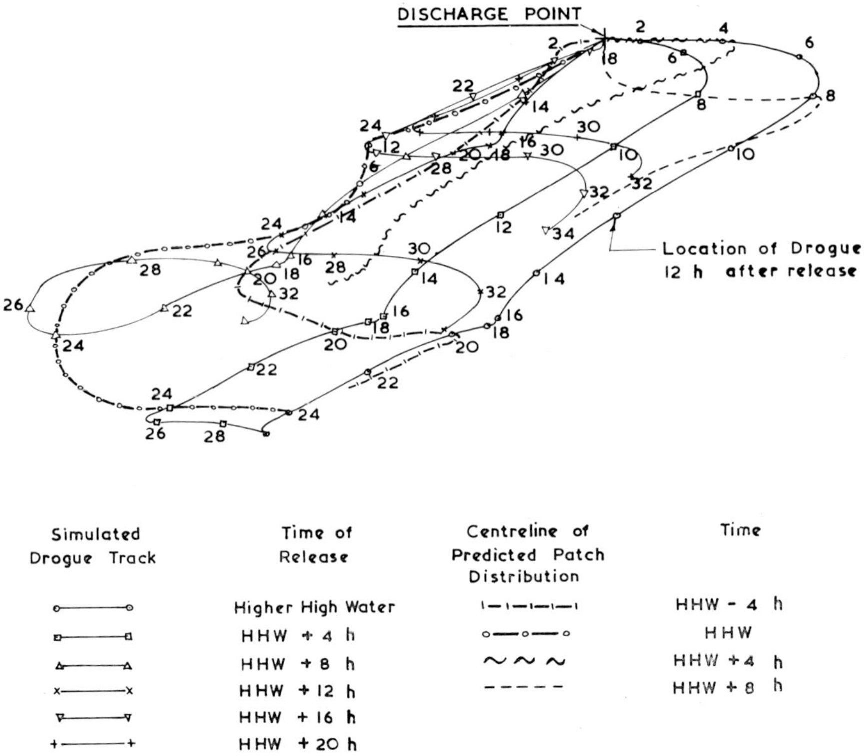

FIG. 4. Predicted drogue tracks and plume centrelines
Neap tide

Physical modelling

The construction of a small-scale physical model which can reproduce large-scale water movement is attractive but requires some compromise in order to conform to model law. There are boundary problems since it is necessary to define the fixed boundaries, that is, the extent of the model in plan and the bed of the sea. There are difficulties in that the surface tension at the air/water boundary cannot be reduced, assuming that water is used in the model, and Coriolis effects are absent from the model.

The application of wind-induced currents to a physical model presents some difficulties, and an advanced form of model has been described by Rance[5]. This model was originally constructed for the Tyneside studies and later adapted for the Teesside investigation where the marine circumstances are similar.

The Tyneside model extended for some 40 km along the coast and 13 km offshore, and embraced discharges from the rivers Tyne, Wear, and Blyth. The field survey, carried out in 1962, included twelve current-meter stations, mostly located on the periphery of the survey area. The model simulated tidal and residual currents and wind-induced currents. For the Tyneside study physical processes were used to identify movement of water in the model. In its adaptation for the Teesside study, a semi-mathematical technique was introduced[9].

Mathematical modelling

The Teesside physical model represents a transient stage between physical and mathematical modelling, and the latter was developed for the study of sewage dispersion off Edinburgh in the Firth of Forth. The Forth Study Area extended for 24 km along the coastline and 13 km offshore, and a brief description has been given of the development of this model[10].

The basic techniques were refined in the model constructed for the South Hampshire Plan Committee for study of effluent disposal into the Solent[11]. The Solent model covered the whole of the enclosed waters lying between the Isle of Wight and the main Hampshire coast, an area extending some 18 km between the coastlines which are about 7 km apart. The residual and tidal currents encountered in the Solent are highly directional and the principal natural shoreward motion is derived from wind-induced currents. Only four main survey stations were used in establishing the data for the Solent model, but much greater use was made of drogue float tracking, which covered the whole of the Study Area.

Comparison of model costs

Because of different circumstances any comparison should be treated with caution, particularly since the extent of field survey and choice of method of synthesis are not entirely dictated by cost considerations. Nevertheless, Table 1 shows the duration of field survey, model technique, and the order of total survey and data evaluation costs (including modelling) for the surveys referred to in this paper. To validate the comparison the costs have been updated to present values.

TABLE 1. Methods and cost of particular field surveys

Study	Synthesis	Duration of field survey (months)	Cost of survey and evaluation (£)
Tyneside	Physical model	12	82 000
Teesside	Physical/mathematical model	6	31 000
Edinburgh	Mathematical model	6	40 000
South Hampshire	Mathematical model	9	45 000
Hong Kong	Intuitive/graphical	17	35 000

Solids Movement

Although in many cases most concern is with the liquid fraction of polluting discharges, the behaviour of settleable solids will often be of importance. One difficulty is to determine how suspended matter will behave after discharge, and experiments of the type described in connection with sludge dispersal experiments for the Greater London Council[12] and for the Department of the Environment[13] will be of considerable value. It does not, of course, follow that settled solids remain in the same condition, and more evidence of the type gathered some forty years ago by the Water Pollution Research Laboratory[14] in studies of the River Mersey is desirable.

Assuming, however, that the settled solids can be related in terms of size distribution, specific gravity, and cohesiveness to naturally occurring materials, it is possible to carry out field experimentation to assess the likely behaviour of such settled solids from a given discharge. If such a relationship can be established, by far the most satisfactory method is to examine the behaviour of natural materials in the vicinity. Since concern will be with long-term behaviour and possible accumulation, short-term experiments are unlikely to be of great value, but a comprehensive study of bed movement in a marine environment is not a light undertaking.

Occasionally, advantage can be taken of studies conducted for other purposes. Thus, in the case of the Tyneside scheme the many years of work undertaken locally by the Dove Marine Laboratory in the vicinity were of relevance, and Buchanan was able to contribute an appreciation from which broad conclusions could be drawn with some confidence. Similarly, with studies in the Solent, studies of sediment drift by Dyer were relevant to the problem, and again enabled conclusions to be drawn as to the ultimate disposal of settleable material[15].

In general terms, it seems reasonable to conclude that, in areas where the sea bed comprises clean rock or coarse grain sediments, accretions of the light flocculent material of the type normally discharged from sewage outfalls is unlikely, and that these organic particles are more likely to accumulate in areas where muds and silts already are found.

It is, of course, relevant that the quantity of solids discharged is frequently small compared with natural sources. At Tyneside the total suspended matter in the sewage from the 1 300 000 equivalent population of that conurbation represented only a small fraction, about 1 per cent, of the suspended solid load discharged by the river, while a similar calculation for the Mersey indicated that the Merseyside sewage and industrial waste contributed less than 0·1 per cent of the total suspended load in the estuary.

To supplement or replace such broadly based studies direct methods of determining the behaviour of suspended matter are necessary. By far the most satisfactory method is the labelling of identical material, and radioactive tracers have been extensively used for this purpose[13].

The interpretation of such experiments over a long term is self evident but short-term experiments need correlation with meteorological conditions and may not give such satisfactory results. Tracer experiments of this nature are expensive and their cost, together with the residual radioactivity derived from each individual release, limits the number of repetitions so that coverage of seasonal and tidal effects, as well as different starter points, is somewhat difficult. For this reason, the cheaper alternative of a sea-bed drifter has been adopted in many cases[16].

It must be assumed that drifter behaviour is comparable to settling matter but this is problematical, both in respect of the scale of the particles and their difference in character, thus, whereas small particles may be re-suspended and dispersed by storm and shoreline turbulence, the larger sea-bed drifters may not. Similarly, sludge particles may cohere and flocculate so that there will be a difference between the respective scouring and movement velocities.

A further disadvantage of sea-bed drifters is that, with few exceptions, only the start and end points are known and the intermediate movement can only be inferred. There is thus difficulty in correlating the drifter performance with physical conditions[12,13].

In the Solent investigation the indication given by the Woodhead drifters as to the desirable point of discharge for sludge disposal was appreciably different from that derived from the examination of sediment drift. The interpretation of data on bed movement is difficult and, although important, is perhaps one of the less satisfactory processes in assessing the movement and dispersal of discharges.

The mechanics of settling-solids movement is of course dependent on water movements close to the sea bed. Currents in the lower levels are normally recorded in field surveys and may be analysed rather more simply than the near-surface water movement. The duration and extent of fixed-station current metering is frequently insufficient to cover longer-term effects and because it usually relates to levels about 1·5 m above the bed, is not entirely representative. Nevertheless such analysis is of assistance in assessing the results of tracer or drifter experiments.

Dispersion

The preceding section on solids movement embraced aspects both of movement and dispersion, while the previous section on water movement was concerned with the path a discharge took and not with its dispersion.

A number of mathematical representations of the dispersion processes have been suggested[17] which in many cases recognize the influence of the instantaneous current velocity, direction, and gradient at each point within the effluent field. Discussion of these models is outside the scope of this paper, but in all cases they depend in practice on the determination of empirical constants and there are, as yet, inadequate data from which the value of such coefficients can be determined for any given set of conditions.

For this reason, it is highly desirable at the present stage of knowledge that dispersion experiments be carried out at proposed discharge locations, and these normally involve some form of tracer of the radioactive, bacterial, chemical, or dye type. It is well known that experiments of this type require considerable effort and result in appreciable cost, and it is general experience that it is not easy to interpret the limited data so expensively obtained.

Because the experiments are necessarily on a limited scale, it cannot be assumed that the observed dispersion will apply uniformly even under prototype conditions. Thus, although it is tempting to assume that the extent of the pollution field will be directly indicated by the dispersion of the tracer, given the same site conditions, such intuitive interpretation is inaccurate at best, and at worst misleading. The more satisfactory approach is to assume a mathematical model and find which set of coefficients would result in a best fit to the experimental data. Such attempts have been made for Sunderland, Edinburgh, South Hampshire, Hong Kong, and other studies.

Where there is an existing discharge, then the possibility exists of using tracers in a direct fashion to show both movement and dispersion under the conditions of the experiment. Much of the work of the Water Pollution Research Laboratory has been of this type, and the difficulties encountered have been described[6].

In a similar manner, surveys of the spread of existing discharges can be used to determine the densimetric flow from massive discharges. In the Tyneside investigation salinity was used to determine the rate of dispersion of the fresh water emerging from the River Tyne and rates of densimetric spread were obtained by combining these data with drogue-float observations[7].

Conclusions

Comprehensive investigation of marine discharges takes much time and effort, and it is important that the data so painfully and expensively acquired are put to best use. It is hoped that this brief review of experience and the interpretation of field data will underline the importance of:

(a) *Obtaining the right data.* It is of vital importance that the data be obtained after and not before determining the method of interpretation. There is nothing worse than to conduct a haphazard field experiment and then to sit down to sort out the result. It is particularly important in the construction of models of water movement and in determining empirical coefficients for mathematical models of dispersion that observation be made at the right places and under the right conditions.

(b) *Obtaining accurate data.* It is hardly necessary to say that the deductions drawn will be as valuable as the data on which they rely, and yet the conditions under which field experimentation is undertaken are often not conducive to accurate recording. There must be great tendency, after tossing about under storm conditions on a small boat for some hours under conditions of physical and perhaps emotional stress, to take and record a reading irrespective as to whether it is right or wrong, and it would be interesting to experiment on the inaccuracies inherent in taking different types of readings with different types of instruments under different conditions. All that can be said here is that it is wise to question the accuracy of data, and to be ruthless in rejecting suspect data before attempting interpretation.

(c) *Proving the data.* It is important to demonstrate the validity of any interpretation or modelling by showing that the model can reproduce observed results. One needs, of course, to be aware of the danger of using the same result to construct and prove the model, and it is perhaps wise to arrange entirely separate proving experiments by which the eventual model can be checked. It is easy to delude oneself into the accuracy of predictions developed by sophisticated mathematical processes. Pictures of developed sewage fields from hypothetical discharges and diagrams of dispersions and dilutions are all too convincing and sometimes the validity of the data on which they have been based is open to serious question.

The other broad conclusion that must be drawn is that there is a minimum requirement in terms of cost, time, and effort to give a meaningful result in a complex situation. The investigation of sea discharges, in terms of survey, evaluation, and assessment, cannot be made on the cheap, and if adequate effort cannot be spared then one might as well throw in one's rotten orange and trust to fortune.

REFERENCES

1. Ministry of Housing and Local Government and Welsh Office. Taken for granted. Report of Working Party on Sewage Disposal. H.M. Stationery Office, London, 1970, pp. 31 & 32.
2. Environmental study of Port Philip Bay: Report on Phase I, 1968–71. Melbourne and Metropolitan Board of Works, and Fisheries and Wildlife Department of Victoria. Melbourne, 1973, p. 230.
3. Eurostaff S.p.A., and Dagh Watson. Studio sull'inquinamento del Golfo di Napoli. Milan, May 1973.
4. J. D. & D. M. Watson. Marine investigations into sewage discharges. Government of Hong Kong, July 1971.
5. Rance, P. J. Investigation of wind-induced currents and their effect on the performance of sea outfalls. *Proc. Instn civ. Engrs*, 1966, **33**, 231.
6. Wheatland, A. B., Agg, A. R., and Bruce, A. M. Some observations on the dispersion of sewage from sea outfalls. *J. Proc. Inst. Sew. Purif.*, 1965, 291.
7. Oakley, H. R., and Dyer, E. A. Investigation of sea outfalls for Tyneside sewage disposal. *Proc. Instn civ. Engrs*, 1966, **33**, 201.
8. Hansen, D. V., and Festa, J. F. Inlet circulation induced by mixing of stratified water masses. *Rapp. P.-v. Réun. Cons. int. Explor. Mer*, 1975, **167**, 163–170.
9. Moir, I. W. M., and Williams, J. M. Physical and mathematical model studies of surface currents. *Proc. 5th int. Conf. Wat. Pollut. Res., San Francisco and Hawaii, 1970*, Vol 2, Paper HA-3.
10. Haverson, D. Mathematical model techniques for coastal waters. *Proc. 5th int. Conf. Wat. Pollut. Res., San Francisco and Hawaii, 1970*, Vol 2, Paper HA-2.
11. J. D. & D. M. Watson. Marine disposal of sewage and sewage sludge. South Hampshire Plan Advisory Committee, May 1972.
12. Pearson, R. F., Finn, E. V., and Miller, D. R. Study for disposal of sewage sludge from the Greater London sewerage area into the North Sea by pipeline. *Proc. Instn civ. Engrs*, 1971, **48**, 375.
13. Department of the Environment. Out of sight out of mind. Report of Working Party on Sludge Disposal in Liverpool Bay. Vol 2. H.M. Stationery Office, London, 1972.
14. Department of Scientific and Industrial Research. Estuary of the River Mersey. The effect of the discharge of crude sewage into the estuary of the River Mersey on the amount and hardness of the deposit in the estuary. Water Pollution Research Technical Paper No. 7. H.M. Stationery Office, London, 1938 (reprinted 1961).

15. Dyer, K. R. The distribution and movement of sediment in the Solent, S. England. *Mar. Geol.*, 1971, **11**, 175.
16. Woodhead, P. M. J., and Lee, A. J. A new instrument for measuring residual currents near the sea bed. I.C.E.S. Hydrographic Committee No. 12, 1960.
17. Okubo, A. Some speculations on oceanic diffusion diagrams. *Rapp. P.–v. Réun. Cons. int. Explor. Mer*, 1975, **167**, 77–85.

Mr. Oakley, in his opening remarks, said the paper had been written from an engineering viewpoint; engineers were concerned more with the end than the means, but recognition of the means would facilitate both economy and expedition. Consideration of marine waste disposal required assessment of present conditions and prediction of future changes, but there might be a wide gulf between observation and prediction. One purpose of the paper was to emphasize the time and cost of data analysis; another was to show the imperfections of present methods of relating fact and theory.

Physical and chemical parameters of water quality were relatively easy to measure, but assessment of biological conditions was still a major problem. Although much progress had been made, by means of mathematical models, in predicting water movements, it should not be assumed that this was the right approach to every problem. Some alternative methods were outlined in the paper and some comparisons of time and cost had been attempted; in Table 1 the cost of survey had not been separated from that of analysis, but in the Authors' experience a ratio of 2:1 was general.

Other important topics were the measurement and prediction of the movement of bed solids and of the rate of dispersion of suspended or dissolved matter. The most satisfactory methods of field examination were expensive; less sophisticated ones were difficult to interpret and sometimes showed disturbing inconsistencies.

The paper concluded with some self-evident but relevant remarks on the need to obtain accurate data and to validate models. Scatter of field observations was unavoidable, but the accuracy of data obtained in unpleasant field conditions might also be questioned, and it was likely that the calibration of current meters, direct-reading dissolved-oxygen meters, and other instruments was not carried out frequently enough. The true test of a model would be comparison of predictions with subsequent observations—such details were all too rarely published.

DISCUSSION

Mr. P. Ackers (U.K.) said the Authors' down-to-earth approach contrasted with the rather esoteric approaches that had been described in some of the earlier papers. One had to bear in mind the uncertainty of the empirical coefficient values incorporated in any mathematical model to describe the dispersion process and the physical, chemical, and biological changes occurring as an effluent moved towards beaches or other critical areas. Previous Authors had frankly admitted this and had commented on order-of-magnitude uncertainties. The range of hydraulic conditions (even at a single site), when combined with these various uncertainties, suggested that there was a long way to go before the best techniques for particular cases could be identified and their reliability ensured.

The Authors had given the only data to the Symposium on costs of surveys and models—ranging from £30 000 to £80 000—and he understood these to cover predictions of water movement alone. Survey and model costs of £100 000 could easily occur, perhaps far more if development of methodology were chargeable. When designing an outfall for a city of a million inhabitants the criteria for judging what was necessary in terms of field surveys and modelling were clearly very different from those for a coastal town with a population of, say, 10 000. Mr. Ackers asked if the Authors could offer any yardstick by which the extent and cost of surveying and modelling could be related to the scale of the project or the local environmental factors. He agreed with the implications of the final paragraph of their paper; at present costs, £20 000–£30 000 might be the minimum cost of feasibility studies, and small outfalls might not justify more than their 'rotten orange' approach. Some simple design criteria were needed for small outfalls, say for populations under 10 000. Such criteria might be a still-water dilution not less than 50, and a nominal travel time to the shore at the worst 10 per cent wind condition (assuming water velocity to be some nominal percentage of wind velocity) not less than a nominal average T_{90} value; provided also that tidal currents at the outfalls had a median exceeding say 0·06 m/s. A month's readings from moored current meters at possible alternative sites, plus analysis of local long-term wind records, would be all the special field work needed. He asked what could be done at reasonable cost to satisfy all interested parties without such 'rules of thumb'.

REPLY TO DISCUSSION

The Authors agreed with much of Mr. Ackers's thoughtful contribution. With regard to the cost of models, it should be noted that the mathematical models listed in Table 1 did provide for estimation of dispersion and degradation, and the cost of incorporating this facility in the model was largely that of the field data required to establish the relevant coefficients. The cost of operating a model to estimate water quality parameters was also slightly greater than that needed for water movement only.

There was no simple answer to the question as to what cost was justifiable in predicting the consequences of marine discharges. Clearly the scale and cost of the scheme and the sensitivity of the environment were all relevant factors, and the extent of the investigation had to be determined by judgement based on experience. For the marine studies described in the paper, costs were generally less than 5 per cent of the estimated value of the capital works, a figure comparable with site investigation and design costs for conventional major engineering works. It did not follow that improvements to existing outfalls justified themselves, since an assessment of benefit relative to the cost of alternative courses of action was still necessary.

They agreed that simple rule-of-thumb guides were appropriate in some instances; advice to the client for one of the studies described in the paper had been that a suitable guideline would be: an initial dilution of not less than 50, conditions which would provide for subsequent dilution of not less than 200, and a travel time of not less than 3 h to recreational waters or beaches. These criteria were not very different from those suggested by Mr. Ackers, but account had to be taken of the background circumstances in each instance, and the use and character of adjacent waters and coastline were clearly of overriding importance. For this reason, the suggestion that information relating to such a guideline could be obtained from current-meter readings at fixed sites, plus an analysis of wind records, was questionable. Small outfalls were presumably relatively close to the shore where current patterns were affected by sea-bed topography, so that current-meter readings at the discharge point would give inadequate information from which to assess the path and travel time following discharge. Given limited time and money, a combination of such readings with float tracking was considered more productive than reliance on one or other of these methods.

While not seeking to justify 'extraordinarily comprehensive surveys', the Authors considered there to be a need for undertaking investigations to acquire and develop sufficient data to demonstrate the efficiency of the adopted solution. Any engineer's reliance on analogous cases and past experience would involve subjective elements not easily defended in the face of the modern critical lay 'conservationist'.

Paper No. **39**

The Hague Outfall

G. C. van Dam

Rijkswaterstaat, The Hague

Several two-dimensional model computations have been made during the past 12 years in connection with the design and operation of the Hague sewage outfall.

In the design period (early sixties), the model was relatively simple and computation times were very long, so that concentrations could be calculated only at a limited number of points, e.g. along the coast. Moreover, great uncertainty existed as to the average life of coliform bacteria in sea water under prevailing climatic conditions.

In later years, during the construction of the outfall and after its coming into operation (1968), the model was gradually improved. Large concentration fields can now be computed and drawn in fairly short times, and more complicated current functions can be simulated. Extensive surveys by the State Institute for Water Purification (RIZA) have provided better data on bacterial lifetimes.

However, observed coliform distributions are often so strange and complicated that it is difficult to simulate and thus 'explain' them with the model. On the other hand it should be mentioned that dye concentrations along the beach, in experiments with continuous dye release, agree better with model simulations than (average) coliform concentrations. One might infer that this is due to a mistaken estimate of bacterial lifetime, based upon a limited set of data, but the two-dimensional surveys suggest some other mechanism (not observed in dye experiments so far) by which, as it were, a certain preference for movement towards the coast exists. This can, to a certain extent, be simulated by the model, *after* it has been observed, but this does not give a physical explanation of the effect, except for the fact that the model can be used to show that such an effect, if it exists, has more influence upon concentrations of bacteria with their relatively short lifetimes than upon concentrations of a conservative substance. Taking this into consideration, it would seem that field surveys of bacterial concentrations are a better tool for testing the model than is the use of conservative substances. This only holds true for a limited range of relatively small scales of 'macro-turbulence' because the bacteria die before they reach the stage in which the larger scales become effective as a dispersing agent.

The explanation for a certain preferential movement towards the coast may be sought in three-dimensional effects, connected with density stratification. The most important stratifications are those arising from the river discharges, 20 km south of the outfall, averaging about 2500 m^3/s. Salinity distributions, such as those shown in Figs. 1 and 2 illustrate that these discharges have an obvious influence in the vicinity of the outfall, in a horizontal as well as a vertical sense. It must be emphasized, moreover, that the surveys concerned were always made under calm weather conditions. Natural stratification is probably less pronounced or even absent with stronger winds, especially from westerly and north-westerly directions. In calm weather periods, winds from the east or south-east are fairly common. Natural density stratification keeps the sewage field submerged during the first few hours after release so that it can move to the beach by the return current while the wind blows the top layers in a seaward direction. Under rougher conditions, the situation is probably nearly neutral, although it is possible that in this case some artificial stratification is induced by the sewage itself (mainly at the turn of the tidal current) and the resulting fields might be transported to the beaches by westerly winds. However, in general the very effective diffusers[1] will prevent artificial stratification of any stability.

In the following section, a short description will be given of the mathematical model and its possibilities in its present state.

In the next section, the lifetimes of coliform bacteria in the area are discussed in the light of the available data and their interpretation. The mathematical model appears to be an important tool for obtaining better estimates of bacterial lifetimes.

In the third section, current characteristics and diffusion parameters are briefly discussed in the light of data from measurements and experiments along the Dutch coast.

In the fourth section, the results are given of a number of computations of bacterial concentrations along the coast based upon various combinations of the parameter values, thus illustrating the influence

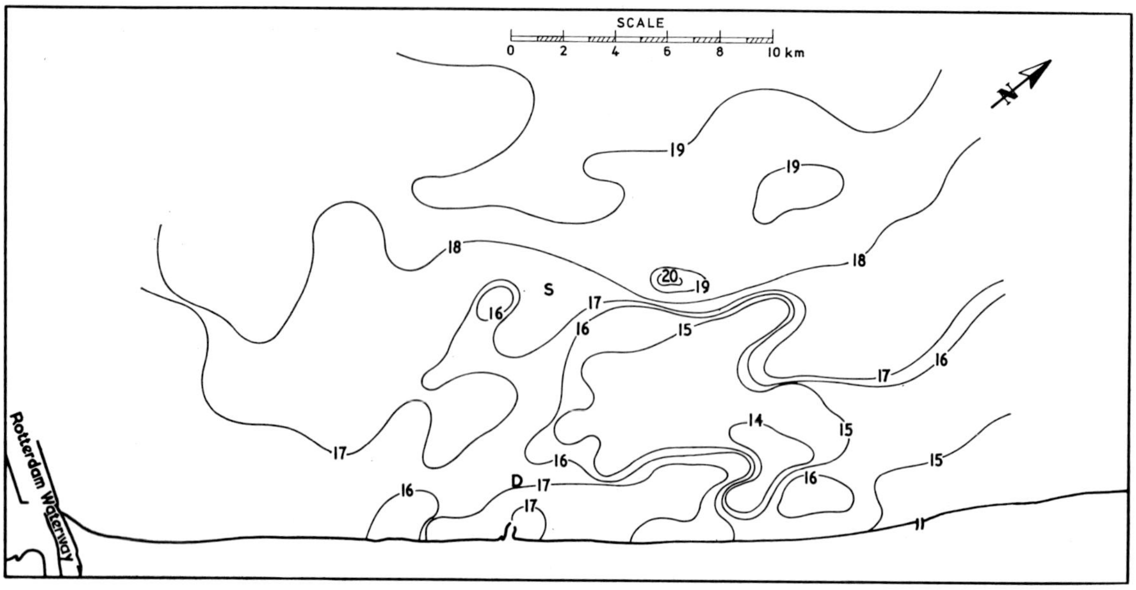

FIG. 1. Surface chloride concentrations (g/l.) near Scheveningen (The Hague) on 1 August 1972
Positions of sewage outfall diffuser (D) *and sludge outlet* (S) *are indicated*

of the various factors. These results are also compared with monthly averages of empirical data from beach sampling.

In the last section, a few cases are considered on the basis of two-dimensional field surveys and computed two-dimensional concentration fields.

In conclusion, it can be said that the applicability of the model is limited, mainly because of the occurrence of effects which can only be described three-dimensionally.

On the other hand, apart from its modest role in the design phase, the model has been quite useful in deepening our insight into relatively complex processes, and in distinguishing various influences and the importance of the role they play. Some of the activities in which the model has had an auxiliary role will be continued in the future. The insights obtained, with the model as one of the tools, may be used in the design of improvements or alterations in the waste discharge and can also be a guide for similar projects that might be considered in the future.

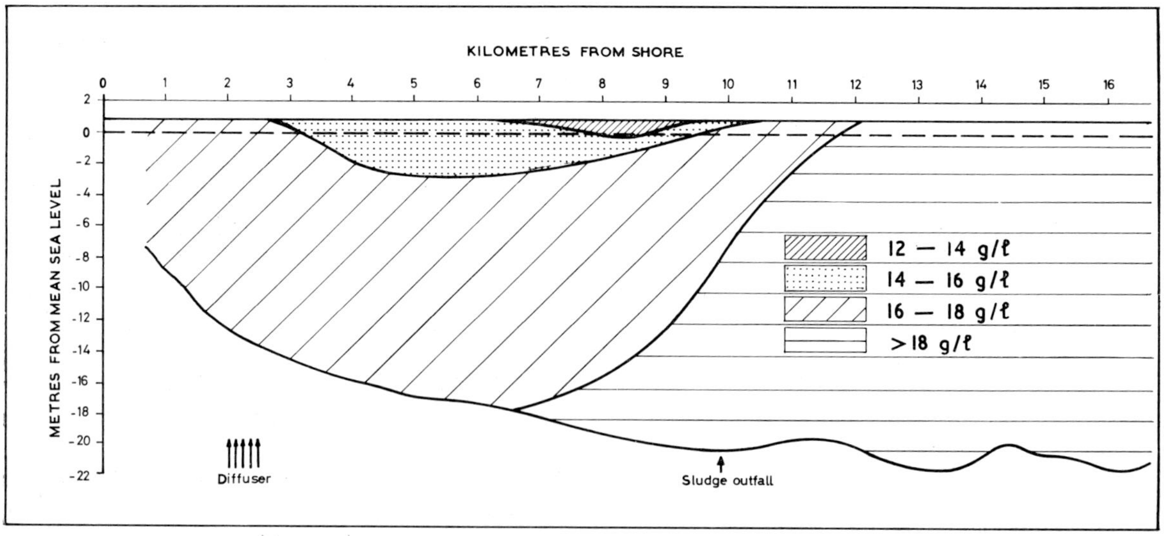

FIG. 2. Vertical cross-section of chloride distribution near outfalls at 0830 on 1 August 1972

Furthermore, the model is used for many other problems, and the insights obtained into its sensitivity to the various factors and into its limitations in scope that have been obtained from this study will be quite profitable for other applications.

The Mathematical Model

The model concerned is a two-dimensional mathematical model based upon the so-called superposition principle. Under this name, the principle was first described by Abraham and van Dam[2] in 1970 but earlier descriptions (1964) were given by Schönfeld[3] and Harremoës[4], and the general ideas are much older. In the present case, the elements which are superimposed are analytical functions symmetric in x- and y-directions (resulting in elliptical patterns), which implies, of course, a simplification of the more whimsical distributions found in dispersion experiments and also implies a more or less homogeneous area with limited depth differences. Rigid boundaries can be taken into account by using a reflection procedure, provided that these boundaries can be considered to be straight lines. This latter condition is met rather well by the very smooth coastline near The Hague. An analysis of computational results and results from dye experiments has revealed that coastline and bottom topography near The Hague do not constitute a serious objection for application of the model. The most serious objection has proved to be the presence of stratification, mainly due to flow from the nearby mouth of the Rotterdam Waterway (Rhine and Meuse). Such conditions disturb the two-dimensional character of dispersion and mixing.

Dispersion from a continuous source is simulated in the model by decomposing the discharge into a sequence of short injections, each injection resulting into an individual 'cloud'. All these clouds constitute a series of elements which are superimposed, taking into account their displacements and their extension, both being dependent upon their 'age'. In this context the extension is considered the result of a dispersion process, often regarded as turbulent diffusion (in a generalized sense). Furthermore, in the present model, size and shape of the clouds are assumed to be functions of 'age' only. The displacement, of course, is not only a function of age, but also dependent on the current conditions encountered during the lifetime of the cloud. This means that the displacements also depend on spatial variations in current velocities, but in its present state, the model can only be fed with velocities (in x- and y-directions) as a function of time. This is not quite as serious an objection as it would seem; because of the relatively simple character of the path along which all centres of the elementary clouds are 'continuously' distributed, each situation (i.e. every arrangement of the cloud centres in the x-y-plane) can in fact be obtained by a proper choice of the respective velocities as a function of time.

The influence of bacterial die-off is easily taken into account by considering the age of each individual cloud. This provides one of the great attractions of the model: the possibility of investigating the influence of decay in a continuous release situation under various conditions.

Lifetime of Coliform Bacteria in the Local Sea Area

When various possible average lifetimes are fed into the model with all other parameters and conditions fixed, the considerable influence of die-off upon concentrations at some distance from the source comes out strikingly. This was already quite clear when the first computations with the earliest version of the model were made[5] in 1962 and 1963. While the total number of coliform bacteria in the entire field is simply proportional to the average lifetime τ, differences in τ of a factor of two may result in differences in (average) concentration of a factor of ten or more at distances of a few kilometres, especially in the direction perpendicular to the prevailing current, because in that direction the 'diffusive' transport, which is relatively slow, prevails. An illustration of this effect was also given by Abraham and van Dam in the forementioned publication[2].

While the Hague outfall was being designed, the actual lifespan of coliform bacteria under the local conditions was not well known. A wide range of figures is reported in the literature, along with indications of strong influences by climatic and even weather conditions (sunshine). From the earliest calculations[5] it could be concluded that coastal concentrations rapidly decrease with decreasing lifetime and secondly (in fact as a logical consequence of the first conclusion) that the influence of outfall length is rather small in the case of long average lifetime and becomes greater and greater as lifetime decreases.

For slow decay rates, and for outfall lengths ranging up to about 2·5 km, a distinct advantage with respect to decrease of average coastal concentrations was shown with increasing length. Therefore, it was decided to choose an outfall length of 2·5 km, considering that if die-off would indeed be slow, the necessary improvement could not be obtained by increasing the length but that additional measures (purification) would then have to be applied.

One might ask whether the existing coastal concentrations resulting from the old release point (400 m from the shore) could not give sufficient information about bacterial survival. In the first place, too little was known about diffusive and advective transport in such a narrow strip (400 m) along the shore. Secondly, the great sensitivity of local concentrations for relatively small variations in die-off time makes results from beach sampling alone an inaccurate tool for determining this factor, certainly in cases in which the physical spread is not well known or variable. Since these considerations are also valid in the new situation, RIZA decided to organize a series of special surveys of the entire area in order to obtain an estimate of the total number of bacteria present in the field, and, at the same time to provide an insight into the patterns of bacterial spread. If the bacterial discharge, as a function of time, in the period preceding the survey is also known, τ can be evaluated from the integral equation

$$\int_{-\infty}^{t_0} q(t) \exp\left(-\frac{t_0 - t}{\tau}\right) dt = N(t_0) \tag{1}$$

where $q(t)$ is the bacterial discharge rate and $N(t_0)$ the total number of bacteria present at time t_0. If $q(t)$ can be considered constant, the simple equation

$$q\tau = N \tag{2}$$

results (N does not even depend upon t_0 as long as q remains constant).

In cases where we need Equation 1 instead of 2, there may be some doubt as to the correctness of the formula, because in such cases the assumption of exponential decay comes into play. The validity of this assumption may be questioned[6], and the same problem arises when computations are made with the mathematical model.

The first six surveys by RIZA have been described by Tromp[7]. His calculation of τ is based on Equation 2. This allows for exponential as well as non-exponential decay if τ is simply defined as average lifetime. Tromp assumed that the amplitudes and frequencies of variations in bacterial discharge were such that the use of an average value was permissible.

Although with this procedure certain inaccuracies must have been incorporated, the differences in τ values found in the various experiments are so large that it must be assumed that at least the order of magnitude of these differences is correct (Table 1). Whatever inaccuracies are present in these results, in predictive computations it will be necessary to take into account the possibility of the larger values that apparently may occur under certain conditions. It is important to observe that the largest values of τ are the most reliable ones, since they are based on very extensive fields of relatively high concentrations so that large errors in the estimate of N (caused by local stratification effects or by a relatively coarse grid of sampling points) are unlikely.

A further series of three surveys was carried out in 1972. The horizontal distributions of coliform bacteria for the upper layers, as resulting from surface sampling, are shown for the first of these surveys in Fig. 3. The assumption that the bacteria are vertically uniformly distributed in all parts of the field leads to an estimate of τ in the order of 5 h (T_{90} about 12 h) for each of the three different surveys. However, if the extent of each of the various equi-concentration contours is compared with the extent of contours computed with the model, the lower concentration contours correspond with τ values which are about twice as large: $\tau \approx 10$ h, $T_{90} \approx 23$ h. (This was worked out in detail for the survey of 20 September 1972; for the other days this tendency is probably even stronger.) This could be explained by assuming that the concentrations in the centre of the field were underestimated. Occasional samples

TABLE 1. Derivation of τ from Equation 2 (from Tromp[7])
$T_{90} = 2{\cdot}30\,\tau$

Outfall	Date	q (10^{10}/s)	N (10^{16})	τ (h)	T_{90} (h)
Old	9 Nov. 1965	(unknown)	2	(9)	(20)
	20 June 1967	60	5	23	54
New	4 Oct. 1968	140	38	76	175
	12 June 1969	210	2	2·2	5
	5 Aug. 1969	385	3	2·6	6
	18 Sept. 1969	465	16	9·6	22

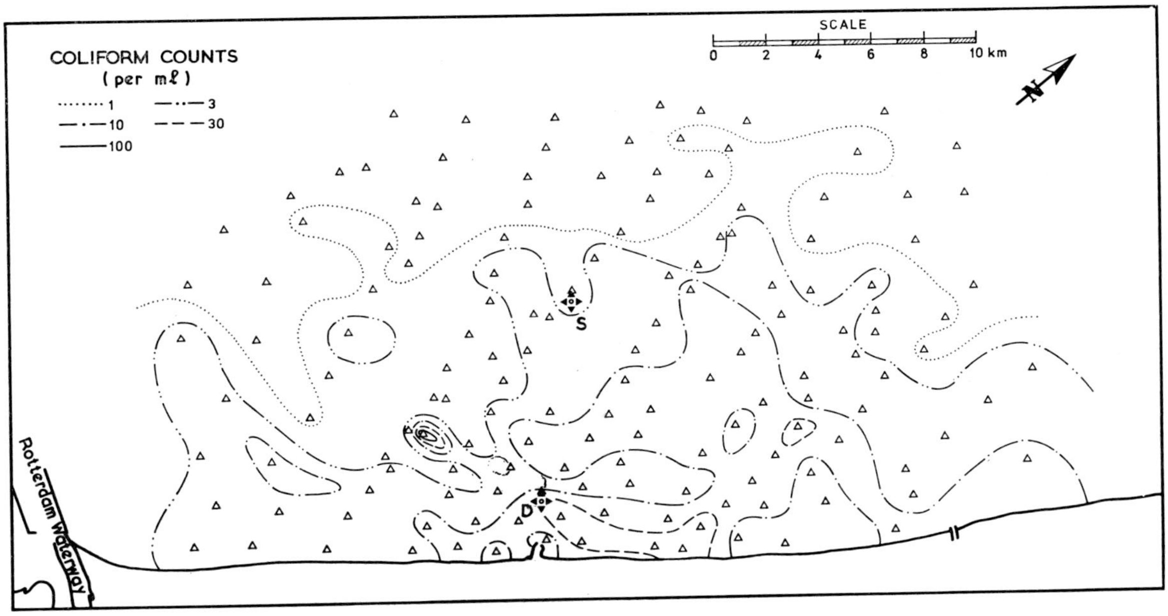

FIG. 3. Surface distributions of coliform bacteria near Scheveningen on 1 August 1972
Data obtained simultaneously with those of Fig. 1

over the entire depth do indeed often show lower values in the surface layers. A dye experiment in May 1969 showed that in a fairly extensive area around the outfall the sewage field was submerged due to natural stratification during almost the entire week of the experiment.

Current Characteristics and Diffusion Parameters

After die-off, the next most important factor in determining concentration distributions and local concentrations is the dispersive action of currents and turbulence. A large number of dye experiments in the region of the outfall and at other places along the Netherlands' coast, has provided a fairly good insight into the parameter values and their variations. The results of these experiments have been published elsewhere[8–11]. Tidal currents parallel to the coast are sufficiently well known; net currents vary but are of little or no interest for substances with decay times such as those found for the coliform bacteria. Current components perpendicular to the coast are variable and may cause large variations in the concentrations along the beaches.

In the following section, variations in concentration along the coast, arising from variations in the relevant parameters within margins derived from the measurements, will be investigated.

Computation of Bacterial Concentrations along the Coast

It is beyond the scope of the present paper to investigate the influence of all those factors which affect the complete two-dimensional distribution of coliform bacteria. Consequently, in this section, only concentrations along the coast will be considered. It is also impracticable to consider the effect of various combinations of factors. Therefore, except for the single case of all deviations working in the same direction (a most unlikely extreme), deviations due to one factor at a time will be shown as compared to one 'standard' case.

For the standard case we will take:

1. Die-off time, 10 h ($T_{90} = 23$ h).
2. Amplitude of tidal current parallel to coast, 0·60 m/s.
3. Net current, 0·03 m/s.
4. Amplitude of current perpendicular to coast, 15 per cent of tidal current amplitude.
5. Diffusion rate, 'average' behaviour.
6. Anisotropy factor of diffusion, 2:1.

In this section the influence of the sludge discharge at 10 km from the shore will be neglected. The initial dispersion by the diffuser of 500 m length is taken into account. It would take too long to discuss

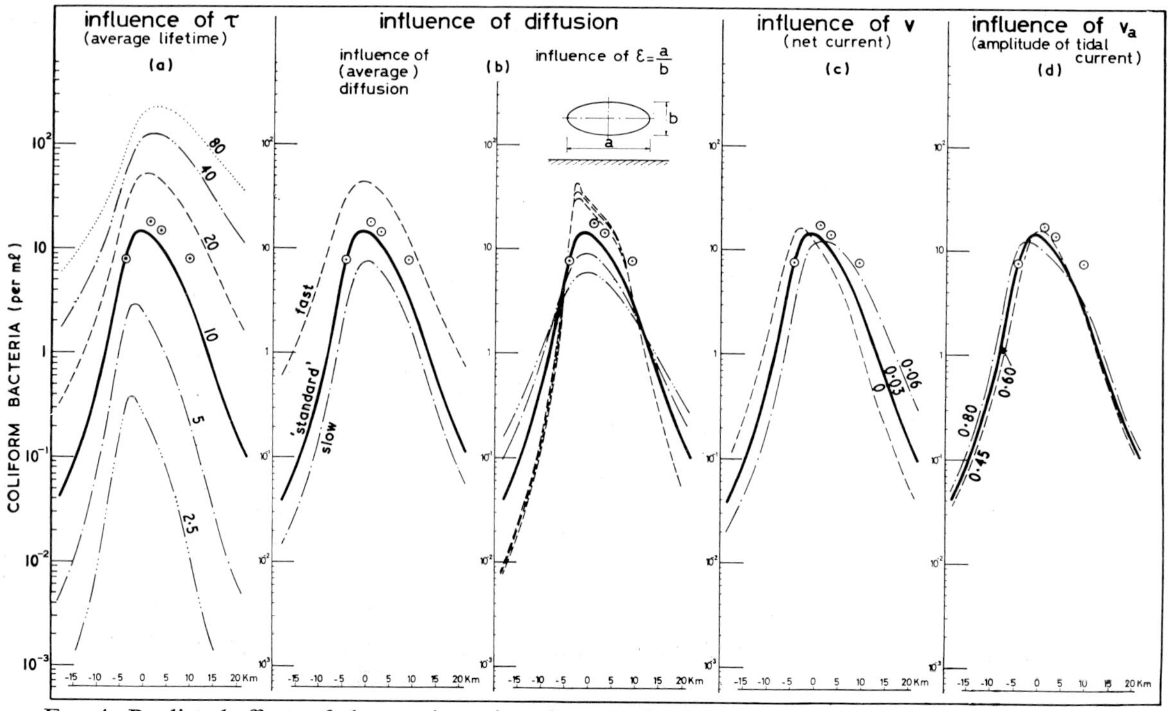

FIG. 4. Predicted effects of changes in various factors affecting coliform concentrations at the shore

Computed 'standard' case indicated by thick continuous curves

Encircled points are average observed values for November 1969 to June 1970 (from Tromp[7])

(*a*) Numbers show values of τ in hours

(*b*) Curves relating to ϵ are (from top to bottom at left) for values of $\frac{1}{4}$, $\frac{1}{2}$, 1, 2, 3, and 4 respectively

(*c*) and (*d*) Numbers show v and v_a in m/s

the way that this is done in any detail. The centre of the diffuser is at a distance of 2250 m from the shore.

In Fig. 4 (a–e) the successive separate influences are shown. Figure 4(f) shows the combined effects of: a long die-off time (τ = 40 and 80 h), a small tidal amplitude (0·45 m/s), a net current of zero, an advective movement towards the coast during the last day preceding the presented situation, the fastest diffusion rate derived from the experiments, and an anisotropy factor (such as might be caused by winds normal to the shoreline) of 0·25 (tending to 1 after a few days).

In all cases it was assumed that the injection had been going on for an infinite time at a constant rate until the moment at which the distribution is presented. For this point in time, a more or less arbitrary phase of the tidal current has been chosen. In other phases of the tide, the distribution will merely be shifted to the right or left over a distance of a few kilometres.

Two-Dimensional Surveys and Computation

The set-up of the mathematical model and the computation rate for the program concerned allow for computing complete two-dimensional distributions; the program can produce a magnetic tape for an automatic drawing machine whereby a topographical presentation of the distribution is obtained.

These facilities are useful for several applications. Comparison of computed and calculated two-dimensional distributions may indicate irregular release rates and deviations from an exponential form of the decay function. Certain irregularities in the measured distribution, such as steep gradients at a considerable distance from the source or the absence of high concentrations in areas where they are expected according to the model, may be indications of three-dimensional effects (stratifications) which cannot be reproduced by the two-dimensional model. Because of the rather large inaccuracies in the laboratory determination of bacterial concentrations, and the fact that no detailed information on bacterial discharge rates (only daily averages) is available for each of the RIZA-surveys, it is not possible to make a direct comparison between details of measured and computed distributions. Available current data are also an insufficient basis for such a direct comparison. It may, however, be useful to

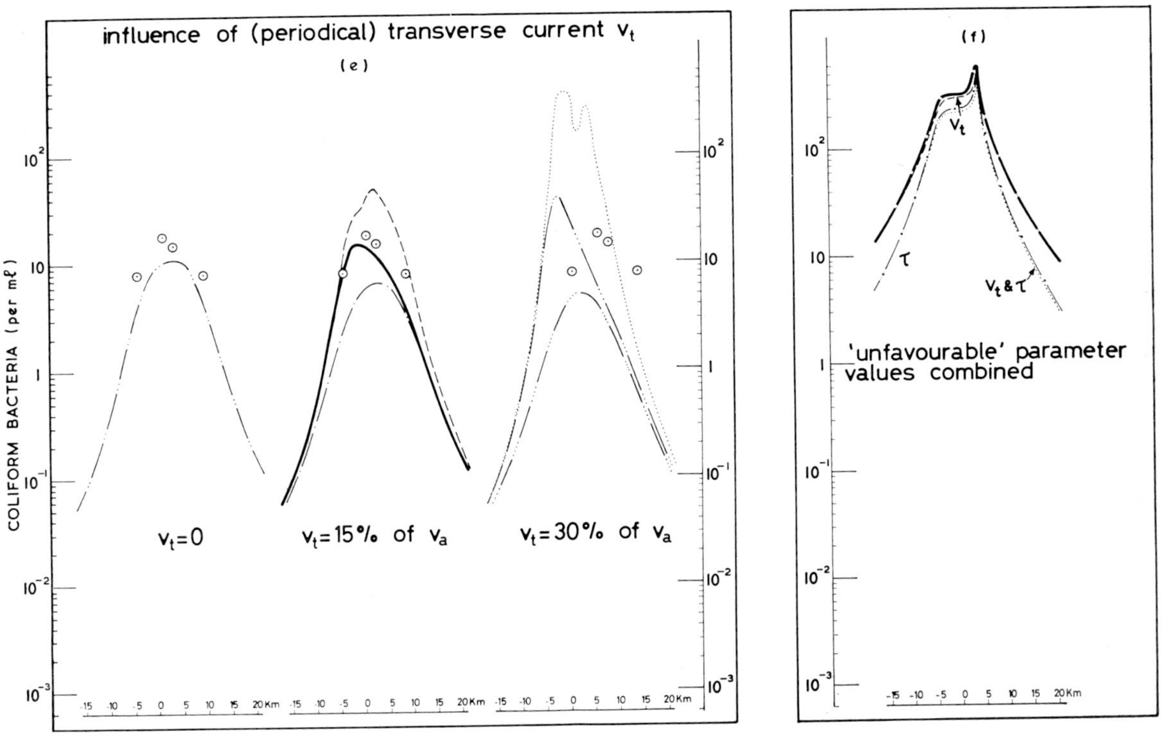

FIG. 4 (continued)

(*e*) Upper, middle, and lower curves for phase with extreme landward, intermediate, and extreme seaward positions

(*f*) Symbols on curves show parameters whose values have been reduced to half the 'extreme' values to which thick curve relates

investigate the extent to which the rather complicated *character* of the measured distributions can be reproduced by the model, keeping the input within the limits of known variations in discharge rates, current directions, and so on. In this way, the artificial two-dimensional distribution presented in Fig. 5 has been produced. Bacterial discharge rates have been varied irregularly within a range of a factor of 9, according to general data and the opinions of experts. Some initial submergence has been simulated by

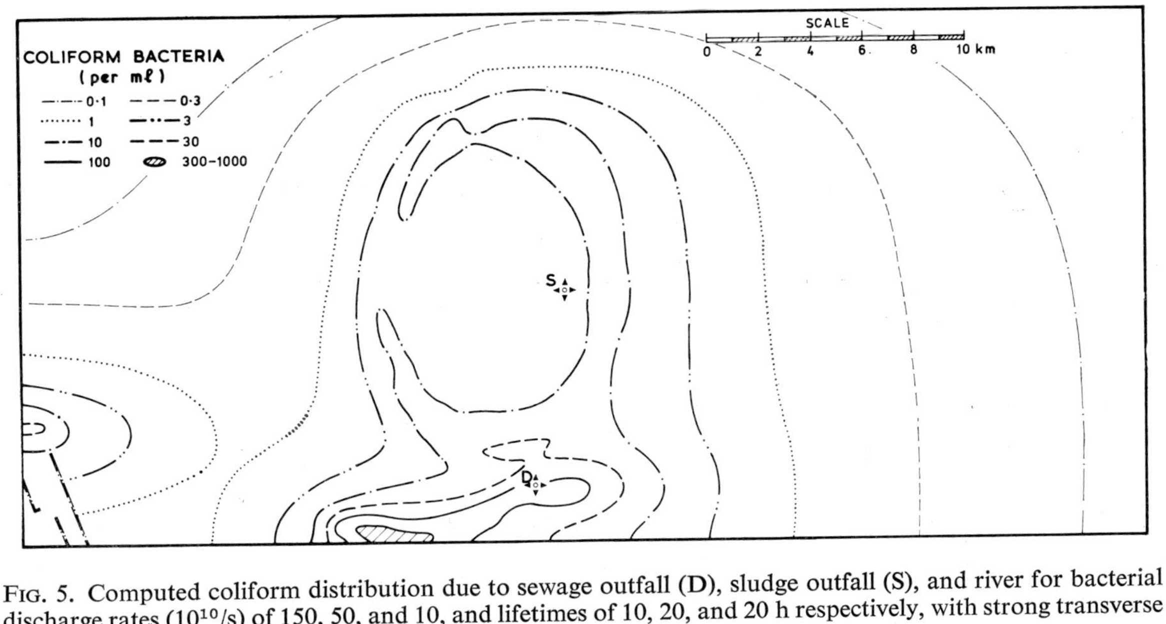

FIG. 5. Computed coliform distribution due to sewage outfall (D), sludge outfall (S), and river for bacterial discharge rates (10^{10}/s) of 150, 50, and 10, and lifetimes of 10, 20, and 20 h respectively, with strong transverse current

At instant shown, transverse current near D *has just transported patch with largest concentration towards coast*

depression of the discharge function in the time interval corresponding with recent release. The sludge outfall and the Rotterdam Waterway have been taken into account. For the sewage sludge, in addition to some initial submergence, a slower decay due to a protective effect by the sludge has been assumed.

It is seen that reasonably analogous results are obtained, to such an extent that certain peculiarities which are only present in the measured distributions might be blamed on the method of determination (inaccuracies in laboratory determination of bacterial numbers; deformations due to inaccuracies in the shift-corrections made to account for advective displacements during the sampling interval of about 4 h).

Of course, the significance of this result is limited. To some extent it gives an illustration of the restricted possibilities of the model. On the other hand, it must be said that a better insight might be obtained by a more extensive experiment with vertical sampling, current measurements at more points, and a detailed record of bacterial discharge rates during the last few days preceding the survey.

REFERENCES

1. ABRAHAM, G. Initiële menging [Initial mixing]. *Water*, 1967, **51**, 170–176 (in Dutch; English summary).
2. ABRAHAM, G., and DAM, G. C. VAN. On the predictability of waste concentrations. In 'Marine pollution and sea life' (ed. M. Ruivo). Fishing News (Books) Ltd, West Byfleet, England, 1972, pp. 135–140.
3. SCHÖNFELD, J. C. Turbulente diffusie: lozing vanuit een bron in zee [Turbulent diffusion: release from a source in the sea]. Rijkswaterstaat, Directie Waterhuishouding en Waterbeweging, MFA 6411, 1964 (in Dutch).
4. HARREMOËS, P. Prediction of pollution from planned sewage outlets. *Teknill. Kem. Aikak.*, 1964, **21**, 251–260.
5. DAM, G. C. VAN. Voorlopige resultaten van eerste metingen en berekeningen aangaande diffusieve verspreiding in de Noordzee op korte afstanden tot de Hollandse kust [Preliminary results of first measurements and calculations with regard to diffusive spread in the North Sea at short distances from the Dutch coast]. Rijkswaterstaat, Directie Waterhuishouding en Waterbeweging, FA-1963-1, 1963 (in Dutch).
6. WON, W. D., and ROSS, H. Persistence of virus and bacteria in seawater. *J. envir. Engng Div. Am. Soc. civ. Engrs*, 1973, **99** (EE3), 205–211.
7. TROMP, D. Rapport over het bacteriologisch onderzoek van het zeewater nabij Scheveningen [Report on the bacteriological investigations of the sea water near Scheveningen]. Rijksinstituut voor Zuivering van Afvalwater, The Hague, 1971 (in Dutch).
8. DAM, G. C. VAN. Horizontal diffusion in the North Sea near the Netherlands coast in connection with waste disposal. International Association for Hydraulic Research, 11th International Congress, Leningrad, 1965.
9. DAM, G. C. VAN, and DAVIDS, J. A. G. Radioactive waste disposal and investigations on turbulent diffusion in the Netherlands' coastal areas. In 'Proc. Symp. Disposal of radioactive wastes into seas, oceans and surface waters'. International Atomic Energy Agency, Vienna, 1966, pp. 233–248.
10. DAM, G. C. VAN. Dispersie van opgeloste en zwevende stoffen in zee gebracht ter hoogte van Wijk aan Zee op 3 km uit de kust [Dispersion of dissolved and suspended matter brought into the sea at 3 km from the shore near Wijk aan Zee]. Rijkswaterstaat, Directie Waterhuishouding en Waterbeweging, MFA 6812, 1968 (in Dutch; extensive English summary).
11. DAM, G. C. VAN, SYDOW, J. S., and WESTHOFF, J. W. Een diffusie-experiment op 10 km uit de Nederlandse kust ter hoogte van Ter Heyde [A diffusion experiment near Ter Heyde at 10 km from the Dutch coast]. Rijkswaterstaat, Directie Waterhuishouding en Waterbeweging, MFA 7003, 1970 (in Dutch).

Introducing his paper Mr. van Dam said that the chosen length of the Hague outfall had been based on the consideration that if T_{90} were not more than about 12 h the operation of the outfall would be satisfactory without additional measures. If T_{90} should appear to be longer, additional measures would be more economical than lengthening the pipeline, because the gain in dilution per kilometre of increase in distance offshore decreased rapidly with the offshore distance.

The values of T_{90} reported in the paper were relatively long, even for the local climate. A possible explanation for this was the very low intensity of solar radiation within the sewage field, owing to the high turbidity in the area and to the initial submergence of the sewage field in most circumstances. Additionally, some of the substances carried by the Rhine might be more harmful to certain natural enemies of coliform bacteria than to the bacteria themselves.

The mathematical model (of the superposition type) had been of great help in interpreting the field data and in determining T_{90} while accounting for the initial submergence. It gave somewhat less complicated two-dimensional distributions than were found in the field, but (as shown in Fig. 4) the computed values for the coast agreed well with the measured results, and also explained the large variations among shoreline samples and gave an insight into the part played by the various factors causing these variations.

DISCUSSION

Mr. J. A. Wakefield (U.K.) said that he had recently observed that the quality of the waters at Scheveningen beach was noticeably worse than some 10 years previously. He asked if Mr. van Dam considered this deterioration was due to the advent of the Hague outfall or the Delta flood barrier.

Mr. R. G. Ludwig (Brazil) asked if the relatively high values of T_{90} found at The Hague were general for the area (or if they related only to local conditions), whether T_{90} had been determined for adjacent areas, and whether the outfall was, in fact, in the optimum place.

REPLY TO DISCUSSION

Mr. van Dam said it was most unlikely that the deterioration Mr. Wakefield had observed was due to the Hague outfall, since (in the period between Mr. Wakefield's observations) the outfall length had been increased from 400 m to 2·25 km for the effluent and 10 km for the sludge. If there had been a systematic deterioration, it could have resulted from decreased quality of river water and increased dumping of materials dredged from the Rotterdam harbours.

In reply to Mr. Ludwig, he said that since the T_{90} values had been obtained from the total bacterial count over the entire area (which was very large) they already represented an average for an extensive region. Separate studies had not been made for adjacent areas. To obtain clearly different conditions—less influence from rivers—it would probably be necessary to go at least 20 km to the north.

Paper No. **40**

Conceptual Design of Marine Waste Disposal Systems

Erman A. Pearson

Professor of Sanitary Engineering, University of California, Berkeley

Increasing public concern about environmental quality coupled with limited factual information about waste discharge effects and a general lack of awareness of the objectives and performance of conventional waste treatment systems all are leading to increased confusion in the development of rational coastal waste management systems. There appears to be a general belief in the minds of the public, conservationists, and even some scientists and regulatory agencies, that all waste treatment systems accomplish the same objectives, that waste treatment is uniformly good depending only upon the degree of treatment, and the higher the degree of treatment for a given discharge location the better the results. Unfortunately, generalizations of the type indicated may lead to system designs that are inappropriate on a conceptual basis and that might result in irreparable ecological damage.

The estuaries and coastal regions of nations are the ultimate recipients of the major fraction of the conservative (non-decayable) waste materials discharged to the inland lakes and rivers directly connected to the sea. In addition there is a substantial additional waste load of both conservative and non-conservative pollutants discharged directly to the estuaries from the metropolitan areas located on the periphery of estuaries. Obviously the conservative pollutants as well as a fraction of the non-conservative pollutants reach the coastal area in a relatively short time. The fraction of non-conservative materials reaching the coast is determined by the waste residence time in the estuary (a function of discharge location, advective flow, mixing, and tidal exchange) and the decay rate of the pollutant.

Unfortunately, the subject of estuarine and marine waste disposal has not received as much attention as inland waste disposal practices and the advantages and disadvantages of traditional inland waste disposal practices in the marine area have not been elucidated clearly. Of the effort expended on estuarine problems, the major fraction has been devoted to physical-hydraulic aspects of the problem rather than to the bio-chemical aspects of pollutant decay and effects on the ecology of both conservative and non-conservative pollutants.

Estuary, Bay, or Open Coast Discharge Dilemma

A major problem in coastal water quality management is determination of the appropriate degree of wastewater treatment and the preferred location of the treated wastewater discharge. Consider an estuary or bay with a significant number of municipal and industrial wastewater discharges located around the periphery of the water body. Historically, the general trend in wastewater management has been to invest heavily in treatment processes—frequently as much as can be financed, and to pay little attention to the location and type of dispersion system for the treated effluent. In the past the choice has been between minimal treatment and primary, or primary and secondary (biological) treatment; however, in the U.S.A. today, the choice appears to be between secondary (the politically determined minimum treatment) and some form of advanced (tertiary) wastewater treatment (AWT).

The fundamental question remains, however: is it better to provide advanced (secondary) treatment and discharge the effluent directly to the estuary or bay, or is it preferable to employ a lesser degree of treatment (primary) and discharge the waste to a distant area that has high dilution capabilities (i.e. $\geqslant 100{:}1$)? The latter alternative would use the incremental cost between primary and secondary wastewater treatment to transport the effluent seaward, or to the open coast, to a disposal location less sensitive ecologically and where much greater volumes of diluting water are available.

Unfortunately, there appears to be a developing attitude of regulatory agencies, at least in the U.S.A., to favor increasing the degree of treatment and to give little if any consideration to the location of the ultimate discharge, the degree of initial as well as subsequent dilution of the waste, and the decay of the pollutants in the receiving water. This trend appears to be supported or encouraged by many, if not most, of the consulting engineers. If the foregoing concept becomes established and widely used, there will be little if any economic or ecological incentive to use the incremental cost between various levels of treatment to transport the lesser treated effluent to a disposal site where there are maximum diluting capabilities for the effluent and/or where the effluent will have the least adverse effect on the ecology.

If one is concerned about pollutant concentrations in the environment and their effects, one must give serious consideration to determination of waste discharge locations where the residual pollutant concentration will have minimal ecological (including human) impact. To accomplish this, it should be obvious that quantitative information must be available on pollutant mass-emission rates and concentrations, on the efficacy of pollutant removal processes, and on the physical, chemical, and biological characteristics of alternative disposal sites. The practical facts are that such information is not generally available to permit rational assessment of alternative treatment and dilution combinations. However, this lack of such information in no way justifies the absence of rational qualitative assessment of the consequences and costs of the various alternatives.

The available initial dilution for treated municipal and industrial wastewaters located around the periphery of a bay or estuary and discharged directly thereto ranges from near zero at the head of a neutral estuary (advective flow $\simeq 0$) to about 20–30 for combined mid-bay or estuary discharges, to possibly as much as 30–50 for a well designed outfall discharge to a large bay. Obviously, the actual dilution of wastewaters in a given restricted body of water depends upon the waste flow rate, the advective flow, and effective tidal/turbulent exchange. For example, if all the estimated 1990 municipal and industrial wastewaters generated around the periphery of San Francisco Bay were collected and discharged in the Central Bay in front of the Golden Gate, the average dilution of the wastewaters would be in the order of only 30–40; and San Francisco Bay is a large estuary with appreciable advective flow (principally the Sacramento River) and tidal exchange.

However, disposal systems located along open coasts (such as along the California Coast including the San Francisco Bay Area) can be designed for essentially any likely magnitude of discharge (up to at least 2×10^6 m^3/day) to achieve immediate initial dilutions in the order of 150–200. It should be noted that conventional secondary waste treatment systems (except for disinfection processes) achieve as a maximum, average pollutant removals of 90–95 per cent (equivalent to dilutions of 10–20) and that primary treatment achieves as a maximum about 65 per cent pollutant removal (equivalent to a dilution of 3). If one considers a terminal dilution of 150–200 with the primary treatment option and about 10 with the secondary system, the total equivalent dilution (removal plus terminal dilution) for the primary option is 450–600 compared to 100–200 for the secondary system. It should be obvious that the resultant effect on the receiving water will be less for the primary treatment (high terminal dilution) system than for the secondary treatment (low terminal dilution) alternative. The foregoing simplified comparison relates to pollutants subject to removal by the processes used. How about the possible effects from pollutants whose removal by the treatment processes used is minimal? Obviously, the high-dilution alternative provides the greatest degree of environmental protection (i.e. lowest pollutant concentrations).

Advantages of Coastal Alternatives

Consider comparison of a typical treatment-disposal system discharging to an estuary/bay with an open-coast treatment-disposal system using one level of treatment lower than that used for the estuary/bay system. For an equal total cost system, the incremental cost between the two levels of treatment can be used to construct an interceptor sewer to the open coast. In general (some exceptions) the coastal alternative produces the lower receiving-water pollutant concentration and hence the least ecological effects. Moreover, the coastal discharge location can be selected so that the area presumed to be affected to some degree is one of lesser ecological significance than would be found in the estuary/bay system.

Typical estuarine-coastal example

Figure 1 and Table 1 present an idealized but representative example of the possible choices in estuarine waste management with respect to disposal location, likely initial waste dilutions obtainable, and pollutant concentrations at each location dependent upon the degree of treatment employed. It is of interest to note that the average waste concentrations in the receiving water at the immediate point of discharge are lowest for the case with the least degree of treatment. Similarly, the degree of treatment (percentage removal) required at each of the discharge locations to produce the same pollutant concentration near the source is shown in the final line of Table 1. To achieve a pollutant concentration of $C_0/150$ in the receiving water requires no treatment at Location C (open ocean), 80 per cent removal at B (mid-estuary), and 93·3 per cent removal at A (landward end of estuary).

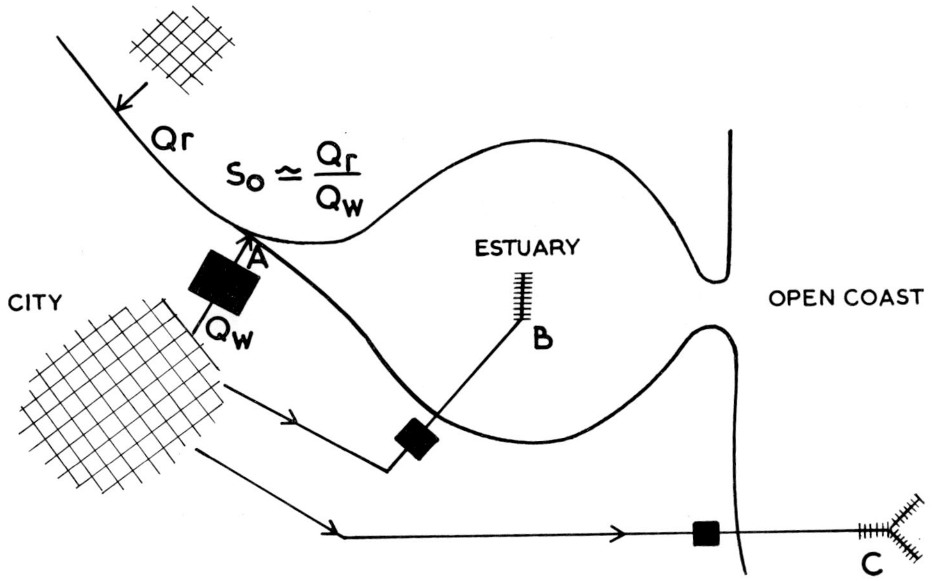

FIG. 1. Idealized estuarine-coastal disposal alternatives

TABLE 1. Worked example for situations shown in Fig. 1

Discharge location	A	B	C
Initial waste dilution, S_0 (at minimum river flow)	10	30	150
Approximate pollutant concentration near source, at diffuser, with no treatment (C_0 = concentration in untreated waste)	$C_0/10$	$C_0/30$	$C_0/150$
Treatment	*Secondary*	*Intermediate*	*Primary*
(Assumed percentage removal)	(90)	(75)	(33)
Approximate pollutant concentration near source	$C_0/100$	$C_0/120$	$C_0/225$
Treatment (percentage removal) required for equal pollutant concentration, $C = C_0/150$, at each location	93·3	80	0

Two points need to be emphasized. First, other than for disinfection processes, there are no conventional waste treatment processes that remove, on the average, 93·3 per cent of BOD, COD, suspended solids, toxicity, or other significant pollutants; secondly, for open-ocean disposal (average initial dilution of ≥ 150) no treatment would be required. Obviously, this kind of system would not be tolerated generally. Thus, it would appear from just the standpoint of initial waste concentrations in the environment, the ocean disposal system has substantial advantages. These advantages become particularly important if the ocean disposal system can be built and operated at about equal or lesser costs than either of the estuarial disposal schemes.

Advantages of Coastal Alternative

1. For the conventional levels of treatment commonly used, the coastal alternative produces the lowest pollutant concentrations in the receiving waters.
2. Generally, the coastal discharge location can be selected in an area of non-critical or of limited (lesser) ecological significance in the area; whereas the estuarine region is one of the most important and critical areas with respect to both the anadromous and marine fishery as well as the local ecology.
3. The residence time of pollutants in the estuary, the critical region, is reduced; hence, all effects of pollutants are minimized.
4. If pollutants of unknown character or origin occur, for which the treatment process may or may not be adequate, the coastal system affects substantial reduction in the pollutant concentration by dilution, thereby providing a safeguard to minimize the effects of any changes in the future.
5. If the treatment process should fail completely or the operation deteriorate as the result of many

factors, the effect of such malfunction is markedly reduced by the substantial dilution available for the pollutants of significance.

6. Concern is often raised about the effect of conservative toxic materials such as DDT, PCB's, etc. which also are not removed to a significant degree by conventional treatment plants built to date—and to an unknown degree by proposed advanced waste treatment systems. Obviously, the best management of such materials is proper source control. However, should such materials gain entrance to the sewage system the coastal disposal system has marked advantages. The high dilution ($\geqslant 150$) afforded by the open coastal dispersion system reduces the initial (as well as subsequent) pollutant concentrations in the receiving water environment. Correspondingly, the lower pollutant concentration reduces the rate at which the pollutant can be incorporated into the biota or food chain. Thus, the biological effects can be expected to be less, as a function of time, which should allow time for correction before irreversible damage is done.
7. It should be recognized that upstream development (beyond the immediate periphery of the estuary/bay) of essentially all river systems will inevitably contribute significant pollutants (waste residuals) to the estuary/bay. Experience to date with many estuaries has shown that, in the long term, the upstream pollutant load alone will impose substantial stress on the estuarine region. Thus, generally it has become a practical and ecological necessity to collect all or most of the locally generated estuarine/bay wastes and transport these wastes, after proper treatment, out of the estuarine system to a point of maximum terminal dilution (the open coast). Only in this way will it bc feasible and possible to protect the estuarine bay region from serious ecological damage. It is of interest to note that recent (November 1973) proposed California water quality control policy states [1] that the objective is "ultimate elimination of municipal and industrial waste discharges (other than cooling water discharges) to bays and estuaries, other than the San Francisco Bay–Delta system."
8. One occasionally encounters the hypothesis put forth by advocates of advanced or complex treatment processes that the transport-dilution system to open-coast disposal sites is an unwarranted expenditure because it detracts from the possible economic feasibility of wastewater reclamation. This is indeed a weak, if not completely erroneous, hypothesis for the following reason. If wastewater reclamation is to be practiced, the net result is the generation of more, not less, pollutants for disposal. This should be obvious if wastewater reclamation practices are to provide waters safe for the user or beneficial uses. What happens to the residual (non-reclaimable) pollutants? Some can be recovered and disposed of as semi-solid wastes on land. Yet, a significant fraction of the non-reclaimables must be disposed of to the environment after suitable treatment. Indeed, a well-designed transport-dispersal system to discharge locations with high diluting capacity is an adjunct, if not a necessity, for economic and ecologically feasible wastewater reclamation systems.
9. An additional advantage of the coastal system is its flexibility and economy to meet future changes in water quality requirements to deal with problems either non-existent or unrecognized at present. Improvements in waste treatment technology can be applied when they become available and are needed to deal with new pollutants or newly discovered pollutant effects. And it will not result in the abandonment of expensive traditional treatment systems not relevant with respect to current technology and problems. An example is the current need for toxicity removal processes as contrasted to traditional emphasis on building plants to upgrade (from 90 to 95 per cent or 95 to 98–99 per cent) removal of BOD and suspended solids; when, as in the case of open coastal disposal of municipal wastewaters, BOD and suspended solids removals at these levels are *not* of *any* practical significance.

The Open Coastal Problem

The conceptual planning of waste disposal systems for dischargers located along an open coastline is very similar to that facing dischargers located along an estuarine or bay system. Assuming that arbitrary and irrational regulatory requirements do not determine the treatment option with little if any regard to disposal location or dilution possibilities available, the problem is one of determining the optimum combination of wastewater treatment and disposal/dispersion system that will meet rational and realistic water quality objectives.

Hypothetical example

Figure 2 presents an idealized set of waste disposal alternatives for a municipality located on the open coast. Three alternative systems are compared as follows: A, conventional secondary treatment, 90 per

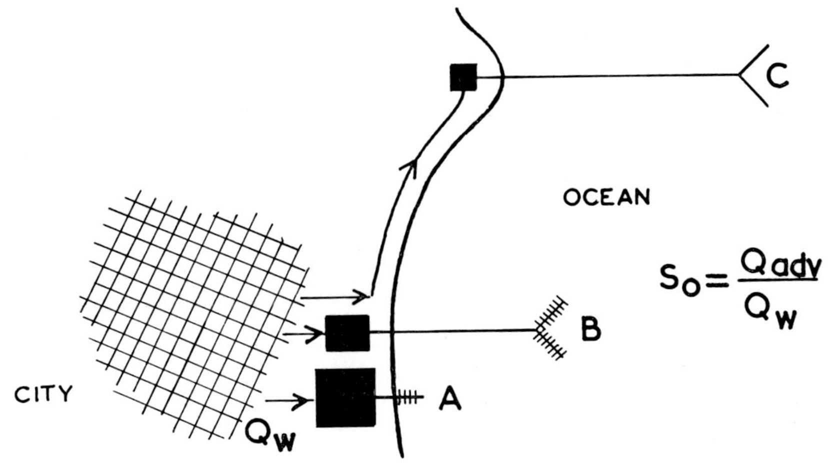

FIG. 2. Idealized open-coast disposal alternatives

TABLE 2. Worked example for situations shown in Fig. 2

Discharge location	A	B	C
Initial dilution at source, S_0 (Value assumed)	5–20 (10)	60–80 (70)	100–200 (150)
Pollution concentration at source, conservative substance (C_0 = concentration in untreated waste)	$C_0/10$	$C_0/70$	$C_0/150$
Treatment (percentage removal)	90	75	50
Assumed transport time to beach (h)	0	12	24
Pollutant concentration at beach			
Conservative substance	$C_0/100$	$C_0/336$	$C_0/450$
Non-conservative substance			
$T_{90} = 4$ h ($k = 0{\cdot}58$ h^{-1})	$C_0/100$	$C_0/3{\cdot}4 \times 10^5$	$C_0/4{\cdot}5 \times 10^8$
$T_{90} = 1$ h ($k = 2{\cdot}3$ h^{-1})	$C_0/100$	$C_0/3{\cdot}4 \times 10^{14}$	$C_0/4{\cdot}5 \times 10^{26}$
$T_{90} = 12$ h ($k = 0{\cdot}19$ h^{-1})	$C_0/100$	$C_0/3{\cdot}4 \times 10^3$	$C_0/4{\cdot}5 \times 10^4$
$T_{90} = 240$ h ($k = 0{\cdot}23$ day^{-1})	$C_0/100$	$C_0/378$	$C_0/567$

cent removal of pollutant, discharge near shore, average dilution 10; B, intermediate treatment, 75 per cent removal of pollutant; discharge some distance from shore, average dilution 70; C, primary treatment, 50 per cent removal of pollutant; discharge offshore with high average dilution 150.

The waste concentrations near shore are compared for each alternative, in terms of the untreated waste pollutant concentration, C_0, in Table 2. The residual waste concentrations consider removal by the various treatments, the initial dilution at the source, and the combined effects of treatment, initial dilution, and dilution during transport. It is apparent that System C, with the minimum treatment (50 per cent removal) and high initial dilution, produces the minimum waste concentrations considering only treatment and dilution processes such as would affect the concentration of conservative (non-decayable) pollutants. It should be noted that the pollutant concentration at the source for Alternative C is $C_0/300$ (50 per cent of $C_0/150$) which is a lower concentration than for Alternative A, $C_0/100$ (10 per cent of $C_0/10$). Thus it appears that Alternative C is the preferred solution with respect to pollutant concentrations in the environment.

If one is concerned with non-conservative (decayable) pollutants in the waste discharge, and this historically has been the focus of all biological treatment processes to date, the obvious preference of Alternative C over A and B is shown in Table 2 for the case where T_{90} is 4 h (i.e. 90 per cent disappearance in 4 h, exponential rate constant $k = 0{\cdot}58$ h^{-1}). Compare a shoreline pollutant concentration $C_0/4{\cdot}5 \times 10^8$ for System C with those of $C_0/3{\cdot}4 \times 10^5$ for System B and $C_0/100$ for System A, the secondary waste treatment system. The foregoing decay rate is a typical design value for coliform bacteria in sea water and it has been used in most of the recently designed marine waste disposal systems in California.

With the recent adoption in the U.S.A. of the federal E.P.A. requirements of secondary treatment for all public waste discharges and the inclusion of a fecal coliform effluent requirement of 200 MPN/100 ml, the impact of this requirement on marine waste disposal practice remains to be seen. Most of the recent marine waste disposal systems in California have been based upon the discharge of non-disinfected effluent and they have relied on coliform disappearance rates equivalent to a T_{90} of approximately 4 h to meet the coliform standards in body-contact areas. The justification of an effluent-disinfection requirement, at least in the U.S.A., considering the potential increase in effluent toxicity associated with disinfection, remains to be demonstrated. If the effluent coliform requirement is enforced, it is expected to have marked impact on outfall design practice in California. It will likely result in considerably shorter and much less effective outfall dispersion systems than are used at present. Whether or not this will have a beneficial effect on marine waste disposal practice remains to be demonstrated. Unless substantial dilution requirements ($\geqslant$100:1) remain in force as a special requirement, there is no doubt in the writer's judgement that sole reliance on arbitrary effluent criteria of the type indicated will be a step backward in marine waste disposal practice, at least in California. Also it will likely have adverse if not disastrous ecological effects.

Treatment, Transport, and Dispersion Systems

To evaluate properly the role and economic justification of the optimum combination of treatment, wastewater transport to a suitable site, and the appropriate dilution-dispersion system, it is necessary first to consider the unit costs of each component of the system.

Treatment

Table 3 presents a summary of 1973 California unit costs for various levels of treatment and size of plant. These values include all costs: capital (based on a 20-year life and 5 per cent interest rate), operation, and maintenance. The design flows shown are equivalent to 1, 10, 100, and 1000 million U.S. gallons per day. It is seen that for 10 mgd (37 800 m³/day) and 100 mgd (378 000 m³/day) the 1973 unit cost increments between primary and secondary treatment with filtration are about 4·1 and 1·9 cents/m³ of waste treated, respectively. Similarly the unit cost increments between secondary and advanced waste treatment (physical, chemical, biological—Alternative B) for the same size plants are about 5·2 and 3·3 cents/m³ respectively. Conceivably the incremental unit cost between the two levels of treatment could be used to transport the waste to a disposal site where the dispersion system could handle without impairment the lesser treated effluent.

TABLE 3. Estimated total unit costs (U.S. cents per m³) of wastewater treatment in California during 1973 Costs include capital costs, based on 20-year life and 5 per cent interest rate, and disposal of waste residuals

Treatment system	Plant design flow (m³/day)			
	3 780	37 800	378 000	3 780 000
Primary	7·4	3·3	1·8	1·5
Intermediate Chemical, filtration, chlorination	13·0	7·4	3·7	2·6
Secondary Activated sludge, filtration, chlorination	13·0	7·4	3·7	2·6
Advanced A. Chemical, filtration, activated carbon, chlorination	24·0	11·4	6·5	4·6
B. Activated sludge, chemical filtration, activated carbon, chlorination	25·9	12·6	7·0	5·3

Transport

The cost of interceptor sewer/transport systems depends primarily upon the surface and sub-surface conditions along the route. Table 4 presents 1973 construction cost estimates for dry-trench construction of interceptor sewers of suitable size to handle the three levels of flow considered.

TABLE 4. Estimated unit costs of interceptor sewers, dry-trench construction, in 1973

Design flow		Sewer size		Construction cost $/m
m³/day	(mgd)	cm	(in)	
3 780	(1)	25·4	(10)	60
37 800	(10)	91·5	(36)	206
378 000	(100)	245	(96)	672

Outfall

The costs of outfall construction vary considerably because of differences in surf and bottom conditions. The cost of the onshore-surf section is high where the outfall must be buried for protection of the pipe; hence the average cost depends upon the length of the outfall and the ratio of the offshore to onshore sections. Table 5 presents reasonable estimates of the average construction costs per unit length of outfall for the sizes of outfall suitable for the three ranges of flows considered. The values reported are in the middle ranges of construction costs (adjusted to 1973 prices) for outfalls built along the Pacific coast during the past 20 years.

TABLE 5. Estimated unit costs of submarine outfalls, California construction practice, in 1973

Design flow		Sewer size		Construction cost $/m
m³/day	(mgd)	cm	(in)	
3 780	(1)	15	(6)	330
37 800	(10)	61	(24)	1 395
378 000	(100)	194	(76)	2 720

Treatment/Transport/Outfall Tradeoffs

In the conceptual design of a wastewater disposal system, one should consider various levels of treatment or pollutant removals, possible transport distances/disposal sites, and outfall-diffuser lengths to produce the minimal impact on the environment. The incremental unit costs between the various levels of treatment can be considered as available for waste transport or outfall construction. Thus, one can compare the effect of high levels of treatment and discharge with minimal dilution close to the source of waste generation with reduced level of treatment and disposal in an area of lesser ecological significance with high initial dilution. Obviously, a rational comparison entails quantitative comparison of both sites with respect to both dilution possibilities and ecological characteristics.

The rational comparison of alternative disposal sites and treatment levels is infrequently done, in part because of the lack of definitive data on dilution capabilities and ecological characteristics of the alternative sites as well as on pollutant parameters and process-removal characteristics. The latter is especially true for the developing parameters such as toxicity and biostimulation. However, a simple economic comparison of different levels of treatment with incremental costs between levels used to estimate break-even land transport distances (interceptor sewers) or lengths of submarine outfalls can be very helpful.

Table 6 presents a summary computation of the distance the treated wastewater can be transported in interceptors or taken to sea in outfalls (such as are constructed on the Pacific coast) for the same cost as the incremental cost between two levels of treatment—in this case between primary treatment and secondary treatment including effluent filtration.

TABLE 6. Lengths of interceptor sewer or submarine outfall that can be constructed for incremental cost between primary and secondaıy waste treatment

Design flow		Incremental unit cost, primary to secondary[a]	Total annual incremental cost, primary to secondary	Equivalent interceptor[b,d]		Equivalent outfall[c,d]	
m³/day	(mgd)	(cents/m³)	($)	Unit cost ($/km)	km	Unit cost ($/km)	km
3 780	(1)	5·6	76 650	3 250	23·6	19 130	4·00
37 800	(10)	4·1	565 750	11 300	50·1	81 400	6·96
378 000	(100)	1·9	2 555 000	36 900	69·3	159 000	16·1

[a] See Table 3 for process description
[b] Based on useful life of 50 years
[c] Based on useful life of 40 years
[d] Assuming interest rate of 5 per cent; excluding pumping, operational, and maintenance costs

It can be noted that for a flow of 37 800 m³/day a secondary plant has the same unit cost ($/m³) as a primary treatment plant with 50·1 km of submarine outfall interceptors or with 6·96 km of open-coast submarine outfall. Obviously, a similar break-even cost solution would be a primary plant with 25 km of interceptors and 3·48 km of submarine outfall. For a 378 000-m³/day plant the incremental cost of going from primary to secondary would permit building 69·3 km of interceptors or 16·1 km of submarine outfall or appropriate combinations of the two to reach a disposal area that would permit or be enhanced by a properly discharged primary effluent. Obviously, if suitable disposal sites were closer to the city than these estimated distances, the primary treatment, transport, and outfall alternative would be more economical than the secondary treatment system.

It should be noted the computed lengths of interceptor sewer and outfall are in addition to the intercepter and outfall lengths required with the secondary treatment solution.

The analysis in Table 6 does not include pumping or interceptor/outfall operation and maintenance costs. In a practical comparison, these costs should be included if pumping is required—the example assumes gravity flow. Inclusion of these costs does not change significantly the break-even lengths, at least for the two largest flows—it reduces the equivalent lengths about 10 per cent. For the flow of 3780 m³/day the length may be reduced as much as 20 per cent.

Certainly the data presented in Table 6 indicate that a reasonable investigation of possible disposal sites is warranted before secondary treatment systems are designed to discharge effluents to the local environment. Also, it should be noted that the preceding analysis is based upon a comparison of primary treatment and secondary waste-treatment removals and costs. Obviously, where regulations do not require primary (or any) treatment as a minimum, a similar comparison can and should be made for the alternative costs and consequences between no or minimal treatment (screening) and primary treatment systems.

Conclusions

Rational analysis of coastal waste disposal systems requires consideration of the efficacy of several treatment levels with respect to pollutant removal and costs, the cost and consequences of transporting adequately treated wastewaters to disposal sites with high diluting capabilities and/or low levels of ecological significance, and the effects of the resulting pollutant concentrations in the receiving water environment.

Increasing levels of wastewater treatment with disposal to local, high-value receiving waters may be the most expensive solution and also the one with the greatest ecological hazard. The possibility of treatment process malfunction or failure and unknown or unremoved pollutants may also have serious ecological consequences.

One should consider the incremental costs between the several possible levels of wastewater treatment, and the use of the incremental cost for transport systems or outfalls to convey the wastewater with the lower level of treatment to a disposal site of high wastewater diluting capacity and/or a low level of ecological significance.

In an example computation, it is shown that for the incremental cost for a 10 mgd (37 800 m³/day) plant between primary treatment and secondary treatment, including filtration, an interceptor sewer 31·1 miles (50·1 km) in length or a marine outfall some 4·32 miles (6·96 km) in length can be constructed at 1973 prices. Obviously, the break-even cost solution would entail some appropriate combination of interceptor and outfall. Moreover, for most alternative solutions, primary treatment with transport to an area of high dilution capability would result in the lowest pollutant concentration in the environment. Thus, it would be reasonable to conclude that this solution would have the least ecological consequences, certainly for non-conservative pollutants. It would also be true for conservative pollutants or any pollutant not significantly removed by conventional treatment processes.

Ecologically sound solutions to waste treatment/disposal problems require the analytical evaluation of pollutant discharges, treatment efficiencies and costs, the consequences and costs of land transport and marine outfalls to convey wastes to areas of high dilution and low levels of ecological significance, the resulting concentrations of pollutants in the receiving waters and their effects. Only in this way can marine waste disposal solutions be developed on a rational and economic basis.

REFERENCE

1. California State Water Pollution Control Board. Proposed water quality control policy for the enclosed bays and estuaries of California. Sacramento, Calif., November 1973.

Prof. Pearson summarized his paper by saying that the design of estuarine and marine waste disposal systems had not received as much attention as inland waste disposal; consequently the applicability of traditional inland waste disposal practices to the marine area had not been elucidated clearly. The consequences and costs of transport, treatment, and dilution (outfall–diffuser) systems in various combinations for discharge into estuaries or to the open coast were analysed in detail and compared on a quantitative basis. Increasing levels of treatment with discharge to local high-value receiving waters might be the most expensive solution and also the one with the greatest ecological hazard. The possibility of malfunction or failure of treatment processes, or of unremoved and/or 'unknown' pollutants, might have serious ecological significance.

It was shown that for the incremental cost of a 37 800-m^3/day plant, between primary and secondary treatment (including filtration), 50·1 km of interceptor sewer or 6·96 km of marine outfall could be constructed at 1973 California prices. Obviously, the appropriate break-even cost solution would entail some appropriate combination of interceptor and outfall. For most situations, primary treatment systems with transport to an area of high dilution capability would result in the lowest pollutant concentrations in the environment. Thus, this solution would have the least ecological consequences, certainly for non-conservative pollutants. It would also be true for conservative pollutants or any pollutant not significantly removed by conventional treatment processes.

Ecologically sound solutions to waste treatment and disposal problems required the analytical evaluation of pollutant discharges, treatment efficiencies and costs, the consequences of costs of land transport and marine outfalls to convey wastes to areas of high dilution and low levels of ecological significance, the resulting concentrations of pollutants in the receiving waters, and their effects. Only in this way could marine waste disposal solutions be developed on a rational and economic basis.

DISCUSSION

Mr. B. Grover (World Bank) said that, in his opinion, Prof. Pearson's paper was one of the most sensible presented at the Symposium. Three particular aspects merited consideration by anybody responsible for making decisions on sea outfall schemes: his attention to concepts which had general application, his scorn for stupid environmental legislation, and his attention to economics (in a rational and common-sense manner) complete with a hypothetical but realistic example. In Mr. Grover's view, the economic aspects should be considered even further by planning groups not bound to irrational criteria. Not only should alternative systems be investigated to determine the least-cost solution to meeting a single water quality objective, but the exercise should also be repeated for a range of objectives; information thus obtained would permit the decision makers to choose rationally—fully aware of all available options—when selecting the optimum combination of treatment and outfalls. He appreciated that such planning involved more time and money than did a simplistic approach; it was a matter for good judgement to determine whether such costs were justifiable.

Dr. H. A. Cole (U.K.) said that he had made considerable efforts over recent years to promote a balanced outlook on marine pollution problems, both nationally and internationally, with the aim of preventing unrealistic legislation and inclusion of unreasonable demands in international conventions to which the U.K. adhered. The impression had been given that the U.K. took an easy-going tolerant view of marine pollution, but this was not correct—the attitude was practical and realistic rather than easy-going. An important omission from Prof. Pearson's paper was that only a very short list of substances (two metals and a couple of groups of persistent organics) should be excluded altogether from wastes discharged to the sea; if this were recognized, the problems of marine waste disposal would be vastly simplified and costs greatly reduced.

Mr. J. A. Wakefield (U.K.) congratulated Prof. Pearson on his paper which, he suspected, would be quoted when River Authorities were attempting to justify sea outfalls rather than biological treatment. He was glad to note, however, that the paper also included material on which environmentalists could argue for the construction of longer outfalls.

Dr. M. Waldichuk (Canada) wished to consider a component not shown in Fig. 1, namely a channel separated from the open ocean by islands. Such a situation existed on the east coast of Vancouver Island where a pulp-mill effluent could not be discharged into the adjacent Osborn Bay because of the presence of a series of barrier islands, barely 1 km offshore, with numerous oyster leases. Consequently, a pipeline with diffuser, discharging into 20 m of water beyond the islands into Stuart Channel, was installed. Although initial dilution seemed satisfactory, the oysters did not appear to have fared too well, and it was currently suspected that the assimilative capacity of the channel had been exceeded—particularly for the refractory long-lived materials like resin acids present in kraft pulp-mill effluent. He asked if those involved in outfall design or mathematical modelling had considered such systems in tandem to avoid serious pollution of one of the 'downstream' basins.

Mr. D. W. Mackay (U.K.) said that one point of view which had not been put forward at the Symposium was that of the regulatory agencies who had to decide what was acceptable in terms of the pollution resulting from discharge of sewage from sea outfalls. Mrs. Jeger (p. xiii) had indicated that this had to be (in the final event) a political decision. As a marine pollution control officer his masters were the Clyde River Purification Board which comprised two-thirds locally elected politicians and one-third members appointed directly by national government (representing such interests as fisheries, agriculture, and industry). Their job was to control pollution along some 500 km of Scottish coastline, and they did this by licensing every discharge, controlling the siting of outfalls, and in many cases carrying out the design surveys. If they were to sit in judgement of the proposals, they had to know at least as much of the relevant circumstances as did the proposers, and to that end they had a multi-disciplinary approach and their survey team was in the field on about 200 days per year.

When considering the discharge of sewage to the sea, at least the following factors had to be taken into account: the nature, volume, and composition of the sewage; the variability of the flow, and the projected change in flow within the lifetime of the system; the characteristics of the receiving waters in terms of density, stratification, and basic circulation, and the degree of pollution already present; the effects of wind-induced currents which might be far greater than basic tidal effects; the topography of the sea bed and adjacent coastline; the proximity of shell-fisheries and other biologically vulnerable areas such as spawning grounds; the degree of use of the receiving waters for bathing or other recreational activities; the proximity of other outfalls or of polluted rivers; the consciousness of the local and visiting populations to aesthetic values; the alternative methods and costs of achieving the desired end, and the finance available. In practice, for any given situation, a desk study of these constraints narrowed the possibilities to such an extent that, after an initial siting survey, a detailed feasibility study of quite a small area was all that was needed.

He was personally very pleased that no catch-all standards based on one or two parameters had emerged from the Symposium. The variability between sites was such that one would not consider automatically imposing the same restrictions on different outfalls. Once standards were set, they could be used in justifying discharges to completely unspoilt areas.

Dr. V. J. Cabelli (U.S.A.) asked Prof. Pearson how he reconciled his conclusions and opinions on secondary treatment with Rittenberg's findings[2] on the relative coliform densities in the sediments around the Hyperion (secondary) as opposed to the White's Point and Long Beach outfalls. With regard to Prof. Pearson's comment on malfunctions of sewage treatment plant, he asked if one should not consider this in relation to the probability of untoward meteorological and hydrographic conditions driving the material onto the shore, even from a 'long' outfall. Referring to Prof. Pearson's comment that standards for chemicals in sewage or receiving waters could be more stringent than those for drinking water, Dr. Cabelli thought this could very properly occur if, in terms of the effective concentration, ecological effects came before human health effects. Finally, he observed that there was a very legitimate political aspect to the development of standards—once the scientists had provided the data the politicians would have to assign an 'acceptable risk' in order to translate these data (criteria) into standards.

REPLY TO DISCUSSION

Prof. Pearson thanked Mr. Grover for his kind remarks, and said that in California there were limitations on the concentrations of at least five substances (total chromium, arsenic, silver, cyanide, and chlorinated hydrocarbons) in the waste-water discharges to the Pacific Ocean which were more restrictive than the allowable concentrations in the public water supplies. It was recognized that it was theoretically possible to have a rational technical justification for such restrictive levels; unfortunately, such rationalizations had not yet been provided.

He generally agreed with Dr. Cole's remarks, but believed that it would be very difficult to get many scientists to agree on the 'short list' of substances to be excluded from marine discharges. His own experience indicated that the consensus might be a rather long 'laundry list' of materials—based more upon feelings than upon facts.

In reply to Mr. Wakefield, he said he considered that the most cost-effective portion of a marine disposal system was the outfall–diffuser component. As indicated in his paper, the economic trade-off in differential costs between incremental levels of conventional treatment would build very long outfalls. The last place to cut back was in the outfall length which, incidentally, was the major ecological safeguard when treatment systems performed poorly or failed as they had been known to do.

Prof. Pearson thought the situation cited by Dr. Waldichuk was unique—he knew of no comparable situation where such estuarine waste-water disposal systems had been designed and constructed. However, the proposed submarine outfall beyond the archipelago at Gothenburg[3] was somewhat similar, though it was questionable whether this outfall would ever be built. To consider that problem in estuarine design one had to examine carefully the initial mixing, subsequent dilution, and the decay rate of the pollutants of concern—apparently toxic substances in this case. It was also necessary to obtain information of the estuary circulation, exchange characteristics, and pollutant residence-time distribution for the discharge location to make a realistic assessment of the pollutant concentration and potential effects on the oyster beds. In short, it was a very difficult problem.

In reply to Dr. Cabelli's first question, he said he did not recall specifically the findings of Rittenberg on coliforms in the sediments around the Hyperion and White's Point outfalls—other than there had been (and still were) lots of them at both locations. In fact, this had to be expected as most California workers in the field had found that one of the major, if not the most significant, mechanisms responsible for coliform disappearance in the water column was that of flocculation and sedimentation. The Hyperion treatment plant had been operated solely as a secondary plant for only about 5 or 6 years (in the early 1950s); excessive coliform concentrations and oil and grease on the beach resulted in redesign and rebuilding—basically back to a primary treatment plant (currently discharging one-third secondary and two-thirds primary effluent) with a fivefold increase in outfall length to 9 km. This was something that would be impossible in the present regulatory climate. The new disposal system had met all the water quality criteria ever since the early 1960s—something the secondary plant had failed to do as early as 5 years after construction.

With regard to Dr. Cabelli's second question, Prof. Pearson's experience was that the probability of treatment-plant malfunction was orders of magnitude greater than that of 'untoward meteorological and hydrographic conditions'. Moreover, plant malfunction was likely to occur during the beach-use season, whereas there should be little risk to human health during 'untoward ocean phenomena' because of the low level of beach usage in such periods.

ADDITIONAL REFERENCES

2. Rittenberg, S. C., Mittwer, T., and Ivler, D. Coliform bacteria in sediments around three marine sewage outfalls. *Limnol. Oceanogr.*, 1958, **3**, 101.

3. Rindgård, E., and Hallquist, S. Proposed plans for sewage discharge from the Gothenberg region, Sweden. *Proc. 5th int. Conf. Wat. Pollut. Res., San Francisco and Hawaii, 1970*, Vol 2, Paper III-5.

Paper No. **41**

Discharge of Sewage from Sea Outfalls

Charles G. Gunnerson

*Marine Ecosystems Analysis Program, Environmental Research Laboratories, National Oceanic and Atmospheric Administration, Boulder, Colorado 80302**

What this Symposium is all about is the point at which we lose engineering control of sewage and industrial waste discharged to the sea. There is abundant and increasing evidence that the fate and effects of these wastes create problems ranging from local to international scales. Present and projected concentrations of people and industry along coastal and estuarine waters emphasize the need for planning and managing the environment in which they live.

There is unhappiness with the current state of knowledge. Nevertheless, most of the information on the viability of enteric organisms in receiving waters, theories of initial dilution, and the relative costs of large collection, treatment, and disposal systems which can be used in engineering design of works discharging to surface waters has been developed by specialists working in estuarine and coastal waters.

A major conclusion which has been made throughout this Symposium is the need for greater efforts in data analysis and utilization and a corresponding reduction in professional and institutional preoccupations with data collection. Information on this need is largely qualitative, but there are several converging lines of evidence showing that from one-third to one-half the total costs of environmental research, monitoring, or control programs should be devoted to data interpretation and use. This emphasis has led from the usual format of a summary paper to the arrangement of topics in the following pages.

Interceptors, Treatment Plants, and Outfalls

Interceptor sewers and treatment plants, like highways and airports, are self-fulfilling prophecies whose land-use implications are imperfectly known. The potential benefit from optimum system design is most obvious in the less-developed regions and in new cities whose water supply and sewage systems are now being developed. Formal bases for planning and optimization are being developed and used in the United States in response to the National Environmental Policy Act of 1969 which requires, in part, a detailed statement by the responsible official on—

(i) The environmental impact of the proposed action,
(ii) Any adverse environmental effects which cannot be avoided should the proposal be implemented,
(iii) Alternatives to the proposed action,
(iv) The relationship between local short-term uses of man's environment and the maintenance and enhancement of long-term productivity, and
(v) Any irreversible and irretrievable commitments of resources which would be involved in the proposed action should it be implemented.

The final configuration of sewage collection, treatment, and disposal facilities rests, in part, on choices between: (1) effluent standards versus receiving water standards—the former generally are politically acceptable and relatively simple to administer, monitor, and enforce, and the latter are generally less expensive—and (2) biological versus physical chemical treatment—the former has higher first costs and lower operating costs and overall costs, and the latter may be politically more acceptable when construction costs are provided from federal or state funds.

In any event, concern for the sanitary environment of individual households results in combining a myriad of small local problems into relatively few large ones. Here, a virtue is made of necessity, and the concept of economy of scale is applied first to sewers and in due course to sewage treatment plants. Economies of scale are limited in that they can become diseconomies where a large quantity of effluent is returned to a single point in the environment. This concept favors more small outfalls over a few large ones, especially where diffusers are required. Further evidence of potential environmental diseconomies associated with large systems is found in the disposal of sewage sludge from Chicago as far away as Alabama. Economies of scale also require an essentially homogeneous raw material and, in sewage

* Formerly: Great Lakes Regional Office, International Joint Commission, Windsor, Ontario, Canada.

treatment, may disappear when low but still dangerous concentrations of toxic industrial wastes are found in municipal sewage.

Receiving Waters

Physical factors which determine the rates and limits of effluent dilution by sea water have been identified in many of the papers presented at this Symposium. Rhodamine-B continues to be used as a tracer because of its lower cost, even though its loss by adsorption onto suspended matter may be recognized. Generally unrecognized is the ability of natural pigments to fluoresce and thus interfere with fluorometry, particularly in estuarial waters[1,2].

Mitchell and Chamberlin (Paper 25) summarized the role of sedimentation as a major factor in the flux of coliform and other enteric organisms in receiving waters. The importance of this factor will be indicated later. The apparent lack of emphasis on determining the sedimentation factor for specific discharges is surprising in view of the recognized epidemiological hazards associated with contaminated shellfish whose food rains down from above.

In their discussions of bacteriological standards and epidemiological risks, several of the contributors to this Symposium indicate that while the risks are dependent upon general health, economic and living standards, and sanitary conditions, a major factor is the local incidence of disease. Coliform bacteria are used as indicators and there is no reason to expect this will not continue. However, assurance of uniform risk associated with the use of a given coliform standard requires quantitative evaluation of the local incidence of disease, mitigated insofar as possible by the specific resistance to infection of the reference population to whom the standards are to be applied.

For example, the United States in 1960 had one typhoid case per year for 22 500 000 people[3,4]. Istanbul in 1968 had one case per 10 000 people. Equal protection against the risk of typhoid in Istanbul as compared with the United States could require coliform standards in the former of only 10 000/22 500 000 or 1/2250 of those in the latter. Corresponding ratios can be developed for hepatitis, bacillary dysentery, gastroenteritis, and other waterborne diseases. Theoretically, the 1/2250 ratio should be

TABLE 1. Reported rates of reductions of coliform bacteria due to mortality and sedimention in ocean waters

$$k_e = \ln 10/T_{90}$$

Location	Date	T_{90} (h)	k_e (h^{-1})	Remarks
Raw sewage				
Honolulu[5]	1970	*ca* 0·2	11	T_{90} too low to account for high pollution levels
Titahi Bay, New Zealand[6]	1959–60	0·65	3·5	
Rio de Janeiro[7]	1963	0·95	2·4	
Israel[8]	—	<1·0	>2·3	Poliovirus T_{90} > 24 h
Istanbul[9]	1968	0·8–1·7	1·4–2·9	Mean value 1·1 h
Genofte, Denmark[10]	—	1·2	2·0	
Tema, Ghana[11]	1964	1·3	1·7	
Nice[12]	—	1·5	1·5	
England[13]	1965	0·78–3·5	0·66–2·9	
Manila[14]	1968–69	1·8–3·4	0·67–1.3	Higher T_{90} value may represent mixing with old effluent
England[15]	1969–73	1·4–5·3	0·43–1·6	Median of 11 results 3·2 h. Extreme values of *ca* 0·6 and 9·0 h rejected
Primary effluent				
Ventura, California[16, 17]	1966	1·7	1·4	
Seaside, New Jersey[18]	1966	1·8	1·3	
Orange Co., California[19]	1954–56	1·8–2·1	1·1–1·3	
Santa Barbara, California[17]	1967	2·4	0·96	
Los Angeles, California[19]	1954–56	4·1[a]	0·56	
Secondary effluent				
Los Angeles, California[19]	1954–56	9·6[b]	0·24	
Combined				
The Hague[20]	1968–69	5–175	0·01–0·5	Higher T_{90} values may represent mixing with old effluent, resuspended sediments, or other sources

[a] T_{90} due to mortality 17·8 h, and to sedimentation 5·3 h; higher T_{90} after chlorination

[b] T_{90} due to mortality 17·8 h, and to sedimentation 21·0 h

reduced by a correction factor relating the disease resistance of the reference population and the local population. Since the large-scale exposure experiments necessary to determine this factor are morally and politically sensitive, it seems likely that the ratios will be either arbitrarily adjusted or not changed at all. An immediate consequence of these ratios is that bacteriological standards for vessel wastes (see below) are at least inappropriate and at worst hazardous. Either a single case or an epidemic among crew or passengers will change the ratio by astronomical levels.

In any event, coliform standards and coliform T_{90}'s will continue to be used for design purposes. Values for coliform T_{90}'s which have been presented by several contributors to this Symposium are summarized in Table 1. In general, T_{90} increases with increasing degree of sewage treatment.

Research Needs

The following list includes a few of those needs which are important to the design, operation, and surveillance of sewerage systems discharging to estuarine or coastal waters. Many will be found to apply equally to large lakes. Some of this research reflects the need for verification or extension of earlier findings. It is disconcerting to be reminded by Fischer (Paper 37) that most conclusions on estuarine circulation are based on Pritchard's three-week investigation of the James River 24 years ago[21]. Similarly, discussions of the survival in sea water of coliform bacteria from secondary effluent are based on 20-year-old studies in Santa Monica Bay[19].

Survival and significance of indicator and pathogenic organisms

Studies of bacterial survival characteristics in receiving waters and laboratories, resistance factors, and viruses are essential to setting standards and for corresponding design and operation of treatment and disposal works. In order to permit comparisons with other locations, it is essential that the degree of treatment and the characteristics of the discharge be described. Special laboratory and field studies should be directed to the role of sedimentation, where differences between primary and secondary effluents should be investigated. Particular attention should be paid to fresh water where *in situ* T_{90} values of 3–10 days have been reported[22]. It is not always possible to rely entirely on *in situ* studies. However, where it is necessary to use pure cultures, these should be related to laboratory studies using mixed cultures which, in turn, are related to specific field observations of survival characteristics.

Tracers

Research into the stability of fluorescent, radioactive, or other tracers is needed in most receiving waters. Losses due to adsorption, ingestion, or decomposition in a variety of environments should be determined and included along with calibration data in reports of investigations. Interferences by natural pigments which result in spurious high fluorometer readings have been observed in San Francisco Bay[2], and probably occur elsewhere.

Stirring, mixing, and diffusion

There are needs for valid data on advective stirring (including chopping and pumping) from both coastal and estuarine environments. The theoretical interrelationships of Eulerian and Lagrangian diffusion models should be validated by field observations.

Epidemiological studies

With increasing competition for available financial and other resources, there are world-wide needs for epidemiological investigations on bathing in fresh and salt water so as to establish firm benefit–cost relationships for coastal or estuarine wastewater treatment projects and for development or evaluation of bathing water standards.

Mineralization and eutrophication

Qualitative evaluation of eutrophication in coastal waters is presented in Paper 14. It is surprising that, given the large number of coastal sewage discharges, very little attention has been given to the rates at which their nitrification takes place in the sea. Preliminary studies on mineralization of secondary effluent in Santa Monica Bay indicated that sufficient quantities of nitrogen and phosphorus were converted within one day to nitrate and phosphate to sustain significant algal blooms[23].

Sludge disposal

Studies are now underway which will characterize the fate, effects, and environmental impacts of dumping sewage sludge from New York City into the New York Bight. The scale of these disposal operations is vast, and there is a need for comparative studies of smaller discharges through outfalls or from barges. Evaluations of pathogenicity and toxicity to both human and marine life, biogeochemical changes in the hydrocarbon fraction of the sludge, and physical movement of the sludge on or near the sea floor are necessary in order to develop benefits of alternative sludge-disposal schemes.

Special Problems

Vessel wastes

The potential hazards of wastes from international shipping originating in ports where cholera or other diseases are epidemic or endemic are of special concern in (1) rivers or lakes used for domestic water supplies and (2) fresh or salt waters used for recreation or fisheries. Generally successful quarantine procedures are applied to people, but their sewage is usually overlooked. For example, a ship detained in a quarantine area is likely to be discharging untreated sewage into waters used downstream for domestic, municipal, or recreational water supplies. The resulting hazards depend upon the surveillance and quality assurance of the basic epidemiological data, the quarantine procedures, and the other health measures taken under World Health Organization International Regulations. The interdependency of standards, ratios of disease, and actual risk of infection discussed above emphasize the need for caution in relying on the waste control measures of Annex IV to the 1973 Convention of the International Marine Consultative Organization in high-use areas.

Dilution

A common defense of 'dilution is the solution to pollution' is to note that a 100-fold initial dilution gives the same result as a 99 per cent removal. However, practical upper limits to initial dilution have not been determined, particularly for toxic wastes. Buoyant-plume theory, including blocking (the effect of the thickness of the field upon further dilution), and buoyant-jet theory, including the effects of jet interference, were applied according to Brooks[24]. Analysis of various configurations of jet diameters, spacing and velocities, diffuser lengths, and premixing ratios indicates that while initial dilutions of, say, 50 000 may be realized, capital and operating costs at the higher ratios almost certainly exceed those for treatment and removal of the most toxic materials. Here, environmental controls based on risks of biomagnification converge with economic considerations and probably indicate mandatory treatment and removal of toxic materials at the source.

Optimizing sampling intervals

The relative ease with which data are collected often results in too much data. Time series may be collected as a substitute for space fields, giving rise to problems which are analogous to those of deriving Lagrangian functions from Eulerian observations described by Callaway (Paper 30), and which are unnecessarily complicated by too short a digitizing, averaging, or other filtering interval. A corresponding problem arises in determining resolution requirements for aerial, satellite, or other remote-sensing imagery. For either time series or space fields, statistical methods useful for optimizing data collection include filtering, probability functions, and spectral analysis. For example, spectral analysis of estuarial data[25] shows that the optimum sampling interval is 2–3 h.

Too often, there is an assumption that the ability to feed data into a computer can be substituted for adequate experimental design. Even with control models which accept continuous data from an outfall or other process-control point, the response characteristics of the system preclude the use of data collected at less than half the interval required to react[25].

Analytical quality control

Sample collection and storage, laboratory procedures and reporting, and data handling, analysis, and utilization vary among the governmental, industrial, and university activities represented at this Symposium. Although procedures for internal quality control are sometimes reported in papers of the sort presented here, their results seldom are. Interlaboratory comparisons, if they exist at all, have not been presented. The result is that we cannot be sure that all of our design, operation, or control premises are valid.

Analytical quality control is essential to development and application of water-quality criteria objectives and standards. Closely related to these needs is that of follow-on operational monitoring of sewage disposal works. It has been suggested that responsibility for performance monitoring be included in all contracts for planning, design, and construction.

Epidemiological studies

Uncertainties in epidemiological data derive from the high costs of data collection, unavailability or bias in the data (after all, few people or communities take pride in being sick), and the realities of statistical inference. Prof. Shuval (Paper 10) remarked that people have the right to ask, "Is it safe?" and it follows that the courts have the right to ask "Is it unsafe?" Insistence on rigorous proof of the latter recently led the U.S. Eighth Circuit Court of Appeals to express the judicial philosophy that "unknowns . . . may not be substituted for proof of a demonstrable hazard to public health." This view suggests that pollution abatement be substituted for pollution prevention and that dead bodies may be useful.

Planning

Planning of environmental or remedial programs requires natural-resources inventories. Resources of the sea are poorly known and such evidence as the finding of DDT in Antarctic penguins raises concern that irreversible damage may be brought about by waste disposal. On a smaller scale, there are sufficient data to show that local conditions result in unique holding capacities. (Holding capacity is the quantity of something which can be held in a container without damaging the contents of the container or the container itself. For example, the number of people who can be supported in the Los Angeles basin at a suitable standard is limited by the air, water, and space available. The quality of air in the basin is limited by the numbers of people and their automobiles. The amount of water free from salt-water intrusion along the coast is limited by the number of people and their wells. And so on . . .) Mr. Calvert (Paper 18) emphasized that waste disposal is not the highest land use and went on to show the need for alternative consideration of waste disposal direct to the sea. Ultimately the need is to optimize the geographic, hydrographic, demographic, and social values as well as economic and energy resources in such a way that the maximum benefit accrues to the maximum number of people. To this end the 1974 U.N. Conferences on the Sea (in Caracas) and on World Population (in Bucharest) are providing much of the paper documentation and philosophy needed by those of us who plan projects for waste disposal into the sea.

Communications

Mrs. Jeger's opening paper of this Symposium was devoted to the political realities of research sponsorship and support. The need for the political link between the 'scientific elite' represented by this Symposium and the scientific public who pays the bills was emphasized by her remark that "scientists do not try as hard to tell us what we want to know as we try to understand what they are trying to tell us." That is why science fiction is so successful although scientific fact is more fun. Having accepted the idea of objectivity and discipline in science, the public does not understand the geographic and professional parochialisms that scientists maintain. Many of the scientific contributions presented at this Symposium are important. It has been startling to learn that foreign embassy science attachés and environmental advisors were not aware of our meeting. No reports have appeared in the press, yet we are discussing matters which affect the lives and pocketbooks of everyone who sees or uses British coastal waters and who ultimately supports our research efforts. These things say that we need help, and that we need professionals in specific activities, including:

(1) Technical editors to ensure that our findings are understood by non-technical audiences. As examples, the magic of a 1000:1 dilution ratio was lost when Prof. Shuval (in discussing Paper 10) described it as one cup in a bathtub, and sewage sludge dumped into the New York Bight takes on a new character when it is called 'black mayonnaise'. The point is that clarity need not lose precision if we accept and rely on professional communications support.

(2) Press rooms at conferences like this where reporters can pick up professionally prepared summaries of papers and releases which they can understand. These will result in requests for interviews which should be granted. The risk of being misunderstood is preferable to that of remaining unheard, unfunded, and still misunderstood.

(3) Communications officers who can train us to not make such statements as "I'm not interested in political problems—I'm only interested in medical ones" and "that decision (on sewage disposal) will be made by a politician whose ignorance permits him to have an opinion."

All of these lead to the need for developing integrated environmental information programs which include the whole spectrum of technical documentation, reports, quality control, technical information storage, retrieval, and exchange, media relations (including government), education at all levels, public participation, and institutional support.

Annotated Summary of Symposium Papers

Regional problems

Review articles by Wood (Paper 1), Aubert and Breittmayer (Paper 2), and Berg (Paper 3) point out the increasing levels of pollution near population centers, recent programs or status of outfall and sewage treatment plant, construction, and some of the special problems of industrial wastes discharged into North Sea, Mediterranean, and United States coastal waters.

Recent investigations of municipal sewage disposal from Hong Kong are summarized by Preston (Paper 4a). Svansson's review (Paper 4b) identifies some of the consequences of pulp, paper, and other industrial waste discharged to the Baltic.

Bacterial indicators and standards

State-of-the-art reviews of research on bacterial indicators are presented by Bonde (Paper 5) and on enteric viruses and bacteriophages by Scarpino (Paper 6). Cabelli, Levin, Dufour, and McCabe (Paper 7) summarize current United States federal epidemiological and receiving-water research.

Metcalf (Paper 8) reports that in shellfish, fecal coli are a satisfactory indicator at high levels for salmonella and viruses, but not at lower levels. Coliphage is an unsatisfactory indicator. In considering epidemiological aspects of sewage discharges, Mosley (Paper 9) concludes that while dilution may save swimmers from high risk exposures, concentrations of pathogens by mollusks partly evens the score. The mere occurrence of pathogens is not significant, but their numbers are. "The ultimate criterion is the number of cases that we would prevent on a long-term basis if contamination of the sea were abolished."

A convincing argument for standards is presented by Shuval (Paper 10). This is based on historical successes of standards for drinking water and milk, on the reality that people do swim and do swallow the water, and on limited but nevertheless real evidence of infection from swimming in polluted waters. Moore (Paper 11) contends that there is no good evidence linking coliform or faecal coliform levels to health, that bathing beaches cannot be satisfactorily classified on the basis of coliforms, and that enforcement of standards on public health grounds is not worth the expense.

Chemical pollution of the sea

Jernelöv (Paper 12) points out that the occurrence of mercury in air, water, land, and living things is related to releases from the earth's crust (possibly increased by the deep plowing of land which was common in Europe by the 15th century), to local effects which are clearly traceable to industrial discharges, and to biogeochemical pathways which are imperfectly known. The finding that tuna and swordfish have contained essentially constant levels of mercury for the last 150 years is reassuring. Nevertheless, mercury-related diseases in natives around Minamata Bay, Japan, and in the Wabigoon, English, and Winnipeg River systems in Canada indicate the need for continued vigilance. Public health concern over the movement of lead from gasoline into the sea is based on the known toxicity of lead and the very large tonnages involved. The significance of other metals reaching the sea is less well known. Locally, anaerobic environments may result in many of them being precipitated as insoluble sulfides by processes similar to those in sewage sludge digesters.

Portmann (Paper 13) has found few persistent organic residues in sewage discharges. This indicates that the sources of significant amounts of these materials in the sea are run-off or dust from agricultural lands or industrial waste discharges. However, there are potentially greater effects from less persistent compounds—such as those from petroleum production, transportation, refining, or spills—because they are present in larger amounts.

McIntyre and Johnston (Paper 14) found that nutrients in sewage had little effect in open marine areas. Locally, sludge deposits have disastrous effects on bottom organisms. Enrichment of waters is most noticed near the mouths of streams where areas of, say, 10 km scale have enough residence time to permit development of plankton blooms. Surface films and particulates with a significant size less than 5 mm from near southern California outfalls were studied by Selleck (Paper 15). These contained approximately 20 000 coliforms per mg hexane-extractable material (HEM). An average of 2–10 mg/m^2 of HEM was found directly over the outfalls. Salt-water bathing standards of $\leqslant 10$ coliforms/ml for $\geqslant 80$ per cent of the time would be met only with HEM levels $< 2\ mg/m^2$. Most of the free saturated fatty acids were present in $<0{\cdot}1$-mm microparticulates which salted out within 14 km of the outfall.

Sewage treatment

Effluent chlorination was found by Kott (Paper 16) to be effective against total and fecal coliforms, *Vibrio cholerae*, and *Entamoeba histolytica* cysts. Enteroviruses were unaffected at doses below about 40 mg/l. at normal pH. Noting that the ocean needs maximum protection because it is the ultimate sink of the world, Baalsrud (Paper 17) recommends no release of pollutants from their source or failing that, their removal by treatment. In contrast, Calvert (Paper 18) argues that since conventional treatment merely speeds up things that would happen anyway, conventional economics favor natural treatment by the receiving waters. The argument is laid aside since the real question is not whether sewage should be treated but where, how much, and when. Too often, the question is resolved on purely economic grounds. This has led Maurice Strong to examine conventional wisdom and to warn that "on a purely economic basis, in fact, it would just not pay to save Planet Earth!"[26]

Microbial mortality

Citing field experiments by Foxworthy[17], Harremoës (Paper 19) confirms earlier work that the dye Rhodamine-B does in fact behave as a dye when it is used as a tracer since it disappears by adsorption onto suspended solids in sewage treatment plant effluent at unpredictable rates which affect dilution computations. Fluorimetric determinations of the latter have been found to show occasional interferences by natural fluorescent pigment in the water or plankton. (Note that Rhodamine-WT has more recently become available, at higher cost, to minimize these problems.) In any event, Harremoës concludes that differences in reported T_{90} values are therefore due to sampling artifacts, although there is evidence that different treatment efficiencies produce effluents with different characteristics having different T_{90}'s (see Table 1 of the present paper). He further concludes that a mass-balance approach is best.

From a review of previously reported work conducted at their laboratory and other places, Aubert, Pesando, and Gauthier (Paper 20) conclude that organic antibiotic substances whose characteristics are only partly known provide stability to at least some life processes in the sea.

In other studies of bactericidal properties of sea water, Jones and Cobet (Paper 21) present data to show that the toxicity to *Escherichia coli* and selected marine bacteria of unidentified heavy metal ions was roughly proportional to their availability under laboratory conditions of chelation.

Gameson and Gould (Paper 22), together with others from the same laboratory, have studied the bactericidal effect of solar radiation in light and dark bottles in the laboratory or at different depths in sewage fields. Under these conditions, T_{90} values for coliform reduction were related to light, mostly in the ultraviolet range. *In situ* studies of T_{90}'s of bacteria in screened, comminuted sewage labelled with bromine-82 did not consistently confirm the results of the bottle experiments, possibly because of a bimodal distribution of T_{90}'s and a normal distribution of the radiation parameter, S_{90} (defined as the radiation in cal/cm^2 which accounts for 90 per cent of the coliform reduction after allowing for mortality which would occur in the dark). This is consistent with results from other studies of T_{90}'s which showed the overriding effects of the characteristics of the sewage itself.

Walker and Guarraia (Paper 23) present a qualitative review of factors which act separately or synergistically in bacterial mortality, including salinity, temperature, pressure, petroleum, pesticides, and predation.

Although enteric viruses survive long enough to reach the sea, public health hazards of virus-contaminated water cannot presently be determined from data on virus survival in marine waters. The review by Akin, Hill, and Clarke (Paper 24) shows (1) the need for research in this area because laboratory data demonstrate that viruses are more resistant to disinfection than are bacteria (and, like bacteria, they are adsorbed onto solids and disappear by sedimentation), and that infection can be caused by as little as one virus unit, (2) consistently high endemic levels of infectious hepatitis and occasional outbreaks of non-bacterial gastroenteritis, and (3) the increasing need for direct recycling of waste waters.

In evaluating and modelling factors influencing the reduction of enteric microorganisms in the sea, Mitchell and Chamberlin (Paper 25) formalized the roles of sedimentation, starvation, predation, physico-chemical effects, and (in clear water) solar radiation.

Initial dilution

After examining eleven earlier models of initial dilution, Hayashi and Ito (Paper 26) have developed a twelfth from an extension of work on continuous convection from an isolated source of heat.

White and Agg (Paper 27) summarize a number of design models and examples for single and multiple port systems, some including pre-mixing in order to achieve required initial dilutions. Their equation for initial dilution (C_0/C_m):

$$(C_0/C_m - 1) = \alpha/x^2,$$

where x is the distance from shore and α is an empirically determined coefficient, is identical with Pomeroy's earlier approach[27] providing unit sewage discharge and a constant slope of the sea floor are assumed. Their Fig. 2 presents classical engineering lore wherein intersections of empirically determined parabolic and hyperbolic curves represent maximum short-term benefit/cost ratios which necessarily neglect environmental costs. Other design formulae and curves follow conventional practice based on depth (or distance from shore assuming constant slope), jet diameter, current velocities, sewage flow, and densimetric Froude number. Cost data are not presented.

Hydén and Larsen (Paper 28) describe a theoretical model for spreading from the top of the rising column based upon density difference, sewage flows, and current velocity.

Sterregaard's review (Paper 29) deals empirically with the importance of initial dilution to hygiene, aesthetics, local circulation, and slick formation.

Subsequent dispersion

Computations of eddy diffusivity (E) from a series of current-meter observations made at 10-min intervals in the thermocline along the Oregon coast were made by Callaway (Paper 30). He reports that over a turbulence scale length (l) of 0·1–6 km the diffusivity varied directly with scale. To one accustomed to the $E = kl^{4/3}$ concept, his Figs. 7 and 8 are equally convincing as further evidence for the four-thirds relationship. In any event, Callaway suggests a series of experiments involving both Eulerian and Lagrangian measurements to further resolve the question.

Cederwall, Göransson, and Svensson (Paper 31) review experimental procedures for field studies of dispersion in coastal surface waters. Talbot (Paper 32) reviews recent work on interpretation of horizontal and vertical diffusion observations, particularly the shear-diffusion models. Kullenberg (Paper 33) summarizes effects of environmental factors in the surface layer, pycnocline, deep layer, and bottom layer. These include air-sea interactions, scale effects of diffusion, buoyancy, and transport. Bretschneider (Paper 34) reported that wave heights, periods, directions, and spreading followed special Weibull distributions whose significance in waste disposal remains to be shown.

Predictive modelling

Munro (Paper 35) presents a two-dimensional model for coliform distributions based on uniform depths, uniform vertical mixing throughout tidal cycles, horizontal diffusion calculated according to Joseph and Sendner, bacterial mortality due to sunlight, and no sedimentation. His predicted and observed values are generally within 2 or 3 orders of magnitude.

Carter (Paper 36) develops a dilution model from Rhodamine-WT studies near the mouth of Chesapeake Bay. Fischer (Paper 37) reviews the use of simplified numerical models for predicting dispersion in estuaries. These include gravitational circulation, cross-sectional gradients in a uniform channel (shear-flow dispersion), bathymetrically induced dispersion (chopping and pumping) from tidal flows, and wind action. Only one predictive model can be used at a time so that the dominant mechanism must be selected on the basis of other information, possibly that from hydraulic models. Oakley and Staples (Paper 38) review approaches to the interpretation of field data. Having noted the limitations of conventional data collection programs and results, their conclusion is consistent with mine—that the costs and time for data analysis are usually overlooked in the excitement of collecting new data.

Observations on operation of the Hague outfall are reported by van Dam (Paper 39). He found T_{90} values ranging from 5 to 175 h and observed that the higher values were taken from extensive fields of high concentrations. Most of his values are inordinately high in comparison with others from all over the world; this may be due to new sewage being diluted with old, to high turbidities from the Rhine or from dredging spoil deposited nearby, or to resuspension of sediments.

Conceptual design of marine disposal systems

After drawing attention to general information needs in marine waste disposal, Pearson (Paper 40) identifies the problems of determining the degree of treatment needed for sewage discharged at various locations and depths into estuarine or coastal waters. Predictably the problems are those of economics, environmental impact, and public health, and are particularly sensitive to interceptor and outfall locations, and to operating costs, especially those of increasing energy costs.

REFERENCES

1. BAILEY, T. E., GUNNERSON, C. G., and McCULLOUGH, C. A. Mixing and dispersion studies in San Francisco Bay. *J. sanit. Engng Div. Am. Soc. civ. Engrs*, 1966, **92** (SA5), 23–45.
2. GUNNERSON, C. G., and McCULLOUGH, C. A. Limitations of Rhodamine-B and Pontacyl brilliant pink as tracers in estuarine waters. Proc. Symp. Diffusion in Ocean and Estuarine Waters, Lamont Geological Observatory, Columbia University, New York, 1964, p. 53.
3. DAMOC, a consortium. Master plan and feasibility report for water supply and sewerage for Istanbul region. Daniel, Mann, Johnson, and Mendenhall, Los Angeles, 1971.
4. DAUER, C. C. 1960 summary of disease outbreaks and a 10-year résumé. *Publ. Hlth Rep., Wash.*, 1961, **76**, 915–922.
5. DILLINGHAM ENVIRONMENTAL CO. Report to City and County of Honolulu Department of Public Work. 90% progress report, work areas 6 & 7, oceanographic studies, Part I. March 1971.
6. THOMAS, R. H. Marine dilution and inactivation of sewage. *Proc. 1st int. Conf. Wat. Pollut. Res., London, 1962*, Vol. 3, pp. 57–83.
7. DA SILVA, P. C. M. Discussion on coastal currents. *Proc. 3rd int. Conf. Wat. Pollut. Res., Munich, 1966*, Vol. 3, pp. 284–290.
8. SHUVAL, H. I. The case for microbial standards for bathing beaches. This volume, pp. 95–101.
9. GUNNERSON, C. G., SUNGUR, O. E., BILAL, E., and ÖZTURGUT, E. Sewage disposal in the Turkish Straits. *Wat. Res.*, 1972, **6**, 763–774.
10. HARREMOËS, P. Field determination of bacterial disappearance in seawater. *Wat. Res.*, 1970, **4**, 737–749.
11. PEARSON, E. A., SELLECK, R. E., and STORRS, P. N. Some physical parameters and their significance. In 'Pollution and marine ecology' (ed. T. A. Olson and F. J. Burgess). Wiley-Interscience, New York, 1962, pp. 297–316.
12. AUBERT, M., GUIZERIX, J., AUBERT, J., MOLINARI, J., and DANIEL, S. Etude de la dispersion bactérienne d'eaux residuaires en mer au moyen de traceurs radioactivs. *Revue int. Océanogr. méd.*, 1966, **1**, 56–91.
13. WHEATLAND, A. B., AGG, A. R., and BRUCE, A. M. Some observations on the dispersion of sewage from sea outfalls. *J. Proc. Inst. Sew. Purif.*, 1965, 291–314.
14. SCHMIDT, O. J. Master sewerage system plan for Metropolitan Manila. *J. sanit. Engng Div. Am. Soc. civ. Engrs*, 1972, **98** (SA1), 125–152.
15. GAMESON, A. L. H., and GOULD, D. J. Effects of solar radiation on the mortality of some terrestrial bacteria. This volume, pp. 209–219
16. HARREMOËS, P. *In situ* methods for determination of microbial disappearance in sea water. This volume, pp. 181–190.
17. FOXWORTHY, J. E., and KNEELING, H. R. Eddy diffusion and bacterial reduction in waste fields in the ocean. University of Southern California, Los Angeles, 1969.
18. CARTER, H. H., CARPENTER, J. H., and WHALEY, R. C. The bactericidal effect of seawater under natural conditions. *J. Wat. Pollut. Control Fed.*, 1967, **39**, 1184–1189.
19. GUNNERSON, C. G. Sewage disposal in Santa Monica Bay. *Trans. Am. Soc. civ. Engrs*, 1959, **124**, 823–851.
20. DAM, G. C. v. The Hague outfall. This volume, pp. 393–401.
21. PRITCHARD, D. W. A study of the salt balance in a coastal plain estuary. *J. mar. Res.*, 1954, **13**, 133–144.
22. CANALE, R. P., PATTERSON, R. L., GANNON, J. J., and POWERS, W. F. Water quality models for total coliforms. *J. Wat. Pollut. Control Fed.*, 1973, **45**, 325–336.
23. RITTENBERG, S. C. Marine bacteriology and the problem of mineralization. In 'Marine microbiology' (ed. C. H. Oppenheimer). C. C. Thomas, Springfield, Ill., 1963, pp. 48–60.
24. BROOKS, N. H. Dispersion in hydrologic and coastal environments. California Institute of Technology, Rep. No. KH-R-29, 1972.
25. GUNNERSON, C. G. Optimizing sampling intervals. Proc. IBM Scientific Computing Symposium on Water and Air Resources Management, White Plains, New York, 1968, pp. 115–140.
26. STRONG, M. One year after Stockholm. *Foreign Affairs*, 1973, **51**, 690–707.
27. POMEROY, R. D. Empirical approach for determining required length of an ocean outfall. *Proc. 1st int. Conf. Waste Disposal mar. Envir.*, Pergamon Press, London, 1960, pp. 268–278.

Mr. Gunnerson was given the opportunity of revising his preprinted paper to reflect the discussions and conclusions of the Symposium. The additional material includes his verbal presentation.

DISCUSSION

Dr. A. J. Lacey (U.K.), in a contribution which would have been more appropriate to the discussion of Papers 17 and 18 and which should thus be considered in conjunction with pp. 177–179, said that in his view as an educator in environmental pollution science the case for treatment was overwhelming, because the risks were too great and the experience in other phases of the environment pointed the lesson that dispersal was not equivalent to disposal. It was his opinion that no authority, however small, should be allowed to discharge sewage into the sea. Small seaside towns in the U.K. should at once be given government grants to build treatment plants. Purely residential towns should be advised to dispose of their sludge to agricultural land and their treated effluent to fresh water—this was particularly important in towns which were likely to increase in size. Towns or villages in high amenity areas should be given additional assistance to reach high standards of effluent to safeguard a national resource. Any industrial establishment discharging directly to the sea should be obliged to incorporate the 'best practicable means' approach to remove from its effluent any substance known to be a pollutant. Criticism could be levelled at this view on the grounds of cost, but (just as in air pollution) the community paid whether it controlled or failed to control pollution. The fallacy that dispersal was equivalent to disposal persisted even into the last papers of the Symposium. Dilution did nothing to diminish the quantity of persistent chemicals released into the environment—in fact it increased the area of the environment (including the food chains) at risk in a way similar to that in which high chimneys increased the population at risk to fall-out.

Mr. B. Grover (World Bank) said that, as an engineer, he had been intrigued by the fundamental scientific aspects that had emerged from the discussions on sea outfalls, though he had been confused and unsettled by the candour with which the various authors had admitted their uncertainties. This was neither the first nor the last time that learned discussions would take place on the topic. It was admitted that too little was known, and recognized that more wastes would be generated in coastal areas in the future. Consequently, he had attempted to evaluate the Symposium, searching for constructive criticism, in the context of the classical scientific approach to the solution of problems—identify the problem, analyse it, consider alternative solutions, select the optimum, and (for feedback to improve future decisions) monitor the results. He thought a few topics might have been missed that should be considered in future discussions:

1. Critical review of many existing outfalls in various environments, to assess their good and bad points. The information gained by collective experience should be shared.
2. A session to discuss the state-of-the-art of outfall construction, since the relevant technology was changing no less than the understanding of the basic principles affecting sea outfalls.
3. Greater efforts to quantify costs. It was not enough to get upset because a few fish got cancer. The problems themselves had to be translated into economic terms. Expenditure on water pollution control had to compete with expenditure in many other areas, so that the case needed to be presented well.

Finally, lest they were to leave the Symposium with the wrong impression, he assured the participants that the World Bank was not a collection of capitalistic Philistines concerned only with myopic penny-pinching. Its task was economic development and its clients were all poor countries. Of course it was aware of the need to prevent pollution and clean up the environment, but it was also aware of the fantastic competition for scarce resources [see also Mr. Grover's contribution on p. 178]. Perspectives had to be kept when planning environmental projects—particularly in developing countries. Otherwise the professionals who could help would lose the credibility they deserved and needed. Compromises were necessary, perfection being the enemy of the good.

Mr. R. G. Ludwig (Brazil), as Director of the Oahu Water Quality Study at Honolulu during 1970–71, took the opportunity to clarify several matters and to correct some misinterpretations of the study data. Clarification was needed because those who had referred to the work did not list as a reference the eleven volumes of reports[28] presenting the results of the studies for which the city paid nearly $1 million. First, regarding coliform counts as reported by Prof. Berg (pp. 21–22) and the pollution levels suggested by Mr. Gunnerson in Table 1: high counts were indeed found in Honolulu Harbor and at the mouths of the drainage channels, but the reports showed these values to be related to materials discharged directly to the affected areas—the Honolulu outfall contributed in no way to these conditions. Secondly, the T_{90} value for coliforms (at the surface of the sea and during the day) in the warm waters at Honolulu was certainly considerably less than an hour, and most probably about 30 min. This result was based not only on several specific T_{90} experiments but also on analysis of coliform and salinity data collected throughout the studies all around the outfall. Finally, for those who might not believe that T_{90} could be so short, he wished to suppress any fears regarding the future of Honolulu beaches, by pointing out that the requirements placed on the design by aesthetic considerations resulted in the magnitude of T_{90} being of academic interest—even if it were several hours the design would not be affected.

Mr. G. C. van Dam (Netherlands) asked why the T_{90} values from Paper 39 had been omitted from Table 1 of the preprint of Mr. Gunnerson's paper. He understood that Mr. Gunnerson (who subsequently included them in the table) had considered them unreliable because they were so different from those found elsewhere in the world. The explanation that they might be due to new sewage being diluted with old was untenable. The basis of the method was given by Equation 1 on p. 396 which deliberately took account of both old and new sewage. Mr. A. L. H. Gameson (U.K.) added that this was the only method reported which integrated the effects of mortality over day and night. If mortality was fairly slow during the day—as would be expected if the depth of the sewage field was great in comparison with the depth of penetration of the lethal component of solar radiation—then the even slower nocturnal mortality would lead to high values of T_{90}.

Mr. J. A. Wakefield (U.K.) said scientists seemed unable to communicate with the public; they failed to divulge basic rather than abstruse information. As an example, the governmental report[29] on disposal of sewage into the sea was almost entirely statistical and of little or no interest to the general public. It had been left to the Coastal Anti-Pollution League (an amenity society) to publish a series of maps giving the exact location of each outfall—a form of information which the public could understand and of particular interest where sewage was discharged onto the beach. The League's problem was the dearth of funds available to pay for the initial printing of these maps.

Dr. E. W. Akin (U.S.A.), on the subject of communication within the technical community, reported that the U.S. Environmental Protection Agency had recently established a policy requiring a final formal report on all projects funded by the E.P.A. (both in-house and extramural) within a few months of completion of the project. These reports would then be made available for distribution, and should certainly help in keeping data out of the filing cabinet and putting it into the hands of the scientific and technical community.

REPLY TO DISCUSSION

Mr. Gunnerson said that Mr. Grover's list of the World Bank's information needs, and his warnings of competition for money and other resources, emphasized his own comments on the need for improved communications. He appreciated Mr. Ludwig's clarification of the Honolulu studies, and maintained that his remarks and the supporting data should have been summarized and published in a technical journal.

He appreciated Mr. Wakefield's problem of obtaining funds to make full use of the information obtained on U.K. coastal discharges. The responsibility for ensuring public availability and utilization of information should lie with those responsible for each project and should be programmed into it. Communication between scientists and engineers was not enough—the results had to be shared with bureaucrats, politicians, and the public.

ADDITIONAL REFERENCES

28. Engineering Science Inc., Sunn, Low, Tom & Hara Inc., and Dillingham Environmental Co. Water quality program for Oahu with special emphasis on waste disposal. City and County of Honolulu, 1971–72, 11 vols.

29. Department of the Environment and Welsh Office. Report of a survey of the discharges of foul sewage to the coastal waters of England and Wales. H.M. Stationery Office, London, 1973.

List of Participants

AUSTRALIA

D. Comar	East Melbourne Laboratories Pty Ltd, 425 Canterbury Road, Surrey Hills, Victoria 3127.
H. L. Jones	Australian Government, Department of Housing and Construction, Box 280 7AA, Melbourne.
D. J. Lane	Chief Chemist, Engineering and Water Supply Department, State Administration Centre, Adelaide, South Australia 5000.
A. Leitch	Chief Engineer (Design), Geelong Waterworks and Sewerage Trust, P.O. Box 659, Geelong 3220.
M. A. Simmonds	Consulting Chemical Engineer, 82 Central Avenue, St. Lucia, Queensland 4067.
A. G. Strom	Gutteridge, Haskins & Davey, 380 Lonsdale Street, Melbourne 3000.
R. W. Whiteside	Chairman, Geelong Waterworks and Sewerage Trust, 65 Ryrie Street, Geelong 3220.

BRAZIL

P. S. Alem	Av. Prof. Frederico Hermann Jr., 465, CETESB, Cep 05459, São Paulo.
M. L. Hess	Planidro, Engenheiros Consultores, Av. Angelica 1854, São Paulo.
Russell G. Ludwig	Caixa Postal 5086, Atlantica ZC-37, Rio de Janeiro Gb.
Prof. A. G. Occhipinti	Director Téchnico, Hidroconsult, Av. Brig. Luiz Antonio, 2637-10° Andar, São Paulo.
C. A. Pessoa	Hidroconsult, Av. Brig. Luiz Antonio, 2637-10° Andar, São Paulo.
A. S. C. Rameh	Av. Prof. Frederico Hermann Jr. 465, CETESB, Cep 05459, São Paulo.
R. T. Tanaka	*Not known.*

CANADA

R. P. Côté	Environment Canada, P.O. Box 2406, Halifax, Nova Scotia, B3J 3E4.
Charles G. Gunnerson	*See* U.S.A.
Dr. D. J. Lawrence	Bedford Institute of Oceanography, Dartmouth, Nova Scotia.
Dr. J. J. Sharp	Chairman of Civil Programmes, Faculty of Engineering, Memorial University of Newfoundland, St. John's, Newfoundland.
Dr. Michael Waldichuk	Department of the Environment, Fisheries and Marine Service, Pacific Environment Institute, 4160 Marine Drive, West Vancouver, B.C.
H. H. Watson	Department of the Environment, Fisheries and Marine Service, Fontaine Building, 11th Floor, Ottawa, Ontario, K1A 0H3.

DENMARK

Prof. G. J. Bonde	Hygiejnisk Institut, Aarhus Universitet, Universitetsparken, 8000 Århus C.
S. Genders	Isotopcentralen, 2 Skelbaekgade, 1717 Copenhagen V.
Dr. K. Grunnet	Hygiejnisk Institut, Aarhus Universitet, Universitetsparken, 8000 Århus C.
Prof. Poul Harremoës	Department of Sanitary Engineering, Technical University of Denmark, Building 115, 2800 Lyngby.
Dr. Gunnar E. B. Kullenberg	Institute of Physical Oceanography, Haraldsgate 6, 2200 Copenhagen N.
Dr. J. H. Nielsen	Cowiconsult Engineers and Planners, Jaegersborg allé 14, 2920 Charlottenlund.
H. Schrøder	Danish Hydraulic Institute, Øster Valdgade 10, 1350 Copenhagen K.
Benny Sterregaard	Nordjyllands Amtskommune, Teknisk Forvaltning. Amtsvandvæsenet, Digtervejen 1, 9200 Skalborg.

FRANCE

Dr. J.-Ph. Breittmayer	C.E.R.B.O.M. (I.N.S.E.R.M.), Parc de la Côte, 1 Avenue Jean-Lorrain, 06300 Nice.
Dr. M. J. Gauthier	*As above*

GERMANY, Federal Republic of

A. Lohmeyer	Institut Wasserbau III an der Universität Karlsruhe, 75 Karlsruhe, Kaiserstrasse 12.
Prof. Gertrud Müller	Direktor, Bundesgesundheltsamt, 1 Berlin 33, Correnplatz 1.

HONG KONG	
P. D. Che	Public Works Department Hong Kong, Civil Engineering Office, Murray Building, Garden Road.
ICELAND	
E. V. Ingimundarson	National Institute of Environmental Health, Post Box 505, Reykjavik.
IRISH REPUBLIC	
G. A. Clemins	Dublin Corporation, 28 Castle Street, Dublin 2.
K. T. Harding	Assistant Chief Engineering Adviser, Department of Local Government, Custom House, Dublin 1.
P. Lynch	Dublin County Council, 11 Parnell Square, Dublin 1.
P. C. MacIntyre	P. H. McCarthy Son & Partners, Iveagh Court, Harcourt Road, Dublin 2.
T. McManus	Institute for Industrial Research and Standards, Glasnerin, Dublin.
A. Whelan	Dublin Corporation, 28 Castle Street, Dublin 2.
IRAN	
M. Namazi	Tehran Regional Water Board, P.O. Box 5, Tehran.
ISRAEL	
Naphtali Cohen	Ministry of Health, Tel-Aviv.
Prof. Yehuda Kott	Environmental Engineering Laboratories, Technion City, Haifa 32000.
Dr. Uri Marinov	Director, Environmental Protection Service, Prime Minister's Office, Hakirya Building 3, Jerusalem 91000.
Prof. Hillel I. Shuval	Environmental Health Laboratory, Hadassah Medical School, The Hebrew University, P.O.B. 1172, Jerusalem.
ITALY	
Dott. Ing. A. Bottino	Comune di Genova, c.so Monte Grappa 34 A-8, 16137 Genoa.
Dott. Ing. J. Giorgio	Comune di Genova, Via don Minetti 8/44 ,Genoa.
Prof. L. Majori	Università Degli Studi di Trieste, Instituto d'Igiene, Via A. Valerio 34, 34127 Trieste.
JAPAN	
Prof. Taizo Hayashi	Department of Civil Engineering, Chuo University, Kasuga 1-chome 13, Bunkyo-ku, Tokyo.
NETHERLANDS	
Dr. G. Abraham	Density Currents Branch, Delft Hydraulics Laboratory, Rotterdamseweg 185, Delft.
Dr. P. K. Baay	Rijks Institute Huivering Afvalwater, Westeinde 3a, Voorburg.
G. C. van Dam	Rijkswaterstaat Directie Waterhuishouding en Waterbeweging, 's-Gravenhage, Hooftskade 1.
J. W. Gunster	Direktie Noordzee, Rijkwaterstaat, Sir Winston Churchill—1 ann 362, Rijswijk 2H.
NEW ZEALAND	
Dr. O. Dinah Brockett	Botany Department, University of Auckland, Auckland.
H. M. Cotton	Chief Engineer, Hutt Valley Drainage Board, Private Bag, Lower Hutt.
C. C. Davidson	Consulting Engineer, P.O. Box 256, Blenheim.
R. M. Hedgland	Hutt Valley Drainage Board, Private Bag, Lower Hutt.
G. B. McBride	Ministry of Works and Development, Wellington.
NORWAY	
Kjell Baalsrud	Norsk Institutt for Vannforskning, Gaustadalléen 25, P.O. Box 333, Blindern, Oslo 3.
—. Bergmann-Paulsen	Statens Forurensningstilsyn, Oslo Dep, Oslo 1.
G. Mosevoll	River and Harbour Laboratory, Technical University of Norway, Klæbuveien 153, 7000 Trondheim.
A. Thendrup	*As above*
PORTUGAL	
M. A. M. M. Pinto	Drena, Av. da Liberdade 177-2°E, Lisbon.

SAUDI ARABIA

A. G. Banks	Watson Saudi Arabia, P.O. Box 2023, Jeddah.

SOUTH AFRICA

Dr. J. J. Barnard	Environmental Health Services Branch, Department of Health, Private Bag X88, Pretoria 0001.

SPAIN

Dr. Josefina P. Castellvi	Instituto de Investigaciones Pesqueras, Paseo Nacional S/N, Barcelona-3.

SWEDEN

Dr. Hans O. Bouveng	Swedish Air and Water Pollution Research Laboratory, Drottning Kristinas Väg 47 D, 114 28 Stockholm 5.
Dr. K. Cederwall	Division of Hydraulics, Chalmers University of Technology, Fack, 402 20 Göteborg 5.
C-G. Göransson	*As above*
Dr. Hans Hydén	Vattenbyggnadsbyrån, Box 5038, 102 41 Stockholm 5.
Mrs. A. L. Martin	Swedish Air and Water Pollution Research Laboratory, Drottning Kristinas Väg 47 D, 114 28 Stockholm 5.
Dr. Artur Svansson	Hydrographic Department, Fishery Board of Sweden, 400 40 Göteborg 4.

TURKEY

Nihat Baş	Iller Bankasi Genel Müdürlüğü (General Directorate of Provincial Bank), Su ve Kanalizasyon Daires, Ankara.
Azmi Erkaslan	*As above*
Yildirim Oktay	*As above*
Reşid Yalvaç	*As above*

UNITED KINGDOM

P. Ackers	Binnie & Partners, Artillery House, Artillery Row, London, SW1P 1RX.
A. R. Agg	Water Research Centre—Stevenage Laboratory, Elder Way, Stevenage, Herts., SG1 1TH.
J. S. Alabaster	*As above*
Dr. R. G. Allen *OBE*	Director, Water Research Centre, Medmenham Laboratory, Marlow, Bucks., SL7 2HD.
Dr. G. K. Anderson	Department of Civil Engineering, University of Newcastle upon Tyne, Cassie Building, Claremont Road, Newcastle upon Tyne, NE1 7RU.
Miss A. Anson	Sabrina Project, Environmental Science Laboratory, University of Bristol, 19 Woodland Road, Bristol, BS8 1TE.
R. A. Apostol	Department of Civil Engineering, University of Dundee, Dundee, DD1 4HN.
J. C. T. Arnold	Harvey, McGill & Hayes, Northernhay House, Northernhay, Exeter, EX4 3RY.
J. L. Arnold	Engineering Inspector, Welsh Office, Cathays Park, Cardiff, CF1 3NQ.
B. Atkinson	Wessex Water Authority, Somerset Rivers Division, The Watergate, West Quay, Bridgwater, Somerset, TA6 3HN.
P. Ayres	Ministry of Agriculture, Fisheries & Food, Fisheries Laboratory, Remembrance Avenue, Burnham-on-Crouch, Essex, CM0 8HA.
D. Battersby	Director of Resource Planning, South West Water Authority, 3–5 Barnfield Road, Exeter, EX1 1RE.
W. Beattie	Water Quality Branch, Department of Industrial and Forensic Science, 5/11 Verner Street, Belfast, BT7 2AD.
R. Best	Directorate General Water Engineering, Department of the Environment, 2 Marsham Street, London, SW1P 3EB.
A. T. Bielak	Water Research Centre—Stevenage Laboratory, Elder Way, Stevenage, Herts., SG1 1TH.
R. J. E. Braybrooks	Directorate General Water Engineering, Department of the Environment, 2 Marsham Street, London, SW1P 3EB.
A. W. J. Bufton	Water Research Centre—Stevenage Laboratory, Elder Way, Stevenage, Herts., SG1 1TH.
John T. Calvert *CBE*	John Taylor & Sons, Artillery House, Artillery Row, Victoria Street, London, SW1P 1RY.

Ian C. Camp	J. D. & D. M. Watson, Terriers House, 201 Amersham Road, High Wycombe, Bucks., HP13 5AJ.
D. E. Cannon	Howard Humphreys & Sons, Westminster House, West Street, Epsom, Surrey, KT18 7RH.
S. Carr	Waterhouse & Partners, Newcastle House, High Spen, Rowlands Gill, Co. Durham, NE39 2BL.
Dr. L. J. Carter	Imperial Chemical Industries Ltd, Brixham Laboratory, Freshwater Quarry, Brixham, Devon, TQ5 8BA.
Dr. J. A. Charlton	Department of Civil Engineering, University of Dundee, Dundee, DD1 4HN.
Dr. H. A. Cole	Garth, Normanston Drive, Lowestoft, Suffolk.
R. I. Collinson	Akins Research & Development, 4 Dorking Road, Epsom, Surrey, KT18 7LX.
J. D. Cooke	Water Quality Branch, Department of Industrial and Forensic Science, 5/11 Verner Street, Belfast, BT7 2AD.
R. J. Cousins	Water Research Centre—Stevenage Laboratory, Elder Way, Stevenage, Herts., SG1 1TH.
R. Cowen	South West Water Authority, Headquarters Design Group, Crownhill Waterworks, Tavistock Road, Plymouth, PL6 5DA.
T. E. Croally	Sandford, Fawcett, Wilton & Bell, 21 Inner Park Road, London, SW19 6ED.
H. Cronshaw	Chief Engineer (Wales), Welsh Office, Cathays Park, Cardiff, CF1 3NQ.
R. S. Darling	Binnie & Partners, Artillery House, Artillery Row, London, SW1P 1RX.
D. G. Davis	*Not known*
J. B. Dawson	A. H. S. Waters & Partners, Canterbury House, 85 Newhall Street, Birmingham, B3 1LZ.
R. G. D. Defriez	c/o Rofe, Kennard & Lapworth, 28/31 St. Mary Street, Cardiff, CF1 2AB.
Dr. A. L. Downing	Binnie & Partners, Artillery House, Artillery Row, London, SW1P 1RX.
P. K. Eastwood	Northumbrian Water Authority, Eldon House, Regent Centre, Gosforth, Newcastle upon Tyne, NE3 3PX.
G. E. Eden	Assistant Director, Water Research Centre—Stevenage Laboratory, Elder Way, Stevenage, Herts., SG1 1TH.
Miss M. S. Eglinton	Manager Analytical Division, Wimpey Laboratories Ltd, Beaconsfield Road, Hayes, Middx.
M. J. Essex	South West Water Authority, Headquarters Design Group, Manley House, Kestrel Way, Exeter, EX2 7LQ.
Mrs. Lilian M. Evison	University of Newcastle upon Tyne, Newcastle upon Tyne, NE1 7RU.
E. W. Flaxman	Binnie & Partners, Artillery House, Artillery Row, London, SW1P 1RX.
K. J. Flemons	Sir Frederick Snow & Partners, Ross House, 144 Southwark Street, London, SE1 0SZ.
I. R. Fozzard	Forth River Purification Board, Causewayhead School, Alloa Road, Stirling, FK9 5PN.
A. L. H. Gameson	Water Research Centre—Stevenage Laboratory, Elder Way, Stevenage, Herts., SG1 1TH.
D. J. Gould	*As above*
A. S. Grey	Edinburgh Corporation, City Chambers, 19 Market Street, Edinburgh, EH1 1BL.
D. J. Griffin	Wallace Evans & Partners, Plymouth House, Penarth, Glam., CF6 2YF.
W. Halcrow	Lothians River Purification Board, Colinton Dell House, West Mill Road, Colinton, Edinburgh, EH13 0PH.
A. W. G. Harley	Scottish Development Department, Room J1/280, Government Buildings, Broomhouse Drive, Saughton, Edinburgh, EH11 3UX.
K. I. M. Henry	Babtie, Shaw & Morton, 95 Bothwell Street, Glasgow, G2 7HX.
E. Hodges	Wessex Water Authority, Techno House, Redcliffe Way, Bristol, BS1 6NY.
J. F. Jackson	Howard Humphreys & Sons, Westminster House, West Street, Epsom, Surrey, KT18 7RH.
Arthur E. James	Glaxo Laboratories Ltd., Ulverston, Cumbria, LA12 9DR.
Mrs Lena Jeger ***MP***	House of Commons, London, SW1A 0AA.
H. H. Jones	Thames Water Authority, New River Head Laboratories, 177 Rosebery Avenue, London, EC1R 4TP.
R. E. Jones	Water Research Centre—Stevenage Laboratory, Elder Way, Stevenage, Herts., SG1 1TH.
R. M. Jones	Edinburgh Corporation, City Chambers, 19 Market Street, Edinburgh, EH1 1BL.
F. M. W. King	Divisional Manager, Wessex Water Authority, Avon and Dorset Recovery Division, 2 Nuffield Road, Poole, Dorset, BH17 7RL.

R. J. King	Water Research Centre—Stevenage Laboratory, Elder Way, Stevenage, Herts., SG1 1TH.
J. B. Kirby	A. H. S. Waters & Partners, Canterbury House, 85 Newhall Street, Birmingham, B3 1LZ.
Dr. A. J. Lacey	Biology Department, Brunel University, Kingston Lane, Uxbridge, Middx., UB8 3PH.
J. S. Lang	D. A. Donald & Wishart, 64 Kelvingrove Street, Glasgow, G3 7SA.
T. J. Leonard	Howard Humphreys & Sons, Westminster House, West Street, Epsom, Surrey, KT18 7RH.
Dr. R. E. Lewis	Imperial Chemical Industries Ltd, Brixham Laboratory, Freshwater Quarry, Brixham, Devon, TQ5 8BA.
W. J. Loneskie	Ayrshire River Purification Board, 2 Alloway Place, Ayr, KA7 2AA.
D. W. Mackay	Clyde River Purification Board, Rivers House, Murray Road, East Kilbride, Glasgow, G75 0LA.
Dr. J. F. Macqueen	Central Electricity Research Laboratory, Kelvin Avenue, Leatherhead, Surrey, KT22 7SE.
J. MacRae	'Municipal Engineering', 178–202 Great Portland Street, London W1N 6NA.
Dr. D. D. Mara	Department of Civil Engineering, University of Dundee, Dundee, DD1 4HN.
Miss A. M. Martin	Water Research Centre—Stevenage Laboratory, Elder Way, Stevenage, Herts., SG1 1TH.
P. R. McFadlean-Ferguson	'Civil Engineer', 26–34 Old Street, London EC1V 9AD.
Dr. A. D. McIntyre	Department of Agriculture and Fisheries for Scotland, Marine Laboratory, P.O. Box 101, Victoria Road, Aberdeen, AB9 8DB.
B. M. Mollowney	Water Research Centre—Stevenage Laboratory, Elder Way, Stevenage, Herts., SG1 1TH.
Dr. H. A. C. Montgomery	Southern Water Authority, Guildbourne House, Chatsworth Road, Worthing, Sussex, BN11 1LD.
Dr. B. Moore	Public Health Laboratory, Church Lane, Heavitree Lane, Heavitree, Exeter, EX2 5AD.
W. D. Mugford	South West Water Authority, 3–5 Barnfield Road, Exeter, EX1 1RE.
D. Munro	Water Research Centre—Stevenage Laboratory, Elder Way, Stevenage, Herts., SG1 1TH.
Dr. B. S. Nau	British Hydromechanics Research Association Fluid Engineering, Cranfield, Bedford, MK43 0AJ.
J. Naughton	'Surveyor', Dorset House, Stamford Street, London, SE1 9LU.
W. B. Norgrove	Northumbrian Water Authority, Directorate of Operations & Works, Eldon House, Regent Centre, Gosforth, Newcastle upon Tyne, NE3 3PX.
H. Roy Oakley	J. D. & D. M. Watson, Terriers House, 201 Amersham Road, High Wycombe, Bucks., HP13 5AJ.
O. C. O'Neill	Industrial Pollution Abatement Ltd, 21 Inner Park Road, London, SW19 6ED.
A. J. O'Sullivan	Akins Research & Development, 4 Dorking Road, Epsom, Surrey, KT18 7LX.
L. N. Payne	Southern Water Authority, Guildborne House, Chatsworth Road, Worthing, Sussex, BN11 1LD.
J. M. A. Perrott	Sir M. MacDonald & Partners, Demeter House, Cambridge.
J. W. Philips	Directorate General Water Engineering, Department of the Environment, 2 Marsham Street, London, SW1P 3EB.
J. Pitcaithley	Babtie, Shaw & Morton, 95 Bothwell Street, Glasgow, G2 7HX.
Dr. J. E. Portmann	Ministry of Agriculture, Fisheries & Food, Fisheries Laboratory, Remembrance Avenue, Burnham-on-Crouch, Essex, CM0 8HA.
Dr. L. J. Powell	Welsh Office, Cathays Park, Cardiff, CF1 3NQ.
Jack R. Preston	J. D. & D. M. Watson, Terriers House, 201 Amersham Road, High Wycombe, Bucks., HP13 5AJ.
Miss K. R. Reynolds	Hydraulic Division, Wimpey Laboratories Ltd, Beaconsfield Road, Hayes, Middx., UB4 0LS.
Miss J. P. Rhodes	Water Research Centre—Stevenage Laboratory, Elder Way, Stevenage, Herts., SG1 1TH.
D. G. M. Roberts	John Taylor & Sons, Artillery House, Artillery Row, Victoria Street, London, SW1P 1RY.
Miss J. A. Robinson	Water Research Centre—Stevenage Laboratory, Elder Way, Stevenage, Herts., SG1 1TH.
S. R. Salt	Divisional Civil Engineer, Welsh National Water Development Authority, Morgannwg Sewage Division, 8 Museum Place, Cardiff, CF1 3BG.

J. R. Saxon	Wessex Water Authority, Techno House, Redcliffe Way, Bristol, BS1 6NY.
D. R. Sayers	Anglian Water Authority, Diploma House, Grammar School Walk, Huntingdon, PE18 6NZ.
Dr. T. L. Shaw	Sabrina Project, Environmental Science Laboratory, University of Bristol, 19 Woodland Road, Bristol, BS8 1TE.
M. T. Shilston	Sir William Halcrow & Partners, Newcombe House, 45 Notting Hill Gate, London, W11 3JX.
Dr. E. A. Simpson	Directorate General Water Engineering, Department of the Environment, 2 Marsham Street, London, SW1P 3EB.
P. H. Smith	Edinburgh Corporation, City Chambers, 19 Market Street, Edinburgh, EH1 1BL.
G. Stanfield	Water Research Centre—Stevenage Laboratory, Elder Way, Stevenage, Herts., SG1 1TH.
Kenneth D. Staples	J. D. & D. M. Watson, Terriers House, 201 Amersham Road, High Wycombe, Bucks., HP13 5AJ.
R. G. Stead	Northumbrian Water Authority, Eldon House, Regent Centre, Gosforth, Newcastle upon Tyne, NE3 3PX.
Dr. J. H. Stoner	Welsh National Water Development Authority, The Barracks, Brecon, Powys.
Dr. L. Streatfield	Senior Consultant, Cremer & Warner, 140 Buckingham Palace Road, London, SW1W 9SQ.
J. W. Talbot	Ministry of Agriculture, Fisheries & Food, Fisheries Laboratory, Lowestoft, Suffolk, NR33 0HT.
J. Tannahill	Clyde River Purification Board, Rivers House, Murray Road, East Kilbride, Glasgow, G75 0LA.
Dr. D. Taylor	Imperial Chemical Industries Ltd, Brixham Laboratory, Freshwater Quarry, Brixham, Devon, TQ5 8BA.
Dr. M. R. G. Taylor	D. Balfour & Sons, Yeoman House, 63 Croydon Road, Penge, London, SE20 7TW.
N. C. Taylor	J. D. & D. M. Watson, Terriers House, 201 Amersham Road, High Wycombe, Bucks., HP13 5AJ.
R. J. Taylor	Ward, Ashcroft & Parkman, Cunard Building, Liverpool, L3 1ES.
J. A. Wakefield	Chairman, The Coastal Anti-Pollution League, Alverstoke, Greenway Lane, Bath, BA2 4LM.
B. E. Waters	Lewis & Duvivier, 4 Deans Yard, London, SW1P 3NP.
E. McG. Weir	Dee and Don River Board, 14 North Silver Street, Aberdeen, AB1 1RL.
S. F. White	Directorate General Water Engineering, Department of the Environment, 2 Marsham Street, London, SW1P 3EB.
Dr. W. R. White	Hydraulics Research Station, Howbery Park, Wallingford, Oxon., OX10 8BA.
E. P. Wilmot-Morgan	Directorate General Water Engineering, Department of the Environment, 2 Marsham Street, London, SW1P 3EB.
J. A. Wilson	Water Quality Branch, Department of Industrial and Forensic Science, 5/11 Verner Street, Belfast, BT7 2AD.
P. C. Wood	Ministry of Agriculture, Fisheries & Food, Fisheries Laboratory, Remembrance Avenue, Burnham-on-Crouch, Essex, CM0 8HA.
P. Worthington	Department of the Environment, 2 Marsham Street, London, SW1P 3EB.

UNITED STATES OF AMERICA

Dr. Elmer W. Akin	United States Environmental Protection Agency, National Environmental Research Center, Cincinnati, Ohio 45268.
Prof. Gerald Berg	*As above*
Prof. Charles L. Bretschneider	Department of Ocean Engineering, University of Hawaii, 2656 The Mall, Honolulu, Hawaii 96822.
Dr. Victor J. Cabelli	National Marine Water Quality Laboratory, Liberty Lane, West Kingston, Rhode Island 02892.
Dr. R. J. Callaway	National Coastal Pollution Research Program, Pacific Northwest Environmental Research Laboratory, Corvallis, Oregon 97330.
Harry H. Carter	Chesapeake Bay Institute, The Johns Hopkins University, Baltimore, Maryland 21218.
Charles Chamberlin	Division of Engineering and Applied Physics, Harvard University, Cambridge, Mass. 02138.
Prof. Rita R. Colwell	Professor of Microbiology, University of Maryland, College Park, Maryland 20742.

Prof. Hugo B. Fischer	Professor of Civil Engineering, 412 O'Brien Hall, University of California, Berkeley, Calif. 94720.
Prof. James E. Foxworthy	College of Science and Engineering, Loyola Marymount University, Loyola Boulevard, Los Angeles, Calif. 90045.
Dr. J. A. French	Camp, Dresser & McKee International, 1 Center Plaza, Boston, Mass. 02108.
Dr. M. Grant Gross	Director, Chesapeake Bay Institute, The Johns Hopkins University, Baltimore, Maryland 21218.
Dr. Leonard J. Guarraia	Office of Water and Hazardous Materials, Environmental Protection Agency, Washington, D.C. 20460.
Charles G. Gunnerson	Marine Ecosystems Analysis Program, Environmental Research Laboratories, National Oceanic and Atmospheric Administration, Boulder, Colorado 80302.
Prof. Galen E. Jones	Professor of Microbiology, University of New Hampshire, Spalding Life Science Building, Durham, New Hampshire 03824.
Prof. T. G. Metcalf	Chairman of Microbiology, University of New Hampshire, Durham, New Hampshire 03824.
Prof. Ralph Mitchell	Division of Engineering and Applied Physics, Pierce Hall, Harvard University, Cambridge, Mass. 02138.
Prof. James W. Mosley	School of Medicine, University of Southern California, 2826 South Hope Street, Los Angeles, Calif. 90007.
Prof. Erman A. Pearson	Room 365, Davis Hall, University of California, Berkeley, Calif. 94720.
Prof. Pasquale V. Scarpino	Professor in Environmental Engineering, Department of Civil and Environmental Engineering, University of Cincinnati, Cincinnati, Ohio 45221.
Prof. Robert E. Selleck	Division of Hydraulic and Sanitary Engineering, University of California, Berkeley, Calif. 94720.

WORLD BANK

Brian Grover	Sanitary Engineer, World Bank, 1818 H Street N.W., Washington, D.C. 20433, U.S.A.

WORLD HEALTH ORGANIZATION

Dr. Michael J. Suess	Regional Officer for Environmental Pollution Control, WHO Regional Office for Europe, Scherfigsvej 8, 2100 Copenhagen Ø, Denmark.
Ian J. Waddington	Regional Officer for Comprehensive Environmental Programmes, WHO Regional Office for Europe, Scherfigsvej 8, 2100 Copenhagen Ø, Denmark.

Name Index

Papers presented at the Symposium are shown by **bold** numerals. Publications referred to in text or discussions are indicated thus: 5/70—Reference 70 in Paper 5. Other references to publications are shown as page numbers in *italics*. References are in approximately chronological order of publication, those separated by commas being to the same paper. Other references to speakers, etc. are page numbers. Addresses of those names preceded by asterisks will be found in the List of Participants.

Subject Index